Outstanding Topics in Ocean Optics

Outstanding Topics in Ocean Optics

Special Issue Editors

Dariusz Stramski
Hubert Loisel

MDPI • Basel • Beijing • Wuhan • Barcelona • Belgrade

MDPI

Special Issue Editors

Dariusz Stramski
University of California San Diego
USA

Hubert Loisel
Université du Littoral Côte d'Opale
France

Editorial Office
MDPI
St. Alban-Anlage 66
4052 Basel, Switzerland

This is a reprint of articles from the Special Issue published online in the open access journal *Applied Sciences* (ISSN 2076-3417) in 2018 (available at: https://www.mdpi.com/journal/applsci/special_issues/ocean_optics)

For citation purposes, cite each article independently as indicated on the article page online and as indicated below:

LastName, A.A.; LastName, B.B.; LastName, C.C. Article Title. *Journal Name* **Year**, *Article Number, Page Range.*

ISBN 978-3-03897-704-9 (Pbk)
ISBN 978-3-03897-705-6 (PDF)

Cover image courtesy of Shea Cheatham.

Contents

About the Special Issue Editors

Dariusz Stramski, Ph.D., is a Professor of Oceanography in the Marine Physical Laboratory division at Scripps Institution of Oceanography, University of California San Diego, U.S.A. He founded the Ocean Optics Research Laboratory at Scripps in 1997. His research covers a wide range of topics in the areas of ocean optics, optical remote sensing of the ocean, and applications of optical methods in oceanography. Example topics include the optical properties of various types of marine particles, radiative transfer in the ocean, underwater light fields, the use of optical measurements including ocean color remote sensing for estimating organic carbon pools, phytoplankton diversity, seawater optical properties and particle characteristics, advances in measurement methodologies in ocean optics, and the development of novel instrumentation. He has published numerous influential contributions on these topics in more than 100 papers in peer-reviewed journals. In recent years he led the development of a new patented technology for counting and sizing submicrometer particles, which was commercialized and has broad applications in various research areas and industrial sectors. Professsor Stramski earned his Ph.D. in Earth Sciences (1985) from the University of Gdansk, Poland. He also received a Degree of Habilitation in Oceanography from the Polish Academy of Sciences (2002) and a title of Professor in Earth Sciences presented by the President of Poland (2014). Prior to joining Scripps Institution of Oceanography (1997), he conducted research at the Institute of Oceanology, Polish Academy of Sciences, Poland (1978–1988), Laboratoire d'Océanographie in Villefranche-sur-Mer, France (1986–88 & 1996), Université Laval in Québec, Canada (1988–1989), and University of Southern California in Los Angeles, U.S.A. (1989–1997). Professor Stramski is the recipient of multiple awards, including the Maurycy Rudzki Prize from the Polish Academy of Sciences for studies of light fluctuations in the sea, and was elected a Distinguished Fellow of the Kosciuszko Foundation Collegium of Eminent Scientists and a Foreign Member of the Polish Academy of Sciences. He has led numerous research projects, participated in many oceanographic expeditions in various regions of the world's ocean, and has been actively involved in teaching and various professional activities, including service as an Associate Editor of the journal Limnology and Oceanography and a member of expert panels, working groups, and science teams.

Hubert Loisel, Ph.D., is a Professor of Oceanography in the Laboratory of Oceanology and Geosciences (LOG) at the Université du Littoral Côte d'Opale in France (ULCO). Professor Loisel earned his Ph.D. in Earth Sciences (1999) from the University of Pierre and Marie Curie (Paris 6, France), and a Degree of Habilitation in Physical Oceanography from ULCO in 2006. He became full professor at ULCO in 2007 and has been teaching courses at undergraduate and graduate levels. He conducted research at the Scripps Institution of Oceanography in San Diego, U.S.A. (1999–2000), Laboratory of Atmospheric Optics in Lille, France (2001–2003), and the Space Technology Institute of the Vietnam Academy of Sciences and Technology (VAST) in Hanoi, Vietnam (2012–2016). He was the leader of the team "Physical oceanography, transport, and remote sensing" at LOG (2007–2012) and is expected to take the position of director of LOG on January 1, 2020. He received the Medal of VAST and the Vietnam Ministry of Science and Technology for his research in Vietnam. He was a member of the CNES Scientific Committee (TOSCA) from 2007 to 2013. Since 2015 he has been a member of the National Committee of Universities (CNU) in France and the Committee of the International Ocean Color Coordinating Group. He was PI and co-PI of numerous projects in marine optics and ocean color remote sensing funded by different space agencies (CNES, ESA, EUMETSAT). His research includes experimental and theoretical studies of inherent optical properties of natural waters, radiative transfer studies in the ocean, development of inverse methods to estimate bio-optical parameters from ocean color remote sensing, and analyses of the spatial/temporal variations of space-derived data products at regional and basin scales. He is particularly interested in the dynamics of suspended organic and mineral marine particles in open-ocean and coastal waters.

Preface to "Outstanding Topics in Ocean Optics"

The transfer of radiant solar energy has vital implications for life and climate on Earth. The interactive nature between radiative transfer and various dissolved and particulate constituents of seawater is at the core of ocean optics science and applications. By the late 1970s, the fundamentals of modern optical oceanography were essentially identified. At that time, the era of optical remote sensing of the ocean also began with the launch of the first proof-of-concept satellite ocean color sensor in 1978. In the past two decades, several satellite missions with ocean color measurements have been launched, providing a new uninterrupted view of the global ocean. In particular, these achievements in optical remote sensing have been reshaping the research landscape towards a better understanding of ocean biology, biogeochemistry, and ecosystems, and their roles in the Earth's system processes. Recent years have also seen an increase in the use of optical technologies on autonomous in situ oceanographic platforms. As a result of these advances, optical theory and measurements are now firmly embedded in studies of a great variety of ocean science and engineering questions.

The optical applications in oceanography rely on intricate linkages between the optical variability in the ocean and a combination of many physical, biological, chemical, and geological processes that control the sources and fates of various optically-significant constituents of seawater. These analyses are, however, incomplete without a firm understanding of radiative transfer and interactions between light and the highly complex and variable constituent composition of seawater. In spite of decades of research, there are still unexplored areas of research and unrealized potential applications of ocean optics which are largely associated with challenges to further advance our understanding of interactions of light with the complex optical medium of seawater at a higher level of detail than in past approaches. Future progress in optical remote sensing and other optical applications in oceanography demand such further advances. While it is true that over the past few decades the community of scientists who use optical techniques has grown considerably and the oceanographic literature has seen a surge of papers that have been catalyzed by and related to satellite ocean color observations, it is also significant to note that available resources and the community's efforts to pursue the outstanding and potentially impactful basic topics in ocean optics which are "out-of-ordinary" and not directly related to ocean color science have been disproportionally low. As a result, fundamental scientific understanding has not always kept pace with technological advances and demand resulting from potential applications.

The goal of this Special Issue entitled "Outstanding Topics in Ocean Optics" has been to draw together a series of papers across a wide range of topics in the field of ocean optics, which generally share a common denominator of frontier research topics that are unique, uncommon, or outstanding in the literature. The intention of this volume is to present the reader with a sampler of exciting research topics that are underway and to provide a balanced view of the extraordinary breadth of research in ocean optics. Leading experts in topics as diverse as measurements and modeling of radiative transfer, light fields, light scattering and polarization, ocean color, benthic optical properties, and the use of optical techniques and approaches for characterizing seawater constituents have contributed to this volume, which makes it special in both its scope and authority. Most of these contributions are indicative of the richly multidisciplinary aspect of modern research in optical oceanography.

Recent progress in the radiative transfer modeling of a coupled atmosphere—snow/ice—ocean system is reviewed by Stamnes and his co-workers. The theme of underwater light field characterization is central to a few papers. Gleason and co-workers present both theoretical and experimental data of polarization of upwelling radiance in the surface ocean, and Li and co-workers present results from a unique set of measurements of downwelling and upwelling (both plane and scalar) irradiances in the ocean euphotic layer. The paper by Massicotte and co-workers addresses the estimation of in-water light field below spatially heterogeneous sea ice cover. Several papers in the Special Issue exhibit a particular focus on light scattering in the ocean. Zhang and Hu discuss light scattering by pure seawater as a function of water temperature. Sun and co-workers review the formalism of Mueller scattering matrix for realistic non-spherical particles suspended in ocean waters. Koestner and co-workers report on measurements of the volume scattering function and degree of linear polarization of light scattered by marine particulate assemblages. The papers by Lain and Bernard and Duforêt-Gaurier and co-workers are focused on the light-scattering properties of phytoplankton, including potential implications for the assessment of phytoplankton communities from satellite ocean color observations. The paper by Twardowski and Tonizzo describes an approach for modeling ocean color, which explicitly incorporates the angularly-resolved light scattering of seawater. Two papers are devoted to algorithms for estimating ocean data products from global satellite observations of ocean color. Specifically, Aurin and co-workers address the estimation of absorption parameters of colored dissolved organic matter and the concentration of dissolved organic carbon, and Wang and co-workers the estimation of concentrations of multiple phytoplankton pigments. Topics associated with benthic optical properties and optical remote sensing of sea floor are covered in three papers. Fournier and co-workers describe an approach for modeling the reflectance of a sea bottom covered with mineral compounds and vegetation. Hedley and co-workers present a model for examining the effects of the three-dimensional structure of coral reefs on benthic reflectance and water-leaving light, and Ackleson and co-workers examine the effects of sensor noise on the optical remote sensing of shallow coral reefs. Last but not least, the utility of field optical measurements for characterizing the particulate and dissolved constituents of seawater is demonstrated in three papers. Agagliate and co-workers describe the use of flow cytometry measurements of discrete seawater samples as a means of estimating the concentrations of suspended particulate matter, particulate organic carbon, and chlorophyll-a. Boss and co-workers examine the in situ measurements of different inherent optical properties as a tool for studying the characteristics of suspended sediment in a bottom boundary layer. Zielinski and co-workers describe the capabilities of an in situ technique of excitation—emission matrix spectroscopy for assessing the fluorescent dissolved organic matter in natural waters. We note that the papers in this volume are ordered according to the date of submission, and not according to subject areas.

We foresee that this collection of papers will be of interest and useful to a broad audience of professional ocean scientists, engineers and advanced students with an interest in ocean optics and applications of optical methods in oceanography. We also hope that the broad perspective and delightful variety of ideas presented in these papers will help to inspire and motivate researchers around the globe to conduct different types of studies using state-of-the-art science and methodologies, which will lead to new discoveries in ocean optics and the further enhancement of optical applications in oceanography. We wish to express our gratitude and appreciation to all authors who provided contributions to this Special Issue. We also thank the staff personnel of *Applied Sciences* for their assistance in the production of this Special Issue.

Dariusz Stramski, Hubert Loisel
Special Issue Editors

applied
sciences

MDPI

Article

Estimation of Suspended Matter, Organic Carbon, and Chlorophyll-a Concentrations from Particle Size and Refractive Index Distributions

Jacopo Agagliate [1],*, Rüdiger Röttgers [2], Kerstin Heymann [2] and David McKee [1]

[1] Department of Physics, University of Strathclyde, 107 Rottenrow, Glasgow G4 ONG, UK; david.mckee@strath.ac.uk

[2] Helmholtz-Zentrum Geesthacht, Max-Planck-Straße 1, 21502 Geesthacht, Germany; rroettgers@hzg.de (R.R.); kerstin.heymann@hzg.de (K.H.)

* Correspondence: jacopo.agagliate@strath.ac.uk

Received: 1 February 2018; Accepted: 23 July 2018; Published: 19 December 2018

Abstract: Models of particle density and of organic carbon and chlorophyll-a intraparticle concentration were applied to particle size distributions and particle real refractive index distributions determined from flow cytometry measurements of natural seawater samples from a range of UK coastal waters. The models allowed for the estimation of suspended particulate matter, organic suspended matter, inorganic suspended matter, particulate organic carbon, and chlorophyll-a concentrations. These were then compared with independent measurements of each of these parameters. Particle density models were initially applied to a simple spherical model of particle volume, but generally overestimated independently measured values, sometimes by over two orders of magnitude. However, when the same density models were applied to a fractal model of particle volume, successful agreement was reached for suspended particulate matter and both inorganic and organic suspended matter values (RMS%E: 57.4%, 148.5%, and 83.1% respectively). Non-linear organic carbon and chlorophyll-a volume scaling models were also applied to a spherical model of particle volume, and after an optimization procedure achieved successful agreement with independent measurements of particulate organic carbon and chlorophyll-a concentrations (RMS%E: 45.6% and 51.8% respectively). Refractive index-based models of carbon and chlorophyll-a intraparticle concentration were similarly tested, and were also found to require a fractal model of particle volume to achieve successful agreement with independent measurements, producing RMS%E values of 50.2% and 45.2% respectively after an optimization procedure. It is further shown that the non-linear exponents of the volume scaling models are mathematically equivalent to the fractal dimensionality coefficients that link cell volume to mass concentration, reflecting the impact of non-uniform distribution of intracellular carbon within cells. Fractal models of particle volume are thus found to be essential to successful closure between results provided by models of particle mass, intraparticle carbon and chlorophyll content, and bulk measurements of suspended mass and total particulate carbon and chlorophyll when natural mixed particle populations are concerned. The results also further confirm the value of determining both size and refractive index distributions of natural particle populations using flow cytometry.

Keywords: forward modeling; suspended matter; marine particles; fractal structure; organic carbon; chlorophyll-a

---

1. Introduction

The determination of suspended particulate mass concentrations (minerogenic mass, biomass, chlorophyll content) in marine particle populations is a matter of particular interest to ocean sciences, and one of the key aspects of the characterization of the properties of marine particles. The relationship between particulate mass properties and optical properties of seawater is important for understanding the formation of optical remote sensing signals and their interpretation.

In previous work carried out on a set of seawater samples collected in UK coastal waters (UKCW dataset), a Mie-based flow cytometric method (FC method) was developed to determine particle size distributions (PSDs) and real refractive index distributions (PRIDs), and its results used as inputs for Mie theory forward modelling to reconstruct not only bulk inherent optical properties (IOPs), but also individual fractions and optical contributions from inorganic, organic, and fluorescent particle subpopulations [1,2]. Although flow cytometric determination of particle physical properties and the subsequent modelling of IOPs have some precedent [3–5], no true effort has been devoted to extending the procedure to the reconstruction of particulate mass concentrations from flow cytometric data. Indeed, coupled with models of particle density and carbon intraparticle concentration, particle size, and real refractive index distributions offer the chance to explore the biogeochemical properties of a particle population from a new perspective.

In this study, models of organic and inorganic particle density are adapted from literature and applied to the UKCW dataset to produce modelled values of suspended particulate matter (SPM), organic suspended matter (OSM), and inorganic suspended matter (ISM). Furthermore, cell volume scaling models and refractive index-based models are also adapted from literature to allow estimation of intraparticle carbon (C_i) and intraparticle chlorophyll-a (Chl_i) content, ultimately producing modelled values of particulate organic carbon (POC) and chlorophyll-a concentrations (ChlA) from the particle data of the UKCW dataset. The modelled values thus obtained are then compared against the results of actual biogeochemical measurements, and the parameters used to assess carbon and chlorophyll concentrations are optimized on a dataset-wide basis to explore the physiology of the cells encountered during the He442 research cruise. A summary of the abbreviations and notations used throughout the study is given in Table 1.

Table 1. Abbreviations and notations used in this study.

Notation	Definition
ChlA	Chlorophyll-a concentration, mg m^{-3}
FC	Flow cytometer; flow cytometry
IOP	Inherent optical property
ISM	Inorganic suspended matter, g m^{-3}
OSM	Organic suspended matter, g m^{-3}
SPM	Suspended particulate matter, g m^{-3}
POC	Particulate organic carbon, mg m^{-3}
PRID	Particle real refractive index distribution
PSD	Particle size distribution
RMS%E	Root-mean-square percentage error
RMSE	Root-mean-square error
UKCW	UK coastal waters (dataset)
a, a_C, a_{chl}	Slopes of the refractive index-based models and of the of the C_i and Chl_i optimized refractive index-based models respectively, kg m^{-3}
b	y-intercepts of the refractive index-based models, kg m^{-3}
$F(r)$	Fractal dimension, dimensionless
h_1, h_2	Power law exponents of the optimized POC and ChlA volume scaling functions respectively, dimensionless
k	Particle size distribution scaling coefficient, mL^{-1}
k_1, k_2	Scaling coefficients of the optimized POC and ChlA volume scaling functions respectively, pg µm^{-3}
m_{tot}	Total particle mass, mg
C_i	Intraparticle carbon concentration, kg m^{-3}
Chl_i	Intraparticle chlorophyll-a concentration, kg m^{-3}

Table 1. *Cont.*

Notation	Definition
$N(D)$	Number concentration of particles within particle size bin corresponding to particle diameter D, mL^{-1}
$N'(D)$	Density function of the particle size distribution, $mL^{-1} \mu m^{-1}$
N_r	Number of particles within particle size bin corresponding to particle radius r, dimensionless
$N_{D,nr}$	Number of particles within particle bin corresponding to particle diameter D and real refractive index n_r, dimensionless
n_0	Real refractive index of the dry matter fraction of the particle, dimensionless
n_r	Real refractive index of the particle, dimensionless
n_i	Imaginary refractive index of the particle, dimensionless
$\bar{n}_{r,1}, \bar{n}_{r,2}$	Average real refractive indices at the upper and lower extremes of the particle size distribution respectively, dimensionless
r, D	Particle radius and particle diameter, μm
r_0	Primary particle radius, μm
V_D	Particle volume, μm^3
V_o	Volume of the dry matter fraction of the particle, μm^3
$y(r)$	Volume scaling function
B	Fractal dimension exponent, dimensionless
γ	Power law slope, dimensionless
P	Particle density, g/m^3
ρ_{nr}	Density of a particle with real refractive index n_r, g/m^3
ρ_0	Density of the dry matter fraction of the particle, g/m^3

2. Materials and Methods

2.1. Theory

2.1.1. Particle Mass Modelling from Apparent Density of Hydrated Matter

Calculations for modelled values of SPM, OSM, and ISM were made following the technique presented by Zhang et al. [6]. Building on the approach presented by Morel & Ahn [7] and Babin et al. [8] the technique estimates a density value for the particulate matter which is dependent on the real part of the refractive index of the particles and is designed to account for their water content. Since this value is neither the value of the dry matter fraction of the particle nor that of water, but rather a combination of the two, this global density is also known as "apparent" density. The equation takes the form

$$\rho = \rho_o V_o = \rho_o \frac{n_r - 1}{n_0 - 1}, \tag{1}$$

where n_r is the real refractive index of the whole particle and n_o, ρ_o, and V_o are respectively the real refractive index, density, and fractional volume of the dry matter fraction of the particle. All refractive index values are given relative to water.

Values for the $\rho_o/(n_o - 1)$ ratio were defined following those employed by Zhang et al. [6]. For organic particles (defined as the fraction of the particle population with $n_r < 1.1$) the mean value of the ratio was set at $(8.56 \pm 1.1) \times 10^6$ g/m³. These are particles with high water content, as high as ~80 ± 10% for algal cells [9]. For mineral particles (defined as the fraction of the particle population with $n_r \geq 1.1$) the mean value of the ratio was instead set at $(15.52 \pm 1.84) \times 10^6$ g/m³. These particles have low water content; when the fractional volume of dry matter reaches unity (i.e., water content within the particles is zero) the apparent density of the particle becomes equal to the density of the dry mineral matter and can be calculated accordingly. Zhang et al. [6] find $n_r = 1.16$ as the threshold above which $V_o = 1$, and give $\rho = [(6.42 \pm 0.85)n_r - (4.86 \pm 0.99)] \times 10^6$ g/m³ as the corresponding density based on a linear regression of literature values of density and refractive index for a number of mineral species. Overall, the final expression of Equation (1) used in practice was

$$\rho = \begin{cases} 8.56 \times 10^6 (n_r - 1) & n_r < 1.1 \\ 15.52 \times 10^6 (n_r - 1) & 1.1 \leq n_r < 1.16 \\ 6.42 \times 10^6 n_r - 4.86 \times 10^6 & n_r \geq 1.16 \end{cases} \quad (2)$$

with all density values given as g/m^3.

2.1.2. Particulate Organic Carbon and Chlorophyll-a Cell Volume Scaling

Organic carbon and chlorophyll-a concentrations within a cell are not linear functions of the cell volume (also defined as biovolume by some authors); C_i and Chl_i values can be instead derived using empirical relationships defined by volume scaling exponents, which can be then summed over organic and fluorescent PSDs to obtain POC and ChlA values respectively, i.e.,

$$C = \sum_r y(r) N_r \quad (3)$$

where C represents either POC or ChlA, $y(r)$ is the corresponding size-dependent total carbon or chlorophyll concentration per cell and N_r is the number of particles within each size bin. A number of these empirical conversion relationships can be found in the literature for the modelling described here: four sets of parameters for carbon [10–12] and two sets of parameters for chlorophyll-a [11,13] were employed. These are presented in Table 2.

Table 2. Particulate organic carbon and chlorophyll-a cell volume scaling models used in this study

POC & ChlA Cell Volume Scaling	
Particulate Organic Carbon	**Source**
$y(r) = 0.433 V(r)^{0.863}$	[10]
$y(r) = 0.109 V(r)^{0.991}$	[11]
$y(r) = 0.288 V(r)^{0.811}$	[12] (diatoms)
$y(r) = 0.216 V(r)^{0.939}$	[12] (non-diatom mixed protists)
Chlorophyll-a	**Source**
$y(r) = 0.00429 V(r)^{0.917}$	[11]
$y(r) = 0.0398 V(r)^{0.863}$	[13]

2.1.3. Refractive Index-Based Estimation of Particulate Organic Carbon and Chlorophyll-a

Research carried out in the 1990s demonstrated that cell volume is not the only parameter that can be used to estimate C_i and Chl_i values. In a series of studies [14–18], a number of empirical relationships were established for various phytoplankton species between the real refractive index n_r and C_i and between the imaginary refractive index (n_i) and Chl_i. Expanding on this premise, Stramski [19] established refractive index-based linear models for the estimation of C_i and Chl_i based on data from two phytoplankton species

$$C_i = 3441.055 n_r (660 \, \text{nm}) - 3404.99 \quad (4)$$

$$Chl_i = 996.86 n_i (675 \, \text{nm}) + 1.17. \quad (5)$$

A follow-up work by DuRand et al. [20] established slightly modified relationships with the inclusion of data from additional phytoplankton species

$$C_i = 3946 n_r (650 \, \text{nm}) - 3922 \quad (6)$$

$$Chl_i = 1244 n_i (675 \, \text{nm}) - 0.32. \tag{7}$$

These can then be associated with a particle volume model and with particle sizes as provided by PSDs to determine POC and ChlA values. Both sets of equations were employed in this work. The n_r values contained in the PRIDs were originally determined by the FC method for $\lambda = 488$ nm, i.e., the wavelength of the laser source used within the flow cytometer [1]; however, n_r values are only weakly dependent on the wavelength (e.g., [21]), and Equations (4) and (6) are thus likely to be usable as is for the UKCW PRIDs as well. The FC method does not provide any information on the n_i values; as will be described in Section 2.2.4, these were derived from literature. The relevant n_i value (i.e., for organic particles at $\lambda = 675$ nm) was adapted from Babin et al. [8] as $n_i = 1.620 \times 10^{-3}$.

2.2. Methods

The particle density, carbon and chlorophyll cell volume scaling, and C_i and Chl_i refractive-index based estimation models were applied to the PSDs and PRIDs of the UKCW dataset as determined by the FC method, which can be found described in detail in [1]. A description of the dataset and of the measurement protocols (particularly those relative to SPM, ISM, OSM, POC, and ChlA measurements) is summarised below. Resulting mass concentrations obtained by modelling from FC data were then compared with corresponding suspended matter, organic carbon, and chlorophyll concentration values determined from traditional sample analysis. Cumulative contributions from different size classes were also calculated for SPM, ISM, OSM, and POC.

2.2.1. UK Coastal Waters (UKCW) Dataset

The UKCW dataset consists of natural water samples obtained during the He442 research cruise in UK waters (4–21 April 2015) on board the R/V Heincke (Alfred-Wegener-Institute, Bremerhaven, Germany). Sixty-two stations were sampled across a variety of optical water conditions around the coast of the UK (Figure 1), supplying a total of 50 samples with complete sets of FC data and matching data from other instruments and independent measurements. This included SPM, ISM, OSM, POC, and ChlA values obtained from lab analysis of the water samples retrieved during the research cruise. Wind conditions were favourable throughout, ranging from calm to moderate gale, and did not hamper the measurement process at any point during the cruise. Day-to-day weather ranged widely from clear sky conditions to heavy rain, although good weather was generally prevalent. Of particular note was the very high particle load found in Bristol Channel waters, which resulted in particularly large values of SPM, ISM, and OSM, as will be described in the following.

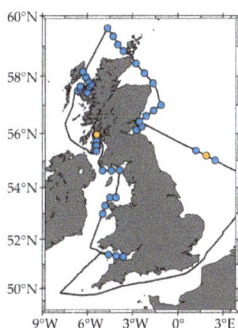

Figure 1. Track of the He442 research cruise, which took place in April 2015 in UK coastal waters aboard R/V Heincke. Out of the 62 measurement stations visited a total of 50 complete sets of data were retrieved, matching flow cytometric data, and ancillary measurements (blue circles). Yellow circles denote stations where two samples were taken. The figure was adapted and modified with permission from Figure 1 of Agagliate et al. [2].

2.2.2. Depth Profiling

Main depth profiling was done via an instrument frame equipped with Niskin bottles for sample retrieval. The frame was lowered through the water column at each of the stations, kept at a maximum depth for a first round of sampling, then raised to near surface depth to retrieve further samples. The samples were taken from the Niskin bottles on the frame as quickly as possible after the frame was back on deck and filled into 10-L plastic containers. In waters with high turbidity the Niskin bottles were flushed twice to avoid settling out of particulate matter. Forty-eight out of the 50 samples of the UKCW dataset are surface samples (depth: 5–7 m), with further two samples taken from bottom depths instead. The prevalence of surface samples within the dataset is due to the focus of the cruise, which was on developing data sets in support of the Sentinel remote sensing missions.

2.2.3. Flow Cytometry Measurement Protocol

All samples were measured by a CytoSense flow cytometer (CytoBuoy b.v., Woerden, The Netherlands) once for each of four sensitivity settings of the side scattering photomultiplier tube (50, 60, 70, 80), for 6 min and at a flow rate of 0.5 μL/s. Side scattering was used as the trigger channel in all cases. The reader is directed to [1] for a detailed description of the CytoSense flow cytometer and its operation, particularly in the context of the UKCW dataset. Additional measurements of standard polymer beads necessary for calibration of the FC method were taken daily across the whole sampling period. A detailed description of the FC method and of the procedure followed to reconstruct PSDs and PRIDs can also be found in [1], where the application of the method to the UKCW dataset is also discussed specifically.

2.2.4. PSD Extrapolations

The mass concentration modelling will require the entire optically relevant particle distribution to be included as the input, or the output will not be comparable with independently measured mass concentration values. The FC method was found to reliably retrieve particle diameters between ~0.5–10 μm [1]. This range covers a large fraction of the contribution to scattering and backscattering, but the whole optically relevant range spans from tens of nanometers to a few millimeters [22,23]. The undetectable fraction of the particle population has to be accounted for using an approximation of the PSD to extend the range of the distribution over the whole relevant range.

Ever since pioneering work in the '60s and '70s found that the number of particles suspended in the ocean increased continuously and monotonically towards smaller scales [24,25], power law distributions of the type used by Junge [26] for aerosols have been the most common form of approximation for natural seawater particle populations [27,28]. The PSDs determined by the FC method for the UKCW dataset broadly conformed to this model, and consequently, following in the steps of Green et al. [5], a least squares best fit of the UKCW PSDs through power law distributions as defined by

$$N(D) = N'(D)dD = kD^{-\gamma}dD \tag{8}$$

was used to extend the range of the measured PSDs (Figure 2). Here $N(D)$ is the number concentration of particles within the size bin corresponding to diameter D, $N'(D)$ the density function of the PSD, dD the width of the size bin, k the scaling coefficient of the PSD, and γ its slope. The form given in Equation (8) is necessary because the FC PSDs have bin-like nature; accordingly, the extrapolations need to be bin-like as well.

Values for the real refractive index n_r in the Junge extensions must also be accounted for using some approximation of the PRID to extend the range of known refractive indices; this was done by averaging n_r at the extremes of the measured PSD fraction (last four bins on either side) and using these averaged values on the respective arms of the extension (Figure 2). Since the FC method does not offer any information on the imaginary part of the refractive indices, n_i values are unknown both in the available FC PSDs and in their extrapolations. Typical values for n_i were therefore adapted from literature (Figure 8 of Babin et al. [8]), for both organic and inorganic particles. These were then assigned to the particles according to the value of the real refractive index of each bin, both directly determined by the FC method and extrapolated.

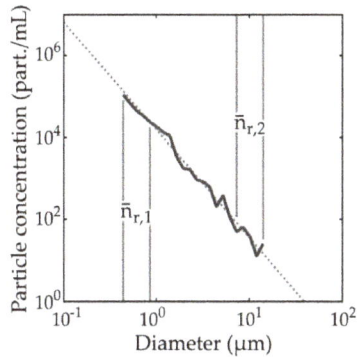

Figure 2. Power law best fit and real refractive index approximation in a typical natural particle population sample. Independent n_r values obtained by averaging the n_r of particles at the extremes of the PSD ($\bar{n}_{r,1}, \bar{n}_{r,2}$) were used to approximate the real refractive index within the respective ends of the PSD extrapolation (dotted line). The figure was adapted and modified with permission from Figure 3 of Agagliate et al. [2].

2.2.5. Suspended Particulate Matter and Inorganic/Organic Suspended Matter

Suspended particulate matter (SPM) was obtained from each sample following procedures detailed by Röttgers et al. [29]. The sample was filtered through filter pads under low vacuum (47-mm Whatman GF/F glass-fiber filters), then immediately placed in a petri dish after filtration and put to dry in a vacuum desiccator: SPM values were subsequently obtained by weighing the mass of dried sample and dividing it by the sample volume used. SPM values ranged between 45.519 and 0.113 g/m^3, with highest values found in the Bristol Channel and lowest values found in the North Sea, although variance in this latter area was large. Further separation into organic suspended matter and inorganic suspended matter fractions (OSM and ISM, also found in literature respectively as particulate organic matter and particulate inorganic matter, POM and PIM) was obtained by volatization of organics at 500 °C. ISM values ranged between 38.373 and 0.038 g/m^3, and followed a geographic pattern similar to that of SPM. OSM values ranged between 4.487 and 0.053 g/m^3, and were found to be more evenly distributed across the cruise track. The lowest value was found in the Irish Sea, while the highest was found once again in the Bristol Channel, due to the very large particle load of its turbid waters.

2.2.6. Particulate Organic Carbon

Particulate organic carbon (POC) concentrations were determined by catalytic combustion using a Vario TOC Cube instrument (Elementar, Langenselbold, Germany). Between 0.5 and 1 L of collected water for each sample were initially filtered onto 25-mm, combusted Whatman GF/F glass-fiber filters. The filters were then frozen and transported to the home laboratory. Once there, the filters were dried at 55 °C, packed into tin capsules and analyzed for their carbon content. Separation or removal of inorganic carbon was not considered, as concentrations of inorganic carbon (calcite or carbonates) are assumed low for these waters. Calibration of the POC/TOC analyzer was done regularly using sulphanilamide as the calibration standard. POC values were found to range between 2.307×10^3 and 6.783×10^1 mg/m^3, and followed a geographic pattern similar to that of OSM. Lowest values were found around Skye and the Hebrides, and highest values in the Bristol Channel.

2.2.7. Chlorophyll-a

Chlorophyll-a concentration (ChlA) was determined using high-performance liquid chromatography (HPLC) following Zapata et al. [30]: specifically, ChlA was determined from fluorescence values using excitation at 440 nm and emission at 650 nm and by comparison with standards of known chlorophyll concentration. ChlA values ranged between 7.620 and 0.096 mg/m³, with lowest values found in the Firth of Clyde and highest values found in the North Sea.

3. Results

3.1. PSDs and PRIDs

The PSDs retrieved by the FC method for the UKCW dataset were found to broadly follow power law distributions, with the main difference between stations being the overall concentration of the particle population (Figure 3a). Two obvious outliers are present, corresponding to samples from the turbid waters of the Bristol Channel; close inspection reveals structures that may be closer in nature to models such as the double gamma distribution proposed by Risović [31], and that indeed may be identified to a lesser degree in the other samples as well. Nonetheless, the power law approach remains a reasonable approximation for a large majority of the dataset, and was used for PSD extrapolations accordingly for all samples including the Bristol Channel ones.

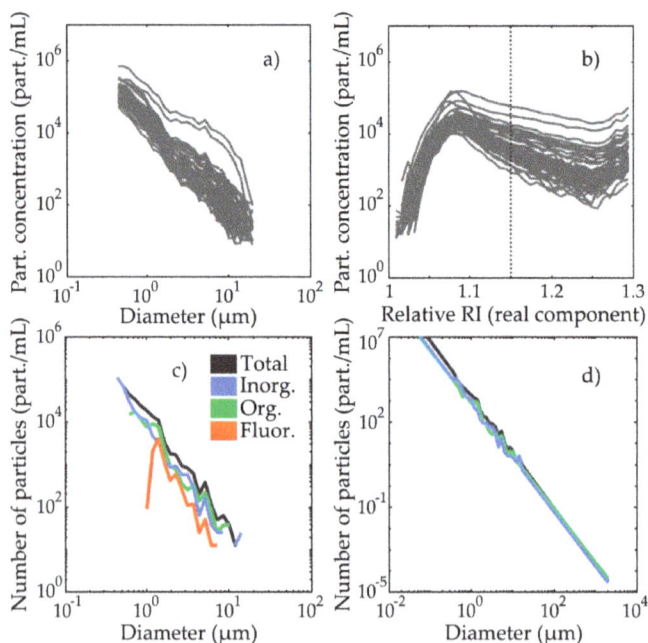

Figure 3. Collective view of (**a**) all 50 UKCW PSDs and (**b**) UKCW PRIDs produced by the FC method. Note that real refractive index values above 1.15 are not precise, but still indicate high refractive indices [1]. (**c**) Total, organic, inorganic, and fluorescent PSDs for a typical sample of the UKCW dataset and (**d**) power law extension of the total, organic and inorganic PSDs. Note that the extended organic and inorganic PSDs intersect the extended total PSD; therefore the sum of the extended organic and inorganic PSDs is not exactly equal to the extended total SPM. To evaluate the error thus introduced, SPM values are modelled both from the total PSD and by summing model ISM and OSM values. Panels (**a**,**b**) of the figure were adapted and modified with permission from Figure 2 of Agagliate et al. [2].

Particle refractive index distributions were found to be fairly homogeneous across all samples (Figure 3b), with distribution peaks found between 1.05–1.15 and within expectations for the real refractive index of the most common components of marine particle populations [9,32]. By exploiting the particle composition information given by the PRIDs, the total PSD of each UKCW sample was further separated into an inorganic PSD, an organic PSD, and a fluorescent fraction, which is itself a sub-fraction of the organic PSD (Figure 3c). These were used to calculate ISM, OSM & POC, and ChlA respectively, while SPM was calculated both from the total PSD and as the sum of the model ISM and OSM. To account for the fraction of the particle population undetected by the FC method, the total organic and inorganic PSDs were extended between 0.05 and 2000 μm as suggested by Davies et al. [23], following Equation (8) (Figure 3d); note that since the PSD extrapolations only approximate the particle populations outside the FC method detection range, the sum of the extended organic and inorganic PSDs generally is not exactly equal to the extended total PSD, making the dual calculation of SPM both from total PSD and as the sum of model ISM and OSM values a useful check of the error so introduced.

3.2. Particle Mass Modelling

Keeping with the assumption of particle sphericity used in the Mie-based models employed by the FC method, a first attempt of total particle mass calculation for the total, organic and inorganic fractions of the particle population was made as a simple bin-by-bin summation of spherical masses, i.e.,

$$m_{tot} = \sum_{D,nr} \rho_{nr} V_D N_{D,nr}, \tag{9}$$

where ρ_{nr} is the apparent density of a particle with real refractive index n_r as defined in Equation (2), V_D is the volume of a sphere of diameter D, D and n_r are the diameter and real refractive index corresponding to each bin, and $N_{D,nr}$ is the number concentration of particles within each bin. Given Equation (9) SPM, ISM, and OSM can then be obtained respectively by summing over the entire range of refractive indices or by limiting the summation to real refractive index values above or below the $n_r = 1.1$ threshold. The model SPM, ISM, and OSM values produced using this simple particle volume model however grossly overestimated the corresponding UKCW measurements, in certain cases by over two orders of magnitude (Figure 4).

A second calculation attempt was therefore carried out using a slightly modified version of the particle volume and total mass model employed in Zhang et al. [6]. This model is designed to account for the fractal nature of some marine particles, which can exist as aggregates of smaller units rather than as individuals exclusively, and takes the form

$$m_{tot} = \sum_{r,nr} \frac{4\pi}{3} \left(\frac{r}{r_0} \right)^{F(r)} r_0^3 \rho_{nr} N_{r,nr}, \tag{10}$$

where

$$F(r) = 3 \left(\frac{r}{r_0} \right)^{\beta}. \tag{11}$$

Here it is assumed that the flow cytometer has measured the radius r of an aggregate particle, which is itself constructed from primary particles of radius r_0. $F(r)$ is known as the fractal dimension of the aggregate. The value of r_0 and of exponent β are given as 0.5 μm and -0.0533 respectively [33], and for $r < r_0$ the value of F is fixed at 3. Crucially, the implementation of the model used here substitutes summation for the original integration to reflect the nature of the FC PSDs, extends the original range of 0.25–1000 μm to the 0.05–2000 μm range suggested by Davies et al. [23] and implements the effective radius r directly as half the particle diameter D determined by the FC method (rather than as the geometric formulation $4/3 \times V/A$ used by Zhang et al. [6], where A is the average projected area). This usage of Mie-derived parameters within a fractal model of mass highlights the dual nature of such modelling procedure: Mie theory is initially employed within the FC method to retrieve an optical size for the particles; the fractal model then reconciles this value with their physical size.

Over the extended size range the value of F was found to vary between 3 (its maximum possible value) and 1.998. The SPM, ISM, and OSM values produced using this fractal procedure were found to model the corresponding UKCW measurements much better than those produced using the simple spherical model, with RMS%E values 57.4%, 148.5%, and 83.1% for SPM, ISM, and OSM respectively (Figure 5). SPM values obtained as the sum of ISM and OSM were found to be close to those derived from the total PSD (RMS%E value 65.2%), indicating that the error introduced by the PSD extension is small. The two Bristol Channel samples, which deviated most obviously from the power law PSD model, produced clear outliers on all three accounts and were not included in the analysis. Median cumulative distributions of SPM, ISM, and OSM were also produced, showing that in a majority of samples 90% of the contribution to all three parameters is from particles between 0.2 and 200 µm (Figure 6).

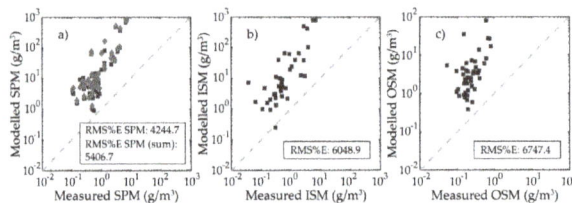

Figure 4. Comparison of modelled vs. measured (**a**) SPM, (**b**) ISM, and (**c**) OSM values for a simple spherical volume model. SPM values derived from the total PSD are represented as dark grey squares, while SPM values calculated as the sum of ISM and OSM are represented as light grey diamonds. The dashed grey lines indicate the 1:1 relationship.

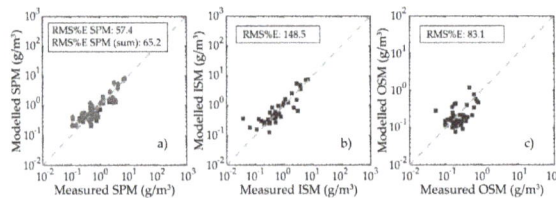

Figure 5. Comparison of modelled vs. measured (**a**) SPM, (**b**) ISM, and (**c**) OSM values for the fractal volume model. SPM values derived from the total PSD are represented as dark grey squares, while SPM values calculated as the sum of ISM and OSM are represented as light grey diamonds. The dashed grey lines indicate the 1:1 relationship.

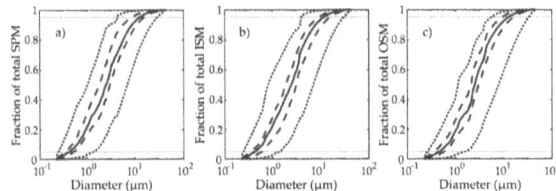

Figure 6. Cumulative distributions of modelled (**a**) SPM, (**b**) ISM, and (**c**) OSM values for the fractal volume model. The SPM curves refer to SPM values calculated from the total PSD. Solid, dashed, and dotted lines represent median, upper/lower quartiles, and maximum/minimum values respectively. The light grey horizontal lines mark the middle 90% of the contribution (i.e., from 5% to 95%) to the total value of each parameter.

3.3. Particulate Organic Carbon and Chlorophyll-a Concentration Modelling

For POC and ChlA modelling, the particle volume V was once again defined using a simple spherical model. The first to be applied were the cell volume scaling models (Table 2). Of the four models used to calculate POC values, only the diatom model given by Menden-Deuer and Lessard [12] produced results compatible with POC measurements (RMS%E: 92.9%, Figure 7a). This possibly reflects the taxonomical composition of the algal populations encountered during the He442 research cruise being mainly composed of diatom species typically associated with the spring bloom. The median cumulative distribution of POC for the diatom model shows an almost linear contribution from all size classes in a majority of samples, although results are shown to range widely from cases where the contribution is dominated by small particles to cases where, oppositely, the largest particles contributed the most (Figure 7b). This is likely the result of the interaction between the model parameters and the slope of the PSDs, and may also indicate that the parameters of the model work well for a majority, but not all of the samples. The two chlorophyll-a models both produced unsatisfactory results, with one data set underestimating ChlA, and the other over-estimating ChlA (Figure 8). The RMS%E values for the two sets were found to be 64.2% and 212.3% respectively. Although the overall quality of the match-up was low, the underlying structure of the data suggests that the form of the relationship might be useful subject to appropriate optimization.

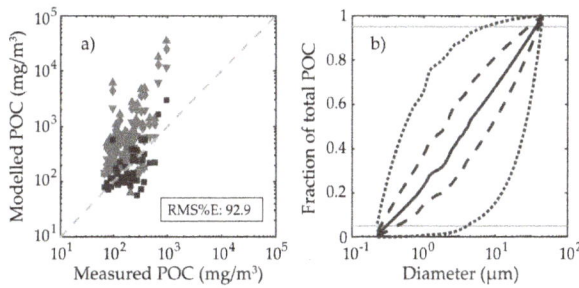

Figure 7. (a) Comparison of modelled vs. measured POC. POC values calculated using the diatom model (Menden-Deuer & Lessard, 2000) are represented by dark grey squares; the RMS%E value refers to these. POC values calculated using the other three models are represented by light grey diamonds and triangles; (b) Cumulative distribution of modelled POC for the diatom model. Solid, dashed, and dotted lines represent median, upper/lower quartiles, and max./min. values respectively. The light grey horizontal lines mark the middle 90% of the contribution (i.e., from 5% to 95%) to the total POC value. The dashed grey line indicates the 1:1 relationship.

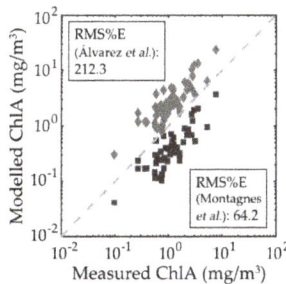

Figure 8. Comparison of modelled vs. measured ChlA. ChlA values calculated using the Montagnes et al. [11] model and the Álvarez et al. [13] model are represented by dark grey squares and light grey diamonds respectively. The dashed grey line indicates the 1:1 relationship.

When the refractive index-based C_i and Chl_i estimation models (Equations (4)–(7)) were applied to POC and ChlA modelling, the comparison between modelled and measured values of POC produced results (Figure 9a) which echo those found for SPM, ISM, and OSM when simple spherical particle volumes are used (see Figure 4). As was the case for particle mass values, the model POC values grossly overestimated the corresponding UKCW measurements, in certain cases by over two orders of magnitude. In contrast, the comparison between modelled and measured values of ChlA (Figure 9b) produced results which are similar to those found with the Montagnes et al. [11] cell volume scaling model (see Figure 8): both the Stramski and the DuRand models i.e., Equations (5) and (7) underestimated the measured ChlA values, producing RMS%E values of 63.3% and 76.5% respectively. Following the successful application of a fractal model of particle volume to the modelling of SPM, ISM, and OSM values, the same fractal model was applied to the refractive index-based C_i and Chl_i estimation as well by simple substitution of the ρ term of Equation (10) with either C_i or Chl_i. The POC values produced using the fractal procedure were found to model the corresponding UKCW measurements much better than those produced using the simple spherical model, with RMS%E values of 51.4% and 49.2% for the Stramski [19] and Durand et al. [20] models respectively (Figure 9c). However, the modelled values of ChlA were driven to further underestimate the measured values by the adoption of fractal volumes. RMS%E values for this new ChlA comparison were found to be 80.7% and 88.1% for the Stramski [19] and Durand et al. [20] models respectively (Figure 9d).

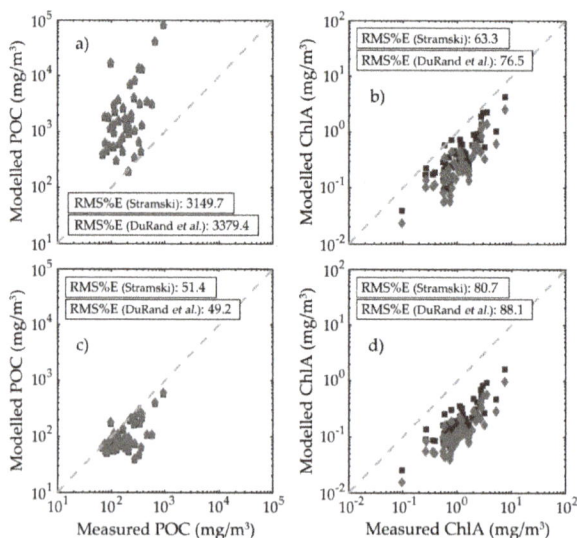

Figure 9. Comparison of (**a**) modelled vs. measured POC and (**b**) modelled vs. measured ChlA when a spherical model of particle volume is employed, and comparison of (**c**) modelled vs. measured POC and (**d**) modelled vs. measured ChlA when a fractal model of particle volume is employed instead. POC and ChlA values calculated using the Stramski [19] and Durand et al. [20] models are represented by dark grey squares and light grey diamonds respectively. The dashed grey lines indicate the 1:1 relationship.

3.4. Particulate Organic Carbon and Chlorophyll-a Concentration Modelling Optimization

A simple inversion of the procedure used to calculate POC and ChlA values allows for the empirical optimization of the parameter pairs used in the cell volume scaling models. Maintaining the general form

$$y(r) = kV(r)^h, \tag{12}$$

arrays of values for the parameters k and h can be generated, combined, and substituted in Equation (12), and the results compared and fitted against available measurements to identify the best parameter combinations, respectively (h_1, k_1) and (h_2, k_2) for POC and chlorophyll-a intraparticle concentrations. The optimization was initially applied to the whole UKCW dataset. Three parameter arrays were generated, one shared by exponents h_1 and h_2 plus one each for factors k_1 and k_2. The ranges were designed to encompass the parameter values of the models used thus far: specifically, 201 linearly spaced values for exponents h_1 and h_2 in a 0.6–1 range, 301 linearly spaced values for factor k_1 in a 0.05–0.65 range and 461 linearly spaced values for factor k_2 in a 0.004–0.05 range. Each (h_1, k_1) and (h_2, k_2) combination was then applied dataset-wide, compared against measured POC and ChlA and evaluated using the goodness-of-fit of a forced linear fit of the 1:1 line. The best parameter combinations were selected as those that minimized the RMSE values of the forced fit. The best POC model for the UKCW dataset was found as

$$y(r) = 0.442V(r)^{0.720} \tag{13}$$

while the best chlorophyll-a model was found as

$$y(r) = 0.029V(r)^{0.736}. \tag{14}$$

Comparisons of the optimized model results with the measured POC and ChlA values of the UKCW dataset are shown in Figure 10. RMS%E values for the comparisons were found to be 45.6% and 51.8% for POC and ChlA respectively.

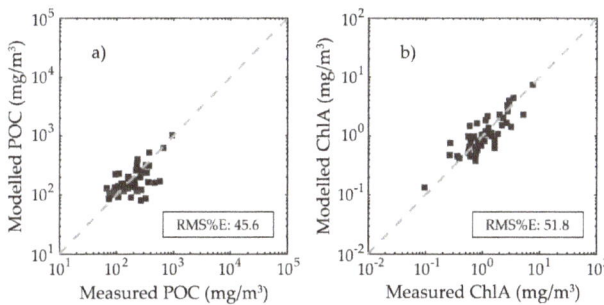

Figure 10. Comparison of (**a**) POC and (**b**) ChlA values as determined by the optimized models of Equations (13) and (14) vs. their respective measured values. The dashed grey lines indicate the 1:1 relationship.

An analogous optimization procedure can be applied to the refractive index-based C_i and Chl_i estimation models. Maintaining the use of fractal particle volumes and the general form

$$C_x = an_x - b \tag{15}$$

common to both models as described by Equations (4)–(7), where C_x is either C_i or Chl_i and n_x is respectively either n_r or n_i, arrays of values for the parameters a and b can be generated, combined and substituted in Equation (15), and the results compared and fitted against available measurements to identify the best parameter combinations. Interestingly, for both POC and ChlA the procedure could not identify single parameter combinations that minimized the RMSE; indeed, in both cases what was found was instead a continuum of parameter pairs which all minimized the RMSE to very similar values across the entire parameter ranges. We interpret this to be indicative of an excess of degrees of freedom in the relationship described by Equation (15) for the models to be properly constrained.

Returning to the physical basis of these relationships given in Stramski [19], the empirical linear relationship between C_i and n_r reflects a physical realization that increasing carbon content will generally increase real refractive index. Here we consider the implication of a natural boundary condition for this relationship: when $n_r = 1$ then $C_i = 0$. Including this constraint results in a slightly modified form of relationship

$$C_i = a_C(n_r(660\,\mathrm{nm}) - 1) \tag{16}$$

We note that this is equivalent to forcing the Stramski [19] regression (Figure 2 therein) through $(n_r, C_i) = (1, 0)$. Given the experimental uncertainties noted in Stramski [19] and the similarity of the a and b components of the original regressions, we believe that the form proposed by Equation (16) is broadly comparable but better reflects the physical relationship under investigation. Similarly, in the case of Chl_i and n_i Stramski [19] provides a clear physical basis for a simple linear relationship based on earlier work by Morel and Bricaud [34]. A small offset is found in the resulting best-fit regressions, which is either small and positive (see Equation (5) here, Stramski [19]), or small and negative (see Equation (7) here, Durand et al. [20]). The initial model suggests that $n_i(675\,\mathrm{nm})$ ought to be a simple linear function of Chl_i, but in practice the small offset could represent residual absorption by pigments other than ChlA (Stramski [19]), or it could be a statistical artefact associated with limitations in data quality (the discrepancy in sign between the two aforementioned studies is possibly significant). In either case, a slightly modified version of the n_i and Chl_i relationship can be given as

$$Chl_i = a_{chl}n_i(675\,\mathrm{nm}) + const. \tag{17}$$

However, it must be noted that the information contained in our data does not provide an indication of the appropriate value for the constant offset in Equation (17): indeed, the absence of a well-determined pair of best parameters when optimization is executed on both a and b terms of the linear model shows that minimal error for the dataset may be reached for any value of the offset. Therefore, the only likely way forward is to use the single well-defined scenario available i.e., the assumption of negligible residual absorption by pigments other than Chl_i. Under this assumption, the constant offset equals zero.

Once applied to Equations (16) and (17), the optimization procedure thus defined identified minimal RMSE values for relationships

$$C_i = 6880(n_r(660\,\mathrm{nm}) - 1) \tag{18}$$

$$Chl_i = 8320n_i(675\,\mathrm{nm}). \tag{19}$$

Comparisons of the optimized model results with the measured POC and ChlA values of the UKCW dataset are shown in Figure 11. RMS%E values for the comparisons were found to be 50.2% and 45.2% for POC and ChlA respectively.

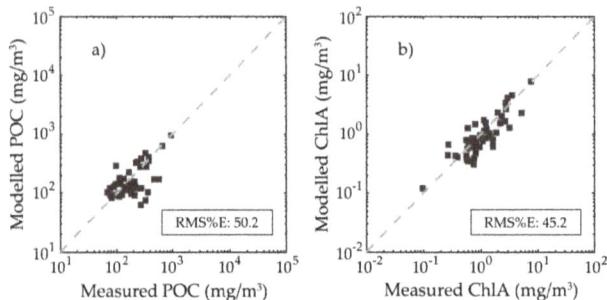

Figure 11. Comparison of (a) POC and (b) ChlA values as determined by the optimized models of Equations (17) and (18) vs. their respective measured values. The dashed grey lines indicate the 1:1 relationship.

4. Discussion

The modelling of suspended matter parameters for the samples of the UKCW dataset produced an interesting result: PSDs and PRIDs generated using a sphere-based, Mie-derived methodology produced SPM, ISM, and OSM values which compared poorly with their corresponding measured parameters when paired with an equally simple spherical volume model (Figure 4). However, the same PSDs and PRIDs produced comparable results when used directly as inputs for a fractal model of particle volume instead (Figure 5). This reflects two important aspects that should be considered carefully. The first is that the particle diameters found by the FC method are equivalent diameters rather than a direct measure of the physical dimension of the particles strictly [2]. The second is that natural particle populations are not solid spheres, so it is necessary to consider the impact of volume scaling and shape effects on apparent densities. The fractal model effectively rescales apparent particle densities across the size range. This is a feature of dealing with polydisperse natural particle populations that may not be encountered when dealing with effectively monodisperse algal cultures.

Application of volume scaling models (Figures 7a and 8) was moderately successful in determining POC and ChlA concentrations, although the variable rate of success between models mirrors the lack of consensus on a single model to accurately represent size–physiology relationships in marine phytoplankton. Indeed, a single set of model parameters is likely to be insufficient to adequately represent the metabolic complexity of all algal organisms [35]. This seems to substantiate the concerns expressed by Stramski [19] in noting the large variability between results obtained using different cell volume scaling models. Nevertheless, the further application of an empirical optimization procedure produced good match-ups between modeled and measured POC and ChlA values across the UKCW dataset (Figure 10).

Refractive index-based models of C_i and Chl_i were found to require a fractal model of particle volume to produce modelled values comparable with measured ones, particularly in the case of POC (Figure 9). This mirrors the results found for the particle density models for SPM, ISM, and OSM, and suggests that fractal models of volume are necessary to reconcile linear formulations of particle density and intraparticle carbon content with corresponding bulk measurements when natural particle populations with a wide size range and complex composition are involved. The matter is less clear in the case of ChlA, for which the refractive index-based models underestimated measured values in the case of spherical volumes, and even more so in the case of fractal volumes. Since the particle volume model must be consistent between POC and ChlA, we interpret this not as an issue with the fractal approach but rather as a sign that other elements within the overall procedure may be problematic. For example, it is important to remember that the imaginary refractive indices used in the Chl_i estimation represent an approximation based on a single n_i value derived from the literature. A simple optimization procedure shows that the modelled ChlA values of Figure 9b,d are reconciled with the corresponding measured values if $n_i(675 \text{ nm}) = 12.3 \times 10^{-3}$ and $n_i(675 \text{ nm}) = 11.1 \times 10^{-3}$ in the fractal volume case and if $n_i(675 \text{ nm}) = 4.2 \times 10^{-3}$ and $n_i(675 \text{ nm}) = 4.6 \times 10^{-3}$ in the spherical volume case (Stramski [19] and Durand et al. [20] models respectively in both cases). All four values are within the range shown in Table 1 of Stramski et al. [36], suggesting that a more dynamic range of n_i values may in fact improve the agreement between measurements and prediction. Furthermore, the fluorescent PSDs used in the ChlA calculations are not always easily reconciled with a Junge-like size distribution, and thus cannot be extended beyond the FC functional size range in a straightforward manner (Figure 3c). This can easily lead to unobserved particle fractions (especially for particle sizes above ~10 μm) and thus to underestimation of the actual total fluorescence.

This underestimation has a direct effect on the parameter values of the optimized refractive index-based models, which were found to be very large compared to those found in the literature (Equations (4)–(7), (18), and (19)). However, it should be noted that while the large value of the parameter a_{chl} can be explained by the issues mentioned above, the large value of the parameter a_C may instead be explained by the very different context of the model's application. The original C_i model was developed from the analysis of a small number of cultured phytoplankton species, which compared to the varied composition and extended size range of the He442 samples are essentially monotypic and probably log-normally distributed in their size. For both parameters, the good agreement obtained here (Figure 11) further confirms that fractal models of particle volume are appropriate and essential for successful closure between the results provided by linear formulations of biogeochemical parameters and their corresponding bulk measurements when natural mixed particle populations are concerned.

Given the results obtained for SPM, ISM, and OSM and for refractive index-derived values of POC and ChlA using a fractal model of particle volume, at a first glance the application of POC and ChlA volume scaling models to simple spherical volumes appears incongruous. However, close inspection of Equations (10)–(12) reveals that the POC and ChlA empirical volume scaling relationships hide proportionalities which are very close to those of the SPM, ISM, and OSM models. Specifically, from Equations (10) and (12) and for a single particle

$$m = r_0^{3-F(r)} \rho \left(\frac{4\pi}{3}\right)^{1-\frac{F(r)}{3}} \left(\frac{4}{3}\pi r^3\right)^{\frac{F(r)}{3}} =$$
$$= u(r) V^{\frac{F(r)}{3}} \propto y = kV^h. \tag{20}$$

Of particular interest then is the comparison between the exponents h and $F(r)/3$. Over the 0.05–2000 μm size range considered here, the value of $F(r)$ was found to vary between 3 and ~2, producing values of $F(r)/3$ between 1 and 2/3. These not only cover the range of values of exponent h in literature-derived and UKCW optimized volume scaling models for both POC and ChlA, but also echo general results found in literature for the volume scaling coefficients of chlorophyll and organic carbon. A physical interpretation of these results is that intracellular carbon is not uniformly distributed within cells, and this non-uniform distribution of intracellular carbon causes the observed non-linear relationship with cell volume. Based on the analysis above, we have found that the observed exponent in the cell volume model reflects the non-uniform distribution of carbon within the cell and can be used to predict the subsequent impact on the fractal dimensionality of the particle. As shown earlier in the paper, the fractal dimension of the particles controls the relationship between particle size distribution and observed mass concentrations (Figure 5).

Non-uniform distribution of intracellular carbon is also closely linked with cellular metabolism. Metabolic rate under optimal growth conditions is seen to scale with volume following a 3/4 exponent for a large number of organisms in what is known as the 3/4 rule or Kleiber's rule [37], and in phytoplankton this relationship is directly tied to the photosynthetic rate, and ultimately to the intracellular chlorophyll-a concentration. In general, phytoplankton cells regulate their pigment concentration in response to environmental irradiance changes [37]. Values for the volume scaling coefficient are then variably predicted to range between 3/4 and 2/3 for optimal growth and light limited conditions respectively [37], or to reach ~1 when nutrients are abundant [35]. Álvarez et al. [13] report values between 3/4 and 1, and Mei et al. [38] modelled cellular growth rate scaling exponents using chlorophyll-a intraparticle concentration scaling exponents ranging between 2/3 and 1. Similarly, empirical estimates of the scaling between cell volume and particulate carbon also vary. Some authors find carbon intraparticle concentration to decrease proportionally with increasing cell size i.e., $h < 1$ [10,12], while other find it to be isometric to cell size i.e., h ~1 [11].

Ultimately, the results presented in this study seem to suggest that, while cellular metabolic rate can be logically expected to play an important part in defining chlorophyll and organic carbon concentrations within organic particles, structural characteristics of the organic particles as described by fractal models can also offer a complementary interpretation for the proportionalities observed in nature and described in existing microbiology literature. This opens up interesting avenues for future research. For example, a simple attempt at repeating the optimization procedure for parameters k_1 and k_2 while substituting $F(r)/3$ for volume scaling coefficient h produced values $k_1 = 0.232$ for POC and $k_2 = 0.100$ for ChlA. The latter value is much larger than corresponding parameters found within the literature-derived ChlA volume scaling models employed in this work; furthermore, in both cases, modelled values of POC and ChlA resulted in a larger RMS%E when compared with their respective measured values from the UKCW dataset, respectively 61.2% and 93.6% for POC and ChlA. This may suggest that the application of fractal volumes to organic carbon and chlorophyll-a concentrations determination via volume scaling models will require a different set of values for the primary radius r_0 and/or exponent β compared to those used in the estimation of suspended matter concentrations. Coincidentally, this might also be further explanation for the underestimation found in modelled POC and especially ChlA values derived from refractive index based models of C_i and Chl_i when fractal volumes are applied. Most certainly, further in-depth research will be needed to answer these questions.

5. Conclusions

The results obtained by the FC method for the UKCW dataset were combined with models of particle density and of organic carbon and chlorophyll-a intraparticle concentrations to investigate the biogeochemical properties of the particle populations. The success of the resulting SPM, ISM, OSM and (after empirical optimization) POC and ChlA estimations lends further credibility to the PSD and PRID determination capabilities of the FC method, and further supports the usefulness of flow cytometry and of the FC method as a tool to complement other established techniques.

The fact that a fractal model of particle structure was key to ensure the quality of the SPM, ISM, OSM and refractive index-derived POC and ChlA match-ups suggests that the FC method observes particles as equivalent spheres, and has therefore the potential to be to some extent resilient both to particles which violate Mie-compatible aspect ratios and to the break-up of flocs and aggregates which derives from the flow cytometric measurement technique. Furthermore, this characteristic makes FC method results more readily comparable with those of other more common marine optics instruments, which for the most part observe bulk seawater IOPs and PSDs. It should be also noted that while the dataset was chiefly composed of surface water samples, no particular limitation exists a priori for the application of the FC method in its present form to samples from any water depth.

The volume scaling models used to calculate POC and ChlA were successfully applied to simple spherical particles instead. This may appear incongruous at first; however, these models too are revealed to hide proportionalities analogous to those caused by fractal structures, shining interesting new light on the volume scaling coefficients described in marine microbiology literature. We have shown here that the volume scaling models provide a route to better understand the impact of carbon distribution within particles and resulting fractal dimensionality, while the successful optimization of refractive index-based models reinforces previous findings that cell composition in the form of intracellular carbon concentration is well represented by corresponding changes in real refractive index. Finally, the size discrimination offered by FC also allows for a better understanding of the contribution of different size classes to the bulk biogeochemical properties, as it previously did for the IOPs of natural particle populations [2]. Taken in combination, these results represent a significant demonstration of the quality of size and refractive index information that is provided by FC data.

Author Contributions: J.A. and D.M. conceived and outlined the concept for this study; R.R. coordinated work during the HE442 research cruise; D.M. and R.R. supervised the retrieval of water samples; J.A. carried out flow cytometry operations; R.R. and K.H. carried out SPM, ISM, OSM, POC, and ChlA measurements on the water samples; J.A. analyzed the data; R.R. provided the Methods paragraphs relative to SPM, ISM, OSM, POC, and ChlA measurements; J.A. wrote the paper.

Funding: The HE442 cruise with RV Heincke was conducted under funding by the Alfred Wegener Institute Helmholtz Centre for Polar and Marine Research (AWI), grant AWI-HE442. This research was further funded by the Scottish Funding Council (SFC) (grant HR09011) via Marine Alliance for Science and Technology for Scotland (MASTS). This work was originally conceived thanks to work conducted under award of NERC grant NE/H021493/1 to McKee and co-investigators.

Acknowledgments: Agagliate and McKee gratefully acknowledge financial support from the MASTS pooling initiative. The authors wish to thank the captain and the crew of RV Heincke for their support and help during the HE442 research cruise. The authors also duly thank D. Stramski and two anonymous reviewers, who all helped improve this manuscript with their comments and suggestions.

Conflicts of Interest: The authors declare no conflict of interest.

References

1. Agagliate, J.; Röttgers, R.; Twardowski, M.S.; McKee, D. Evaluation of a flow cytometry method to determine size and real refractive index distributions in natural marine particle populations. *Appl. Opt.* **2018**, *57*, 1705–1716. [CrossRef] [PubMed]
2. Agagliate, J.; Lefering, I.; McKee, D. Forward modelling of inherent optical properties from flow cytometry estimates of particle size and refractive index. *Appl. Opt.* **2018**, *57*, 1777–1788. [CrossRef] [PubMed]
3. Ackleson, S.G.; Spinrad, R.W. Size and refractive index of individual marine particulates: A flow cytometric approach. *Appl. Opt.* **1988**, *27*, 1270–1277. [CrossRef] [PubMed]
4. Green, R.E.; Sosik, H.M.; Olson, R.J.; DuRand, M.D. Flow cytometric determination of size and complex refractive index for marine particles: Comparison with independent and bulk estimates. *Appl. Opt.* **2003**, *42*, 526–541. [CrossRef] [PubMed]
5. Green, R.E.; Sosik, H.M.; Olson, R.J. Contributions of phytoplankton and other particles to inherent optical properties in New England continental shelf waters. *Limnol. Oceanogr.* **2003**, *48*, 2377–2391. [CrossRef]
6. Zhang, X.; Stavn, R.H.; Falster, A.U.; Gray, D.; Gould, R.W., Jr. New insight into particulate mineral and organic matter in coastal ocean waters through optical inversion. *Estuar. Coast. Shelf Sci.* **2014**, *149*, 1–12. [CrossRef]
7. Morel, A.; Ahn, Y.-H. Optical efficiency factors of free-living marine bacteria: Influence of bacterioplankton upon the optical properties and particulate organic carbon in oceanic waters. *J. Mar. Res.* **1990**, *48*, 145–175. [CrossRef]
8. Babin, M.; Morel, A.; Fournier-Sicre, V.; Fell, F.; Stramski, D. Light scattering properties of marine particles in coastal and open ocean waters as related to the particle mass concentration. *Limnol. Oceanogr.* **2003**, *48*, 843–859. [CrossRef]
9. Aas, E. Refractive index of phytoplankton derived from its metabolite composition. *J. Plankton Res.* **1996**, *18*, 2223–2249. [CrossRef]
10. Verity, P.G.; Robertson, C.Y.; Tronzo, C.R.; Andrews, M.G.; Nelson, J.R.; Sieracki, M.E. Relationships between cell volume and the carbon and nitrogen content of marine photosynthetic nanoplankton. *Limnol. Oceanogr.* **1992**, *37*, 1434–1446. [CrossRef]
11. Montagnes, D.J.S.; Berges, J.A.; Harrison, P.J.; Taylor, F.J.R. Estimating carbon, nitrogen, protein, and chlorophyll *a* from volume in marine phytoplankton. *Limnol. Oceanogr.* **1994**, *39*, 1044–1060. [CrossRef]
12. Menden-Deuer, S.; Lessard, E.J. Carbon to volume relationships for dinoflagellates, diatoms, and other protist plankton. *Limnol. Oceanogr.* **2000**, *45*, 569–579. [CrossRef]
13. Álvarez, E.; Nogueira, E.; López-Urrutia, Á. In Vivo Single-Cell Fluorescence and Size Scaling of Phytoplankton Chlorophyll Content. *Appl. Environ. Microbiol.* **2017**, *83*, e03317-16. [CrossRef] [PubMed]
14. Stramski, D.; Morel, A. Optical properties of photosynthetic picoplankton in different physiological states as affected by growth irradiance. *Deep Sea Res. Part A Oceanogr. Res. Pap.* **1990**, *37*, 245–266. [CrossRef]
15. Stramski, D.; Reynolds, R.A. Diel variations in the optical properties of a marine diatom. *Limnol. Oceanogr.* **1993**, *38*, 1347–1364. [CrossRef]
16. Stramski, D.; Shalapyonok, A.; Reynolds, R.A. Optical characterization of the oceanic unicellular cyanobacterium *Synechococcus* grown under a day-night cycle in natural irradiance. *J. Geophys. Res.* **1995**, *100*, 13295–13307. [CrossRef]
17. Reynolds, R.A.; Stramski, D.; Kiefer, D.A. The effect of nitrogen limitation on the absorption and scattering properties of the marine diatom *Thalassiosira pseudonana*. *Limnol. Oceanogr.* **1997**, *42*, 881–892. [CrossRef]
18. Durand, M.D.; Olson, R.J. Diel patterns in optical properties of the chlorophyte *Nannochloris* sp.: Relating individual-cell to bulk measurements. *Limnol. Oceanogr.* **1998**, *43*, 1107–1118. [CrossRef]
19. Stramski, D. Refractive index of planktonic cells as a measure of cellular carbon and chlorophyll a content. *Deep Sea Res. Part I Oceanogr. Res. Pap.* **1999**, *46*, 335–351. [CrossRef]
20. Durand, M.D.; Green, R.E.; Sosik, H.M.; Olson, R.J. Diel Variations in Optical Properties of *Micromonas Pusilla* (Prasinophyceae). *J. Phycol.* **2002**, *38*, 1132–1142. [CrossRef]
21. Stramski, D.; Morel, A.; Bricaud, A. Modeling the light attenuation and scattering by spherical phytoplanktonic cells: A retrieval of the bulk refractive index. *Appl. Opt.* **1988**, *27*, 3954–3956. [CrossRef] [PubMed]
22. Reynolds, R.A.; Stramski, D.; Wright, V.M.; Woźniak, S.B. Measurements and characterization of particle size distributions in coastal waters. *J. Geophys. Res.* **2010**, *115*. [CrossRef]

23. Davies, E.J.; McKee, D.; Bowers, D.; Graham, G.W.; Nimmo-Smith, W.A.M. Optically significant particle sizes in seawater. *Appl. Opt.* **2014**, *53*, 1067. [CrossRef] [PubMed]

24. Bader, H. The hyperbolic distribution of particle sizes. *J. Geophys. Res.* **1970**, *75*, 2822–2830. [CrossRef]

25. Sheldon, R.W.; Prakash, A.; Sutcliffe, W.H., Jr. The size distribution of particles in the ocean. *Limnol. Oceanogr.* **1972**, *17*, 327–340. [CrossRef]

26. Junge, C.E. *Air Chemistry and Radioactivity*; Academic Press: New York, NY, USA, 1963; ISBN-13 9780123921505.

27. Stramski, D.; Kiefer, D.A. Light scattering by microorganisms in the open ocean. *Prog. Oceanogr.* **1991**, *28*, 343–383. [CrossRef]

28. Ulloa, O.; Sathyendranath, S.; Platt, T. Effect of the particle-size distribution on the backscattering ratio in seawater. *Appl. Opt.* **1994**, *33*, 7070. [CrossRef] [PubMed]

29. Röttgers, R.; Heymann, K.; Krasemann, H. Suspended matter concentrations in coastal waters: Methodological improvements to quantify individual measurement uncertainty. *Estuar. Coast. Shelf Sci.* **2014**, *151*, 148–155. [CrossRef]

30. Zapata, M.; Rodríguez, F.; Garrido, J.L. Separation of chlorophylls and carotenoids from marine phytoplankton: A new HPLC method using a reversed phase C8 column and pyridine-containing mobile phases. *Mar. Ecol. Prog. Ser.* **2000**, *195*, 29–45. [CrossRef]

31. Risović, D. Two-component model of sea particle size distribution. *Deep Sea Res. Part I Oceanogr. Res. Pap.* **1993**, *40*, 1459–1473. [CrossRef]

32. Twardowski, M.S.; Boss, E.; Macdonald, J.B.; Pegau, W.S.; Barnard, A.H.; Zaneveld, J.R.V. A model for estimating bulk refractive index from the optical backscattering ratio and the implications for understanding particle composition in case I and case II waters. *J. Geophys. Res.* **2001**, *106*, 14129–14142. [CrossRef]

33. Khelifa, A.; Hill, P.S. Models for effective density and settling velocity of flocs. *J. Hydraul. Res.* **2006**, *44*, 390–401. [CrossRef]

34. Morel, A.; Bricaud, A. Theoretical results concerning light absorption in a discrete medium, and application to specific absorption by phytoplankton. *Deep Sea Res. Part A Oceanogr. Res. Pap.* **1981**, *28*, 1375–1393. [CrossRef]

35. Marañón, E.; Cermeño, P.; Rodríguez, J.; Zubkov, M.V.; Harris, R.P. Scaling of phytoplankton photosynthesis and cell size in the ocean. *Limnol. Oceanogr.* **2007**, *52*, 2190–2198. [CrossRef]

36. Stramski, D.; Bricaud, A.; Morel, A. Modeling the inherent optical properties of the ocean based on the detailed composition of the planktonic community. *Appl. Opt.* **2001**, *40*, 2929–2945. [CrossRef] [PubMed]

37. Finkel, Z.V.; Irwin, A.J.; Schofield, O. Resource limitation alters the 3/4 size scaling of metabolic rates in phytoplankton. *Mar. Ecol. Prog. Ser.* **2004**, *273*, 269–279. [CrossRef]

38. Mei, Z.P.; Finkel, Z.V.; Irwin, A.J. Light and nutrient availability affect the size-scaling of growth in phytoplankton. *J. Theor. Biol.* **2009**, *259*, 582–588. [CrossRef] [PubMed]

applied sciences

MDPI

Article

Characterization of the Light Field and Apparent Optical Properties in the Ocean Euphotic Layer Based on Hyperspectral Measurements of Irradiance Quartet

Linhai Li [1], Dariusz Stramski [1,*] and Mirosław Darecki [2]

[1] Scripps Institution of Oceanography, University of California San Diego, La Jolla, CA 92093-0238, USA; lil032@ucsd.edu
[2] Institute of Oceanology, Polish Academy of Sciences, 81-712 Sopot, Poland; darecki@iopan.gda.pl
* Correspondence: dstramski@ucsd.edu; Tel.: +1-858-534-3353

Received: 23 February 2018; Accepted: 1 June 2018; Published: 19 December 2018

Abstract: Although the light fields and apparent optical properties (AOPs) within the ocean euphotic layer have been studied for many decades through extensive measurements and theoretical modeling, there is virtually a lack of simultaneous high spectral resolution measurements of plane and scalar downwelling and upwelling irradiances (the so-called irradiance quartet). We describe a unique dataset of hyperspectral irradiance quartet, which was acquired under a broad range of environmental conditions within the water column from the near-surface depths to about 80 m in the Gulf of California. This dataset enabled the characterization of a comprehensive suite of AOPs for realistic non-uniform vertical distributions of seawater inherent optical properties (IOPs) and chlorophyll-*a* concentration (*Chl*) in the common presence of inelastic radiative processes within the water column, in particular Raman scattering by water molecules and chlorophyll-*a* fluorescence. In the blue and green spectral regions, the vertical patterns of AOPs are driven primarily by IOPs of seawater with weak or no discernible effects of inelastic processes. In the red, the light field and AOPs are strongly affected or totally dominated by inelastic processes of Raman scattering by water molecules, and additionally by chlorophyll-*a* fluorescence within the fluorescence emission band. The strongest effects occur in the chlorophyll-*a* fluorescence band within the chlorophyll-*a* maximum layer, where the average cosines of the light field approach the values of uniform light field, irradiance reflectance is exceptionally high approaching 1, and the diffuse attenuation coefficients for various irradiances are exceptionally low, including the negative values for the attenuation of upwelling plane and scalar irradiances. We established the empirical relationships describing the vertical patterns of some AOPs in the red spectral region as well as the relationships between some AOPs which can be useful in common experimental situations when only the downwelling plane irradiance measurements are available. We also demonstrated the applicability of irradiance quartet data in conjunction with Gershun's equation for estimating the absorption coefficient of seawater in the blue-green spectral region, in which the effects of inelastic processes are weak or negligible.

Keywords: oceanic light field; irradiance quartet; apparent optical properties; inelastic processes; Gershun equation; ocean euphotic zone

---

1. Introduction

The ocean epipelagic zone that extends from the surface to approximately 200 m depth is extremely important for ocean-atmosphere interactions with implications to climate and supporting life on Earth. Most solar radiation incident on the ocean is absorbed within this layer [1,2]. In the euphotic layer that overlaps with the upper portion or the entire epipelagic zone in very clear ocean waters, there is enough light to support the process of photosynthesis, which contributes nearly half of the world's total biological primary production [3]. The studies of light propagation and light field characteristics are crucial for understanding many physical and biological processes in the upper ocean, which are driven by or depend on solar radiation. For example, for studying the heating rate, the radiometric quantities of spectral downward plane irradiance, $E_d(z, \lambda)$, and the spectral upward plane irradiance, $E_u(z, \lambda)$, are essential [4]. The symbols z (units of m) and λ (units of nm) stand for depth in the ocean and light wavelength in vacuum, respectively. In photosynthesis studies the key quantity of photosynthetically available radiation (PAR) is best represented on the basis of radiometric quantity of spectral scalar irradiance, $E_o(z, \lambda)$, and its integration over the spectral range of 350–700 nm or 400–700 nm with conversion from energy units to quantum units [5–9]. This measure of PAR is hereafter referred to as the PAR quantum scalar irradiance, E_{oPAR}. In the past, however, the PAR estimates in the ocean have been often based on the measurements of downward plane irradiance [10], so this measure of PAR will be referred to as PAR quantum downward irradiance, E_{dPAR}. The PAR estimate in terms of E_{oPAR} is superior to E_{dPAR}, mainly because it does not ignore the contribution of upwelling light to photosynthetically available radiation and the measurement of E_o with a spherical collector gives equal weight to quanta arriving at a point from all possible directions, in contrast to the cosine weighting in the measurement of E_d with a plane collector. For simplicity, the explicit dependence of optical quantities on z and λ is omitted hereafter unless causing ambiguity (symbols and definitions are summarized in Table 1, see also [9] for terminology and definitions used in hydrologic optics).

Table 1. Symbols of basic variables used in this study.

Symbol	Description	Units
λ	Light wavelength in vacuum	nm
z	Depth in water	m
θ_s	Solar zenith angle	degree
a	Absorption coefficient	m^{-1}
b	Scattering coefficient	m^{-1}
c	Beam attenuation coefficient (sum of a and b)	m^{-1}
L_u	Spectral upwelling radiance at zenith direction	$W\ m^{-2}\ sr^{-1}\ nm^{-1}$
E_d, E_u	Spectral downwelling and upwelling plane irradiances	$W\ m^{-2}\ nm^{-1}$
E_o, E_{od}, E_{ou}	Spectral total, downwelling, and upwelling scalar irradiances	$W\ m^{-2}\ nm^{-1}$
K_x	Diffuse attenuation coefficients for irradiance or radiance x	m^{-1}
$\bar{\mu}, \bar{\mu}_d, \bar{\mu}_u$	Average cosines of total, down- and upwelling light fields	dimensionless
R	Irradiance reflectance	dimensionless
Chl	Chlorophyll-a concentration	$mg\ m^{-3}$
Subscripts		
w	Water	
p	Suspended particulate matter	
g	Colored dissolved organic matter (CDOM)	

Radiometric measurements in the ocean started in the early 1930s and a summary of pioneering work can be found in [11]. Example early studies include measurements of spectral irradiance, e.g., [12,13] and the angular distribution of radiance, e.g., [14–16], and determinations of some apparent optical properties (AOPs) of the ocean from light measurements, e.g., [12,17,18]. Early reports on radiometric and other optical measurements in the ocean can be also found in the Russian literature, e.g., [19].

More recent studies have considerably expanded these early measurements by using advanced instrumentation and various deployment approaches, e.g., [20–32]. Much research interest in recent decades has been concentrated on ocean reflectance in relation to applications of ocean color remote sensing. For these applications, many radiometric measurements have been collected for spectral upwelling (zenith) radiance, L_u, and E_d, and optionally also for E_u [32–34]. Note that the zenith radiance indicates that the measured light travels toward zenith. The analysis of radiometric measurements driven by the interest in ocean color has been typically focused on near-surface layer, approximately the top 10–20 m. With regard to AOPs, most experimental efforts have been placed on the determinations of (i) the spectral diffuse attenuation coefficients K_d, K_u, and K_{Lu} for E_d, E_u, and L_u, respectively; (ii) the spectral irradiance reflectance, $R = E_u/E_d$; and (iii) the spectral remote-sensing reflectance just above the sea surface, $R_{rs}(z = 0^+) = L_w(z = 0^+)/E_d(z = 0^+)$ where L_w is the spectral water-leaving radiance and $z = 0^+$ is just above the sea surface, e.g., [32,34–43]. Note also that the measurements of both E_d and E_u allow for determination of the net irradiance, $E = E_d - E_u$, and the diffuse attenuation coefficient for net irradiance, K_E.

In contrast to the plane irradiances E_d and E_u, the measurements of spectral scalar irradiance, E_o, including its downward, E_{od}, and upward, E_{ou}, components have seldom been conducted. In particular, simultaneous measurements of the irradiance quartet that includes E_d, E_u, E_{od}, and E_{ou} have been very rare [44] and we are unaware of explicit presentation of experimental data of the complete irradiance quartet in the literature. This greatly limits the availability of experimental data of the average cosines that provide a simple way of specifying the angular structure of the light field. The average cosines include: (i) the average cosine for the downwelling light field, $\bar{\mu}_d = E_d/E_{od}$; (ii) the average cosine for the upwelling light field, $\bar{\mu}_u = E_u/E_{ou}$; and (iii) the average cosine for the entire light field, $\bar{\mu} = (E_d - E_u)/(E_{od} + E_{ou}) \equiv E/E_o$ [8,9]. We note that according to the original concept and definition of AOPs introduced by Preisendorfer [45,46], the average cosines (and the reciprocal of the average cosines referred to as the distribution functions) can be considered AOPs. Preisendorfer [46] provides the following definition of AOPs: "The apparent optical properties of a natural hydrosol are those radiometrically determined scattering- and absorbing-induced quantities which generally depend on the geometrical structure of the light field (i.e., whether the light field is more or less collimated or diffuse) but which have enough regular features and enough stability to be entitled to the appellation optical property". In some studies, however, for example in the work of Kirk [10,47], the average cosines are considered just as simple parameters that specify the angular structure of the light field, and not AOPs. In this paper, we choose to refer to the average cosines as AOPs, which is consistent with the traditional view [9,46,48] and simplifies the narrative structure of the presentation of our results for the three classes of parameters (i.e., K-coefficients, reflectances, and average cosines) which are derived from radiometric quantities. The issue of whether or not we refer to the average cosines as AOPs has no impact on the presented results.

The scarcity of simultaneous measurements of E and E_o has also greatly limited the potential use of the so-called Gershun equation [49] for estimating the absorption coefficient of seawater, a, from radiometric measurements or AOPs. This equation can be derived directly from the 1-D scalar (i.e., depth dependence only and no consideration of polarization of light) radiative transfer equation (RTE) for radiance, and takes a simple form $a = K_E \cdot \bar{\mu}$ when the inelastic radiative processes and internal sources are ignored [9]. The inelastic processes in the ocean include Raman scattering by water molecules, fluorescence by phytoplankton pigments, especially chlorophyll-a, and fluorescence by colored dissolved organic matter (CDOM). Few experimental studies have incorporated measurements of both spectral plane and scalar irradiances [18,44] or derived the irradiance quartet from radiance measurements limited to a single or a few wavebands [20,29,31,50] to examine the application of Gershun's equation. Højerslev [51] demonstrated that the irradiances involved in Gershun's equation, E and E_o, can be measured with two irradiance sensors equipped with spherical collectors oriented in upward and downward looking directions and masked so that they each collect light over a hemispherical portion of the collector. This latter approach was also used by Spitzer and Wernand [52] to determine the absorption coefficient of seawater from Gershun's equation.

In principle, theoretical simulations of radiative transfer (RT) in the ocean enable the analysis of essentially all radiometric quantities of underwater light field and AOPs for various pre-defined scenarios of input data of inherent optical properties (IOPs) of seawater and boundary conditions at the sea surface and ocean bottom. There is a large body of literature in this area with a great majority of studies focusing on the so-called inverse problem or developing methods for estimating

IOPs from radiometric measurements of natural light field or AOPs in a water body (see [53] for a review of this topic). In general, the limitations of such RT models are set by the assumptions used in their development, which include the assumptions regarding the vertical structure of IOPs and inelastic processes. The RT simulations have usually assumed a homogeneous water column and ignored all or some inelastic processes [47,54–60]. If included, the vertical stratification normally requires some idealized parameterization of the depth profile of IOPs, e.g., [60–64]. The models based on such simplifying assumptions can be satisfactory within certain limitations for the purposes of their development, for example the retrieval of average IOPs in the near-surface ocean layer. Notwithstanding the simplifying assumptions about the ocean environment, the RT modeling has provided insights into the complexity of the effects of the vertical inhomogeneity of the water column [60,62,63,65,66] and inelastic processes [60,67–75] on in-water and water-leaving light fields as well as AOPs. However, a comprehensive theoretical analysis of all (or nearly all) radiometric quantities (including plane and scalar irradiances) as well as the three AOP classes (K-coefficients, reflectances, and average cosines) throughout the water column for a variety of complex environmental scenarios is not easily tractable, especially for various naturally-occurring vertical profiles of IOPs within the upper ocean. Such analysis has usually remained beyond the scope of RT modeling; rare examples are found in Morel and Gentili [60], which focused on the well-lit upper ocean layer, and Li et al. [64], which focused on the dimly-lit mesopelagic zone ($z > 200$ m). Morel and Gentili [60] demonstrated, for example, the intricate patterns of the influences of many factors, including the sun position, seawater IOPs (as driven primarily by chlorophyll-a concentration, Chl, and co-varying water constituents), and Raman scattering, on the various AOPs within a hypothetically homogeneous ocean.

In spite of decades of experimental and theoretical studies of optical radiometry of the ocean, there exist some gaps or interesting questions, which deserve further attention. This study has been motivated by a few such questions, in particular (i) how the spectral and vertical patterns of light field characteristics and AOPs are affected by the interplay of inelastic processes (i.e., Raman scattering and chlorophyll-a fluorescence) and the actual non-uniform vertical distributions of Chl (and hence the vertically varying IOPs) within the euphotic layer; (ii) what relationships can be established between different AOPs using the depth-resolved data within the euphotic layer; and (iii) what is the quality and applicability of estimation of seawater absorption coefficient from measured AOPs and Gershun's equation under real conditions of non-uniform vertical distributions of Chl and IOPs, and inelastic processes within the euphotic layer. To address these questions, we collected simultaneous hyperspectral measurements of the irradiance quartet (E_d, E_u, E_{od}, and E_{ou}) and the zenith radiance L_u from the surface to depths of 60–80 m in the Gulf of California. This dataset is unique because it allows conducting a case study of a comprehensive set of radiometric quantities and all essential AOPs for a broad range of specific optical and biogeochemical conditions within the water column as described by the actual vertical profiles of Chl and IOPs.

2. Methods

2.1. Study Area

The field data were collected in the Gulf of California onboard the R/V New Horizon in June 2010 and June–July 2011 (Figure 1). A total of 11 stations were investigated in 2010 and 14 stations in 2011. These stations were located within three regions of the Gulf of California, Guaymas Basin, Carmen Basin, and Farallon Basin, which differ significantly in terms of water optical properties. The Guaymas Basin is the central part of the Gulf of California, Farallon Basin is the southernmost part of the Gulf of California, and Carmen Basin is located between the Guaymas and Farallon Basins. The Gulf of California has unique oceanographic and ecosystem characteristics [76,77]. According to Jerlov's optical classification of natural water bodies based on near-surface values of $K_d(\lambda)$ [11], the waters in the Guaymas Basin range typically between the oceanic Type IB and Type II, the Carmen Basin waters between the oceanic Type I and Type IA, and the Farallon Basin waters between Type II and coastal Type 1.

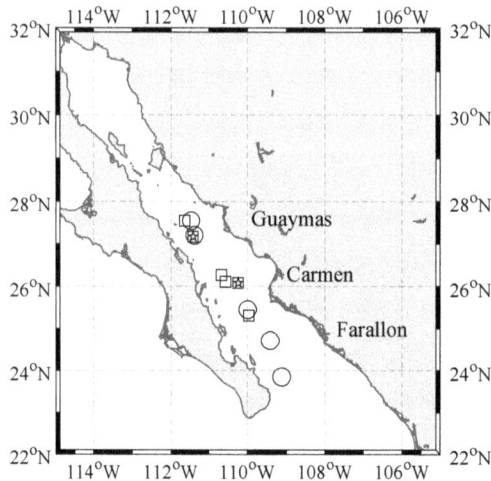

Figure 1. Location of stations in the Gulf of California where underwater radiometric and ancillary measurements were made. Circles indicate the station sites in 2010 and squares in 2011. Optical measurements were made at four, three, and four stations in the Guaymas, Carmen, and Farallon Basins, respectively. Stars indicate two stations where the absorption and beam attenuation measurements with an ac-9 instrument were made in 2011.

2.2. CTD Measurements

At each station, a conductivity-temperature-depth (CTD) package equipped with Niskin bottles was deployed to determine the depth profiles of seawater physical parameters (temperature T, salinity S, and density anomaly σ), chlorophyll-*a* fluorescence, and non-water optical beam attenuation coefficient at 660 nm, $c_{nw}(z, 660)$ (i.e., the total beam attenuation of seawater after subtraction of pure seawater contribution). Water samples were also collected at discrete depths with the CTD-Rosette for subsequent determinations of *Chl* in the laboratory.

The CTD package was deployed from water surface to ~500 m depth. The acquired raw data were processed using a manufacturer's software, SBEDataProcessing v7.23.2 (Sea-Bird Electronics Inc., Bellevue, WA, USA), and converted to values in physical units with calibration files. The processing software automatically aligned the measurements of each sensor to the common depth vector. The calibrated data were subsequently split into down- and up-casts. Preliminary inspection of the data showed that the up-casts of CTD measurements often exhibited less noise and thus were chosen for further processing. Depth profiles of each parameter were despiked using a running median method and smoothed using a Savitzky-Golay filter in Matlab R2015b (MathWorks, Natick, MA, USA). The profiles were then binned to 0.5 m depth resolution to yield the final data of T, S, σ, chlorophyll-*a* fluorescence, and $c_{nw}(660)$. The $c_{nw}(660)$ was considered to be equivalent to the particulate attenuation coefficient at 660 nm, $c_p(660)$, because the absorption of dissolved material at 660 nm was assumed negligible.

2.3. Chlorophyll-a Concentration

The chlorophyll-*a* concentration, *Chl*, was determined for discrete water samples obtained from the Niskin bottles deployed during the CTD-Rosette casts. The particulate material was collected by filtration onto glass-fiber filters (GF/F) and stored in liquid nitrogen for post-cruise analysis. The samples collected in 2010 were analyzed with High-Performance Liquid Chromatography (HPLC) using the method described in [78]. The total *Chl*, which represents the summed contributions of monovinyl chlorophyll-*a*, divinyl chlorophyll-*a*, chlorophyllide-*a*, and the allomeric and epimeric forms of chlorophyll-*a*, was determined. In 2010, *Chl* was usually determined for water samples collected at two depths, 1–3 m below the sea surface and around the depth of chlorophyll-*a* fluorescence maximum, z_{Chlmax}. In 2011, water samples were also collected below z_{Chlmax}, resulting in three water samples at each station. These samples were analyzed spectrophotometrically to determine *Chl*. The absorbance spectra of acetone extracts of pigments were measured inside an integrating sphere of a spectrophotometer (Perkin-Elmer Lambda 18 equipped with a 15-cm sphere) from 290 to 860 nm with 1-nm interval. The *Chl* concentrations were computed from absorbance values at 630, 647, 665, and 691 nm using the equation of Ritchie [79].

Based on *Chl* determinations at discrete depths, depth profiles of *Chl* were derived by calibrating the chlorophyll-*a* fluorescence profiles from the CTD cast. The values of fluorescence at depths of *Chl* determination were used to establish a relationship between the fluorescence signal and *Chl*. The squared correlation coefficient r^2 of this relationship was 0.942 and 0.938 for the data collected in 2010 and 2011, respectively. With this relationship, the depth profiles of *Chl* were constructed using the depth profiles of measured fluorescence at all stations (see Figure 2 for an example). We note that the analysis and discussion of our results do not, however, rely strongly on exact values of *Chl*, but rather on relative changes in *Chl*.

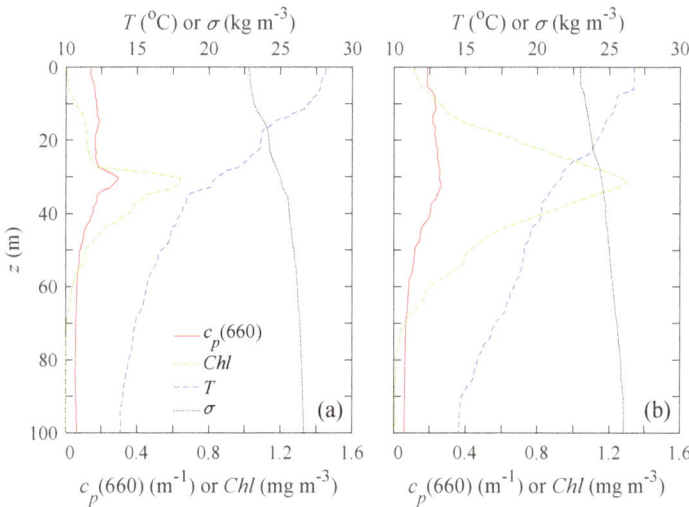

Figure 2. Vertical profiles of water temperature *T*, density anomaly σ, chlorophyll-*a* concentration *Chl*, and beam attenuation coefficient of particles at 660 nm $c_p(660)$ measured at two contrasting stations. (**a**) Station in the Guaymas Basin (27.54° N, 111.64° W) with lower *Chl* and c_p and smaller solar zenith angle of ~5° and (**b**) station in the Farallon Basin (25.28° N, 109.58° W) with higher *Chl* and c_p and larger solar zenith angle of ~57°.

2.4. Radiometric Quantities and Apparent Optical Properties

Underwater radiometric measurements, including the irradiance quartet, E_d, E_u, E_{od}, and E_{ou} (for all stations) and upwelling zenith radiance L_u (for a few stations in 2011), were collected using an instrument package equipped with five RAMSES sensors (TriOS Mess- und Datentechnik GmbH, Rastede, Germany). Concurrent in-air downwelling plane irradiance, $E_s(\lambda)$, was also measured with an additional sensor mounted on board the ship. These sensors typically acquired signal over the approximate spectral range 320–950 nm (with slight difference among sensors) with ~3.3 nm spectral interval across the spectrum. The radiance sensor had a 20° field of view in air. All five in-water sensors were mounted on a common frame and aligned to the exact same depth level (Figure 3). The in-water package was lowered into water from the sun-exposed side of the vessel at a distance of ~15 m from the vessel to minimize ship-induced light field perturbations [80,81]. Vertical profile measurements were taken at several discrete depths from a few meters below the surface to depths of 50 to 80 m depending on the station. The package was stopped and held every 3–10 m to acquire 5–10 spectral scans at each measurement depth. The integration time of each sensor was automatically adjusted to the light intensity at each measurement depth to ensure an adequate signal to noise ratio. The depth resolution of discrete measurements was highest near the surface and decreased to 10 m below the 20 m depth. Generally, at least one measurement was made near the depth of the *Chl* maximum as indicated by the chlorophyll-*a* fluorescence profile from the CTD cast.

Figure 3. The instrument package used to measure the irradiance quartet and upwelling radiance. As shown, the five radiometric sensors are aligned to the same depth level.

The raw radiometric data acquired in the field were converted to physical units of spectral irradiance (W m^{-2} nm^{-1}) and spectral radiance (W m^{-2} sr^{-1} nm^{-1}). For the underwater measurements, the correction for immersion effect was made for each sensor with the coefficients provided by the manufacturer, which were verified and revised in the laboratory experiment [82]. All measurements were matched to each other by their time stamp and assigned depth. Each spectrum was filtered using the Savitzky-Golay filter in Matlab R2015b to minimize noise, and the linear interpolation was then applied to obtain the final spectral data with a 1 nm spectral interval. The quality of radiometric measurements was best when the irradiance magnitude was higher than the detection limit of 10^{-6} W m^{-2} nm^{-1}. The measurements of L_u below ~30–40 m were often not possible because of insufficient sensitivity of the sensor. All measurements with a signal below the detection limit were excluded from the analysis. As a result, at some stations the greatest depth of measurements of upwelling irradiance in the red spectral region was somewhat smaller (by ~10–20 m) compared with measurements of upwelling irradiance in the blue-green spectral region or measurements of downwelling irradiances. The underwater irradiance and radiance data that passed quality criteria were adjusted to eliminate the effects of temporal changes in sky conditions during vertical profiling with the underwater system. This was accomplished by appropriate normalization of underwater data acquired at different depths using simultaneously acquired data of E_s.

The data of irradiance quartet were used to derive a comprehensive set of AOPs at each discrete depth, including, $\bar{\mu}_d$, $\bar{\mu}_u$, $\bar{\mu}$, R (all dimensionless), and various diffuse attenuation coefficients, collectively denoted with a symbol K (units of m^{-1}). The K coefficients for various irradiances were derived from

$$K_x(z_1 + \Delta z, \lambda) = -\frac{\ln E_x(z_2, \lambda) - \ln E_x(z_1, \lambda)}{z_2 - z_1}, \tag{1}$$

where x represents d, u, od, and ou, z_1 is a depth corresponding to the upper boundary of the layer over which K_x is determined, z_2 is the lower boundary of this layer, and $\Delta z = (z_2 - z_1)/2$. Similarly, other K_x coefficients were also calculated, specifically the diffuse attenuation coefficient of net irradiance, K_E, by replacing E_x with net irradiance, $E = E_d - E_u$, and the diffuse attenuation coefficient of upwelling zenith radiance, K_{Lu}, by replacing E_x with L_u. PAR quantum scalar irradiance, E_{oPAR} (in units of µmol quanta m^{-2} s^{-1}), was calculated by the conversion of spectral E_o from energy units to quantum units and integration between 400 and 700 nm. Similarly, PAR quantum downward irradiance, E_{dPAR}, was calculated from spectral E_d. The corresponding diffuse attenuation coefficients, K_{oPAR} and K_{dPAR}, were calculated by replacing E_x with E_{oPAR} and E_{dPAR}, respectively, in Equation (1).

The spectra of all AOPs were inspected to ensure the quality of data. Because at each depth multiple radiometric measurements were taken, AOPs were initially determined from each radiometric measurement. The unrealistic AOPs, for example $R > 1$ or average cosines > 1, or obvious outliers were then discarded. The abnormal data were often observed near the surface, most likely owing to the effects of wave-induced light fluctuations in E_d and E_{od} [83,84]. After this quality control, the accepted spectra of radiometric quantities were averaged to generate final data of irradiances and radiance for each depth of measurement. The average spectra of irradiances and radiance were then used to compute the final AOPs. Such final data from 11 stations (five in 2010 and six in 2011; see Figure 1) that included valid measurements at more than four depths were selected for further analysis in this study.

2.5. Absorption and Beam Attenuation Coefficients

The non-water absorption coefficient, $a_{nw}(z, \lambda) = a(z, \lambda) - a_w(\lambda)$, and non-water beam attenuation coefficient, $c_{nw}(z, \lambda) = c(z, \lambda) - c_w(\lambda)$, were measured at five stations in 2011 with an ac-9 instrument equipped with nine spectral bands; 412, 440, 488, 510, 532, 555, 650, 676, and 715 nm (WET Labs, Inc., Philomath, OR, USA). $a(z, \lambda)$ and $c(z, \lambda)$ are the total absorption and beam attenuation coefficients of seawater and $a_w(\lambda)$ and $c_w(\lambda)$ represent the pure seawater contributions to these coefficients, respectively. Note that $a_{nw}(z, \lambda)$ is often written as a sum of $a_p(z, \lambda)$ and $a_g(z, \lambda)$ where the subscripts p and g represent the contributions associated with suspended particulate matter and CDOM, respectively. The coefficient $c_{nw}(z, \lambda)$ can be written as a sum of $a_{nw}(z, \lambda)$ and $b_p(z, \lambda)$, where $b_p(z, \lambda)$ represents the particulate component of the total scattering coefficient of seawater, $b(z, \lambda)$.

Vertical profiles of ac-9 data were obtained from the surface to ~500 m depth. The first step of data processing included quality checking of downcast and upcast measurements and rejecting doubtful and excessively noisy data. Because data were often missing for the upcast profiles within the top ~30–50 m layer, only the downcast data were used for further processing, which included temperature and salinity correction [85], subtraction of a baseline determined with Milli-Q water before the cruise, and binning to 0.5 m depth resolution. The $a_{nw}(z, \lambda)$ values were additionally corrected for the scattering error by subtracting the values at 715 nm at all ac-9 wavelengths. After quality control, only two stations (indicated in Figure 1) provided ac-9 data with concurrent radiometric measurements described in Section 2.4.

2.6. Regression Analysis

The presented results of regression analysis are mostly based on the Model I regression method because our objective was to formulate a functional relationship for estimating the ordinate variable y from the abscissa variable x [86]. The ordinary least squares method was applied for all linear relationships and the nonlinear least squares method was applied to all non-linear relationships. The regression calculations were performed using standard fitting methods in Matlab R2015b.

The ordinary (untransformed) variables x and y were used as input to the fitting procedure. The fitting process in Matlab R2015b also outputs root mean squared error, *RSME*, that is calculated as:

$$RMSE = \sqrt{\frac{\sum_{i=1}^{N} (y_i - x_i)^2}{N - m}},$$

$$(2)$$

where N is the total number of data points and m is the number of coefficients fitted via the regression analysis.

In one case of our data analysis, we applied the Model II regression method in which the relationship was estimated using the slope of major (principal) axis of the bivariate dataset (Section 3.3). In this specific analysis we compared the estimates of the absorption coefficient obtained with two different experimental methods based on irradiance quartet and ac-9 measurements.

3. Results and Discussion

3.1. Contrasting Examples of Light Field Characteristics and AOPs

Our measurements show that the bio-optical properties within the Gulf of California exhibited a large range of variability, which is summarized in terms of *Chl* and $c_p(660)$ data in Table 2. The examined water column extended from the surface to a maximum measurement depth of 80 m. The depth of euphotic layer, z_{eu}, was determined using a conventional criterion by estimating a depth at which the surface PAR (i.e., E_{oPAR}) was reduced to 1%. The depth z_{eu} ranged from about 30 to 60 m depending on time and location of measurement. Typically, our measurements (at least in the blue-green spectral region) extended ~10–30 m below z_{eu}, and these maximum measurement depths corresponded to ~0.1–0.5% of surface E_{oPAR}. Because phytoplankton have specific energetic demand rather than demand for percentage of light, it is reasonable to expect that the compensation depth of phytoplankton photosynthesis for specific latitude, season, and optical water types investigated in our study was greater than z_{eu}, probably closer to 0.1% of surface PAR [87]. Therefore, we refer to the entire investigated water column as the euphotic layer.

Table 2. Summary of data characterizing the shallowest and greatest depths of radiometric measurements, the depth of chlorophyll-*a* maximum, and the corresponding chlorophyll-*a* concentration, *Chl*, and particulate beam attenuation coefficient at 660 nm, $c_p(660)$, for these depths based on all stations during the 2010 and 2011 cruises in the Gulf of California. For each of these variables the average value ± standard deviation with the minimum and maximum values in parenthesis are given. In addition, the minimum and maximum values of the solar zenith angle θ_s and the data characterizing the euphotic depth z_{eu} (the average value ± standard deviation with the minimum and maximum values in parenthesis) are shown for all stations.

Depth of Measurement	Shallowest	Greatest	*Chl* Maximum
z (m)	5.4 ± 2.9 (1.2, 10.5)	64.9 ± 12.1 (47.6, 80.7)	33.0 ± 9.9 (20.4, 50.5)
Chl (mg m^{-3})	0.067 ± 0.067 (0.005, 0.24)	0.13 ± 0.15 (0.004, 0.42)	0.81 ± 0.54 (0.25, 2.17)
$c_p(660)$ (m^{-1})	0.173 ± 0.035 (0.103, 0.238)	0.089 ± 0.048 (0.043, 0.202)	0.239 ± 0.076 (0.125, 0.405)
Solar zenith angle θ_s (degrees)	0.5–57.1		
Euphotic depth z_{eu} (m)	40.5 ± 8.9 (29.3, 60.0)		

Within the examined water column, *Chl* varied from extremely low levels on the order of 10^{-3} to more than 2 mg m^{-3} with the highest values observed within the *Chl* maximum (*Chl$_{max}$*) layer and the lowest values either near the surface or at the deepest measurement depths (~60–80 m). The shape of the *Chl(z)* vertical profile also changed significantly within the study area. The depth z_{Chlmax} of *Chl$_{max}$* ranged from about 20 m to 50 m. The full width half maximum (FWHM) of the *Chl$_{max}$* layer, when fitted to a Gaussian function, ranged from 5 m to 30 m. Significant variations spanning one order of magnitude from about 0.04 to 0.4 m^{-1} were also observed in $c_p(z, 660)$. In addition, our measurements were conducted under a broad range of the solar zenith angle, θ_s, varying between 0.5° (i.e., the sun nearly at zenith) to 57.1°. Such broad range of environmental conditions is evident in terms of diverse scenarios in the measured light field characteristics and AOPs.

In this section we discuss two contrasting scenarios of light fields and AOPs within the euphotic layer, one was located in the Guaymas Basin which was observed in 2011 (Figure 2a) and the other in the Farallon Basin which was observed in 2010 (Figure 2b). Figure 2 compares the vertical profiles of *T*, σ, *Chl*, and $c_p(660)$ at both stations. These two stations have highly contrasting *Chl* profiles and also exhibit noticeable differences in the $c_p(660)$ profiles. Although both stations had a similar depth of *Chl$_{max}$* (~30 m) the magnitude and FWHM of the *Chl$_{max}$* layer were very different. Specifically, the magnitude of *Chl$_{max}$* in the Farallon Basin (1.3 mg m^{-3}) was about twice as high as compared with that observed in the Guaymas Basin (0.64 mg m^{-3}). The FWHM was also considerably larger in the Farallon Basin, i.e., 25 m vs. 10 m in the Guaymas Basin. In general, the magnitudes of $c_p(660)$ for these two stations did not differ significantly but the vertical distributions were different. In particular, there was a noticeable local maximum of $c_p(660)$ near the depth of *Chl$_{max}$* in the Guaymas Basin, and no similar feature was observed in the Farallon Basin. The observed optical differences between the stations could have been caused by differences in the characteristics of suspended particulate matter such as organic and mineral composition. In addition, whereas the measurements were made under clear skies, the solar zenith angle was dramatically different at these two stations, 5.5° in the Guaymas Basin (the measurement was taken around noon at 12:45 p.m. local time) and 57.1° in the Farallon Basin (the measurement taken late afternoon at 4:35 p.m.). Thus, the comparison of these two stations provides an illustration of differences in the light field and AOPs, which were affected not only by water optical properties but also solar zenith angle. We will first present the data of irradiance quartet (Figure 4) and AOPs (Figures 5 and 6) for the station in the Guaymas Basin, and then the AOPs for the station in the Farallon Basin (Figure 7).

The spectra of irradiances, E_d, E_u, E_{od}, and E_{ou}, at several discrete depths within the top 70 m of the water column are presented in Figure 4a–d. These results illustrate the well-known general pattern of the vertical change in the irradiance spectra in the upper ocean, including a decrease in the magnitude and narrowing of the spectral distribution with an increase in depth, e.g., [11]. In the investigated waters, all four irradiances exhibited well-pronounced maximum at a wavelength of about 495 nm at the deepest measurement depth of 67 m, but the spectral distributions of downwelling light, E_d and E_{od}, were clearly broader with relatively more light in the green compared with the distributions for the upwelling light, E_u and E_{ou}. A remarkable feature associated with vertically non-uniform fluorescence of chlorophyll-*a* was also evident in the irradiance spectra. At a depth of 29 m that coincides with z_{Chlmax} the spectra of E_d and E_{od} showed a distinct maximum in the spectral band of maximum fluorescence centered around 683 nm. This maximum was even more pronounced in the spectra of E_u and E_{ou}. This feature weakened greatly for E_d and E_{od} at depths shallower or deeper than z_{Chlmax}, but was preserved in the spectra of E_u and E_{ou} throughout the water column, which was expected as long as *Chl* did not drop to a very low level.

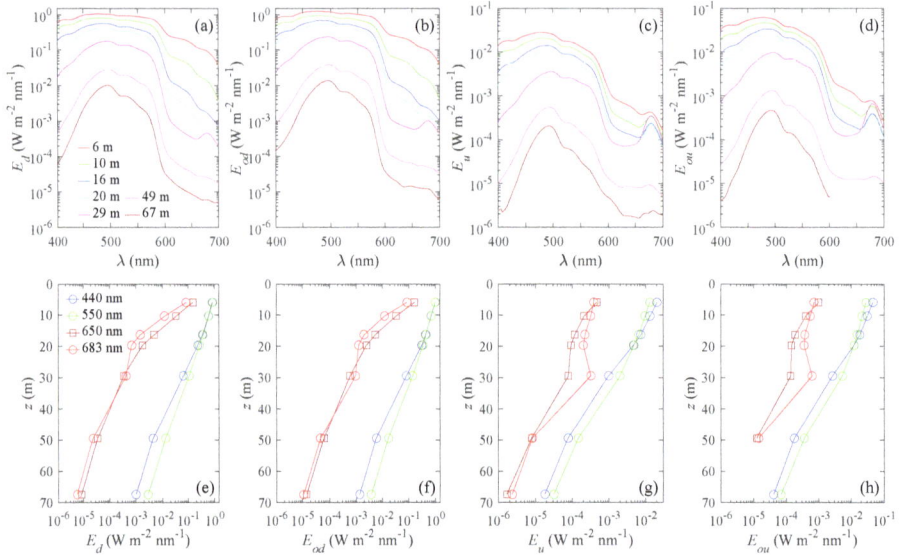

Figure 4. Spectra and vertical profiles of the irradiance quartet, E_d (**a,e**), E_{od} (**b,f**), E_u (**c,g**), and E_{ou} (**d,h**), measured at the station in the Guaymas Basin shown in Figure 2a. In panels (**a–d**) different colors represent the measurement depths as indicated in (**a**). In (**e–h**) different colors and symbols represent selected light wavelengths as indicated in (**e**).

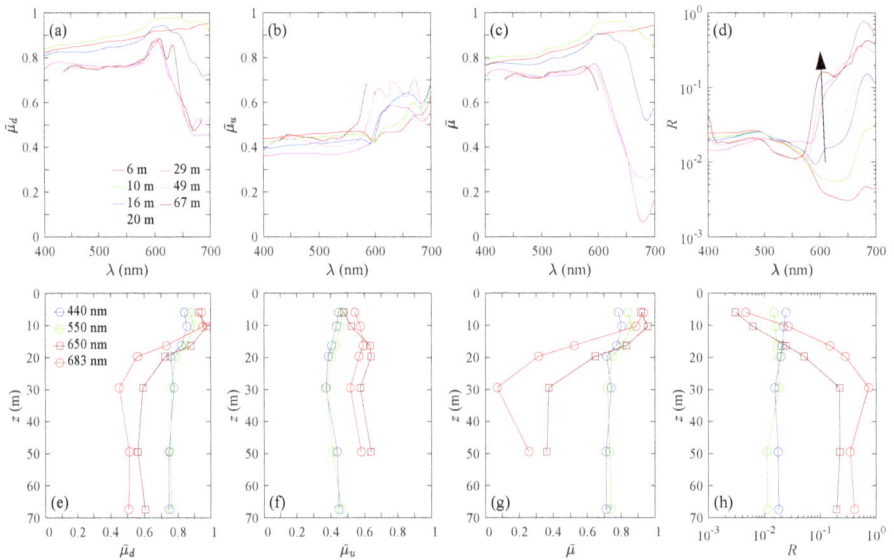

Figure 5. Spectra and vertical profiles of average cosines of underwater light field, $\bar{\mu}_d$ (**a,e**), $\bar{\mu}_u$ (**b,f**) and $\bar{\mu}$ (**c,g**) as well as irradiance reflectance, R (**d,h**), derived from the irradiance quartet for Guaymas Basin shown in Figure 4. In panels (**a–d**) different colors represent depths as indicated in (**a**). The black arrow in (**d**) indicates a spectral feature caused by Raman scattering of water molecules. In (**e–h**) different colors and symbols represent light wavelengths as indicated in (**e**).

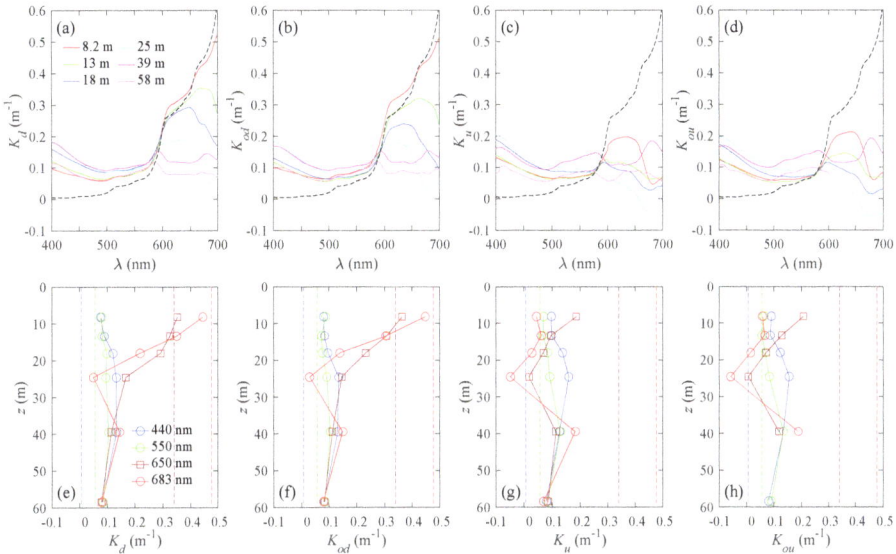

Figure 6. Spectra and vertical profiles of diffuse attenuation coefficients, K_d (**a**,**e**), K_{od} (**b**,**f**), K_u (**c**,**g**), and K_{ou} (**d**,**h**), derived from the irradiance quartet for Guaymas Basin in Figure 4. In panels (**a**–**d**) different colors represent the mid-point depths of the layers within which the K coefficients were determined as indicated in panel (**a**). Dashed lines represent the spectrum of pure water absorption coefficient, which is a theoretical minimum of K for a hypothetical case when no inelastic processes and true emission sources are present. In (**e**–**h**) different colors and symbols represent the light wavelengths as indicated in panel (**e**) and dashed lines represent pure water absorption at respective wavelengths.

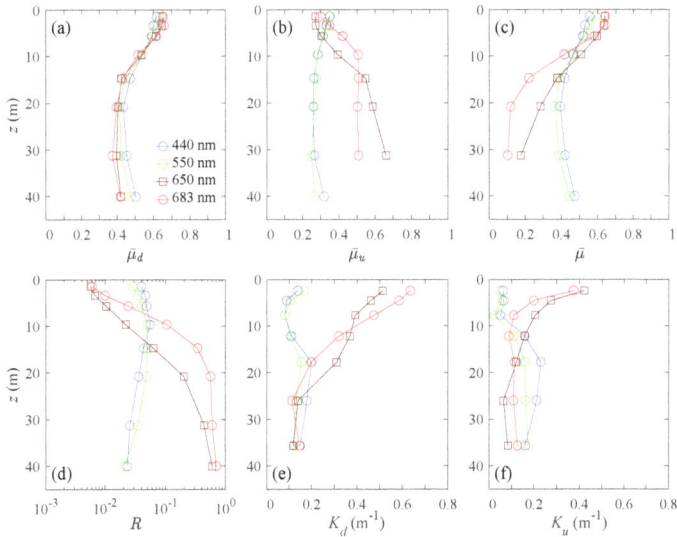

Figure 7. Vertical profiles of average cosines, irradiance reflectance, and diffuse attenuation coefficients for the station in the Farallon Basin shown in Figure 2b. Data for $\bar{\mu}_d$ panel (**a**); $\bar{\mu}_u$ (**b**); $\bar{\mu}$ (**c**); R (**d**); K_d (**e**); and K_u (**f**) are shown for selected light wavelengths as indicated in panel (**a**).

The effects of vertically non-uniform IOPs and chlorophyll-*a* fluorescence were also clearly revealed in the depth profiles of the irradiance quartet at four selected wavelengths (Figure 4e–h). Because the unscattered and elastically scattered light in the red spectral region was efficiently removed within the near-surface layer owing to strong absorption by water molecules, the vertical profiles in the red (650 and 683 nm) were affected by inelastic processes (Raman scattering and chlorophyll-*a* fluorescence) to much greater extent compared with shorter wavelengths, and hence differed significantly from the profiles in the blue and green (440 and 550 nm). The slope of the profiles in the blue and green showed relatively small variations with depth, which were driven primarily by vertical inhomogeneity in IOPs associated with the strength of absorption and elastic scattering processes. For example, at 440 nm all four irradiances showed steeper profiles (i.e., higher attenuation rate with depth) within the Chl_{max} layer, which can be attributed largely to enhanced absorption of blue light by phytoplankton. In contrast, the slope of irradiance profiles in the red varied within the water column to much greater extent compared with shorter wavelengths. For example, in the 650 nm band the profiles of E_d and E_{od} became gradually less steep with depth, which can be attributed largely to the increasing contribution of Raman scattering. In the 683 nm band where the contribution of chlorophyll-*a* fluorescence was strongest, E_d and E_{od} remained nearly constant with increasing depth within the upper portion of the Chl_{max} layer (note that no data is available to resolve the depth variations throughout the lower portion of the Chl_{max} layer as the next data point at 50 m was near the base of the Chl_{max} layer). The effects of inelastic processes were, however, most pronounced for E_u and E_{ou} in the red spectral region. At 683 nm the effect of chlorophyll-*a* fluorescence on the upwelling light field was so strong that both E_u and E_{ou} increased with depth within the upper portion of the Chl_{max} layer. Although Raman scattering is known to have very important contribution to the upwelling light in the red [9,60,64,71,88], our data show that the fluorescence effect in the 683 nm band can be dominant in the presence of Chl_{max} layer. At 650 nm E_u and E_{ou} remain nearly constant within the upper portion of the Chl_{max} layer. This suggests that the fluorescence fingerprint in the upwelling irradiances is still noticeable at this wavelength, which can be attributed to relatively broad emission band of chlorophyll-*a* fluorescence (FWHM of ~20–25 nm) [89,90].

The vertical variations in the spectral AOPs derived from the irradiance quartet measurements reveal features that were clearly associated with the effects of inelastic processes in the presence of non-uniform profiles of *Chl* and IOPs (Figures 5 and 6). The spectra and depth profiles of the average cosines, $\bar{\mu}_d$, $\bar{\mu}_u$, and $\bar{\mu}$, were relatively weakly variable within the blue-green spectral region where the effects of inelastic processes are much weaker compared with the long-wavelength portion of the spectrum (Figure 5a–c,e–g). In the spectral range of 400–580 nm, the values of $\bar{\mu}_d$ and $\bar{\mu}$ were in the range 0.7–0.9 with higher values observed at shallower depths and related to the high solar elevation in the clear sky during the measurement. Thus, the downwelling light field or the entire light field was characterized by the predominant downward direction of light propagation within 45° of the vertical. The values of $\bar{\mu}_u$ in the blue-green were around 0.4 suggesting more uniform angular distribution of upwelling light field. Our data of average cosines in the blue-green are in the range of theoretical and experimental values reported in other studies [9,10,29,52,60,73]. In the red spectral region, the average cosines exhibited large variations owing to the effects of inelastic processes within the non-uniform water column. We observed a large rapid decrease in $\bar{\mu}_d$ and $\bar{\mu}$ below the near-surface layer (below 10 m) to about 0.5 and 0.05, respectively, at depths $\geq$20 m. For $\bar{\mu}_u$ in the red we observed a variation between about 0.45 and 0.65. The observed values of average cosines in the red differ significantly from those which would have occurred if inelastic processes were absent [9,73]. In the presence of inelastic processes, the light field in the red approached a more uniform distribution with increasing depth quite rapidly, and could achieve a nearly isotropic distribution within moderately strong Chl_{max} layer at relatively shallow depths, such as ~30 m in the examined case (Figure 5e–g). Note also that, unlike Raman scattering that leads gradually to a more uniform light field in the red with increasing depth, the vertically non-uniform chlorophyll-*a* fluorescence caused additional depth variations in the angular distribution of light field within the 683 nm band. In contrast, in the blue-green region where the effects of inelastic processes are much weaker, the light field remained highly directional throughout the examined euphotic layer (Figure 5e–g).

Similar to the average cosines, the spectral irradiance reflectance R (Figure 5d,h) showed relatively little variation with depth in the blue-green spectral region as the R values ranged between 0.01 and 0.025. In contrast, R exhibited very large vertical variations at wavelengths longer than about 580 nm. In the red R varied from about 0.003 at 6 m to 0.8 at z_{Chlmax} = 29 m, and relatively high values (0.2–0.4) persisted below the Chl_{max} layer (Figure 5h). Similar observations were previously reported by Dirks and Spitzer [37]. This result was caused by inelastic processes, including a dominant effect of chlorophyll-*a* fluorescence that produces a maximum in the R spectra within the fluorescence emission band centered at 683 nm (Figure 5d). At shorter wavelengths within the approximate range 580–650 nm that was away from significant effect of chlorophyll-*a* fluorescence, Raman scattering appeared to exert a dominant influence on R throughout much of the examined water column. Owing to this inelastic process a shoulder-like feature was seen in the R spectra around 600 nm at intermediate measurement depths (16–49 m), which developed into a distinct maximum with further increase of depth to 67 m (Figure 5d). This pattern is consistent with an increasing role of Raman scattering with depth [37,69,70,73]. These results also suggest that the effects of Raman scattering were more easily detectable or distinguishable from the effects of chlorophyll-*a* fluorescence in the data of AOPs than in the irradiance data.

Figure 6 shows the attenuation coefficients for the four irradiances, specifically K_d, K_u, K_{od}, and K_{ou}, as derived from the measurements of irradiance quartet, E_d, E_u, E_{od}, and E_{ou}, respectively. Similar to other AOPs discussed above, the K-coefficients were less variable in the blue-green than in the red portion of the spectrum, both in terms of the spectral shape (Figure 6a–d) and vertical profiles (Figure 6e–h). One of the most remarkable features was that the K-coefficients in the red (more specifically for $\lambda > 580$–600 nm) were generally smaller than the pure water absorption coefficient, a_w. This feature resulted from inelastic processes and was more pronounced for K_u and K_{ou} than K_d and K_{od}. In the chlorophyll-*a* fluorescence band, K_d and K_{od} approached zero in the upper portion of the Chl_{max} layer (Figure 6e,f), which is consistent with an earlier observation that E_d and E_{od} are nearly constant within that layer (Figure 4e,g). A stronger effect within the 683 nm band was observed for the K_u and K_{ou} coefficients that assume negative values in the upper portion of the Chl_{max} layer (Figure 6g,h). This is again consistent with the measurements of E_u and E_{ou} which both increased with depth within that layer (Figure 4f,h). Similar increase of E_u in the red band (685 nm) with depth within the Chl_{max} layer was previously observed by Dirks and Spitzer [37]. The presence of negative values for the attenuation of upwelling irradiance was also demonstrated through RT simulations of light field within highly scattering near-surface layer of bubble clouds entrained by wave breaking [91]. Other interesting features of the long-wavelength portion of K-spectra include a maximum at 683 nm below z_{Chlmax} (for all K-coefficients, see Figure 6a–d) or a maximum that could peak at wavelengths longer than about 600 nm but shorter than 683 nm at depths above z_{Chlmax} (especially for K_d and K_{od}, see Figure 6a,b). These features arise from differential spectral effects of inelastic processes as a function of increasing wavelength. It is also interesting to note that a distinct feature at these intermediate depths was observed in the spectra of $\bar{\mu}_d$ and $\bar{\mu}$ (Figure 5b,d). This can be explained by higher attenuation coefficient leading to more light concentrated around the vertical direction.

In contrast to the red portion of the spectrum, in the blue-green spectral region where the effects of inelastic processes are much weaker, the K-coefficients were higher than $a_w(\lambda)$. At these wavelengths, especially at 440 nm, the vertical profiles of K-coefficients exhibited maximum values within the Chl_{max} layer, and hence generally resembled the shape of $Chl(z)$ profile. It is also noteworthy that the K-coefficients at all different wavelengths became nearly identical at the greatest depth of about 60 m used in these determinations (Figure 6e–h). This was primarily because the effect of Raman scattering generally led to a gradual decrease of K-coefficients in the red, for example at 650 nm, with increasing depth until approaching the level of K-coefficients in the blue and green. This pattern that eventually led to relatively flat K-coefficients has been demonstrated previously in experimental data and theoretical results from RT simulations [37,64,73]. It is, however, important to emphasize that we do not suggest that the light field in the examined case has reached an asymptotic regime within the deepest portion of the examined layer (i.e., ~60–70 m) where we observe such similarity of K-coefficients and relatively constant AOPs with depth. While such patterns can be associated with the asymptotic regime, previous deep-sea RT simulations demonstrated that a nearly asymptotic light field could generally be reached only at significantly greater depths within the ocean mesopelagic zone [64,73].

For comparison with the results discussed above in the Guaymas Basin, Figure 7 shows the vertical profiles of AOPs obtained from measurements at a station in the Farallon Basin where the Chl_{max} layer was significantly broader with a two-fold higher Chl_{max} and the solar elevation was significantly lower (see Figure 2). As a result of these differences in the environmental conditions, which could also include differences in the IOPs, some features of AOPs measured in the Farallon Basin were different compared to those in the Guaymas Basin. We note, however, that the data from the Farallon Basin do not extend throughout the entire Chl_{max} layer but end at 40 m which is about 10 m below z_{Chlmax}. The values of the average cosines in the blue-green spectral region in the Farallon Basin were much lower (Figure 7a–c) than those in the Guaymas Basin (Figure 5f–h). The lower values of $\bar{\mu}_d$ and $\bar{\mu}$ in the blue and green can be explained by the higher solar zenith angle and slightly higher IOPs as indicated by the $c_p(z, 660)$ profiles (Figure 2b). This result is consistent with previous experimental and modeling results [29,60,92–94]. The observation of very low values (0.25–0.35) of $\bar{\mu}_u$ in the blue and green (Figure 7b) will be discussed is some detail in Section 3.2.1, but it is interesting to note that the values less than 0.35 were previously reported on the basis of RT simulations of highly scattering environment within near-surface bubble clouds [91]. In the red where the effects of inelastic processes are strong, the vertical profiles of average cosines exhibited similar features to those described earlier for the station in the Guaymas Basin. Overall, our data of the average cosines support the notion that sun position in a clear sky sets the values of $\bar{\mu}_d$ and $\bar{\mu}$ near the ocean surface, which then change with depth in a manner dependent on the interplay of the vertical structure of IOPs and inelastic processes, and $\bar{\mu}_u$ may vary within a broader range than commonly assumed. It is thus important to emphasize that the average cosines are sensitive to surface illumination, especially for the downwelling and total light in the near-surface layer, so these average cosines may potentially serve as AOPs or useful descriptors of near-surface water bodies only under specified surface boundary conditions.

The vertical profiles of R in the Farallon Basin (Figure 7d) showed patterns that were similar to those observed in the Guaymas Basin (Figure 5e), but the R values were generally higher because of higher Chl and water turbidity at the station in the Farallon Basin. The properties of particulate assemblages, including the composition and size distribution, could have also been different at these stations. The K_d and K_u values within the Chl_{max} layer in the Farallon Basin (Figure 7e,f) were not as low as in the examined case from the Guaymas Basin (Figure 6e,g). This observation also applies to K_{od} and K_{ou} (not shown in Figure 7). In particular, unlike in the Guaymas Basin case, the chlorophyll-a fluorescence within the much stronger Chl_{max} layer in the Faralllon Basin did not produce a minimum of K_u with negative values around the depth of z_{Chlmax} (Figure 7f). This lack of distinct negative minimum of K_u can be attributed to the weakening of the effects of inelastic processes when the water column becomes more turbid, as previously demonstrated by RT simulations [60]. In addition, it is also likely that the influence of Chl_{max} is less profound when the FWHM of the Chl_{max} layer is large, which is the case for the station in the Farallon Basin.

3.2. Overall Variability and Relationships Involving the Apparent Optical Properties

In this section we examine the overall variability of AOPs and the relationships involving AOPs and some environmental factors (depth, solar zenith angle, Chl) using the entire dataset collected in the Gulf of California.

3.2.1. Average Cosines

As indicated earlier the average cosines serve as simple proxies of the angular structure of the light field and they have or have not been treated as AOPs by different investigators in the past [9,10,46,48]. Based on our measurements of irradiance quartet the average cosine of underwater light field, $\bar{\mu}$, ranged from about 0.38 to 0.88 in the blue and green spectral regions. In the red, the range was much larger, from 0.07 to 0.94, owing to strong effects of inelastic processes. The average cosine for downwelling light field, $\bar{\mu}_d$, varied from 0.41 to 0.91 in the blue and green and 0.37 to 0.98 in the red. The approximate range for the average cosine for the upwelling light field, $\bar{\mu}_u$, was 0.2–0.5 in the blue and green and 0.27–0.75 in the red. Our data are consistent but generally exhibit a broader range of values compared with relatively few experimental data of average cosines reported in literature. Højerslev [18] reported on measurements of E_o, E_d, and E_u in the Mediterranean Sea and Norwegian fjord. Using his data, we calculated that $\bar{\mu}$ was in the range of 0.47–0.95 in the blue-green spectral region (427, 477, and 532 nm) between the sea surface and 70 m depth, and 0.7–0.97 in the red (633 nm) within top 20 m layer. Spitzer and Wernand [52] reported on measurements of $\bar{\mu}$ in the range 0.7–0.9 at wavelengths shorter than 600 nm and depths between 24 and 55 m. Lewis et al. [29] measured the full angular distribution of radiance at a single wavelength of 555 nm in surface waters of the Pacific Ocean off Hawaii Islands, Santa Barbara Channel, and in Bedford Basin, and used these measurements to determine the ranges for $\bar{\mu}$, $\bar{\mu}_d$, and $\bar{\mu}_u$, which were approximately 0.41–0.82, 0.5–0.94, and 0.2–0.48, respectively. Our determinations in the blue and green are generally consistent with the data from literature. The comparison for the red spectral region is, however, more limited because only few fragmentary data in the red have been previously reported. Our measurements of the average cosines in the red exhibited a very wide range of variability within the water column, providing a more comprehensive characterization of the average cosines that are strongly affected by inelastic processes in this spectral region.

It has long been recognized that at near-surface depths in relatively clear ocean waters, the downwelling radiance distribution, and hence $\bar{\mu}_d$, are strongly dependent on the sun position in clear sky or more generally on sky conditions, for example clear skies vs. overcast skies [11]. Such sensitivity of the downwelling average cosine to surface illumination conditions cautions against its indiscriminate use as a general descriptor (in the sense of AOPs) of water bodies, especially in the surface layers, and emphasizes a need for specifying surface boundary conditions. Our data collected within the top 10 m layer in the Gulf of California support the general trend of a decrease in near-surface $\bar{\mu}_d$ with increasing solar zenith angle regardless of light wavelength (Figure 8a). For comparison, the best fit lines to experimental data of Aas and Højerslev [94] obtained at a 5 m depth (Mediterranean Sea) in the blue (465 and 474 nm) and experimental data of Lewis et al. [29] at a 5 m depth (open ocean waters in the Pacific) in the green (555 nm) are also shown in Figure 8a. In contrast, our data of $\bar{\mu}_u$ at near-surface depths did not exhibit a clear dependence on the solar zenith angle (Figure 8b). We note, however, that Aas and Højerslev [94] and Lewis et al. [29] suggested a very weak dependence based on their data, as indicted in Figure 8b.

The increased role of inelastic processes, especially Raman scattering, with increasing depth is illustrated for the average cosines in the red part of the spectrum in Figure 9. This effect was particularly well-pronounced for the downwelling light field in terms of a well-behaved trend of a decrease in $\bar{\mu}_d$ with increasing depth towards $\bar{\mu}_d = 0.5$, which is clearly seen in our data at the two examined wavelengths in the red, 650 nm and 683 nm (Figure 9a). The best-fit exponential functions to the data are:

$$\bar{\mu}_d(650) = 0.5058 \exp(-0.0460\,z) + 0.5140\ (r^2 = 0.828, RMSE = 0.061, N = 65); \quad (3a)$$

$$\bar{\mu}_d(683) = 0.6512 \exp(-0.0715\,z) + 0.4659\ (r^2 = 0.855, RMSE = 0.067, N = 65), \quad (3b)$$

where r^2 is the determination coefficient and N the number of observations. A distinct group of outlying data marked in Figure 9a was collected at significantly larger solar zenith angle ($\theta_s = 57.1°$)

than the remaining data. These outliers were not included in the regression analysis. Whereas at 650 nm Raman scattering is expected to have a dominant effect, at 683 nm chlorophyll-*a* fluorescence can have an additional significant or even dominant effect at certain depths, depending on chlorophyll-*a* concentration and its variation within the water column. The combined effects of Raman scattering and chlorophyll-*a* fluorescence at 683 nm result in faster decrease of $\bar{\mu}_d(683)$ with depth from the near surface through the lower portion of the *Chl*$_{max}$ layer compared with $\bar{\mu}_d(650)$. In addition, in our dataset the interplay of vertically variable chlorophyll-*a* fluorescence and Raman scattering throughout the water column does not seem to have a noticeable deteriorating effect on the strength of the relationship $\bar{\mu}_d(683)$ vs. z as compared with $\bar{\mu}_d(650)$ vs. z.

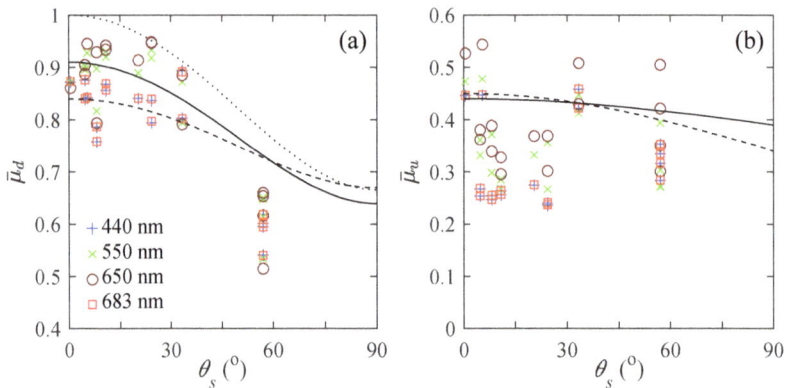

Figure 8. (a) Relationship between $\bar{\mu}_d$ at selected light wavelengths as indicated and solar zenith angle θ_s. (b) Same as (a) but for $\bar{\mu}_u$. The presented data were collected at all stations within the top 10 m layer. For comparison, solid lines in (**a**,**b**) represent the relationship reported by Lewis et al. [29] and dashed lines by Aas and Højerslev [94]. Dotted line in (**a**) represents $\bar{\mu}_d = \cos(\theta_{sw})$ where θ_{sw} is the zenith angle of the refracted solar beam just beneath the ocean surface obtained from Snell's law.

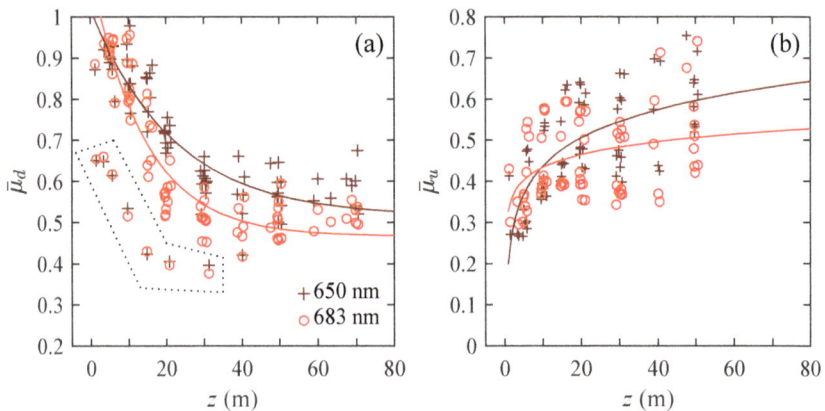

Figure 9. Average cosines in the red spectral bands (650 and 683 nm) plotted as a function of depth z using data from all stations and depths. (**a**) $\bar{\mu}_d(z, 650)$ (crosses) and $\bar{\mu}_d(z, 683)$ (open circles) vs. z. (**b**) Same as (**a**) but for $\bar{\mu}_u$. Data points surrounded by dotted lines in (**a**) are for the station in the Farallon Basin with the solar zenith angle significantly larger ($\sim57°$) than at other stations. Solid lines show the best-fit regression functions for the average cosines vs. depth.

In contrast to $\bar{\mu}_d$, $\bar{\mu}_u$ in the red did not exhibit such strong dependence on depth although it is clear that $\bar{\mu}_u$ assumed generally lower values near the surface (Figure 9b). The r^2 values for the $\bar{\mu}_u$ vs. z data dropped to 0.488 at 650 nm and 0.147 at 683 nm. The differences in the patterns of depth dependencies for $\bar{\mu}_d$ and $\bar{\mu}_u$ can be attributed to relatively smaller role played by inelastic processes at near-surface depths for the downwelling light field compared with the upwelling light field in the red spectral region. It is also important to point out that in comparison with our dataset the dependence of $\bar{\mu}_d$ on depth is expected to be reduced for datasets collected in turbid waters, at large solar zenith angles, and/or heavily overcast skies. Under such conditions the near-surface values of $\bar{\mu}_d$ are expected to be lower (i.e., closer to 0.5) compared with most of our near-surface data, and hence the expected changes of $\bar{\mu}_d$ with depth would not be as well pronounced. Thus, the strong sensitivity of near surface $\bar{\mu}_d$ to illumination conditions at the sea surface suggests that no single relationship can describe a change of $\bar{\mu}_d$ with depth. This is illustrated in Figure 9a which includes an outlying case of data points collected at significantly larger θ_s than the remaining data, emphasizing a need to pay particular attention to the possible effect of solar zenith angle. In addition, it is important to emphasize that in the blue-green spectral region where the effects of inelastic processes are weaker, the average cosines exhibited much less distinct dependency on depth (see Figures 5e–g and 7a–c).

Our analysis also shows that regardless of light wavelength, none of the average cosines were related significantly to *Chl* in our dataset. The highest r^2 of 0.42 was found between $\bar{\mu}_d(z, 650)$ and *Chl*(z) (not shown). For $\bar{\mu}_d$ in the blue-green where inelastic processes are less important than in the red, and for $\bar{\mu}_u$ at all examined spectral bands, the relation was much weaker or essentially indiscernible. This lack of relation with *Chl* is not surprising because the average cosine within the water column, if not dominated by inelastic processes, depends on the interplay of the inherent absorption and scattering properties of seawater in rather complex way [85], and *Chl* cannot adequately represent such effects.

Similar patterns observed in our data of $\bar{\mu}$ and $\bar{\mu}_d$ (e.g., Figure 5) indicate that these two average cosines were well correlated. The relationships between $\bar{\mu}$ and $\bar{\mu}_d$ obtained with our data are presented in Figure 10a. Specifically, the linear relationship for the blue-green spectral region as determined from the data collected at various depths and two wavelengths, 440 nm and 550 nm, is:

$$\bar{\mu} = 0.996\,\bar{\mu}_d - 0.0352 \quad \left(r^2 = 0.978, RMSE = 0.042, N = 158\right). \tag{4a}$$

On the basis of data at two wavelengths in the red, 650 and 683 nm, the relationship is:

$$\bar{\mu} = 1.51\,\bar{\mu}_d - 0.472 \quad \left(r^2 = 0.929, RMSE = 0.167, N = 134\right). \tag{4b}$$

These relationships indicate that reasonably good estimates of $\bar{\mu}$ can be obtained from downwelling irradiance measurements. As seen, $\bar{\mu}$ and $\bar{\mu}_d$ in the blue-green were typically close to one another within the examined euphotic layer. In the red, the correlation is marginally lower and still very strong but $\bar{\mu}$ tends to be somewhat smaller than $\bar{\mu}_d$ over a nearly complete data range with the exception of the highest values.

In contrast to $\bar{\mu}$ vs. $\bar{\mu}_d$ relationship, $\bar{\mu}_u$ and $\bar{\mu}_d$ were not correlated or very weakly correlated (Figure 10b). The r^2 values for the blue-green and red spectral regions are 0.026 and 0.109, respectively. Therefore, $\bar{\mu}_u$ cannot be estimated from $\bar{\mu}_d$ or vice versa, which means that a complete irradiance quartet or radiance angular distribution is required for investigating both $\bar{\mu}_u$ and $\bar{\mu}_d$. Figure 10b also reveals that the $\bar{\mu}_u$ data in the blue-green appeared to be grouped in at least two distinct clusters, one with $\bar{\mu}_u > \sim 0.3$ and the other with $\bar{\mu}_u < \sim 0.3$. A large number of $\bar{\mu}_u$ measurements fells in the range from 0.2 to ~0.3, suggesting that much of the upwelling light in these cases propagated relatively close (within ~10°–20°) to the horizontal direction. This is not consistent with a common assumption that the angular distribution of upwelling light field is more isotropic. Nonetheless, such low values of $\bar{\mu}_u$ were also observed by Lewis et al. [29] and reported on the basis of RT simulations within highly-scattering bubble clouds at near-surface depths [91] and for certain combinations of absorption coefficient, scattering coefficient, and scattering phase function in the water column (C. Mobley, personal communication). A closer inspection of our data indicates that the group with $\bar{\mu}_u < {\sim}0.3$ was generally characterized by relatively larger values of $c_p(z, 660)$, especially within the top 30–40 m, compared to the group with $\bar{\mu}_u > {\sim}0.3$. This is seen in the example data shown in Figures 3, 5b,f and 7b, supporting the notion that certain suites of IOPs can play an important role for producing low values of $\bar{\mu}_u$.

Although the behavior of average cosines in natural waters can be complex, these quantities are usually assumed to be constant or vertically uniform when used in the context of ocean color remote sensing applications, bio-optical modeling, and/or estimation of water constituents, e.g., [39,95–99]. The assumption that $\bar{\mu}$ obtained from simulations for a homogeneous water column is representative of a stratified water column was also used [100,101]. The present study demonstrates that these simplifying assumptions are not valid. Further studies to determine how such assumptions lead to errors in data products of interest are needed. For example, Sathyendranath and Platt [102] reported that oceanic primary productivity could be systematically underestimated by 5 to 13% when ignoring the angular distribution of the underwater light field, for which the average cosines serve as simple yet useful proxies.

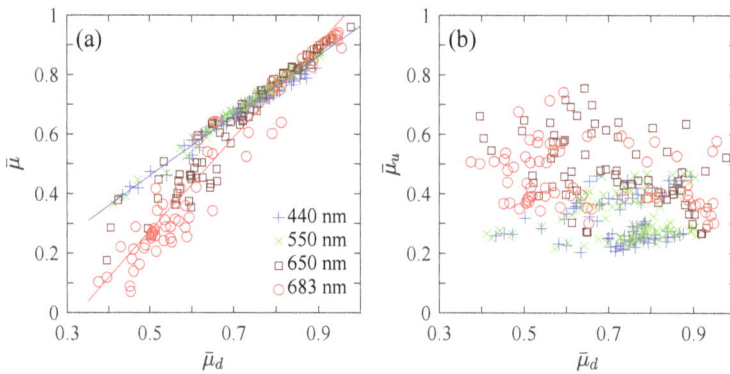

Figure 10. Relationships between average cosines based on measurements from all stations and depths. Data for (**a**) $\bar{\mu}$ vs. $\bar{\mu}_d$ and (**b**) $\bar{\mu}_u$ vs. $\bar{\mu}_d$ are shown for selected light wavelengths as indicated in panel (**a**). In (**a**) data points for the blue and green (440 nm and 550 nm) were combined to determine the best-fit regression function (blue line) and data points for the red (650 nm and 683 nm) were combined to determine the best-fit regression function (red line).

3.2.2. Irradiance Reflectance

Similar to the average cosines, the variations in the irradiance reflectance, R, throughout the euphotic layer are characterized by a much broader range in the red spectral region where the effects of inelastic processes are very strong compared with the blue-green part of the spectrum where these effects are much weaker. Our data of R in the blue-green ranged from 0.006 to 0.056, whereas in the red from 0.0015 near the surface to 0.75 around z_{chlmax}. The observed range in the blue-green is similar to that of 0.006–0.063 based on E_d and E_u measurements reported by Højerslev [18] for 427, 477, 532, and 572 nm between the surface and ~70 m. The R values at 633 nm in Højerslev's dataset range from 0.001 to 0.004, but these values were determined only at shallow depths, typically less than 5 m. Our data in the red extend to much higher values within the euphotic layer owing to strong inelastic effects in this spectral region, which is consistent with a few measurements from undisclosed locations reported by Dirks and Spitzer [37].

As emphasized before, the increasing role of inelastic processes with increasing depth is most significant in the long-wavelength part of the spectrum. This effect is clearly seen in our data of R

plotted as a function of depth for wavelengths of 650 and 683 nm (Figure 11). Both $R(650)$ and $R(683)$ showed gradual increase from the values of the order of 10^{-3} near the surface to more than 0.1 within the mid- to deep portions of the examined water column. These data suggest that depth can serve as a useful proxy for the vertical changes in R in the red part of the spectrum. The best-fit exponential functions to the data are:

$$R(650) = 10^{0.7633 \ln(z) - 3.6633} \quad (r^2 = 0.820, RMSE = 0.331, N = 74) \tag{5a}$$

$$R(683) = 10^{0.7690 \ln(z) - 3.2582} \quad (r^2 = 0.794, RMSE = 0.362, N = 74). \tag{5b}$$

Whereas these vertical trends in R reflect primarily the increasing effect of Raman scattering with depth, the effects of chlorophyll-*a* fluorescence were also evident within the 683 nm band. The values of $R(683)$ were comparable to $R(650)$ near the surface but increased faster with depth owing to increasing *Chl* and associated fluorescence until the Chl_{max} was reached. Consequently, $R(683)$ tended to have the highest values at depths close to z_{Chlmax}, which was about 30 m for most investigated stations in this study.

The dominant role of Raman scattering and additional complexities caused by chlorophyll-*a* fluorescence emission in the red spectral region indicate that this part of the spectrum is not well suited for establishing a relationship between R and *Chl* within the euphotic layer. The analysis of our entire dataset including all measurement depths confirmed the lack of relationship between R in the red and *Chl* (not shown). For example, a very low determination coefficient (r^2 in the range of ~0.091–0.155) was found between the log-transformed data of red (650 or 683 nm)-to-green (550 nm) reflectance ratios and *Chl*. In contrast, our data of the blue-to-green reflectance ratio, $R(440)/R(550)$, and *Chl* collected at all measurement depths exhibited a relationship with significantly higher determination coefficient of 0.608 (not shown). Although this may not seem surprising given the common use of the blue-to-green ratio of ocean surface reflectance (typically the remote-sensing reflectance rather than irradiance reflectance) for estimating surface *Chl* in the context of remote sensing applications, e.g., [103,104], it is important to emphasize a distinct difference which is that our data represent irradiance reflectance and *Chl* measured at different discrete depths within the entire euphotic layer down to depths as high as ~80 m.

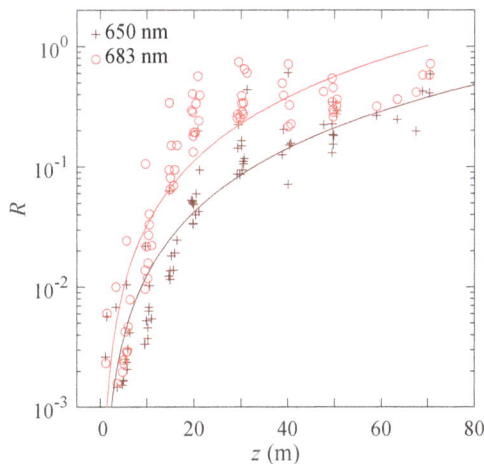

Figure 11. Irradiance reflectance R in the red spectral bands (650 and 683 nm as indicated) plotted as a function of depth z using data from all stations and depths. Solid lines represent the best-fit regression functions.

3.2.3. Diffuse Attenuation Coefficients

The diffuse attenuation coefficients K in our dataset span a relatively wide range of values. For example, the ranges for K_d at 440, 550, 650, and 683 nm are 0.031–0.246, 0.054–0.168, 0.072–0.514, and 0.044–0.635 m^{-1}, respectively. The ranges for K_u at the same wavelengths are 0.041–0.253, 0.021–0.192, 0.018–0.421, and −0.095–0.374 m^{-1}, respectively. The lowest values of K_u at longer wavelengths (green and red) were significantly smaller than those for K_d and can even include the negative values because the inelastic processes exerted a stronger reducing effect on K_u than K_d. The possibility of negative values of the K-coefficients at 685 nm within the Chl_{max} layer were also demonstrated through earlier observations and modeling [37].

Because E_d has been by far the most commonly measured radiometric quantity in the ocean, K_d in the surface layer is relatively well documented in the literature. An overview of early data of K_d, including the classification of optical water types based on the spectral K_d within the top 10 m layer of the ocean, is given in Jerlov [11]. The range of K_d values observed in our study in the blue spectral region indicates that the investigated waters in the Gulf of California fall between the oceanic water Types IB or II and the coastal water Type 1 according to Jerlov's classification. Following an increased interest in ocean color remote sensing in early 1970s and the introduction of the concept of bio-optical state of ocean waters representing a measure of the effect of biological processes on ocean optical properties [105], several studies examined experimental data of K_d in the context of the relationship with chlorophyll-*a* concentration over a broad range of these variables measured in various oceanic waters [35,38,39,106]. In those studies, K_d was typically estimated to represent the top layer from the ocean surface to one attenuation depth ($1/K_d$) or euphotic depth (1% of surface PAR). The lowest limiting values of spectral K_d for the clearest natural waters were also estimated, for example 0.017 m^{-1} or 0.00885 m^{-1} at 440 nm, 0.0648 m^{-1} or 0.05746 m^{-1} at 550 nm, and 0.35 m^{-1} or 0.34052 m^{-1} at 650 nm by Morel and Maritorena [39] and Smith and Baker [107], respectively. These estimates were generally consistent with the lowest measured values in the datasets presented in [11,38,39] although, to our knowledge, the lowest reported measurements of K_d in the blue are 0.012 m^{-1} at 420 nm [26]. The lowest K_d values in the blue in our dataset are thus 1.8 to 3.5 higher than these previous estimates for the clearest waters. In the red, our data of K_d collected within the top 10 m layer (Figures 6e and 7e) are similar or somewhat higher than the previous determinations for the clearest waters. At larger depths, K_d in the red was reduced owing to the increasing effect of inelastic processes with depth. This effect on the vertical changes in both K_d and K_u in the red spectral region is illustrated for our dataset in Figure 12. Both K coefficients decrease considerably with depth. This reduction is more pronounced at 683 nm than 650 nm. K_u decreased more rapidly than K_d and approached a relatively stable level of values at shallower depths (i.e., about 15–20 m) than K_d. The best-fit exponential functions to the data in Figure 12 are:

$$K_d(650) = 0.4792 \exp(-0.0451\,z) + 0.0387 \quad (r^2 = 0.925, RMSE = 0.033\ \text{m}^{-1}, N = 66) \qquad (6\text{a})$$

$$K_d(683) = 0.7229 \exp(-0.0947\,z) + 0.0846 \quad (r^2 = 0.898, RMSE = 0.050\ \text{m}^{-1}, N = 66) \qquad (6\text{b})$$

$$K_u(650) = 0.5614 \exp(-0.2258\,z) + 0.0730 \quad (r^2 = 0.806, RMSE = 0.032\ \text{m}^{-1}, N = 64) \qquad (6\text{c})$$

$$K_u(683) = 1.0000 \exp(-0.4711\,z) + 0.0641 \quad (r^2 = 0.391, RMSE = 0.062\ \text{m}^{-1}, N = 64). \qquad (6\text{d})$$

The relationship for $K_u(683)$ is greatly inferior because at this wavelength the effect of Raman scattering is reinforced by chlorophyll-*a* fluorescence emission, which resulted in a very steep decrease of $K_u(683)$ in the top 20 m (Figure 12b).

In contrast to K_d, other K coefficients are poorly documented in the literature. It is therefore useful to examine the relationships between K_d and other K coefficients using our dataset containing depth-resolved measurements within the euphotic layer. The results from this analysis are shown in Figure 13. Not shown in the figure are results that indicated a very good linear relationship with close agreement between the K_d and K_{od} values ($r^2 = 0.985$) and between K_u and K_{ou} ($r^2 = 0.953$).

This agreement was generally observed regardless of light wavelength and depth of measurement within our dataset.

The relationship between K_u and K_d is reasonably good in the blue and green spectral regions (Figure 13a). For example, at 440, 490, and 550 nm the best-fit linear relationships are:

$$K_u(z, 440) = 1.0115\, K_d(z, 440) + 9.47 \times 10^{-3} \quad (r^2 = 0.828, RMSE = 0.023\ \mathrm{m}^{-1}, N = 66) \qquad (7a)$$

$$K_u(z, 490) = 1.0220\, K_d(z, 490) + 1.71 \times 10^{-3} \quad (r^2 = 0.691, RMSE = 0.021\ \mathrm{m}^{-1}, N = 66) \qquad (7b)$$

$$K_u(z, 550) = 1.4699\, K_d(z, 550) - 4.83 \times 10^{-2} \quad (r^2 = 0.672, RMSE = 0.023\ \mathrm{m}^{-1}, N = 66). \qquad (7c)$$

As the wavelength increases into the red spectral region the relationship weakens considerably or nearly vanishes as suggested by the large scatter of data points for 650 nm and 683 nm in Figure 13a. For these wavelengths, the r^2 values drop to 0.444 and 0.079, respectively. This result can be attributed to increased effects of inelastic processes with increasing wavelength, which act differentially on the downwelling and upwelling light attenuation. It is noteworthy that one possible application of the relationships between K_u and K_d in the blue and green is to use the K_d measurements for extrapolating E_u measurements to deeper depths where upwelling light is no longer detectable with the E_u sensor but is still detectable with the E_d sensor. However, a special caution must be exercised for situations with large solar zenith angles, as evidenced by one outlying data point in our dataset which was obtained for relatively large θ_s of 57.1° compared with the remaining data. While it is known that at large θ_s K_d tends to increase significantly in the surface layer [11,17], the effect of solar angle is expected to be less pronounced for K_u. Such result is seen in our dataset in Figure 6e,f for the near-surface measurements taken at the example station in the Farallon Basin.

Our analysis shows that the linear relationship can be also used for estimating the diffuse attenuation coefficient of net irradiance, K_E, from K_d using the depth-resolved data in the euphotic layer (Figure 13b). For the blue-green spectral region the best-fit relationship is nearly a perfect 1:1 line with the correlation coefficient $r^2 > 0.998$. Specifically, the slope and offset parameters of the linear fit are: 1.0001 and -1.67×10^{-4} m^{-1} for 440 nm, 0.9999 and 1.66×10^{-5} m^{-1} for 490 nm, and 1.0015 and 1.78×10^{-5} m^{-1} for 550 nm ($N = 66$ for each wavelength). For the red spectral region, the best fit parameters are still very close to the 1:1 line, specifically 0.9746 and 1.26×10^{-2} m^{-1} for 650 nm and 0.9857 and 1.92×10^{-2} m^{-1} for 683 nm ($N = 62$ for each wavelength). The r^2 values at these wavelengths are only slightly lower than in the blue-green region and remained above 0.965. These results indicate that the magnitudes and behavior of K_E throughout the water column were consistent with those of K_d regardless of the spectral region including the long-wavelength portion of the spectrum where the effects of inelastic processes are strongest. Therefore, the use of a single relationship between K_E and K_d may be satisfactory regardless of wavelength. Such relationship determined from the combined data collected at wavelengths of 440, 550, and 650 nm is:

$$K_E(z) = 1.0093\, K_d(z) + 7.88 \times 10^{-4} \quad (r^2 = 0.996, RMSE = 0.006\ \mathrm{m}^{-1}, N = 194). \qquad (8)$$

This relationship is plotted in Figure 13b. It can be useful when K_E is required, for example for the analysis of Gershun's equation, when E_d measurements are available but are not accompanied by E_u measurements.

Similar to K_E vs. K_d, the linear relationship between the K_o and K_d values measured throughout the water column is also quite robust regardless of light wavelength (Figure 13c). The plotted regression line corresponds to the best fit obtained from the combined data collected at 440, 550, and 650 nm:

$$K_o(z) = 0.9480\, K_d(z) - 1.24 \times 10^{-3} \quad (r^2 = 0.984, RMSE = 0.015\ \mathrm{m}^{-1}, N = 192). \qquad (9)$$

This type of relationship can be useful in the absence of scalar irradiance measurements throughout the water column which is a common situation, but when the measurement or estimate of spectral E_o

just below the surface is available. In such case, E_o can be propagated throughout the water column on the basis of vertical measurements of spectral E_d using the relationship between K_o and K_d. The resulting depth profile of spectral E_o can, in turn, be subject to conversion to quantum units and spectral integration to obtain the depth-resolved PAR quantum scalar irradiance, E_{oPAR}.

Assuming the availability of E_{oPAR} just below the surface, e.g., [108], another relationship between the K coefficients can be useful for estimating E_{oPAR} at different depths from vertical measurements of spectral E_d. This relationship of K_{oPAR} vs. K_d on the basis of data collected throughout the water column is shown in Figure 13d. In this analysis we ignored, however, the data collected at near-surface depths (<5 m) because K_{oPAR} decreases rapidly with depth within the near-surface layer owing to high absorption by water molecules within the long-wavelength portion of the spectrum [10,109–111]. Figure 13d demonstrates that there is a reasonably good linear relationship between K_{oPAR} and K_d in the blue and green spectral regions, as illustrated for 440 nm, 490 nm, and 550 nm. The best-fit relationships are:

$$K_{oPAR}(z) = 0.5643\, K_d(z, 440) + 3.07 \times 10^{-2} \quad (r^2 = 0.911,\ RMSE = 0.029\ \mathrm{m}^{-1},\ N = 54) \tag{10a}$$

$$K_{oPAR}(z) = 0.8504\, K_d(z, 490) + 2.41 \times 10^{-2} \quad (r^2 = 0.923,\ RMSE = 0.015\ \mathrm{m}^{-1},\ N = 54) \tag{10b}$$

$$K_{oPAR}(z) = 1.2256\, K_d(z, 550) - 2.52 \times 10^{-2} \quad (r^2 = 0.856,\ RMSE = 0.012\ \mathrm{m}^{-1},\ N = 54). \tag{10c}$$

A closer examination also showed that $K_{PAR}(z) \approx [K_d(z, 490) + K_d(z, 550)]/2$:

$$K_{oPAR}(z) = 1.0182[K_d(z, 490) + K_d(z, 550)]/2 - 2.61 \times 10^{-3}$$
$$(r^2 = 0.908,\ RMSE = 0.010\ \mathrm{m}^{-1},\ N = 54). \tag{10d}$$

Being very close to the 1:1 line, this relationship provides a convenient approximation to $K_{oPAR}(z)$ based on $K_d(z)$ and is plotted in Figure 13d.

Because of unavailability or scarcity of scalar irradiance measurements, the measurements of downward plane irradiance have been typically used in the past for estimating photosynthetically available radiation in terms of PAR quantum downward irradiance, E_{dPAR} [11,105,109–111]. Kirk [10] provides a comprehensive overview of K_{dPAR} values for various oceanic, coastal, estuarine, and inland aquatic environments. Our dataset shows a strong linear relationship between K_{oPAR} and K_{dPAR} throughout the examined water column (Figure 13e):

$$K_{oPAR}(z) = 0.9516\, K_{dPAR}(z) - 3.12 \times 10^{-3} \quad (r^2 = 0.937,\ RMSE = 0.012\ \mathrm{m}^{-1},\ N = 54). \tag{11}$$

This relationship provides an alternative to Equation (10d) for the use in a situation when E_{oPAR} just below the surface and vertical measurements of spectral E_d are available.

It is noteworthy to mention that because of widespread interest in the determinations of remote-sensing reflectance the contemporary underwater radiometric systems are often configured to include only two sensors, one for the spectral measurements of E_d and another for the spectral measurements of upwelling zenith radiance, L_u. Because such systems lack the measurement of E_u, it is useful to examine the relationship between K_u and K_{Lu}. The analysis of our dataset indicates that a single linear relationship provides reasonably good estimates of K_u from K_{Lu} regardless of light wavelength and depth (Figure 13f):

$$K_u = 0.9838\, K_{Lu} - 0.0010 \quad (r^2 = 0.905,\ RMSE = 0.016\ \mathrm{m}^{-1},\ N = 52). \tag{12}$$

This relationship was obtained from data collected at three wavelengths, 440, 550, and 650 nm. Note that the slope coefficient of the regression is close to 1 but many data points of K_u in the blue and green spectral bands are higher than K_{Lu}. This tendency is not seen in the red band, however.

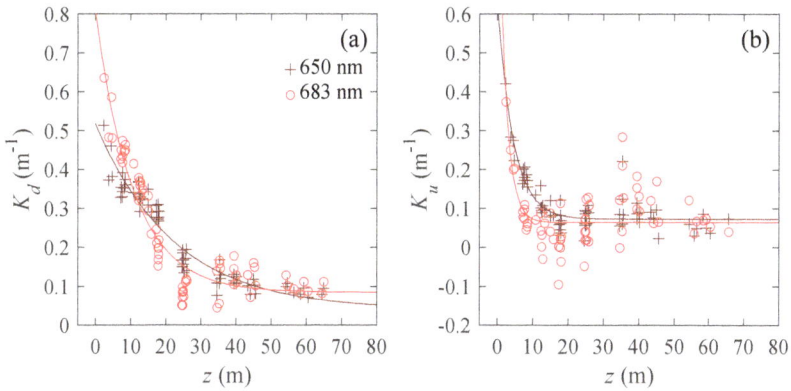

Figure 12. Diffuse attenuation coefficients in the red spectral bands (650 and 683 nm) plotted as a function of depth z using data from all stations and depths. (**a**) $K_d(z, 650)$ (crosses) and $K_d(z, 683)$ (open circles) vs. z. (**b**) Same as (**a**) but for K_u. Solid lines represent the best-fit regression functions.

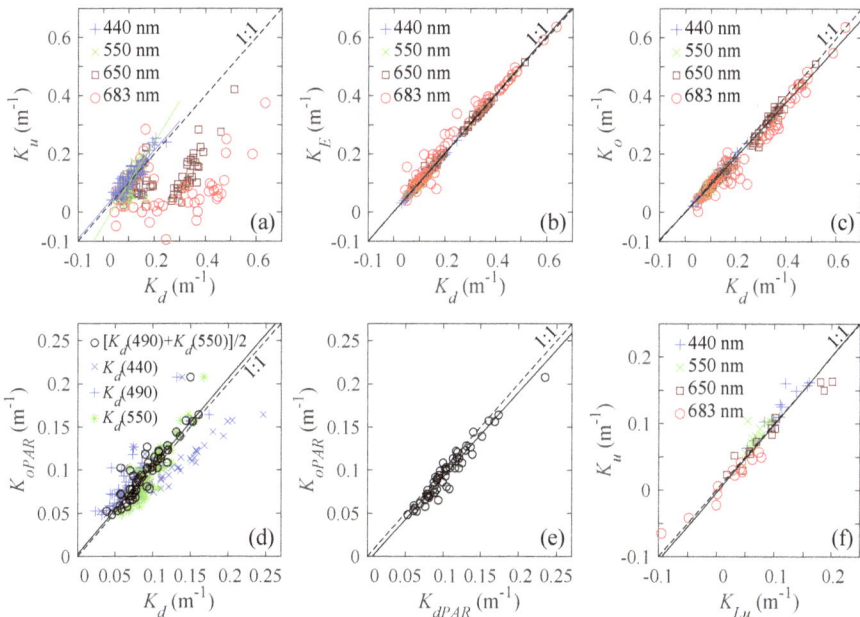

Figure 13. Relationships between diffuse attenuation coefficients based on measurements from all stations and depths. (**a**) K_u vs. K_d; (**b**) K_E vs. K_d; (**c**) K_o vs. K_d. (**d**) K_{dPAR} vs. K_d; (**e**) K_{oPAR} vs. K_{dPAR}; and (**f**) K_u vs. K_{Lu}. In (**a**–**c**,**f**), different symbols represent data at different light wavelengths as indicated. In (**d**) symbols indicate the relationships between $K_{oPAR}(z)$ and different K-coefficients as indicated, including the relationship between $K_{oPAR}(z)$ and $[K_d(z, 490) + K_d(z, 550)]/2$. The best fit regression functions (solid lines) are shown for (**a**) data at 440 nm and 550 nm analyzed separately, (**b**,**c**) combined data at 440 nm, 550 nm, and 650 nm, (**d**) data of $K_{oPAR}(z)$ vs. $[K_d(z, 490) + K_d(z, 550)]/2$, and (**e**,**f**) all data displayed in these panels. Dashed lines represent the 1:1 agreement between the variables.

3.3. Application of Gershun's Equation

The underwater radiometric measurements in conjunction with the use of Gershun's equation provide a method for determining the absorption coefficient of seawater, $a(\lambda, z)$, as a product of $K_E(\lambda, z)$ and $\overline{\mu}(\lambda, z)$. The main limitation of this radiometric method is that Gershun's equation was derived from a simplified radiative transfer equation that ignores inelastic processes of Raman scattering and fluorescence and internal sources (i.e., true light emission) such as bioluminescence, e.g., [9,11,49]. Therefore, in principle, this method can work satisfactorily only for cases in which the contributions of inelastic processes and internal sources are negligible. Because bioluminescence is typically intermittent and discrete under natural conditions in the ocean, the primary limitation of the applicability of Gershun's equation in the upper ocean or euphotic layer is associated with the presence of inelastic processes. In spite of these limitations, this radiometric method offers a unique benefit resulting from the estimation of the absorption coefficient that is representative of relatively large volume of seawater. This is because the radiometric measurements and AOPs are representative of much larger volumes of water compared with small volumes involved in direct IOP measurements including in situ absorption meters.

We examined the applicability of the radiometric method using our data collected at two stations, one in the Carmen Basin and another in the Farallon Basin (see Figure 1), where concurrent radiometric and ac-9 measurements were made. In this analysis, we compare the values of the absorption coefficient, $a_{ac9}(z, \lambda)$, measured with the ac-9 instrument with those derived from Gershun's equation, i.e., $a_{AOP}(z, \lambda) = K_E(z, \lambda) \cdot \overline{\mu}(z, \lambda)$. In order to enable such comparison, the data of $a_{ac9}(z, \lambda)$ and $\overline{\mu}(z, \lambda)$ were adjusted to common depths corresponding to the determinations of $K_E(z, \lambda)$. Specifically, $\overline{\mu}(z, \lambda)$ was linearly interpolated to obtain the values for depths at which $K_E(z, \lambda)$ was determined, i.e., the mid-points between the discrete depths (see Equation (1)) where the radiometric measurements, and hence $\overline{\mu}(z, \lambda)$, were made. The high depth resolution ac-9 data required only minor interpolation to these mid-point depths. In addition, the high spectral resolution data of $K_E(z, \lambda)$ and $\overline{\mu}(z, \lambda)$ were linearly interpolated to ac-9 wavelengths. Our comparative analysis of $a_{AOP}(z, \lambda)$ and $a_{ac9}(z, \lambda)$ is focused on five ac-9 wavelengths, 440, 488, 555, 650, and 676 nm.

In addition to the effects of inelastic processes and mismatch of spatial scales of AOP and ac-9 measurements, it is noteworthy to mention other sources of uncertainty that can affect the comparisons of $a_{ac9}(z, \lambda)$ and $a_{AOP}(z, \lambda)$. The accurate direct in situ measurements of $a(z, \lambda)$ with instruments such as ac-9 are difficult, mainly because of scattering error and stringent calibration requirements [112–114]. The determinations of $K_E(z, \lambda)$ and $\overline{\mu}(z, \lambda)$ require measurements with four (E_d, E_u, E_{od}, E_{ou}) or at least three (E_d, E_u, E_o) radiometric sensors, so these determinations are critically dependent on accurate calibrations of multiple sensors. The accuracy of determinations of $K_E(z, \lambda)$ for specific depths can be also affected by relatively low depth resolution of radiometric measurements that were taken at several discrete depths in our study.

In spite of multiple sources of uncertainty, we found reasonably good agreement between a_{AOP} and a_{ac9} throughout the euphotic layer for the blue and green spectral regions where the effects of inelastic processes were relatively weak (Figure 14a–c). As shown for the example data collected in the Farallon Basin, the applicability of Gershun's equation within the examined water column extends from the short-wavelength portion of the spectrum to about 580 nm, as indicated by the vertical dotted line in Figure 14a. At longer wavelengths this application failed owing to stronger effects of Raman scattering and chlorophyll-*a* fluorescence, even at the shallowest depth of 7.2 m examined in this case (Figure 14a,b). As the Raman scattering effects continue to accumulate with increasing depth the short-wavelength boundary of the affected spectral region is expected to shift towards shorter wavelengths [60,64,73]. Therefore, one can expect that at depths greater than those examined in our study the application of Gershun's equation may fail for wavelengths shorter than ~580 nm. Figure 14b also shows that somewhat larger discrepancy between a_{AOP} and a_{ac9} at 440 nm was observed within the Chl_{max} layer around the depth of z_{Chlmax}. This may be associated with the limitation that the radiometric measurements were taken at discrete depths with relatively low depth resolution (i.e., 10 m below $z = 20$ m), as opposed to nearly continuous depth profiles which would be required for better determinations of $K_E(z, \lambda)$ within the optically heterogeneous layer.

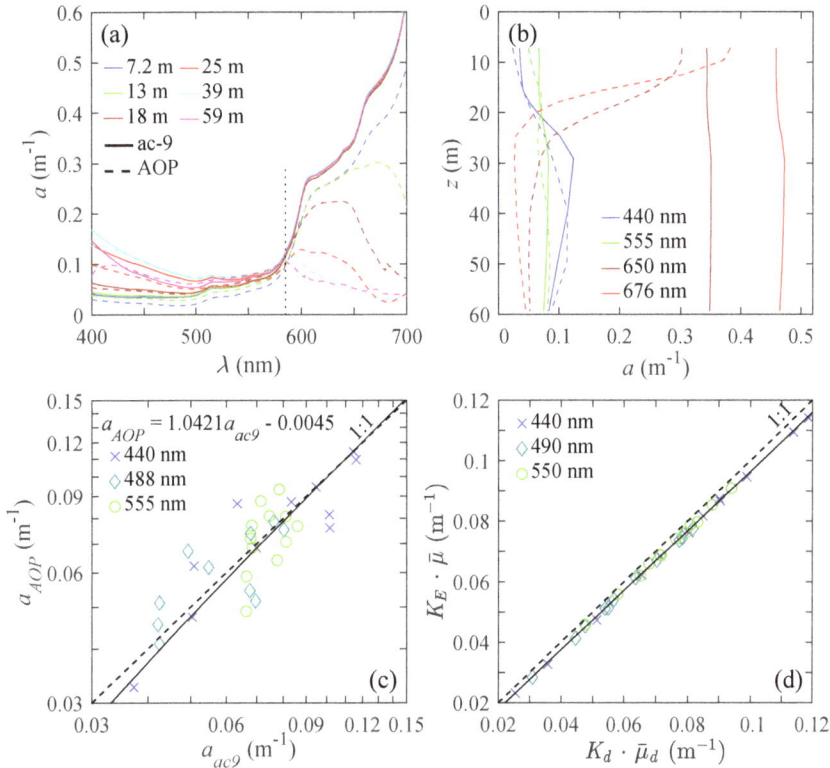

Figure 14. Comparison of the absorption coefficient of seawater measured with an ac-9 instrument (a_{ac9}) and estimated from AOPs using the irradiance quartet measurements and Gershun's equation (a_{AOP}). (**a**) Spectra of absorption coefficient at depths for which the determinations of K_E were made on the basis of radiometric measurements taken at one station in the Carmen Basin where ac-9 measurements were made during the 2011 cruise. The vertical dotted line indicates the transition region to long-wavelength portion of the spectrum where the use of Gershun's is inadequate because of the effects of inelastic processes. (**b**) Vertical profiles of absorption coefficient measured with the ac-9 instrument (solid lines) and estimated from AOPs (dashed lines) at selected light wavelengths as indicated. Data were collected at the same station as in panel (**a**). (**c**) Direct comparison between a_{AOP} and a_{ac9} at selected blue and green spectral bands as indicated. Data were collected at one station in the Guaymas Basin and one station in the Carmen Basin where ac-9 measurements were made during the 2011 cruise. (**d**) Relationship between $K_E \cdot \bar{\mu}$ and $K_d \cdot \bar{\mu}_d$ for the blue-green spectral regions based on data at the three selected light wavelengths as indicated. Solid lines in (**c**,**d**) represent the best-fit regression functions and dashed lines the 1:1 agreement between the variables.

Figure 14c shows that a_{AOP} and a_{ac9} compare favorably for the combined dataset from the two stations, which includes data points for the three wavelengths from the blue-green spectral region (440, 488, and 555 nm) and all discrete depths included in this analysis (i.e., where K_E was determined). The best linear fit to the data is very close to the 1:1 line and the correlation is strong ($r^2 = 0.750$). This was the only case in this study when Model II regression analysis was applied (see Section 2.6). If the ac-9-measured values are tentatively assumed to represent true values, then the statistical indicators suggest a very small bias of a_{AOP} estimates as the median value for the ratio of a_{AOP}/a_{ac9} is very close to 1 (0.989) and the mean difference between a_{AOP} and a_{ac9} is very small (-0.0016 m^{-1}). The median absolute percent difference between a_{AOP} and a_{ac9} is 12.4% and the root mean square deviation is 0.0116 m^{-1}.

Because of tight relationships $\bar{\mu}$ vs. $\bar{\mu}_d$ (Figure 10a) and K_E vs. K_d (Figure 13b), it is natural to expect a good relationship between the products $K_E \cdot \bar{\mu}$ and $K_d \cdot \bar{\mu}_d$. This is indeed the case as shown for our data combining the three wavelengths from the blue-green spectral region (440, 490, and 550 nm) and all examined discrete depths at all stations (Figure 14d). The best-fit linear fit to the data is:

$$K_E \cdot \bar{\mu} = 0.980 \, K_d \cdot \bar{\mu}_d - 0.0017 \left(r^2 = 0.999, RMSE = 0.003 \, m^{-1}, N = 36 \right). \tag{13}$$

By indicating that $K_E \cdot \bar{\mu}$ can be quite accurately estimated from $K_d \cdot \bar{\mu}_d$, this relationship has important practical ramifications. Specifically, the measurements of three (E_d, E_u, E_0) or four (E_d, E_u, E_{od}, E_{ou}) irradiances required for determining the absorption coefficient from Gershun's equation can be replaced with just two downwelling irradiance measurements (E_d and E_{od}), thus significantly simplifying the experimental requirements. We note that the use of two AOPs based on the downwelling light field, K_d and $\bar{\mu}_d$, has been previously introduced in the context of estimation of absorption by phytoplankton and CDOM, as well as chlorophyll-*a* concentration from measurements of E_d alone [96,115]. The modeling formalism based on E_d alone required assumptions about $\bar{\mu}_d$ and the backscattering coefficient of seawater.

4. Summary

Although quantitative radiometric measurements in the ocean have a long-recorded history with the beginning in the early 1930s, the measurements of scalar irradiance have been rare and simultaneous measurements of downward and upward plane and scalar irradiances (E_d, E_u, E_{od}, and E_{ou}), the so-called irradiance quartet, are virtually lacking. To our knowledge, this is the first-ever study to report data of high spectral resolution measurements of irradiance quartet within the ocean euphotic layer and the apparent optical properties (AOPs) derivable from this quartet, including a complete set of average cosines characterizing the angular structure of underwater light field, i.e., the downwelling ($\bar{\mu}_d$), upwelling ($\bar{\mu}_u$), and total ($\bar{\mu}$) light field. Because the experimental irradiance data collected in the past have been fragmentary in terms of incomplete set of irradiances or limited spectral coverage in rare experiments when E_d, E_u, and E_0 were measured directly or derived from radiance measurements, currently available descriptions and understanding of underwater light field and AOPs associated with a complete irradiance quartet have relied mostly on theoretical simulations of radiative transfer in the ocean. Such simulations unavoidably involve some idealized assumptions about the real environment. Thus, the experimental data presented in this paper provide a unique and revealing resource on the patterns of irradiance quartet and associated suite of AOPs within a real water column that was characterized by the interplay of non-uniform vertical distribution of inherent optical properties (IOPs) of seawater and the effects of inelastic radiative processes such as Raman scattering by water molecules and chlorophyll-*a* fluorescence.

Our dataset was collected in the Gulf of California within the euphotic layer from the near-surface depths to ~80 m and covered a broad range of IOPs, sky conditions, and solar zenith angle. Our data show that in the blue and green spectral regions, the vertical patterns of AOPs were driven primarily by IOPs of seawater with weak or no discernible effects of inelastic radiative processes. In contrast, in the red spectral region the radiometric variables characterizing the light field and the AOPs were strongly affected or totally dominated by inelastic processes of Raman scattering, and additionally by chlorophyll-*a* fluorescence within the fluorescence emission band, the latter being particularly well-pronounced within the chlorophyll-*a* maximum layer. Some of the most notable features caused by inelastic processes in the red include the values of average cosines which approached those of the uniform light field, exceptionally high values of irradiance reflectance R approaching

1, and exceptionally low values (i.e., smaller than the seawater absorption coefficient) of the diffuse attenuation coefficients, K, for various irradiances, including the negative values for the attenuation of upwelling irradiances in the chlorophyll-*a* fluorescence emission band within the chlorophyll-*a* maximum layer.

Whereas details of the vertical patterns in the light field characteristics and AOPs depended on the interplay of inelastic processes and vertically non-uniform IOPs including the chlorophyll-*a* concentration, there were general vertical patterns associated with an increase in the contribution of inelastically produced light with increasing depth. We observed, for example, that the vertical patterns of some AOPs in the red, such as $\bar{\mu}_d$, $\bar{\mu}$, R, K_d, and K_u, can be approximated by relatively simple functions of depth. We also established empirical relationships between some AOPs, which can be useful in common experimental situations when only data on downwelling plane irradiance are available. For example, we proposed single relationships for estimating $K_E(z)$ and $K_0(z)$ from $K_d(z)$ at any depth z within the examined depth range regardless of light wavelength. Similarly, we also determined that a single relationship can be used for estimating $K_u(z)$ from $K_d(z)$ within the blue-green spectral region, in which the effects of inelastic processes are generally weak. Our results also demonstrated that below the near-surface layer (i.e., z approximately greater than 5 m in our dataset) the values of $K_{oPAR}(z)$ can be estimated from measurements of $K_d(z)$ in the blue-green bands using a single relationship. Typically, the empirical relationships are expected to be applicable within a range of environmental conditions consistent with the dataset used in the development of the relationships. This limitation also applies to the relationships established in this study, in which we covered a fairly broad range of conditions (Table 2).

We tested the use of irradiance quartet measurements in conjunction with Gershun's equation for estimating the absorption coefficient of seawater and found that in the blue-green spectral region such estimates agree reasonably well with direct determinations of the absorption coefficient from measurements with the ac-9 instrument. As the Gershun equation is inadequate in situations when inelastic processes are significant, no agreement was observed at light wavelengths longer than about 580 nm. Specifically, in the long-wavelength portion of the spectrum the absorption coefficients estimated from Gershun's equation were greatly underestimated due to the effects of inelastic processes. The analysis of our data from the entire examined depth range also showed that the product of two AOPs in the blue-green spectral region, $K_E \cdot \bar{\mu}$, which is used for estimating the absorption coefficient from Gershun's equation, can be adequately estimated from the product $K_d \cdot \bar{\mu}_d$. This is potentially useful because the determinations of $K_E \cdot \bar{\mu}$ require the measurements of irradiance quartet or at least three irradiances $E_d(z)$, $E_u(z)$, and $E_0(z)$, whereas the determinations of $K_d \cdot \bar{\mu}_d$ require a simpler experimental design with only two irradiances, E_d and E_{od}.

Author Contributions: L.L. and D.S. wrote the paper; M.D. collected the field data; L.L. and M.D. processed the data; L.L. analyzed the data; and all authors made intellectual contributions and reviewed the manuscript.

Acknowledgments: This study was supported by the U.S. Office of Naval Research (grant N00014-09-1-1053) as part of the Department of Defense Multidisciplinary University Research Initiative (MURI) Program. Partial support for Linhai Li was provided by the Scripps Institution of Oceanography. The scientists, crew members, and officers of R/V New Horizon are acknowledged for providing logistical support and help during the field work. In particular, we thank Alison Sweeney for providing data from CTD measurements and Yakir Gagnon for ac-9 data. We also thank Rick A. Reynolds and anonymous reviewers for providing comments on the manuscript.

Conflicts of Interest: The authors declare no conflict of interest.

References

1. Lewis, M.R.; Carr, M.E.; Feldman, G.C.; Esaias, W.; McClain, C. Influence of penetrating solar-radiation on the heat-budget of the equatorial Pacific Ocean. *Nature* **1990**, *347*, 543–545. [CrossRef]
2. Lee, Z.P.; Du, K.P.; Arnone, R.; Liew, S.C.; Penta, B. Penetration of solar radiation in the upper ocean: A numerical model for oceanic and coastal waters. *J. Geophys. Res.* **2005**, *110*, C09019. [CrossRef]
3. Field, C.B.; Behrenfeld, M.J.; Randerson, J.T.; Falkowski, P. Primary production of the biosphere: Integrating terrestrial and oceanic components. *Science* **1998**, *281*, 237–240. [CrossRef] [PubMed]

4. Mobley, C.D.; Chai, F.; Xiu, P.; Sundman, L.K. Impact of improved light calculations on predicted phytoplankton growth and heating in an idealized upwelling-downwelling channel geometry. *J. Geophys. Res. Oceans* **2015**, *120*, 875–892. [CrossRef]
5. Smith, R.C; Wilson, W.H., Jr. Photon scalar irradiance. *Appl. Opt.* **1972**, *11*, 934–938. [CrossRef] [PubMed]
6. Booth, C.R. The design and evaluation of a measurement system for photosynthertically active quantum scalar irradiance. *Limnol. Oceanogr.* **1975**, *21*, 326–336. [CrossRef]
7. Højerslev, N.K. Daylight measurements appropriate for photosynthetic studies in natural sea waters. *J. Cons. Int. Explor. Mer* **1978**, *38*, 131–146. [CrossRef]
8. Morel, A. Available, usable, and stored radiant energy in relation to marine photosynthesis. *Deep-Sea Res.* **1978**, *25*, 673–688. [CrossRef]
9. Mobley, C.D. *Light and Water: Radiative Transfer in Natural Waters*; Academic Press: San Diego, CA, USA, 1994.
10. Kirk, J.T. *Light and Photosynthesis in Aquatic Ecosystems*; Cambridge University Press: Cambridge, UK, 1994.
11. Jerlov, N.G. *Marine Optics*; Elsevier Scientific Publishing Company: Amsterdam, The Netherlands, 1976.
12. Jerlov, N.G. Optical studies of ocean water. *Rep. Swed. Deep-Sea Exped.* **1951**, *3*, 1–59.
13. Tyler, J.E.; Smith, R.C. *Measurements of Spectral Irradiance Underwater*; Gordon & Breach Publishing Group: Philadelphia, PA, USA, 1970.
14. Jerlov, N.G.; Liljequist, G.H. On the angular distribution of submarine daylight and on the total submarine illumination. *Svenska Hydrografisk-Biologiska Kommisionens Skrifter Ny Serie Hydrografi* **1938**, *14*, 1–15.
15. Whitney, L.V. The angular distribution of characteristic diffuse light in natural waters. *J. Mar. Res.* **1941**, *4*, 122–131.
16. Smith, R.C. Structure of solar radiation in the upper layers of the sea. In *Optical Aspects of Oceanography*; Jerlov, N.G., Steemann, E., Eds.; Academic Press: San Diego, CA, USA, 1974; pp. 95–117.
17. Lundgren, B.; Højerslev, N.K. *Daylight Measurements in the Sargasso Sea: Results from the "Dana" Expedition January–April 1966. Report 14*; University of Copenhagen, Institution of Physical Oceanography: Copenhagen, Denmark, 1971.
18. Højerslev, N.K. *Inherent and Apparent Optical Properties of the Western Mediterranean and the Hardangerfjord. Report 21*; University of Copenhagen, Institution of Physical Oceanography: Copenhagen, Denmark, 1973.
19. Ochakovskiy, Y.Y.; Kopelevich, O.V.; Voytov, V.I. *Light in the Sea*; (Edited translation from Russian) FTD-HT-24-105-72; Foreign Technology Division, Air Force System Command, US Air Force; National Technical Information Service, U.S. Department of Commerce: Springfieldm, VA, USA, 1972.
20. Voss, K.J. Electro-optic camera system for measurement of the underwater radiance distribution. *Opt. Eng.* **1989**, *28*, 384–387. [CrossRef]
21. Dickey, T.; Frye, D.; Jannasch, H.; Boyle, E.; Manov, D.; Sigurdson, D.; McNeil, J.; Stramska, M.; Michaels, A.; Nelson, N.; et al. Initial results from the Bermuda testbed mooring program. *Deep Sea Res. Part I* **1998**, *45*, 771–794. [CrossRef]
22. Hooker, S.B.; Maritorena, S. An evaluation of oceanographic radiometers and deployment methodologies. *J. Atmos. Ocean. Technol.* **2000**, *17*, 811–830. [CrossRef]
23. Clark, D.K.; Feinholz, M.; Yarbrough, M.; Johnson, B.C.; Brown, S.W.; Kim, Y.S.; Barnes, R.A. Overview of the radiometric calibration of MOBY. In *Earth Observing Systems VI*; Barres, W.L., Ed.; SPIE Proceedings; SPIE: Bellingham, WA, USA, 2002; Volume 4483, pp. 64–76. [CrossRef]
24. Bulgarelli, B.; Zibordi, G.; Berthon, J.F. Measured and modeled radiometric quantities in coastal waters: Toward a closure. *Appl. Opt.* **2003**, *42*, 5365–5381. [CrossRef] [PubMed]
25. Voss, K.J.; Chapin, A. Upwelling radiance distribution camera system, NURADS. *Opt. Express* **2005**, *13*, 4250–4262. [CrossRef] [PubMed]
26. Morel, A.; Gentili, B.; Claustre, H.; Babin, M.; Bricaud, A.; Ras, J.; Tieche, F. Optical properties of the "clearest" natural waters. *Limnol. Oceanogr.* **2007**, *52*, 217–229. [CrossRef]
27. Antoine, D.; d'Ortenzio, F.; Hooker, S.B.; Bécu, G.; Gentili, B.; Tailliez, D.; Scott, A.J. Assessment of uncertainty in the ocean reflectance determined by three satellite ocean color sensors (MERIS, SeaWiFS and MODIS-A) at an offshore site in the Mediterranean Sea (BOUSSOLE project). *J. Geophys. Res.* **2008**, *113*, C07013. [CrossRef]
28. Chang, G.C.; Dickey, T.D. Interdisciplinary sampling strategies for detection and characterization of harmful algal blooms. In *Real-Time Observation Systems for Ecosystem Dynamics and Harmful Algal Blooms*; Babin, M., Roesler, C.S., Cullen, J.J., Eds.; UNESCO: Paris, France, 2008; pp. 43–84.

29. Lewis, M.R.; Wei, J.; Van Dommelen, R.; Voss, K.J. Quantitative estimation of the underwater radiance distribution. *J. Geophys. Res.* **2011**, *116*, C00H06. [CrossRef]
30. Wei, J.; Dommelen, R.V.; Lewis, M.R.; McLean, S.; Voss, K.J. A new instrument for measuring the high dynamic range radiance distribution in near surface sea water. *Opt. Express* **2012**, *20*, 27024–27038. [CrossRef] [PubMed]
31. Antoine, D.; Morel, A.; Leymarie, E.; Houyou, A.; Gentili, B.; Victori, S.; Buis, J.-P.; Buis, N.; Meunier, S.; Canini, M.; et al. Underwater radiance distributions measured with miniaturized multispectral radiance cameras. *J. Atmos. Ocean. Technol.* **2013**, *30*, 74–95. [CrossRef]
32. Cunningham, A.; McKee, D. Measurement of hyperspectral underwater light fields. In *Subsea Optics and Imaging*; Watson, J.E., Zielinski, O., Eds.; Woodhead Publishing: Philadelphia, PA, USA, 2013; pp. 83–97.
33. Lewis, M.R. Measurement of apparent optical properties for diagonosis of harmful algal blooms. In *Real-Time Observation Systems for Ecosystem Dynamics and Harmful Algal Blooms*; Babin, M., Roesler, C.S., Cullen, J.J., Eds.; UNESCO: Paris, France, 2008; pp. 207–236.
34. Zibordi, G.; Voss, K.J. In situ optical radiometry in the visible and near infrared. In *Optical Radiometry for Ocean Climate Measurements*; Zibordi, G., Donlon, C.J., Parr, A.C., Eds.; Academic Press: San Diego, CA, USA, 2014; pp. 248–305.
35. Smith, R.C.; Baker, K.S. The bio-optical state of ocean waters and remote sensing. *Limnol. Oceanogr.* **1978**, *23*, 247–259. [CrossRef]
36. Siegel, D.A.; Dickey, T.D. Observations of the vertical structure of the diffuse attenuation coefficient spectrum. *Deep-Sea Res.* **1987**, *34*, 547–563. [CrossRef]
37. Dirks, R.W.J.; Spitzer, D. On the radiative transfer in the sea, including fluorescence and stratification effects. *Limnol. Oceanogr.* **1987**, *32*, 942–953. [CrossRef]
38. Morel, A. Optical modeling of the upper ocean in relation to its biogenous matter content (case I waters). *J. Geophys. Res.* **1988**, *93*, 10749–10768. [CrossRef]
39. Morel, A.; Maritorena, S. Bio-optical properties of oceanic waters: A reappraisal. *J. Geophys. Res.* **2001**, *106*, 7163–7180. [CrossRef]
40. Zibordi, G.; D'Alimonte, D.; Berthon, J.-F. An evaluation of depth resolution requirements for optical profiling in coastal waters. *J. Atmos. Ocean. Technol.* **2004**, *21*, 1059–1073. [CrossRef]
41. Yarbrough, M.A.; Houlihan, T.; Feinholz, M.; Flora, S.; Johnson, B.C.; Kim, Y.S.; Murphy, M.Y.; Ondrusek, M.; Clark, D. Results in coastal waters with high resolution in-situ spectral radiometry: The Marine Optical System ROV. In *Coastal Ocean Remote Sensing*; Frouin, R.J., Ed.; SPIE Proceedings; SPIE: Bellingham, WA, USA, 2007; Volume 6680, p. 66800I.
42. Antoine, D.; Hooker, S.B.; Bélanger, S.; Matsuoka, A.; Babin, M. Apparent optical properties of the Canadian Beaufort Sea—Part 1: Observational overview and water column relationships. *Biogeosciences* **2013**, *10*, 4493–4509. [CrossRef]
43. Lee, Z; Shaoling, S; Stavn, R.H. AOPs are not additive: On the biogeo-optical modeling of the diffuse attenuation coefficient. *Front. Mar. Sci.* **2018**, *5*, Article 8. [CrossRef]
44. Pegau, W.S.; Cleveland, J.S.; Doss, W.; Kennedy, C.D.; Maffione, R.A.; Mueller, J.L.; Stone, R.; Trees, C.C.; Weidemann, A.D.; Wells, W.H.; et al. A comparison of methods for the measurement of the absorption coefficient in natural waters. *J. Geophys. Res.* **1995**, *100*, 13201–13220. [CrossRef]
45. Preisendorfer, R.W. Application of radiative transfer theory to light measurements in the sea. *Int. Union Geod. Geophys. Monogr. Symp. Radiant Energy Sea* **1961**, *10*, 11–30.
46. Preisendorfer, R.W. *Hydrologic Optics. Volume I. Introduction*; US Department of Commerce, National Oceanic and Atmospheric Administration, Environmental Research Laboratories: Honolulu, HI, USA, 1976.
47. Kirk, J.T. Volume scattering function, average cosines, and the underwater light field. *Limnol. Oceanogr.* **1991**, *36*, 455–467. [CrossRef]
48. Gordon, H.R. Modeling and simulating radiative transfer in the ocean . In *Ocean Optics*; Spinrad, R.W., Carder, K.L., Perry, M.J., Eds.; Oxford University Press: New York, NY, USA, 1994; pp. 3–39.
49. Gershun, A. The light field. *J. Math. Phys.* **1939**, *18*, 51–151. [CrossRef]
50. Tyler, J.E.; Richardson, W.H.; Holmes, R.W. Method for obtaining the optical properties of large bodies of water. *J. Geophys. Res.* **1959**, *64*, 667–673. [CrossRef]
51. Højerslev, N.K. A spectral light absorption meter for measurements in the sea. *Limnol. Oceanogr.* **1975**, *20*, 1024–1034. [CrossRef]

52. Spitzer, D.; Wernand, M.R. In situ measurements of absorption spectra in the sea. *Deep-Sea Res.* **1981**, *28A*, 165–174. [CrossRef]
53. Gordon, H.R. Inverse methods in hydrologic optics. *Oceanologia* **2002**, *44*, 9–58.
54. Gordon, H.R.; Brown, O.B.; Jacobs, M.M. Computed relationships between the inherent and apparent optical properties of a flat homogeneous ocean. *Appl. Opt.* **1975**, *14*, 417–427. [CrossRef] [PubMed]
55. Kirk, J.T. Monte Carlo study of the nature of the underwater light field in, and of the relationship between optical properties of, turbid yellow waters. *Aust. J. Mar. Freshw. Res.* **1981**, *32*, 517–532. [CrossRef]
56. Gordon, H.R. Absorption and scattering estimates from irradiance measurements: Monte Carlo simulations. *Limnol. Oceanogr.* **1991**, *36*, 769–777. [CrossRef]
57. Sathyendranath, S.; Platt, T. Angular distribution of the submarine light field: Modification by multiple scattering. *Proc. R. Soc. Lond. A* **1991**, *433*, 287–297. [CrossRef]
58. Stramska, M.; Stramski, D.; Mitchell, B.G.; Mobley, C.D. Estimation of the absorption and backscattering coefficients from in-water radiometric measurements. *Limnol. Oceanogr.* **2000**, *45*, 628–641. [CrossRef]
59. Loisel, H.; Stramski, D. Estimation of inherent optical properties of natural waters from the irradiance attenuation coefficient and reflectance in the presence of Raman scattering. *Appl. Opt.* **2000**, *39*, 3001–3011. [CrossRef] [PubMed]
60. Morel, A.; Gentili, B. Radiation transport within oceanic (case 1) water. *J. Geophys. Res.* **2004**, *109*, C06008. [CrossRef]
61. Lewis, M.R.; Cullen, J.J.; Platt, T. Phytoplankton and thermal structure of the ocean: Consequences of nonuniformity in the chlorophyll profile. *J. Geophys. Res.* **1983**, *88*, 2565–2570. [CrossRef]
62. Gordon, H.R. Diffuse reflectance of the ocean: Influence of nonuniform phytoplankton pigment profile. *Appl. Opt.* **1992**, *31*, 2116–2129. [CrossRef] [PubMed]
63. Stramska, M.; Stramski, D. Effects of a nonuniform vertical profile of chlorophyll concentration on remote-sensing reflectance of the ocean. *Appl. Opt.* **2005**, *44*, 1735–1747. [CrossRef] [PubMed]
64. Li, L.; Stramski, D.; Reynolds, R.A. Characterization of the solar light field within the ocean mesopelagic zone based on radiative transfer simulations. *Deep Sea Res. Part I* **2014**, *87*, 53–69. [CrossRef]
65. Gordon, H. Remote sensing of optical properties in continuously stratified waters. *Appl. Opt.* **1978**, *17*, 1893–1897. [CrossRef] [PubMed]
66. Yang, Q.; Stramski, D.; He, M.-X. Modeling the effects of near-surface plumes of suspended particulate matter on remote-sensing reflectance of coastal waters. *Appl. Opt.* **2013**, *52*, 359–374. [CrossRef] [PubMed]
67. Stavn, R.H.; Weidemann, A.D. Optical modeling of clear ocean light fields: Raman scattering effects. *Appl. Opt.* **1988**, *27*, 4002–4011. [CrossRef] [PubMed]
68. Marshall, B.R.; Smith, R.C. Raman scattering and in-water ocean optical properties. *Appl. Opt.* **1990**, *29*, 71–84. [CrossRef] [PubMed]
69. Kattawar, G.W.; Xu, X. Filling in of Fraunhofer lines in the ocean by Raman scattering. *Appl. Opt.* **1992**, *31*, 6491–6500. [CrossRef] [PubMed]
70. Ge, Y.; Gordon, H.R.; Voss, K.J. Simulation of inelastic-scattering contributions to the irradiance field in the ocean: Variation in Fraunhofer line depths. *Appl. Opt.* **1993**, *32*, 4028–4036. [CrossRef] [PubMed]
71. Stavn, R.H. Effects of Raman scattering across the visible spectrum in clear ocean water: A Monte Carlo study. *Appl. Opt.* **1993**, *32*, 6853–6863. [CrossRef] [PubMed]
72. Hu, C.; Voss, K.J. In situ measurements of Raman scattering in clear ocean water. *Appl. Opt.* **1997**, *36*, 6962–6967. [CrossRef] [PubMed]
73. Berwald, J.; Stramski, D.; Mobley, C.D.; Kiefer, D.A. The effect of Raman scattering on the average cosine and the diffuse attenuation coefficient of irradiance in the ocean. *Limnol. Oceanogr.* **1998**, *43*, 564–576. [CrossRef]
74. Gordon, H.R. Contribution of Raman scattering to water-leaving radiance: A reexamination. *Appl. Opt.* **1999**, *38*, 3166–3174. [CrossRef] [PubMed]
75. Li, L.; Stramski, D.; Reynolds, R.A. Effects of inelastic radiative processes on the determination of water-leaving spectral radiance from extrapolation of underwater near-surface measurements. *Appl. Opt.* **2016**, *55*, 7050–7067. [CrossRef] [PubMed]
76. Lavín, M.F.; Marinone, S.G. An overview of the physical oceanography of the Gulf of California. In *Nonlinear Processes in Geophysical Fluid Dynamics*; Velasco Fuentes, O.U., Sheinbaum, J., Ochoa, J., Eds.; Springer: Dordrecht, The Netherlands, 2003; pp. 173–204.

77. Lluch-Cota, S.E.; Aragon-Noriega, E.A.; Arreguín-Sánchez, F.; Aurioles-Gamboa, D.; Bautista-Romero, J.J.; Brusca, R.C.; Cervantes-Duarte, R.; Cortés-Altamirano, R.; Del-Monte-Luna, P.; Esquivel-Herrera, A.; et al. The Gulf of California: Review of ecosystem status and sustainability challenges. *Prog. Oceanogr.* **2007**, *73*, 1–26. [CrossRef]

78. Ras, J.; Claustre, H.; Uitz, J. Spatial variability of phytoplankton pigment distributions in the Subtropical South Pacific Ocean: Comparison between in situ and predicted data. *Biogeosciences* **2008**, *5*, 353–369. [CrossRef]

79. Ritchie, R.J. Universal chlorophyll equations for estimating chlorophylls a, b, c, and d and total chlorophylls in natural assemblages of photosynthetic organisms using acetone, methanol, or ethanol solvents. *Photosynthetica* **2008**, *46*, 115–126. [CrossRef]

80. Gordon, H.R. Ship perturbation of irradiance measurements at sea 1: Monte Carlo simulations. *Appl. Opt.* **1985**, *24*, 4172–4182. [CrossRef] [PubMed]

81. Waters, K.J.; Smith, R.C.; Lewis, M.R. Avoiding ship-induced light-field perturbation in the determination of oceanic optical properties. *Oceanography* **1990**, *3*, 18–21. [CrossRef]

82. Zibordi, G.; Darecki, M. Immersion factors for the RAMSES series of hyper-spectral underwater radiometers. *J. Opt. A Pure Appl. Opt.* **2006**, *8*, 252–258. [CrossRef]

83. Stramski, D. Fluctuations of solar irradiance induced by surface waves in the Baltic. *Bull. Pol. Acad. Sci. Earth Sci.* **1986**, *34*, 333–344.

84. Darecki, M.; Stramski, D.; Sokólski, M. Measurements of high-frequency light fluctuations induced by sea surface waves with an Underwater Porcupine Radiometer System. *J. Geophys. Res.* **2011**, *116*, C00H09. [CrossRef]

85. Sullivan, J.M.; Twardowski, M.S.; Zaneveld, J.R.V.; Moore, C.M.; Barnard, A.H.; Donaghay, P.L.; Rhoades, B. Hyperspectral temperature and salt dependencies of absorption by water and heavy water in the 400–750 nm spectral range. *Appl. Opt.* **2006**, *45*, 5294–5309. [CrossRef] [PubMed]

86. Sokal, R.R.; Rohlf, F.J. *Biometry: The Principles and Practice of Statistics in Biological Research*, 3rd ed.; W.H. Freeman: New York, NY, USA, 1995.

87. Banse, K. Should we continue to use the 1% light depth convention for estimating the compensation depth of phytoplankton for another 70 years. *Limnol. Oceanogr. Bull.* **2004**, *13*, 49–52. [CrossRef]

88. Sugihara, S.; Kishino, M.; Okami, N. Contribution of Raman scattering to upward irradiance in the sea. *J. Oceanogr. Soc. Jpn.* **1984**, *40*, 397–404. [CrossRef]

89. Bristow, M.; Nielsen, D.; Bundy, D.; Furtek, R. Use of water Raman emission to correct airborne laser fluorosensor data for effects of water optical attenuation. *Appl. Opt.* **1981**, *20*, 2889–2906. [CrossRef] [PubMed]

90. Hoge, F.E.; Swift, R.N. Airborne simultaneous spectroscopic detection of laser-induced water Raman backscatter and fluorescence from chlorophyll a and other naturally occurring pigments. *Appl. Opt.* **1981**, *20*, 3197–3205. [CrossRef] [PubMed]

91. Stramski, D.; Tęgowski, J. Effects of intermittent entrainment of air bubbles by breaking wind waves on ocean reflectance and underwater light field. *J. Geophys. Res.* **2001**, *106*, 31345–31360. [CrossRef]

92. Bannister, T.T. Model of the mean cosine of underwater radiance and estimation of underwater scalar irradiance. *Limnol. Oceanogr.* **1992**, *37*, 773–780. [CrossRef]

93. Berwald, J.; Stramski, D.; Mobley, C.D.; Kiefer, D.A. Influences of absorption and scattering on vertical changes in the average cosine of the underwater light field. *Limnol. Oceanogr.* **1995**, *40*, 1347–1357. [CrossRef]

94. Aas, E.; Højerslev, N.K. Analysis of underwater radiance observations: Apparent optical properties and analytic functions describing the angular radiance distribution. *J. Geophys. Res.* **1999**, *104*, 8015–8024. [CrossRef]

95. Sathyendranath, S.; Platt, T. Ocean-color model incorporating transspectral processes. *Appl. Opt.* **1998**, *37*, 2216–2227. [CrossRef] [PubMed]

96. Nahorniak, J.S.; Abbott, M.R.; Letelier, R.M.; Pegau, W.S. Analysis of a method to estimate chlorophyll-a concentration from irradiance measurements at varying depths. *J. Atmos. Ocean. Technol.* **2001**, *18*, 2063–2073. [CrossRef]

97. Brown, C.A.; Huot, Y.; Purcell, M.J.; Cullen, J.J.; Lewis, M.R. Mapping coastal optical and biogeochemical variability using an autonomous underwater vehicle and a new bio-optical inversion algorithm. *Limnol. Oceanogr. Methods* **2004**, *2*, 262–281. [CrossRef]

98. Xing, X.; Morel, A.; Claustre, H.; d'Ortenzio, F.; Poteau, A. Combined processing and mutual interpretation of radiometry and fluorometry from autonomous profiling Bio-Argo floats: 2. Colored dissolved organic matter absorption retrieval. *J. Geophys. Res.* **2012**, *117*, C04022. [CrossRef]
99. Westberry, T.K.; Boss, E.; Lee, Z. Influence of Raman scattering on ocean color inversion models. *Appl. Opt.* **2013**, *52*, 5552–5561. [CrossRef] [PubMed]
100. Liu, C.C.; Carder, K.L.; Miller, R.L.; Ivey, J.E. Fast and accurate model of underwater scalar irradiance. *Appl. Opt.* **2002**, *41*, 4962–4974. [CrossRef] [PubMed]
101. Liu, C.C.; Miller, R.L.; Carder, K.L.; Lee, Z.; D'Sa, E.J.; Ivey, J.E. Estimating the underwater light field from remote sensing of ocean color. *J. Oceanogr.* **2006**, *62*, 235–248. [CrossRef]
102. Sathyendranath, S.; Platt, T. Computation of aquatic primary production: Extended formalism to include effect of angular and spectral distribution of light. *Limnol. Oceanogr.* **1989**, *34*, 188–198. [CrossRef]
103. O'Reilly, J.E.; Maritorena, S.; Mitchell, B.G.; Siegel, D.A.; Carder, K.L.; Garver, S.A.; Kahru, M.; McClain, C. Ocean color chlorophyll algorithms for SeaWiFS. *J. Geophys. Res.* **1998**, *103*, 24937–24953. [CrossRef]
104. O'Reilly, J.E.; Maritorena, S.; Siegel, D.A.; O'Brien, M.C.; Hooker, S.B.; Smith, R.; Menzies, D.; Mueller, J.L.; Kahru, M.; Toole, D.; et al. *SeaWiFS Postlaunch Calibration and Validation Analyses, Part. 3, NASA/TM-2000-206892*; Hooker, S.B., Firestone, E.R., Eds.; NASA Goddard Space Flight Center: Greenbelt, MD, USA, 2000; Volume 11.
105. Smith, R.C.; Baker, K.S. Optical classification of natural waters. *Limnol. Oceanogr.* **1978**, *23*, 260–267. [CrossRef]
106. Gordon, H.R.; Morel, A. *Remote Assessment of Ocean Color for Interpretation of Satellite Visible Imagery: A Review*; Springer: New York, NY, USA, 1983.
107. Smith, R.C.; Baker, K.S. Optical properties of the clearest natural waters (200–800 nm). *Appl. Opt.* **1981**, *20*, 177–184. [CrossRef] [PubMed]
108. Frouin, R.; Lingner, D.W.; Gautier, C.; Baker, K.S.; Smith, R.C. A simple analytical formula to compute clear sky total and photosynthetically available solar irradiance at the ocean surface. *J. Geophys. Res.* **1989**, *94*, 9731–9742. [CrossRef]
109. Morel, A. Optical properties and radiant energy in the waters of the Guinea dome and the Mauritanian upwelling area in relation to primary production. *J. Cons. Int. Explor. Mer* **1982**, *180*, 94–107.
110. Siegel, D.A.; Dickey, T.D. On the parameterization of irradiance for open ocean photoprocesses. *J. Geophys. Res.* **1987**, *92*, 14648–14662. [CrossRef]
111. Smith, R.C.; Marra, J.; Perry, M.J.; Baker, K.S.; Swift, E.; Buskey, E.; Kiefer, D.A. Estimation of a photon budget for the upper ocean in the Sargasso Sea. *Limnol. Oceanogr.* **1989**, *34*, 1673–1693. [CrossRef]
112. Moore, C.C.; Bruce, E.J.; Pegau, W.S.; Weidemann, A.D. WET Labs ac-9: Field calibration protocol, deployment techniques, data processing, and design improvements. In *Ocean Optics XIII*; SPIE Proceedings; SPIE: Bellingham, WA, USA, 1997; Volume 2963, pp. 725–731. [CrossRef]
113. Twardowski, M.S.; Sullivan, J.M.; Donaghay, P.L.; Zaneveld, J.R.V. Microscale quantification of the absorption by dissolved and particulate material in coastal waters with an ac-9. *J. Atmos. Ocean. Technol.* **1999**, *16*, 691–707. [CrossRef]
114. Röttgers, R.; McKee, D.; Woźniak, S.B. Evaluation of scatter corrections for ac-9 absorption measurements in coastal waters. *Methods Oceanogr.* **2013**, *7*, 21–39. [CrossRef]
115. Sathyendranath, S.; Platt, T. The spectral irradiance field at the surface and in the interior of the ocean: A model for applications in oceanography and remote sensing. *J. Geophys. Res.* **1988**, *93*, 9270–9280. [CrossRef]

applied
sciences

MDPI

Article

Concentrations of Multiple Phytoplankton Pigments in the Global Oceans Obtained from Satellite Ocean Color Measurements with MERIS

Guoqing Wang [1],*, Zhongping Lee [1] and Colleen B. Mouw [2]

[1] School for the Environment, University of Massachusetts Boston, Boston, MA 02125, USA; zhongping.lee@umb.edu

[2] Graduate School of Oceanography, University of Rhode Island, Narragansett, RI 02882, USA; cmouw@uri.edu

* Correspondence: guoqing.wang001@umb.edu or gqwang18@gmail.com; Tel.: +1-860-771-0330

Received: 4 May 2018; Accepted: 6 September 2018; Published: 19 December 2018

Abstract: The remote sensing of chlorophyll *a* concentration from ocean color satellites has been an essential variable quantifying phytoplankton in the past decades, yet estimation of accessory pigments from ocean color remote sensing data has remained largely elusive. In this study, we validated the concentrations of multiple pigments (Cpigs) retrieved from in situ and MEdium Resolution Imaging Spectrometer (MERIS) measured remote sensing reflectance ($R_{rs}(\lambda)$) in the global oceans. A multi-pigment inversion model (MuPI) was used to semi-analytically retrieve Cpigs from $R_{rs}(\lambda)$. With a set of globally optimized parameters, the accuracy of the retrievals obtained with MuPI is quite promising. Compared with High-Performance Liquid Chromatography (HPLC) measurements near Bermuda, the concentrations of chlorophyll *a*, *b*, *c* ([Chl-a], [Chl-b], [Chl-c]), photoprotective carotenoids ([PPC]), and photosynthetic carotenoids ([PSC]) can be retrieved from MERIS data with a mean unbiased absolute percentage difference of 38%, 78%, 65%, 36%, and 47%, respectively. The advantage of the MuPI approach is the simultaneous retrievals of [Chl-a] and the accessory pigments [Chl-b], [Chl-c], [PPC], [PSC] from MERIS $R_{rs}(\lambda)$ based on a closure between the input and output $R_{rs}(\lambda)$ spectra. These results can greatly expand scientific studies of ocean biology and biogeochemistry of the global oceans that are not possible when the only available information is [Chl-a].

Keywords: phytoplankton pigments; ocean color; remote sensing; MERIS; global oceans

---

1. Introduction

Ocean color remote sensing has been focused on phytoplankton due to the important role that they play in the global biogeochemical cycles and ocean food webs [1,2]. With the development of remote sensing technology, a variety of approaches have been developed to remotely obtain information about phytoplankton, such as their chlorophyll concentration [3–5], functional groups, and size classes [6–11]. The most widely used satellite-based product of phytoplankton is chlorophyll *a* concentration ([Chl-a], mg·m^{-3}) [3–5,12,13]. Satellite retrieved [Chl-a] has been utilized in estimation of phytoplankton biomass, primary production, and detection of harmful algal blooms [14,15]. However, many studies have indicated that [Chl-a] alone is not a good indicator of phytoplankton biomass or physiological status [16–20]. Some accessory pigments have been recognized as biomarkers for phytoplankton groups or species [18,21–24]. These accessory pigments provide better estimation of the biomass of particular phytoplankton groups or species, such as phycocyanin (PC) for cyanobacteria [25,26]. The variation in the accessory pigment composition has been widely used in estimating different phytoplankton functional groups [27–42], and physiological status of the phytoplankton [43,44].

In an effort to obtain pigment concentrations beyond [Chl-a] from remote sensing reflectance ($R_{rs}(\lambda)$, sr^{-1}), phycocyanin concentration, instead of [Chl-a], has been retrieved and used as a better index for cyanobacteria biomass and potential toxicity for cyanobacteria bloom waters [25,26]. To obtain phycocyanin and [Chl-a], empirical and semi-analytical methods have been proposed and good results obtained in their application to bloom detection and monitoring [13,25,26]. Either empirical or analytical, these methods are based upon relationships between bio-optical information and one or two pigment concentrations.

Empirical approaches have been the most widely used to obtain information for two or more pigments. Similar to the empirical relationships used by NASA for the estimation of [Chl-a] from $R_{rs}(\lambda)$ [4], Pan et al. [45] developed empirical relationships between High-Performance Liquid Chromatography (HPLC) measured pigment concentrations and $R_{rs}(\lambda)$ for coastal waters in the northeast coast of the United States. Moissan et al. [46] directly used satellite-derived [Chl-a] as model input to retrieve other pigments in the Atlantic Ocean off the east coast of United States with the underlining assumption that all accessory pigments co-vary with chlorophyll *a*.

Semi-analytical models, which are based on mechanistic relationships derived from radiative transfer, allow the estimation of inherent optical properties (absorption and backscattering) from $R_{rs}(\lambda)$ measured by any radiometer [47–62]. These semi-analytical models make it possible to obtain optical properties of the water components simultaneously from measured $R_{rs}(\lambda)$. Taking advantage of this property of semi-analytical algorithms, Wang et al. [63] incorporated the Gaussian decomposition method proposed by Hoepffner and Sathyendranath [64] into a semi-analytical model, termed as multi-pigment inversion model (MuPI), and demonstrated the potential of obtaining Gaussian peak heights representing the absorption coefficients from various pigments. Chase et al. [65] also adopted a similar scheme and applied it to hyperspectral in situ $R_{rs}(\lambda)$ measurements from the open ocean for the estimation of accessory pigments. However, as demonstrated in many studies [37,39,64–66], the assumption that each Gaussian amplitude represents the light absorption of one specific pigment is not always feasible. This is further shown in Chase et al. [65] where mixed results were obtained when a Gaussian peak height was linked with a single pigment. In this study, with in situ data from the global oceans, a thorough examination between the Gaussian peak heights and pigment concentrations was conducted.

The purpose of this study is twofold: (1) to evaluate the updated MuPI in retrieving concentrations of multiple phytoplankton pigments across the global ocean from MEdium Resolution Imaging Spectrometer (MERIS) measurements, and (2) to present the spatial distributions of accessory pigments across the global ocean that were previously not available. Model parameters were updated, and its performance was evaluated with different datasets covering a large dynamic range of ocean water conditions. The model was then applied to satellite remote sensing data from MERIS to obtain the global distribution and variation of different pigment concentrations. Finally, limitations and future developments of the MuPI model are discussed.

2. Data and Methods

2.1. Datasets and Study Sites

The datasets used in this study can be broadly classified into six different categories: (a) phytoplankton absorption coefficients ($a_{ph}(\lambda)$) from the global oceans; (b) simultaneously collected $a_{ph}(\lambda)$ and HPLC; (c) simultaneously measured $R_{rs}(\lambda)$, $a_{ph}(\lambda)$ and/or HPLC; (d) HPLC time series; (e) $R_{rs}(\lambda)$ from MERIS imagery; and (f) HydroLight simulated $R_{rs}(\lambda)$, $a_{ph}(\lambda)$, particulate backscattering coefficients ($b_{bp}(\lambda)$) and absorption coefficients of colored dissolved and detrital matters ($a_{dg}(\lambda)$) (International Ocean-Color Coordinating Group (IOCCG) dataset [50]). Table 1 provides an overview of the different datasets, time, size, variables, [Chl-a] range, and the main usage. The description of each dataset and the data measurements are included in the following paragraphs.

Table 1. Datasets, time, variables, size and their usages in this study. Cpigs: pigment concentrations, N: the number of samples, NA: not applicable.

Datasets/Cruises	Time	Size (N)	Measurements	Chl-a (mg·m^{-3})	Usage
SeaBASS	2001–2012	1619	$a_{ph}(\lambda)$	NA	Gaussian curves
	1991–2007	430	$a_{ph}(\lambda)$, HPLC	0.02–13.2	$a_{Gau}(\lambda)$ vs. Cpigs relationships
IOCCG	NA	500	$R_{rs}(\lambda)$, $a_{ph}(\lambda)$, $a_{dg}(\lambda)$, $b_{bp}(\lambda)$	0.03–30	
Tara Oceans expedition	2010–2012	23	$R_{rs}(\lambda)$, $a_{ph}(\lambda)$, HPLC	0.02–0.95	$a_{Gau}(\lambda)$ and Cpigs validation
VIIRS cal/val	2014–2015	21	$R_{rs}(\lambda)$, $a_{ph}(\lambda)$, HPLC	0.15–1.5	
BIOSOPE	2004	31	$R_{rs}(\lambda)$, $a_{ph}(\lambda)$, HPLC	0.00036–3.06	
BATS	2002–2012	148	HPLC	0.002–0.486	Cpigs variation
MERIS	2002–2012	148	$R_{rs}(\lambda)$	0.037–0.325	

A series of $a_{ph}(\lambda)$ spectra measured with the quantitative filter technique (QFT [67]) were used to find the globally optimized Gaussian parameters and the relationships among them. This dataset was obtained by searching the SeaWiFS Bio-optical Archive and Storage System (SeaBASS), which covers 1619 stations across the global oceans observed during 2001–2011. A set of 430 observations that had $a_{ph}(\lambda)$ and HPLC measurements coincidently observed were obtained from this dataset and were further randomly separated into two equal subsets (N = 215). The Subset_1 was used for regression analysis between Gaussian peak height ($a_{Gau}(\lambda)$) and pigment concentrations (Cpigs) to obtain the relationships among them; and the Subset_2 together with the $a_{ph}(\lambda)$ and HPLC from Tara Oceans, BIOSOPE, and VIIRS cal/val cruises were used to validate the relationships obtained from Subset_1.

The Tara Oceans expedition contains 23 match-ups of $R_{rs}(\lambda)$, $a_{ph}(\lambda)$ and HPLC around the global ocean. The BIOSOPE dataset includes 31 match-ups of $R_{rs}(\lambda)$, $a_{ph}(\lambda)$ and HPLC which were collected in the southeastern Pacific Ocean (obtained from: http://www.obs-vlfr.fr/proof/vt/op/ec/biosope/bio.htm). The VIIRS cal/val dataset is composed of 21 $R_{rs}(\lambda)$, $a_{ph}(\lambda)$ and HPLC measurements obtained from cruises covering the coastal oceans in North Atlantic Ocean off the United States east coast. The BATS (Bermuda Atlantic Time-Series Study) dataset is composed of HPLC time series from 2002 to 2012 and was obtained from the Bermuda Atlantic Time-Series Study (near Bermuda) (http://bats.bios.edu/bats_measurements.html). All of the in situ measurements used are from the surface, defined as a depth ≤5 m. The sampling locations of all these measurements are shown in Figure 1.

The IOCCG dataset was simulated using HydroLight software version 5.1 [68]. It was designed to cover the dynamic range observed across the global ocean but is biased to coastal waters (http://www.ioccg.org/data/synthetic.html). The $R_{rs}(\lambda)$, $a_{ph}(\lambda)$, $a_{dg}(\lambda)$ and $b_{bp}(\lambda)$ obtained from this dataset were used to validate the MuPI retrievals.

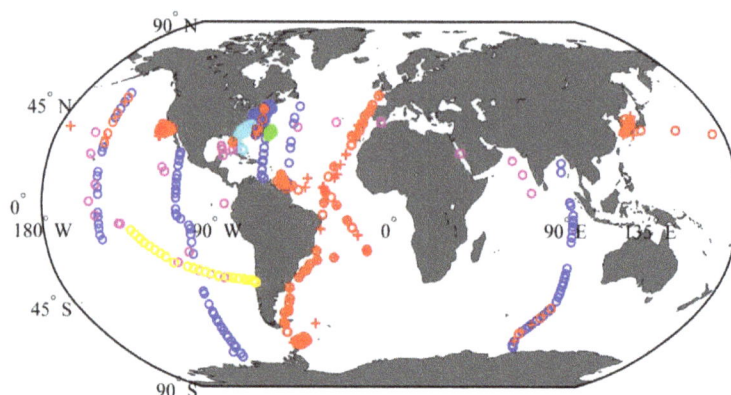

Figure 1. In situ data distribution, the (o) are the stations for quantitative filter technique (QFT) $a_{ph}(\lambda)$ from SeaBASS, (o) and (+) are the subset_1 and subset_2 stations of matchups of $a_{ph}(\lambda)$ and HPLC from SeaBASS, (o) is the HPLC location for BATS (Bermuda Atlantic Time-Series Study), (o) are the BIOSOPE $R_{rs}(\lambda)$, $a_{ph}(\lambda)$ and HPLC locations, (o) are locations of the $R_{rs}(\lambda)$, $a_{ph}(\lambda)$ and HPLC from VIIRS val/cal cruises in 2014 and 2015, and (o) are the locations of $R_{rs}(\lambda)$, $a_{ph}(\lambda)$ and HPLC from Tara Oceans expedition.

2.2. Radiometric Measurements

The in situ remote sensing reflectance, $R_{rs}(\lambda)$, was calculated based on the measurements of radiance and irradiance sampled with the Radiometer Incorporating the Skylight-Blocked Apparatus (RISBA) [69], Hyper Spectral Radiometer HyperPro free-fall profiler (Satlantic, Inc. Halifax, Nova Scotia, Canada), or above water radiometers [70]. The $R_{rs}(\lambda)$ spectra from 350–800 nm with different spectral increments were interpolated to 1 nm resolution.

Standard Level 3 MERIS $R_{rs}(\lambda)$ was acquired from the National Aeronautics and Space Administration (NASA) ocean color website (https://oceancolor.gsfc.nasa.gov). The HPLC data from BATS were matched to Level 3 MERIS 8-day products, at 4 km resolution and plus or minus one pixel (3 × 3 window). This criterion, although less restricting than NASA's 3-h window for data and algorithm validation [71], was adopted to maximize the number of match-ups.

The MuPI model was applied to MERIS $R_{rs}(\lambda)$ imagery from 2002–2012 to obtain the seasonal variation of chlorophyll *a*, *b*, *c*, photoprotective and photosynthetic carotenoids concentrations ([Chl-a], [Chl-b], [Chl-c], [PPC], [PSC]) near Bermuda. As examples, global maps of these five different pigments were also obtained from MERIS $R_{rs}(\lambda)$ imagery of 2007. The ratios of these concentrations to [Chl-a] are also presented to highlight their independence from chlorophyll *a*.

2.3. Absorption Measurements

Water samples for absorption and HPLC measurements were filtered onto a GF/F filter and stored in liquid nitrogen before laboratory measurements. Spectrophotometers were used to measure the absorbance and then to calculate the absorption coefficient of particles (a_p) and detrital matter (a_d). The phytoplankton absorption coefficient ($a_{ph}(\lambda)$) were obtained by subtracting a_d from a_p following NASA Ocean Optics Protocols, Revision 4, Volume IV protocol [72]. These $a_{ph}(\lambda)$ spectra generally cover 400–800 nm with spectral resolution around 3 nm. They were interpolated into 1 nm resolution for studies here.

Following Hoepffner and Sathyendranath [37,64] and Wang et al. [63,73], the phytoplankton absorption coefficients were decomposed into 13 Gaussian curves using the least square curve fitting technique provided in MATLAB and Statistics Toolbox (Release 2016a, MathWorks, Inc. Natick, MA, USA):

$$a_{\mathrm{ph}}(\lambda) = \sum_{i=1}^{n} a_{\mathrm{Gau}}(\lambda_i) \exp\left[-0.5\left(\frac{\lambda - \lambda_i}{\sigma_i}\right)^2\right] \tag{1}$$

where σ_i and $a_{\mathrm{Gau}}(\lambda_i)$ are the width and peak magnitude of the i-th Gaussian curve at peak center (λ_i) as shown in Table 2. The obtained $a_{\mathrm{Gau}}(\lambda)$ are used as ground truth to validate the inversion results from $R_{\mathrm{rs}}(\lambda)$.

Table 2. The 12 Gaussian curves corresponding to the phytoplankton pigment absorption coefficients, with Peak_loc as the center location of each pigment absorption peak and width as the full width at half maximum (FWHM). The relationships indicate the power-law relationships used to estimate the Gaussian peak amplitudes from the two independent variables: x_1: $a_{\mathrm{Gau}}(434)$ and x_2: $a_{\mathrm{Gau}}(492)$. Chl-a: chlorophyll a, Chl-b: chlorophyll b, Chl-c: chlorophyll c, PPC: photo-protective carotenoids, PSC: photosynthetic carotenoids, PE: phycoerythrin and PC: phycocyanin.

Peak	Pigments	Peak_loc (nm)	Width(FWHM) (nm)	Relationships	R^2
1	Chl-a	406	16	$1.13x_1^{1.01}$	0.98
2	Chl-a	434	12	x_1	–
3	Chl-c	453	12	$0.60x_1^{0.95}$	0.99
4	Chl-b	470	13	$0.51x_1^{0.97}$	0.98
5	PPC	492	16	x_2	–
6	PSC	523	14	$0.87x_2^{1.17}$	0.99
7	PE	550	14	$0.79x_2^{1.27}$	0.96
8	Chl-c	584	16	$0.40x_2^{1.17}$	0.96
9	PC	617	13	$0.34x_1^{1.14}$	0.93
10	Chl-c	638	11	$0.47x_2^{1.19}$	0.96
11	Chl-b	660	11	$0.30x_2^{1.11}$	0.94
12	Chl-a	675	10	$0.86x_1^{1.11}$	0.98

2.4. Pigment Concentrations

All the HPLC analyses were carried out according to the method following or adapted from Van Heukelem and Thomas [23]. The concentrations of chlorophyll a, b, c, photo-protective carotenoids (PPC) and photosynthetic carotenoids (PSC) were estimated from HPLC measurements as:

(A) Total chlorophyll a (Chl-a) = chlorophyll a + divinyl chlorophyll a + chlorophyllide a;
(B) Total chlorophyll b (Chl-b) = chlorophyll b + divinyl chlorophyll b;
(C) chlorophyll c (Chl-c) = chlorophyll $c1$ + chlorophyll $c2$;
(D) PPC = α-carotene + β-carotene + zeaxanthin + alloxanthin + diadinoxanthin;
(E) PSC = 19′-hexanoyloxyfucoxanthin + fucoxanthin + 19′-butanoyloxyfucoxanthin + peridinin.

MERIS [Chl-a] was estimated from the Level 3 $R_{\mathrm{rs}}(\lambda)$ following the standard algorithm OC4E provided by NASA [4]. Details about this algorithm can be found on the following webpage: https://oceancolor.gsfc.nasa.gov/atbd/chlor_a/.

2.5. Pigment Retrieval from $R_{\mathrm{rs}}(\lambda)$

2.5.1. $a_{\mathrm{Gau}}(\lambda)$ from $R_{\mathrm{rs}}(\lambda)$

The multi-pigment inversion model (MuPI) was used to retrieve $a_{\mathrm{Gau}}(\lambda)$ from $R_{\mathrm{rs}}(\lambda)$. Wang et al. [63,73] developed this semi-analytical inversion model (MuPI) to retrieve $a_{\mathrm{Gau}}(\lambda)$ from hyper- or multi-spectral $R_{\mathrm{rs}}(\lambda)$. A brief description of MuPI is presented here. The functional relationship between $R_{rs}(\lambda)$ and inherent optical properties (IOPs) is taken from Gordon et al. [47] and Lee et al. [49]:

$$R_{rs}(\lambda) = 0.52 \sum_{i=1}^{2} g_i \left[\frac{b_b(\lambda)}{a(\lambda) + b_b(\lambda)} \right]^i / \left\{ 1 - 1.7 \sum_{i=1}^{2} g_i \left[\frac{b_b(\lambda)}{a(\lambda) + b_b(\lambda)} \right]^i \right\} \tag{2}$$

where g_1 (sr^{-1}) and g_2 (sr^{-1}) are fixed to 0.089 and 0.125 sr^{-1}. The IOP spectra, $a(\lambda)$ and $b_b(\lambda)$, are partitioned into relevant components

$$b_b(\lambda) = b_{bw}(\lambda) + b_{bp}(\lambda) \tag{3}$$

$$a(\lambda) = a_w(\lambda) + a_{ph}(\lambda) + a_{dg}(\lambda) \tag{4}$$

with $b_{bw}(\lambda)$ for seawater backscattering coefficient [74] and $a_w(\lambda)$ for seawater absorption coefficient [75,76]. Phytoplankton absorption coefficient ($a_{ph}(\lambda)$) is modeled following Equation (1); $b_{bp}(\lambda)$, particulate backscattering coefficient, is modeled following Equations (5) and (6) [49]; and the combined dissolved and detrital particulate absorption coefficient $a_{dg}(\lambda)$ is modeled using Equation (7) [52,77,78].

$$b_{bp} = b_{bp}(\lambda_0) \left(\frac{\lambda_0}{\lambda} \right)^\eta \tag{5}$$

$$\eta = 2 \left(1 - 1.2 \exp \left(-0.9 \frac{R_{rs}(440)}{R_{rs}(550)} \right) \right) \tag{6}$$

$$a_{dg}(\lambda) = a_{dg}(\lambda_0) \exp(-S(\lambda - \lambda_0)) \tag{7}$$

where λ_0 is a reference wavelength (nearest to 440 nm), S is the spectral decay constant for absorption of detrital and dissolved materials and kept as an unknown within 0.007 to 0.02 nm^{-1} [52,77,78]. η is the power-law exponent for the particulate backscattering coefficient calculated from the $R_{rs}(440)$ to $R_{rs}(550)$ ratio following Lee et al. [49].

In the determination of Gaussian parameters (σ_i and λ_i) for $a_{ph}(\lambda)$ in the global scale, we also tested various combinations of parameters using data published in the literature [63–65]. The existing parameters were not successful at obtaining satisfactory results for every data range due to various reasons, including the fact that the initial datasets used to obtain the parameters had a small dynamic range unable to cover varied conditions such as coastal regions and non-bloom natural oceanic waters. Thus, a refinement of the parameters σ_i and λ_i was conducted using the $a_{ph}(\lambda)$ dataset obtained from SeaBASS to improve the overall performance of the MuPI model for global oceans. The non-linear least square fitting procedure in MATLAB was used to solve Equation (1). A set of refined Gaussian parameters for σ_i and λ_i were obtained and are presented in Table 2. For oceanic waters, as the absorption coefficient from water molecules contributes >80% of the total absorption coefficient for wavelengths >550 nm, it is difficult to obtain accurate $a_{Gau}(\lambda)$ by directly inverting $R_{rs}(\lambda)$ in the longer wavelengths. On the other hand, since the Gaussian peaks at 434 and 492 nm cover the main absorption features of the different pigments, the two $a_{Gau}(\lambda)$ at 434 and 492 nm were chosen as the independent variables in this effort.

Following Wang et al. [73], the implementation of this model used two Gaussian peak heights [$a_{Gau}(\lambda_1)$ and $a_{Gau}(\lambda_2)$] to reconstruct $a_{ph}(\lambda)$, in which empirical relationships as shown in Table 2 were used. These relationships between $a_{Gau}(\lambda)$ were obtained by regression analysis with the purpose of reducing the unknowns in the $R_{rs}(\lambda)$ inversion procedure [63,73]. With this design, there will be five unknowns [$a_{\mathrm{Gau}}(\lambda_1)$, $a_{\mathrm{Gau}}(\lambda_2)$, $b_{\mathrm{bp}}(\lambda_0)$, $a_{\mathrm{dg}}(\lambda_0)$ and S] to be retrieved from a $R_{\mathrm{rs}}(\lambda)$ spectrum, which is obtained by a minimization of the cost function (Equation (8)):

$$\delta = \frac{\sqrt{\frac{1}{N_\lambda}\sum_{i=1}^{N_\lambda}\left(\hat{R}_{\mathrm{rs}}(\lambda_i) - R_{\mathrm{rs}}(\lambda_i)\right)^2}}{\frac{1}{N_\lambda}\sum_{i=1}^{N_\lambda}R_{\mathrm{rs}}(\lambda_i)} \tag{8}$$

with N_λ as the wavelength number, $R_{\mathrm{rs}}(\lambda)$ as the measured, and $\hat{R}_{\mathrm{rs}}(\lambda)$ the modeled spectrum, respectively. Basically, δ value provides a measure of the relative difference between the input and output R_{rs} spectra. The generalized reduced gradient (GRG) nonlinear optimization procedure [79] was used to solve Equation (8).

The statistical indices used to estimate the agreement between the two values ($\hat{R}$ and R) was the unbiased absolute percentage difference (UAPD), defined as Equation (9) and root mean square error (RMSE, Equation (10)) with N as the number of samples.

$$\mathrm{UAPD} = \frac{|\hat{R} - R|}{0.5(\hat{R} + R)} \times 100\% \tag{9}$$

$$\mathrm{RMSE} = \sqrt{\frac{1}{N}(\hat{R} - R)^2} \tag{10}$$

2.5.2. $a_{\mathrm{Gau}}(\lambda)$ Versus Cpigs

Hoepffner and Sathyendranath [64] indicated that each Gaussian curve represents the absorption contributed by one or multiple pigments. However, attempts to obtain the concentration of a specific pigment from a single Gaussian curve is not always successful [39,65,66]. For a better understanding of the Gaussian curves and their relationships with Cpigs, a series of regression analyses were applied to relate $a_{\mathrm{Gau}}(\lambda)$ with Cpigs for data from SeaBASS. The t-statistics and p-value were calculated to test the significance of the parameters. Using $p < 0.05$ as the criteria, the significant contributors to each Gaussian peak and the corresponding R^2 of these parameters were obtained (see Table 3). The possible existence of other pigments that are not detectable with current HPLC techniques, such as phycoerythrin (PE) and phycocyanin (PC), likely explains the relatively lower R^2 values for Peaks 7, 8, 9 and 10.

After a series of multivariable regression analyses, it was found that Cpigs could be estimated from $a_{\mathrm{Gau}}(\lambda)$ following the function:

$$\log_{10}(\mathrm{Cpigs}) = a_0 + \sum_{i=1}^{n} a_i \log_{10}(a_{\mathrm{Gau}}(\lambda_i)) \tag{11}$$

The corresponding $a_{\mathrm{Gau}}(\lambda)$, parameters, and the R^2 value are shown in Table 4. Further, it was found that the estimated Cpigs agree with the measured values very well throughout the concentration range when the relationships were applied to the validation dataset (with data points scattered closely to the 1:1 line; Figure 2).

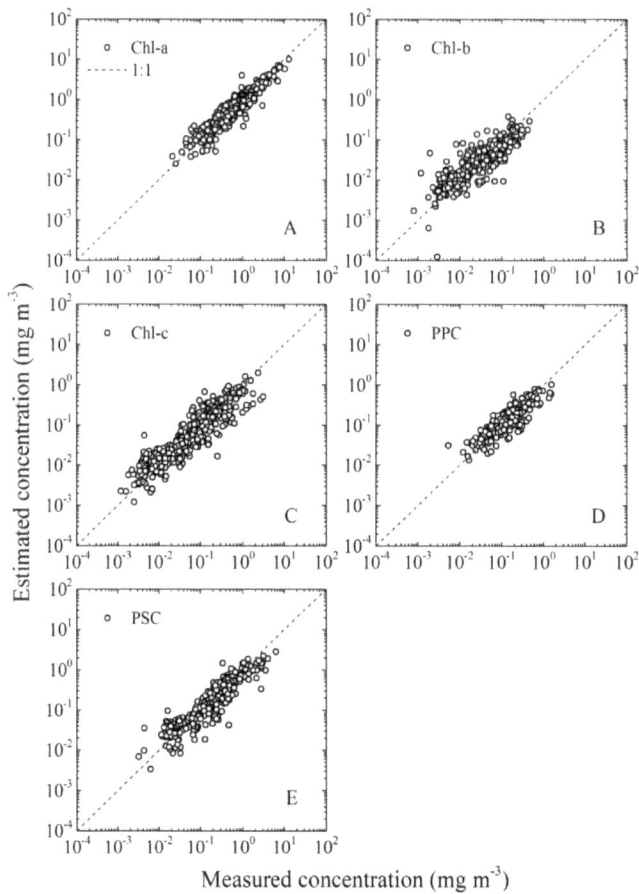

Figure 2. $a_{Gau}(\lambda)$ estimated pigment concentrations versus the measured concentrations from HPLC using the $a_{ph}(\lambda)$ and HPLC from SeaBASS: **A**: chlorophyll *a* (Chl-a), **B**: chlorophyll b (Chl-b), **C**: chlorophyll c (Chl-c), **D**: photoprotective carotenoids (PPC), and **E**: photosynthetic carotenoids (PSC).

Table 3. The *p*-value and R^2 from the t-statistics, with $a_{Gau}(\lambda)$ at 12 different wavelengths: 406, 434, 453, 470, 492, 523, 550, 584, 617, 638, 660, and 675 nm, and Chl-a: chlorophyll *a*, Chl-b: chlorophyll *b*, Chl-c: chlorophyll *c*, PPC: photo-protective carotenoids, PSC: photosynthetic carotenoids.

p-Value	Peak 406	Peak 434	Peak 453	Peak 470	Peak 492	Peak 523	Peak 550	Peak 584	Peak 617	Peak 638	Peak 660	Peak 675
Chl-a	0.01	0.00	0.00			0.01	0.03	0.00	0.00	0.00	0.00	0.00
Chl-b		0.03		0.00								
Chl-c							0.04	0.03			0.00	0.04
PPC	0.00		0.00	0.00	0.00							
PSC				0.00	0.02	0.01						
R^2	0.80	0.87	0.83	0.87	0.83	0.78	0.64	0.68	0.76	0.73	0.81	0.91

Table 4. Parameters for estimation of pigment concentrations: the pigment-specific $a_{Gau}(\lambda)$, coefficients and R^2. Chl-a: chlorophyll *a*, Chl-b: chlorophyll *b*, Chl-c: chlorophyll *c*, PPC: photo-protective carotenoids, PSC: photosynthetic carotenoids.

Pigments	$a_{Gau}(\lambda)$	Parameters ($a_0, a_1, \ldots, a_i$)	R^2
Chl-a	675	1.804, 0.975	0.89
Chl-b	434, 453, 470	−0.066, 2.470, −3.073, 1.379	0.72
Chl-c	470, 492, 523, 675	1.334, 2.022, −3.125, 0.745, 1.119	0.83
PPC	453, 470	0.734, 1.311, −0.416	0.76
PSC	470, 492, 523	1.67, 3.034, −2.670, 0.725	0.84

3. Results

3.1. Retrievals from $R_{rs}(\lambda)$

3.1.1. $a_{Gau}(\lambda)$ Validation

The MuPI model was first tested with datasets that contained different levels of chlorophyll *a* concentration from the IOCCG synthesized data and different cruises in the global ocean. The main purpose of this test was to evaluate, and validate, the implementation of the MuPI approach for a wide range of environments. A mean UAPD of 36% was obtained between $a_{Gau}(\lambda)$ from $R_{rs}(\lambda)$ inversion and $a_{Gau}(\lambda)$ from water samples throughout the data range for different datasets (see Figure 3). The differences in statistical results, noted in Table 5 for $a_{Gau}(\lambda)$ retrieval from different datasets, are strongly influenced by their different dynamic ranges and characteristics as implied in Section 2.1.

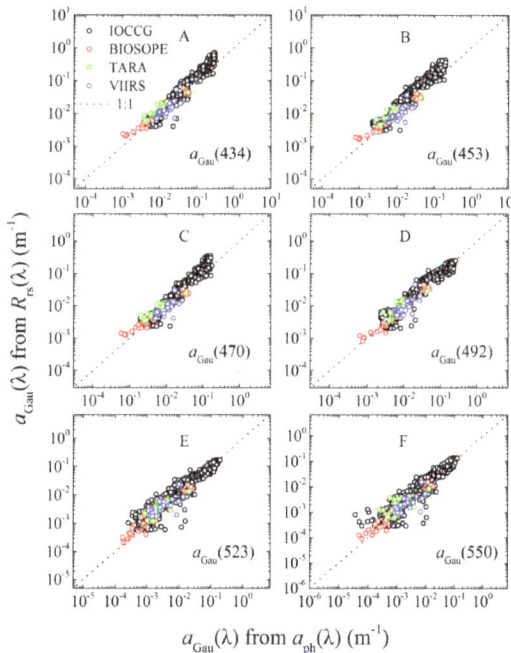

Figure 3. $a_{Gau}(\lambda)$ heights retrieved from $R_{rs}(\lambda)$ at MEdium Resolution Imaging Spectrometer (MERIS) bands versus measured Gaussian peak (decomposed from $a_{ph}(\lambda)$) for the data from different datasets (shown in different colors).

Table 5. Mean (Mea.) and median (Med.) of the unbiased percentage errors for $a_{Gau}(\lambda)$ retrieval from $R_{rs}(\lambda)$ for different datasets.

Peak Center	IOCCG		Tara Oceans		BIOSOPE		VIIRS Cruises	
	Mea.	Med.	Mea.	Med.	Mea.	Med.	Mea.	Med.
406	45	45	34	27	34	28	28	20
434	37	36	34	28	26	25	28	13
453	47	49	28	24	23	18	27	15
470	35	34	29	25	30	27	31	18
492	34	31	26	21	22	18	29	18
523	44	34	38	28	34	29	48	44
550	45	35	53	41	37	34	41	35
584	55	48	48	37	38	36	53	57
617	51	45	47	38	36	29	37	40
638	54	42	66	68	41	35	41	35
660	52	48	35	23	32	29	43	34
675	46	40	30	26	60	56	32	21

3.1.2. $b_{bp}(\lambda)$ and $a_{dg}(\lambda)$ Validation

The backscattering coefficients of particles and absorption coefficients of detrital and dissolved materials retrieved from $R_{rs}(\lambda)$ by MuPI have also been validated with the IOCCG dataset (with values at 440 nm as examples). The $b_{bp}(440)$ showed very high accuracy with mean UAPD of 4.8%, and all samples showed no bias in the entire data range as presented in Figure 4A. The estimated $a_{dg}(440)$ also showed very good agreement with the simulated values, with the mean UAPD of 21.3%. No inter-comparisons were made for the products $a_{dg}(\lambda)$ and $b_{bp}(\lambda)$ for other datasets because of lacking corresponding measured data.

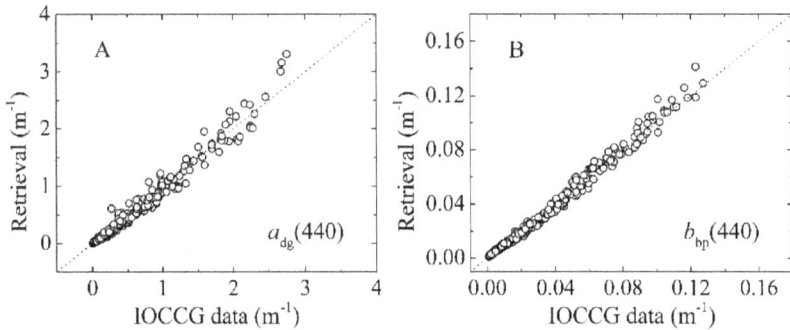

Figure 4. MuPI retrieved $a_{dg}(440)$ and $b_{bp}(440)$ versus those from the International Ocean-Colour Coordinating Group (IOCCG) dataset.

3.2. Cpigs from Satellite Remote Sensing

3.2.1. Cpigs Validation and Their Seasonal Variation

The ability of MuPI to capture the magnitudes of Cpigs and their seasonal variability from satellite $R_{rs}(\lambda)$ was then validated using a time series of HPLC measurements at BATS from the years 2002 to 2012. During this period of time, Cpigs varied in these ranges: [Chl-a]: 0.016–0.486 mg·m^{-3}, [Chl-b]: 0.001–0.108 mg·m^{-3}, [Chl-c]: 0.001–0.206 mg·m^{-3}, [PPC]: 0.004–0.147 mg·m^{-3}, and [PSC]: 0.003–0.106 mg·m^{-3}.

To obtain Cpigs from satellite $R_{rs}(\lambda)$, $a_{Gau}(\lambda)$ were inverted first from $R_{rs}(\lambda)$ using MuPI, then Equation (11) was applied to convert the retrieved $a_{Gau}(\lambda)$ to Cpigs. As shown in Figure 5, it is found that there are good matches in the magnitudes and the seasonal cycles for the five pigments, with mean UAPD values as 38%, 78%, 65%, 36%, and 47% (and the medians are 34%, 79%, 64%, 30% and 55%) for [Chl-a], [Chl-b], [Chl-c], [PPC] and [PSC], respectively. The [Chl-a] accuracy is comparable with the NASA adopted standard [Chl-a] algorithms, for which the color index (CI) algorithm [5] showed a mean UAPD of 38.6% and OC4E of 46.7%. There are many reasons for the relatively low accuracy in the retrieval of [Chl-b] and [Chl-c], which include very low concentrations (e.g., in situ [Chl-b] and [Chl-c] were close to the HPLC detection minimum), as well as uncertainties in satellite measured $R_{rs}(\lambda)$, and the derived $a_{Gau}(\lambda)$ from $R_{rs}(\lambda)$.

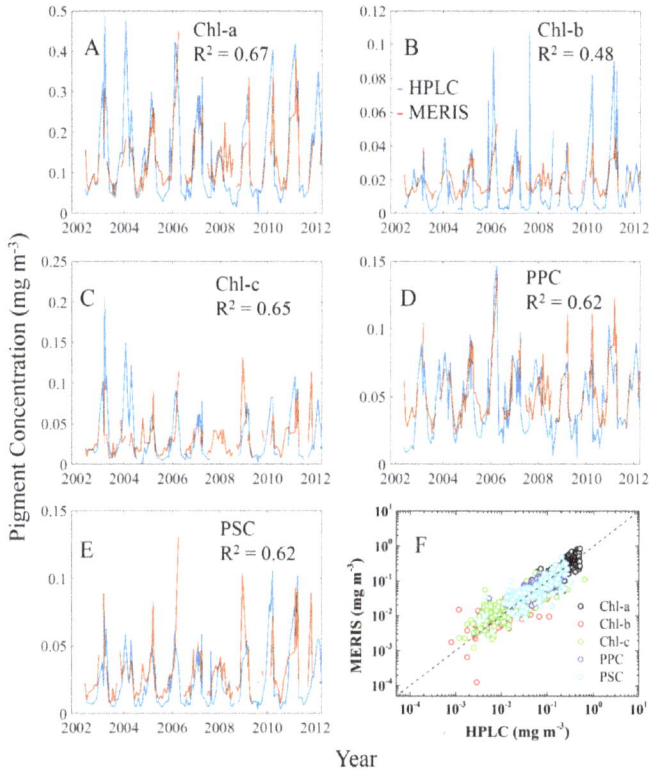

Figure 5. Time series of pigment concentrations from BATS HPLC and MERIS $R_{rs}(\lambda)$, and the determination coefficients (R^2). **A**: Chl-a: chlorophyll *a*, **B**: Chl-b: chlorophyll *b*, **C**: Chl-c: chlorophyll *c*, **D**: PPC: photoprotective carotenoids, **E**: PSC: photosynthetic carotenoids, **F**: the scatterplot of estimated versus in situ pigment concentrations.

Beyond the seasonal cycles in Cpig magnitudes, variation in pigment composition over time implied in the change of pigment ratios was also noticed in Figure 6A. On further examination, we found the ratios derived from MERIS $R_{rs}(\lambda)$ using MuPI can pick up the variation in the [Chl-b], [Chl-c], [PPC], [PSC] to [Chl-a] ratios observed from HPLC measurements very well with the mean UAPD of 50%, 47%, 25%, 37%, and median of 38%, 39%, 19% and 29% respectively (Figure 6B). Since these phytoplankton pigment ratios do not co-vary with the [Chl-a] product and cannot be empirically estimated from [Chl-a] alone, the Gaussian peaks and multiple pigments retrieved here provide a valuable glimpse into potential applications of these ratios in ocean changes that can be studied at large spatial and high temporal scales.

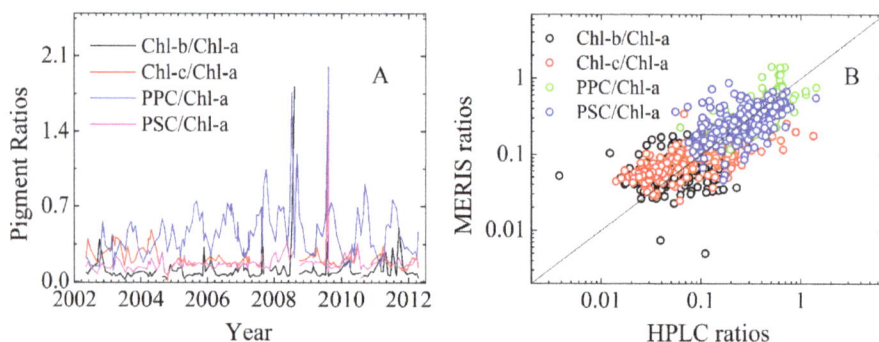

Figure 6. A: Time series of phytoplankton pigment to chlorophyll *a* (Chl-a) ratios at BATS from HPLC measurements. **B**: Chl-b, Chl-c, PPC and PSC to Chl-a ratios from HPLC versus from MERIS measured $R_{rs}(\lambda)$ using MuPI.

3.2.2. Global Distribution of Cpigs

As an example, MuPI was applied to MERIS L3 global annual (2007) average $R_{rs}(\lambda)$ to obtain the global distribution of the five different pigment concentrations (Figures 7 and 8) and their ratios to [Chl-a] (Figure 9). The global patterns of [Chl-a], [Chl-b], [Chl-c], [PPC], and [PSC] mimic the major gyre systems and other large-scale circulation features of the world ocean. High values of Cpigs are found within regions of persistent large-scale upwelling (e.g., subarctic gyres, equatorial divergences, eastern boundary currents, etc.), while low values are observed where large-scale downwelling is observed (e.g., subtropical gyres).

The basin-scale [Chl-a] distribution follows the pattern reported in the literature [4,5]. In comparing the global map of [Chl-a] derived in this study with the standard product from NASA ocean color website (Figure 7), some differences were noticed. In the previous section, when using HPLC data from BATS for validation, the retrieved [Chl-a] showed higher accuracy via MuPI than that from the standard OC4E algorithm. To better understand the differences in [Chl-a] distribution at the global scale, a validation dataset from SeaBASS was used. This dataset was obtained by searching match-ups of in situ measured [Chl-a] with those from MERIS $R_{rs}(\lambda)$. A dataset containing 608 pairs of [Chl-a] and MERIS $R_{rs}(\lambda)$ were obtained in which [Chl-a] concentrations ranged from ~0.017 to ~40.3 mg·m^{-3} (locations shown in Figure 7A). In comparing the estimated [Chl-a] from two different methods with in situ measurements (Figure 7C), MuPI and OC4E showed comparable results with mean UAPD of 48.8% and RMSE of 4.51 mg·m^{-3} for OC4E and mean UAPD of 49.3% and RMSE of 4.05 mg·m^{-3} for MuPI. As shown in Figure 7C, the [Chl-a] estimated from OC4E is biased slightly high (~10%) in the range of 1–10 mg·m^{-3} compared with results from MuPI. This range of [Chl-a] (1–10 mg·m^{-3}) is mainly from coastal and inland waters for which the influences from colored dissolved and detrital matter result in lower accuracy in band-ratio estimated [Chl-a] [80]. For several samples, the [Chl-a] values from MuPI are biased low (~70%) as shown in Figure 7C. There are two possible reasons for this: (1) bad input MERIS $R_{rs}(\lambda)$, not only the values but also the spectral shape, especially at the blue bands that are susceptible to poor atmospheric correction, where negative values are often observed for coastal waters [81]; (2) the limitation of the algorithm as a result of the empirical parameters used to reduce the unknowns in MuPI, and the low contribution of $a_{ph}(\lambda)$ to the total absorption coefficients in the 400–750 nm range, which will be further discussed in Section 4.

The pigment ratios to [Chl-a] showed complicated patterns (Figure 9). In high [Chl-a] regions, [Chl-b]/[Chl-a] and [PPC]/[Chl-a] are low. In some low [Chl-a] regions, the ratios of [Chl-c]/[Chl-a] and [PSC]/[Chl-a] are relatively high, such as in the East Pacific Ocean. These results agree with previous findings about the global distribution of phytoplankton groups and pigment ratios, as lower [PPC]/[Chl-a] ratios correspond to high [Chl-a] and larger particle size [9,17,82–84]. As recorded in the literature, in different regions of the global ocean, the quality and quantity of light and nutrient, as well as temperature, is highly variable [85]. These highly unpredictable and rapid changes of the environment usually result in phytoplankton taxonomic composition variation (long-term adaptation) or physiological acclimation (short-term acclimation) [24]. The variation in pigment ratios obtained from MuPI can directly reflect these changes in phytoplankton and provide valuable information for phytoplankton studies in large spatial and high temporal scales.

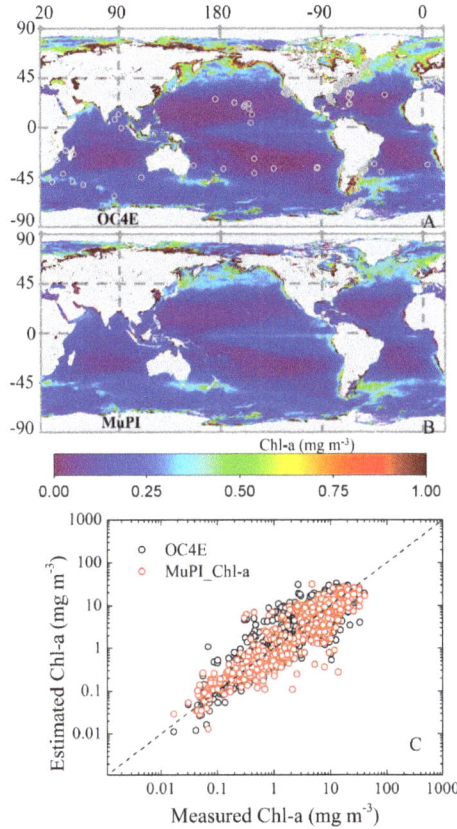

Figure 7. Global distributions of chlorophyll *a* concentration estimated from 2007 MERIS L3 $R_{rs}(\lambda)$ imagery using NASA standard algorithm OC4E (**A**) and MuPI model (**B**). The locations (**o**) of in situ Chl-a and MERIS $R_{rs}(\lambda)$ matchups for further comparison of OC4E and MuPI were plotted on the OC4E Chl-a map. **C**: Chlorophyll *a* concentration (Chl-a) from in situ measurements and from those estimated from matchups of MERIS $R_{rs}(\lambda)$ using OC4E and MuPI algorithms with mean UAPD of 48.8% and RMSE of 4.51 mg·m^{-3} for OC4E and mean UAPD of 49.3% and RMSE of 4.05 mg·m^{-3} for MuPI.

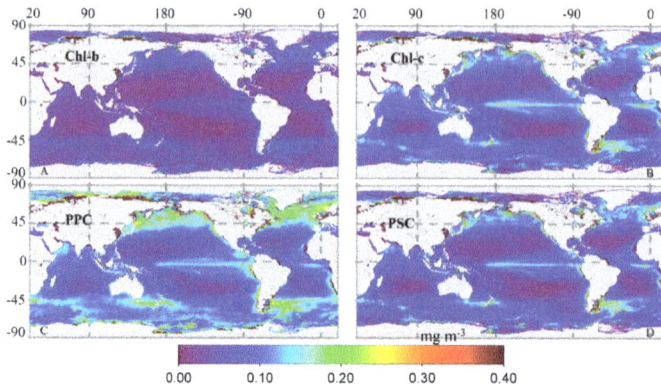

Figure 8. Global distributions of chlorophyll *b* (Chl-a), chlorophyll *c* (Chl-c), photoprotective carotenoids (PPC) and photosynthetic carotenoids (PSC) from 2007 L3 annual MERIS $R_{rs}(\lambda)$ imagery.

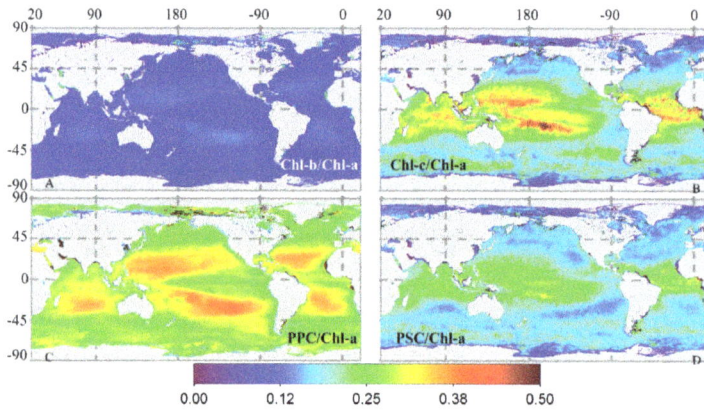

Figure 9. Global distributions of the accessory pigment to chlorophyll *a* ratios: ratio of concentrations of chlorophyll *b* (Chl-b/Chl-a), chlorophyll *c* (Chl-c/Chl-a), photoprotective carotenoids (PPC/Chl-a), and photosynthetic carotenoids to chlorophyll *a* (PSC/Chl-a).

4. Discussion

Based on the initial model of Wang et al. [63] that was developed for bloom waters, we have demonstrated that using a set of refined parameters for the Gaussian curves for global waters (Table 2), the MuPI model demonstrates consistent performance in $a_{Gau}(\lambda)$ retrievals on a global scale, as shown in Figures 3 and 4 and Table 5. Compared with HPLC measurements, the estimates of [Chl-a], [Chl-b], [Chl-c], [PPC] and [PSC] from satellite $R_{rs}(\lambda)$ showed reasonable accuracies, with the mean UAPD of 38%, 78%, 65%, 36%, and 47% respectively (Figure 5). Further, the distribution of these pigments and their ratios to [Chl-a], were obtained from MERIS measurements on the global scale (Figures 8 and 9).

The phytoplankton pigment ratios are critical indicators of the variation in phytoplankton groups and species due to their physiological adaptation to changes in nutrients, temperature, and light availability over time and space. The HPLC measured pigment concentrations and ratios have been widely used for determining the phytoplankton taxonomic composition and estimating the biomass of different groups, such as in CHEMTAX [27] and PFT analyses [10]. However, lacking effective methods, the estimation of phytoplankton pigments from satellite remote sensing has been limited to only [Chl-a] in the past decades [3–5], and the efforts made to obtain the accessory pigments have been more or less based on empirical relationships with Chl-a [45,46]. Thus, these products provided limited ability to capture the variation in the physiological status of phytoplankton. However, these unknown physiological variations in phytoplankton pigments have been one of the main uncertainties in traditional phytoplankton remote sensing models that use [Chl-a] as input to represent the phytoplankton biomass, such as the traditional [Chl-a] based primary productivity [14,20]. MuPI, as shown in this study, fills in this gap by obtaining not only the accessory pigment concentrations but also their ratios to [Chl-a] on the global scale, and reasonable accuracy has been obtained in validation with in situ data.

[Chl-a] is by far the easiest quantity to validate as it is routinely measured. An independent evaluation of the model has been conducted with the MERIS match-up dataset (Figure 7). This dataset contains nearly simultaneous in situ [Chl-a] measurements and MERIS $R_{rs}(\lambda)$ at coincident locations. To evaluate other pigments, an independent evaluation was conducted with MERIS $R_{rs}(\lambda)$ and in situ HPLC match-ups from 2002 to 2012 near Bermuda (Figure 5). The match-up dataset contains data from different seasons over a decadal scale. Use of the model with the match-up dataset from BATS confirms good overall behavior of the MuPI model for pigment concentration and pigment ratio retrievals, demonstrating the ability of the model to obtain accurate information from satellite ocean color imagery. However, because of the limitation of in situ pigment concentrations, the potential of obtaining PE and PC concentrations from satellite remote sensing data was not addressed in this study.

The main difficulty in making the model more applicable with any waterbody comes from the parameterization of the Gaussian curves, particularly the empirical relationships among $a_{Gau}(\lambda)$. Although it is reasonably straightforward to optimize the parameters with each dataset to obtain better retrievals, it would be extremely difficult (if not impossible) to do so when in situ measurements from the target location are hard to obtain. Instead, a set of globally optimized parameters were obtained using a dataset that covers a large dynamic range of the global ocean. Another challenge for $a_{Gau}(\lambda)$ retrieval in the open ocean lies in the low contribution of pigment absorption to the total absorption coefficient around 550–650 nm. Thus, it is difficult to directly invert $R_{rs}(\lambda)$ to obtain pigment absorption coefficients at the longer wavelengths (>550 nm).

This version of the MuPI model should be considered interim because the model could be further updated when more global data become available. In its present form, the model is optimized to work with $R_{rs}(\lambda)$ data from the first nine MERIS bands. As a first step, several components of the Gaussian model were deliberately formulated by use of empirical relationships to limit the number of unknowns to be solved via the spectral optimization procedure. This is particularly true for estimation of 13 $a_{Gau}(\lambda)$ from two independent Gaussian curves, which significantly reduced the unknowns. Instead of the Gaussian scheme, another potential way to obtain different pigment information from remote sensing data is through the specific absorption coefficients as adopted by many studies [46,86]. However, the specific absorption coefficients have significant limitations, such as the variation of the coefficients in different waters and the lack of routine measurements of some pigments, such as PE and PC [46,86,87].

With the information of accessory pigments obtained from MuPI, different biogeochemical studies could be conducted: 1. Remotely sensed PE and PC concentrations could be validated and applied to the estimation of cyanobacteria on a global scale. 2. The pigment ratios could be used as a direct indicator for estimation of phytoplankton functional types or functional traits, and phytoplankton physiological variation over space and time. 3. The pigment absorption coefficients (photoprotective and photosynthetic) could be estimated from satellite remote sensing data and incorporated into models for more accurate estimation of primary productivity.

5. Conclusions

The multi-pigment inversion model, namely MuPI, which semi-analytically obtains concentrations of multiple pigments from remote sensing reflectance, has been validated and applied to MERIS $R_{rs}(\lambda)$ imagery to obtain not only [Chl-a], but also [Chl-b], [Chl-c], [PPC], [PSC] (and subsequently their ratios to [Chl-a]) in the global oceans. The obtained pigment concentrations and the pigment ratios showed good agreement with in situ HPLC data, with the mean UAPD of 38%, 78%, 65%, 36%, and 47% respectively. Further, at the global scale, the MuPI obtained [Chl-a] from MERIS showed comparable results with those estimated from the widely used OC4E algorithm with mean UAPD of 48.8% and RMSE of 4.51 mg·m^{-3} for OC4E and mean UAPD of 49.3% and RMSE of 4.05 mg·m^{-3} for MuPI. However, unlike OC4E, MuPI as a semi-analytical model provided reasonable retrievals of several parameters {[Chl-a], [Chl-b], [Chl-c], [PPC], [PSC], a_{dg}(440) and b_{bp}(440)} simultaneously from satellite obtained remote sensing reflectance. The information of these accessory pigments would extend the application of satellite ocean color data in global biogeochemical studies that was previously limited due to [Chl-a] as the only available pigment.

Author Contributions: G.W. conceived and designed the analysis, collected the data and wrote the paper; Z.L. and C.B.M. supervised the project, analyzed results and wrote the paper.

Funding: This research was funded by National Aeronautics and Space Administration (NASA) Ocean Biology and Biogeochemistry and Water and Energy Cycle Programs, the National Oceanic and Atmospheric Administration (NOAA) JPSS VIIRS Ocean Color Cal/Val Project.

Acknowledgments: We are grateful for the two anonymous reviewers who have provided constructive suggestions to improve the quality of this manuscript. We thank Charles W. Kovach very much for the careful reading of the manuscript and all the helps he provided in improving the language. We thank Jianwei Wei and Junfang Lin in helping to take in situ measurements of remote sensing reflectance, Chuanmin Hu's group for sharing the phytoplankton absorption coefficients, and National Aeronautics and Space Administration (NASA) for sharing the HPLC measurements from 2014–2015 VIIRS Cal/Val cruises. We really appreciate those who shared the data from cruises of the Tara Oceans expedition and BIOSOPE and made it publicly accessible. We thank ESA for providing the MERIS imagery and NASA for SeaWiFS Bio-optical Archive and Storage System (SeaBASS) dataset, the Bermuda Institute of Ocean Science for the BATS dataset, and the International Ocean Color Coordinating Group for the IOCCG dataset. Support from the National Aeronautics and Space Administration (NASA) Ocean Biology and Biogeochemistry and Water and Energy Cycle Programs, the National Oceanic and Atmospheric Administration (NOAA) JPSS VIIRS Ocean Color Cal/Val Project and cruises are greatly appreciated.

Conflicts of Interest: The authors declare no conflict of interest.

References

1. Falkowski, P.G. The role of phytoplankton photosynthesis in global biogeochemical cycles. *Photosynth. Res.* **1994**, *39*, 235–258. [CrossRef] [PubMed]

2. Kiørboe, T. Turbulence, phytoplankton cell size, and the structure of pelagic food webs. In *Advances in Marine Biology*; Academic Press: Cambridge, MA, USA, 1993; Volume 29, pp. 1–72.

3. Gordon, H.R.; Clark, D.K.; Brown, J.W.; Brown, O.B.; Evans, R.H.; Broenkow, W.W. Phytoplankton pigment concentrations in the Middle Atlantic Bight: Comparison of ship determinations and CZCS estimates. *Appl. Opt.* **1983**, *22*, 20–36. [CrossRef] [PubMed]

4. O'Reilly, J.E.S.; Maritorena, B.G.; Mitchell, D.A.; Siegel, K.L.; Carder, S.A.; Garver, M.; Kahru, C.R. McClain, Ocean color chlorophyll algorithms for SeaWiFS. *J. Geophys. Res.* **1998**, *103*, 24937–24953. [CrossRef]

5. Hu, C.; Lee, Z.; Franz, B. Chlorophyll algorithms for oligotrophic oceans: A novel approach based on three-band reflectance difference. *J. Geophys. Res.* **2012**, *117*. [CrossRef]

6. Alvain, S.; Moulin, C.; Dandonneau, Y.; Bréon, F.M. Remote sensing of phytoplankton groups in case 1 waters from global SeaWiFS imagery. *Deep Sea Res. Part I Oceanogr. Res. Pap.* **2005**, *52*, 1989–2004. [CrossRef]

7. Bracher, A.; Vountas, M.; Dinter, T.; Burrows, J.P.; Röttgers, R.; Peeken, I. Quantitative observation of cyanobacteria and diatoms from space using PhytoDOAS on SCIAMACHY data. *Biogeosciences* **2009**, *6*, 751–764. [CrossRef]

8. Ciotti, A.M.; Bricaud, A. Retrievals of a size parameter for phytoplankton and spectral light absorption by colored detrital matter from water-leaving radiances at SeaWiFS channels in a continental shelf region off Brazil. *Limnol. Oceanogr. Methods* **2006**, *4*, 237–253. [CrossRef]

9. Brewin, R.J.; Sathyendranath, S.; Hirata, T.; Lavender, S.J.; Barciela, R.M.; Hardman-Mountford, N.J. A three-component model of phytoplankton size class for the Atlantic Ocean. *Ecol. Model.* **2010**, *221*, 1472–1483. [CrossRef]

10. Hirata, T.; Hardman-Mountford, N.J.; Brewin, R.J.W.; Aiken, J.; Barlow, R.; Suzuki, K.; Yamanaka, Y. Synoptic relationships between surface Chlorophyll-a and diagnostic pigments specific to phytoplankton functional types. *Biogeosciences* **2011**, *8*, 311–327. [CrossRef]

11. Mouw, C.B.; Yoder, J.A. Optical determination of phytoplankton size composition from global SeaWiFS imagery. *J. Geophys. Res. Oceans* **2010**, *115*. [CrossRef]

12. Sathyendranath, S.; Cota, G.; Stuart, V.; Maass, H.; Platt, T. Remote sensing of phytoplankton pigments: A comparison of empirical and theoretical approaches. *Int. J. Remote Sens.* **2001**, *22*, 249–273. [CrossRef]

13. Gitelson, A.A.; Schalles, J.F.; Hladik, C.M. Remote chlorophyll-a retrieval in turbid, productive estuaries: Chesapeake Bay case study. *Remote Sens. Environ.* **2007**, *109*, 464–472. [CrossRef]

14. Behrenfeld, M.J.; Falkowski, P.G. Photosynthetic rates derived from satellite-based chlorophyll concentration. *Limnol. Oceanogr.* **1997**, *42*, 1–20. [CrossRef]

15. Stumpf, R.P.; Culver, M.E.; Tester, P.A.; Tomlinson, M.; Kirkpatrick, G.J.; Pederson, B.A.; Soracco, M. Monitoring Karenia brevis blooms in the Gulf of Mexico using satellite ocean color imagery and other data. *Harmful Algae* **2003**, *2*, 147–160. [CrossRef]

16. Lehman, P.W. Comparison of chlorophyll a and carotenoid pigments as predictors of phytoplankton biomass. *Mar. Biol.* **1981**, *65*, 237–244. [CrossRef]

17. Schitüter, L.; Riemann, B.; Søndergaard, M. Nutrient limitation in relation to phytoplankton carotenoid/chlorophyll a ratios in freshwater mesocosms. *J. Plankton Res.* **1997**, *19*, 891–906. [CrossRef]

18. Breton, E.; Brunet, C.; Sautour, B.; Brylinski, J.M. Annual variations of phytoplankton biomass in the Eastern English Channel: Comparison by pigment signatures and microscopic counts. *J. Plankton Res.* **2000**, *22*, 1423–1440. [CrossRef]

19. Kruskopf, M.; Flynn, K.J. Chlorophyll content and fluorescence responses cannot be used to gauge reliably phytoplankton biomass, nutrient status or growth rate. *New Phytol.* **2006**, *169*, 525–536. [CrossRef] [PubMed]

20. Behrenfeld, M.J.; O'Malley, R.T.; Boss, E.S.; Westberry, T.K.; Graff, J.R.; Halsey, K.H.; Brown, M.B. Revaluating ocean warming impacts on global phytoplankton. *Nat. Clim. Chang.* **2016**, *6*, 323. [CrossRef]

21. Bidigare, R.R.; Morrow, J.H.; Kiefer, D.A. Derivative analysis of spectral absorption by photosynthetic pigments in the western Sargasso Sea. *J. Mar. Res.* **1989**, *47*, 323–341. [CrossRef]

22. Jeffrey, S.W.; Vesk, M. Introduction to marine phytoplankton and their pigment signature. In *Phytoplankton Pigments in Oceanography*; UNESCO Publishing: Paris, France, 1997; p. 3784.

23. Kirkpatrick, G.J.; Millie, D.F.; Moline, M.A.; Schofield, O. Optical discrimination of a phytoplankton species in natural mixed populations. *Limnol. Oceanogr.* **2000**, *45*, 467–471. [CrossRef]

24. Van Heukelem, L.; Thomas, C.S. Computer-assisted high-performance liquid chromatography method development with applications to the isolation and analysis of phytoplankton pigments. *J. Chromatogr. A* **2001**, *910*, 31–49. [CrossRef]

25. Roy, S.; Llewellyn, C.A.; Egeland, E.S.; Johnsen, G. (Eds.) *Phytoplankton Pigments: Characterization, Chemotaxonomy and Applications in Oceanography*; Cambridge University Press: Cambridge, UK, 2011.

26. Simis, S.G.; Peters, S.W.; Gons, H.J. Remote sensing of the cyanobacterial pigment phycocyanin in turbid inland water. *Limnol. Oceanogr.* **2005**, *50*, 237–245. [CrossRef]

27. Wynne, T.; Stumpf, R.; Tomlinson, M.; Warner, R.; Tester, P.; Dyble, J.; Fahnenstiel, G. Relating spectral shape to cyanobacterial blooms in the Laurentian Great Lakes. *Int. J. Remote Sens.* **2008**, *29*, 3665–3672. [CrossRef]

28. Mackey, M.D.; Mackey, D.J.; Higgins, H.W.; Wright, S.W. CHEMTAX—A program for estimating class abundances from chemical markers: Application to HPLC measurements of phytoplankton. *Mar. Ecol. Prog. Ser.* **1996**, *144*, 265–283. [CrossRef]

29. Vidussi, F.; Claustre, H.; Manca, B.B.; Luchetta, A.; Marty, J.C. Phytoplankton pigment distribution in relation to upper thermocline circulation in the eastern Mediterranean Sea during winter. *J. Geophys. Res. Oceans* **2001**, *106*, 19939–19956. [CrossRef]

30. Uitz, J.; Claustre, H.; Morel, A.; Hooker, S.B. Vertical distribution of phytoplankton communities in open ocean: An assessment based on surface chlorophyll. *J. Geophys. Res. Oceans* **2006**, *111*. [CrossRef]

31. Sathyendranath, S.; Aiken, J.; Alvain, S.; Barlow, R.; Bouman, H.; Bracher, A.; Clementson, L.A. Phytoplankton functional types from Space. In *Reports of the International Ocean-Colour Coordinating Group (IOCCG)*; International Ocean-Colour Coordinating Group: Dartmouth, Canada, 2014; pp. 1–156.

32. Uitz, J.; Stramski, D.; Reynolds, R.A.; Dubranna, J. Assessing phytoplankton community composition from hyperspectral measurements of phytoplankton absorption coefficient and remote-sensing reflectance in open-ocean environments. *Remote Sens. Environ.* **2015**, *171*, 58–74. [CrossRef]

33. Catlett, D.; Siegel, D.A. Phytoplankton pigment communities can be modeled using unique relationships with spectral absorption signatures in a dynamic coastal environment. *J. Geophys. Res. Oceans* **2018**, *123*, 246–264. [CrossRef]

34. Bracher, A.; Bouman, H.A.; Brewin, R.J.; Bricaud, A.; Brotas, V.; Ciotti, A.M.; Hardman-Mountford, N.J. Obtaining phytoplankton diversity from ocean color: A scientific roadmap for future development. *Front. Mar. Sci.* **2017**, *4*, 55. [CrossRef]

35. Bricaud, A.; Mejia, C.; Blondeau-Patissier, D.; Claustre, H.; Crepon, M.; Thiria, S. Retrieval of pigment concentrations and size structure of algal populations from their absorption spectra using multilayered perceptrons. *Appl. Opt.* **2007**, *46*, 1251–1260. [CrossRef] [PubMed]

36. Ciotti, A.M.; Lewis, M.R.; Cullen, J.J. Assessment of the relationships between dominant cell size in natural phytoplankton communities and the spectral shape of the absorption coefficient. *Limnol. Oceanogr.* **2002**, *47*, 404–417. [CrossRef]

37. Devred, E.; Sathyendranath, S.; Stuart, V.; Platt, T. A three component classification of phytoplankton absorption spectra: Application to ocean-color data. *Remote Sens. Environ.* **2011**, *115*, 2255–2266. [CrossRef]

38. Hoepffner, N.; Sathyendranath, S. Determination of the major groups of phytoplankton pigments from the absorption spectra of total particulate matter. *J. Geophys. Res. Oceans* **1993**, *98*, 22789–22803. [CrossRef]

39. Hirata, T.; Aiken, J.; Hardman-Mountford, N.; Smyth, T.J.; Barlow, R.G. An absorption model to determine phytoplankton size classes from satellite ocean colour. *Remote Sens. Environ.* **2008**, *112*, 3153–3159. [CrossRef]

40. Lohrenz, S.E.; Weidemann, A.D.; Tuel, M. Phytoplankton spectral absorption as influenced by community size structure and pigment composition. *J. Plankton Res.* **2003**, *25*, 35–61. [CrossRef]

41. Moisan, J.R.; Moisan, T.A.; Linkswiler, M.A. An inverse modeling approach to estimating phytoplankton pigment concentrations from phytoplankton absorption spectra. *J. Geophys. Res. Oceans* **2011**, *116*. [CrossRef]

42. Organelli, E.; Bricaud, A.; Antoine, D.; Uitz, J. Multivariate approach for the retrieval of phytoplankton size structure from measured light absorption spectra in the Mediterranean Sea (BOUSSOLE site). *Appl. Opt.* **2013**, *52*, 2257–2273. [CrossRef] [PubMed]

43. Mouw, C.B.; Hardman-Mountford, N.J.; Alvain, S.; Bracher, A.; Brewin, R.J.; Bricaud, A.; Hirawake, T. A consumer's guide to satellite remote sensing of multiple phytoplankton groups in the global ocean. *Front. Mar. Sci.* **2017**, *4*, 41. [CrossRef]

44. Jensen, A.; Sakshaug, E. Studies on the phytoplankton ecology of the trondheemsfjord. II. Chloroplast pigments in relation to abundance and physiological state of the phytoplankton. *J. Exp. Mar. Biol. Ecol.* **1973**, *11*, 137–155. [CrossRef]

45. Suggett, D.J.; Moore, C.M.; Hickman, A.E.; Geider, R.J. Interpretation of fast repetition rate (FRR) fluorescence: Signatures of phytoplankton community structure versus physiological state. *Mar. Ecol. Prog. Ser.* **2009**, *376*, 1–19. [CrossRef]

46. Pan, X.; Mannino, A.; Russ, M.E.; Hooker, S.B.; Harding, L.W., Jr. Remote sensing of phytoplankton pigment distribution in the United States northeast coast. *Remote Sens. Environ.* **2010**, *114*, 2403–2416. [CrossRef]

47. Moisan, T.A.; Rufty, K.M.; Moisan, J.R.; Linkswiler, M.A. Satellite observations of phytoplankton functional type spatial distributions, phenology, diversity, and ecotones. *Front. Mar. Sci.* **2017**, *4*, 189. [CrossRef]

48. Gordon, H.R.; Brown, O.B.; Evans, R.H.; Brown, J.W.; Smith, R.C.; Baker, K.S.; Clark, D.K. A semianalytic radiance model of ocean color. *J. Geophys. Res. Atmos.* **1988**, *93*, 10909–10924. [CrossRef]

49. Lee, Z.; Carder, K.L.; Mobley, C.D.; Steward, R.G.; Patch, J.S. Hyperspectral remote sensing for shallow waters. I. A semianalytical model. *Appl. Opt.* **1998**, *37*, 6329–6338. [CrossRef] [PubMed]

50. Lee, Z.; Carder, K.L.; Arnone, R.A. Deriving inherent optical properties from water color: A multiband quasi-analytical algorithm for optically deep waters. *Appl. Opt.* **2002**, *41*, 5755–5772. [CrossRef] [PubMed]

51. Lee, Z. *Remote Sensing of Inherent Optical Properties: Fundamentals, Tests of Algorithms, and Applications*; International Ocean-Colour Coordinating Group: Dartmouth, Canada, 2006; Volume 5.

52. Maritorena, S.; Siegel, D.A.; Peterson, A.R. Optimization of a semianalytical ocean color model for global-scale applications. *Appl. Opt.* **2002**, *41*, 2705–2714. [CrossRef] [PubMed]

53. Werdell, P.J.; Franz, B.A.; Bailey, S.W.; Feldman, G.C.; Boss, E.; Brando, V.E.; Mangin, A. Generalized ocean color inversion model for retrieving marine inherent optical properties. *Appl. Opt.* **2013**, *52*, 2019–2037. [CrossRef] [PubMed]

54. Werdell, P.J.; McKinna, L.I.; Boss, E.; Ackleson, S.G.; Craig, S.E.; Gregg, W.W.; Stramski, D. An overview of approaches and challenges for retrieving marine inherent optical properties from ocean color remote sensing. *Prog. Oceanogr.* **2018**. [CrossRef]

55. Brando, V.E.; Dekker, A.G.; Park, Y.J.; Schroeder, T. Adaptive semianalytical inversion of ocean color radiometry in optically complex waters. *Appl. Opt.* **2012**, *51*, 2808–2833. [CrossRef] [PubMed]

56. Brewin, R.J.; Raitsos, D.E.; Dall'Olmo, G.; Zarokanellos, N.; Jackson, T.; Racault, M.F.; Hoteit, I. Regional ocean-colour chlorophyll algorithms for the Red Sea. *Remote Sens. Environ.* **2015**, *165*, 64–85. [CrossRef]

57. Bukata, R.P.; Jerome, J.H.; Kondratyev, A.S.; Pozdnyakov, D.V. *Optical Properties and Remote Sensing of Inland and Coastal Waters*; CRC Press: Boca Raton, FL, USA, 2018.

58. Devred, E.; Sathyendranath, S.; Stuart, V.; Maass, H.; Ulloa, O.; Platt, T. A two-component model of phytoplankton absorption in the open ocean: Theory and applications. *J. Geophys. Res. Oceans* **2006**, *111*. [CrossRef]

59. Garver, S.A.; Siegel, D.A. Inherent optical property inversion of ocean color spectra and its biogeochemical interpretation: 1. Time series from the Sargasso Sea. *J. Geophys. Res. Oceans* **1997**, *102*, 18607–18625. [CrossRef]

60. Hoge, F.E.; Lyon, P.E. Satellite retrieval of inherent optical properties by linear matrix inversion of oceanic radiance models: An analysis of model and radiance measurement errors. *J. Geophys. Res. Oceans* **1996**, *101*, 16631–16648. [CrossRef]

61. Roesler, C.S.; Perry, M.J. In situ phytoplankton absorption, fluorescence emission, and particulate backscattering spectra determined from reflectance. *J. Geophys. Res. Oceans* **1995**, *100*, 13279–13294. [CrossRef]

62. Wang, P.; Boss, E.S.; Roesler, C. Uncertainties of inherent optical properties obtained from semianalytical inversions of ocean color. *Appl. Opt.* **2005**, *44*, 4074–4085. [CrossRef] [PubMed]

63. Loisel, H.; Stramski, D.; Dessailly, D.; Jamet, C.; Li, L.; Reynolds, R.A. An Inverse Model for Estimating the Optical Absorption and Backscattering Coefficients of Seawater from Remote-Sensing Reflectance over a Broad Range of Oceanic and Coastal Marine Environments. *J. Geophys. Res. Oceans* **2018**, *123*, 2141–2171. [CrossRef]

64. Wang, G.; Lee, Z.; Mishra, D.R.; Ma, R. Retrieving absorption coefficients of multiple phytoplankton pigments from hyperspectral remote sensing reflectance measured over cyanobacteria bloom waters. *Limnol. Oceanogr. Methods* **2016**, *14*, 432–447. [CrossRef]

65. Hoepffner, N.; Sathyendranath, S. Effect of pigment composition on absorption properties of phytoplankton. *Mar. Ecol. Prog. Ser.* **1991**, *73*, 1–23. [CrossRef]

66. Chase, A.P.; Boss, E.; Cetinić, I.; Slade, W. Estimation of phytoplankton accessory pigments from hyperspectral reflectance spectra: Toward a global algorithm. *J. Geophys. Res. Oceans* **2017**. [CrossRef]

67. Lutz, V.A.; Sathyendranath, S.; Head, E.J.H. Absorption coefficient of phytoplankton: Regional variations in the North Atlantic. *Mar. Ecol. Prog. Ser.* **1996**, 197–213. [CrossRef]

68. Mitchell, B.G. Algorithms for determining the absorption coefficient for aquatic particulates using the quantitative filter technique. In *Ocean Optics X*; International Society for Optics and Photonics: Bellingham, WA, USA, 1990; Volume 1302, pp. 137–149.

69. Mobley, C.D. *Light and Water: Radiative Transfer in Natural Waters*; Academic Press: Cambridge, MA, USA, 1994.

70. Lee, Z.; Pahlevan, N.; Ahn, Y.H.; Greb, S.; O'Donnell, D. Robust approach to directly measuring water-leaving radiance in the field. *Appl. Opt.* **2013**, *52*, 1693–1701. [CrossRef] [PubMed]

71. Mueller, J.L. *Ocean Optics Protocols for Satellite Ocean Color Sensor Validation, Revision 4: Radiometric Measurements and Data Analysis Protocols*; Goddard Space Flight Center: Greenbelt, MD, USA, 2003; Volume 3.

72. Bailey, S.W.; Werdell, P.J. A multi-sensor approach for the on-orbit validation of ocean color satellite data products. *Remote Sens. Environ.* **2006**, *102*, 12–23. [CrossRef]

73. Mitchell, B.G.; Kahru, M.; Wieland, J.; Stramska, M. Determination of spectral absorption coefficients of particles, dissolved material and phytoplankton for discrete water samples. In *Ocean Optics Protocols for Satellite Ocean Color Sensor Validation, Revision 4, Volume IV: Inherent Optical Properties: Instruments, Characterizations, Field Measurements and Data Analysis Protocols*; NASA/TM-2003-211621; Mueller, J.L., Fargion, G.S., McClain, C.R., Eds.; NASA Goddard Space Flight Center: Greenbelt, MD, USA, 2003; pp. 39–64.
74. Wang, G.; Lee, Z.; Mouw, C. Multi-spectral remote sensing of phytoplankton pigment absorption properties in cyanobacteria bloom waters: A regional example in the western basin of Lake Erie. *Remote Sens.* **2017**, *9*, 1309. [CrossRef]
75. Morel, A. Optical properties of pure water and pure sea water. *Opt. Asp. Oceanogr.* **1974**, *1*, 22.
76. Pope, R.M.; Fry, E.S. Absorption spectrum (380–700 nm) of pure water. II. Integrating cavity measurements. *Appl. Opt.* **1997**, *36*, 8710–8723. [CrossRef] [PubMed]
77. Lee, Z.; Wei, J.; Voss, K.; Lewis, M.; Bricaud, A.; Huot, Y. Hyperspectral absorption coefficient of "pure" seawater in the range of 350–550 nm inverted from remote sensing reflectance. *Appl. Opt.* **2015**, *54*, 546–558. [CrossRef]
78. Nelson, N.B.; Siegel, D.A.; Michaels, A.F. Seasonal dynamics of colored dissolved material in the Sargasso Sea. *Deep Sea Res. Part I Oceanogr. Res. Pap.* **1998**, *45*, 931–957. [CrossRef]
79. Babin, M.; Stramski, D.; Ferrari, G.M.; Claustre, H.; Bricaud, A.; Obolensky, G.; Hoepffner, N. Variations in the light absorption coefficients of phytoplankton, nonalgal particles, and dissolved organic matter in coastal waters around Europe. *J. Geophys. Res. Oceans* **2003**, *108*. [CrossRef]
80. Lasdon, L.S.; Waren, A.D.; Jain, A.; Ratner, M. Design and testing of a generalized reduced gradient code for nonlinear programming. *ACM Trans. Math. Softw. (TOMS)* **1978**, *4*, 34–50. [CrossRef]
81. Gilerson, A.A.; Gitelson, A.A.; Zhou, J.; Gurlin, D.; Moses, W.; Ioannou, I.; Ahmed, S.A. Algorithms for remote estimation of chlorophyll-a in coastal and inland waters using red and near infrared bands. *Opt. Express* **2010**, *18*, 24109–24125. [CrossRef] [PubMed]
82. Ruddick, K.; Park, Y.; Astoreca, R.; Neukermans, G.; Van Mol, B. Validation of MERIS water products in the Southern North Sea. In *Proceedings of the 2nd MERIS—(A) ATSR Workshop*; ESA Publications Office Frascati: Frascati, Spain, 2008.
83. Claustre, H. The trophic status of various oceanic provinces as revealed by phytoplankton pigment signatures. *Limnol. Oceanogr.* **1994**, *39*, 1206–1210. [CrossRef]
84. Aiken, J.; Pradhan, Y.; Barlow, R.; Lavender, S.; Poulton, A.; Holligan, P.; Hardman-Mountford, N. Phytoplankton pigments and functional types in the Atlantic Ocean: A decadal assessment, 1995–2005. *Deep Sea Res. Part II Top. Stud. Oceanogr.* **2009**, *56*, 899–917. [CrossRef]
85. Descy, J.P.; Sarmento, H.; Higgins, H.W. Variability of phytoplankton pigment ratios across aquatic environments. *Eur. J. Phycol.* **2009**, *44*, 319–330. [CrossRef]
86. Behrenfeld, M.J.; O'Malley, R.T.; Siegel, D.A.; McClain, C.R.; Sarmiento, J.L.; Feldman, G.C.; Boss, E.S. Climate-driven trends in contemporary ocean productivity. *Nature* **2006**, *444*, 752. [CrossRef] [PubMed]
87. Bricaud, A.; Claustre, H.; Ras, J.; Oubelkheir, K. Natural variability of phytoplanktonic absorption in oceanic waters: Influence of the size structure of algal populations. *J. Geophys. Res. Oceans* **2004**, *109*. [CrossRef]

applied
sciences

MDPI

Article

Anomalous Light Scattering by Pure Seawater

Xiaodong Zhang [1,*] and Lianbo Hu [1,2]

[1] Department of Earth System Science and Policy, University of North Dakota, Grand Forks, ND 58202, USA; hulb@ouc.edu.cn
[2] Ocean Remote Sensing Institute, Ocean University of China, Qingdao 266001, China
* Correspondence: zhang@aero.und.edu; Tel.: +1-701-777-6087

Received: 17 May 2018; Accepted: 19 July 2018; Published: 19 December 2018

Abstract: The latest model for light scattering by pure seawater was used to investigate the anomalous behavior of pure water. The results showed that water exhibits a minimum scattering at 24.6 °C, as compared to the previously reported values of minimum scattering at 22 °C or maximum scattering at 15 °C. The temperature corresponding to the minimum scattering also increases with the salinity, reaching 27.5 °C for $S = 40$ psu.

Keywords: light scattering; light scattering by pure water; light scattering by pure seawater; anomalous properties of water

1. Introduction

Light scattering by pure water or pure seawater is a fundamental quantity in aquatic optics. Because of hydrogen bonding, many bulk properties of water exhibit anomalous behavior with temperature that is unlike any other liquids [1,2]. For example, liquid water has a maximum density near 4 °C [3], a minimum isothermal compressibility near 46 °C [4] and a maximum refractive index near 0 °C [5]. The scattering seems to behave "anomalously" too [6]. Cohen and Eisenberg [6] measured the scattering at 436 and 546 nm by pure water at temperatures from 5 to 65 °C, and found a scattering minimum at approximately 22 °C that is consistent with their theoretical estimate using the Einstein–Smoluchowski equation and the temperature variation of the isothermal compressibility. Using the same Einstein-Smoluchowski equation, with inputs re-evaluated using newer experimental results, Buiteveld et al. [7] improved the estimate of light scattering by pure water, which showed a better agreement with the spectral values measured by Morel [8]. However, their model predicts a maximum scattering at 15 °C, which differs from Cohen and Eisenberg [6] not only in value but in the behavior as well. To the best of our knowledge, few other studies have explored the temperature behavior of scattering by water. In addition, it is still unknown whether and how this temperature dependence of scattering by water would vary in the presence of sea salts.

2. Methods

Recently, Zhang and coworkers refined the models for light scattering by pure water [9], by pure seawater [10,11], and by simple sea salt solutions [12] using the improved measurements of the key thermodynamic parameters, and their models agree with the spectral scattering measurements [13,14] for both pure water and seawater within the experimental errors (2%). Localized fluctuation in density for pure water, as well as additional fluctuations in the mixing ratio of salt ions and water for pure seawater, lead to microscopic inhomogeneities in the refractive index (n) [15], which in turn cause scattering of light. Since the fluctuations in density and mixing ratio are independent, the scattering coefficient of seawater, b (m^{-1}) can be expressed as

$$b = b_d + b_c, \tag{1}$$

where b_d represents the scattering due to density fluctuation, and b_c the scattering due to fluctuation of mixing ratio (concentration). Following Zhang and Hu [9],

$$b_d = \frac{8\pi^3}{\lambda^4} (\rho \frac{\partial n^2}{\partial \rho})_T^2 kT\beta_T h(\delta), \tag{2}$$

and following Zhang et al. [11]

$$b_c = \frac{8\pi^3}{\lambda^4 N_A} (\frac{\partial n^2}{\partial S})^2 \frac{M_w}{\rho} \frac{S}{-\partial \ln a_w/\partial S} h(\delta), \tag{3}$$

where, respectively, λ, k (=1.38064852 $\times$ 10^{-23} m^2·kg·s^{-2}·K^{-1}), and N_A (=6.022 $\times$ 10^{23} mol^{-1}) are the wavelength of light, the Boltzmann constant, and Avogadro's number; ρ, n, T, β_T, S, and δ are the density, the absolute refractive index, the absolute temperature, the isothermal compressibility, the mass concentration of salts, and the depolarization ratio of the seawater; and a_w and M_w (=18.01528 g mol^{-1}) are the activity and molecular weight of pure water. Also, $h(\delta) = (2 + \delta)/(6 - 7\delta)$.

In Equation (1), b_c vanishes for pure water, and the scattering of light is due entirely to density fluctuation. Replacing the density derivative in Equation (2) with pressure derivative, i.e., $(\rho \frac{\partial n^2}{\partial \rho})_T = \frac{2n}{\beta_T} (\frac{\partial n}{\partial P})_T$, Equation (2) becomes the Einstein–Smoluchowski equation

$$b_d = \frac{32\pi^3}{\lambda^4} \frac{n^2}{\beta_T} (\frac{\partial n}{\partial P})_T^2 kTh(\delta) \tag{4}$$

which was used by Cohen and Eisenberg [6] and Buiteveld et al. [7] in evaluating the temperature dependence of scattering by pure water. Historically, Equation (4) was often used because the isothermal piezo-optic coefficient $(\partial n/\partial P)_T$ was relatively easier to measure, even though the uncertainty was high as compared to $n(T)$, $n(\lambda)$, or $n(S)$ [16]. However, recent theoretical development [17] has greatly improved our knowledge in $(\rho\partial n/\partial \rho)_T$. This, together with the development of Equation (3) to explicitly account for the effect of salinity on scattering [11], has advanced our capability for modeling scattering by seawater [9–11], and our confidence in using Equations (1)–(3) to evaluate its temperature effect. The formulae used in the equations to estimate n, ρ, β_T, and a_w can be found in Zhang and Hu [9] and Zhang et al. [11], and the Matlab code for the model can be accessed at https://goo.gl/jKAZgT. Light scattering by seawater is a function of salinity, temperature, and pressure. In this study, we focus on the temperature and salinity ranges of 0–60 °C and 0–40 psu under one atmospheric pressure, which cover the majority of natural inland, coastal, and oceanic surface water bodies. The presence of sea salts is expected to modify the value of δ through two contrasting effects: isotropic ions would decrease δ; whereas their electrostatic field would increase anisotropy, and hence the value of δ [18,19]. Both effects have been observed in pure salt solutions: δ for KNO$_3$ solution increases and δ for KCl solution decreases, with their respective concentrations [20]. To the best of our knowledge, however, no studies have been reported on how the δ of seawater would vary with salinity. For this study, we assumed a constant value of 0.039 for the depolarization ratio δ for pure water [21] and for seawater [8,9,22,23].

3. Results and Discussion

Light scattering by pure water at 436 and 546 nm was estimated using Equation (2) for temperatures 0–60 °C, and the values normalized to the scattering at 25 °C were compared with the measurements by Cohen and Eisenberg [6] in Figure 1. Between the two wavelengths, the temperature variations of the scattering are almost identical, showing a minimum scattering at 24.7 °C $\pm$ 0.2%. The scattering increases by 4.3% towards 0 °C and increases by 4.7% towards 60 °C at both wavelengths. The root mean square difference between our model and the measurements by Cohen and Eisenberg [6] is approximately 1.3% at both wavelengths. The refractive index model [24] used in Equations (2) and (3) were developed using the Austin and Halikas [16] measurements, which had a temperature precision of 0.1 °C. Also, the Cohen and Eisenberg [6] data we used for comparison in Figure 1 had a temperature precision of 0.1 °C as well. Therefore, we report the temperature in this study at a precision of 0.1 °C. We denote the temperature at which the scattering reaches the minimum as T_{min} hereafter. The predicted values of T_{min} are close to the value of 22 °C measured by Cohen and Eidenberg [6], but differ significantly in both value and trend from the Buiteveld, et al. [7] model, which predicts a maximum near 15 °C. We believe the difference is largely due to the uncertainty in modeling $(\partial n/\partial P)_T$ in Equation (4) that was used by Buiteveld, et al. [7]. Austin and Halikas [16] pointed out that the measurements of the refractive index of water as a function of the pressure, i.e., $n(P)$, were of worse quality when compared to those of $n(T)$, $n(\lambda)$, or $n(S)$. Also, it is well-known that to numerically approximate a derivative, such as $\partial n/\partial P$, as a ratio of measured values is very sensitive to the uncertainties in the measurements of $n(P)$. In addition, Buiteveld, et al. [7] derived the temperature dependency of $(\partial n/\partial P)_T$ by fitting the measurements [25] between 5 and 35 °C, which also explains the relatively large deviation as shown in Figure 1 when extrapolating their model beyond 35 °C.

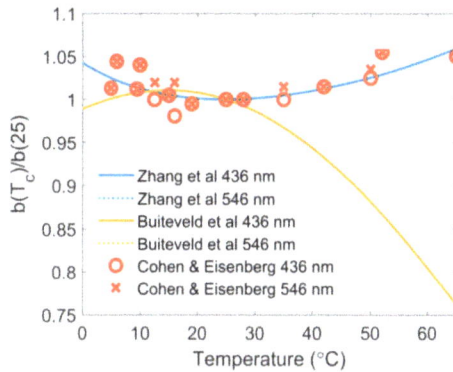

Figure 1. The temperature variations of light scattering by pure water, calculated using the Zhang and Hu [9] model (i.e., Equation (2)) at 436 and 546 nm and normalized by their respective values at 25 °C, are compared with the estimates using the Buiteveld, et al. [7] model and with the measurements by Cohen and Eisenberg [6]. Note that the normalized variations estimated by the Zhang and Hu model overlap with each other at the two wavelengths.

Several bulk properties of pure water needed to estimate the scattering coefficient behave "anomalously": density has the maximum near 4 °C [3]; isothermal compressibility has the minimum near 46 °C [4]; and the refractive index has the maximum near 0 °C [5]. Also, less apparent but indirectly relevant is that $(\partial n/\partial P)_T$, as in Equation (4), has its minimum near 50 °C [5]. Clearly, anomalous light scattering by pure water results from the combination of all of these anomalous properties, as well as its direct proportionality with the temperature (Equation (2) or (4)). Even though scattering by pure water varies strongly with wavelength, with a spectral slope of −4.28 [9], the anomalous temperature behavior of scattering varies little with wavelength (Figure 1).

The scattering coefficient at 546 nm as a function of temperature for different salinities is shown in Figure 2a for b_d (due to density fluctuation) and Figure 2b for b_c (due to concentration fluctuation). Both b_d and b_c vary with temperature in the same anomalous way, all exhibiting a minimum. Also, both T_{min} for b_d and T_{min} for b_c change with salinity—however, with differing patterns. T_{min} for b_d decreases about 20% from 24.6 °C to 19.1 °C for salinity varying from 0 to 40 psu, whereas over the same salinity range T_{min} for b_c increases slightly by ~3%, from 32.2 °C to 33.2 °C. In terms of absolute magnitude, b_d is about 2–10 times greater than b_c (Figure 2a), but in terms of change with respect to salinity, b_c is about 10 times greater than b_d (Figure 2b). As a result, the change of T_{min} for the total scattering coefficient, b is dominated by b_c, and increases from 24.6 °C to 27.5 °C for S from 0–40 psu (Figure 3). It is well known that T_{max} for density [26] and T_{min} for isothermal compressibility [27] decrease with the salinity. Here, we show for the first time that T_{min} for light scattering increases with salinity, which is largely due to the temperature variation of scattering introduced by sea salts. Table 1 lists the variations of T_{min} for b_d, b_c, and b at different salinities.

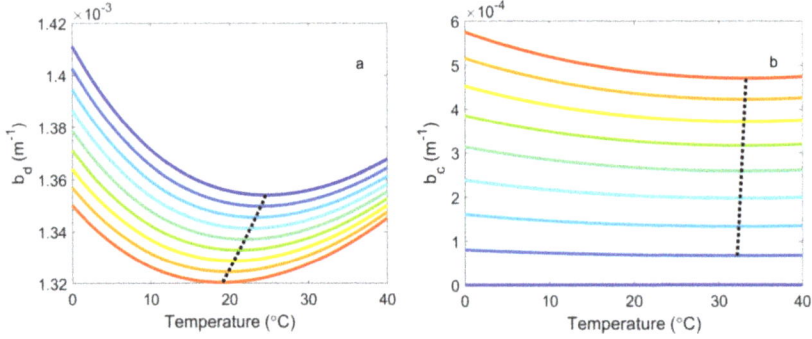

Figure 2. Light scattering by pure seawater at 546 nm as a function of temperature and salinity. (**a**) b_d, the scattering due to density fluctuation; and (**b**) b_c, the scattering due to concentration fluctuation. Lines of progressive colors from blue to red correspond to different salinities from 0 to 40 psu, at 5 psu increments. The dotted line in each plot connects T_{min} at different salinities.

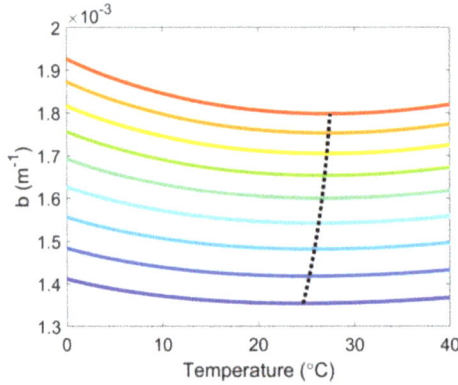

Figure 3. Total scattering coefficient by pure seawater at 546 nm as a function of temperature and salinity. Lines of progressive colors from blue to red correspond to different salinities from 0 to 40 psu at 5 psu increments. The dotted line connects T_{min} at different salinities.

Table 1. Temperatures (T_{min} in °C) at which the scattering of light at 546 nm by pure seawater due to density fluctuations (b_d), concentration fluctuations (b_c), and their total (b) reach the minimum for various salinities (S).

S (psu)	0	5	10	15	20	25	30	35	40
b_d (m^{-1})	24.6	24.0	23.4	22.7	22.0	21.2	20.5	19.8	19.1
b_c (m^{-1})		32.2	32.3	32.5	32.7	32.8	33.0	33.1	33.2
b (m^{-1})	24.6	25.3	25.9	26.3	26.6	26.9	27.2	27.3	27.5

In conclusion, using the latest scattering models for water and seawater, we find that water exhibits an anomalous light scattering behavior, with a minimum occurring at 24.6 °C for pure water, and that this minimum increases with the salinity, reaching 27.5 °C at 40 psu. This temperature behavior changes little spectrally. Caution should be exercised when using the Buiteveld, et al. [7] model, which predicts a temperature behavior of scattering that is inconsistent with the measurements [6] or the results of this study.

Author Contributions: X.Z. and L.H. conceived and designed the experiments; X.Z. wrote the paper.

Funding: National Aeronautics and Space Administration (NASA) (NNX13AN72G, NNX15AC85G) and National Science Foundation (NSF) (1355466, 1459168).

Acknowledgments: We thank three anonymous reviewers and Dariusz Stramski for their comments, which have improved the manuscript.

Conflicts of Interest: The authors declare no conflict of interest. The funding sponsors had no role in the design of the study; in the collection, analyses, or interpretation of data; in the writing of the manuscript; and in the decision to publish the results.

References

1. Brovchenko, I.; Oleinikova, A. Multiple Phases of Liquid Water. *ChemPhysChem* **2008**, *9*, 2660–2675. [CrossRef] [PubMed]
2. Ball, P. Water—An enduring mystery. *Nature* **2008**, *452*, 291. [CrossRef] [PubMed]
3. Vedamuthu, M.; Singh, S.; Robinson, G.W. Properties of Liquid Water: Origin of the Density Anomalies. *J. Phys. Chem.* **1994**, *98*, 2222–2230. [CrossRef]
4. Vedamuthu, M.; Singh, S.; Robinson, G.W. Properties of Liquid Water. 4. The Isothermal Compressibility Minimum near 50 °C. *J. Phys. Chem.* **1995**, *99*, 9263–9267. [CrossRef]
5. Cho, C.H.; Urquidi, J.; Gellene, G.I.; Robinson, G.W. Mixture model description of the *T*-, *P* dependence of the refractive index of water. *J. Chem. Phys.* **2001**, *114*, 3157–3162. [CrossRef]
6. Cohen, G.; Eisenberg, H. Light Scattering of Water, Deuterium Oxide, and Other Pure Liquids. *J. Chem. Phys.* **1965**, *43*, 3881–3887. [CrossRef]
7. Buiteveld, H.; Hakvoort, J.H.M.; Donze, M. The optical properties of pure water. *SPIE* **1994**, *2258*, 174–183.
8. Twardowski, M.S.; Claustre, H.; Freeman, S.A.; Stramski, D.; Huot, Y. Optical backscattering properties of the "clearest" natural waters. *Biogeosciences* **2007**, *4*, 1041–1058. [CrossRef]
9. Zhang, X.; Hu, L. Estimating scattering of pure water from density fluctuation of the refractive index. *Opt. Express* **2009**, *17*, 1671–1678. [CrossRef] [PubMed]
10. Zhang, X.; Hu, L. Scattering by pure seawater at high salinity. *Opt. Express* **2009**, *17*, 12685–12691. [CrossRef] [PubMed]
11. Zhang, X.; Hu, L.; He, M.-X. Scattering by pure seawater: Effect of salinity. *Opt. Express* **2009**, *17*, 5698–5710. [CrossRef] [PubMed]
12. Zhang, X.; Hu, L.; Twardowski, M.S.; Sullivan, J.M. Scattering by solutions of major sea salts. *Opt. Express* **2009**, *17*, 19580–19585. [CrossRef] [PubMed]
13. Morel, A. Etude Experimentale de la diffusion de la lumiere par l'eau, les solutions de chlorure de sodium et l'eau de mer optiquement pures. *J. Chim. Phys.* **1966**, *10*, 1359–1366. [CrossRef]
14. Morel, A. Note au sujet des constantes de diffusion de la lumiere pour l'eau et l'eau de mer optiquement pures. *Cahiers Oceanogr.* **1968**, *20*, 157–162.
15. Einstein, A. Theorie der Opaleszenz von homogenen Flüssigkeiten und Flüssigkeitsgemischen in der Nähe des kritischen Zustandes. *Annalen der Physik* **1910**, *338*, 1275–1298. [CrossRef]
16. Austin, R.W.; Halikas, G. *The Index of Refraction of Seawater*; SIO Ref. No. 76-1; Scripps Institute of Oceanography: La Jolla, CA, USA, 1976; p. 121.
17. Proutiere, A.; Megnassan, E.; Hucteau, H. Refractive index and density variations in pure liquids: A new theoretical relation. *J. Phys. Chem.* **1992**, *96*, 3485–3489. [CrossRef]
18. Morel, A. Optical Properties of Pure Water and Pure Sea Water. In *Optical Aspects of Oceanography*; Jerlov, N.G., Nielsen, E.S., Eds.; Academic Press: New York, NY, USA, 1974; pp. 1–24.
19. Shifrin, K.S. *Physical Optics of Ocean Water*; American Institute of Physics: New York, NY, USA, 1988; p. 285.

20. Pethica, B.A.; Smart, C. Light scattering of electrolyte solutions. *Trans. Faraday Soc.* **1966**, *62*, 1890–1899. [CrossRef]
21. Farinato, R.S.; Rowell, R.L. New values of the light scattering depolarization and anisotropy of water. *J. Chem. Phys.* **1976**, *65*, 593–595. [CrossRef]
22. Zhang, X. Molecular Light Scattering by Pure Seawater. In *Light Scattering Reviews 7*; Kokhanovsky, A., Ed.; Springer: Heidelberg, Germany, 2013; pp. 225–243.
23. Werdell, P.J.; McKinna, L.I.; Boss, E.; Ackleson, S.G.; Craig, S.E.; Gregg, W.W.; Lee, Z.; Maritorena, S.; Roesler, C.S.; Rousseaux, C.S.; et al. An overview of approaches and challenges for retrieving marine inherent optical properties from ocean color remote sensing. *Prog. Oceanogr.* **2018**, *160*, 186–212. [CrossRef]
24. Quan, X.; Fry, E.S. Empirical equation for the index of refraction of seawater. *Appl. Opt.* **1995**, *34*, 3477–3480. [CrossRef] [PubMed]
25. O'Connor, C.L.; Schlupf, J.P. Brillouin Scattering in Water: The Landau—Placzek Ratio. *J. Chem. Phys.* **1967**, *47*, 31–38. [CrossRef]
26. Millero, F.J.; Chen, C.-T.; Bradshaw, A.; Schleicher, K. A new high pressure equation of state for seawater. *Deep-Sea Res.* **1980**, *27*, 255–264. [CrossRef]
27. Lepple, F.K.; Millero, F.J. The isothermal compressibility of seawater near one atmosphere. *Deep-Sea Res.* **1971**, *18*, 1233–1254. [CrossRef]

![applied sciences logo] *applied sciences*

MDPI

Article

Modeling Sea Bottom Hyperspectral Reflectance

Georges Fournier *, Jean-Pierre Ardouin and Martin Levesque

DRDC Valcartier Research Centre, Québec, QC G3J1X5, Canada;
jean-pierre.ardouin@drdc-rddc.gc.ca (J.-P.A.); martin.levesque@drdc-rddc.gc.ca (M.L.)
* Correspondence: grfournier1@gmail.com; Tel.: +1-418-844-4000 (ext. 4313)

Received: 5 June 2018; Accepted: 5 September 2018; Published: 19 December 2018

Featured Application: Hyperspectral Bathymetry and near-shore bottom mapping. Retrieving both depth and bottom types from hyperspectral remote-sensing reflectance requires inverting the remote-sensing reflectance profile to fit both the inherent optical properties of the water column and the bottom spectral reflectance profile. In order to obtain a robust fit, the number of parameters required to characterize the bottom reflectance spectrum must be kept to a minimum. The model which we have developed allows one to model a good approximation to bottom spectra by using at most three parameters.

Abstract: Over the near-ultraviolet (UV) and visible spectrum the reflectance from mineral compounds and vegetation is predominantly due to absorption and scattering in the bulk material. Except for a factor of scale, the radiative transfer mechanism is similar to that seen in murky optically complex waters. We therefore adapted a semi-empirical algebraic irradiance model developed by Albert and Mobley to calculate the irradiance reflectance from both mineral compounds and vegetation commonly found on the sea bottom. This approach can be used to accurately predict the immersed reflectance spectra given the reflectance measured in air. When applied to mineral-based compounds or various types of marine vegetation, we obtain a simple two-parameter fit that accurately describes the key features of the reflectance spectra. The non-linear spectral combination effect as a function of the thickness of vegetation growing on a mineral substrate is then accounted for by a third parameter.

Keywords: remote-sensing reflectance; bathymetry; hyperspectral; bottom mapping; radiative transfer

---

1. Introduction

The application that supplied the primary impetus for the present work was bathymetry and near shore bottom mapping. Both problems require inverting the remote-sensing reflectance profile to simultaneously fit both the inherent optical properties of the water column and the bottom spectral reflectance profile. In order to obtain robust and reliable results, the number of fitted parameters must be kept to a minimum. The parameters required to model the water column are already well known from numerous and extensive remote-sensing reflectance studies and detailed in the semi-empirical algebraic irradiance model developed by Albert and Mobley [1] that we use as a basis for our work. The model we are proposing here produces a good approximation to bottom reflectance spectra by using only three parameters.

In the near-ultraviolet (UV) and visible, the reflectance from mineral compounds and vegetation is predominantly due to absorption and backscattering in the bulk material. For most inorganic liquids or solids such as minerals the absorption comes from the broadened far wing of electronic transitions in the deep UV [2–4] and the backscattering is dominated by reflections at the interface between the crystalline grains of the material. For vegetation the absorption is primarily due to the chlorophyll-a and accessory pigments contained in the plant chloroplasts while the backscattering is due to reflections at the membranes of the cells and their inner components. The radiative transfer processes in both minerals and vegetation, even though occurring on a much smaller scale, are very similar to that occurring in murky waters. This prompted us to adapt a semi-empirical model for murky type II waters due to Albert and Mobley [1] and generalize its results with another model due to Aas [5].

Except for the obvious size scale factor, the key difference between the radiative transfer that occurs in minerals and vegetation against that found in murky waters is due to the physics of the backscattering term. We assume the backscattering term comes from the reflection of the interfaces between the structural elements of the solid. The surfaces of the interfaces are modeled to be rough and randomly oriented. The formulas for this type of backscattering are identical to those derived for randomly oriented particles with rough surfaces [6,7]. The formulas scale as a function of the relative index of refraction of the solid grains and the material of the gap. If the original reflectance was measured for dry samples, the gaps contain air. If the sample is immersed the gaps are water filled and the relative index is smaller. This occurs at or just below the surface of solid rocks and depends on the porosity and on the state and time of immersion. The same effect occurs to an even greater depth when the mineral is in powdered form such as sand We have used this effect to predict the immersed reflectance spectra given the reflectance measured in air. This new model allows one to use the vast library of spectral reflectance signatures measured in air to the underwater environment. We have also used the model in our bathymetric work by measuring the hyperspectral signature of the coastline and modifying it to use as bottom reflectance. We have found this approach to be particularly effective with sand beaches. The only parameter that needs to be estimated is the mean index of refraction of the sand grains which is very close to either silica or in some cases calcite.

To properly model vegetation absorption several effects must be accounted for. The absorption spectrum of chlorophyll-a and accessory pigments at low concentrations is modified by saturation of the absorption through the chloroplasts as the concentration increases. This is known as the package effect and has been extensively studied for spherical chloroplast by Morel and Bricaud [8]. We extend this work to include disk-shaped chloroplast. We then use the resulting formulas to fit with a single parameter the measured phytoplankton absorption spectra as a function of concentration [9,10]. The backscattering cellular interfaces are assumed to be composed of cellulose and the reflectance spectra are computed for several types of algae and underwater vegetation.

The spectra show that, as is well known, vegetation is actually translucent which means that when it grows over a mineral substrate the reflectance spectra changes significantly as a non-linear function of the thickness. We use a normalized version of the Albert and Mobley model for finite depth [1] to evaluate this effect. The reflectance from the mineral substrate replaces the bottom reflectivity in the model and the water column absorption and backscatter properties are replaced by those of the vegetation. If the vegetation cover is complete over one pixel, the complete vegetation model depends on three parameters: the concentration of chlorophyll-a, the chloroplast absorption saturation parameters and the thickness.

The aim of the present work is to help limit the number of fit parameters in order to better constrain the water depth value. This is particularly significant in conditions were there is little or no a priori knowledge of the bottom type. For convenience, Table 1 lists the symbols we use and their definitions and units.

Table 1. List of abbreviations, symbols, definitions and units.

Symbol or Abbreviation	Definition, Units
$a(\lambda)$	Absorption coefficient, m^{-1}
$a_{cl}(\lambda)$	Cellulose absorption coefficient, m^{-1}
$a_w(\lambda)$	Pure water absorption coefficient, m^{-1}
$a^*(\lambda)$	Extended Bricaud specific absorption coefficient, m^2/mg

Table 1. *Cont.*

Symbol or Abbreviation	Definition, Units
$a_o^*(\lambda)$	Specific absorption coefficient at low concentration, m^2/mg
$a_r^*(\lambda)$	Specific absorption coefficient at the reference concentration
$a_v^*(\lambda)$	Specific absorption coefficient at any concentration, m^2/mg
A_1, A_2	Coefficients for the finite thickness translucent model
α_o	Amplitude coefficient for the mineral fit, units $\lambda^{-\nu}$
$b_b(\lambda)$	Backscattering coefficient in air, m^{-1}
$b_{bw}(\lambda)$	Backscattering coefficient in water, m^{-1}
<cos>	Mean scattering cosine
<d>	Mean diameter of the scattering structures, m^{-1}
$\delta(\lambda)$	Bottom reflectance attenuation coefficient, m^{-1}
f	Fitting parameter for alternate $R_{spc-\infty}$ formula
f_{cl}	Mass fraction of cellulose in a vegetation cell
f_{vp}	Fraction of vegetation cover per pixel
$\gamma(\lambda)$	Translucent substance irradiance attenuation coefficient, m^{-1}
$\kappa_0, \kappa_{1W}, \kappa_{2W}, \kappa_{1b}, \kappa_{2b}$	Coefficients for the finite thickness translucent model
λ	Wavelength in air, microns
λ_0	Wavelength coefficient for the mineral fit, microns
ν	Power coefficient for the mineral fit, dimensionless
$\mu_{(d)}$	Mean value of the cell size, microns
n	Real Index of refraction in air
n_{cw}	Index of refraction of cell walls
n_w	Real Index of refraction in water
n_c	Number of cells per unit volume, m^{-3}
n_{cp}	Number of chloroplasts per cell
N	Number of scattering elements per unit volume, m^3
$p_1, p_2, p_3, p_4, p_5,$	Coefficients of the irradiance reflectance model
$p(\theta, \lambda)$	Total scattering phase function.
Q_a	Absorption efficiency, dimensionless
R_∞	Irradiance reflectance with no bottom contribution
$R_{spc-\infty}$	Spectralon reference normalized irradiance reflectance
R_b	Bottom irradiance reflectance
R_m	Mixed pixel irradiance reflectance
R_t	Irradiance reflectance for translucent materials
ρ_{chl}	Chlorophyll-a mass density, mg/m^3
ρ_{cp}	Chlorophyll-a mass density inside the chloroplasts, mg/m^3
ρ_r	Chlorophyll-a mass density concentration reference, mg/m^3
σ_g	Geometric cross-section, m^2
$\sigma_b(\lambda)$	Backscattering cross section, m^2
$\sigma_{(d)}$	Standard deviation of the cell size, microns
$\sigma_r(\%)$	Standard deviation of the relative error, units %
$\theta_{\bar{s}}$	Sun zenith angle in water
τ_{cp}	Thickness of the disk shaped chloroplasts, m
u_b	Backscattering coefficient times z_b, dimensionless
u_{cp}	$\rho_{cp}\tau_{cp}$, units, mg/m^2
u_r	u_{cp} at the reference chlorophyll-a concentration ρ_r
V_c	Volume of vegetation cell, m^3
V_{cp}	Volume of chloroplast, m^3
V_m	Volume of vegetation filled by cells, m^3
$x(\lambda)$	Backscattering albedo in air, dimensionless, range 0 to 1
$x_{ba}(\lambda)$	Backscattering albedo in air, dimensionless, range 0 to 1
$x_{bw}(\lambda)$	Backscattering albedo in water, dimensionless, range 0 to 1
z_b	Translucent material layer thickness, m^{-1}
ω_b	Backscattering reflection coefficient for random orientation
ω_t	Reflection coefficient for random orientation, range 0 to 1

2. Materials and Methods

2.1. Basic Model

The key parameter in any radiative transfer model of reflectance is a parameter we will refer to in this paper as the backscattering albedo $x(\lambda)$. This defined as the ratio of the total scattering in the back hemisphere to the sum of the absorption and total backscattering.

$$x(\lambda) = \frac{b_b(\lambda)}{a(\lambda) + b_b(\lambda)} , \tag{1}$$

In the above expression $a(\lambda)$ is the absorption coefficient while $b_b(\lambda)$ is the backscattering coefficient. The backscattering coefficient is defined in standard form by the following expression.

$$b_b(\lambda) = 2\pi \int_{\pi/2}^{\pi} p(\theta, \lambda) \sin \theta \, d\theta , \tag{2}$$

In the expression above $p(\theta, \lambda)$ is the scattering phase function. The main aim of our work from now on is to obtain expressions for the various contributions to both $b_b(\lambda)$ and $a(\lambda)$. Figure 1 shows graphically the various mechanisms discussed above and will serve as a guide in this task.

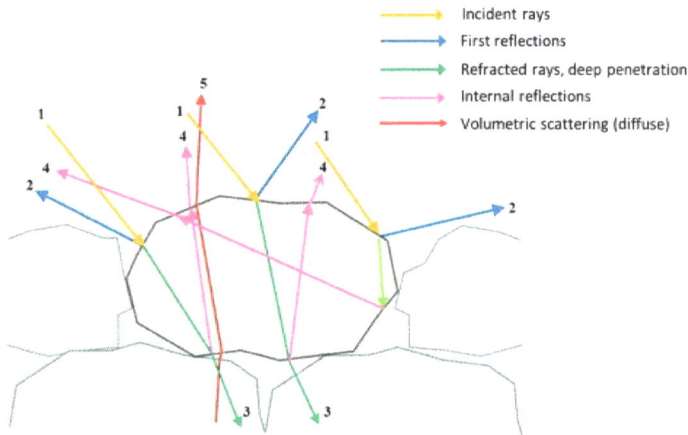

Figure 1. This figure is a schematic of the microstructure elements relevant to scattering and absorption for both minerals (grains) and vegetation (cells). The incident light rays (1) are reflected (2) and transmitted at the first surface (3). The rays transmitted through the first surface are subsequently both reflected back from the inner surfaces of the grains (4) and absorbed. The rays that penetrate deeper (5) are multiply scattered before coming back to the surface and have a near Lambertian (uniform) scattering distribution.

The materials of interest to us, minerals and vegetation, absorb little per grain or cell and the light ray will encounter many inner surfaces before being scattered back out. In our evaluation of $b_b(\lambda)$ we will treat the reflection from the first surface boundary as similar to that of the inner deeper boundaries. The formulas we will use for the backscattering from the inner surfaces of minerals are those which describe reflection from rough surface elements with random orientation. For vegetation we will use the formulas for backscattering from smooth surfaces which is a more appropriate representation. This model has been used recently to describe the backscattering from complex naturally occurring structures such as coccoliths [6,7]. The backscattering cross-section function for the reflection from a randomly oriented set of uniform (Lambertian) diffusers is given by:

$$\sigma_b(\lambda) = \sigma_g \, \omega_t(n) 2\pi \int_{\pi/2}^{\pi} \frac{4}{3\pi} (\sin\theta - \theta \cos\theta) \sin\theta d\theta = \frac{5}{6} \sigma_g \, \omega_t(n) \,, \tag{3}$$

σ_g is the geometric cross-section of an individual scattering structure and $\omega_t(n)$ is the Fresnel reflectance integrated over a set of randomly oriented surfaces of relative index of refraction n that together compose the surface of those structures. The total backscattering coefficient of the ensemble of the scattering elements is by definition:

$$b_b(\lambda) = \frac{5}{6} \omega_t(n) \, N \, \sigma_g \,, \tag{4}$$

Assuming that the number density of the scattering structures N is such that the sum of their geometric cross-sections is equal to the area of the material normal to the impinging light, we obtain the following formula for the backscattering coefficient of the material.

$$b_b(\lambda) = \frac{5}{6} \frac{\omega_t(n)}{d} \,, \tag{5}$$

where $<d>$ is the mean diameter of the scattering structures. $\omega(n)$ for unpolarised light is given by the following formulas [3,4].

$$\omega_t = \left(\frac{\omega_\perp + \omega_\parallel}{2} \right) \,, \tag{6}$$

$$\omega_\perp = \frac{(3n+1)(n-1)}{3(n+1)^2} \,, \tag{7}$$

$$\omega_\parallel = \frac{1}{(n^2+1)^3(n^2-1)^2} \left\{ (n^4-1)(n^6 - 4n^5 - 7n^4 + 4n^3 - n^2 - 1) \right.$$
$$\left. + 2n^2 \left[(n^2-1)^4 \ln\left(\frac{n-1}{n+1}\right) + 8n^2(n^4+1) \ln(n) \right] \right\} \,, \tag{8}$$

Corresponding formulas for smooth surfaces are:

$$\omega_{b\perp} = \frac{3n^4 - 16n^3 + 12n^2 - 1 + 2(2n^2-1)^{3/2}}{6(n^2-1)^2} \,, \tag{9}$$

$$\omega_{b\parallel} = \omega_{b\perp} \left[(3 - \ln 16) + \frac{37}{40}\left(\frac{n-1}{n+1}\right) \right] \,, \tag{10}$$

$$\omega_b = \left(\frac{\omega_{b\perp} + \omega_{b\parallel}}{2} \right) \tag{11}$$

$$b_b(\lambda) = \frac{\omega_b(n)}{<d>} \,, \tag{12}$$

The formulas above were derived assuming the same Fresnel coefficients for both the entrance and exit faces of the scattering structures. We do this because for most randomly oriented convex objects the outgoing light ray has a nearly symmetrical angular relationship with the incoming light ray which implies close to identical surface reflectivity. This symmetrical relationship is strictly true for the extreme cases of spherical, cylindrical and flat plate shapes. Given the near universality of the relationship we expect that in almost all cases of interest to us any deviation from it will be small and to first order can be neglected.

2.2. Dry to Wet Reflectance Ratio

Note that the wavelength dependence of the backscattering coefficient is a direct consequence of the wavelength dependence of the relative index of refraction. One important consequence of this dependence on the relative index of refraction is the reduction in $b_b(\lambda)$ when the interfaces between the grains are filled with water instead of air. This effect is the source of the lowering of the irradiance reflectance of materials and vegetation immersed in water. Because the interstitial gaps are small the grain structure and spacing <*d*> is the same in both cases, we can estimate the water to air ratio directly.

$$\frac{b_{bw}(\lambda)}{b_b(\lambda)} = \frac{\omega_t(n/n_w)}{\omega_t(n)}, \tag{13}$$

Since the absorption does not change, we can directly estimate the ratio of backscattering albedo.

$$\frac{x_{bw}(\lambda)}{x_{ba}(\lambda)} = \frac{\frac{a(\lambda)}{b_b(\lambda)} + 1}{\left[\frac{\omega_t(n)}{\omega_t(n/n_w)}\right]\frac{a(\lambda)}{b_b(\lambda)} + 1}, \tag{14}$$

$x_{bw}(\lambda)$ is the backscattering albedo in water while $x_{ba}(\lambda)$ is the corresponding backscattering albedo in air.

We can at this time estimate the wet to dry reflectivity factors for three of the most important and frequently found components of materials and vegetation, crystalline quartz, calcite and cellulose.

These are shown in Figure 2. The detailed formulas as a function of wavelength for these important indices are given in Appendix A. These indices can be found in references [11–15].

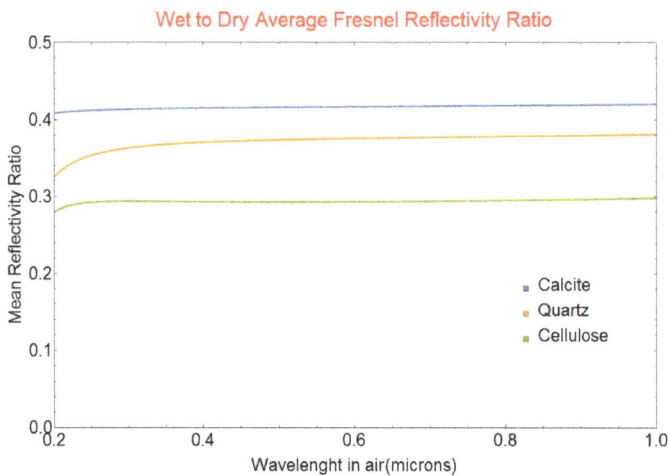

Figure 2. Wet to dry angularly averaged Fresnel reflectivity factors as a function of wavelength for important components of materials and vegetation: crystalline quartz, calcite and cellulose.

2.3. Basic Irradiance Reflectance Model

To estimate the irradiance reflectance from the backscattering albedo we use the Albert and Mobley algebraic radiance model valid for the infinitely deep medium case. The model is based on a careful analysis of solutions of the radiative transfer equation by the Hydrolight code for over 177,000 cases that encompass the full range of parameters for optically complex waters [1].

$$R_\infty = p_1 x \left(1 + p_2 x + p_3 x^2 + p_4 x^3\right)\left(1 + p_5\frac{1}{\cos\theta_s}\right), \tag{15}$$

In this formula R_∞ is the irradiance reflectance over waters deep enough that there is no contribution from the bottom reflectance. θ_s is the sun angle just below the surface of the scattering medium. Standard irradiance reflectance measurements are carried out by comparing the signal from a high-quality diffuse reflecting surface (Spectralon) that fills the field of view of the portable spectrometer to the signal from the substance to be measured under the same illumination conditions. Therefore, we must normalize the original Albert and Mobley expression to have $R_\infty = 1$ when $x = 1$.

$$R_{spc-\infty} = \frac{p_1 x (1 + p_2 x + p_3 x^2 + p_4 x^3)}{p_1 (1 + p_2 + p_3 + p_4)} , \tag{16}$$

We have used the notation $R_{spc-\infty}$ in Equation (16) to clearly note that we are referring to the calibrated Spectralon normalized irradiance reflectance but from now on we will simply assume that all reflectances have been properly normalized. Table 2 gives the coefficients of Equations (15) and (16).

Table 2. Coefficients of the Albert-Mobley model for the infinite medium depth case.

Coefficient	R_∞
p_1	0.1034
p_2	3.3586
p_3	−6.5358
p_4	4.6638
p_5	2.4121

The expression that Albert and Mobley use is based on an extensive empirical survey carried out with an exact radiative code. The results of this survey are fitted as a fourth order polynomial which is an inconvenient form to use if we need in some cases to reverse the process and, for instance, estimate x from $R_{spc-\infty}$. Aas [5] developed a two-stream radiative model and obtained approximate solutions for the irradiance reflectance from an infinite depth medium. We found that we could closely match the result of Albert and Mobley by parametrizing the formulas given by Aas. This approach yields simpler more general formulas that can easily be inverted as desired.

$$R_{spc-\infty} = \frac{(1+f^2) - \sqrt{(1+f^2)^2 - 4f^2 x^2}}{2 f^2 x} , \tag{17}$$

f is an empirical parameter that varies from 0 to 1. A very close fit to the results of Albert and Mobley is obtained with = 0.79. We have used the notation $R_{spc-\infty}$ in Equations (16) and (17) to clearly note that we are referring to the calibrated Spectralon normalized irradiance reflectance but from now on we will simply assume that all reflectance have been properly normalized to unity. Formula (17) is easily inverted to obtain x as a function of R if required.

$$x = \left(\frac{1+f^2}{f}\right)\left(\frac{fR}{1+fR^2}\right) , \tag{18}$$

The value of f is correlated with but not equal to the mean cosine of the total scattering function <cos> which is defined as follows.

$$<\cos> = 2\pi \int_0^\pi p(\theta, \lambda) \cos \theta \sin \theta \, d\theta , \tag{19}$$

when $f = 0$, the single scattering is nearly isotropic and the irradiance reflectance is equal to the backscattering albedo. When f approaches 1 the single scattering becomes highly forward peaked. Note that $p(\theta, \lambda)$ is the total scattering function and it includes both the reflected and the transmitted part of the radiation. The transmitted part is controlled by refraction and diffraction which dominate scattering in the forward hemisphere for grains or cells much larger that the wavelength. In the cases that concern us in this work the grains or cells are large enough that the transmitted part controls the value of the mean cosine.

2.4. Irradiance Reflectance Model for Translucent Subtances

There is one more common case we have to concern ourselves with: translucent organic materials growing on a mineral substrate. In order to model this situation we use the irradiance reflection model for finite bottom depth of Albert and Mobley [1]. The irradiance reflectance of the underlying material is used as a bottom irradiance reflectance R_b in this case. The irradiance reflectance of the combination of translucent overlay of reflectance R_∞ and thickness z_b with a substrate of reflectance R_b is modeled by the following equations.

$$R_t = R_\infty \left[1 - A_1 e^{-\gamma(\lambda)z_b} \right] + R_b A_2 e^{-\delta(\lambda)z_b} , \tag{20}$$

with:

$$\delta(\lambda) = \left[\kappa_0 + (1 + x(\lambda))^{\kappa_{1w}} (1 + \kappa_{2w}) \right] \left(\frac{b_b(\lambda)}{x(\lambda)} \right) , \tag{21}$$

$$\gamma(\lambda) = \left[\kappa_0 + (1 + x(\lambda))^{\kappa_{1b}} (1 + \kappa_{2b}) \right] \left(\frac{b_b(\lambda)}{x(\lambda)} \right) , \tag{22}$$

Table 3 gives the coefficients of Equations (20)–(22).

Table 3. Coefficients of the Albert-Mobley model for the finite medium depth case.

Coefficient	R_∞
A_1	1.0000
κ_0	1.0546
κ_{1W}	1.9991
κ_{2W}	0.2995
A_2	1.0000
κ_{1b}	1.2441
κ_{2b}	0.5182

The new parameter that controls the behavior of the combined solution is $u_b = b_b(\lambda)z_b$. Assuming the overlaying vegetation completely covers the substrate then, as the thickness of the overlay z_b increases, the combined reflectance R_t goes to the reflectance of the overlay R_∞ while when z_b becomes small the combined reflectance approaches R_b as expected. What the model above shows is that the usual approach of linearly combining the separate reflectance signatures of the mineral substrate and the organic cover according to the weights of their relative areas only works in the limit where the cover is thick enough. For thin organic covers there is an exponential transfer of signature from substrate to cover which is a strong function of wavelength through the backscattering albedo of the translucent overlay $x(\lambda)$.

3. Results

3.1. Specific Properties of Minerals

As mentioned briefly in the introduction, in the near-UV and visible the reflectance from mineral compounds is predominantly due to absorption and backscattering in the bulk material. For minerals this absorption comes from the broadened far wing of the lowest energy electronic transitions in the deep UV [2–4] and the backscattering is dominated by reflections at the interface between the crystalline grains of the material. The usual method for obtaining the absorption spectrum of mineral compounds is to measure the transmission loss through a sample of known thickness made from mineral powder that has been pressed and sintered. This is a time-consuming process that requires great care to obtain sufficiently low backscatter. Using our model opens up the possibility of obtaining the relative absorption spectra in the visible near-infrared (IR) region by simply measuring their irradiance reflectance.

Given the irradiance reflectance R we first obtain the backscattering albedo x from Equation (18). Using the definition of the backscattering albedo (1) and the Formula (5) we derived for $b_b(\lambda)$ we obtain the following expression for the absorption spectrum:

$$<d> a(\lambda) = \frac{5}{6} \omega_t(n) \frac{(1-x)}{x} ,$$ (23)

Except for the scale factor of the mean crystalline grain size $<d>$ we can now directly obtain the absorption spectrum for any substance for which we have measured an irradiance reflectance. We will use a simple approximate empirical functional form for the far wing absorption spectrum of an electronic transition which includes the cases of broadening due to internal collision and Van der Wall like interactions in the bulk of the material.

$$<d> a(\lambda) = \frac{\alpha_o}{(\lambda - \lambda_o)^{\nu}} ,$$ (24)

Note that there is still a considerable amount of physical meaning to the parameters in the formula above. λ_o is an estimate of the central wavelength of the lowest energy electronic transition. The value of ν is a function of the shape of the interaction potential of the molecular components of the crystalline grain. In the limiting case of an abrupt delta function like interaction potential $\nu = 2$ and Equation (24) becomes a far wing Lorentzian profile [4] which is the standard abrupt collison lineshape. In the case of a sample of Trenton Limestone measured on the shore of Lake Ontario we obtain a very good fit of $<d> a(\lambda)$ using the $\nu = 1$ solution. This solution is indicative of a smoothly varying interaction potential similar to that of a linear spring.

$$<d> a(\lambda) = \frac{0.081}{(\lambda - 0.183)} ,$$ (25)

Many other values of ν are obviously possible and depend on the form of the interaction potential. Using expression (25) we can reconstruct the reflectance spectrum. Figure 3 shows a graph of the fit between the original reflectance and the one computed using our formulas.

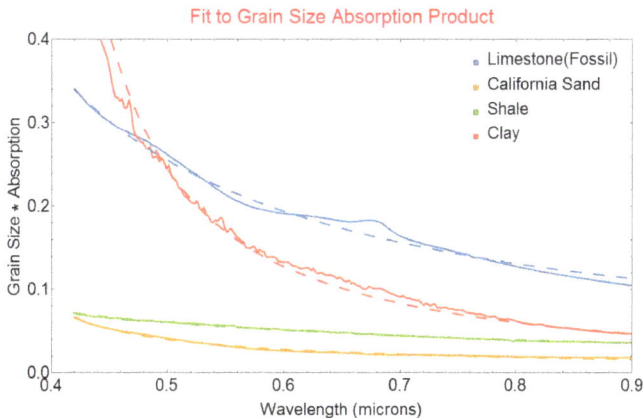

Figure 3. Comparison of the dimensionless parameter grain size times absorption coefficient estimated using Formula (23) for various mineral compounds (solid lines) with the fit (dashed lines) obtained using Equation (25). The fit parameters are given in the corresponding entries of Table 4. The spectral features seen in the experimental reflectance of the Trenton limestone sample are due to an interstitial chlorophyll-a residue lying on top of the limestone.

The fits are quite accurate over the visible spectrum from 0.35 to 0.90 micron and we expect similar accuracy for materials whose absorption is not dominated by inclusions containing color centers. The results presented in Table 4 demonstrate that this is indeed the case.

Table 4 is the result of the fit for a set of materials of interest that could possibly be found on the bottom of the water column. The fit was constrained to a region from 0.42 to 0.90 microns. This wavelength zone was chosen to avoid the reflectance measurement accuracy problems that notoriously plague the near-UV and deep blue region of the spectrum. The standard deviation of the relative error in percent between the model and the data $\sigma_r(\%)$ is given in the last column.

Table 4. This is a table of the functional fits to various minerals according to the formula.

Substance	a_o	ν	λ_o	$\sigma_r(\%)$
California Sand	0.01055	0.708	0.346	5.0
Hawaii Sand	0.04631	1.438	0.102	4.8
Greenland Sand	0.03323	0.453	0.350	4.9
Limestone (Trenton)	0.08401	1.217	0.102	3.8
Limestone (Fossil)	0.01662	0.688	0.348	6.6
Clay	0.02180	1.273	0.350	5.9
Sandy Loam	0.01087	1.882	0.195	4.1
Gray Silty Loam	0.00850	1.636	0.298	3.3
Brown Loam	0.01292	1.595	0.286	3.4
Dark Loam	0.02345	1.395	0.350	4.5
Granite	0.02102	0.737	0.304	8.6
Schist	0.11558	0.134	0.338	0.9
Shale	0.02988	0.738	0.113	1.5
Shale	0.04426	1.165	0.101	6.5
Shale	0.02364	0.219	0.344	2.8
Shale	0.11558	0.368	0.350	3.3
Siltstone	0.02669	0.229	0.345	3.9
Siltstone	0.01803	0.377	0.343	4.2

The low standard deviation of the relative error shows that the fits are very close and are in several cases within the instrumental reflectance measurement variation. Formula (24) can, therefore, serve to fit experimentally measured reflectances. We originally hoped that in the limit, the values of the parameters λ_o, ν and α_o could even be used as markers to identify an unknown material. The results given in Table 3 are not encouraging in this respect as there is a great deal of variability even for similar materials. The situation is, however, not hopeless as we have noted that several of the signatures are affected by the presence of absorbtion by organic compounds and by the colour centers of mineral inclusions. Whether these effects can be properly adressed will require further studies. We begin to address the problem of the presence vegetation in the following sections.

3.2. Specific Properties of Vegetation

Absorption in vegetation is controlled by the absorption of the chlorophyll-a filled chloroplasts in the cell. As the concentration of chlorophyll-a and/or the size of the chloroplasts increases the absorption through the cell increases until the chloroplast absorbs more of the light at a given wavelength until in the limit of large concentrations and/or size it becomes a dark spot masking all the light its surface intercepts at this wavelength. This absorption saturation effect was first extensively studied by Morel and Bricaud [8] who called it the package effect. This is the factor that dominantes the variability in the absorption spectrum for different types of vegetation.

To compute this effect first we need formulas for the absorption efficiency Q_a of the chloroplasts. These are derived in Appendix B for both the original model that asssumed a spherical shape for the chloroplasts and for a new model that assumes disk-like chloroplasts.

As mentioned in Appendix B, the exact formulas can be approximated to a suffcent accuracy by a simpler exponential model. From now on we will use the more realistic disk-like shape to model the absorption saturation effect.

$$a_v^*(\lambda) = \left(\frac{1}{2\,u_{cp}}\right)\left(1 - e^{-a_o^*(\lambda)2\,u_{cp}}\right), \tag{26}$$

With:

$$u_{cp} = \rho_{cp}\tau_{cp}, \tag{27}$$

$a_o^*(\lambda)$ is the specific mass absorption coefficient of chlorophyll-a at low concentration in units of m^2 gr^{-1}. ρ_{cp} is the chlorophyll-a mass density inside the chloroplast in gr m^{-3} and τ_{cp} is the thickness of the chloroplast disk in meters. The mean thickness of a randomly oriented set of disks is 2 τ_{cp} which explains the factor of 2 seen in Equation (26).

We verified the validity of this model by first comparing the theory for disks given by Equation (26) with the Bricaud et al. [9] empirical formula for chlorophyll-a absorption in type I waters which is based on in-depth analysis of a compilation of most of the available datasets. To do this we rewrite Equation (26) as a specific absorption gain function:

$$\frac{a_v^*(\lambda)}{a_o^*(\lambda)} = \left(\frac{1}{a_o^*(\lambda)\,2\,u_{cp}}\right)\left(1 - e^{-a_o^*(\lambda)2\,u_{cp}}\right), \tag{28}$$

As can be seen in Figure 4, the overall behavior of the absorption is captured by the gain formula and this over three orders of magnitude in chlorophyll-a density. We note that the hysteresis seen in the empirical curves is due to an additional wavelength shift as a function of chlorophyll-a density. This effect was in fact observed by Gitelson [16]. The results shown in Figure 4 are a strong indication that the dominant effect in the spectral variation as a function of chlorophyll-a density is the absorption saturation effect. There was a large amount of variability in the original experimental data sets on which the empirical formulas are based so the discrepancies are not surprising. However, in the case that concerns us, which is the absorption in vegetation itself, the number density of phytoplankton which is the main uncontrolled empirical variable becomes severely constrained. The bulk chlorophyll-a mass density ρ_{chl} is given by:

$$\rho_{chl} = n_c\,n_{cp}\,\rho_{cp}V_{cp}, \tag{29}$$

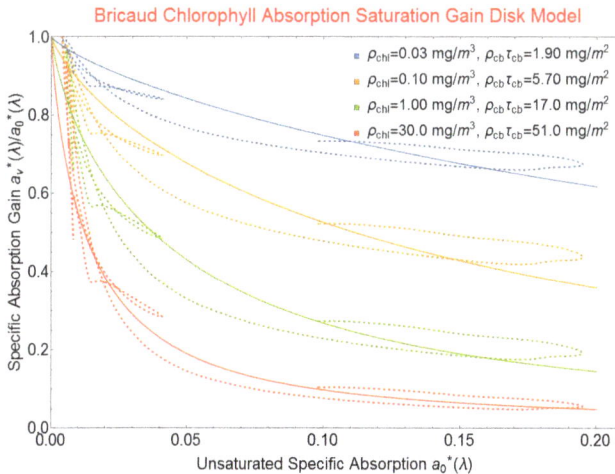

Figure 4. Graph of the ratio of the specific absorption gain to the unsaturated absorption gain. The dotted lines are from the empirical formula of Bricaud et al. [9] for chlorophyll-a. The solid lines are from Equation (21) with the parameters noted.

The number density of cells is n_c, the number of chloroplasts per cell is n_{cp} and the volume of each chloroplast is V_{cp}. In the open ocean, n_c may be weakly correlated with the other parameters while in a continuous block of cells as is the case in vegetation there is a very strongly constrained relationship. We can see this as follows. For a volume of vegetation V_m filled by cells with a volume V_c we have:

$$\rho_{chl} V_m = \left(\frac{V_m}{V_c}\right) \rho_{cp} \, n_{cp} \, V_{cp} = V_m \, \rho_{cp} \left(\frac{n_{cp} \, V_{cp}}{V_c}\right), \tag{30}$$

$$\rho_{chl} = \rho_{cp} \left(\frac{n_{cp} \, V_{cp}}{V_c}\right), \tag{31}$$

We expect the ratio of the total volume of chloroplasts $n_{cp} \, V_{cp}$ to the cell volume V_c to be almost constant. The variability induced by n_c for open water has disappeared and the bulk chlorophyll-a concentration is now as expected simply proportional to the chlorophyll-a concentration inside the chloroplasts that we use to estimate the absorption saturation.

The main implication of the discussion above is that we expect to be able to model the spectral shape of the absorption spectrum of chlorophyll-a with a single fitting parameter u_{cp}. Before this becomes feasible, there are, however, several significant hurdles which have to be overcome. First, the Bricaud formula for chlorophyll-a absorption can be scaled to any concentration no matter how small even for ranges that lie well outside the zone of the data used for the original fit. This creates a problem when trying to determine $a_o^*(\lambda)$ as a limiting value for low chlorophyll-a concentrations as we can extrapolate back to unphysically small values of concentration. Ciotti et al. [10] used a different approach to model the chlorophyll-a absorption from naturally occurring populations of organisms. They determine the absorption spectra for two limiting populations of organisms, the nano population and micro population. For low concentrations of chlorophyll-a the absorption spectrum of the nano population applies while for high concentrations the spectrum of the micro population is the appropriate one to use. As the concentration of chlorophyll-a increases, the spectrum evolves as a linear combination of both these extreme cases. We first attempted to use the nano population spectrum from Ciotti et al. [10] as the limiting case $a_o^*(\lambda)$ for low chlorophyll-a concentrations. Unsurprisingly, we found that the difference between our model and the Bricaud form diverged significantly at the higher concentrations.

This is problematic since for vegetation, which is the case of interest for us, the chlorophyll-a concentrations are expected to be large. In fact, they exceed the range of validity of the Bricaud formulas. To handle these extreme cases with a reasonable expectation of accuracy we decided to take a different approach. The technique is based on using a Bricaud spectrum at a given reference value with sufficiently high chlorophyll-a concentration but in a zone where the fit is still valid and extending the range from that point using the gain saturation equations. The rationale to do this is based on the fact that the concentration exceeds the measurement range and until data is available there is no other valid approach. This extension method proceeds as follows. Defining u_r as the value of u_{cp} at a reference bulk concentration ρ_r we have:

$$a_r^*(\lambda)2u_r = \left(1 - e^{-a_o(\lambda)2u_r}\right), \tag{32}$$

$$a_o(\lambda)2u_r = \ln[1/(1 - a_r^*(\lambda)2u_r)], \tag{33}$$

$$a_o(\lambda) = \left(\frac{1}{2u_r}\right) \ln[1/(1 - a_r^*(\lambda)2u_r)], \tag{34}$$

This $a_o(\lambda)$ is completely determined by the reference spectrum $a_r^*(\lambda)$ and the value we choose for u_r. Note that for $a_o(\lambda)$ to stay finite at all wavelengths there is a maximum value that u_r can take:

$$\mathrm{max}u_r = \left(\frac{1}{\mathrm{max}[2\,a_r^*(\lambda)]}\right), \tag{35}$$

Using (34) we can write that:

$$a_o(\lambda) 2\, u_{cp} = \left(\frac{u_{cp}}{u_r}\right) \ln[1/(1 - a_r^*(\lambda) 2 u_r)] ,\tag{36}$$

Finally, we obtain the following general expression for an extended Bricaud absorption spectrum that can be used at chlorophyll-a densities appropriate for vegetation:

$$a^*(\lambda) = \left(\frac{1}{2\, u_{cp}}\right)\left(1 - e^{-a_o(\lambda) 2 u_{cp}}\right),\tag{37}$$

As a final practical step we need to determine what value of ρ_r we will use a reference spectrum and what value of u_r leads to the most reliable extrapolation. To do this we first choose the Bricaud spectrum for 5.0 mg/m^3 which is a value at the high end of the bulk concentration range but still well below the 20.0 mg/m^3 extreme limit of the data on which the formula was based. To determine the best value of u_r we varied that parameter until we obtained the best fit to the Bricaud spectra at 1.0, 3.0 and 10.0 mg/m^3. In all these cases we found that the optimum reference u_r asymptotically approached max$u_r(\lambda)$. In practice, therefore, we recommend using a value of 0.99 max$u_r(\lambda)$.

To completely model the absorption due to vegetation we need to include the absorption of water and of the cellulose that makes up the walls and internal structures of the cell. The absorption spectrum of water is taken from the data of Pope and Fry [17] for the zone from 0.38 to 0.70 microns and from the data of Kou [18] normalized to the data of Pope and Fry in their wavelength overlap zone for the 0.65 to 2.5 microns range.

The specific absorption spectrum of naturally occurring lignin cellulose from 0.4 to 2.5 microns is given in [19]. The spectrum of crystalline cellulose from 0.2 to 0.5 micron in arbitrary units can be found in reference [20]. We used the overlap zone from 0.4 to 0.5 microns with the calibrated spectrum in [16] to transform the UV-visible spectrum given in [20] to specific absorption in units of m^2·gr^{-1}. We will use the pure water, chlorophyll-a and cellulose absorption spectra to model the irradiance reflectance spectra of algae and other marine vegetation.

3.3. Modeling Algae

We are now in a position to analyse the spectral signature of algae and other marine vegetation. The spectra we will be using were collected on the shores of Janvrin Island in Nova Scotia. These calibrated reflectance spectra range from 0.35 to 2.5 microns. This range extends beyond the wavelength band over which chlorophyll-a absorption has a significant amplitude. This is fortunate in as much as we can use the reflectance measured in the wavelength range over 0.90 micron to obtain a measurement of <d>. This is because the components which dominate absorption in that wavelength range are water and cellulose and their absolute values and relative abundances are well known. In that spectral band, therefore, we have:

$$<d> = \left(\frac{1 - x(\lambda)}{x(\lambda)}\right) \frac{\omega_b(n_{cw}/n_w)}{[a_w(\lambda)(1 - f_{cl}) + f_{cl}\, a_{cl}(\lambda)]} ,\tag{38}$$

f_{cl} is the mass fraction of cellulose in the cell and n_{cw} is the index of refaction of the walls of the cell and its subcomponents and n_w is the index of refraction of water. This cell wall index has been estimated for both the mesophyll and antidermal cell walls by Baranoski [12]. Since both are quite close to one another we use their average value as an estimate for n_{cw}.

We can use the fact that the mean spacing between backscattering layers <d> should be independent of wavelength to estimate f_{cl}. To do this, we vary f_{cl} to minimize the variance in the estimate of <d> as a function of wavelength computed with Equation (39). We use the wavelength range from 0.90 to 1.35 micron. We need to be above 0.90 microns to ensure that we are completely out of the zone where there could be remaining absorption by chlorophyll-a and other pigments in the algae. We also must stay below 1.35 micron to remain below the large water absorption band which reduces the irradiance signal to levels where instrument noise totally dominates. Figure 5 and Table 5 show the result of such a fit to the reflectance spectra of wet *Fucus* sp. and a drying mixture of *Fucus* sp. and *F. serratus* from Janvrin Island in Nova Scotia. These samples were chosen because they represent the extreme values of the reflectance spectra we measured. The results show the potential of this approach to estimate the status of the vegetation. In this case the *Fucus* sp. is much more saturated with water than the drying mixture sample while the mean backscattering feature size of the mixture is larger. The ratio of the standard deviation to the mean value of <d> is of the order of 5% in both cases which shows that a constant mean value <d> is a good model for the data.

Table 5. Functional fits to the mean spacing of backscatter layers <d>.

Substance	f_{cl}	<d>(*microns*)	$\sigma_{<d>}$	$\sigma_{<d>}/\mu_{<d>}$
Fucus sp.	0.043	3.29	0.15	0.048
Fucus sp. & *F. serratus*	0.164	4.07	0.21	0.054

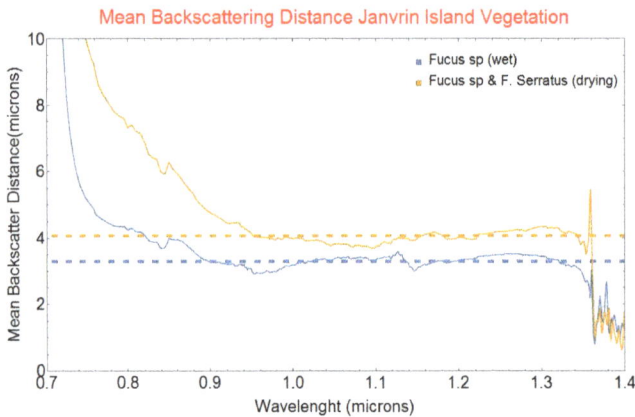

Figure 5. Fit of the mean backscatter distance derived from the irradiance reflectance of wet *Fucus* sp. and a drying mixture of *Fucus* sp. and *F. serratus* from Janvrin Island in Nova Scotia. In the zone below 0.90 microns the absorption of the chlorophyll-a and various pigments starts to dominate while in the zone above 1.35 microns the absorption of water becomes large enough that the resulting irradiance reflectance signal is dominated by noise. The full parameters of the fit are given in Table 5.

Once we have obtained the value of <d> we can use it in the wavelength range where the absorption of cholorophyll is dominant.

$$a_v(\lambda) = \rho_{chl} a_v^*(\lambda) = \left(\frac{1}{<d>}\right)\left(\frac{1-x(\lambda)}{x(\lambda)}\right) w_b(n_{cw}/n_w) - [a_w(\lambda)(1-f_{cl}) + f_{cl}\, a_{cl}(\lambda)] , \qquad (39)$$

As a final step we can now estimate the absorption saturation parameter u_{cp} and the bulk cholorophyll concentration ρ_{chl} by using Equation (39) and performing a non-linear least squares fit on the ratio of the experimental absorption obtained with the procedure described above to the low chlorophyll-a concentration limit $a_o^*(\lambda)$.

$$\left(\frac{\rho_{chl}}{2\,u_{cp}}\right)\left(1 - e^{-a_o^*(\lambda)2\,u_{cp}}\right) = \left(\frac{1}{<d>}\right)\left(\frac{1-x(\lambda)}{x(\lambda)}\right) w_b(n_{cl}/n_w) - [a_w(\lambda)(1-f_{cl}) + f_{cl}\, a_{cl}(\lambda)] , \qquad (40)$$

Figure 6 shows the result of such a fit for a reflectance spectrum of wet *Fucus* sp. and a drying mixture of *Fucus* sp. and *F. serratus*. Note the significant noise increase in the short wavelength region.

Appl. Sci. **2018**, *8*, 2680

This is due to the signal to noise of the reflectance measuring instrument in the blue and near UV. This significant spectral variation of the signal to noise forces us to use of an appropriate weighing function in performing the fit.

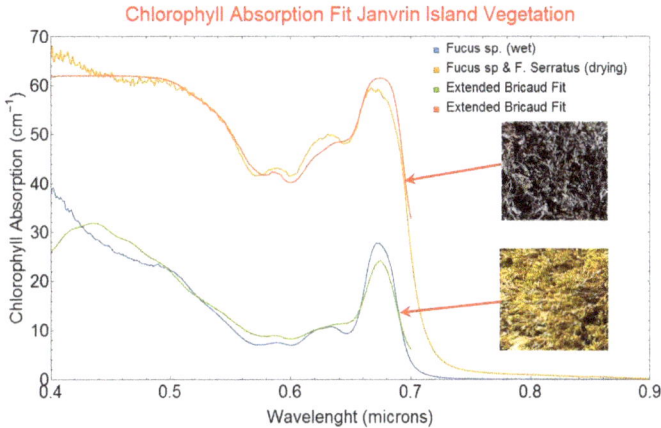

Figure 6. Fit of the absorption spectrum of wet *Fucus* sp. and a drying mixture of *Fucus* sp. and *F. serratus* with the extended Bricaud model. Note the noise due to the instrumental signal to noise degradation in the blue wavelength range. This effect was compensated by weighing the fit function inversely proportional to the S/N. The blue and yellow curves are the derived spectrum from the reflectance measurements and Equation (29). The green and red curves are the fit using Equation (37).

Once this weighing is applied we can see that the resulting modeled absorption spectrum approaches the experimental results.

In order to further verify the accuracy of the predictions of the model we have used the parameters of the fit to compute directly the predicted irradiance reflectance spectra and compare them to the original measured spectra. The results are shown in Figure 7 below.

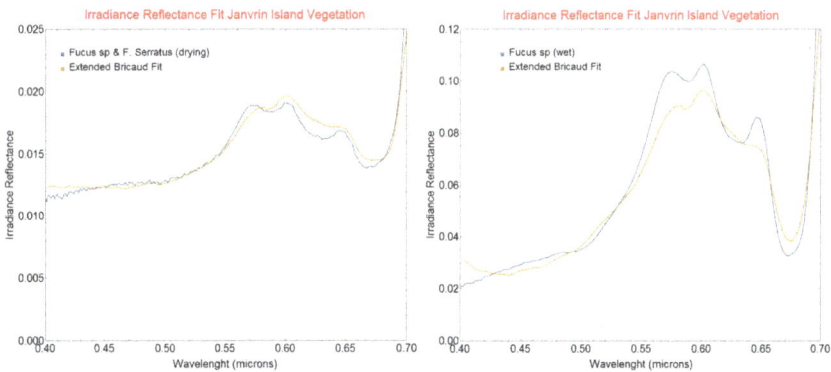

Figure 7. Comparison of modeled irradiance reflectance spectrum of wet *Fucus* sp. and a drying mixture of *Fucus* sp. and *F. serratus* (yellow curves) with the experimental measurements (blue curves). Note the noise degradation of the instrumental signal in the blue wavelength range. This reduced sensitivity may explain part of the incipient discrepancy in that spectral region.

The error between the modeled and measured reflectance signatures is of the same relative magnitude as the corresponding error in the absorption fit shown in Figure 6. Given the simplicity and generality of the model the overall precision of the fit is sufficient to satisfy our original purpose of hyperspectral bathymetry in uncharted waters.

One should note that the present model does not explicitly involve accessory pigments such as fucoxanthin which is known to be present in *Fucus*. It is based on the transformation due to absorption saturation in the chloroplasts (package effect) of the spectrum of phytoplankton. This phytoplankton spectrum is taken here as an archetype of a naturally occuring assemblage of various pigments dominated by chlorophyll-a. The absorption saturation effect shifts the resulting reflectance spectrum to the yellow and red which accounts for the relative closeness of the fit even without specific contributions from the accessory pigments. In the bathymetry application which most concerns us, the overall spectral shift and absolute level of the absorption are the key parameters needed to obtain reliable estimates of the depth. Given the exact pigment composition one could obviously improve the fit to the reflectance spectrum. However, this would defeat the purpose of obtaining the depth and the bottom spectrum without any a priori information other than the reflectance spectrum of the shoreline. This approach can, however, be improved by an iterative technique as we shall see in the discussion.

3.4. Non-Linear Effects of Vegetation Cover

We can now compute the effect of translucent vegetation growing over a mineral substrate. Because we have already obtained the mean cell size of the algae *<d>* we can directly compute its $b_b(\lambda)$ from Equation (5). This allows us to evaluate all the terms in Equation (16) and solve for the reflectance spectrum as a function of the actual thickness of the vegetation layer.

Figure 8 shows the variation in the spectral reflectance signature as a function of thickness for fucus over Trenton limestone. In the near IR, the spectra evolve from a high reflectance translucent signature for the pure *Fucus* to the low reflectivity of the wet limestone while in the visible zone that trend is reversed and the spectra go from the low reflectance of pure *Fucus* to the higher reflectance of wet limestone.

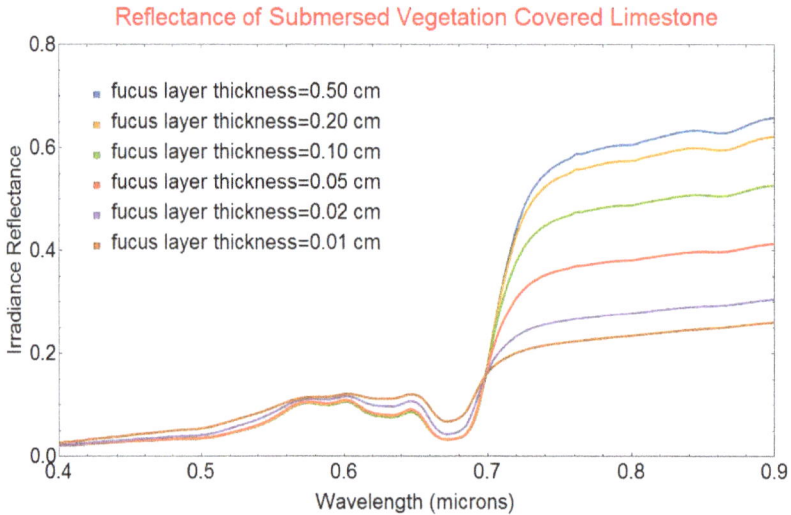

Figure 8. Computed variation of the spectral signature of translucent fucus vegetation over Trenton limestone as a function of the thickness of the layer. The mean spacing between scattering surfaces *<d>* is 3.3 microns.

Note that in cases where we don't have a separate estimate of $b_b(\lambda)$ we simply need to use directly the parameter $u_b = b_b(\lambda)\, z_b$. Given that we know the limiting spectra for the pure vegetation and for the mineral substrate u_b is the only parameter required to define the reflectance spectrum of their combination.

4. Discussion

The simple model presented in the previous sections leads to several important insights into the behavior of the irradiance reflectance spectra of minerals and vegetation in the underwater environment. The first significant result is that we are now able to estimate the ratio of the reflectance of materials immersed in water to their dry state. Figure 9 shows the ratio of irradiance reflectance for limestone and for beach sand that can be computed using Equations (14) and (17) from our model.

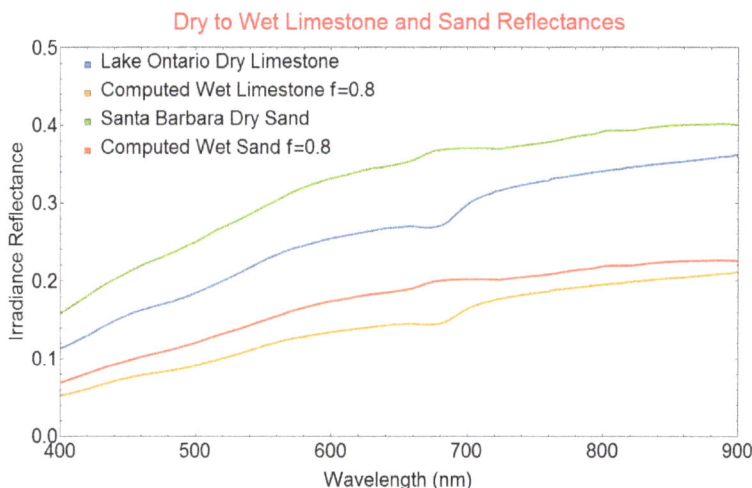

Figure 9. Computed irradiance reflectance for wet limestone and beach sand. The spectral signature for dry limestone comes from the shore of Lake Ontario. The signature of dry sand comes from a beach in Santa Barbara.

The ability to transfer reflectance spectra measured in air to their in-water equivalent is of great importance in practice as there are many comprehensive sources of spectral signatures measured in air while very few data are available under water due to the obvious difficulties in measurement. These underwater reflectance spectra are the backbone of all near shore shallow water hyperspectral surveys and the accuracy of any depth or bottom cover composition depends directly on their estimates.

Using the algebraic radiative transfer model, we have shown that we can estimate directly from reflectance measurement the relative absorption spectrum $<d>\, a(\lambda)$. Using these spectra, we have managed to obtain a simple and accurate fitting function whose structure is nevertheless based on fundamental physical considerations in far wing line broadening of the absorption from the lowest energy electronic transition in the material. Note that, strictly speaking, this fitting function is only appropriate for dielectric materials since the presence of the conduction bands in metals is not accounted for. The existence of this simple function valid over the range of wavelength of relevance to underwater hyperspectral measurements opens the possibility of identifying the material by a direct fit to the absorption parameters obtained from an inversion of the irradiance reflectance measurements of the water column using the algebraic radiative transfer model. The variance of the ν and λ_o parameters seen in Table 4 which control the spectral shape of the material may be indicative of a fundamental difficulty in obtaining directly bottom-type identification from the measured airborne hyperspectral reflectance. Addressing the scope and precise nature of this problem will be the subject of a further study.

We have also obtained a similarly simple four-parameter fitting function to the reflectance spectra of vegetation. The first parameter is the mean size of the vegetative cells $<d>$, the second parameter is the chlorophyll-a absorption saturation factor u_{cp}, the third parameter is the mass fraction of cellulose contained in a cell f_{cl}, and the fourth parameter is the bulk chlorophyll-a mass concentration ρ_{chl}. The backscattering term is controlled by the relative index of cellulose in water and the mean size of the cells while the absorption is the weighed sum of the absorption of water, cellulose and chlorophyll-a. The cell absorption is composed the absorption of pure water, cellulose and chlorophyll-a. The shape of the absorption of chlorophyll-a is controlled by the package effect through the absorption saturation parameter u_{cp} and its magnitude is controlled by the chlorophyll-a bulk density ρ_{chl}. The parameters of the model are interrelated. Relationships such as the one given in Equation (27) open up the possibility of obtaining estimates of parameters such as the ratio of the total volume of the chloroplasts to the total cell volume which could be used as an indicator of cell health.

We have shown that the reflectance of a mix of vegetation and minerals is not just a simple relative area coverage problem. When vegetation grows on top of a mineral substrate there results a combined spectrum which depends in a highly non-linear fashion on the product of thickness of the vegetation times its backscattering coefficient u_b. The overall effect for the reflectance spectra $R_m(\lambda)$ of pixels which are partially covered in vegetation is a combination of this non-linear mixing and area coverage factor.

$$R_m(\lambda) = \left[R_b(\lambda)\left(1 - f_{vp}\right) + f_{vp}\, R_t(\lambda) \right], \tag{41}$$

f_{vp} is the fractional per pixel vegetation cover. Given that the vegetation is generally expected to be of the same type and in substantially the same state of health over areas larger than a pixel it will in many cases be possible to separate the area coverage factor f_{vp} from the backscattering thickness factor u_b. This new information has the potential to increase significantly the level of knowledge about the ecologically relevant status and distribution of the near shore underwater vegetation.

In summary, the model we proposed here helps limit the number of parameters that need to be fitted for an analysis of the marine environment with hyperspectral irradiance reflectance spectra. This is an important factor because of the restricted wavelength band available when working in the underwater environment. The added complexity of the overlying water column absorption and scattering spectrum renders extremely difficult and unstable any inversion directly based on fitting linear combinations of bottom reflectance spectra. The low reflectance values and the low signal-to-noise ratios as depth increases severely affect the detectable level of spectral variation. The spectral angle is often near or within the noise band so the only hope for reasonable depth and type of bottom estimates and identifications are to use general parameters in low numbers. This is the case with our model. Furthermore, all the parameters in the model have a physical basis and are amenable to being further constrained in their fitting range by any information available from other sources such as the size and shape of chloroplasts, size of mineral grains and size of the spacing for near-surface fragmentation and porosity of rocks and sand.

We are currently using as a default reference the specific absorption spectrum of Bricaud et al. [9]. Given the extensive work in relation to coral reefs on the end member spectra and their variability [21] it may be possible in future to derive reference spectra better suited to modeling vegetation that also include a better balanced and more comprehensive mix of accessory pigments. As we have seen, a substantial part of the vegetation reflectance signature differences and spectral variability may be explained by the choloroplast absorption saturation effect. The remaining differences could, therefore, be less significant than appear at first glance, thus potentially reducing the number of distinct spectral absorption compositions required to model the end member signatures set. The other potential contribution of the approach we have taken of modeling the reflectance signatures by a radiative transfer model is that the variability in the spectral signatures clearly outlined for instance by Hochberg et al. [21] can be related explicitly to several parameters of interest in the study of corals such as the thickness of the thin translucent organic cover z_b over the mineral substrate (Equation (20)), the size of the scattering features of the cells, and their chloroplast pigment concentration.

We must remember that bathymetry in unknown waters is one of the main drivers for restricting the number of variables to optimize to obtain an estimate of the water column depth. The standard approach of using a combination of linear mixes is problematic when there is no ground truth or a priori knowledge of the bottom to restrict the space of end members for bottom reflectance. However, there is a way to use the best features of our model and the linear mixing method. We first solve for depth using the generic bottom reflectance model proposed here. Once the depth is estimated, we can use the measured water surface irradiance reflectance to derive the bottom spectrum that would produce that measured surface irradiance reflectance. Given this bottom spectrum we can then use the standard linear mixing method to determine the bottom vegetation and mineral types that compose it, thus extracting valuable information about the bottom type. Given this new information we can recompute the depth and correct for any error in the original approximate model, therefore maximizing the benefits of both approaches. This mixed method will be the subject of future investigations.

Author Contributions: G.F., J.-P.A. and M.L. conceptualized the study together. G.F. developed the model and wrote up the work. J.-P.A. led and was responsible for all the trials that gathered the hyperspectral data from airborne and ground-based detectors. He also analyzed the results to extract the irradiance reflectance spectra. M.L. validated the reflectance spectra using a precision polarized bi-directional reflectance function measurement system he designed and developed.

Funding: This research received no external funding.

Acknowledgments: The totality of the funding for this work and its publication costs was from Defence Research and Development Canada (DRDC). The authors would like to thank Dorte Krause-Jensen and Birgit Olesen for kindly using their expertise and taking the time to identify the vegetation species in the sample used in Section 3.3. The authors would also like to thanks both reviewers for very detailed and constructive comments which helped significantly improve this paper.

Conflicts of Interest: The authors declare no conflict of interest.

Appendix A

The index of refraction equations for water, calcite, quartz and cellulose used in this paper are given below. Calcite and crystalline quartz are birefringent materials with two orientation dependent indices of refraction, the extraordinary index for propagation along the direction of the optical axis and the ordinary index for propagation orthogonal to the optical axis. These were measured by Gosh [8].

$$n_{co}^2 - 1 = 0.73358749 + \frac{0.96464345 \, \lambda^2}{\lambda^2 - 0.0194325203} + \frac{1.8283145 \, \lambda^2}{\lambda^2 - 120.} , \tag{A1}$$

$$n_{ce}^2 - 1 = 0.35859695 + \frac{0.82427830 \, \lambda^2}{\lambda^2 - 0.0106689543} + \frac{0.14429128 \, \lambda^2}{\lambda^2 - 120.} , \tag{A2}$$

$$n_{qo}^2 - 1 = 0.28604141 + \frac{1.07044083 \, \lambda^2}{\lambda^2 - 0.0100585997} + \frac{1.10202242 \, \lambda^2}{\lambda^2 - 100.} , \tag{A3}$$

$$n_{qe}^2 - 1 = 0.28851804 + \frac{1.09509924 \lambda^2}{\lambda^2 - 0.0102101864} + \frac{1.15662475 \, \lambda^2}{\lambda^2 - 100.} , \tag{A4}$$

$$n_{cl}^2 - 1 = \frac{1.124 \, \lambda^2}{\lambda^2 - 0.011087} , \tag{A5}$$

$$n_{wq} = 1.31405 - 2.02 \times 10^{-6} T_c^2 + \frac{0.01586 - 4.23 \times 10^{-6} T_c}{\lambda} - \frac{0.004382}{\lambda^2} + \frac{0.0011455}{\lambda^3} , \tag{A6}$$

The wavelength in the expressions above is in microns. n_{co} and n_{ce} are, respectively, the ordinary and extraordinary index of calcite. n_{qo} and n_{qe} are the ordinary and extraordinary index of crystalline quartz. n_{cl} is the index of pure solid cellulose measured by Sultanova et al. [9]. n_{wq} is the index of pure water measured by Quan and Fry [10] which is strictly only valid to its full accuracy between 0.4 to 0.7

microns. In that last expression, T_c is the temperature in degrees centigrade. However, we need for our approach an expression for the index of water that is valid in the near IR. Schriebener et al. [11] have proposed such an expression valid from 0.2 to 2.5 microns. We have verified that it does match with the available data and the Quan and Fry formula over its range of applicability.

$$\frac{(n_{ws}^2-1)}{(n_{ws}^2+2)}\left(\frac{1}{\bar{\rho}}\right) = a_0 + a_1\bar{\rho} + a_2\bar{T} + a_3\bar{\lambda}\,\bar{T} + \frac{a_4}{\bar{\lambda}^2} + \frac{a_5}{\left(\bar{\lambda}^2-\bar{\lambda}_{uv}^2\right)} + \frac{a_6}{\left(\bar{\lambda}^2-\bar{\lambda}_{ir}^2\right)} + a_7\bar{\rho}^2 \tag{A7}$$

In this expression we have:

$$\bar{T} = \frac{T_k}{273.15}$$

$$\bar{\lambda} = \frac{\lambda}{0.589}$$

$$\bar{\rho} = \frac{\rho}{1\text{ g cm}^3}$$

Table A1. Coefficients of the water index of refraction Formula (A7).

Coefficient	Coefficient
$a_0 = 0.244257733$	$a_4 = 1.58920570 \times 10^{-3}$
$a_1 = 9.74634476 \times 10^{-3}$	$a_5 = 2.45934259 \times 10^{-3}$
$a_2 = -3.73234996 \times 10^{-3}$	$a_6 = 0.900704920$
$a_3 = 2.68678472 \times 10^{-4}$	$a_7 = -1.66626219 \times 10^{-2}$
$\bar{\lambda}_{uv} = 0.2292020$	$\bar{\lambda}_{ir} = 5.432937$

Formula (A7) is the one we use in this paper because of the extended range we require. We have verified that it matches to one part in a thousand the index formula given by Quan and Fry and that it tracks closely the available experimental data on water index in the near IR and UV.

If we assume that the orientation of the optical axis of the calcite and quartz grains is random, we need to compute the resulting average index as follows. The ordinary index n_o is the same no matter the angular orientation of the incoming ray with the optical axis of the crystal. The extraordinary index n_e varies as a function of the angles with respect to the optical axis $n_e(\varphi, \theta)$. The shape of the variation is this spheroid defined by:

$$\frac{k_x^2}{n_e^2} + \frac{k_y^2}{n_e^2} + \frac{k_z^2}{n_0^2} = \frac{\omega^2}{c^2} \tag{A8}$$

The light-wave propagation vector is k and its angular frequency is ω with the speed of light being given by c. The optical axis is along the z direction. Transforming to cylindrical coordinates the spheroid is symmetrical about the angle φ and elliptical in θ. If we assume that the distribution of the optical axis is random we can derive an expression for the mean extraordinary index.

$$n_e(\theta) = \sqrt{n_0^2 \cos\theta^2 + n_e^2 \sin\theta^2}\,, \tag{A9}$$

$$<n_e> = \frac{\int_0^{\pi/2} n_e(\theta) \sin\theta\, d\theta}{\int_0^{\pi/2} \sin\theta\, d\theta}\,, \tag{A10}$$

The result of the integral is:

$$<n_e> = \frac{n_0}{2}\left\{1 + \frac{n_e^2}{n_0^2}\frac{1}{\sqrt{1-n_e^2/n_0^2}} + \ln\left[\frac{n_0}{n_e} + \sqrt{1-n_e^2/n_0^2}\right]\right\}\,, \tag{A11}$$

The final result for the mean index for random birefringent crystal orientation is:

$$n_e(\theta) = <n> = \frac{n_o + n_e}{2} \, , \tag{A12}$$

We use these formulae to compute the mean index of both calcite and quartz crystals.

Appendix B

To compute the absorption saturation effect, first we need formulas for the absorption efficiency Q_a of the chloroplasts. The original model assumed that the chloroplasts were spherical and that their absorption efficiency can be modeled using the anomalous diffraction theory which is applicable since there is almost no difference in the real part of the index of refraction for the chloroplasts and the surrounding cell medium.

For spherical chloroplasts, the absorption efficiency is given by:

$$Q_{a-sph}(z) = 2\left[\frac{1}{2} + \frac{e^{-z}}{z} + \frac{(e^{-z}-1)}{z^2}\right] \, , \tag{A13}$$

$$z = a_o^*(\lambda)u \, ,$$

$$u = \rho_{cp}d_{cp} \, ,$$

$a_o^*(\lambda)$ is the specific mass absorption coefficient of chlorophyll-a at low concentration in units of $m^2 \cdot gr^{-1}$. ρ_{cp} is the chlorophyll-a mass density inside the chloroplast in $gr \cdot m^{-3}$ and d_{cp} is the diameter of the chloroplast in meters.

Note that in the limit of small z we have:

$$Q_{a-sph}(z) = \frac{2z}{3} \, , \tag{A14}$$

Since in the limit of small concentrations the chloroplast absorption will be unsaturated and equal to the limiting absorption, we can write the absorption saturation gain function for a spherical chloroplast as:

$$G_{a-sph}(z) = \left(\frac{3}{2z}\right)Q_{a-sph}(z) \, , \tag{A15}$$

The saturated absorption spectrum for spherical chloroplasts can, therefore, be computed as:

$$a_v^*(\lambda) = \left(\frac{3}{2z}\right)Q_{a-sph}(z)a_o^*(\lambda) \, , \tag{A16}$$

It is interesting to estimate what the effect of chloroplast shape maybe in the estimate of this packaging effect. Chloroplasts are often disk shaped and we will use the form for the absorption efficiency for randomly oriented disks.

$$Q_{a-dsk}(z_d) = 2 - E_3(z_d) \, , \tag{A17}$$

With,

$$z_d = a_o^*(\lambda)\rho_{cp}\tau_{cp} = a_o^*(\lambda)\,u_{cp}, \tag{A18}$$

τ_{cp} is the thickness of the disk. We also define f as the ratio of the thickness τ_{cp} of the disk to its diameter d_{cp}. $E_3(z)$ is the exponential integral function of order 3 which is defined as:

$$E_n(z) = \int_1^\infty \frac{e^{-z\,t}}{t^n}dt \, ,$$

In the limit of small z we have:

$$Q_{a-dsk}(z_d) = 2\,z_d\,, \tag{A19}$$

The gain function for randomly oriented disks becomes:

$$G_{a-dsk}(z_d) = \left(\frac{1}{2\,z_d}\right)Q_{a-dsk}(z_d)\,, \tag{A20}$$

Finally the saturated absorption spectrum for disks can be computed as:

$$a_v^*(\lambda) = \left(\frac{1}{2\,z_d}\right)Q_{a-dsk}(z_d)a_o^*(\lambda)\,, \tag{A21}$$

The exact formulas given above can be approximated to within a 10% relative error by the following simple exponential forms.

$$a_v^*(\lambda) = \left(\frac{3}{2\,z}\right)\left(1 - e^{-2z/3}\right)a_o^*(\lambda)\,, \tag{A22}$$

$$a_v^*(\lambda) = \left(\frac{1}{2\,z_d}\right)\left(1 - e^{-2z_d}\right)a_o^*(\lambda)\,, \tag{A23}$$

References

1. Albert, A.; Mobley, C. An analytical model for subsurface irradiance and remote sensing reflectance in deep and shallow case-2 waters. *Opt. Express* **2003**, *11*, 2873–2890. [CrossRef] [PubMed]
2. Jonasz, M.; Fournier, G. *Light Scattering by Particles in Water: Theoretical and Experimental Foundations*; Academic Press: New York, NY, USA, 2007; pp. 39–42, 77–80, ISBN 10: 0-12-388751-8.
3. Szudy, J.; Bayliss, W.E. Uniform Frank-Condon treatment of pressure broadening of spectral lines. *J. Quant. Spectrosc. Radiat. Transf.* **1975**, *15*, 641–668. [CrossRef]
4. Wooten, F. *Optical Properties of Solids*; Academic Press: New York, NY, USA, 1972; pp. 42–52, ISBN 9781483220765.
5. Aas, E. Two-stream irradiance model for deep waters. *Appl. Opt.* **1987**, *26*, 2095–2101. [PubMed]
6. Jonasz, M.; Fournier, G. *Light Scattering by Particles in Water: Theoretical and Experimental Foundations*; Academic Press: New York, NY, USA, 2007; pp. 119–120, ISBN 10: 0-12-388751-8.
7. Fournier, G.; Neukermans, G. An Analytical Model for Light Backscattering by Coccoliths and Coccospheres of Emiliania Huxleyi. *Opt. Express* **2017**, *25*, 14996–15009. [CrossRef] [PubMed]
8. Morel, A.; Bricaud, A. Theoretical results concerning light absorption in a discrete medium, and application to specific absorption of phytoplankton. *Deep Sea Res.* **1981**, *28*, 1375–1393. [CrossRef]
9. Bricaud, A.; Morel, A.; Babin, M.; Allali, K.; Claustre, H. Variations of light absorption by suspended particles with the chlorophyll-a a concentration in oceanic (Case 1) waters: Analysis and implications for bio-optical models. *J. Geophy. Res.* **1998**, *103*, 31033–31044. [CrossRef]
10. Ciotti, A.M.; Lewis, M.R.; Cullen, J.J. Assessment of the relationships between dominant cell size in natural phytoplankton communities and the spectral shape of the absorption coefficient. *Limnol. Oceanogr.* **2002**, *47*, 404–417. [CrossRef]
11. Ghosh, G. Dispersion-equation coefficients for the refractive index and birefringence of calcite and quartz crystals. *Opt. Commun.* **1999**, *163*, 95–102. [CrossRef]
12. Sultanova, N.; Kasarova, S.; Nikolov, I. Dispersion properties of optical polymers. *Acta Phys. Pol. A* **2009**, *116*, 585–587. [CrossRef]
13. Quan, X.; Fry, E.S. Empirical equation for the index of refraction of seawater. *Appl. Opt.* **1995**, *34*, 3477–3480. [CrossRef] [PubMed]
14. Schiebener, P.; Straub, J.; Levelt Sengers, J.M.H.; Gallagher, J.S. Refractive Index of Water and Steam as Function of Wavelength, Temperature and Density. *J. Phys. Chem. Ref. Data* **1990**, *19*, 677–717. [CrossRef]
15. Baranoski, G.V.G. Modeling the interaction of infrared radiation (750 to 2500 nm) with bifacial and unifacial plant leaves. *Remote Sens. Environ.* **2006**, *100*, 335–347. [CrossRef]

16. Gitelson, A. The peak near 700 nm on radiance spectra of algae and water: Relationship of its magnitude and position with chlorophyll-a concentration. *Int. J. Remote Sens.* **1992**, *13*, 3367–3373. [CrossRef]

17. Pope, R.M.; Fry, E.S. Absorption spectrum (380–700 nm) of pure water. II. Integrating cavity measurements. *Appl. Opt.* **1997**, *36*, 8710–8723.

18. Kou, L.; Labrie, D.; Chýlek, P. Refractive indices of water and ice the 0.65 to 2.5 m spectral range. *Appl. Opt.* **1993**, *32*, 3531–3540. [CrossRef] [PubMed]

19. Jacquemoud, S.; Ustin, S.L.; Verdebout, J.; Schmuck, J.; Andreoli, G.; Hosgood, B. Estimating leaf biochemistry using the PROSPECT leaf optical properties model. *Remote Sens. Environ.* **1996**, *56*, 194–202. [CrossRef]

20. Adolfo, A.; Martin, P.; UV-Visible NIR Microspectroscopy of Nanocrystalline cellulose. *CRAIC Technol.* **2013**. Available online: http://www.warsash.com.au/news/articles/craic-application-paper.pdf (accessed on 10 May 2018).

21. Hochberg, E.J.; Atkinson, M.J.; Andrefouet, S. Spectral reflectance of coral reef bottom-types worldwide and implications for coral reef remote sensing. *Remote Sens. Environ.* **2003**, *85*, 159–173. [CrossRef]

applied sciences

MDPI

Article

The Fundamental Contribution of Phytoplankton Spectral Scattering to Ocean Colour: Implications for Satellite Detection of Phytoplankton Community Structure

Lisl Robertson Lain [1,*] and Stewart Bernard [1,2]

1   Department of Oceanography, University of Cape Town, Cape Town 7700, South Africa; sbernard@csir.co.za
2   NRE, Centre for Scientific and Industrial Research (CSIR), Cape Town 7700, South Africa
*   Correspondence: lislrobertson@gmail.com; Tel.: +27-(0)72-200-6369

Received: 14 June 2018; Accepted: 28 October 2018; Published: 19 December 2018

Abstract: There is increasing interdisciplinary interest in phytoplankton community dynamics as the growing environmental problems of water quality (particularly eutrophication) and climate change demand attention. This has led to a pressing need for improved biophysical and causal understanding of Phytoplankton Functional Type (PFT) optical signals, in order for satellite radiometry to be used to detect ecologically relevant phytoplankton assemblage changes. Biophysically and biogeochemically consistent phytoplankton Inherent Optical Property (IOP) models play an important role in achieving this understanding, as the optical effects of phytoplankton assemblage changes can be examined systematically in relation to the bulk optical water-leaving signal. The Equivalent Algal Populations (EAP) model is used here to investigate the source and magnitude of size- and pigment-driven PFT signals in the water-leaving reflectance, as well as the potential to detect these using satellite radiometry. This model places emphasis on the determination of biophysically consistent phytoplankton IOPs, with both absorption and scattering determined by mathematically cogent relationships to the particle complex refractive indices. All IOPs are integrated over an entire size distribution. A distinctive attribute is the model's comprehensive handling of the spectral and angular character of phytoplankton scattering. Selected case studies and sensitivity analyses reveal that phytoplankton spectral scattering is most useful and the least ambiguous driver of the PFT signal. Key findings are that there is the most sensitivity in phytoplankton backscatter ($b_{b\phi}$) in the 1–6 µm size range; the backscattering-driven signal in the 520 to 570 nm region is the critical PFT identifier at marginal biomass, and that, while PFT information does appear at blue wavelengths, absorption-driven signals are compromised by ambiguity due to biomass and non-algal absorption. Low signal in the red, due primarily to absorption by water, inhibits PFT detection here. The study highlights the need to quantitatively understand the constraints imposed by phytoplankton biomass and the IOP budget on the assemblage-related signal. A proportional phytoplankton contribution of approximately 40% to the total b_b appears to a reasonable minimum threshold in terms of yielding a detectable optical change in R_{rs}. We hope these findings will provide considerable insight into the next generation of PFT algorithms.

Keywords: phytoplankton; PFT; ocean colour; satellite radiometry; radiative transfer; optical modelling

---

1. Introduction

Phytoplankton across the world's oceans represent about half of all primary production on our planet [1,2]. Their growth and function are fundamental to sustaining life: they constitute the foundation of the aquatic food web, and serve critical roles in the recycling of essential elements such as carbon and nitrogen, as well as in remineralisation [3–5]. Being so responsive to nutrient availability and water temperature, these tiny organisms are key indicators of ecosystem change, and understanding their community dynamics is key to answering some of the most challenging earth science questions of our time about the impacts of climate change on local, regional and global scale aquatic systems and the carbon cycle. The widespread distribution and integral role of phytoplankton in global marine ecosystems means that these fields of study depend heavily on modelling together with satellite data for any large scale analysis. In situ data collection is indispensable for local scale investigations and for ground truthing of satellite and model data, but simultaneous large scale direct measurements are logistically impossible. Optical measurements in natural waters are challenging: they are expensive and logistically difficult, technically complex due to large dynamic ranges of the signal, and overall require delicate, rigorously calibrated instrumentation with precise knowledge of sources of error. Remote sensing and moored in situ instrumentation are the only feasible ways to acquire continuous data series, but these largely involve bulk measurements of the total optical signal. Isolating the respective optical components for laboratory assessment is a significant further undertaking. In situ and laboratory measurements are consequently extremely valuable, and appropriate bio-optical models provide essential tools for the analysis and understanding of these bulk measurements, whether above- or sub-surface.

It has long been appreciated that phytoplankton have a direct effect on the observable colour of the ocean, and broad scale biomass estimates based on Chl *a* concentrations derived from satellite radiometry are widely relied upon despite persistent uncertainty in the accuracy of information derived from satellite imagery [6,7]. Recently, there has been considerable interest in more detailed information on phytoplankton assemblage characteristics [8–11], but it has not been widely ascertained to what degree Phytoplankton Functional Type (PFT) information can be gleaned from satellite data, and at what level of confidence. Furthermore, descriptions of PFTs differ with context—and the potential for identifying relationships between the ecological roles of phytoplankton and their optical properties must also be considered. Understanding the causal effect of biophysical phytoplankton characteristics on the optical water-leaving signal is at the heart of addressing these questions, and this is undoubtedly an outstanding topic in ocean optics.

Any useable radiometric PFT-related signal results directly from the interaction of phytoplankton with their light environment, but the physical basis of this interaction is not well understood in terms of observed variability across the wide diversity of aquatic environments and phytoplankton assemblages [12,13]. Generally, in oceanic waters, it is the strong absorption by phytoplankton which dominates the phytoplankton contribution to the ocean colour signature, and has therefore been identified as a promising signal in terms of PFT identification. However, distinguishing the effects of variable phytoplankton absorption due to biomass changes from the effects due to functional type changes (and further from changes induced by photoacclimation and photoprotective pigments) is not straightforward. This ambiguity in the phytoplankton community signal is at the core of the PFT problem. It is then overlaid with further complexity, given that a potential PFT signal from the phytoplankton component of a water body's optical constituents must be considered in the context of the other components in the water, recognising the contributions from the non-algal sources of optical variability: absorption due to CDOM (Coloured Dissolved Organic Matter) and detrital particles, i.e., $a_{gd}(\lambda)$, and non-algal backscatter i.e., $b_{bnap}(\lambda)$ [13]. The blue spectral region of maximum phytoplankton absorption is also the region most affected by CDOM and detrital absorption. (It should also be noted that, in the context of satellite radiometry, blue spectral bands display the largest absolute measurement uncertainties [7,14]. While the blue water-leaving signal may be large in oceanic regions, resulting in a small relative uncertainty, this is when the signal is overwhelmingly dominated by the backscattering of water, decreasing confidence in the lesser contributions of $a_{gd}(\lambda)$ and a_ϕ. Generally, a_{gd} product retrievals from the satellite tend to be less robust than those of other IOPs [15,16].)

A comprehensive guide to PFT approaches is given by Mouw et al. [17], dividing them into four categories: abundance-based e.g., Hirata et al. [18] and Brewin et al. [19,20]; radiance based e.g., the PHYSAT method: Alvain et al. [21,22]; absorption based e.g., Devred et al. [23], Ciotti and Bricaud [24], and PhytoDOAS: Bracher et al. [25]; and scattering based e.g., Kostadinov et al. [10,26] (all references in [17]). Existing scattering-based approaches [10,11] assume a Jungian (exponential) particle size distribution and rely on Mie modelling, which does not adequately represent phytoplankton angular scattering [27], and there are consequently high uncertainties in PSD retrieval where the particle size distribution slope is low, i.e., highly productive and coastal areas dominated by relatively large cells [10]. Low biomass (Chl $a < 1$ mg·m^3) oceanic conditions with an absorption-dominated phytoplankton component of the water-leaving signal can exhibit good relationships with differential pigment absorption e.g., the diagnostic pigment approach used on satellite R_{rs} in Uitz et al. [28]. It follows that, in the context of additional non-algal absorption, differentiated spectra show better similarity than non-differentiated [29], and also that high spectral resolution measurements show better potential for retrieving phytoplankton assemblage information than multi-spectral R_{rs} [29] when retrieving diagnostic features of the first derivative of R_{rs}. However, other methods using the fourth derivative of pigment absorption and R_{rs} to identify fine-scale phytoplankton absorption features have found that objective discrimination of pigment groups from hyperspectral R_{rs} may not be feasible at low biomass < 1 mg·m^3 due to the high similarities in the derivative spectra [30]. This suggests that the primary phytoplankton signal in R_{rs} is due to biomass (Chl a) rather than the accessory pigments, and that with the exception of uniquely diagnostic pigment absorption outside the spectral regions of that of Chl a, phytoplankton information cannot be retrieved without assumptions about PFT relationships with biomass. The PhytoDOAS method also employs a fourth derivative analysis [25,31] but is performed on hyperspectral top-of-atmosphere satellite measurements, avoiding the uncertainties associated with poor atmospheric correction, but the sensitivity of this high spectral resolution approach to biomass and both algal and non-algal scattering contributions to the IOP budget has yet to be determined.

Generally, phytoplankton absorption- and abundance-based methods rely on empirical relationships between biomass, functional type and CDOM. Where these quantities co-vary predictably or are exactly known, empirical PFT algorithms may be successful. However, Brewin et al. [6] acknowledges that both the abundance-based approaches as well as approaches relying on differential pigment absorption break down in environments that do not conform to the generalised relationships between community structure and biomass upon which these approaches are based, usually in elevated biomass comprised of small cells. The relative contributions of phytoplankton absorption and scatter to light emerging from seawater change with biomass, size and other functional type traits, and as the $a_{gd}(\lambda)$ and $b_{bnap}(\lambda)$ components vary (see Stramski et al. [32] and references therein). The total water-leaving signal is a delicate balance of the frequently opposing optical effects of biomass and phytoplankton assemblage variability such as size, pigments and ultrastructure, together with the optical effects of the non-algal in-water constituents. An interactive webpage demonstrating the first order effect of variability in these parameters on Rrs is available in the Supplementary Material. It was observed by Brown et al. [13] that backscatter anomaly maps (i.e., backscatter independent of variability due to biomass) correlate approximately with PFT distribution maps calculated from optical anomalies which were initially attributed to differences in phytoplankton accessory pigments [21]. This leads to the suggestion that radiance-based methods, e.g., the Alvain (PHYSAT) criteria used to distinguish PFTs, are in fact primarily due to backscattering characteristics [9,13], indicating that phytoplankton groups either directly determine, or perhaps are simply associated with, backscattering variability around the mean.

Brown et al. [13] conclude that these relationships can only be fully explored if a method is applied where the phytoplankton groups are causally linked to the optical conditions. The Equivalent Algal Populations (EAP) model provides exactly such a method, and is used here to investigate the impact of size- and pigment-based PFT variability on the optical signal, and to confirm the assertion that biomass drives the largest part of observed variability in the water-leaving signal, and that the radiometric signal in the blue is ambiguous due to the effects of $a_{gd}(\lambda)$, and the additional effects of $b_{bnap}(\lambda)$ [13,33].

The EAP model is a fully physics-based two-layered spherical model, which calculates, from first principles, biophysically linked phytoplankton absorption and scattering characteristics from particle refractive indices reflecting the primary light-harvesting pigments of various phytoplankton groups. IOPs are calculated at high spectral resolution between 400 and 900 nm and are integrated over an entire equivalent size distribution [34,35], simulating the dominant optical characteristics of natural phytoplankton assemblages. The EAP is used here only as a forward model: the intention of this study is to isolate the biophysical driver(s) of PFT optical signals and determine the associated implications for detecting PFT changes from satellite radiometry.

In this study, the term "Phytoplankton Functional Type" is used in a broad sense of the dominant characteristics of a phytoplankton assemblage, with respect to both cell size and accessory pigments, from an optical perspective.

Study Objectives and Outline

The aim of this work is to investigate the magnitude and spectral location of optical water-leaving signals resulting from phytoplankton assemblage changes; to determine how these signals respond to changes in biomass and functional type; to evaluate their optical ambiguity in the context of the optical effects of other in-water constituents; and to assess their robustness against measurement uncertainties in satellite radiometry. This work does not present a PFT detection method, but instead aims to identify the reasonable limits of PFT detection from satellite, inferred from appropriate illuminative case studies and sensitivity analyses demonstrating the source and magnitude of PFT signals in terms of both cell size (assemblage D_{eff}) and accessory pigments.

To give context to the discussion on the case studies, an analysis is first made of the contribution of the phytoplankton-driven signal to the bulk R_{rs}, and how this relates to the proportional contribution of phytoplankton to the IOP budget. A Southern Ocean based case study then demonstrates the optical impact on the R_{rs} of transitioning assemblages in terms of both biomass and D_{eff} changes. This discussion is developed further with a Benguela-like example more representative of productive upwelling systems investigating the relative magnitude of pigment-driven PFT changes. A sensitivity analysis then shows the spectral position and magnitude of the accessible phytoplankton optical signal in R_{rs} as biomass and D_{eff} vary. The source of these signals is traced back to phytoplankton backscatter and its relationship with biomass, and ambiguity associated with non algal variability is evaluated.

It is clear that the case studies reflect simplified representative examples of much wider pigment- and size-related variability in nature, but the described dependence of absorption-driven pigment signals versus scattering-driven cell size signals on biomass holds across assemblage types. Optical PFT effects are most easily identified in relatively high biomass environments (Chl $a > 1$ mg·m^3) [36–38], and where the IOP budget is dominated by phytoplankton [37,39], and so the case studies deal with these water types. However, as the sensitivity analysis shows, together with the contextual discussion around ambiguity and uncertainty in satellite R_{rs}, the conclusions of this study have implications for the identification of PFT changes from satellite R_{rs} across all water types.

2. Methods: Modelling Approach

2.1. The Requirement for a Biophysically Consistent PFT Optical Model

The EAP model was developed to understand the causality-driven impact of different phytoplankton assemblages on the water-leaving optical signal. Optical variability in phytoplankton is known to be driven by particle size (effective diameter D_{eff}) [32,40,41], pigment quantity and type, cellular material, shape and internal structure, fine-scale morphology, and aggregation [42–45].

The model focuses primarily on particle size as given by the D_{eff} parameter, which is of fundamental importance both optically and ecologically [10,46].

Due to immense species diversity and variability in distribution, the Phytoplankton Functional Type (PFT) approach (e.g., Sathyendranath et al. [8], Alvain et al. [21], Ciotti and Bricaud [24], Bouman [47]) groups phytoplankton species according to their biogeochemical function and attempts to relate this to their biophysical characteristics, with size as a major consideration [10,46,48]. This approach is important for oceanic waters, characterised by widespread but low biomass, which contribute the largest proportion of global oceanic primary production [1]. Cell size governs many biological traits [49]; smaller phytoplankton are ubiquitous and play an important role in nutrient recycling, while larger phytoplankton often display the highest growth rates [49]. The dynamics of phytoplankton ecology have profound and intricate influence not only on oceanic biogeochemistry (e.g., acidification, and its effects on both CO_2 uptake and on marine life) but also at higher trophic levels e.g., on fish ecology, as certain phytoplankton environments promote the development of different fish populations [48]. A size-based PFT approach is particularly meaningful in the context of carbon sequestration [46], as particle size determines sinking rates for a large part.

However, phytoplankton ecology is complex, and modelling PFTs with adequate parameterisation in a biogeochemical context is consequently extremely challenging [12]. Following the EAP's conceptual intent to understand the impact of D_{eff} as the primary optical determinant once the effect of biomass has been accounted for, other sources of bio-optical variability are intentionally constrained. PFTs can therefore, to the first order, be approached from a size-based perspective, and the EAP model consequently lends itself extremely well to PFT sensitivity studies in terms of its ability to isolate small differences in reflectance resulting only from variability in assemblage size distribution [37]. The model does additionally provide scope for varying other biophysical attributes within a population (such as the pigment-determined spectral refractive indices, the shape of the size distribution itself, the ratio of core to shell sphere volumes, and the cellular Chlorophyll *a* density of the cells in the distribution), as required. It should be noted, however, that the model is not intended as a full representation of phytoplankton optical complexities, and there is certainly ecologically significant natural variability in phytoplankton IOPs e.g., dependent on their growth state [50], in response to growth irradiance, nutrient availability and water temperature [51–53] and diel cycles [54,55]. These effects can be a large (e.g., 80% increase in the phytoplankton scattering cross section between sunrise and sunset [54]), and while they are not explicitly addressed here, they serve to add further uncertainty to PFT retrievals from the optical water-leaving signal.

Empirically based phytoplankton abundance-type approaches, following observed relationships between phytoplankton assemblage taxonomic information (e.g., pigments) and biomass, show good results in low biomass conditions (i.e., where phytoplankton absorption dominates the phytoplankton IOP contribution), and where the covariability of the phytoplankton optical contribution with that of other in-water constituents generally holds [56], but do not address the sources of second order variability or optical causality [13], or the likelihood that these empirical relationships will not withstand the ecological shifts resulting from changing climatic conditions [6]. A biophysical approach to PFTs not only allows improved analysis of sensitivity and causality but is likely to have greater validity in a future ocean (see also [57]).

The optical impact of a phytoplankton assemblage interacting with its aquatic environment is by no means straightforward, and a rigorous IOP model such as the EAP can systematically vary phytoplankton biogeophysical attributes in the context of likely additional non-algal absorption and scatter, and can examine the resulting effects on the light field when used in combination with a Radiative Transfer (RT) model. The value of this reductionist approach has been demonstrated [58,59] (and furthermore by Stramski et al. [32]) for separating and understanding the effects of various phytoplankton groups and accompanying in-water constituents on the oceanic light field and emergent R_{rs}. There is a bulk effect attributable simply to biomass, for which Chlorophyll *a* (Chl *a*) is used as a proxy, and which for the most part dominates the phytoplankton-related signal in Case 1 waters [60]

(It is acknowledged that Chl *a* concentration and biomass are not equivalent, as biomass includes non-pigmented biological matter in quantities which may not be proportional to pigmented matter. However, for the purposes of this study, biomass and Chl *a* concentration are used interchangeably, as this work is approached from a purely optical perspective and ignores non-pigmented biological matter.). PFT characteristics generally result in optical effects secondary to those of Chl *a*: accessory pigments dominate assemblage absorption characteristics [61], and particle size is usually the primary determinant of phytoplankton scattering characteristics [62] (excepting the influence of ultrastructure in certain species, e.g., highly scattering liths or vacuoles). Natural waters are also subject to non-algal absorption, the dissolved part of which is frequently referred to as Coloured Dissolved Organic Matter (CDOM) or gelbstoff, but which may also have a particulate component in addition to non-algal scatter that can include scatter by detrital matter, sediment, bacteria, and/or bubbles. These quantities absorb and scatter light with spectral signatures distinct from those of phytoplankton, and their subsequent optical interactions and resulting effect on the total water-leaving signal are highly complex. Understanding the interaction between cells' biophysical characteristics and the light field in the presence of these additional optically active constituents is central to determining which parts of the optical signal are useable for PFT diagnostics, and, likewise, where signal ambiguity is prohibitive.

2.2. Equivalent Algal Populations Model: Principal Attributes

The EAP model has been used for a variety of applications [35,37,63,64]. It can be assumed that a model demonstrated as successful in phytoplankton-dominated waters [65] addresses the phytoplankton component accurately. Models designed for low biomass, with simplistic and absorption-decoupled phytoplankton scattering models, tend to underperform in higher biomass conditions when phytoplankton IOPs dominate [65]. It follows that the phytoplankton component of the combined optical properties is not generally well represented in these models. Following a reductionist approach, good information on the phytoplankton component is a prerequisite for any quantitative comment on the optical contribution of respective PFTs, or identifying changes in the bulk optical properties of seawater as dominant PFTs change. Only when representing the detailed nature of phytoplankton optics, with absorption and scattering biophysically consistent—as they are in nature—is a causal understanding of their interactive effect on the optical signal possible [32,66].

The EAP model exhibits a two-layered sphere particle and equivalent size-based community structure [27], which enables the calculation of phytoplankton IOPs from first principles, presenting a valuable opportunity for furthering the understanding of causal relationships between phytoplankton physiology and their optical characteristics based on quantified community structure. It is emphasised that this is not an empirical model and its use here is not to provide optical closure, but rather to identify and understand the biophysical drivers of phytoplankton optics and their contribution to an observable signal in the context of different water types.

At the core of the model are the phytoplankton particle refractive indices, with the imaginary part of the refractive index approximately representing that portion of light that is absorbed by the cell, and the real part of the refractive index representing that portion of light which is scattered. The imaginary and real parts of the refractive index spectra are numerically linked through the Kramers–Kronig relations [67], whereby the real part of the refractive index $n(\lambda)$ is calculated as the imaginary part of a Hilbert transform of the imaginary refractive index, originally derived from cellular absorption measurements. It should be noted that the imaginary refractive index characterises the absorption of the intracellular material and has no dependency on cell size. This has implications for the applicability of the model to a wide range of cell sizes, and is discussed further in Appendix A.1.

With a real refractive index of 1.12 for the 'chloroplast' sphere, and as 1.02 for the 'cytoplasm' sphere, this yields an overall particle spectral real refractive index of between 1.03 and 1.04 for phytoplankton cells (see also Stramski et al. [32], Aas [68]). Full details of the refractive index calculations can be found in Bernard et al. [69].

In this model, the imaginary part of the refractive index is also numerically linked to the specified intracellular Chl *a* concentration [27,52,54,55,70]. For eukaryotic particles, a core sphere represents the cytoplasm (which contains approximately 80% water, and is almost colourless), while an outer sphere represents the more refractive chloroplast, where the pigmented material (generally Chl *a* in the largest part) is also strongly absorbing.

A critical feature of the model is that Chl *a*-specific absorption (a*ϕ) is constrained at 675 nm to reflect the theoretical maximum absorption by unpackaged phytoplankton of 0.027 mg/m^2 as per Johnsen et al. [71]. This is incorporated into the calculation of the imaginary refractive index of the chloroplast layer n'_{chlor} (outer sphere), based on the assumption that the cytoplasm layer (inner sphere) has no signficant absorption at 675 nm:

$$n'_{chlor}(675) = \frac{675}{n_{media}} \frac{\pi c_i a^*_{sol}(675)}{4Vv},$$

(1)

where $n_{media} = 1.334$ and Vv is the relative chloroplast volume, c_i is the intracellular Chl *a*, and $a^*_{sol}(675)$ is the Chl *a*-specific absorption at 675 nm of that pigment in solution, i.e., unpackaged [27].

The effect of constraining the unpackaged absorption in this way is to establish a quantitative relationship between the intracellular Chl *a* and the cell volume; a relationship that is biophysically consistent as the cell size varies [27]. This results in an effectively decreasing Chl *a*-specific absorption with increasing size, observable in the resulting optics as the "package effect" [40,72].

When coupled with a radiative transfer model—here, Hydrolight-Ecolight (Numerical Optics, Ltd., Devon, UK) is used—the interactions of phytoplankton IOPs (in combination with those of other in-water constituents) with the surrounding light field can be examined systematically. A full physics-based model such as this has the additional advantage of providing not only biophysically interrelated particle absorption, scattering and backscattering, but IOPs for assemblages that are integrated over the entire assemblage size distribution, and which are fully angularly resolved. This presents the unique opportunity of closely examining simulated phytoplankton phase functions, which are notoriously difficult to measure, and whose behaviour in terms of variability in particle size and wavelength is poorly understood. With no decoupling of absorption and backscattering, and IOPs integrated over the entire size distribution, the model provides an unprecedented opportunity to examine the drivers of variability in phytoplankton optical signals systematically.

2.3. Case Study Methods

The complex optical interactions of D_{eff} and biomass, and the question of whether they can be separated into a useable PFT signal from a background environment of further non-algal optical complexity, is best addressed by investigating specific ecological events of interest to the remote sensing community.

The case studies outlined in the *Introduction* consider phytoplankton from two groups—a Chl *a*-carotenoid phytoplankton group, representing phytoplankton dominated by Chl *a*, and fucoxanthin and/or peridinin; and a Chl *a*- and phycoerythrin-containing group [27]. The former group is chosen as representative of a wide range of phytoplankton across size classes, and the latter for the unique absorption characteristics of phycoerythrin-associated phytoplankton species. This selection is intentionally kept limited in order to assess the relative magnitude of particle size- vs. accessory pigment-related optical signals in likely ecological scenarios, in the context of changing biomass.

Refractive indices for the chloroplast spheres are derived from measurements of cells from blooms in the Benguela—dinoflagellate and diatoms, dominated by Chl *a* and the carotenoid pigments fucoxanthin and peridinin—as well as for a phycoerythrin-associated cryptophyte group (based on a *Mesodinium rubrum/Myrionecta rubra*—dominated assemblage [27]). A justification for using these derived refractive indices across wide size ranges of modelled phytoplankton assemblages is included in Appendix A.1. Phytoplankton assemblages are modelled using a Standard Normal size distribution with a nominal effective variance of 0.6, recognising that while Jungian (exponential) distributions are frequently used for bulk particulate in oceanic conditions, the former is more appropriate for representing the increased species monospecificity associated with elevated biomass ([34]), and the case studies refer mainly to biomass > 1 mg·m^3. Assemblages are modelled with appropriate effective diameters to represent the effective diameters of measured size distributions in the case studies. The resulting IOPs are presented, with explanatory notes, in Appendix A.2. (A cyanobacterial group with substantially altered geometry to represent vacuolated cells has also been developed [63]).

Phytoplankton IOPs are combined, in various proportions as indicated, with appropriate non-algal optical constituents as detailed for each experiment. The phytoplankton-related optical signal is assessed against variability in the non-algal contributions (detailed in Appendices A.3 and A.4), so their absolute magnitude is not critical. Both $a_{gd}(\lambda)$ and $b_{bnap}(\lambda)$ do, however, assume a smooth spectral shape with predictable spectral structure. The potential for additional spectral features in these contributions is not addressed here, and would add further complexity (and hence ambiguity) to resolving phytoplankton scattering characteristics. Water types are considered homogenous with depth (i.e., IOPs constant with depth), generic atmospheric and geographic conditions, and the full radiative transfer solution is calculated by Hydrolight at a spectral resolution of 5 nm. Given the technical challenges with using EAP phase functions for modelling high resolution spectra [35], a Fournier Forand phase function chosen for the backscatter fraction of the combined particulate IOPs is used at each wavelength throughout these experiments. A basic fluorescence efficiency model is included for completeness (detailed in Appendix B.2), but modelling this spectral region accurately is challenging and outside of the scope of this work, so the features of this spectral region are not discussed in terms of PFT sensitivity.

2.3.1. Southern Ocean Case Study: Separating the Effects of Biomass From the Effects of D_{eff} Change

As shown in Lain et al. [65] and Lain et al. [35], where the water-leaving signal is phytoplankton-dominated (e.g., in the Benguela system), it is quite reasonable to expect that some PFT information may be derived from the bulk radiometric signal. However, the challenge for the ocean colour community is determining the PFT signal in low biomass oceanic conditions, for example in the Southern Ocean.

Phytoplankton dynamics in the Southern Ocean are particularly important for their role in uptake of anthropogenic CO_2 (around half of all oceanic uptake), and hence carbon sequestration [3,4]. Variability in phytoplankton ecology is directly linked to mineral and nutrient cycles: assemblages of large diatoms drive primary productivity and carbon export, while assemblages of small phytoplankton play a significant role in nutrient recycling although the net productivity is very low [73].

The third Southern Ocean Seasonal Cycle Experiment (SOSCEx III) undertaken on the SANAE 55 cruise (austral winter 2015) provides the phytoplankton size distribution and Chl *a* data for this experiment [74]. Assemblage D_{eff} were calculated from Coulter Counter measurements, and Chl *a* determined by fluorometric analysis [5]. The additional $a_{gd}(\lambda)$ and $b_{bnap}(\lambda)$ components were estimated guided by observations in [75,76] respectively, noting that these are simply used to approximate the bulk R_{rs} and do not influence any of the other results, as they are discussed in terms of likely variability rather than absolute magnitudes. EAP phytoplankton IOPs with generalised Chl *a*-carotenoid eukaryotic refractive indices were calculated according to the measured D_{eff} and Chl *a* concentrations, and were combined with these estimates and run through Hydrolight to produce the modelled R_{rs}.

Given that the refractive indices used to model the EAP IOPs for this example are from the generalised Chl *a*-carotenoid group suitable for diatom and dinoflagellate species, the likelihood of encountering *Phaeocystis* sp. in the Southern Ocean must be addressed. Given the oceanographic context, as the D_{eff} of 16 µm is reached, it can reasonably be assumed that the assemblage comprises both diatoms and *Phaeocystis*. The main accessory pigment in *Phaeocystis* is 19-hexanolyoxyfucoxanthin, a derivative of fucoxanthin, a dominant light harvesting pigment in diatoms, and so it may be reasonable to model the intracellular absorption properties of individual cells with the generalised eukaryote refractive indices, but this species forms large floating colonies which result in quite different optical effects, and this cannot currently be addressed with the model. Thus, while the likely presence of *Phaeocystis* is acknowledged, it is not explicitly catered for in the modelling. This does not affect the observations on identifying changes in D_{eff} in the discussion below.

2.3.2. Benguela-Like Case Study: Addressing Pigment Variability

The assemblages modelled in the first case study address optical changes due only to biomass (i.e., concentration of Chl *a* pigment) and size (assemblage D_{eff}), as the same set of generalised Chl *a*-carotenoid refractive indices is used for all phytoplankton particles represented. However, this approach addresses only a small subset of important changes in phytoplankton assemblage type, and in the presence of variability in dominant accessory pigments, the EAP model can be set to incorporate different refractive indices as appropriate for phytoplankton displaying accessory pigments other than carotenoids.

To illustrate the effects of pigment variability, this case study simulates a transition from a high biomass *Myrionecta rubra*-dominated assemblage, to a high biomass peridinin (carotenoid)-containing dinoflagellate-dominated assemblage. *M. rubra* is a fascinating but troublesome ciliate species, and enjoys an endosymbiontic relationship with cryptophytes containing the diagnostic pigment phycoerythrin [77], and so "borrows" their characteristic red colour. *M. rubra* blooms can reach extraordinary biomass, resulting in darkly pigmented 'red tide' waters that have negative impacts both ecologically (depletion of nutrients, and the potential for anoxia as the bloom dies), as well as on the recreational use of coastal waters [77]. Again, assemblages are modelled using a Standard normal size distribution and the same values of ϵ are used as for the carotenoids. This ensures that, in this example, all assemblage changes observed are due only to pigment-related differences.

It should be made clear that the modelled transition is not intended to represent a likely ecological succession (except possibly a Lagrangian one, if a dinoflagellate bloom is advected into a previously *M. rubra*-dominated region), but rather to test what biomass and pigment differences are required for the detection of distinct optical conditions, particularly in the context of remote sensing.

2.3.3. Spectral Shape and Sensitivity Analyses

To test the sensitivity of the EAP model, a general allometric approximation of changing D_{eff} from 2 to 8 μm was chosen for this analysis, which ranges from 0.1 to 10 mg/m^3. It is recognised that this scenario does not represent all possible ecological changes, but is a reasonable approximation for a mid-range biomass diatom and dinoflagellate-dominated environment where there may be a detectable PFT signal.

3. Results and Discussion

3.1. Quantifying the Contribution of Phytoplankton to the R_{rs} Signal

Remembering that the Remote Sensing Reflectance (R_{rs}) is grossly proportional to b_b/a [78], it should be noted that, for a given D_{eff} and phytoplankton group, $b_{b\phi}/a_\phi$ will be constant for any given concentration of Chl *a* because the package effect observed with increasing Chl *a* concentration is implicit in the model (see the comparison of Bricaud and EAP a_ϕ^* for varying Chl *a* concentrations in Lain et al. [65]). However, the contribution of the phytoplankton IOPs to the total, i.e., $b_{b\phi}/a_\phi$ as a percentage of total b_b/a, will vary. The EAP model, used together with Hydrolight, allows the inspection of any component optical quantity of interest, and here, the contribution of the phytoplankton IOPs to the total IOP budget is investigated. EAP phytoplankton IOPs are used with Hydrolight to calculate a full radiative transfer solution resulting in a new theoretical quantity, $R_{rs}\phi$. This quantity is introduced as an approximate quantification of the phytoplankton contribution to the bulk R_{rs}, in order to more intuitively understand the relative optical contributions in terms of remote sensing. While acknowledging that R_{rs} is not an additive quantity, $R_{rs}\phi$ is the calculation of reflectance with only water and phytoplankton IOPs. It does not account for any optical interaction between the phytoplankton and other in-water constituents likely to be present in natural waters, such as CDOM or detrital and mineral particles. These interactions are assumed to be secondary to the contribution of phytoplankton, but have not been quantified. It is anticipated that trans-spectral effects are most likely to suffer from this type of subtractive approach, but a full photon tracing model (such as a Monte Carlo model) would be needed to ascertain this. By modelling the phytoplankton contribution to the water-leaving signal, we can assess the availability of signal for PFT retrieval.

Being able to identify the spectral regions sensitive to changes in phytoplankton assemblage (focusing on those due to change in assemblage D_{eff}) is valuable, especially to identify spectral regions which might be sufficiently independent from the ambiguity introduced by other in-water constituents. This allows the quantification of the phytoplankton signal with confidence, even where these other constituents are not well characterised. The spectral regions of maximum proportional phytoplankton signal are the ones which hold potential for detecting PFT changes from an in-water perspective, as these represent the regions of the largest phytoplankton-related signal variability as the assemblage changes.

The resulting contribution of phytoplankton to the total R_{rs} is shown in Figure 1, for typical Case 1 waters as a simple illustrative example. In this example, a_{gd} covaries with Chl a while b_{bnap} is constant. This represents the combined a_{cdom}, known to approximately covary with Chl a, and detrital absorption a_{det}, which is assumed to be small with respect to a_{cdom} [79] and approximately constant (a_{det} is neglected altogether in some IOP models e.g., Alvain et al. [9]). As CDOM does not scatter, b_{bnap} represents the scatter of only the detrital component of the non-algal constituents. Thus, it is assumed that biomass increases together with a_{cdom}, against a relatively unchanging background detrital population whose IOPs are dominated by backscatter. As the phytoplankton contribution to the IOP budget increases (i.e., generally, as biomass increases), the impact of the other constituents is proportionally less in the R_{rs}. This varies with D_{eff}, and is observable in Figure 1 to a greater degree in the R_{rs} with a smaller (nominal) D_{eff} of 2 μm as compared with a larger (nominal) D_{eff} of 12 μm to show the higher level of phytoplankton backscatter of small cells contributing to brighter R_{rs}, which is less sensitive to the addition of scattering from other sources.

For each D_{eff}, it is evident that the phytoplankton percentage contribution to the bulk R_{rs} increases with biomass. However, it can be seen that there is a dependency on D_{eff} which, when considered in the context of transitioning assemblages, is not straightforward. This observation indicates a requirement to go beyond the Case 1/Case 2 water type distinction for PFT signal analysis and applications: the differential in phytoplankton scatter as D_{eff} varies in both water types must be considered as well as variable b_{bnap} in Case 2. When it comes to retrieving information about the phytoplankton IOPs, their proportional contribution to the bulk water-leaving signal (or the total IOPs) should be considered. Figure 2 demonstrates the proportional IOP contributions for the 0.1, 1 and 10 mg/m^3 cases.

3.2. Case Study 1: Separating the Effects of Biomass from the Effects of D_{eff} Change

Figure 3 presents two distinct events which illustrate the interdependency of the size and biomass signals. Modelled R_{rs} are shown for selected adjacent stations (20 to 21 is marked A; 12 to 13 is marked B) where a nominal threshold of change detectable by satellite is reached in the blue and green spectral regions, in other words, where a change in R_{rs} would be evident on a satellite image. Given the ambiguity in the causality of the phytoplankton signal, assessing the magnitude of changes to the water-leaving signal as the in-water constituents vary will give an indication of whether there may be enough radiometric signal at TOA to even detect the change. A threshold in situ measurement resolution of 1×10^{-4} sr^{-1} [80] is taken as an indication of sensitivity to detecting change in R_{rs} by direct measurement. Given an average estimated uncertainty in satellite R_{rs} of $\pm 0.6 \times 10^{-3}$ sr^{-1} across the spectrum [81], here a conservative 1×10^{-3} sr^{-1} is used to indicate a potentially detectable change in water-leaving signal from satellite. These thresholds are not definitive and are used purely for the purpose of contextualising the discussion. Both examples in Figure 3 display large changes in R_{rs}, but these are causally distinct: (A) represents a large change in Chl *a* concentration *and* in D_{eff}, while (B) represents a large change in Chl *a* concentration but a negligible change in D_{eff}.

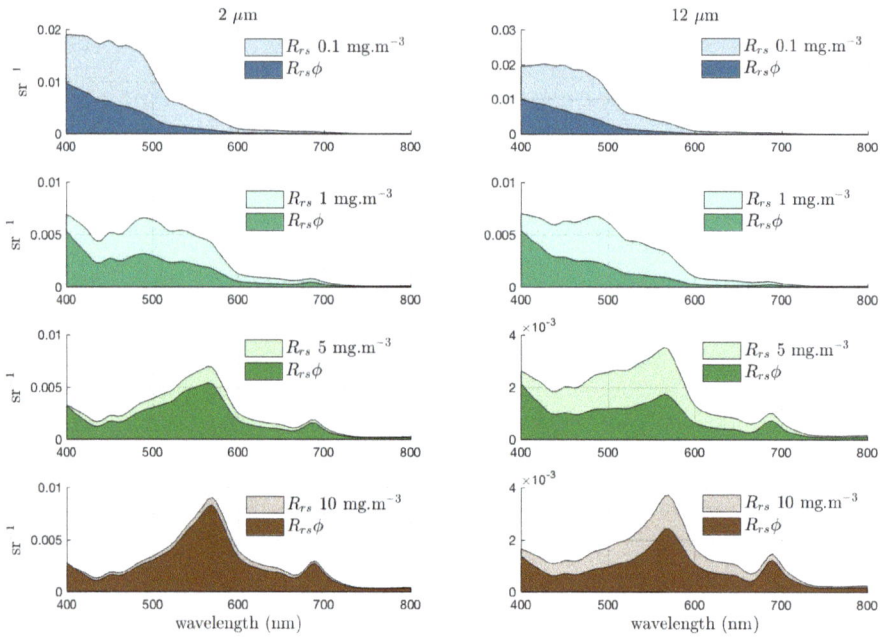

Figure 1. Relative contribution of phytoplankton to total R_{rs} (with $a_{gd}(400) = 0.07 \cdot [Chl_a]^{0.75}$, and $b_{bnap}(550) = 0.005$ m^{-1}) for increasing biomass with $D_{eff} = 2$ and 12 µm. These populations are idealised examples and not intended to represent any observed relationship between Chl *a* concentration and D_{eff}.

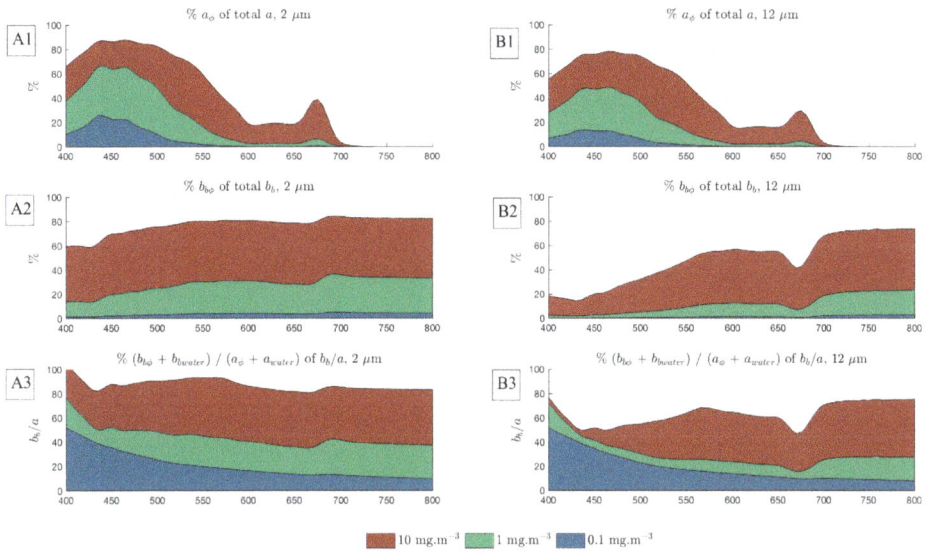

Figure 2. Example proportional phytoplankton to total Inherent Optical Property (IOP) contributions for Case 1 waters, for idealised eukaryote assemblages of 2 and 12 μm.

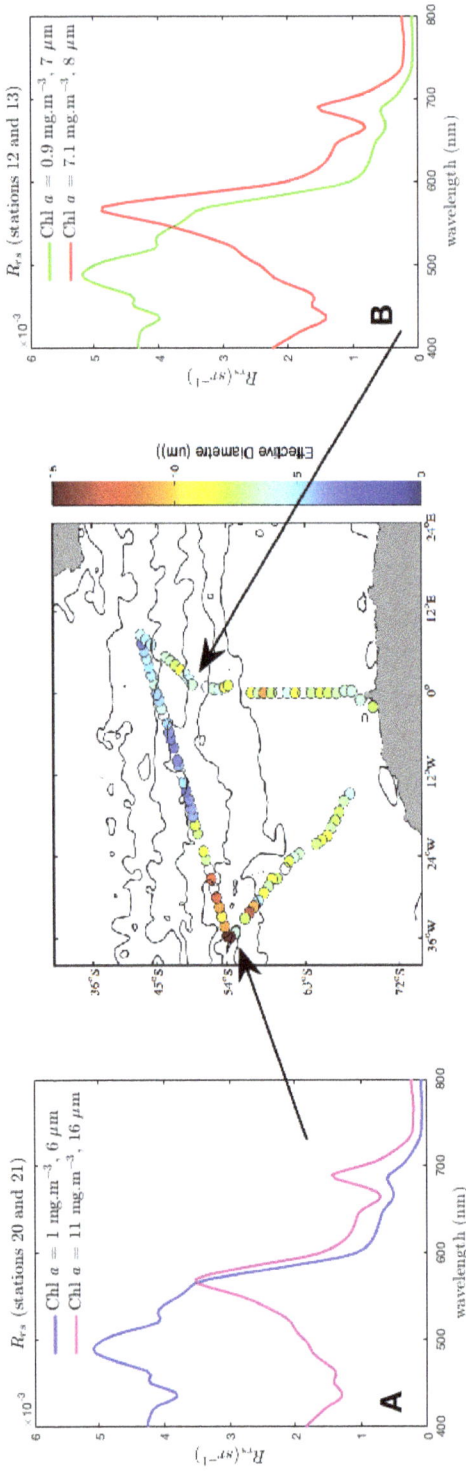

Figure 3. Modelled R_{rs} for stations 20, 21, 12 and 13 of SOSCEX III. The modelled bulk R_{rs} are calculated using Equivalent Algal Populations (EAP) generalised Chl *a*-carotenoid refractive indices and measured Chl *a* concentrations for the phytoplankton component, and include estimated $a_{gd}(\lambda)$ and $b_{bmap}(\lambda)$ contributions appropriate for this region [75,76]. Stations 20 to 21 (**A**) represent a large change in both Chl *a* concentration and in D_{eff}. Stations 12 to 13 (**B**) represent a large change in Chl *a* concentration only. The centre panel shows the measured D_{eff} for the cruise track (starting at the ice shelf on the bottom right and continuing in an anticlockwise direction.) The effective diameter image is courtesy of SANAE 55 Report [74].

Station 20 to 21 therefore represents a significant phytoplankton community shift, as large changes in both D_{eff} (from 6 to 16 μm) and Chl a concentration (from 1 to 11 /m^3) were recorded. To isolate this change in phytoplankton signal, the differences in $R_{rs}\phi$ for an assemblage with D_{eff} = 6 μm and an assemblage with D_{eff} = 16 μm are presented in Figure 4A for the measured range of Chl a concentration. Note that the large differences of about 2 × 10^{-3} sr^{-1} in R_{rs} in the blue, observable in Figure 3A, only appears at very low biomass in Figure 4A as this difference in R_{rs} is almost entirely due to the difference in biomass and not a change in D_{eff}. The spectral location of the most promising size-related signal for PFT retrieval is evidently dependent on biomass, and, at low biomass, it is positioned near 435 nm, while, at higher biomass, it is around 570 nm. As this is the phytoplankton-only signal, the question remains to what extent this signal is expressed in the bulk R_{rs}, when the optical impact of the non-algal constituents is also considered.

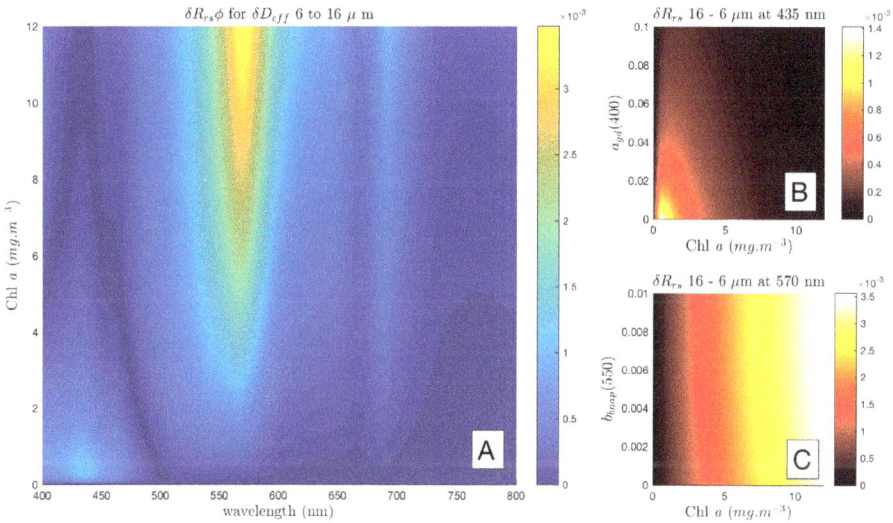

Figure 4. Southern Ocean stations 20 to 21: $\delta R_{rs}\phi$ is shown for δD_{eff} of 6 to 16 μm (**A**). The effect of $a_{gd}(\lambda)$ at 435 nm is shown in (**B**), and $b_{bnap}(\lambda)$ at 570 nm in (**C**). The units of the colour bars are sr^{-1}.

Working with the change in phytoplankton size signal identified at 435 nm (bottom left corner, Figure 4A), $a_{gd}(\lambda)$ is added at increasing concentrations to simulate a range of bulk R_{rs} at 435 nm in Figure 4B, and $b_{bnap}(\lambda)$ is likewise added incrementally at 570 nm in Figure 4C. In these plots, horizontal gradients indicate R_{rs} sensitivity primarily to the constituent on the y axis, while vertical gradients indicate that the change in R_{rs} is driven by the biomass, and is not sensitive to variability on the y axis.

Figure 4B shows that the difference in bulk R_{rs} for the given δD_{eff} is only detectable at the satellite threshold level (shown in yellow) at low biomass under very low $a_{gd}(\lambda)$ conditions. As biomass increases, increasing absorption by phytoplankton as well as by additional $a_{gd}(\lambda)$, reduces the magnitude of the water-leaving signal and renders any δD_{eff} information ambiguous. When additionally considering the brightening effect of $b_{bnap}(\lambda)$ in the blue (not quantified here), it can readily be perceived that the water-leaving signal is too complex at 435 nm to retrieve useful size information.

In Figure 4C, the relationship with $b_{bnap}(\lambda)$ at 570 nm is more straightforward. Change in R_{rs} due to δD_{eff} is detectable in the bulk R_{rs} at the satellite threshold (in red) from about 2.5 mg/m^3 upwards regardless of the $b_{bnap}(\lambda)$ contribution, at least for Case 1 type conditions. The magnitude of this signal is almost entirely biomass driven. (This is in line with the observation made by [13] that the MODIS wavebands at 531 and 551 nm are good indicators of backscatter anomalies because their magnitude is proportional to the addition or removal of particulate backscattering, and the longer wavelength band at 551 nm is less affected by variability in both $a_{gd}(\lambda)$ and phytoplankton absorption [10].)

It should be appreciated, though, that $R_{rs}\phi$ in these figures is representing the change in R_{rs} due to size *at* a particular biomass (i.e., biomass is constant while assemblage characteristics vary), effectively removing the effects of simultaneous biomass changes. Figure 5 simulates a transition from 6 to 16 D_{eff} with biomass 1 to 11 mg/m^3, where the intermediate values of both D_{eff} and Chl a are simply linearly interpolated. The vertical lines highlight 435 and 570 nm which were identified in Figure 4A as being the spectral regions of greatest size-driven signal. In Figure 5, while biomass and size effects combine to form large changes in $R_{rs}\phi$ in the blue, it is the smaller signal around 570 nm that contains the most size-driven change as it is not affected by biomass to the same degree. Figure 4B,C show that the signal at 435 nm is sensitive to the effects of variable a_{gd}, while the phytoplankton signal at 570 nm remains robust against variability in the non-algal optical contributions.

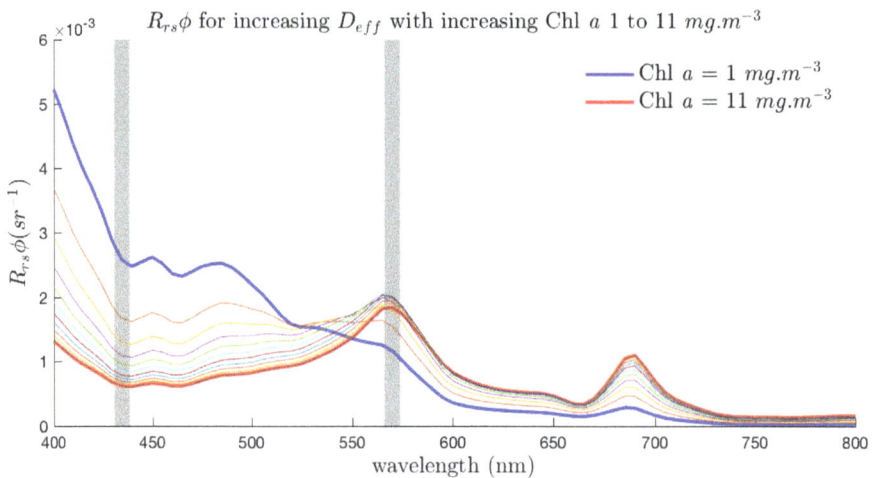

Figure 5. A simulated transition from 6 to 16 D_{eff} with biomass 1 to 11 mg/m^3. Intermediate values of D_{eff} and Chl a are simply linearly interpolated. The lines highlight 435 nm and 570 nm, regions of maximum size signal, which are (at 435 nm) and are not (at 570 nm) sensitive to the effects of additional optical constituents.

By contrast, stations 12 to 13 exhibit a large change in R_{rs}—seen first in Figure 4B; shown again in Figure 6A—with an increase in Chl a from 0.9 to 7.1 mg/m^3 but only a very small change in D_{eff} from 7 to 8 µm. This is likely, given the location in the lee of the South Sandwich Islands, to reflect a diatom bloom associated with island wake effects, due to fertilisation by terrestrial iron [82]. Tracing the signal due to this change in D_{eff} across all Chl a concentrations in this range in Figure 6B shows that there is a size related signal between 550 and 600 nm, but it is of an order of magnitude less than in the previous example, and so does not show potential for detection by satellite radiometry. This is illustrated further in the lower panel (C), showing the location of this signal but also that it is almost all attributable to biomass—as shown by the $R_{rs}\phi$ representing D_{eff} 7 at 7.1 mg/m^3 i.e., what the higher biomass R_{rs} would look like without the increase in effective diameter as the assemblage changes. It can be seen quite clearly from these spectra that a difference in the blue due only to this δD_{eff}, with any variability $a_{gd}(\lambda)$, would not be detectable by any means.

It should be noted that the spectral locations of maximum δD_{eff} features are a direct consequence of the spectral nature of the IOPs used in the modelling, and that both of these examples use the same Chl a-carotenoid refractive indices to generate the phytoplankton IOPs. In other words, as phytoplankton IOPs are adjusted to represent pigment differences, the spectral character of the assemblage change will vary. A slight migration in the exact location of the maximum available δD_{eff} signal is observable with different ranges of D_{eff}, although within the Chl *a*-carotenoid group it remains between 550 and 600 nm for any difference in D_{eff} between 1 and 40 μm. (This is discussed in more detail later with respect to Figure 9, and an additional figure is shown in Appendix.)

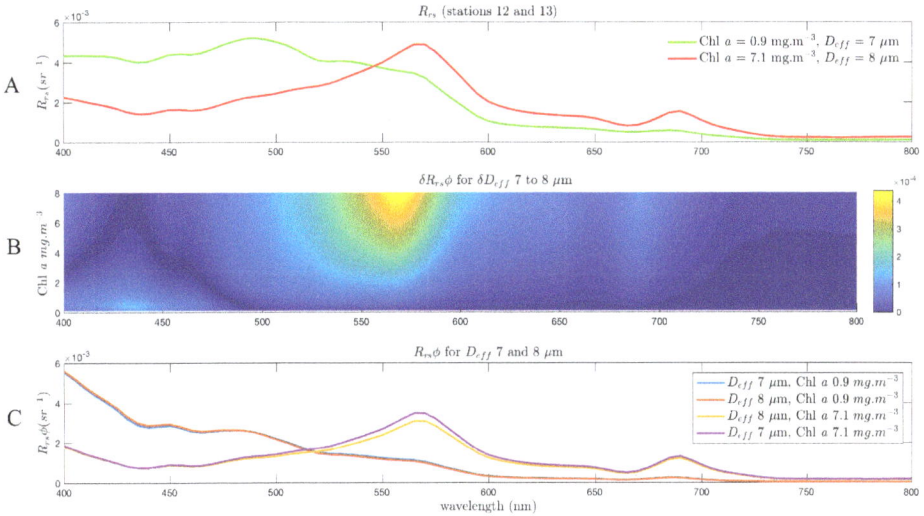

Figure 6. Modelled R_{rs} for Stations 12 and 13 (**A**), with EAP eukaryote phytoplankton IOPs, and $a_{gd}(\lambda)$ and $b_{bnap}(\lambda)$ components estimated guided by observations in [75,76], respectively; (**B**) shows $\delta R_{rs}\phi$ for this large change in Chl *a* concentration (1 to 7 mg/m^3) but a small δD_{eff} of 7–8 μm. The unit of the colour bar is sr^{-1}. Note that the results are one order of magnitude less than in the previous example; (**C**) shows the negligible effect on $R_{rs}\phi$ of a change in D_{eff} from 7 to 8 μm at the measured Chl *a* concentrations.

3.3. Case Study 2: Addressing Pigment Variability

Both the phycoerythrin-containing and peridinin-containing assemblages are modelled here, Figures 7 and 8) with D_{eff} of 12 μm, so the simulated optical changes as the assemblage changes from *M. rubra*-dominated to dinoflagellate-dominated are all due to differences in pigmentation, for any given Chl *a* concentration. From the log-scale R_{rs}, it is evident that the pigment-related differences in R_{rs} become larger as biomass increases. In the very high biomass blooms ($\geq$30 mg·m^{-3}) typical of the Benguela system, it is known that *M. Rubrum*—containing assemblages are identifiable from MERIS satellite imagery [83] due to the effects of the diagnostic phycoerythrin peak (at 565 nm) appearing in the 560:520 nm band ratio.

An analogous study of the sensitivity of the maximum $\delta R_{rs}\phi$ signal to non-algal constituents is made at 570 nm for the pigment-driven feature appearing at high biomass (10 mg·m^3, Figure 9). The sensitivity of pigment-driven differences to non-algal effects is in contrast to the Southern Ocean size example in that it is largely driven by variability in $b_{bnap}(\lambda)$. As biomass increases past 10 mg·m^3, the magnitude of the $\delta R_{rs}\phi$ grows as $b_{bnap}(\lambda)$ increases, showing no impact at all of biomass past 20 mg·m^3. What this means is that while significant biomass is required to detect pigment changes, past a certain upper biomass limit, the magnitude of the pigment differential signal grows proportionally as R_{rs} is augmented by non-algal scatter.

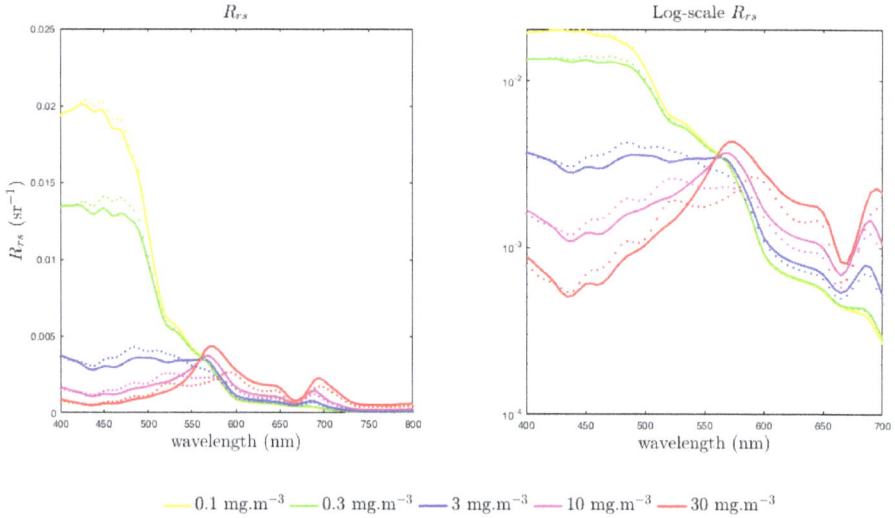

Figure 7. Benguela-like pigment-based experiment: Modelled R_{rs} shown for Chl a-carotenoid pigmented assemblages (solid lines) and phycoerythrin containing assemblages (dotted lines) for identical Chl a concentrations, at 0.1, 0.3, 3, 10 and 30 mg·m^{-3}. There is no change in D_{eff}, both are 12 μm. The non-algal optical constituents are modelled with $a_{gd}(400) = 0.07 * [Chl_a]^{0.75}$, and $b_{bnap}(550) = 0.005$ m^{-1}.

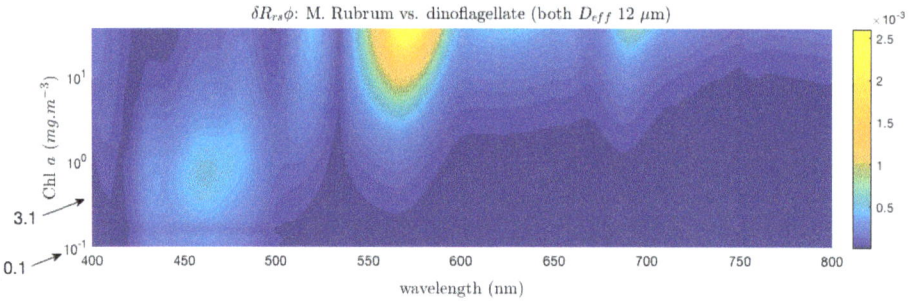

Figure 8. δR_{rs} shown for a change from a high biomass *Myrionecta rubra*-dominated assemblage, to a high biomass peridinin (carotenoid)-containing dinoflagellate-dominated assemblage. There is no change in D_{eff}.

Absorption-based pigment differences are therefore sensitive to scattering variability unless the biomass is very high, and this is particularly relevant in spectral regions affected by scattering variability due to changes in D_{eff} (Figure 4). Noting the log-scale Chl a axis of Figure 9, it can be observed that while the respective magnitudes of the pigment-driven ($D_{eff} = 12$ μm, Figure 9) and size-driven (D_{eff} from 6 to 16 μm, Figure 4) signals are comparable at the point that pigment differences appear (i.e., 10 mg·m^3), the size-driven feature is more sensitive at lower biomass—detectable at around 2 mg·m^3 for the given change in D_{eff}. This sensitivity will be affected by the size range in question and also by pigment concentrations, but it can be inferred that, generally, where size changes of this range (or larger) take place together with pigment changes, it is the size change that drives the variability in the water-leaving signal, and changes in the reflectance due to a substantial change in D_{eff} are observable at lower biomass than those due to pigment changes.

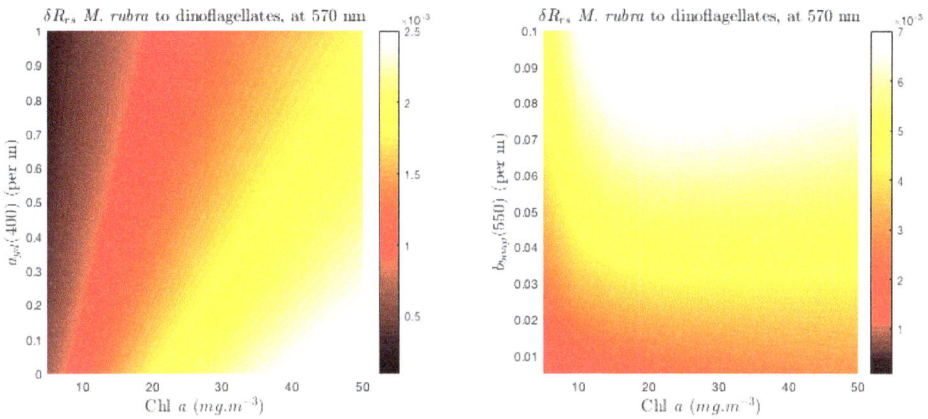

Figure 9. δR_{rs} sensitivity to a_{gd} and b_{bnap} at 570 nm, for a high biomass *Myrionecta rubra*-dominated assemblage, to a high biomass peridinin (carotenoid)-containing dinoflagellate-dominated assemblage.

This applies equally to accessory pigments other than phycoerythrin absorbing in spectral regions affected by phytoplankton scatter and/or by variability in non-algal scatter. This speaks to the importance of both the magnitude of the change in $b_b\phi(\lambda)/a\phi(\lambda)$ and the proportional contribution of $b_b\phi(\lambda)/a\phi(\lambda)$ to the total $b_b(\lambda)/a(\lambda)$ when evaluating the potential for accessing these signals from R_{rs} (see Figure 2).

This result also implies that there are cases where the δD_{eff} signal is augmented by pigment changes, for example when moving from small fucoxanthin- or peridinin-dominated cells to large phycoerythrin-dominated cells, as the optical effect of reduction in backscatter around 560–570 nm by large cells will be enhanced by additional pigment absorption in the δR_{rs} signal. However, there remains a complex optical relationship with biomass, and this effect needs to be properly accounted for in order to accurately detect the augmented signal.

The EAP approach to pigments does not address the extent to which fine spectral resolution accessory pigment absorption features persist in R_{rs}, nor their retrieval from hyperspectral radiometry. The intention with the EAP model is to demonstrate the dependence on biomass to retrieve absorption features, and the inherent signal ambiguity as the contrasting optical effects of $b_b\phi(\lambda)$ and $a\phi(\lambda)$ interact to form the phytoplankton signal within the R_{rs}. It is worth noting that hyperspectral radiometry will not overcome the inherent signal-related constraints identified in this study.

3.4. Radiometric Sensitivity of EAP Size-Based PFT Detection—Magnitude of $\delta R_{rs}\phi$

Having established that at low biomass the PFT signal in the blue is easily overwhelmed by the effects of $a_{gd}(\lambda)$ and $b_{bnap}(\lambda)$, and that pigment effects are generally secondary to those of δD_{eff}, the PFT signal due to phytoplankton scattering in the 500 to 600 nm region can be evaluated for sensitivity in terms of changes in D_{eff} and biomass. To this end, the EAP model is again coupled with Hydrolight to simulate expected variability in R_{rs} due to changes in D_{eff} with the aim of evaluating the sensitivity of the model.

Figure 10A demonstrates how the combined effects of biomass and D_{eff} interact to form the maximum available $\delta R_{rs}\phi$ signal at low biomass and small size ranges. The figure shows that this maximum lies between 520 and 570 nm—the exact wavelength varies with both size difference and biomass. The shifting position of maximum $\delta R_{rs}\phi$ is shown in Figure 10B. Increasing biomass improves the ability to trace the size-related effects, and a D_{eff} change from 2 μm up to at least 8 μm is not detectable at the threshold in oceanic conditions with Chl $a < 1$ mg/m^3.

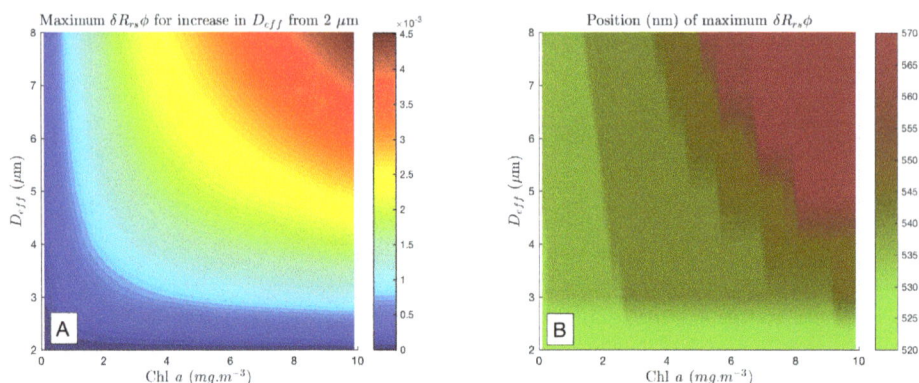

Figure 10. Maximum $\delta R_{rs}\phi$ for δD_{eff} from a starting assemblage with D_{eff} 2 μm, as Chl *a* varies (**A**). Note that the $\delta R_{rs}\phi$ occurs at different wavelengths from 500 to 600 nm (**B**), and this shows the maximum signal, so there is no exact wavelength information in (**A**). Using a difference of 1×10^{-3} sr^{-1} as a threshold for detection by satellite, it can be seen that, while the maximum size change here (2 to 8 μm) is not detectable with Chl *a* < 1 mg/m^3, by 10 mg/m^3, even a small change in D_{eff} results in a detectable change in R_{rs}.

Using 1×10^{-3} sr^{-1} as a threshold for detection by satellite, it can be seen that an ecologically significant shift in D_{eff} from 2 or 3 to 6 μm, such as at the onset of an oceanic bloom, looks potentially detectable from about 2 mg/m^3. By 10 mg/m^3, even a small change in D_{eff} results in a detectable change in $R_{rs}\phi$, but, as biomass falls below this, the change in D_{eff} must be increasingly large to be detected. This is consistent with inversion studies of EAP sensitivity [37]. Note that this experiment addresses only the phytoplankton-related signal, and that when attempting to identify these signals in the bulk R_{rs}, it is necessary to consider the sensitivity of the phytoplankton signal to the optical effects of the non-algal constituents. These results can be considered to show the minimum threshold for potential detection i.e., the signal is further ambiguated by non-algal optics in a real-world context.

The spectrally shifting nature of the $\delta R_{rs}\phi$ signal for oceanic PFT applications provides a strong case for hyperspectral sensors in the 520 to 570 nm wavelength region. The extent to which the $\delta R_{rs}\phi$ signal persists in fixed waveband ratios is investigated in the next section on shape sensitivity.

3.5. Spectral Shape Sensitivity of EAP Size-Based PFT Detection

To further test the sensitivity of the EAP model and the causal IOP variability in terms of identifiable changes in spectral shape from a multi-spectral perspective, $R_{rs}\phi$ ratios for 440:560 nm (blue:green), 560:665 nm (green:red) and 665:710 nm (red:NIR) wavelengths were calculated for a range of D_{eff} and biomass.

These are shown in Figure 11, representing corresponding changes in the $R_{rs}\phi$ and in the phytoplankton backscattering, for these wavelength pairs. The B:G $R_{rs}\phi$ ratio shows a strong biomass dependency and a small sensitivity to size at large sizes, for $0.5 \leq$ Chl $a \leq 4.5$ mg/m^3. The R:NIR ratio shows some sensitivity to larger sizes from about 3 mg/m^3, but this decreases as biomass increases. The G:R ratio shows a significant size-related feature for small sizes (≤ 6 μm) from biomass of about 2 mg/m^3 upwards (encircled in Figure 10). This is where a peak in the corresponding $b_{b\phi}$ ratio appears, suggesting that the large change in magnitude of the Chl *a*-specific backscatter $b_b^*\phi$ between small D_{eff} (Figure 12) is directly responsible for the sensitivity in the $R_{rs}\phi$ G:R ratio seen in Figure 11.

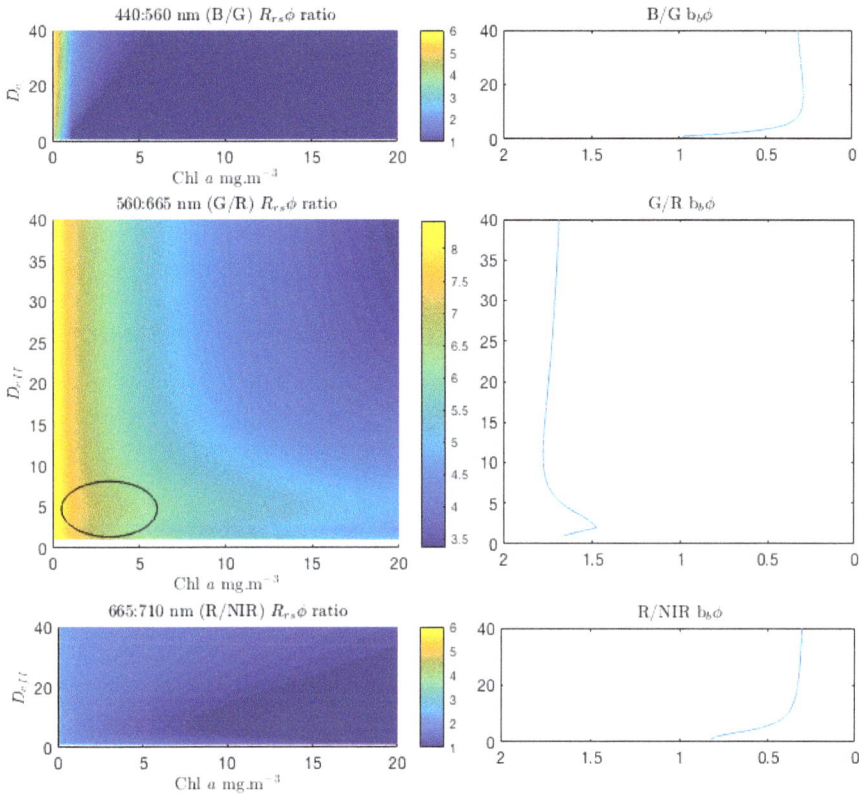

Figure 11. $R_{rs}\phi$ ratios for blue:green, green:red and red:NIR (Near Infra-Red) wavelengths as shown, for Chl a concentrations of 0.1 to 20 mg/m^3 and D_{eff} 1 to 40 μm. The B/G ratio shows a strong biomass dependency and a small sensitivity to size at large sizes, for $0.5 \leq$ Chl $a \leq 4.5$ mg/m^3. The $b_{b\phi}$ ratios all display a strong size signal at 2–4 μm, and the G/R ratio shows a corresponding size-related feature.

This is an important finding. There is a marked size dependency in all of the $b_b^*\phi$ ratios, with the greatest rate of change somewhere between D_{eff} 2 and 8 μm, but it is only in the case of the G:R ratio that the magnitude of the backscatter is sufficient for this signal to be identifiable in the $R_{rs}\phi$. Given that the radiometric signal in the blue is greatly reduced by large phytoplankton absorption and a_{gd}, and the red and NIR wavelengths are similarly affected by the absorption of water, it can be concluded that the main driver of the useable PFT signal in the green and red is phytoplankton backscatter.

Figure 13 shows the rapid increase in the proportional contribution of phytoplankton to total backscatter at 560 and 665 nm. It is known that, for typical diatom/dinoflagellate assemblages, the 560 nm region is more influenced by backscatter than by absorption. The fact that the magnitude of the total backscatter is much lower at 665 than at 560 nm, together with the strong absorption by water in this region, result in a small useable R_{rs} signal. A contribution of approximately 40% of phytoplankton to total b_b at 560 nm corresponds with the limits of detectable $\delta R_{rs}\phi$ (see Figure 2), indicating that this is the proportion at which phytoplankton backscatter starts driving the total water-leaving signal around 560 nm. Consequently, this is the minimum contribution for which some δD_{eff} information may be known. For an oceanic bloom example δD_{eff} from 2–6 μm, this threshold contribution is reached at about 2 mg/m^3, while to detect an example δD_{eff} of 10 to 20 μm in a eukaryotic succession, extremely high biomass is required. The mid-range biomass sensitivity demonstrated here presents opportunities for identifying higher resolution size classes than the 2 to 20 μm and >20 μm categories currently frequently employed [6,8,10,84]. The ability to achieve better resolution within the 2–20 μm size class is particularly desirable for marine ecosystem modelling [6].

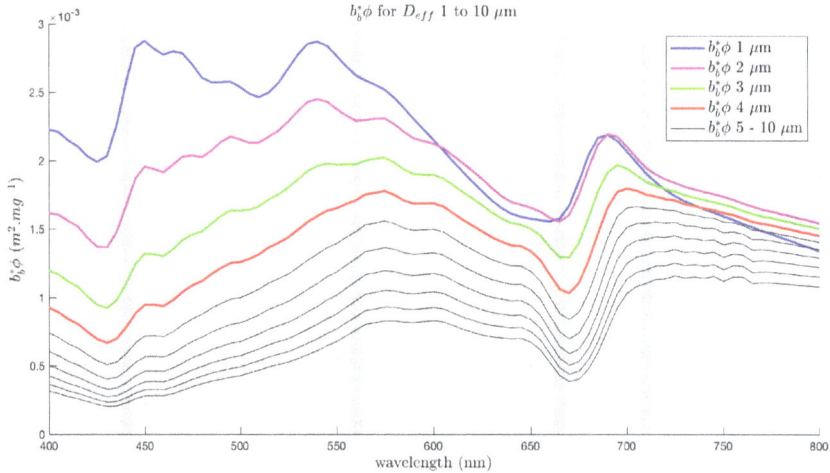

Figure 12. $b_b^* \phi$ shown for D_{eff} 1 to 10 μm. The largest differences in backscatter across the spectrum occur between 1 and 4 μm, with the exception of the overlapping of $b_b^* \phi$ in the red and NIR.

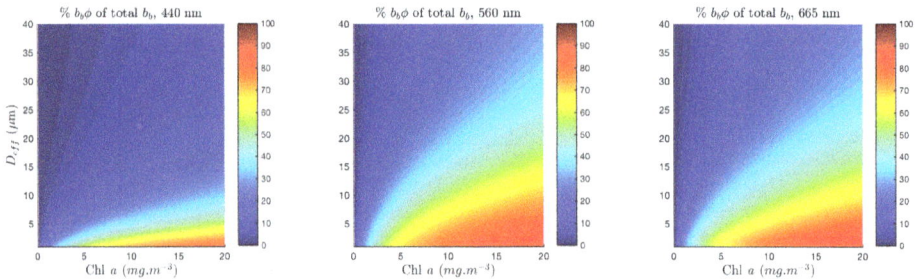

Figure 13. Percentage contribution of phytoplankton to total backscatter (including water, and with nominal $b_{bnap}(550) = 0.005$), shown for D_{eff} 1 to 40 μm and Chl *a* from 0.1 to 20 mg/m³, at 440, 560 and 665 nm.

3.6. Considering Uncertainties

Particularly when considering δR_{rs} retrievals from satellite, it is important and necessary to contextualise the magnitude of the PFT signal with respect to uncertainties on the satellite radiometry. A brief study on model and associated radiometric uncertainty is available in Appendix D. An important observation is that, while the 500 to 600 nm region of the promising PFT signal may be mostly insensitive to the effects of non-algal constituents, it is also where variability in R_{rs} due to the different approaches to phytoplankton phase functions is important, emphasising the critical role of phytoplankton scatter in this signal.

4. Conclusions

The distinct causal optical effects of variations in phytoplankton biomass, mean assemblage size, pigment-related spectral variability and non-algal constituents are not easily identified, with substantial interdependency and spectral ambiguity. Consistent with previous studies [37,39], it can be seen here that ambiguity is critical in attempting to resolve the phytoplankton community structure signals.

The case studies illustrate how most of the R_{rs} signal that is due to phytoplankton is driven by biomass, an expected result. This concept underpins, after all, the primary missions of most ocean colour sensors: resolving variability in phytoplankton biomass. The study shows that quantitative consideration of the constraints of biomass and the phytoplankton contribution to the total IOP budget is required when addressing the PFT question.

The key findings include the assertions that most of the absorption driven phytoplankton signal in R_{rs} in the blue is too ambiguous to use, and that the most useful PFT signal is caused by spectral backscatter. Furthermore, the ability to assess the PFT signal against non-algal optical contributions is largely driven by biomass and the IOP budget.

Overall, spectral scattering properties of natural waters are not well characterised [85,86], and phytoplankton spectral backscattering characteristics are underexploited in terms of their impact on the water-leaving signal. The importance of better representing the angular and spectrally variable nature of phytoplankton scattering has been established [35], and it is clear that phytoplankton backscatter is at the heart of the PFT question.

The size-related PFT signal is driven by phytoplankton scattering, and spectral regions where scattering is at its most sensitive to D_{eff} show the most potential for PFT detection from the total water-leaving signal. There is most sensitivity to size-related changes in $b_{b\phi}$ in the 1–6 µm size range. Phytoplankton size-related features, most likely driven by phytoplankton absorption variablity, appear in R_{rs} around 430 nm at low biomass, and scattering-driven size-related features in the 520 to 570 nm spectral region at elevated biomass. The water-leaving signal in the blue spectral region is highly complex and ambiguous, being the result of varied and contrasting effects of absorbing and scattering characteristics of both algal and non-algal in-water constituents. Consequently, the size-related signal appearing in low biomass waters (<1 mg/m^3) may be useful only when $a_{gd}(\lambda)$ and $b_{bnap}(\lambda)$ are exactly known. Accessory pigment absorption features that persist in R_{rs} in low biomass suffer from the same vulnerability to uncertainty in the non-algal constituents. Satellite measurement uncertainty and $a_{gd}(\lambda)$ retrievals may in future be improved (e.g., with the use of radiometry in the UV), but, given current uncertainties, achieving sufficiently accurate satellite estimates of the non-algal optical components is unlikely for this purpose.

This finding exposes a vulnerability in historical approaches to phytoplankton identification and quantification based on the features of phytoplankton absorption characteristics in the blue. Satellite PFT methods using this approach all suffer from this shortcoming where assemblage-related variability is secondary to biomass effects, and where phytoplankton relationships with $a_{gd}(\lambda)$ and $b_{bnap}(\lambda)$ are not precisely known. These approaches additionally rely on implicit relationships between Chl a and D_{eff} which may not always hold. This work shows that, at low biomass (< 1 mg·m^3), where R_{rs} is absorption-dominated, it is unlikely that there is sufficient size- or pigment-driven PFT signal to be retrieved from satellite radiometry without making these assumptions. (Phytoplankton whose prominent absorption features are at longer wavelengths, such as phycocyanin-containing cyanobacteria, present a different case). Isolating variability in $R_{rs}\phi$ as D_{eff} and biomass vary shows that an example oceanic bloom δD_{eff} from 2 to 6 µm is only detectable at the satellite measurement threshold of 1×10^{-3} sr^{-1} when the biomass reaches about 2 mg/m^3 (Figure 10A).

Consequently, it is the size-related backscatter-driven signal in the 500 to 570 nm region, appearing at substantial biomass, that is the most useful for PFT identification from satellite radiometry as it is sufficiently insensitive to reasonable variability in both $a_{gd}(\lambda)$ and $b_{bnap}(\lambda)$ (when composed of small particles) (Figure 4). Variability in scatter due to non-algal particulate in the same size range as phytoplankton will likely ambiguate the distinctive spectral scatter of PFT changes, and this has not yet been tested. The location of the maximum $\delta R_{rs}\phi$ size feature shifts between 520 and 570 nm (Figure 10B), suggesting strongly that hyperspectral data in this region would add greater capability here. Further analysis is needed to quantify the potential advantages of hyper- over multi-spectral data with respect to this shifting maximum signal, and also with respect to the reduced SNR implicit in narrow waveband measurements.

Understanding the proportional phytoplankton contribution to the total IOP budget and the resulting water-leaving signal is central to the determination of sufficient phytoplankton-driven signal containing PFT information. The proportional 'net' contribution of phytoplankton i.e., $b_{b\phi}/a_\phi$ as a percentage of total b_b/a, has been identified as the driver of PFT sensitivity in the R_{rs}. Given the detectable differences in R_{rs} as size and biomass change, a proportional phytoplankton contribution of approximately 40% to the total b_b appears to a reasonable minimum threshold in terms of yielding a detectable optical change. The proportional contribution always varies with the non-algal optical constituents $a_{gd}(\lambda)$ and $b_{nap}(\lambda)$.

Despite the many sources of model uncertainty and the requirement for model validation in specific regions, these results indicate the necessity of approaching PFTs from a strongly biophysical perspective. There is a great need for better characterisation of phytoplankton community structure and improved handling of the complex spectral and angular nature of phytoplankton scattering.

The EAP model code in Matlab R2018a (The Mathworks Inc., Natick, Mass, United States) or in Python 3.7, as well as the Fortran routine for Hydrolight allowing the choice of discretised EAP phase function based on wavelength rather than backscatter fraction (see Appendix D: Uncertainties), are freely available to the community. Please contact the corresponding author.

Supplementary Materials: The following are available online at http://www.mdpi.com/2076-3417/8/12/2681/s1, Visualisation Tool: Interactive visualisation of spectral R_{rs} with assemblage D_{eff} shown for 2–20 µm, with user-controlled Chl *a* concentration, $a_{gd}(400)$ and $b_{bnap}(550)$.

Author Contributions: Conceptualization, L.R.L. and S.B.; Formal analysis, L.R.L.; Investigation, L.R.L.; Project administration, S.B.; Software, L.R.L.; Supervision, S.B.; Writing—original draft, L.R.L.; Writing—review & editing, S.B.

Funding: Funding that was awarded to Lisl Robertson Lain from the Centre for Scientific and Industrial Research (CSIR) and the University of Cape Town (UCT) PhD Scholarship Programme is gratefully acknowledged, as is funding from the CSIR/DST SWEOS Strategic Research Programme.

Acknowledgments: Thanks to Curtis Mobley for assistance with Hydrolight and the examiners of Lisl Robertson Lain's PhD thesis for their valuable comments. Input from two anonymous reviewers of this manuscript, as well as from Dariusz Stramski, was gratefully received.

Conflicts of Interest: The authors declare no conflict of interest.

Appendix A. Phytoplankton Assemblage Variability in the EAP Model

The successful validation of the model in very high biomass Benguela conditions [65] gives confidence in the representation of the phytoplankton component of the water-leaving signal, as it is known that, in these cases, the R_{rs} is overwhelmingly dominated by phytoplankton. It is concluded in Lain et al. [65] that it is the EAP's detailed handling of phytoplankton spectral backscatter that sets it apart from other IOP models. The core mathematics of the model are fully described in Bernard et al. [27]. A detailed study of EAP phytoplankton angular scattering and phase functions is available in Lain et al. [35].

Appendix A.1. Justification for Using Measurement-Derived Refractive Indices across Wide Size Ranges

The main light-harvesting pigments in typical diatom and dinoflagellate assemblages (fucoxanthin and peridinin, respectively)—while chemotaxonomically distinct—display the typical broad, featureless absorption spectra characteristic of carotenoids, with peaks centered around 500 nm [87] and vary well within the natural variability of phytoplankton absorption (Figure A1). They consequently have similar refractive indices [27] and so these types were combined into a generalised set of diatom/dinoflagellate IOPs, as no significant difference was found between the dinoflagellate and diatom groups in terms of their optics that could not be attributed to the respective particle sizes (see also [38,88]). This group of IOPs should correctly be referred to as Chl *a*-carotenoid IOPs.

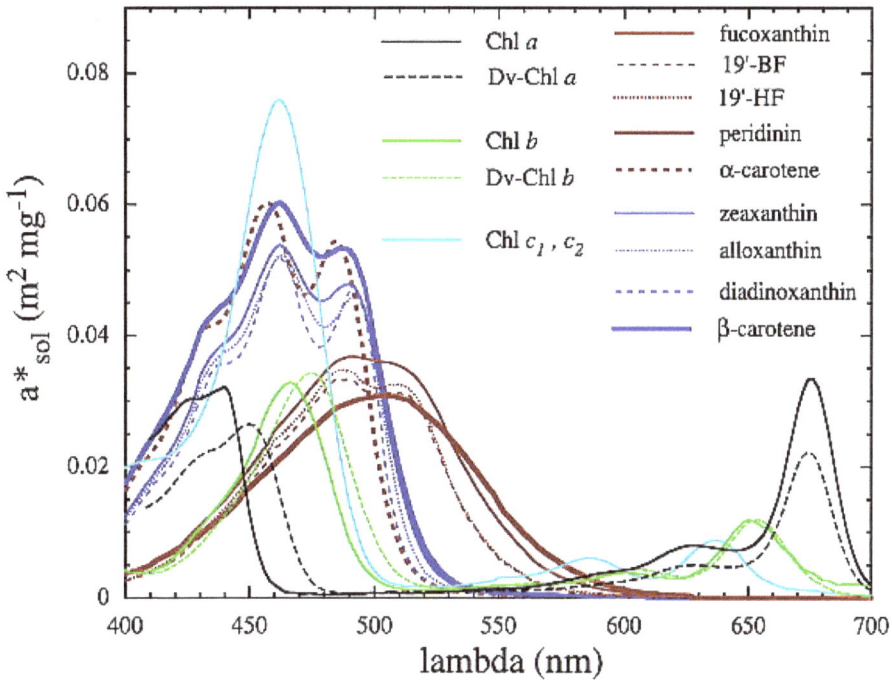

Figure A1. Pigment absorption spectra from Bricaud et al. [87], reprinted with permission from the American Geophysical Union. The broad featureless absorption spectra of fucoxanthin and peridinin peaking at around 500 nm are shown by the thin and thick brown lines, respectively.

While the measurements and refractive index derivations [89] were performed for cells of approximately 12 to 20 μm, it should be made clear that the imaginary refractive index characterises the absorption of the intracellular material and has absolutely no dependency on cell size. It can be inferred therefore that these refractive indices can be used to represent a range of phytoplankton sizes displaying dominant Chl *a* and carotenoid pigments. Southern Ocean nanophytoplankton comprise mainly diatoms and dinoflagellates, but also chlorophytes and haptophytes in smaller proportions [90,91]. The latter two phytoplankton groups are generally dominated by Chl *a* and the fucoxanthin pigment derivatives 19′-hex-fucoxanthin and 19′-but-fucoxanthin [91], which display somewhat elevated absorption peaks located at slightly shorter wavelengths with respect to fucoxanthin [87,92]. The optical influence of these derivative pigments is assumed to be negligible in the context of the case studies for two reasons. Firstly, in the nanophytoplankton group as a whole, the fucoxanthin-derived pigments occur in far lesser concentrations than that of fucoxanthin itself [91], which is represented by the refractive indices. This is reinforced by the derivation of a nanophytoplankton group of refractive indices in non-bloom conditions (D_{eff} = 2 μm), whose impact on the IOPs of small cells was not sufficiently different from the diatom-dinoflagellate group to warrant its routine use for small cells. Secondly, these pigments act to increase phytoplankton absorption around the 450 nm spectral region, identified the spectral region as very vulnerable to small variability in a_{gd}, and to large satellite R_{rs} measurement uncertainty, so the conclusions regarding the spectral regions containing the most useful signatures for PFT detection still hold.

EAP sensitivity testing has indicated that increasing the intracellular Chl *a* density C_i may be appropriate for small cells, but this is not explored here.

Appendix A.2. EAP Phytoplankton IOPss

Phytoplankton-specific IOPs are presented in Figures A5 and A6 for generalised Chl *a*-carotenoid assemblages and for phycoerythrin-containing assemblages, respectively.

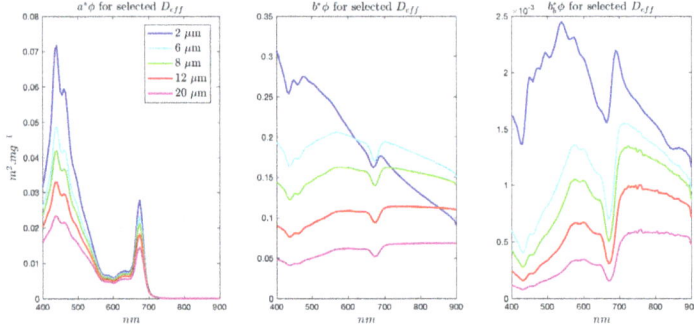

Figure A2. EAP Eukaryote Chl *a*-carotenoid-dominated IOPs for a range of assemblage D_{eff}.

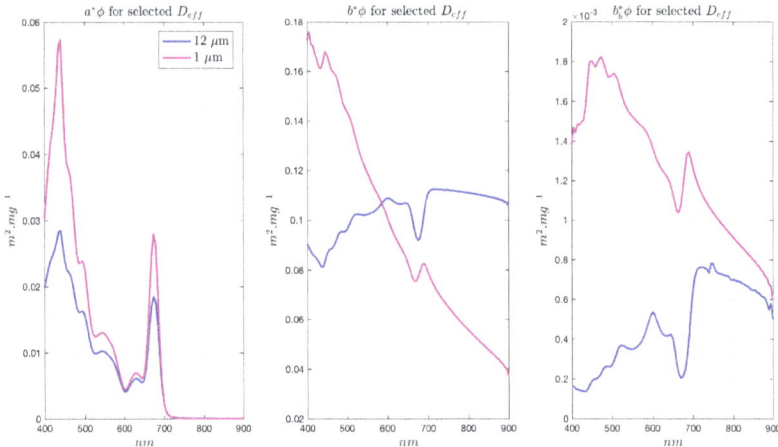

Figure A3. EAP Phycoerythrin-containing IOPs (based on *Myrionecta Rubra*), used for cryptophyte-dominated assemblages in the Benguela, and *Synechococcus* sp. in the Southern Ocean.

Appendix A.3. EAP $a_{gd}(\lambda)$ Parameterisation

A simple exponential combined gelbstoff and detrital absorption term $a_{gd}(\lambda)$ [93,94] is used as a representative of commonly occurring conditions in the Benguela:

$$a_{gd}(\lambda) = a_{gd}(400) \exp[-S(\lambda - 400)]. \tag{A1}$$

The exponential slope factor S is given a constant value of 0.012 [95]. This value, derived for the Benguela system, is not adjusted for the $a_{gd}(\lambda)$ term used in the Southern Ocean Case Studies. This is acknowledged as a source of uncertainty, but supporting literature suggests that values in the range 0.0140 ± 0.0032 nm^{-1} cater adequately for a variety of water types [93].

An observed relationship of

$$a_{gd}(400) = 0.0904 \log[Chl_a] + 0.1287 \tag{A2}$$

from measurements in the Benguela is used to scale the gelbstof/detrital exponential term, and $a_{gd}(750)$ onwards is assumed to be zero. This parameterisation was derived for high biomass environments. At very low biomass ($<$ 1 mg/m^3), the $\log[Chl_a]$ term becomes negative, and so, for the Southern Ocean case studies, this parameterisation was amended to

$$a_{gd}(400) = 0.07 \cdot [Chl_a]^{0.75} \tag{A3}$$

following Alvain et al. [21], noting that the referenced parameterisation is for 440 nm and not 400, but also that the a_{gd} term is used as an approximate measure of total signal sensitivity, and so, in this sense, an absolutely accurate term is not a requirement.

Appendix A.4. EAP $b_{bnap}(\lambda)$ Parameterisation

Non-algal backscattering is modelled after Roesler and Perry [96] who describe a small particle backscattering term represented by a power law relationship (their b_{bs}, referred to as b_{bnap} in the EAP model). It has a constant spectral shape dependent only on wavelength, but variable in magnitude.

$$b_{bnap}(\lambda) = \lambda^{-1.2}. \tag{A4}$$

This is then adjusted to a selected value of $b_{bnap}(550)$, as detailed in the text.

Small particle (non-algal) scatter b_{nap} is approximated as 50 times the b_{bnap} in the Benguela examples and as 100 times the b_{bnap} in the Southern Ocean examples. This yields a non-algal particulate backscattering probability ($\tilde{b}_{bnap}$) of 0.02 (2%) and 0.01 (1%), respectively. This is assumed to be reasonable given that it has been shown that the total particulate backscattering probability $\tilde{b}_b$ varies in the range 1.2 to 3.2 % in coastal waters dominated by non-algal particles (i.e., Case 2) [97], and that generally accepted values for $\tilde{b}_b$ in Case 1 waters is around 1% [98].

Keeping the non-phytoplankton backscattering constant with Chl a results in a dependent but nonlinear relationship, resulting in an overall $\tilde{b}_b$ that decreases as Chl a increases (Figure A4), noting the spectral variability of small phytoplankton at elevated biomass.

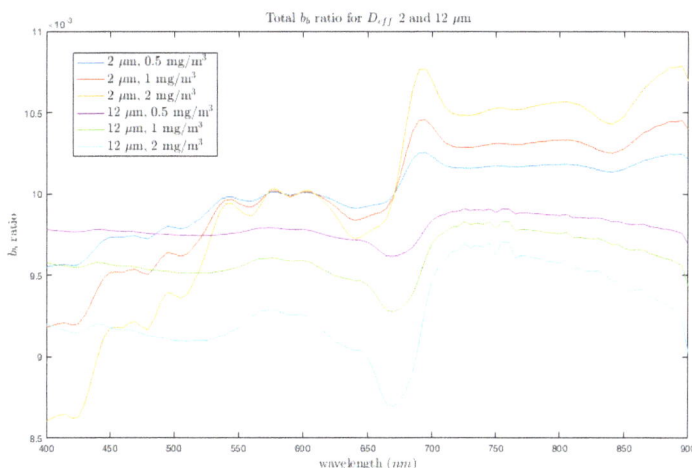

Figure A4. Bulk backscatter ratio shown for D_{eff} 2 and 12 μm, with nominal $b_{bnap}(550) = 0.01$ m^{-1} and b_{nap} as 50 times the b_{bnap}, as for a coastal environment, shown for Chl a of 0.5, 1.0 and 2.0 mg/m^3. The elevated backscatter ratio of coastal environments with respect to the Southern Ocean (where b_{nap} is modelled as 100 times the b_{bnap}) is attributed to the contribution of terrestrial mineral particles with a high refractive index [66,99].

Appendix B. Measurements and Modelling Parameters

Appendix B.1. Chl a

Chl *a* measurements are made using a Turner 10-AU Fluorometer (Turner Designs, San Jose, CA, USA), following Holm-Hansen et al. [100].

Appendix B.2. Model Parameters Used for Hydrolight-Ecolights

For most of the experiments, Ecolight's 2-component IOP model was used to generate $R_{rs}(\lambda)$. The "clearest natural water" IOPs were selected for Component 1 (water). IOPs for component 2 (everything else) were precomputed in Matlab from the EAP phytoplankton IOPs and additional $a_{gd}(\lambda)$ and $b_{bnap}(\lambda)$ contributions as required.

Fluorescence quantum efficiency ϕ was approximated by Chl *a* concentration:

$$< 10 \text{ mg/m}^3 = 1\%$$
$$10\text{--}50 \text{ mg/m}^3 = 0.6\%$$
$$50\text{--}100 \text{ mg/m}^3 = 0.2\%$$
$$> 100 \text{ mg/m}^3 = 0.1\%.$$

These values are based on MODIS ϕ_{sat} climatologies [101], and measurements [102] to characterise the reduction in ϕ as eutrophication increases.

A constant set of generalised atmospheric conditions was selected for all experiments. An annual average for solar irradiance and a solar zenith of 30° was used in lieu of time and location.

Appendix C. Position of Maximum $\delta R_{rs}\phi$

Further to Figure 9B in the main text, Figure A5 shows the position of maximum $\delta R_{rs}\phi$ for assemblage changes from 8 µm and 14 µm, respectively, for Chl *a* concentrations between 10 and 20 mg/m³. It can be seen that, at these high biomass concentrations, there is no spectral migration of the maximum $\delta R_{rs}\phi$ with biomass, and that once the maximum $\delta R_{rs}\phi$ signal reaches 570 nm, its location does not change with increasing D_{eff} for biomass up to 20 mg/m³. This has been tested up to 40 µm (not shown).

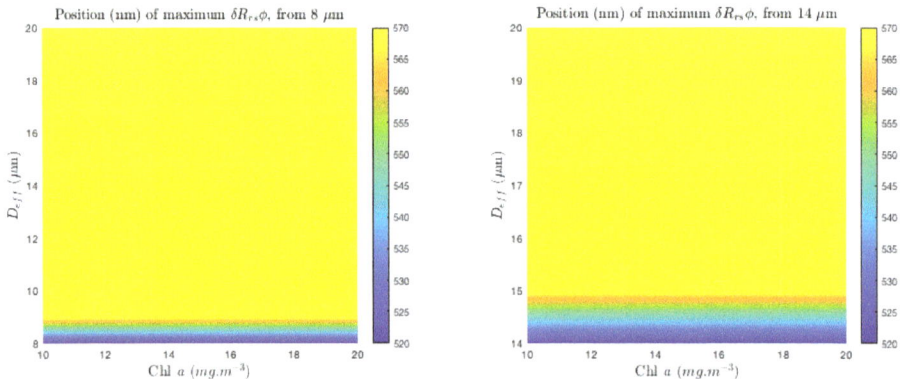

Figure A5. Spectral position of maximum $\delta R_{rs}\phi$ for assemblage changes from 8 µm and 14 µm, respectively

Appendix D. Uncertaintiess

Uncertainties in satellite radiometry are given in the main text, and model error in terms of uncertainty/variability in the phase function is described fully in [35]. There are many additional sources of uncertainty in the model (non-sphericity of phytoplankton, approximations in size distribution, Chlorophyll *a* density, to name a few), and further work is needed to quantify them appropriately. For demonstrative purposes here, given that any retrieval of size properties would be performed with the model itself, the model uncertainty is constrained to just that of the phase function variability, as this has a size implication in itself, as shown in [35].

In Figure A6, the model uncertainty is shown for $R_{rs}\phi$ against a background of total R_{rs} with nominal additional $a_{gd}(\lambda)$ and $b_{nap}(\lambda)$, together with the satellite R_{rs} measurement uncertainty. Despite the small model uncertainty on the phytoplanton signal in the blue, the huge impact of additional $a_{gd}(\lambda)$ and the large satellite radiometric uncertainty clearly show the large degree of ambiguity and potential error in the retrieval of the phytoplankton component, even if the $a_{gd}(\lambda)$ is exactly known. Satellite-derived $a_{gd}(\lambda)$ (and CDOM) products have large uncertainties: r^2 of less than 0.25 for three different $a_{gd}(\lambda)$ algorithms against in situ data [16]—noting that dependence on the atmospheric correction means that a significant level of error is propagated through the algorithms from this source, particularly in the blue.

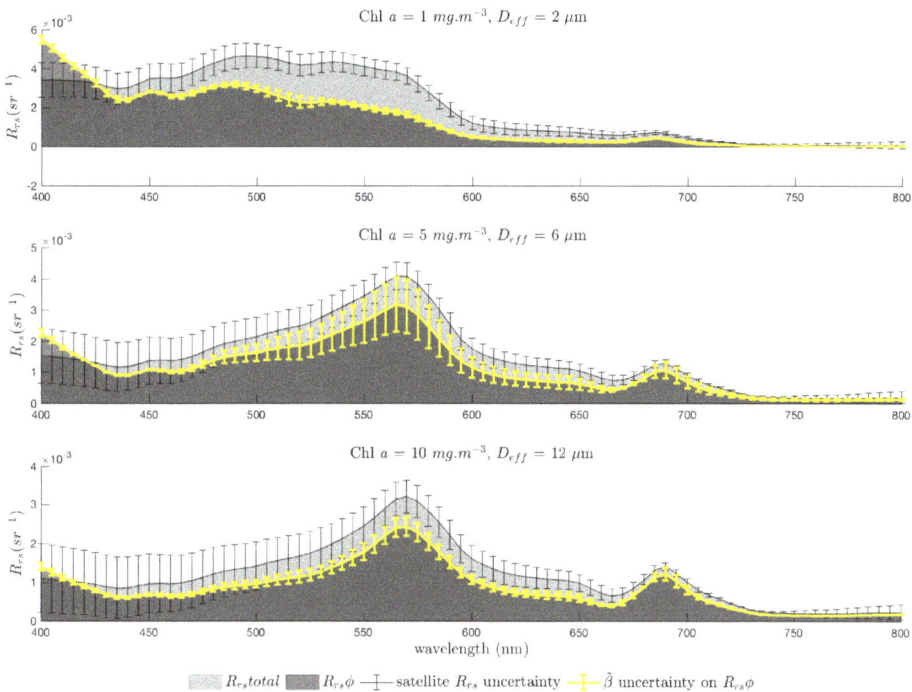

Figure A6. Total R_{rs} with satellite measurement uncertainties in the blue and red bands from [16] and linearly interpolated between them. An indication of model uncertainty on the $R_{rs}\phi$ is calculated by the spectral differences resulting from the use of a combined $b_{bp}(\lambda)$-specific Fournier Forand phase function independent of wavelength, vs. wavelength- and $b_{b\phi}(\lambda)$-dependent EAP phase functions.

The most significant spectral regions of Figure A6 in the context of this study are those where the uncertainty on $R_{rs}\phi$ overlap with the bulk satellite R_{rs} measurement uncertainty in each example. These are the spectral regions where the phytoplankton-specific signal dominates the bulk signal to the point that they are arguably indistinguishable, so these regions are particularly promising in terms of PFT detection from the bulk R_{rs}. It is also encouraging to note that the regions of maximum δD_{eff} previously identified fall within these regions, meaning that particularly close to 570 nm, the bulk signal not only closely reflects the causal phytoplankton signal but is also not very sensitive to reasonable variability in $a_{gd}(\lambda)$ and $b_{nap}(\lambda)$. However, it is an important consideration that these are also regions of large uncertainty in the size signal, and that, as D_{eff} increases, the R_{rs} expression of reduced phytoplankton scatter becomes more vulnerable to variable $b_{nap}(\lambda)$.

At low biomass, the phytoplankton signal falls well outside of the bulk measurement uncertainty, but the question of whether phytoplankton IOPs could be retrieved from the bulk depends on the resulting proportional contribution to the total. With reduced $b_b\phi/a\phi$, even small variability in the non-algal contribution to b_b/a results in signal ambiguity. In this case, the additional $a_{gd}(\lambda)$ and $b_{nap}(\lambda)$ contributions need to be exactly known, in order to be able to retrieve any PFT information.

It can also be observed in Figure A6 that the magnitude of model uncertainty is less, and the proportional contribution of phytoplankton to the bulk IOPs is greater, at wavelengths slightly shorter than those of the maximum δD_{eff} in the case studies. Thus, the spectral location of the largest observable δD_{eff} signal may not necessarily be the most revealing of PFT discrimination in terms of the associated uncertainties. A sophisticated uncertainty model would be necessary to calculate the respective advantages of reduced contribution uncertainties on a smaller signal vs. slightly larger uncertainties on a larger workable signal. It is also worth considering that, even where the bulk and $R_{rs}\phi$ signals are distinct, there are spectral regions where they are parallel i.e., maintain the same shape. It can be concluded that the phytoplankton contribution determines the spectral shape in these regions—although the uncertainty associated with a smooth $b_{bnap}(\lambda)$ curve is also not quantified here. This information could potentially be exploited to investigate PFT signal from the bulk R_{rs}.

Further work on incorporating EAP phase functions into Hydrolight has enabled the $R_{rs}\phi$ presented here to include the fluorescence term, and this is also a spectral region of a large proportional phytoplankton contribution together with small model uncertainty as calculated by the difference in approach to scattering phase functions. This region (around 685 nm) appears in the maximum δR_{rs} plots from the case studies, but has not been discussed as confidence in modelling this spectral region accurately needs to be improved with respect to natural variability in a fluorescence quantum yield and phytoplankton response to the light environment. However, it is known that this region holds further useful information on phytoplankton health [103] as well as size.

Overall, the uncertainties in both measured and modelled quantities should be considered in terms of the proportional contribution by phytoplankton. The highest proportional phytoplankton contribution to the bulk optics, and therefore the most promising signal for PFTs, occurs where elevated scatter due to biomass is complemented by the elevated scatter of small phytoplankton cells. Approaches to modelling the phase functions result in an inherent ambiguity of about 4 μm at very high biomass, but this drops with biomass and as D_{eff} increases.

References

1. Field, C.B.; Behrenfeld, M.J.; Randerson, J.T. Primary production of the biosphere: Integrating terrestrial and oceanic components. *Science* **1998**, *281*, 237–240. [CrossRef] [PubMed]
2. McClain, C.R. A decade of satellite ocean color observations. *Annu. Rev. Mar. Sci.* **2009**, *1*, 19–42. [CrossRef] [PubMed]
3. Swart, S.; Chang, N.; Fauchereau, N.; Joubert, W.; Lucas, M.; Mtshali, T.; Roychoudhury, A.; Tagliabue, A.; Thomalla, S.; Waldron, H.; et al. Southern Ocean Seasonal Cycle Experiment 2012: Seasonal scale climate and carbon cycle links. *S. Afr. J. Sci.* **2012**, *108*, 11–13. [CrossRef]
4. Thomalla, S.; Fauchereau, N.; Swart, S.; Monteiro, P. Regional scale characteristics of the seasonal cycle of chlorophyll in the Southern Ocean. *Biogeosciences* **2011**, *8*, 2849. [CrossRef]
5. Ryan-Keogh, T.J.; Thomalla, S.J.; Mtshali, T.N.; Little, H. Modelled estimates of spatial variability of iron stress in the Atlantic sector of the Southern Ocean. *Biogeosciences* **2017**, *14*, 3883–3897. [CrossRef]
6. Brewin, R.J.W.; Ciavatta, S.; Sathyendranath, S.; Jackson, T.; Tilstone, G.; Curran, K.; Airs, R.L.; Cummings, D.; Brotas, V.; Organelli, E.; et al. Uncertainty in Ocean-Color Estimates of Chlorophyll for Phytoplankton Groups. *Front. Mar. Sci.* **2017**, *4*, 104. [CrossRef]

7. Antoine, D.; d'Ortenzio, F.; Hooker, S.B.; Bécu, G.; Gentili, B.; Tailliez, D.; Scott, A.J. Assessment of uncertainty in the ocean reflectance determined by three satellite ocean color sensors (MERIS, SeaWiFS and MODIS-A) at an offshore site in the Mediterranean Sea (BOUSSOLE project). *J. Geophys. Res. Oceans* **2008**, *113*, 2156–2202. [CrossRef]
8. Sathyendranath, S.; Watts, L.; Devred, E.; Platt, T.; Caverhill, C.; Maass, H. Discrimination of diatoms from other phytoplankton using ocean-colour data. *Mar. Ecol. Prog. Ser.* **2004**, *272*, 59–68. [CrossRef]
9. Alvain, S.; Loisel, H.; Dessailly, D. Theoretical analysis of ocean color radiances anomalies and implications for phytoplankton groups detection in Case 1 waters. *Opt. Express* **2012**, *20*, 1070–1083. [CrossRef]
10. Kostadinov, T.; Siegel, D.; Maritorena, S. Retrieval of the particle size distribution from satellite ocean color observations. *J. Geophys. Res. Oceans* **2009**, *114*. [CrossRef]
11. Kostadinov, T.S. Carbon-based phytoplankton size classes retrieved via ocean color estimates of the particle size distribution. *Ocean Sci.* **2016**, *12*, 561. [CrossRef]
12. Anderson, T.R. Plankton functional type modelling: Running before we can walk? *J. Plankton Res.* **2005**, *27*, 1073–1081. [CrossRef]
13. Brown, C.A.; Huot, Y.; Werdell, P.J.; Gentili, B.; Claustre, H. The origin and global distribution of second order variability in satellite ocean color and its potential applications to algorithm development. *Remote Sens. Environ.* **2008**, *112*, 4186–4203. [CrossRef]
14. MÉlin, F.; Sclep, G.; Jackson, T.; Sathyendranath, S. Uncertainty estimates of remote sensing reflectance derived from comparison of ocean color satellite data sets. *Remote Sens. Environ.* **2016**, *177*, 107–124. [CrossRef]
15. Lee, Z.P. Remote Sensing of Inherent Optical Properties: Fundamentals, Tests of Algorithms and Applications. *Rep. Int. Ocean Colour Coord. Group* **2006**, *5*, 1–122.
16. Mélin, F.; Zibordi, G.; Berthon, J.F. Assessment of satellite ocean color products at a coastal site. *Remote Sens. Environ.* **2007**, *110*, 192–215. [CrossRef]
17. Mouw, C.B.; Hardman-Mountford, N.J.; Alvain, S.; Bracher, A.; Brewin, R.J.W.; Bricaud, A.; Ciotti, A.M.; Devred, E.; Fujiwara, A.; Hirata, T.; et al. A Consumer's Guide to Satellite Remote Sensing of Multiple Phytoplankton Groups in the Global Ocean. *Front. Mar. Sci.* **2017**, *4*, 41. [CrossRef]
18. Hirata, T.; Hardman-Mountford, N.J.; Brewin, R.J.W.; Aiken, J.; Barlow, R.; Suzuki, K.; Isada, T.; Howell, E.; Hashioka, T.; Noguchi-Aita, M.; et al. Synoptic relationships between surface Chlorophyll-a and diagnostic pigments specific to phytoplankton functional types. *Biogeosciences* **2011**, *8*, 311–327. [CrossRef]
19. Brewin, R.J.W.; Hardman-Mountford, N.J.; Lavender, S.J.; Raitsos, D.E.; Hirata, T.; Uitz, J.; Devred, E.; Bricaud, A.; Ciotti, A.; Gentili, B. An intercomparison of bio-optical techniques for detecting dominant phytoplankton size class from satellite remote sensing. *Remote Sens. Environ.* **2011**, *115*, 325–339. [CrossRef]
20. Brewin, R.J.W.; Sathyendranath, S.; Hirata, T.; Lavender, S.J.; Barciela, R.M.; Hardman-Mountford, N.J. A three-component model of phytoplankton size class for the Atlantic Ocean. *Ecol. Modell.* **2010**, *221*, 1472–1483. [CrossRef]
21. Alvain, S.; Moulin, C.; Dandonneau, Y.; Bréon, F.M. Remote sensing of phytoplankton groups in Case 1 waters from global SeaWiFS imagery. *Deep Sea Res. Part I Oceanogr. Res. Pap.* **2005**, *52*, 1989–2004. [CrossRef]
22. Alvain, S.; Moulin, C.; Dandonneau, Y.; Loisel, H. Seasonal distribution and succession of dominant phytoplankton groups in the global ocean: A satellite view. *Glob. Biogeochem. Cycles* **2008**, *22*, 1–15. [CrossRef]
23. Devred, E.; Sathyendranath, S.; Stuart, V.; Platt, T. A three component classification of phytoplankton absorption spectra: Application to ocean-color data. *Remote Sens. Environ.* **2011**, *115*, 2255–2266. [CrossRef]
24. Ciotti, A.M.; Bricaud, A. Retrievals of a size parameter for phytoplankton and spectral light absorption by colored detrital matter from water-leaving radiances at SeaWiFS channels in a continental shelf region off Brazil. *Limnol. Oceanogr. Methods* **2006**, *4*, 237–253. [CrossRef]
25. Bracher, A.; Vountas, M.; Dinter, T.; Burrows, J.; Röttgers, R.; Peeken, I. Quantitative observation of cyanobacteria and diatoms from space using PhytoDOAS on SCIAMACHY data. *Biogeosciences* **2009**, *6*, 751–764. [CrossRef]
26. Kostadinov, T.S.; Siegel, D.A.; Maritorena, S. Global variability of phytoplankton functional types from space: assessment via the particle size distribution. *Biogeosciences* **2010**, *7*, 3239–3257. [CrossRef]
27. Bernard, S.; Probyn, T.A.; Quirantes, A. Simulating the optical properties of phytoplankton cells using a two-layered spherical geometry. *Biogeosci. Discuss.* **2009**, *6*, 1497–1563. [CrossRef]

28. Uitz, J.; Stramski, D.; Reynolds, R.A.; Dubranna, J. Assessing phytoplankton community composition from hyperspectral measurements of phytoplankton absorption coefficient and remote-sensing reflectance in open-ocean environments. *Remote Sens. Environ.* **2015**, *171*, 58–74. [CrossRef]
29. Torrecilla, E.; Stramski, D.; Reynolds, R.A.; Millán-Núñez, E.; Piera, J. Cluster analysis of hyperspectral optical data for discriminating phytoplankton pigment assemblages in the open ocean. *Remote Sens. Environ.* **2011**, *115*, 2578–2593. [CrossRef]
30. Xi, H.; Hieronymi, M.; Röttgers, R.; Krasemann, H.; Qiu, Z. Hyperspectral differentiation of phytoplankton taxonomic groups: A comparison between using remote sensing reflectance and absorption spectra. *Remote Sens.* **2015**, *7*, 14781–14805. [CrossRef]
31. Sadeghi, A.; Dinter, T.; Vountas, M.; Taylor, B.; Altenburg-Soppa, M.; Bracher, A. Remote sensing of coccolithophore blooms in selected oceanic regions using the PhytoDOAS method applied to hyper-spectral satellite data. *Biogeosciences* **2012**, *9*, 2127–2143. [CrossRef]
32. Stramski, D.; Bricaud, A.; Morel, A. Modeling the inherent optical properties of the ocean based on the detailed composition of the planktonic community. *Appl. Opt.* **2001**, *40*, 2929–2945. [CrossRef] [PubMed]
33. Matsuoka, A.; Huot, Y.; Shimada, K.; Saitoh, S.I.; Babin, M. Bio-optical characteristics of the western Arctic Ocean: Implications for ocean color algorithms. *Can. J. Remote Sens.* **2007**, *33*, 503–518. [CrossRef]
34. Bernard, S.; Shillington, F.A.; Probyn, T.A. The use of equivalent size distributions of natural phytoplankton assemblages for optical modeling. *Opt. Express* **2007**, *15*, 1995–2007. [CrossRef] [PubMed]
35. Lain, L.R.; Bernard, S.; Matthews, M.W. Understanding the contribution of phytoplankton phase functions to uncertainties in the water colour signal. *Opt. Express* **2017**, *25*, A151–A165. [CrossRef] [PubMed]
36. Morel, A. Consequences of a Synechococcus bloom upon the optical properties of oceanic (case 1) waters. *Limnol. Oceanogr.* **1997**, *42*, 1746–1754. [CrossRef]
37. Evers-King, H.; Bernard, S.; Lain, L.R.; Probyn, T.A. Sensitivity in reflectance attributed to phytoplankton cell size: Forward and inverse modelling approaches. *Opt. Express* **2014**, *22*, 11536–11551. [CrossRef]
38. Organelli, E.; Nuccio, C.; Lazzara, L.; Uitz, J.; Bricaud, A.; Massi, L. On the discrimination of multiple phytoplankton groups from light absorption spectra of assemblages with mixed taxonomic composition and variable light conditions. *Appl. Opt.* **2017**, *56*, 3952. [CrossRef]
39. Defoin-Platel, M.; Chami, M. How ambiguous is the inverse problem of ocean color in coastal waters? *J. Geophys. Res. Oceans* **2007**, *112*. [CrossRef]
40. Kirk, J. A theoretical analysis of the contribution of algal cells to the attenuation of light within natural waters I. General treatment of suspensions of pigmented cells. *New Phytol.* **1975**, *75*, 11–20. [CrossRef]
41. Morel, A.; Bricaud, A. Theoretical results concerning light absorption in a discrete medium, and application to specific absorption of phytoplankton. *Deep Sea Res.* **1981**, *28*, 1375–1393. [CrossRef]
42. Sathyendranath, S.; Lazzara, L.; Prieur, L. Variations in the spectral values of specific absorption of phytoplankton. *Limnol. Oceanogr.* **1987**, *32*, 403–415. [CrossRef]
43. Bricaud, A.; Bédhomme, A.; Morel, A. Optical properties of diverse phytoplanktonic species: Experimental results and theoretical interpretation. *J. Plankton Res.* **1988**, *10*, 851–873. [CrossRef]
44. Ahn, Y.H.; Bricaud, A.; Morel, A. Light backscattering efficiency and related properties of some phytoplankters. *Deep Sea Res. Part A Oceanogr. Res. Pap.* **1992**, *39*, 1835–1855. [CrossRef]
45. Bricaud, A.; Stramski, D. Spectral absorption coefficients of living phytoplankton and nonalgal biogenous matter: A comparison between the Peru upwelling areaand the Sargasso Sea. *Limnol. Oceanogr.* **1990**, *35*, 562–582. [CrossRef]
46. Le Quéré, C.; Harrison, S.P.; Prentice, I.C.; Buitenhuis, E.T.; Aumont, O.; Bopp, L.; Claustre, H.; Cotrim Da Cunha, L.; Geider, R.; Giraud, X.; et al. Ecosystem dynamics based on plankton functional types for global ocean biogeochemistry models. *Glob. Chang. Biol.* **2005**, *11*, 2016–2040.
47. Nair, A.; Sathyendranath, S.; Platt, T.; Morales, J.; Stuart, V.; Forget, M.H.; Devred, E.; Bouman, H. Remote sensing of phytoplankton functional types. *Remote Sens. Environ.* **2008**, *112*, 3366–3375. [CrossRef]
48. IOCCG. Phytoplankton Functional Types from Space. In *Reports of the International Ocean Colour Coordinating Group*; IOCCG: Dartmouth, NS, Canada, 2014.
49. Dutkiewicz, S.; Hickman, A.; Jahn, O.; Gregg, W.; Mouw, C.; Follows, M. Capturing optically important constituents and properties in a marine biogeochemical and ecosystem model. *Biogeosciences* **2015**, *12*, 4447–4481. [CrossRef]

50. Moutier, W.; Duforet-Gaurier, L.; Thyssen, M.; Loisel, H.; Meriaux, X.; Courcot, L.; Dessailly, D.; Reve, A.H.; Gregori, G.; Alvain, S.; et al. Evolution of the scattering properties of phytoplankton cells from flow cytometry measurements. *PLoS ONE* **2017**, *12*, e0181180. [CrossRef]

51. Stramski, D.; Morel, A. Optical properties of photosynthetic picoplankton in different physiological states as affected by growth irradiance. *Deep Sea Res.* **1990**, *37*, 245–266. [CrossRef]

52. Reynolds, R.A.; Stramski, D.; Kiefer, D.A. The effect of nitrogren limitation on the absorption and scattering properties of the marine diatom Thalassiosira pseudonana. *Limnol. Oceanogr.* **1997**, *42*, 881–892. [CrossRef]

53. Stramski, D.; Sciandra, A.; Claustre, H. Effects of temperature, nitrogen, and light limitation on the optical properties of the marine diatom Thalassiosira pseudonana. *Limnol. Oceanogr.* **2002**, *47*, 392–403. [CrossRef]

54. Stramski, D.; Reynolds, R.A. Diel variations in the optical properties of a marine diatom. *Limnol. Oceanogr.* **1993**, *38*, 1347–1364. [CrossRef]

55. Stramski, D.; Shalapyonok, A.; Reynolds, R.A. Optical characterization of the oceanic unicellular cyanobacterium Synechococcus grown under a day-night cycle in natural irradiance. *J. Geophys. Res. Oceans* **1995**, *100*, 13295–13307. [CrossRef]

56. Sauer, M.J.; Roesler, C.; Werdell, P.; Barnard, A. Under the hood of satellite empirical chlorophyll a algorithms: Revealing the dependencies of maximum band ratio algorithms on inherent optical properties. *Opt. Express* **2012**, *20*, 20920–20933. [CrossRef] [PubMed]

57. Sathyendranath, S.; Brewin, R.J.; Jackson, T.; Mélin, F.; Platt, T. Ocean-colour products for climate-change studies: What are their ideal characteristics? *Remote Sens. Environ.* **2017**, *203*, 125–138. [CrossRef]

58. Mobley, C.D.; Stramski, D. Effects of microbial particles on oceanic optics: Methodology for radiative transfer modeling and example simulations. *Limnol. Oceanogr.* **1997**, *42*, 550–560. [CrossRef]

59. Stramski, D.; Mobley, C.D. Effects of microbial particles on oceanic optics: A database of single-particle optical properties. *Limnol. Oceanogr.* **1997**, *42*, 538–549. [CrossRef]

60. Morel, A.; Prieur, L. Analysis of variations in ocean color. *Limnol. Oceanogr.* **1977**, *22*, 709–722. [CrossRef]

61. Hoepffner, N.; Sathyendranath, S. Effect of pigment composition on absorption properties of phytoplankton. *Mar. Ecol. Prog. Ser.* **1991**, 11–23. [CrossRef]

62. Olson, R.; Zettler, E.; Anderson, O. Discrimination of eukaryotic phytoplankton cell types from light scatter and autofluorescence properties measured by flow cytometry. *Cytom. Part A* **1989**, *10*, 636–643. [CrossRef] [PubMed]

63. Matthews, M.W.; Bernard, S. Using a two-layered sphere model to investigate the impact of gas vacuoles on the inherent optical properties of M. aeruginosa. *Biogeosciences* **2013**, *10*, 8139–8157. [CrossRef]

64. Smith, M.E.; Lain, L.R.; Bernard, S. An optimized Chlorophyll a switching algorithm for MERIS and OLCI in phytoplankton-dominated waters. *Remote Sens. Environ.* **2018**, *215*, 217–227. [CrossRef]

65. Lain, L.R.; Bernard, S.; Evers-King, H. Biophysical modelling of phytoplankton communities from first principles using two-layered spheres: Equivalent Algal Populations (EAP) model. *Opt. Express* **2014**, *22*, 16745–16758. [CrossRef] [PubMed]

66. Stramski, D.; Boss, E.; Bogucki, D.; Voss, K. The role of seawater constituents in light backscattering in the ocean. *Prog. Oceanogr.* **2004**, *61*, 27–56. [CrossRef]

67. Toll, J.S. Causality and the Dispersion Relation: Logical Foundations. *Phys. Rev.* **1956**, *104*, 1760–1770. [CrossRef]

68. Aas, E. Refractive index of phytoplankton derived from its metabolite composition. *J. Plankton Res.* **1996**, *18*, 2223–2249. [CrossRef]

69. Bernard, S.; Probyn, T.; Barlow, R. Measured and modelled optical properties of particulate matter in the southern Benguela. *S. Afr. J. Sci.* **2001**, *97*, 410–420.

70. Stramski, D. Refractive index of planktonic cells as a measure of cellular carbon and chlorophyll a content. *Deep Sea Res. Part I Oceanogr. Res. Pap.* **1999**, *46*, 335–351. [CrossRef]

71. Johnsen, G.; Samset, O.; Granskog, L.; Sakshaug, E. In vivo absorption characteristics in 10 classes of bloom-forming phytoplankton: Taconomic characteristics and responses to photoadaptation by means of discriminant and HPLC analysis. *Mar. Ecol. Prog. Ser.* **1994**, 149–157. [CrossRef]

72. Bricaud, A.; Babin, M.; Morel, A.; Claustre, H. Variability in the chlorophyll-specific absorption coefficients of natural phytoplankton: Analysis and parameterization. *J. Geophys. Res. Oceans* **1995**, *100*, 13321–13332. [CrossRef]

73. Constable, A.J.; Melbourne-Thomas, J.; Corney, S.P.; Arrigo, K.R.; Barbraud, C.; Barnes, D.; Bindoff, N.L.; Boyd, P.W.; Brandt, A.; Costa, D.P.; et al. Climate change and Southern Ocean ecosystems I: How changes in physical habitats directly affect marine biota. *Glob. Chang. Biol.* **2014**, *20*, 3004–3025. [CrossRef] [PubMed]

74. Mtshali, T.N. *SANAE 55 Cruise Report*; South African National Antarctic Programme (SANAP): Cape Town, South Africa, 2016.

75. Del Castillo, C.E.; Miller, R.L. Horizontal and vertical distributions of colored dissolved organic matter during the Southern Ocean Gas Exchange Experiment. *J. Geophys. Res. Oceans* **2011**, *116*. [CrossRef]

76. Reynolds, R.; Stramski, D.; Mitchell, B. A chlorophyll-dependent semianalytical reflectance model derived from field measurements of absorption and backscattering coefficients within the Southern Ocean. *J. Geophys. Res. Oceans* **2001**, *106*, 7125–7138. [CrossRef]

77. Gustafson D.E., Jr.; Stoecker, D.K.; Johnson, M.D.; Van Heukelem, W.F.; Sneider, K. Cryptophyte algae are robbed of their organelles by the marine ciliate Mesodinium rubrum. *Nature* **2000**, *405*, 1049. [CrossRef] [PubMed]

78. Stramska, M.; Stramski, D.; Mitchell, B.G.; Mobley, C.D. Estimation of the absorption and backscattering coefficients from in water radiometric measurements. *Limnol. Oceanogr.* **2000**, *45*, 628–641. [CrossRef]

79. Oubelkheir, K.; Claustre, H.; Bricaud, A.; Babin, M. Partitioning total spectral absorption in phytoplankton and colored detrital material contributions. *Limnol. Oceanogr. Methods* **2007**, *5*, 384–395. [CrossRef]

80. Chami, M.; McKee, D.; Leymarie, E.; Khomenko, G. Influence of the angular shape of the volume-scattering function and multiple scattering on remote sensing reflectance. *Appl. Opt.* **2006**, *45*, 9210–9220. [CrossRef] [PubMed]

81. Gordon, H.R. Atmospheric correction of ocean color imagery in the Earth Observing System era. *J. Geophys. Res. Atmos.* **1997**, *102*, 17081–17106. [CrossRef]

82. Boyd, P.; Ellwood, M. The biogeochemical cycle of iron in the ocean. *Nat. Geosci.* **2010**, *10*, 675–682. [CrossRef]

83. Bernard, S.; Pitcher, G.; Evers-King, H.; Robertson, L.; Matthews, M.; Rabagliati, A.; Balt, C. Ocean Colour Remote Sensing of Harmful Algal Blooms in the Benguela System. In *Remote Sensing of the African Seas*; Springer: Berlin, Germany, 2014; pp. 185–203.

84. Devred, E.; Sathyendranath, S.; Stuart, V.; Maass, H.; Ulloa, O.; Platt, T. A two-component model of phytoplankton absorption in the open ocean: Theory and applications. *J. Geophys. Res. Oceans* **2006**, *111*. [CrossRef]

85. Tan, H.; Oishi, T.; Tanaka, A.; Doerffer, R. Accurate estimation of the backscattering coefficient by light scattering at two backward angles. *Appl. Opt.* **2015**, *54*, 7718–7733. [CrossRef] [PubMed]

86. Harmel, T.; Hieronymi, M.; Slade, W.; Röttgers, R.; Roullier, F.; Chami, M. Laboratory experiments for inter-comparison of three volume scattering meters to measure angular scattering properties of hydrosols. *Opt. Express* **2016**, *24*, 234–256. [CrossRef] [PubMed]

87. Bricaud, A.; Claustre, H.; Ras, J.; Oubelkheir, K. Natural variability of phytoplanktonic absorption in oceanic waters: Influence of the size structure of algal populations. *J. Geophys. Res. Oceans* **2004**, *109*. [CrossRef]

88. Dierssen, H.M.; Kudela, R.M.; Ryan, J.P. Red and black tides: Quantitative analysis of water-leaving radiance and perceived color for phytoplankton, colored dissolved organic matter, and suspended sediments. *Limnol. Oceanogr.* **2006**, *51*, 2646–2659. [CrossRef]

89. Quirantes, A.; Bernard, S. Light-scattering methods for modelling algal particles as a collection of coated and/or nonspherical scatterers. *J. Quant. Spectrosc. Radiat. Transf.* **2006**, *100*, 315–324. [CrossRef]

90. Ishikawa, A.; Wright, S.W.; van den Enden, R.; Davidson, A.T.; Marchant, H.J. Abundance, size structure and community composition of phytoplankton in the Southern Ocean in the austral summer 1999/2000. *Pol. Biosci.* **2002**, *15*, 11–26.

91. Wright, S.; Thomas, D.; Marchant, H.; Higgins, H.; Mackey, M.; Mackey, D. Analysis of phytoplankton of the Australian sector of the Southern Ocean: Comparisons of microscopy and size frequency data with interpretations of pigment HPLC data using the\'CHEMTAX\'matrix factorisation program. *Mar. Ecol. Prog. Ser.* **1996**, *144*, 285–298. [CrossRef]

92. Wright, S.W.; Jeffrey, S. Fucoxanthin pigment markers of marine phytoplankton analysed by HPLC and HPTLC. *Mar. Ecol. Prog. Ser.* **1987**, 259–266. [CrossRef]

93. Bricaud, A.; Morel, A.; Prieur, L. Absorption by dissolved organic matter of the sea (yellow substance) in the UV and visible domains. *Limnol. Oceanogr.* **1981**, *26*, 43–53. [CrossRef]

94. Roesler, C.S.; Perry, M.J.; Carder, K.L. Modeling in situ phytoplankton absorption from total absorption spectra in productive inland marine waters. *Limnol. Oceanogr.* **1989**, *34*, 1510–1523. [CrossRef]

95. Bernard, S.; Probyn, T.A.; Shillington, F.A. Towards the validation of SeaWiFS in southern African waters: The effects of gelbstoff. *S. Afr. J. Mar. Sci.* **1998**, *19*, 15–25. [CrossRef]

96. Roesler, C.S.; Perry, M.J. In situ phytoplankton absorption, fluorescence emission, and particulate backscattering spectra determined from reflectance. *J. Geophys. Res.* **1995**, *100*, 13279–13294. [CrossRef]

97. Chami, M.; Shybanov, E.B.; Khomenko, G.A.; Lee, M.E.G.; Martynov, O.V.; Korotaev, G.K. Spectral variation of the volume scattering function measured over the full range of scattering angles in a coastal environment. *Appl. Opt.* **2006**, *45*, 3605–3619. [CrossRef] [PubMed]

98. Twardowski, M.S.; Boss, E.; Macdonald, J.B.; Pegau, W.S.; Barnard, A.H.; Zaneveld, J.R.V. Model for estimating bulk refractive index from the optical backscattering ratio and the implications for understanding particle composition in case I and case II waters. *J. Geophys. Res* **2001**, *106*, 14129–14142. [CrossRef]

99. Boss, E.; Pegau, W.; Lee, M.; Twardowski, M.; Shybanov, E.; Korotaev, G.; Baratange, F. Particulate backscattering ratio at LEO 15 and its use to study particle composition and distribution. *J. Geophys. Res. Oceans* **2004**, *109*. [CrossRef]

100. Holm-Hansen, O.; Lorenzen, C.J.; Holmes, R.W.; Strickland, J.D. Fluorometric determination of chlorophyll. *ICES J. Mar. Sci.* **1965**, *30*, 3–15. [CrossRef]

101. Behrenfeld, M.J.; Westberry, T.K.; Boss, E. Satellite-detected fluorescence reveals global physiology of ocean phytoplankton. *Biogeosciences* **2009**, 779–794. [CrossRef]

102. Ostrowska, M.; Woźniak, B.; Dera, J. Modelled quantum yields and energy efficiency of fluorescence photosynthesis and heat production by phytoplankton in the World Ocean. *Oceanologia* **2012**, *54*, 565–610. [CrossRef]

103. Greene, R.M.; Geider, R.J.; Kolber, Z.; Falkowski, P.G. Iron-induced changes in light harvesting and photochemical energy conversion processes in eukaryotic marine algae. *Plant Physiol.* **1992**, *100*, 565–575. [CrossRef]

![applied sciences logo] *applied sciences*

MDPI

Article

Progress in Forward-Inverse Modeling Based on Radiative Transfer Tools for Coupled Atmosphere-Snow/Ice-Ocean Systems: A Review and Description of the AccuRT Model

Knut Stamnes [1,*], Børge Hamre [2,†], Snorre Stamnes [3,†], Nan Chen [1,†], Yongzhen Fan [1,†], Wei Li [1,†], Zhenyi Lin [1,†] and Jakob Stamnes [2,†]

1   Stevens Institute of Technology, Hoboken, NJ 07030, USA; nchen@stevens.edu (N.C.); yfan3021@gmail.com (Y.F.); wli4@stevens.edu (W.L.); lzhenyi@stevens.edu (Z.L.)
2   University of Bergen, 5020 Bergen, Norway; Borge.Hamre@uib.no (B.H.); Jakob.Stamnes@uib.no (J.S.)
3   NASA LARC, Hampton, VA 23681, USA; snorre.a.stamnes@nasa.gov
*   Correspondence: Knut.Stamnes@stevens.edu; Tel.: +1-201-216-8194
†   These authors contributed equally to this work.

Received: 14 June 2018; Accepted: 3 September 2018; Published: 19 December 2018

Abstract: A tutorial review is provided of forward and inverse radiative transfer in coupled atmosphere-snow/ice-water systems. The coupled system is assumed to consist of two adjacent horizontal slabs separated by an interface across which the refractive index changes abruptly from its value in air to that in ice/water. A comprehensive review is provided of the inherent optical properties of air and water (including snow and ice). The radiative transfer equation for unpolarized as well as polarized radiation is described and solutions are outlined. Several examples of how to formulate and solve inverse problems encountered in environmental optics involving coupled atmosphere-water systems are discussed in some detail to illustrate how the solutions to the radiative transfer equation can be used as a forward model to solve practical inverse problems.

Keywords: vector radiative transfer; polarization; coupled systems; atmosphere; ocean; forward modeling; inverse problems

---

1. Introduction

Reliable, accurate, and efficient modeling of electromagnetic radiation transport in turbid media has important applications in studies of Earth's climate by remote sensing. For example, such modeling is needed to develop forward-inverse methods used to quantify types and concentrations of aerosol and cloud particles in the atmosphere, as well as dissolved organic and particulate biogeochemical matter in lakes, rivers, coastal water, and open-ocean water, and to simulate the performance of remote sensing detectors deployed on aircraft, balloons, and satellites. Accurate radiative transfer (RT) modeling is also required to compute irradiances and scalar irradiances that are used to compute warming/cooling and photolysis rates in the atmosphere, solar energy deposition in the cryosphere including frozen fresh water (lakes and rivers), sea ice, and glaciers, as well as primary production rates in the water. Finally, RT modeling is needed to compute the Stokes vector describing the polarization state of the radiation field, which is desired in many remote sensing applications.

Accurate, efficient, and easy-to-use radiative transfer (RT) simulation tools are important because they (i) can be used to generate irradiances as well as total and polarized radiances (including degree of polarization) at any location and direction; (ii) will provide accurate results for given input parameters and specified inherent optical properties (IOPs); (iii) will lead to significant progress in research areas such as remote sensing algorithm development, climate research, and other atmospheric and hydrologic applications.

Available tools for atmospheric applications include: (i) SBDART [1], Streamer [2], and LibRadtran (www.libradtran.org), which all apply to the atmosphere only; there is no coupling to an underlying surface consisting of e.g., solid (snow/ice) or liquid water; (ii) Hydrolight, which applies to water only, provides water-leaving radiance, but not top-of-the-atmosphere (TOA) radiance; there is no coupling to the atmosphere (assumed to be a boundary condition). To remedy this situation a new Accurate Radiative Transfer (AccuRT) tool was developed to facilitate well-tested and robust RT simulations in coupled systems consisting of two slabs with different refractive indices. Please note that we here use the word "water" generically to describe the solid phase (i.e., snow and ice) as well as the liquid phase. The AccuRT tool accounts for reflection and transmission at the interface between the two slabs, and allows each slab to be divided into a sufficient number of layers to resolve the variation in the IOPs with depth.

Notation

Radiative transfer practitioners in the atmosphere, liquid water (ocean, lakes, rivers) and cryosphere (snow/ice) communities use different nomenclatures and terminologies. This situation can be confusing and frustrating to students and researchers addressing interdisciplinary problems in environmental optics. In this paper, we will adopt the notation of [3]. Letting z be the vertical position in the plane-parallel medium under consideration, and letting Θ be the scattering angle, we will denote the:

1. the absorption coefficient $[\text{m}^{-1}]$ by the letter $\alpha(z)$;
2. the scattering coefficient $[\text{m}^{-1}]$ by the letter $\beta(z)$;
3. the extinction coefficient $[\text{m}^{-1}]$ by the letter $\gamma(z) = \alpha(z) + \beta(z)$;
4. the single-scattering albedo by $\varpi(z) = \beta(z)/(\alpha(z) + \beta(z))$;
5. the volume scattering function $[\text{m}^{-1}\text{sr}^{-1}]$ by $\text{vsf}(z, \cos\Theta, \phi)$ and the related scattering phase function (dimensionless) by $p(z, \cos\Theta, \phi)$;
6. the scattering phase matrix (dimensionless) by $\mathbf{P}_S(\Theta)$ in the Stokes vector representation $\mathbf{I}_S = [I, Q, U, V]^T$ and by $\mathbf{P}(\Theta)$ in the Stokes vector representation $\mathbf{I} = [I_\parallel, I_\perp, U, V]^T$.

The corresponding notation used in the Ocean Optics community is a instead of α, b instead of β, and c instead of γ. Since α, β, and γ are the three first letters in the Greek alphabet it should be easy to recall the connection with a, b, and c.

This tutorial review is organized as follows. In Section 2 the input parameters needed to describe the coupled system are specified. Then in Section 3 we describe the inherent optical properties (IOPs) of the two adjacent coupled slabs, consisting of absorption and scattering coefficients as well as scattering phase functions. In addition, one needs the scattering phase matrix for polarized radiation. For unpolarized radiation only one element of this matrix, namely the scattering phase function, is required. These IOPs appear in the radiative transfer equation (RTE) described in Section 4, where we also review how to formulate and solve the RTE for unpolarized as well as polarized radiation. In Section 5 we provide several examples of how the solution of the forward problem discussed in Section 4 can be used to solve the corresponding inverse problem. In the forward problem, the IOPs are assumed to be known, so that the solution of the RTE provides the total (and polarized) radiances. To solve the inverse problem we ask: given the measured total (and/or polarized) radiances can we determine the IOPs? The inverse problem is generally much more difficult to solve than the forward problem. It can be formulated as a classical, nonlinear minimization problem, which can be solved in an iterative manner. We will also demonstrate how neural networks can be used to help tackle the inverse problem in a reliable and efficient manner. Finally, in Section 6 we briefly discuss some remaining problems in Ocean Optics, while in Section 7 we provide a brief summary.

2. Input and Output Parameters for the Forward Radiative Transfer Problem

2.1. Input Parameters

The following input parameters must be specified (i) the physical properties of each of the two slabs that constitute the coupled system, (ii) the radiative energy input at the top of the upper slab (top-of-the-atmosphere, TOA), and (iii) the boundary conditions at the bottom of the lower slab (water bottom). Each of the two slabs is assumed to be a plane-parallel, vertically stratified structure in which the scattering and absorption properties, i.e., the IOPs are defined in Section 3. To resolve changes in the IOPs as a function of vertical position z, each slab can be divided into several adjacent layers such that the IOPs are constant within each layer, but allowed to vary from one layer to the next. The impact of a wind-roughened air-water interface is described in Section 5.3.

To specify the IOPs, we will use the concept of materials, which are radiatively significant constituents in the atmosphere-water system. Examples of such materials are atmospheric gases, aerosols, clouds, snow, ice, pure water and water impurities. These materials can be designed to account for the wavelength dependence of the IOPs so that all one needs to do is to decide which of them to include in each layer. More specifically the following input parameters must be specified: (i) solar beam irradiance [W·m^{-2}]; (ii) wavelength range, number of center wavelengths and widths [nm]; (iii) solar zenith angle(s) in degrees; (iv) the number of "discrete ordinate streams" used to solve the radiative transfer equation (RTE) as described in Section 4; (v) IOPs of materials used. Specifications required for the upper slab include (i) layer boundaries, ground-level altitude (sea-level is default); (ii) atmospheric type; (iii) aerosol particle types (e.g., a bimodal lognormal volume distribution) for each layer as described in Section 3.4.2; (iv) cloud particles as described in Section 3.4.3; (v) snow particles as described in Section 3.5. Specifications required for the lower slab include (i) layer boundaries; (ii) ice material as described in Section 3.5; (iii) refractive index as a function of wavelength in the lower slab (water)—set to "one" in the upper slab (atmosphere); (iv) water impurities; (v) water bottom albedo.

2.2. Output Parameters

Once the input parameters above have been specified, the solution of the RTE as described in Section 4, will provide two types of output, namely, (i) irradiances and mean intensities (scalar irradiances in Ocean Optics terminology) at specified vertical positions in the coupled system; (ii) total and polarized radiances in desired directions at specified vertical positions in the coupled system.

3. Inherent Optical Properties (IOPs)

3.1. General Definitions

An *inherent optical property* (IOP) depends only on the medium itself, and is independent of the ambient light field within the medium [4]. An *apparent optical property* (AOP) depends also on the illumination, i.e., on light propagating in particular directions inside and outside the medium (Apparent optical properties (1) depend both on the medium (the IOPs) and on the geometric (directional) structure of the polarized radiance distribution, and (2) display enough regular features and stability to be useful descriptors of a water body [4]. Hence, a radiance or an irradiance would satisfy only the first part of the definition, while a radiance reflectance or irradiance reflectance, obtained by division of the radiance or the upward irradiance by the downward irradiance, would satisfy also the second part of the definition.).

Two important IOPs are the absorption coefficient $\alpha(z)$ and the scattering coefficient $\beta(z)$ defined as [3]

$$\alpha(z) = \frac{1}{I^i}\left(\frac{dI^\alpha}{dz}\right), \qquad \beta(z) = \frac{1}{I^i}\left(\frac{dI^\beta}{dz}\right) \qquad [\text{m}^{-1}]. \tag{1}$$

Here I^i is the incident radiance entering a volume element $dV = dA dz$ of the medium of cross sectional area dA and thickness dz, and $dI^\alpha > 0$ and $dI^\beta > 0$, respectively, are the radiances that are absorbed and scattered in all directions as the light propagates the distance dz along the direction of the incident light. The extinction coefficient is given by $\gamma(z) = \alpha(z) + \beta(z)$, and the single-scattering albedo is defined as $\varpi(z) \equiv \beta(z)/\gamma(z)$.

The angular distribution of the scattered light is given in terms of the *volume scattering function* (vsf), which is defined as

$$\mathrm{vsf}(z, \hat{\Omega}', \hat{\Omega}) = \frac{1}{I^i} \frac{d^2 I^\beta}{dz d\omega} = \frac{1}{I^i} \frac{d}{dz}\left(\frac{dI^\beta}{d\omega}\right) \qquad [\mathrm{m}^{-1}\,\mathrm{sr}^{-1}]. \tag{2}$$

Here $\hat{\Omega}'$ and $\hat{\Omega}$ are unit vectors, and $d^2 I^\beta$ is the radiance scattered from an incident direction $\hat{\Omega}'$ into a cone of solid angle $d\omega$ around the direction $\hat{\Omega}$ as the light propagates the distance dz along the direction $\hat{\Omega}'$ (see Figure 1). The plane spanned by $\hat{\Omega}'$ and $\hat{\Omega}$ is called the *scattering plane*, and the *scattering angle* Θ is given by $\cos\Theta = \hat{\Omega}' \cdot \hat{\Omega}$. Integration of Equation (2) over all scattering directions yields

$$\beta(z) = \frac{1}{I^i} \frac{d}{dz} \int_{4\pi} \left(\frac{dI^\beta}{d\omega}\right) d\omega = \frac{1}{I^i}\left(\frac{dI^\beta}{dz}\right)$$
$$= \int_{4\pi} \mathrm{vsf}(z, \hat{\Omega}', \hat{\Omega}) d\omega = \int_0^{2\pi}\int_0^{\pi} \mathrm{vsf}(z, \cos\Theta, \phi) \sin\Theta d\Theta d\phi \qquad [\mathrm{m}^{-1}] \tag{3}$$

where Θ and ϕ are respectively the polar angle and the azimuth angle in a spherical coordinate system in which the polar axis is along $\hat{\Omega}'$. As indicated in Equation (3), the volume scattering function $[\mathrm{vsf}(z, \cos\Theta, \phi)]$ is generally a function of both Θ and ϕ, but for randomly oriented scatterers one may assume that the scattering potential is spherically symmetric implying that there is no dependence on azimuth, so that $\mathrm{vsf} = \mathrm{vsf}(z, \cos\Theta)$. Then one finds, with $x = \cos\Theta$

$$\beta(z) = 2\pi \int_0^{\pi} \mathrm{vsf}(z, \cos\Theta) \sin\Theta d\Theta = 2\pi \int_{-1}^{1} \mathrm{vsf}(z, x) dx \qquad [\mathrm{m}^{-1}]. \tag{4}$$

A normalized vsf, denoted by $p(z, \cos\Theta)$ and referred to hereafter as the *scattering phase function*, may be defined as follows

$$p(z, \cos\Theta) = 4\pi \frac{\mathrm{vsf}(z, \cos\Theta)}{\int_{4\pi} \mathrm{vsf}(z, \cos\Theta) d\omega} = \frac{\mathrm{vsf}(z, x)}{\frac{1}{2}\int_{-1}^{1}\mathrm{vsf}(z, x) dx} \tag{5}$$

so that

$$\frac{1}{4\pi}\int_{4\pi} p(z, \cos\Theta) d\omega = \frac{1}{2}\int_{-1}^{1} p(z, x) dx = 1. \tag{6}$$

The scattering phase function has the following physical interpretation. Given that a scattering event has occurred, $p(z, \cos\Theta) d\omega/4\pi$ is the probability that a light beam traveling in the direction $\hat{\Omega}'$ is scattered into a cone of solid angle $d\omega$ around the direction $\hat{\Omega}$.

The scattering phase function $[p(z, \cos\Theta)]$ describes the angular distribution of the scattered light, while the scattering coefficient $\beta(z)$ describes the total amount of scattered light integrated over all scattering directions. A convenient measure of the "shape" of the scattering phase function is the average over all scattering directions (weighted by $p(z, \cos\Theta)$) of the cosine of the scattering angle Θ, i.e.,

$$g(z) = \langle\cos\Theta\rangle = \frac{1}{4\pi}\int_{4\pi} p(z, \cos\Theta) \cos\Theta d\omega$$
$$= \frac{1}{2}\int_0^{\pi} p(z, \cos\Theta) \cos\Theta \sin\Theta d\Theta = \frac{1}{2}\int_{-1}^{1} p(z, x) x dx. \tag{7}$$

The average cosine $g(z)$ is called the *asymmetry factor* of the scattering phase function. Equation (7) yields complete forward scattering if $g = 1$, complete backward scattering if $g = -1$, and $g = 0$ if $p(z, \cos \Theta)$ is symmetric about $\Theta = 90°$. Thus, isotropic scattering also gives $g = 0$. Similarly, the probability of scattering into the backward hemisphere, is given by the backscattering ratio (or backscatter fraction) b, defined as

$$b(z) = \frac{1}{2} \int_{\pi/2}^{\pi} p(z, \cos \Theta) \, \sin \Theta \, d\Theta = \frac{1}{2} \int_0^1 p(z, -x) dx. \tag{8}$$

The scattering phase function $p(z, \cos \Theta)$ depends on the refractive index as well as the size and shape of the scattering particles, and will thus depend on the physical situation and the practical application of interest. Two different scattering phase functions, which are useful in practical applications, are discussed below.

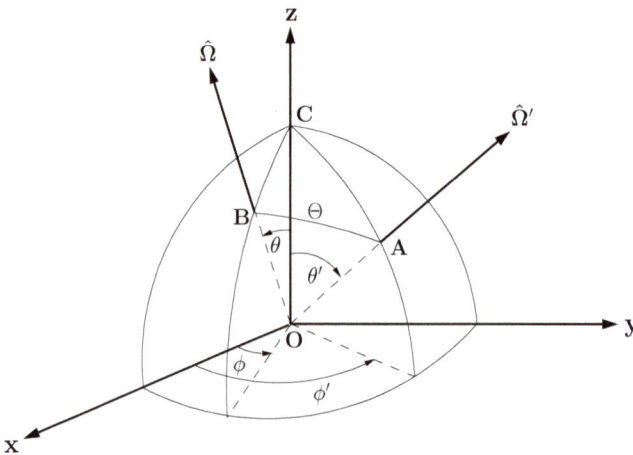

Figure 1. Coordinate system for scattering by a volume element at **O**. The points **C**, **A** and **B** are located on the unit sphere. The incident light beam with Stokes vector $\mathbf{I}_S^{inc}$ is in direction **OA**(θ', ϕ') with unit vector $\hat{\Omega}'$, the scattered beam with Stokes vector $\mathbf{I}_S^{sca}$ is in direction **OB**(θ, ϕ) with unit vector $\hat{\Omega}$ [5].

3.1.1. Rayleigh Scattering Phase Function

When the size d of the scatterers is small compared with the wavelength of light $(d < \frac{1}{10}\lambda)$, the Rayleigh scattering phase function gives a good description of the angular distribution of the scattered light. The Rayleigh scattering phase function for unpolarized light is given by

$$p(\cos \Theta) = \frac{3}{3+f}(1 + f \cos^2 \Theta) \tag{9}$$

where the parameter $f = \frac{1-\rho}{1+\rho}$, and ρ is the depolarization ratio, attributed to the anisotropy of the scatterer (molecule) [6–8]. Originally this scattering phase function was derived for light radiated by an electric dipole [9]. Since the Rayleigh scattering phase function is symmetric about $\Theta = 90°$, the asymmetry factor is $g = 0$. If the Rayleigh scattering phase function is expanded in Legendre polynomials, the expansion coefficients χ_ℓ [see Equation (117) below] are simply given by $\chi_0 = 1$, $\chi_1 = 0$, $\chi_2 = \frac{2f}{5(3+f)}$, and $\chi_\ell = 0$ for $\ell > 2$.

Using $\rho = 0.0286$ for air at 500 nm [10], we get $f = \frac{1-\rho}{1+\rho} = 0.944$, and using $\rho = 0.039$ [11,12] for water, we get $f = \frac{1-\rho}{1+\rho} = 0.925$. Hence, for Rayleigh scattering, the scattering phase function moments become:

- $\chi_0 = 1$, $\chi_1 = 0$, $\chi_2 = 0.0957$ and $\chi_\ell = 0$ for $\ell > 2$ for air, and
- $\chi_0 = 1$, $\chi_1 = 0$, $\chi_2 = 0.0943$, and $\chi_\ell = 0$ for $\ell > 2$ for water.

3.1.2. Henyey-Greenstein Scattering Phase Function

Henyey and Greenstein [13] proposed the one-parameter scattering phase function given by (oppressing the dependence on the position z)

$$p(\cos\Theta) = \frac{1 - g^2}{(1 + g^2 - 2g \cos\Theta)^{3/2}} \tag{10}$$

where the parameter g is the asymmetry factor defined in Equation (7). The Henyey–Greenstein (HG) scattering phase function has no physical basis, but is very useful for describing a highly scattering medium, such as turbid water or sea ice, for which the actual scattering phase function is unknown. The HG scattering phase function is convenient for Monte Carlo simulations and other numerical calculations because it has an analytical form. In deterministic plane-parallel RT models it is also very convenient because the addition theorem of spherical harmonics can be used to expand the scattering phase function in a series of Legendre polynomials [3,14], as reviewed in Section 4.1.1. For the HG scattering phase function, the expansion coefficients χ_ℓ in this series [see Equation (117) below] are simply given by $\chi_\ell = g^\ell$, where $g = \chi_1$ is the asymmetry factor defined in Equation (7). The HG scattering phase function is useful for scatterers with sizes comparable to or larger than the wavelength of light. Although the HG scattering phase function is easy to use, it is not as realistic as the Fournier-Forand scattering phase function discussed in Section 3.6.1.

3.2. Scattering Phase Matrix

The theoretical development of vector radiative transfer theory may start with the Stokes vector representation $\mathbf{I} = [I_\parallel, I_\perp, U, V]^T$, where the superscript ^{T} denotes the transpose. In terms of the complex transverse electric field components of the radiation field $E_\parallel = |E_\parallel|e^{-i\epsilon_1}$ and $E_\perp = |E_\perp|e^{-i\epsilon_2}$, these Stokes vector components are given by:

$$\begin{aligned}
I_\parallel &= E_\parallel E_\parallel^* \\
I_\perp &= E_\perp E_\perp^* \\
U &= 2|E_\parallel||E_\perp|\cos\delta \\
V &= 2|E_\parallel||E_\perp|\sin\delta
\end{aligned} \tag{11}$$

where $\delta = \epsilon_1 - \epsilon_2$. The connection between this Stokes vector representation, $\mathbf{I} = [I_\parallel, I_\perp, U, V]^T$, and the more commonly used representation $\mathbf{I}_S = [I, Q, U, V]^T$, where $I = I_\parallel + I_\perp$ and $Q = I_\parallel - I_\perp$, is given by:

$$\mathbf{I}_S = \mathbf{D}\mathbf{I} \tag{12}$$

where

$$\mathbf{D} = \begin{pmatrix} 1 & 1 & 0 & 0 \\ 1 & -1 & 0 & 0 \\ 0 & 0 & 1 & 0 \\ 0 & 0 & 0 & 1 \end{pmatrix}, \quad \mathbf{D}^{-1} = \frac{1}{2}\begin{pmatrix} 1 & 1 & 0 & 0 \\ 1 & -1 & 0 & 0 \\ 0 & 0 & 2 & 0 \\ 0 & 0 & 0 & 2 \end{pmatrix}. \tag{13}$$

The degree of polarization is defined as

$$p = [Q^2 + U^2 + V^2]^{1/2}/I \tag{14}$$

so that $0 \leq p \leq 1$, where $p = 1$ corresponds to completely polarized light and $p = 0$ to natural (unpolarized) light. The degree of circular polarization is defined as

$$p_c = V/I \tag{15}$$

the degree of linear polarization as

$$p_l = [Q^2 + U^2]^{1/2}/I \tag{16}$$

and alternatively, when $U = 0$ as

$$p_l = \frac{|Q|}{I} = \frac{|I_\perp - I_\parallel|}{I_\perp + I_\parallel}. \tag{17}$$

The transverse electric field vector $[E_\parallel, E_\perp]^T$ of the scattered field can be obtained in terms of the transverse field vector $[E_{\parallel 0}, E_{\perp 0}]^T$ of the incident field by a linear transformation:

$$\begin{pmatrix} E_\parallel \\ E_\perp \end{pmatrix} = \mathbf{A} \begin{pmatrix} E_{\parallel 0} \\ E_{\perp 0} \end{pmatrix}$$

where $\mathbf{A}$ is a 2×2 matrix, referred to as the amplitude scattering matrix, which includes a $1/r$ dependence of the scattered field. The corresponding linear transformation connecting the Stokes vectors of the incident and scattered fields in the scattering plane is called the Mueller matrix (in the case of a single scattering event). For scattering by a small volume containing an ensemble of particles, the ensemble-averaged Mueller matrix is referred to as the Stokes scattering matrix $\mathbf{F}$. Finally, when transforming from the scattering plane to a fixed laboratory frame, the corresponding matrix is referred to as the scattering phase matrix $\mathbf{P}$.

3.2.1. Stokes Vector Representation $\mathbf{I}_S = [I, Q, U, V]^T$

The scattering geometry is illustrated in Figure 1. The plane **AOB**, defined as the scattering plane, is spanned by the directions of propagation of the incident parallel beam with Stokes vector $\mathbf{I}_S^{inc}$ and the scattered parallel beam with Stokes vector $\mathbf{I}_S^{sca}$. Here the subscript S pertains to the Stokes vector representation $\mathbf{I}_S = [I, Q, U, V]^T$. The scattered radiation, represented by the Stokes vector $\mathbf{I}_S^{sca}$, is related to the incident radiation, represented by the Stokes vector $\mathbf{I}_S^{inc}$, by a 4×4 scattering matrix [see Equations (18) and (19) below] and two rotations are required to properly connect the two Stokes vectors as explained below. We describe the Stokes vector of the incident beam in terms of two unit vectors $\hat{\ell}'$ and $\hat{r}'$, which are normal to one another and to the unit vector $\hat{\Omega}' = \hat{r}' \times \hat{\ell}'$ along the propagation direction of the incident beam. Similarly, we describe the Stokes vector of the scattered beam in terms of two unit vectors $\hat{\ell}$ and $\hat{r}$, which are normal to one another and to the unit vector $\hat{\Omega} = \hat{r} \times \hat{\ell}$ along the propagation direction of the scattered beam. The unit vector $\hat{\ell}'$ is along the direction of $\mathbf{E}_\parallel'$ of the incident beam and lies in the meridian plane of that beam, which is defined as the plane **OAC** in Figure 1. Similarly, the unit vector $\hat{\ell}$ is along the direction of $\mathbf{E}_\parallel$ of the scattered beam and lies in the meridian plane of that beam, which is defined as the plane **OBC** in Figure 1. For the incident beam, the unit vector $\hat{\ell}'$, may be defined to be tangent at the point **A** to the unit circle passing through the points **A** and **C** in Figure 1. For the scattered beam, the unit vector $\hat{\ell}$ may be defined to be tangent at the point **B** to the unit circle passing through the points **B** and **C** in Figure 1. For either beam, its meridian plane acts as a plane of reference for the Stokes vector, so that the point **A** in Figure 1 is the starting point for the unit vector $\hat{\Omega}' = \hat{r}' \times \hat{\ell}'$ along the direction of propagation of the incident beam, and the point **B** in Figure 1 is the starting point for the unit vector $\hat{\Omega} = \hat{r} \times \hat{\ell}$ along the direction of propagation of the scattered beam.

As explained above, the Mueller matrix describes scattering by a single particle, and for scattering by a small volume of particles, the ensemble-averaged Mueller matrix is referred to as the Stokes scattering matrix $\mathbf{F}_S$. If any of the following conditions are fulfilled [15] (i) each particle in the volume element has a plane of symmetry, and the particles are randomly oriented, (ii) each volume element contains an equal number of particles and their mirror particles in random orientation, (iii) the particles are much smaller than the wavelength of the incident light, then the Stokes scattering matrix in the $\mathbf{I}_S = [I, Q, U, V]^T$ representation has the following form

$$\mathbf{F}_S(\Theta) = \begin{bmatrix} a_1(\Theta) & b_1(\Theta) & 0 & 0 \\ b_1(\Theta) & a_2(\Theta) & 0 & 0 \\ 0 & 0 & a_3(\Theta) & b_2(\Theta) \\ 0 & 0 & -b_2(\Theta) & a_4(\Theta) \end{bmatrix}. \tag{18}$$

Each of the six independent matrix elements in Equation (18) depends on the scattering angle Θ, and will in general also depend on the position in the medium. For spherical particles, the matrix in Equation (18) simplifies, since $a_1 = a_2$ and $a_3 = a_4$, so that only four independent elements remain.

As already mentioned, two rotations are required to connect the Stokes vector of the scattered radiation to that of the incident radiation. As illustrated in Figure 1, the first rotation is from the meridian plane **OAC**, associated with the Stokes vector $\mathbf{I}_S^{inc}$, into the scattering plane **OAB**, whereas the second rotation is from the scattering plane **OAB** into the meridian plane **OBC**, associated with the Stokes vector $\mathbf{I}_S^{sca}$. Hence, the Stokes vector for the scattered radiation is given by [16]

$$\mathbf{I}_S^{sca} = \mathbf{R}_S(\pi - i_2)\mathbf{F}_S(\Theta)\mathbf{R}_S(-i_1)\mathbf{I}_S^{inc} \equiv \mathbf{P}_S(\Theta)\mathbf{I}_S^{inc}. \tag{19}$$

The matrix $\mathbf{R}_S$ is called the Stokes rotation matrix. It represents a rotation in the clockwise direction with respect to an observer looking into the direction of propagation, and can be written as $(0 \leq \omega \leq 2\pi)$

$$\mathbf{R}_S(\omega) = \begin{bmatrix} 1 & 0 & 0 & 0 \\ 0 & \cos(2\omega) & -\sin(2\omega) & 0 \\ 0 & \sin(2\omega) & \cos(2\omega) & 0 \\ 0 & 0 & 0 & 1 \end{bmatrix}. \tag{20}$$

Hence, according to Equation (19), the scattering phase matrix, which connects the Stokes vector of the scattered radiation to that of the incident radiation, is obtained from the Stokes scattering matrix $\mathbf{F}_S(\Theta)$ in Equation (18) by

$$\mathbf{P}_S(\theta', \phi'; \theta, \phi) = \mathbf{R}_S(\pi - i_2)\mathbf{F}_S(\Theta)\mathbf{R}_S(-i_1) = \mathbf{R}_S(-i_2)\mathbf{F}_S(\Theta)\mathbf{R}_S(-i_1) \tag{21}$$

where $\mathbf{R}_S$ is the rotation matrix described in Equation (20) [16], and $\mathbf{R}_S(\pi - i_2) = \mathbf{R}_S(-i_2)$ since the rotation matrix is periodic with a period π.

According to Equation (19) (see also Figure 1), the Stokes vector $\mathbf{I}_S^{inc}$ of the incident parallel beam must be multiplied by the rotation matrix $\mathbf{R}_S(-i_1)$ before it is multiplied by the Stokes scattering matrix $\mathbf{F}_S(\Theta)$, whereafter it must be multiplied by the rotation matrix $\mathbf{R}_S(\pi - i_2)$. These matrix multiplications are carried out explicitly in some radiative transfer (RT) models including Monte Carlo simulations, while they are implicitly taken care of in other RT models such as the adding-doubling method [17] and the discrete ordinate method [18,19] which use the expansion of the scattering phase matrix in generalized spherical functions [20,21] as discussed in Section 3.2.3.

Carrying out the matrix multiplications in Equation (21) one finds:

$$\mathbf{P}_S(\Theta) = \begin{bmatrix} a_1 & b_1 C_1 & -b_1 S_1 & 0 \\ b_1 C_2 & C_2 a_2 C_1 - S_2 a_3 S_1 & -C_2 a_2 S_1 - S_2 a_3 C_1 & -b_2 S_2 \\ b_1 S_2 & S_2 a_2 C_1 + C_2 a_3 S_1 & -S_2 a_2 S_1 + C_2 a_3 C_1 & -b_2 C_2 \\ 0 & -b_2 S_1 & -b_2 C_1 & a_4 \end{bmatrix} \tag{22}$$

where $a_i = a_i(\Theta), i = 1, \ldots, 4, b_i = b_i(\Theta), i = 1, 2$, and

$$\begin{aligned} C_1 &= \cos 2i_1, & C_2 &= \cos 2i_2 \tag{23} \\ S_1 &= \sin 2i_1, & S_2 &= \sin 2i_2. \tag{24} \end{aligned}$$

A comparison of Equations (18) and (22) shows that only the corner elements of $\mathbf{F}_S(\Theta)$ remain unchanged by the rotations of the reference planes. The (1,1)-element of the scattering phase matrix $\mathbf{P}_S(\Theta)$ (and of the Stokes scattering matrix $\mathbf{F}_S(\Theta)$) is the scattering phase function. Since also the (4,4)-element of the scattering phase matrix remains unchanged by the rotations, the state of circular polarization of the incident light does not affect the intensity of the scattered radiation after one scattering event.

To compute $\mathbf{P}_S(\theta', \phi'; \theta, \phi)$ given by Equation (21) we must relate the angles θ', ϕ', θ, and ϕ on the left side with the angles i_1, i_2, and Θ on the right side. Using spherical geometry, we may apply the cosine rule for Θ, θ, and θ' successively, in Figure 1, to obtain ($u = \cos\theta, u' = \cos\theta'$) [5]

$$\cos\Theta = uu' + (1 - u^2)^{1/2}(1 - u'^2)^{1/2}\cos(\phi' - \phi) \tag{25}$$

$$\cos i_1 = \frac{-u + u'\cos\Theta}{(1 - u'^2)^{1/2}(1 - \cos^2\Theta)^{1/2}} \tag{26}$$

$$\cos i_2 = \frac{-u' + u\cos\Theta}{(1 - u^2)^{1/2}(1 - \cos^2\Theta)^{1/2}}. \tag{27}$$

The trigonometric functions for the double angles can be obtained by using

$$\cos 2i = 2\cos^2 i - 1 \tag{28}$$

and

$$\sin 2i = 2\sin i \cos i \tag{29}$$

or

$$\sin 2i = \begin{cases} 2(1 - \cos^2 i)^{1/2}\cos i & \text{if } 0 < \phi' - \phi < \pi \\ -2(1 - \cos^2 i)^{1/2}\cos i & \text{if } \pi < \phi' - \phi < 2\pi \end{cases} \tag{30}$$

where i is i_1 or i_2. We now have all the information needed to compute the scattering phase matrix [see Equation (21)] as a function of the three variables u, u' and $\phi' - \phi$:

$$\mathbf{P}_S(u', u, \phi' - \phi) = \mathbf{R}_S(-i_2)\mathbf{F}_S(\Theta)\mathbf{R}_S(-i_1).$$

If there is no difference in azimuth (i.e. $\phi' - \phi = 0$), then the meridian planes of the incident and scattered beams in Figure 1 coincide with the scattering plane. Hence there is no need to rotate the reference planes ($\mathbf{R}(-i_2)$ and $\mathbf{R}(-i_1)$ both reduce to the identity matrix), so that

$$\mathbf{P}_S(u', u, 0) = \mathbf{P}_S(u', u, \pi) = \mathbf{F}_S(\Theta). \tag{31}$$

It follows from Equation (25) that the phase matrix is invariant to three basic changes in the polar angles u' and u and azimuthal angles ϕ' and ϕ which leave the scattering angle unaltered: (i) changing the signs of u and u' simultaneously: $\mathbf{P}_S(-u', -u, \phi' - \phi) = \mathbf{P}_S(u', u, \phi' - \phi)$, (ii) interchange of u and

u': $\mathbf{P}_S(u', u, \phi' - \phi) = \mathbf{P}_S(u, u', \phi' - \phi)$ (iii) interchange of ϕ and ϕ': $\mathbf{P}_S(u', u, \phi' - \phi) = \mathbf{P}_S(u', u, \phi - \phi')$. Also, if the b_2-element in Equation (22) is zero, the circular polarization component decouples from the other three components. Then, the Stokes parameter V is scattered independently of the others, according to the phase function $a_4(\Theta)$, and the remaining part of the scattering phase matrix referring to I, Q, and U becomes a 3×3 matrix:

$$\mathbf{P}_S(\Theta) = \begin{bmatrix} a_1 & b_1 C_1 & -b_1 S_1 \\ b_1 C_2 & C_2 a_2 C_1 - S_2 a_3 S_1 & -C_2 a_2 S_1 - S_2 a_3 C_1 \\ b_1 S_2 & S_2 a_2 C_1 + C_2 a_3 S_1 & -S_2 a_2 S_1 + C_2 a_3 C_1 \end{bmatrix}. \tag{32}$$

Finally, in a plane-parallel or slab geometry, there is no azimuth-dependence for light beams traveling in directions perpendicular to the slab (either up or down). Thus, if either the incident or the scattered beam travels in a perpendicular direction, we may use the meridian plane of the other beam as a reference plane for both beams. Since this plane coincides with the scattering plane, Equation (31) applies in this situation too.

3.2.2. Stokes Vector Representation $\mathbf{I} = [I_\parallel, I_\perp, U, V]^T$

The Stokes vector $\mathbf{I} = [I_\parallel, I_\perp, U, V]^T$ is related to $\mathbf{I}_S = [I, Q, U, V]^T$ by

$$\mathbf{I}_S = \mathbf{D}\mathbf{I} \tag{33}$$

where $\mathbf{D}$ is given by Equation (13), so that $I = I_\parallel + I_\perp$, and $Q = I_\parallel - I_\perp$. Denoting the Stokes vector obtained after a rotation by

$$\mathbf{I}'_S = \mathbf{R}_S(\omega)\mathbf{I}_S \tag{34}$$

we find

$$\mathbf{I}' = \mathbf{D}^{-1}\mathbf{I}'_S = \mathbf{D}^{-1}\mathbf{R}_S(\omega)\mathbf{I}_S = \mathbf{D}^{-1}\mathbf{R}_S(\omega)\mathbf{D}\mathbf{I} = \mathbf{R}(\omega)\mathbf{I}. \tag{35}$$

Hence, the rotation matrix for the Stokes vector in the representation $\mathbf{I} = [I_\parallel, I_\perp, U, V]^T$ becomes:

$$\mathbf{R}(\omega) = \mathbf{D}^{-1}\mathbf{R}_S(\omega)\mathbf{D} = \begin{bmatrix} \cos^2 \omega & \sin^2 \omega & -\frac{1}{2}\sin(2\omega) & 0 \\ \sin^2 \omega & \cos^2 \omega & \frac{1}{2}\sin(2\omega) & 0 \\ \sin(2\omega) & -\sin(2\omega) & \cos(2\omega) & 0 \\ 0 & 0 & 0 & 1 \end{bmatrix}. \tag{36}$$

The scattering phase matrix $\mathbf{P}(\Theta)$ in the Stokes vector representation $\mathbf{I} = [I_\parallel, I_\perp, U, V]^T$ is related to scattering phase matrix $\mathbf{P}_S(\Theta)$ in the Stokes vector representation $\mathbf{I}_S = [I, Q, U, V]^T$ by

$$\mathbf{P}(\Theta) = \mathbf{D}^{-1}\mathbf{P}_S(\Theta)\mathbf{D}. \tag{37}$$

Similarly, the Stokes scattering matrix $\mathbf{F}(\Theta)$ associated with the Stokes vector representation $\mathbf{I} = [I_\parallel, I_\perp, U, V]^T$ is related to the Stokes scattering matrix $\mathbf{F}_S(\Theta)$ in Equation (18) by

$$\mathbf{F}(\Theta) = \mathbf{D}^{-1}\mathbf{F}_S(\Theta)\mathbf{D} = \begin{pmatrix} \frac{1}{2}(a_1 + a_2 + 2b_1) & \frac{1}{2}(a_1 - a_2) & 0 & 0 \\ \frac{1}{2}(a_1 - a_2) & \frac{1}{2}(a_1 + a_2 - 2b_1) & 0 & 0 \\ 0 & 0 & a_3 & b_2 \\ 0 & 0 & -b_2 & a_4 \end{pmatrix}. \tag{38}$$

For Rayleigh scattering with parameter $f = \frac{1-\rho}{1+\rho}$, where ρ is the depolarization factor defined in Equation (43), the Stokes scattering matrix in the Stokes vector representation $\mathbf{I}_S = [I, Q, U, V]^T$ is given by [16,22]

$$
F_S(\Theta) = \frac{3}{3+f}
\begin{bmatrix}
1 + f\cos^2\Theta & -f\sin^2\Theta & 0 & 0 \\
-f\sin^2\Theta & f(1+\cos^2\Theta) & 0 & 0 \\
0 & 0 & 2f\cos\Theta & 0 \\
0 & 0 & 0 & (3f-1)\cos\Theta
\end{bmatrix}.
\tag{39}
$$

For the first scattering event of unpolarized light, only the (1,1)-element of Equation (39) matters, and leads to the scattering phase function given by Equation (9).

In the Stokes vector representation $I = [I_{\parallel}, I_{\perp}, U, V]^T$, the corresponding Stokes scattering matrix for Rayleigh scattering becomes (using Equations (38) and (39) [16]):

$$
F(\Theta) = \frac{3}{2(1+2\zeta)}
\begin{pmatrix}
\cos^2\Theta + \zeta\sin^2\Theta & \zeta & 0 & 0 \\
\zeta & 1 & 0 & 0 \\
0 & 0 & (1-\zeta)\cos\Theta & 0 \\
0 & 0 & 0 & (1-3\zeta)\cos\Theta
\end{pmatrix}
\tag{40}
$$

where $\zeta = \rho/(2-\rho) = \frac{1-f}{1+3f}$.

From Equation (40) we see that for an incident beam of natural unpolarized light given by $I^{inc} = [I_{\parallel}^{inc}, I_{\perp}^{inc}, U^{inc}, V^{inc}]^T = [\frac{1}{2}I^{inc}, \frac{1}{2}I^{inc}, 0, 0]^T$, the scattered intensities in the plane parallel and perpendicular to the scattering plane are obtained by carrying out the multiplication $I^{sca} = F(\Theta)I^{inc}$:

$$
I_{\parallel}^{sca} \propto \frac{3}{4(1+2\zeta)}[2\zeta + (1-\zeta)\cos^2\Theta]I^{inc}
\tag{41}
$$

$$
I_{\perp}^{sca} \propto \frac{3}{4(1+2\zeta)}[(1+\zeta)]I^{inc}.
\tag{42}
$$

Thus, for unpolarized incident light, the scattered light at right angles ($\Theta = 90°$) to the direction of incidence defines the depolarization ratio:

$$
\rho \equiv \left(\frac{I_{\parallel}^{sca}}{I_{\perp}^{sca}}\right)_{\Theta=90°} = \frac{2\zeta}{1+\zeta}
\tag{43}
$$

whereas the degree of linear polarization becomes [Equation (17)]:

$$
p_l = \frac{I_{\perp} - I_{\parallel}}{I_{\perp} + I_{\parallel}} = \frac{(1-\zeta)(1-\cos^2\Theta)}{1+3\zeta+(1-\zeta)\cos^2\Theta} \rightarrow \frac{1-\zeta}{1+3\zeta} = \frac{1-\rho}{1+\rho} = f \text{ as } \Theta \rightarrow 90°.
$$

3.2.3. Generalized Spherical Functions—The "Greek Constants"

For unpolarized radiation, only the $a_1(\Theta)$ element of the Stokes scattering matrix Equation (18) is relevant, and this element is the scattering phase function given by Equation (5) in general, and by Equation (9) for Rayleigh scattering. As discussed below, the scattering phase function can be expanded in Legendre polynomials (see Equation (116)), enabling one to express it as a Fourier cosine series (see Equation (115)).

In a similar manner, the scattering phase matrix can be expanded in generalized spherical functions. In the Stokes vector representation $I_S = [I, Q, U, V]^T$, the scattering phase matrix is $P_S(\Theta) = P_S(u', u; \phi' - \phi)$ with $u = \cos\theta$, θ being the polar angle after scattering, and $u' = \cos\theta'$, θ' being the polar angle prior to scattering. Similarly, ϕ and ϕ' are the azimuth angles after and prior to scattering, respectively. To accomplish the expansion in generalized spherical functions, the scattering phase matrix is first expanded in a $(M+1)$-term Fourier series in the azimuth angle difference ($\Delta\phi = \phi' - \phi$):

$$P_S(u', u; \Delta\phi) = \sum_{m=0}^{M} \{P_c^m(u', u)\cos m(\Delta\phi) + P_s^m(u', u)\sin m(\Delta\phi)\} \tag{44}$$

where $P_c^m(u', u)$ and $P_s^m(u', u)$ are the coefficient matrices of the cosine and sine terms, respectively, of the Fourier series.

An addition theorem for the generalized spherical functions can be used to express the Fourier expansion coefficient matrices directly in terms of the expansion coefficients of the Stokes scattering matrix $\mathbf{F}_S(\Theta)$ [see Equation (18)] as follows [20,21,23]:

$$\begin{align}
\mathbf{P}_c^m(u', u) &= \mathbf{A}^m(u', u) + \Delta_{3,4}\mathbf{A}^m(u', u)\Delta_{3,4} \tag{45}\\
\mathbf{P}_s^m(u', u) &= \mathbf{A}^m(u', u)\Delta_{3,4} - \Delta_{3,4}\mathbf{A}^m(u', u) \tag{46}
\end{align}$$

where $\Delta_3 = \text{diag}(1, 1, -1, 1)$. The matrix $\mathbf{A}^m(u', u)$ is given by:

$$\mathbf{A}^m(u', u) = \sum_{\ell=m}^{M} \mathbf{P}_\ell^m(u)\Lambda_\ell\mathbf{P}_\ell^m(u') \tag{47}$$

where

$$\Lambda_\ell = \begin{pmatrix} \alpha_{1,\ell} & \beta_{1,\ell} & 0 & 0 \\ \beta_{1,\ell} & \alpha_{2,\ell} & 0 & 0 \\ 0 & 0 & \alpha_{3,\ell} & \beta_{2,\ell} \\ 0 & 0 & -\beta_{2,\ell} & \alpha_{4,\ell} \end{pmatrix} \tag{48}$$

and

$$a_1(\Theta) = \sum_{\ell=0}^{M} \alpha_{1,\ell}P_\ell^{0,0}(\cos\Theta) \tag{49}$$

$$a_2(\Theta) + a_3(\Theta) = \sum_{\ell=2}^{M} (\alpha_{2,\ell} + \alpha_{3,\ell})P_\ell^{2,2}(\cos\Theta) \tag{50}$$

$$a_2(\Theta) - a_3(\Theta) = \sum_{\ell=2}^{M} (\alpha_{2,\ell} - \alpha_{3,\ell})P_\ell^{2,-2}(\cos\Theta) \tag{51}$$

$$a_4(\Theta) = \sum_{\ell=0}^{M} \alpha_{4,\ell}P_\ell^{0,0}(\cos\Theta) \tag{52}$$

$$b_1(\Theta) = \sum_{\ell=2}^{M} \beta_{1,\ell}P_\ell^{0,2}(\cos\Theta) \tag{53}$$

$$b_2(\Theta) = \sum_{\ell=2}^{M} \beta_{2,\ell}P_\ell^{0,2}(\cos\Theta). \tag{54}$$

Here the so-called "Greek constants" $\alpha_{j,\ell}$ and $\beta_{j,\ell}$ are expansion coefficients, and $a_j(\Theta)$ and $b_j(\Theta)$ are the elements of the Stokes scattering matrix $\mathbf{F}_S(\Theta)$ in Equation (18). An example of Greek constants for Rayleigh scattering is provided in Table 1 (see [24]):

Table 1. Expansion Coefficients for Rayleigh Scattering.

ℓ	$\alpha_{1,\ell}$	$\alpha_{2,\ell}$	$\alpha_{3,\ell}$	$\alpha_{4,\ell}$	$\beta_{1,\ell}$	$\beta_{2,\ell}$
0	1	0	0	0	0	0
1	0	0	0	$3d/2$	0	0
2	$c/2$	$3c$	0	0	$\sqrt{3/2}c$	0

where

$$c = \frac{2(1-\rho)}{2+\rho} \qquad d = \frac{2(1-2\rho)}{2+\rho}$$

and ρ is the depolarization ratio given by Equation (43).

The matrix $\mathbf{P}_\ell^m(u)$ occurring in Equation (47) is defined as:

$$\mathbf{P}_\ell^m(u) = \begin{pmatrix} P_\ell^{m,0}(u) & 0 & 0 & 0 \\ 0 & P_\ell^{m,+}(u) & P_\ell^{m,-}(u) & 0 \\ 0 & P_\ell^{m,-}(u) & P_\ell^{m,+}(u) & 0 \\ 0 & 0 & 0 & P_\ell^{m,0}(u) \end{pmatrix} \tag{55}$$

where

$$P_\ell^{m,\pm}(u) = \frac{1}{2}[P_\ell^{m,-2}(u) \pm P_\ell^{m,2}(u)] \tag{56}$$

and the functions $P_\ell^{m,0}(u)$ and $P_\ell^{m,\pm2}(u)$ are the generalized spherical functions. More details about these functions and how they are computed are available in Appendix B of the book by Hovenier et al. [5].

We note that in the scalar (unpolarized) case all the components of the Stokes scattering matrix $\mathbf{F}_S(\Theta)$ [see Equation (18)] are zero except for $a_1(\Theta)$, and:

$$a_1(\Theta) = \sum_{\ell=0}^{M} \alpha_{1,\ell}(\tau) P_\ell^{0,0}(\cos\Theta) \equiv p(\tau, \cos\Theta) \approx \sum_{\ell=0}^{M} (2\ell+1)\chi_\ell(\tau) P_\ell(\cos\Theta) \tag{57}$$

since $P_\ell^{0,0}(\cos\Theta) \equiv P_\ell(\cos\Theta)$, where $P_\ell(\cos\Theta)$ is the Legendre polynomial of order ℓ, and $\alpha_{1,\ell}(\tau) \equiv (2\ell+1)\chi_\ell(\tau)$. Note also that the expansion coefficients given above are for the scattering phase matrix $\mathbf{P}_S(\Theta)$, which relates the incident and scattered Stokes vectors in the representation $\mathbf{I}_S = [I, Q, U, V]^T$.

3.3. IOPs for a Size Distribution of Particles

Particles encountered in nature consist of a variety of chemical compositions, sizes, and shapes. The chemical composition determines the refractive index of the particle, and unless the composition is the same throughout the particle, the refractive index will depend on location inside the particle. The computation of IOPs for such a collection of particles requires solutions of Maxwell's equations for electromagnetic radiation interacting with an inhomogeneous, non-spherical particle of a given size. Then one needs to integrate over size and shape for particles of a given chemical composition, and finally average over the particle composition. To avoid having to deal with this complexity it is frequently assumed that the particles are homogeneous with a constant refractive index, and that the shape can be taken to be spherical. Even with these assumptions, one still needs to deal with the variety of particles sizes encountered in nature.

For a spherical particle with a specified radius and refractive index, Mie theory (and its numerical implementation) may be used to generate IOPs for a single particle. Thus, if we have computed the IOPs for a single spherical particle with specified refractive index and a given size, we may compute the absorption and scattering coefficients and the scattering phase function for a polydispersion of particles by integrating over the particle size distribution (PSD):

$$\alpha_p(\lambda) = \int_{r_{min}}^{r_{max}} \alpha_n(\lambda, r)n(r)dr = \int_{r_{min}}^{r_{max}} \pi r^2 Q'_\alpha(\lambda, r)n(r)dr \tag{58}$$

$$\beta_p(\lambda) = \int_{r_{min}}^{r_{max}} \beta_n(\lambda, r)n(r)dr = \int_{r_{min}}^{r_{max}} \pi r^2 Q'_\beta(\lambda, r)n(r)dr \tag{59}$$

$$p_p(\lambda, \Theta) = \frac{\int_{r_{min}}^{r_{max}} p(\lambda, \Theta, r)n(r)dr}{\int_{r_{min}}^{r_{max}} n(r)dr} \tag{60}$$

where $n(r)$ is the PSD and $\alpha_n(\lambda, r)$, $\beta_n(\lambda, r)$, and $p(\lambda, \Theta, r)$ are the absorption cross section, the scattering cross section, and the scattering phase function per particle of radius r. The absorption or scattering "efficiency", $Q'_\alpha(r)$ or $Q'_\beta(r)$, is defined as the ratio of the absorption or scattering cross section for a spherical particle of radius r to the geometrical cross section πr^2. The scattering phase function $p(\lambda, \Theta, r)$ in Equation (60) is the $a_1(\Theta)$ element of the Stokes scattering matrix [Equation (18)]. Since a Mie code can be used to compute all elements of the Stokes scattering matrix in Equation (18), we may use an expression analogous to Equation (60) to carry out the integration over the PSD for each of the matrix elements.

IOPs for a Mixture of Different Particle Types

Consider a particle mixture consisting of a total of N particles per unit volume in a layer of thickness Δz, and let $N = \sum_i n_i$ and $f_i = n_i/N$, where n_i is the concentration and f_i the fraction of homogeneous particles (with fixed chemical composition or refractive index) of type labeled i. To compute IOPs for the mixture of particles, we define $\beta_{n,i}$ = scattering cross section, $\alpha_{n,i}$ = absorption cross section, $\gamma_{n,i} = \beta_{n,i} + \alpha_{n,i}$ = extinction cross section, and $\varpi_i = \beta_{n,i} n_i / \gamma_{n,i} n_i = \beta_i / \gamma_i$ = single-scattering albedo, where the subscript i stands for particle type. Weighting by number concentration may be used to create IOPs for the particle mixture. Thus, by combining the absorption and scattering cross sections, and the moments of the scattering phase matrix elements, one obtains the following IOPs for the mixture (subscript m stands for mixture):

$$\Delta \tau_m = \Delta z \sum_i n_i \gamma_{n,i} = \Delta z \sum_i \gamma_i = \gamma_m \Delta z \tag{61}$$

$$\varpi_m = \frac{\beta_m}{\gamma_m} = \frac{\sum_i \beta_{n,i} n_i}{\sum_i \gamma_{n,i} n_i} = \frac{\sum_i \beta_i}{\sum_i \gamma_i} = \frac{\sum_i \varpi_i \gamma_{n,i} f_i}{\sum_i \gamma_{n,i} f_i} \tag{62}$$

$$\chi_{m,\ell} = \frac{\sum_i \beta_{n,i} n_i \chi_{i,\ell}}{\sum_i \beta_{n,i} n_i} = N \frac{\sum_i \beta_{n,i} f_i \chi_{i,\ell}}{\beta_m} = \frac{\sum_i f_i \varpi_i \gamma_{n,i} \chi_{i,\ell}}{\sum_i f_i \beta_{n,i}} \tag{63}$$

where $\Delta \tau_m$ = layer optical depth; β_m = total scattering coefficient; γ_m = total extinction coefficient; ϖ_m = single-scattering albedo; and $\chi_{m,\ell}$ = scattering phase function expansion coefficient for the particle mixture. A mixing rule similar to Equation (63) may be used for each element of the scattering phase matrix.

3.4. Atmosphere IOPs

The stratified vertical structure of the bulk properties of an atmosphere is a consequence of hydrostatic balance. By equating pressure forces and gravitational forces and invoking the ideal gas law, one may derive the barometric law for the pressure $p(z)$ as function of altitude z above the surface z_0 [14]:

$$p(z) = p(z_0) \exp\left[-\int_{z_0}^{z} dz' / H(z') \right] \tag{64}$$

where $H(z) = kT(z)/\bar{M}g$ is the atmospheric scale height, $\bar{M}$ is the mean molecular weight, k is is Boltzmann's constant, g is the acceleration due to gravity, and $T(z)$ is the temperature. The ideal gas law allows one to write similar expressions for the bulk density $\rho(z)$ and the bulk concentration $n(z)$. Clearly, from a knowledge of the surface pressure $p(z_0)$ and the variation of the scale height $H(z)$ with height z, Equation (64) allows us to determine the bulk gas properties at any height. Equation (64) applies to well-mixed gases, but not to short-lived species such as ozone, which is chemically created and destroyed, or water, which undergoes phase changes on short time scales.

3.4.1. Gases in the Earth's Atmosphere

The total number of air molecules in a 1 m^2 wide vertical column extending from sea level to the top of the atmosphere is about 2.15×10^{29}. In comparison, the total amount of ozone (a trace gas) in the same vertical column is about 1.0×10^{23}. Anderson et al. [25] compiled six model atmospheres including, (i) the US Standard atmosphere 1976, (ii) tropical, (iii) midlatitude summer, (iv) midlatitude winter, (v) subarctic summer, and (vi) subarctic winter (see Appendix U of [14] for a numerical tabulation of these models). These atmospheric models contain profiles of temperature, pressure, and the concentrations of the main atmospheric constituents, molecular nitrogen (N_2) and molecular oxygen (O_2). In addition, they contain profiles of the concentrations of several trace gases including water vapor (H_2O), ozone O_3, carbon dioxide (CO_2), methane (CH_4), nitrous oxide (N_2O), and the chlorofluoromethanes (CFCs) in the Earth's atmosphere.

The clear atmosphere (no clouds or aerosols) molecular (Rayleigh) scattering coefficient can be expressed as

$$\sigma_{\text{Ray}}(\lambda, z) \equiv \sigma_{\text{Ray,n}} n(z) = \frac{32\pi^3 (m_r - 1)^2}{3\lambda^4 n(z)} \quad [\text{m}^{-1}] \tag{65}$$

where $n(z)$ is the bulk air concentration (see Equation (64)), m_r is the real part of the refractive index, and $\sigma_{\text{Ray,n}}$ is the Rayleigh scattering cross section. Please note that since m_r depends on wavelength, the Rayleigh scattering coefficient does not have an exact λ^{-4} dependence. For air, a convenient numerical formula for the Rayleigh scattering cross section (accurate to 0.3%) is given by [14]

$$\sigma_{\text{Ray,n}} = \lambda^{-4} \sum_{i=0}^{3} a_i \lambda^{-2i} \times 10^{-28} \quad [\text{cm}^2] \quad (0.205 < \lambda < 1.05 \ \mu\text{m})$$

where the coefficients are $a_0 = 3.9729066$, $a_1 = 4.6547659 \times 10^{-2}$, $a_2 = 4.5055995 \times 10^{-4}$, and $a_3 = 2.3229848 \times 10^{-5}$. The scattering phase function is given by Equation (9).

Computer codes like MODTRAN [26,27] have been developed to provide atmospheric transmittance and thereby absorption coefficients for all important atmospheric trace gases for a large variety of atmospheric conditions. In AccuRT we use a band model based on MODTRAN [28] to generate absorption coefficients and optical depths due to atmospheric gases including O_2, H_2O, CO_2, O_3, CH_4, and NO_2. An example of how to use this approach to deal with gaseous absorption in shortwave near infrared bands for an atmosphere overlying a snow surface is provided in [29].

3.4.2. Aerosol IOPs

If we know the size distribution and the refractive index of the aerosol particles, we may use available aerosol models to generate aerosol IOPs. For example, one may use the aerosol models employed in the Sea-viewing Wide Field-of-view Sensor (SeaWiFS) Database Analysis System (SeaDAS), and described by Ahmad et al. [30]. Alternatively, we may use the OPAC models [31]. For atmospheric correction of ocean color imagery, it is customary to assume a lognormal distribution of aerosol sizes [32]. Based on AERONET data [33,34], Ahmad et al. [30] adopted a bimodal lognormal volume size distribution:

$$v(r) = \frac{dV(r)}{dr} = \frac{1}{r} \frac{dV(\ln r)}{d \ln r} = \sum_{i=1}^{2} \frac{V_i}{\sqrt{2\pi}\sigma_i} \frac{1}{r} \exp\left[-\left(\frac{\ln r - \ln r_{vi}}{\sqrt{2}\sigma_i} \right)^2 \right] \tag{66}$$

where the subscript i represents the mode, V_i is the total volume of particles with mode i, r_{vi} is the mode radius, also called the volume geometric mean radius, and σ_i is the geometric standard deviation. Please note that since the numerator in the exponential of Equation (66), $\ln(r/r_{vi})$, is dimensionless, so is σ_i. Since

$$\int_0^\infty \frac{dr}{\sqrt{2\pi}\sigma} \frac{1}{r} \exp\left[-\left(\frac{\ln r - \ln r_v}{\sqrt{2}\sigma} \right)^2 \right] = 1$$

integration over all sizes for both modes, yields:

$$\int_0^\infty v(r)dr = V_1 + V_2 = V.$$

In terms of the number density (concentration), Equation (66) becomes

$$n(r) = \frac{dN(r)}{dr} = \frac{1}{r}\frac{dN(r)}{d(\ln r)} = \sum_{i=1}^{2} \frac{N_i}{\sqrt{2\pi}\sigma_i}\frac{1}{r}\exp\left[-\left(\frac{\ln r - \ln r_{ni}}{\sqrt{2}\sigma_i}\right)^2\right] \qquad (67)$$

where the number of particles N_i and the mean geometric (or mode) radius r_{ni} are related to V_i and r_{vi} as follows

$$\ln r_{ni} = \ln r_{vi} - 3\sigma_i^2 \qquad (68)$$

$$N_i = \frac{V_i}{\frac{4}{3}\pi r_{ni}^3}\exp(-4.5\sigma_i^2) \qquad (69)$$

and integration over all sizes for both modes, yields:

$$\int_0^\infty n(r)dr = N_1 + N_2 = N.$$

If we use the subscript $i = f$ to denote the fine mode, and the subscript $i = c$ to denote the coarse mode, we have $V = V_f + V_c$, and the volume fraction of fine mode particles becomes $f_v = V_f/V$.

Relationship between Effective Radius and Mode Radius

The particle size distribution may also be characterized by an effective radius

$$r_{\text{eff}} = \frac{\int_{r_{\min}}^{r_{\max}} n(r)r^3 dr}{\int_{r_{\min}}^{r_{\max}} n(r)r^2 dr} \qquad (70)$$

and an effective variance

$$v_{\text{eff}} = \frac{\int_{r_{\min}}^{r_{\max}} (r - r_{\text{eff}})^2 n(r)r^2 dr}{r_{\text{eff}}^2 \int_{r_{\min}}^{r_{\max}} n(r)r^2 dr} \qquad (71)$$

where r_{eff}^2 is included in the denominator of Equation (71) to make v_{eff} dimensionless [35]. The effective radius, r_{eff}, can be used to describe the IOPs in an approximate manner as will be discussed below for cloud as well as snow/ice materials. For a single mode, the lognormal size distribution is given by [see Equation (67)]

$$n(r) = \frac{dN(r)}{dr} = \frac{N}{\sqrt{2\pi}\sigma}\frac{1}{r}\exp\left[-\left(\frac{\ln r - \ln r_n}{\sqrt{2}\sigma}\right)^2\right]$$

where r_n is the mode radius, $n(r)$ is the number density or PSD in units of $[\text{m}^{-3} \cdot \text{m}^{-1}]$ and $N = \int_0^\infty n(r)dr \ [\text{m}^{-3}]$ is the total number of particles per unit volume since

$$\int_0^\infty \frac{dr}{\sqrt{2\pi}\sigma}\frac{1}{r}\exp\left[-\left(\frac{\ln r - \ln r_n}{\sqrt{2}\sigma}\right)^2\right] = 1. \qquad (72)$$

With the change of variable $x = \frac{\ln(r/r_n)}{\sqrt{2}\sigma}$, Equation (72) becomes

$$\frac{1}{\sqrt{\pi}}\int_{-\infty}^{+\infty}\exp(-x^2)dx = 1 \qquad (73)$$

and it can be shown that [35]:

$$r_{\text{eff}} = r_n \exp[2.5\sigma^2], \qquad (74)$$

and

$$v_{\text{eff}} = \exp\left[\sigma^2\right] - 1 \tag{75}$$

(see [3] for details).

Impact of Relative Humidity

A change in the relative humidity (RH) will affect bot the size and refractive index of a particle. The particle radius can be parameterized as a function of RH from the wet-to-dry mass ratio:

$$r(a_w) = r_0 \left[1 + \rho \frac{m_w(a_w)}{m_0}\right]^{1/3} \tag{76}$$

where the water activity a_w of a soluble aerosol at radius r [μm] can be expressed as

$$a_w = \text{RH} \exp\left[\frac{-2\sigma V_m}{R_w T} \frac{1}{r(a_w)}\right]. \tag{77}$$

Here r_0 is the dry particle radius (RH = 0), ρ is the particle density relative to that of water, $m_w(a_w)$ is the mass of condensed water, m_0 is the dry particle mass (RH = 0), σ is the surface tension on the wet surface, V_m is the specific volume of water, R_w is the gas constant for water vapor, and T is the absolute temperature [K] [36]. Similarly, the change in refractive index with RH can be determined from [36]

$$\tilde{m}_c = \tilde{m}_{c,w} + (\tilde{m}_{c,0} - \tilde{m}_{c,w})\left[\frac{r_0}{r_{\text{RH}}}\right]^3 \tag{78}$$

where $\tilde{m}_{c,w}$ and $\tilde{m}_{c,0}$ are the complex refractive indices of water and dry aerosols, respectively, and r_0 and r_{RH} are the radii of the aerosols in the dry state and at the given RH, respectively. From these formulas we note that the magnitude of the particle growth and the change of refractive index with increasing RH depend on the size r_0 of the dry aerosol but also on the type of aerosol through the water uptake [the ratio $m_w(a_w)/m_0$ in Equation (76)] [36–38].

A Mie code [39] is needed to compute the IOPs of aerosol particles [$Q'_\alpha(r), Q'_\beta(r)$, and $p_p(\lambda, \Theta, r)$ in Equations (58)–(60)], and numerical integration is required to evaluate the integrals over the lognormal size distributions to obtain $\alpha_p(\lambda), \beta_p(\lambda)$, and $p_p(\lambda, \Theta)$. For polarized radiation all elements of the scattering phase matrix as well as the Greek constants appearing in Equation (48) must be computed. For a bimodal lognormal volume size distribution [Equation (66)] one must specify the fine mode volume fraction $f_v = V_f/V$, where $V = V_f + V_c$, the volume mode radii r_{vf} and r_{vc} as well as the corresponding standard deviations σ_f and σ_c in addition to the refractive index of the particles relative to air.

In analogy to the liquid water content (see Equation (81) below), we may introduce the aerosol mass content (AMC) for each mode defined as

$$\text{AMC} = \rho_a \int_{r_{\text{min}}}^{r_{\text{max}}} \left(\frac{4\pi}{3}\right) r^3 n(r) dr \equiv \rho_a f_V \quad [\text{kg} \cdot \text{m}^{-3}] \tag{79}$$

where $n(r)$ is the aerosol size distribution [$\text{m}^{-3} \cdot \text{m}^{-1}$], ρ_a is the bulk aerosol density [$\text{kg} \cdot \text{m}^{-3}$], and f_V is the aerosol volume fraction (not to be confused with the fine mode volume fraction, f_v) given by:

$$f_V \equiv \int_{r_{\text{min}}}^{r_{\text{max}}} \left(\frac{4\pi}{3}\right) r^3 n(r) dr = \text{AMC}/\rho_a \quad \text{(dimensionless)}. \tag{80}$$

Typical values of atmospheric aerosol densities are $\rho_a \approx 1\,\text{g} \cdot \text{cm}^{-3} = 1 \times 10^6\,\text{g} \cdot \text{m}^{-3}$. Hence, an AMC value of $10^{-6}\,\text{g} \cdot \text{m}^{-3}$ would yield $f_V = 10^{-12}$.

3.4.3. Cloud IOPs

Clouds consist of liquid water droplets or ice (frozen water) particles. While liquid water droplets can be assumed to have spherical shape, ice crystals can have a variety of non-spherical shapes. If we assume for simplicity that all cloud particles consist of spherical water droplets or spherical ice particles, (For ice crystals, a spherical model may be unrealistic. For a good introduction to this topic, see the textbook by Wendisch and Yang [40].) we can use a Mie code to compute their IOPs because their refractive index is known. Hence, we may use Equations (58)–(60) to compute $\alpha_p(\lambda), \beta_p(\lambda)$, and $p_p(\lambda, \Theta)$.

The real part of the refractive index of pure water needed in the Mie computations may be taken from [41], while the imaginary part $\tilde{m}_{i,w}$ is calculated from the absorption coefficient $(\alpha_w(\lambda) = 4\pi\tilde{m}_{i,w}/\lambda)$ obtained from published data [42–44] for wavelengths between 340 and 700 nm, and from another source [45] for wavelengths between 720 and 900 nm.

It is customary to introduce the liquid water content (LWC) defined as

$$\text{LWC} \equiv \rho_w \int_{r_{\min}}^{r_{\max}} \left(\frac{4\pi}{3}\right) r^3 n(r) dr \equiv \rho_w \, f_V \qquad [\text{kg} \cdot \text{m}^{-3}] \tag{81}$$

where $n(r)$ is the cloud droplet size distribution [$\text{m}^{-3} \cdot \text{m}^{-1}$] and ρ_w is the liquid water mass density [$\text{kg} \cdot \text{m}^{-3}$] and f_V stands for the dimensionless liquid (cloud) particle volume fraction defined in a similar manner as AMC in Equation (80), i.e., $f_V = \text{LWC}/\rho_w$. For a liquid water cloud, a typical value for LWC is about 0.5 g $\cdot$ m^{-3}, implying that $f_V = 5 \times 10^{-7}$ for $\rho_w = 10^3$ kg $\cdot$ m^{-3}. In Equation (70) for the effective radius, the numerator is proportional to the concentration or LWC, while the denominator is related to the scattering coefficient:

$$\beta_c = \int_0^\infty dr (\pi r^2) Q_\beta(r) \, n(r) dr \qquad [\text{m}^{-1}].$$

If the size of the droplet is large compared to the wavelength λ, then $Q_\beta(r) \to 2$. Therefore, in the visible spectral range where $2\pi r/\lambda \gg 1$, we find:

$$\beta_c \approx \frac{3}{2} \frac{1}{\rho_w} \frac{\text{LWC}}{r_{\text{eff}}} = \frac{3}{2} \frac{f_V}{r_{\text{eff}}} \qquad [\text{m}^{-1}]. \tag{82}$$

For ice cloud particles assumed to be spherical in shape a similar expression for the scattering coefficient is obtained with f_V being the ice particle volume fraction. For a liquid water cloud with $f_V = 5 \times 10^{-7}$ and $r_{\text{eff}} = 5 \times 10^{-6}$ m, we get $\beta_c = \frac{3}{2}\frac{f_V}{r_{\text{eff}}} = 0.15\,\text{m}^{-1}$, and hence an optical thickness of 15 for a 100 m thick cloud layer.

3.5. Snow and Ice IOPs

3.5.1. General Approach

Assuming that snow grains and sea ice inclusions have spherical shape, we may obtain their IOPs from Mie computations, which require the refractive index and the size distribution of the particles as input. Then, the IOPs, i.e., the absorption and scattering coefficients and the scattering phase function, $\alpha_p(\lambda), \beta_p(\lambda)$, and $p_p(\lambda, \Theta)$, can be obtained from Equations (58)–(60). This approach leads to computed snow albedo values that agree surprisingly well with available observations [46,47]. The following reasons why large errors are not incurred by assuming spherical shape have been advocated by Craig Bohren as quoted elsewhere [48]: *The orientationally averaged extinction cross section of a convex particle that is large compared with the wavelength is one-half its surface area. The absorption cross section of a large, nearly transparent particle is proportional to its volume almost independent of its shape. The closer the real part of the particle's refractive index is to 1, the more irrelevant the particle shape.*

The asymmetry parameter of a large particle is dominated by near-forward scattering, which does not depend greatly on particle shape.

Hence, we may assume that snow grains and ice inclusions (air bubbles and brine pockets) consist of homogeneous spheres with a single-mode lognormal volume size distribution [see Equation (66)], and use the refractive index data base for ice compiled by [49]. Specifying the effective radius r_{eff} and the width of the distribution σ, one obtains the geometrical mean radius r_n from Equation (74). Since the complex refractive index is prescribed, r_n and σ constitute the only input required for a Mie code (see Section 3.4.2), which can be used to compute absorption and scattering coefficients as well the scattering phase function. One may choose to use only the first moment of the scattering phase function in conjunction with the Henyey-Greenstein scattering phase function because the Mie scattering phase function is unrealistic for non-spherical snow grains and ice inclusions.

3.5.2. Fast, yet Accurate Parameterization of Snow/Ice IOPs

Building on previous work [50–52], Stamnes et al. [53] created a generic tool for computing snow/ice IOPs (τ, ϖ, and g). This tool can be used to generate wavelength-dependent ice/snow IOPs from ice/snow physical parameters: real and imaginary parts of the ice/snow refractive index, brine pocket concentration and effective size (sea ice), air bubble concentration and effective size (sea ice), volume fraction and absorption coefficient of sea ice impurities, asymmetry factors for scattering by snow grains, brine pockets, and air bubbles, and sea ice thickness. We can compute $Q'_\alpha(r)$, $Q'_\beta(r)$, and $p_p(\lambda, \Theta, r)$ using a Mie code, but evaluation of Equations (58)–(60) requires knowledge of the particle size distribution $n(r)$, which is usually unknown. Equations (58)–(60) can be considerably simplified by making the following assumptions [53]:

- The particle distribution is characterized by an effective radius [Equation (70)], which obviates the need for an integration over r.
- The particles are weakly absorbing, so that [51]

$$Q'_\alpha(r) \equiv Q'_\alpha \approx \frac{16\pi \, r_{\text{eff}} \, \tilde{m}_{i,p}}{3\lambda} \frac{1}{m_{\text{rel}}} [m_{\text{rel}}^3 - (m_{\text{rel}}^2 - 1)^{3/2}] \tag{83}$$

where $\tilde{m}_{i,p}$ is the imaginary part of the refractive index of the particle, λ is the wavelength in vacuum, and $m_{\text{rel}} = \tilde{m}_{r,p} / \tilde{m}_{r,\text{med}}$ is the ratio of the real part of the refractive index of the particle ($\tilde{m}_{r,p}$) to that of the surrounding medium ($\tilde{m}_{r,\text{med}}$).

- The particles are large compared to the wavelength ($2\pi r/\lambda \gg 1$) which implies

$$Q'_\beta(r) \equiv Q'_\beta = 2. \tag{84}$$

The scattering phase function may be represented by the one-parameter Henyey-Greenstein scattering phase function [see Equation (10)], which depends only on the asymmetry factor defined in Equation (7).

With these assumptions, Equations (58)–(59) become:

$$\alpha_p(\lambda) = \alpha(\lambda) \frac{1}{m_{\text{rel}}} [1 - (m_{\text{rel}}^2 - 1)^{3/2}] f_V \tag{85}$$

$$\beta_p(\lambda) = \frac{3}{2} \frac{f_V}{r_{\text{eff}}}. \tag{86}$$

Here $\alpha(\lambda) = 4\pi \tilde{m}_{i,p}/\lambda$ is the absorption coefficient of the material of which the particle is composed, and $f_V \equiv \frac{4\pi}{3} \int n(r) r^3 dr \approx \frac{4}{3} \pi r_{\text{eff}}^3 n_e$, where n_e = number of particles per unit volume with radius r_{eff}. Since Equation (86) is identical to Equation (82), it is clear that f_V represents the volume fraction of the particles as defined in Equation (80).

For wavelengths $\lambda \leq 1.2$ µm, the absorption and scattering efficiency for snow grains, brine inclusions in sea ice, and air bubbles in ice may be parameterized by Equations (83) and (84), and the asymmetry factor g can be held constant with wavelength and set equal to 0.85, 0.89, and 0.997 for air bubbles, snow grains, and brine pockets, respectively. To extend the validity to NIR wavelengths, we may use the following modified parameterizations [53]:

$$Q_\alpha = 0.94[1 - \exp(-Q'_\alpha/0.94)]; \qquad Q_\beta = 2 - Q_\alpha; \qquad g = g_0^{(1-Q_\alpha)^{0.6}} \tag{87}$$

where Q'_α is given by Equation (83). Here g is the asymmetry factor of the scattering phase function, and g_0 is the asymmetry factor for non-absorbing particles. For large particles ($r > \sim 50$ µm) g_0 depends only on the real part of the refractive index. For a medium consisting of several absorbing and scattering constituents the total absorption and scattering efficiencies are just the sum of those due to the separate constituents. The optical thickness τ and single-scattering albedo ϖ for a slab of thickness h become [53]:

$$\tau = \pi r_{\text{eff}}^2 Nh(Q_\alpha + Q_\beta); \qquad \varpi = \frac{Q_\beta}{Q_\alpha + Q_\beta} \tag{88}$$

where N is the total number of particles per unit volume, and Q_α and Q_β are the total absorption and scattering efficiencies, each equal to the sum of those due to the separate constituents. These modified parameterizations work well for all wavelengths for Q_α, while for Q_β and g they work well for wavelengths shorter than about 2.8 µm, but deviate significantly from predictions by Mie theory for longer wavelengths. Thus, for wavelengths longer than 2.8 µm one should preferably use results from the computationally less efficient Mie theory. Note that for wavelengths shorter than 2.8 µm, where the parameterizations work well, the variations in $\tilde{m}_{r,p}$ and $\tilde{m}_{i,p}$ are large. Thus, one would expect these parameterizations to be valid for most types of large particles [53].

3.5.3. Impurities, Air Bubbles, Brine Pockets, and Snow

If the volume fraction of impurities within a snow grain or brine pocket is not too large, which is the case for typical situations occurring in nature, scattering by impurities can be ignored, so that their effects can be included by simply adding the imaginary part $\tilde{m}_{i,\text{imp}}$ of the refractive index for impurities to $\tilde{m}_{i,p}$ in Equation (83). For typical impurities in snow and ice, the wavelength dependence of $\tilde{m}_{i,\text{imp}}$ can be parameterized as [53]

$$\tilde{m}_{i,\text{imp}}(\lambda) = \tilde{m}_{i,\text{imp}}(\lambda_0) (\lambda_0/\lambda)^\eta \tag{89}$$

where η would be close to zero for black carbon, but larger for other impurities, and $\tilde{m}_{i,\text{imp}}(\lambda_0 = 440\,\text{nm})$ has values that depend on the type of impurity. Equation (89) is based on the observation that the absorption coefficient α of non-algal impurities tend to have a smooth increase towards shorter wavelengths [54–57], and α is connected to the imaginary part of the refractive index through $\alpha = 4\pi \tilde{m}_{i,\text{imp}}/\lambda$. For snow, the number of snow grain particles per unit volume is $N = \frac{1}{\frac{4}{3}\pi r_{\text{eff}}^3} \frac{\rho_s}{\rho_i}$, where r_{eff} is the effective particle radius, while ρ_s and ρ_i are the mass densities of snow and pure ice, respectively. The optical thickness and the single-scattering albedo can be calculated from Equations (87) and (88), using the refractive indices of pure ice [49] and impurities [Equation (89)].

We assume that sea ice consists of pure ice with embedded brine pockets, air bubbles, and impurities. To include the effects of the embedded components, we first calculate the absorption coefficient α for sea ice [53]

$$\alpha = \pi r_{\text{br}}^2 N_{\text{br}} Q_{\alpha,\text{br}} + \left[1 - \frac{4}{3}\pi r_{\text{br}}^3 N_{\text{br}} - \frac{4}{3}\pi r_{\text{bu}}^3 N_{\text{bu}}\right] \frac{4\pi(\tilde{m}_{i,p} + f_{\text{imp}}\tilde{m}_{i,\text{imp}})}{\lambda} \tag{90}$$

where f_{imp} is the volume fraction of impurities, N_{br} and N_{bu} are the number concentrations of brine pockets and air bubbles, respectively, r_{br} and r_{bu} are the corresponding effective radii, and $Q_{\alpha,\text{br}}$ is the absorption efficiency for brine pockets. The two terms on the right side of Equation (90) represent the absorption coefficients of brine pockets and surrounding ice (including impurities), respectively. In Equation (90), we have used the general relation $\alpha = 4\pi \tilde{m}_{i,p}/\lambda$, where λ is the wavelength in vacuum, and the expression inside the square brackets is the volume fraction of the ice surrounding all brine pockets and bubbles.

The air bubbles were assumed to be non-absorbing ($Q_{a,bu} = 0$), and the impurities were assumed to be uniformly distributed in the ice with $\tilde{m}_{i,p}$ and $\tilde{m}_{i,imp}$ being the imaginary parts of the refractive indices for pure ice and impurities, respectively. For brine pockets, which are in the liquid phase, the refractive index of sea water was used. The volume fraction f_{imp} of impurities typically lies in the range between 1×10^{-7} and 1×10^{-5}. The scattering coefficient β of sea ice is given by [53]

$$\beta = \beta_{br} + \beta_{bu}; \quad \beta_{br} = \pi r_{br}^2 N_{br} Q_{\beta,br}; \quad \beta_{bu} = \pi r_{bu}^2 N_{bu} Q_{\beta,bu} \tag{91}$$

where β_{br} and β_{bu} are the scattering coefficients for brine pockets and air bubbles, respectively, and $Q_{\beta,br}$ and $Q_{\beta,bu}$ are the corresponding scattering efficiencies. Here we have ignored the scattering coefficient for pure sea ice because it is very small compared to either β_{br} or β_{bu}. The optical thickness τ, the single-scattering albedo ϖ, and the asymmetry factor g for sea ice now become

$$\tau = (\alpha + \beta)h; \quad \varpi = \frac{\beta}{\alpha + \beta}; \quad g = \frac{\beta_{br} g_{br} + \beta_{bu} g_{bu}}{\beta_{br} + \beta_{bu}} \tag{92}$$

where h is the sea ice thickness.

The merit of these IOP parameterizations have been provided by comparisons with field measurements and laboratory data [50–53].

3.6. Ocean IOPs—Bio-Optical Models

In open ocean water, it is customary to assume that the IOPs of particulate matter can be parameterized in terms of the chlorophyll concentration. In coastal water, the IOPs will depend on the presence of "impurities" consisting of inorganic (mineral) particles, organic (algae) particles, and Colored Dissolved Organic Matter (CDOM) in addition to pure water. Due to the complexity of coastal water, we introduce three bio-optical models that have been adopted to represent different types of water. The CoastColour Round Robin (CCRR) model [58] is a useful proxy for turbid coastal water frequently observed in estuary areas, where suspended sediment (i.e., mineral) particles have a strong influence on water IOPs. The Santa Barbara Channel (SBC) and Garver-Siegel-Maritorena (GSM) bio-optical models described below provide useful representations of clean to moderately turbid water, where the IOPs are primarily dominated by algae. By varying the slope parameter S that describe the CDOM spectral absorption (see Equation (109) below), the GSM model may be used to represent CDOM dominated water.

As mentioned in Section 3.4.3, for pure water we may adopt the real part of the refractive index of pure water from [41], and we use the absorption coefficient $\alpha_w(\lambda)$ based on published data [42–44] for wavelengths between 340 and 700 nm, and other data [45] for wavelengths between 720 and 900 nm. Pure water scattering coefficients $\beta_w(\lambda)$ are based on published data [7], and the Rayleigh scattering phase function is given by Equation (9) with depolarization ratio $\rho = 0.039$, and thus $f = (1 - \rho)/(1 + \rho) = 0.925$ (see Section 3.1.1).

3.6.1. The CCRR Water Impurity IOPs

Here we first describe a bio-optical model used in the CoastColour Round Robin (CCRR) effort [58]. The CCRR bio-optical model consists of the three input parameters chlorophyll concentration (CHL), mineral concentration (MIN), and $\alpha_{CDOM}(443)$, which are allowed to vary. According to this decomposition into three basic components, the "mineral particle" component can include also non-algae particles whose absorption does not covary with that of the algae particles [58].

Mineral Particle IOPs

The absorption coefficient for mineral particles at 443 nm is given by [59] (Note on units: $\alpha_{MIN}(\lambda)/MIN = 0.041$ has units $[m^2 \cdot g^{-1}]$, so that if MIN has units of $[g \cdot m^{-3}]$, then the units of $\alpha_{MIN}(\lambda)$ will be $[m^{-1}]$.):

$$\alpha_{MIN}(443) = 0.041 \times 0.75 \times MIN$$

and its spectral variation is described by [58,59]:

$$\alpha_{MIN}(\lambda) = \alpha_{MIN}(443)[\exp(-0.0123(\lambda - 443))]. \tag{93}$$

The scattering coefficient at 555 nm is given by [60]

$$\beta_{MIN}(555) = 0.51 \times MIN$$

and the spectral variation of the attenuation coefficient is

$$\gamma_{MIN}(\lambda) = \gamma_{MIN}(555) \times (\lambda/\lambda_0)^{-c}; \qquad c = 0.3749, \qquad \lambda_0 = 555\,\text{nm} \tag{94}$$

where

$$\begin{aligned}
\gamma_{MIN}(555) &= \alpha_{MIN}(555) + \beta_{MIN}(555) \\
&= [0.041 \times 0.75 \exp(-0.0123(555 - 443)) + 0.51] \times MIN \\
&= 0.52 \times MIN.
\end{aligned}$$

The spectral variation of the scattering coefficient for mineral particles follows from

$$\beta_{MIN}(\lambda) = \gamma_{MIN}(\lambda) - \alpha_{MIN}(\lambda). \tag{95}$$

The average Petzold phase function with a backscattering ratio of 0.019 [4], may be used to describe the scattering phase function for mineral particles.

Algae Particle IOPs

The absorption coefficient for *pigmented* particles (algae particles or phytoplankton) can be written [61]:

$$\alpha_{pig}(\lambda) = A_\phi(\lambda) \times [CHL]^{E_\phi(\lambda)} \tag{96}$$

where $A_\phi(\lambda)$ and $E_\phi(\lambda)$ are given by [61], and where CHL is the chlorophyll concentration, which represents the concentration of pigmented particles (algae particles or phytoplankton).

The attenuation coefficient for pigmented particles at 660 nm is given by [62]:

$$\gamma_{pig}(660) = \gamma_0 \times [CHL]^\eta; \qquad \gamma_0 = 0.407; \qquad \eta = 0.795$$

and its spectral variation is taken to be [63]:

$$\gamma_{pig}(\lambda) = \gamma_{pig}(660) \times (\lambda/660)^\nu \tag{97}$$

where

$$\nu = \begin{cases} 0.5 \times [\log_{10} CHL - 0.3] & 0.02 < CHL < 2.0 \\ 0 & CHL > 2.0. \end{cases}$$

The spectral variation of the scattering coefficient for pigmented particles follows from the difference:

$$\beta_{pig}(\lambda) = \gamma_{pig}(\lambda) - \alpha_{pig}(\lambda). \tag{98}$$

The scattering phase function for pigmented particles may be described by the Fournier-Forand phase function (see below) with a backscattering ratio equal to 0.006 [63,64].

CDOM IOPs

The absorption by CDOM is given by [60]:

$$\alpha_{\text{CDOM}}(\lambda) = \alpha_{\text{CDOM}}(443) \times \exp[-S(\lambda - 443)]; \qquad S = 0.0176. \tag{99}$$

The total absorption and scattering coefficients due to water impurities for the CCRR IOP model are given by:

$$\alpha_{\text{tot}}(\lambda) = \alpha_{\text{MIN}}(\lambda) + \alpha_{\text{pig}}(\lambda) + \alpha_{\text{CDOM}}(\lambda) \tag{100}$$

$$\beta_{\text{tot}}(\lambda) \equiv \beta_{\text{p}}(\lambda) = \beta_{\text{MIN}}(\lambda) + \beta_{\text{pig}}(\lambda). \tag{101}$$

Scattering Phase Function for Particles

Measurements have shown that the particle size distribution (PSD) function in oceanic water can be accurately described by an inverse power law (Junge distribution) $F(r) = C_r/r^\xi$, where $F(r)$ is the number of particles per unit volume per unit bin width, and r [μm] is the radius of the assumed spherical particle. C_r [cm$^{-3} \cdot \mu$m$^{\xi-1}$] is the Junge coefficient, and ξ is the PSD slope, which typically varies between 3.0 and 5.0 [65,66]. By assuming an inverse power law (Junge distribution) for the PSD, [67] derived an analytic expression for the scattering phase function of oceanic water (hereafter referred to as the FF scattering phase function), given by [64]

$$
\begin{aligned}
p_{\text{FF}}(\Theta) \;=\; & \frac{1}{4\pi(1-\delta)^2\delta^\nu}\left\{\nu(1-\delta) - (1-\delta^\nu) + \frac{4}{u^2}[\delta(1-\delta^\nu) - \nu(1-\delta)]\right\} \\
& + \frac{1-\delta_{180}^\nu}{16\pi(\delta_{180}-1)\delta_{180}^\nu}[3\cos^2\Theta - 1]
\end{aligned}
\tag{102}
$$

where $\nu = 0.5(3-\xi)$, $u = 2\sin(\Theta/2)$, $\delta \equiv \delta(\Theta) = \frac{u^2}{3(\tilde{m}_r-1)^2}$, $\delta_{180} = \delta(\Theta = 180°) = \frac{4}{3(\tilde{m}_r-1)^2}$, Θ is the scattering angle, and $\tilde{m}_r$ is the real part of the refractive index.

Integrating $p_{\text{FF}}(\Theta)$ over the backward hemisphere (setting $x = \cos\Theta$), one obtains the backscattering ratio or backscatter fraction defined in Equation (8) [64]

$$
\begin{aligned}
b_{\text{FF}} = \frac{1}{2}\int_{\pi/2}^{\pi} p_{\text{FF}}(\cos\Theta)\sin\Theta d\Theta &= \frac{1}{2}\int_0^1 p_{\text{FF}}(-x)dx \\
&= 1 - \frac{1 - \delta_{90}^{\nu+1} - 0.5(1-\delta_{90}^\nu)}{(1-\delta_{90})\delta_{90}^\nu}
\end{aligned}
\tag{103}
$$

where $\delta_{90} = \delta(\Theta = 90°) = \frac{4}{3(\tilde{m}_r-1)^2}\sin^2(45°) = \frac{2}{3(\tilde{m}_r-1)^2}$. Equation (103) can be solved for ν in terms of b_{FF} and δ_{90}, implying that ν and thus ξ can be determined if the real part of the refractive index $\tilde{m}_r$ and the backscatter ratio b_{FF} are specified. As a consequence, the FF scattering phase function can be evaluated from a measured value of b_{FF} if the real part of the refractive index $\tilde{m}_r$ is known.

As already mentioned, in the CCRR bio-optical model, the Petzold scattering phase function with a backscattering ratio of 0.019 is used to represent mineral (non-algal) particles. These scattering phase functions are shown in Figure 2 together with the Rayleigh scattering phase function, which represents scattering by water molecules.

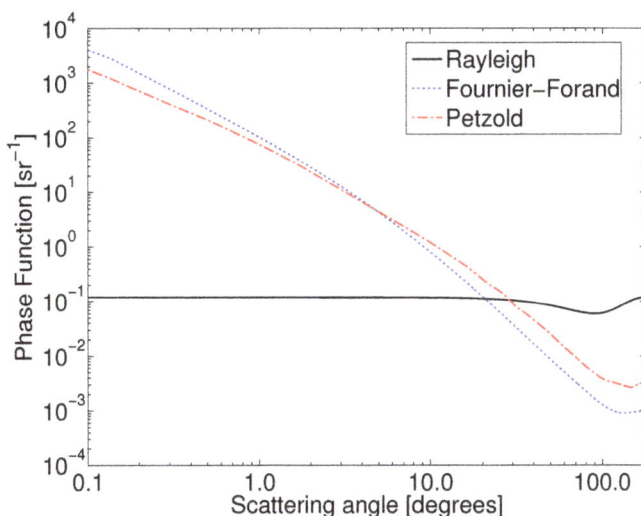

Figure 2. Rayleigh, Fournier-Forand, and Petzold scattering phase functions used to represent scattering by water molecules, pigmented particles, and non-algal particles, respectively, in the CCRR bio-optical model. To generate the FF scattering phase function the values $\xi = 3.38$ and $\tilde{m}_r = 1.068$ were used.

A moment-fitting program [68] may be used to create Legendre expansion coefficients $\chi_{\ell,\mathrm{PET}}$ and $\chi_{\ell,FF}$ for the Petzold and FF scattering phase functions. Hence, the total scattering phase function Legendre expansion coefficients are given by:

$$\chi_\ell = \frac{\beta_{\mathrm{pig}}(\lambda)\chi_{\ell,\mathrm{FF}} + \beta_{\mathrm{MIN}}(\lambda)\chi_{\ell,\mathrm{PET}} + \beta_{\mathrm{w}}(\lambda)\chi_{\ell,\mathrm{water}}}{\beta_{\mathrm{pig}}(\lambda) + \beta_{\mathrm{MIN}}(\lambda) + \beta_{\mathrm{w}}(\lambda)}. \tag{104}$$

Thus, to use the CCRR bio-optical model one must specify the three input parameters CHL, MIN, and $\alpha_{\mathrm{CDOM}}(443)$.

3.6.2. The SBC and GSM Bio-Optical Models

Based on field measurements conducted in the Santa Barbara Channel and compiled in the NOMAD data base [69], Li et al. [70] constructed a local bio-optical model representative for the SBC coastal waters. Another frequently used bio-optical model is the GSM model [71,72], which is included in NASA's SeaDAS software package. GSM is a global model which has the same structure as the SBC model, but with different coefficients.

In the SBC model, the water body, in addition to pure water, is assumed to be described by three parameters that can be varied: (i) the chlorophyll concentration [CHL] (a proxy for the concentration of pigmented particles), the CDOM absorption coefficient at 443 nm [$\alpha_{\mathrm{CDOM}}(443)$], and the total scattering coefficient at 443 nm [$\beta_{\mathrm{tot}}(443)$]. The IOPs of the SBC model are described by:

$$\alpha_{\mathrm{pig}}^{\mathrm{SBC}}(\lambda) = a_1^{\mathrm{SBC}}(\lambda)[\mathrm{CHL}]^{a_2(\lambda)} \tag{105}$$

$$\alpha_{\mathrm{CDOM}}^{\mathrm{SBC}}(\lambda) = \alpha_{\mathrm{CDOM}}^{\mathrm{SBC}}(443)\exp[-S(\lambda - 443)]; \qquad S = 0.012 \tag{106}$$

$$\beta_{\mathrm{tot}}^{\mathrm{SBC}}(\lambda) = \beta_{\mathrm{tot}}^{\mathrm{SBC}}(443)(\lambda/443). \tag{107}$$

Similarly, for the GSM model the IOPs are described by

$$\alpha_{\text{pig}}^{\text{GSM}}(\lambda) = a_1^{\text{GSM}}(\lambda)[\text{CHL}] \tag{108}$$

$$\alpha_{\text{CDOM}}^{\text{GSM}}(\lambda) = a_{\text{CDOM}}^{\text{GSM}}(443) \exp[-S(\lambda - 443)]; \qquad S = 0.0206 \tag{109}$$

$$\beta_{\text{tot}}^{\text{GSM}}(\lambda) = \beta_{\text{tot}}^{\text{GSM}}(443)(\lambda/443)^{1.0337}. \tag{110}$$

Please note that the wavelength dependent factors $a_1^{\text{SBC}}(\lambda)$ and $a_2(\lambda)$ in Equation (105) as well as $a_1^{\text{GSM}}(\lambda)$ in Equation (108) are determined from field measurements compiled in the NOMAD data base. For simplicity we will assume here that $\alpha_{\text{pig}}^{\text{SBC}}(\lambda) = \alpha_{\text{pig}}^{\text{GSM}}(\lambda) \equiv \alpha_{\text{pig}}^{\text{CCRR}}(\lambda)$, so that the difference between the three models lies in the treatment of scattering and CDOM absorption.

For both the SBC and GSM models, CDOM represents a combination of colored dissolved organic matter and mineral particles. Hence, values of $a_{\text{CDOM}}^{\text{SBC}}(443)$ in Equation (106) and $a_{\text{CDOM}}^{\text{GSM}}(443)$ in Equation (109) should be compared to the sum $a_{\text{CDOM}}^{\text{CCRR}}(443) + 0.041 \times 0.75 \text{ MIN}$. The total suspended particle scattering coefficient is $\beta_{\text{tot}}^{\text{SBC}}(\lambda)$ and $\beta_{\text{tot}}^{\text{GSM}}(\lambda)$ in the SBC and GSM model, respectively, each being comparable to $\beta_{\text{tot}}(\lambda) = \beta_{\text{MIN}}(\lambda) + \beta_{\text{pig}}(\lambda)$ in the CCRR model.

The most significant difference between these three models is that the SBC and GSM models do not separately include mineral particles although the total scattering coefficient does include the total suspended particle scattering. Another difference is that the CCRR model is based on three different reference wavelengths, namely 443 nm for α_{MIN}, 555 nm for γ_{MIN}, and 660 nm for α_{pig}, while the SBC and the GSM models are based only on 443 nm as a reference wavelength. In the SBC and GSM models one may use the FF scattering phase function for pigments, but other scattering phase functions, such as the Petzold scattering phase function may work better in coastal areas.

4. Radiative Transfer in Coupled Atmosphere-Water (Including Snow/ice) Systems

4.1. Radiative Transfer Equation—Unpolarized Radiation

In the AccuRT computational tool, one considers a *coupled* system consisting of two adjacent slabs (atmosphere overlying a water body) separated by a plane, horizontal interface. The refractive index changes abruptly across this interface from a value $\tilde{m}_{c,1}$ in the upper slab (hereafter slab$_1$, the atmosphere) to a value $\tilde{m}_{c,2}$ in the lower slab (hereafter slab$_2$, a water body). If the IOPs in each of the two slabs vary only in the vertical direction denoted by z, where z increases upward, the corresponding vertical optical depth, denoted by $\tau(z)$, is defined by

$$\tau(z) = \int_z^\infty [\alpha(z') + \beta(z')]dz' \tag{111}$$

where the absorption and scattering coefficients α and β are defined in Equations (1). Please note that the vertical optical depth is defined to increase downward from $\tau(z = \infty) = 0$ at the top of the atmosphere. In either of the two slabs, assumed to be in local thermodynamic equilibrium so that they emit radiation according to the local temperature $T(\tau(z))$, the diffuse radiance distribution $I(\tau, u, \phi)$ can be described by the radiative transfer equation (RTE)

$$\mu \frac{dI(\tau, u, \phi)}{d\tau} = I(\tau, u, \phi) - S(\tau, u', \phi') \tag{112}$$

where

$$S(\tau, u', \phi') = S^*(\tau, u', \phi') + [1 - \omega(\tau)]B(\tau) + \frac{\omega(\tau)}{4\pi} \int_0^{2\pi} d\phi' \int_{-1}^1 p(\tau, u', \phi'; u, \phi)I(\tau, u', \phi')du'. \tag{113}$$

Here u is the cosine of the polar angle θ, ϕ is the azimuth angle, $\omega(\tau) = \beta(\tau)/[\alpha(\tau) + \beta(\tau)]$ is the single-scattering albedo, $p(\tau, u', \phi'; u, \phi)$ is the scattering phase function defined by Equation (5), and $B(\tau)$ is the thermal radiation field given by the Planck function. The differential vertical optical depth is (see Equation (111))

$$d\tau(z) = -[\alpha(\tau) + \beta(\tau)]dz \tag{114}$$

where the minus sign indicates that τ increases in the downward direction, whereas z increases in the upward direction, as noted above. The scattering angle Θ and the polar and azimuth angles are related by (see Equation (25))

$$\hat{\Omega}' \cdot \hat{\Omega} = \cos\Theta = \cos\theta\cos\theta' + \sin\theta'\sin\theta\cos(\phi' - \phi).$$

By definition, $\theta = 180°$ is directed toward nadir (straight down) and $\theta = 0°$ toward zenith (straight up). Thus, $u = \cos\theta$ varies in the range $[-1, 1]$ (from nadir to zenith). For cases of oblique illumination of the system, $\phi = 180°$ is defined to be the azimuth angle of the incident light.

4.1.1. Isolation of Azimuth Dependence

The azimuth dependence in Equation (112) may be isolated by expanding the scattering phase function in *Legendre polynomials*, $P_\ell(\cos\Theta)$, and making use of the addition theorem for spherical harmonics [14]

$$p(\cos\Theta) = p(u', \phi'; u, \phi) = \sum_{m=0}^{2N-1} (2 - \delta_{0,m})p^m(u', u)\cos m(\phi' - \phi) \tag{115}$$

where $\delta_{0,m}$ is the Kronecker delta, i.e., $\delta_{0,m} = 1$ for $m = 0$ and $\delta_{0,m} = 0$ for $m \neq 0$, and

$$p^m(u', u) = \sum_{\ell=m}^{2N-1} (2l+1)\chi_\ell \Lambda_\ell^m(u')\Lambda_\ell^m(u). \tag{116}$$

Here

$$\chi_\ell = \frac{1}{2}\int_{-1}^{1} P_\ell(\cos\Theta)p(\cos\Theta)d(\cos\Theta) \tag{117}$$

is an expansion coefficient and $\Lambda_\ell^m(u)$ is given by

$$\Lambda_\ell^m(u) \equiv \sqrt{\frac{(\ell - m)!}{(\ell + m)!}}P_\ell^m(u) \tag{118}$$

where $P_\ell^m(u)$ is an associated Legendre polynomial of order m. Expanding the radiance in a similar way,

$$I(\tau, u, \phi) = \sum_{m=0}^{2N-1} I^m(\tau, u)\cos m(\phi - \phi_0) \tag{119}$$

where ϕ_0 is the azimuth angle of the incident light, one finds that each Fourier component satisfies the following RTE (see [14] for details)

$$\mu\frac{dI^m(\tau, u)}{d\tau} = I^m(\tau, u) - \frac{\omega(\tau)}{2}\int_{-1}^{1} p^m(\tau, u', u) I^m(\tau, u)d\mu - S^{*m}(\tau, u) \tag{120}$$

where $m = 0, 1, 2, \ldots, 2N - 1$ and $p^m(\mu', \mu)$ is given by Equation (116).

4.1.2. The Interface between the Two Slabs—Calm (Flat) Water Surface

When a beam of light is incident upon a plane interface between two slabs of different refractive indices, one fraction of the incident light will be reflected and another fraction will be transmitted or refracted. For unpolarized light incident upon the interface between the two slabs, the Fresnel reflectance ρ_F is given by

$$\rho_F = \frac{1}{2}(\rho_\perp + \rho_\|) \tag{121}$$

where $\rho_\perp$ is the reflectance for light polarized with the electric field perpendicular to the plane of incidence, and $\rho_\|$ is the reflectance for light polarized with the electric field parallel to the plane of incidence [3,14,73,74]. Thus, one finds

$$\rho_F = \frac{1}{2}\left[\left|\frac{\mu_1 - m_{rat}\mu_t}{\mu_1 + m_{rat}\mu_2}\right|^2 + \left|\frac{\mu_2 - m_{rat}\mu_1}{\mu_2 + m_{rat}\mu_1}\right|^2\right] \tag{122}$$

where $\mu_1 \equiv \mu_{air} = \cos\theta_1$, θ_1 being the angle of incidence, $\mu_2 \equiv \mu_{ocn} = \cos\theta_2$, θ_2 being the angle of refraction determined by Snell's law ($\tilde{m}_{r,1}\sin\theta_1 = \tilde{m}_{r,2}\sin\theta_2$), and $m_{rat} = \tilde{m}_{c,2}/\tilde{m}_{c,1}$. Similarly, the Fresnel transmittance becomes

$$T_F = 2m_{rel}\mu_i\mu_t\left[\left|\frac{1}{\mu_i + m_{rat}\mu_t}\right|^2 + \left|\frac{1}{\mu_t + m_{rat}\mu_i}\right|^2\right] \tag{123}$$

where $m_{rel} = \tilde{m}_{r,2}/\tilde{m}_{r,1}$.

4.1.3. A Wind-Blown (Rough) Air-Water Interface—Pseudo-Two-Dimensional BRDF Treatment

A calm (flat) atmosphere-water interface occurs only for very low wind speeds. A wind-roughened water surface occurs more frequently and is therefore more realistic.

Consider a Cartesian coordinate system (x, y, z) in which z is the vertical coordinate. To calculate the slope distribution, $p(z_x, z_y)$, we consider a plane wave incident on a rough surface characterized by a Gaussian random height distribution $z = f(x,y) = f(\mathbf{r}_\perp)$ where $f(\mathbf{r}_\perp) = f(x,y)$ with mean height $\langle z \rangle = \langle f(x,y) \rangle \equiv \langle f(\mathbf{r}_\perp) \rangle = 0$. We now focus on a particular tilted surface facet that makes a polar angles θ_n with respect to the vertical direction and a relative azimuth angle α. Let the incident solar radiance I_i be at a zenith angle θ_0, the reflected radiance I_r be at zenith angle θ, and the relative azimuth between I_i and I_r be at angle $\Delta\phi$. Then the slope of the tilted surface facet has components z_x and z_y defined by:

$$z_x = \frac{\partial z}{\partial x} = \frac{\partial f(x,y)}{\partial x} = \sin\alpha\tan\theta_n \qquad z_y = \frac{\partial z}{\partial y} = \frac{\partial f(x,y)}{\partial y} = \cos\alpha\tan\theta_n.$$

For an anisotropic distribution of slope components (dependent on the wind direction), we define new slope components as follows:

$$z'_x = \cos(\chi)z_x + \sin(\chi)z_y \qquad z'_y = -\sin(\chi)z_x + \cos(\chi)z_y$$

where $\chi = \phi_s - \phi_W = $ rotation from the sun-observation system (x, y, z) and $\phi_W = $ the wind direction. The slope distribution can be written as a Gram-Charlier series [75]:

$$p(z'_x, z'_y) = \frac{1}{2\pi\sigma_x\sigma_y}\exp\left[-\frac{1}{2}\left(\frac{z'^2_x}{\sigma^2_x} + \frac{z'^2_y}{\sigma^2_y}\right)\right]\left[1 - \Delta(\xi, \eta)\right] \tag{124}$$

where $\xi = \frac{z'_x}{\sigma_x}$, $\eta = \frac{z'_y}{\sigma_y}$, σ^2_x and σ^2_y are variances of z'_x and z'_y, and the function $\Delta(\xi, \eta)$ represents the departure of the slope distribution from a strict two-dimensional (2D) Gaussian due to skewness and peakedness.

Please note that in the absence of skewness and peakedness ($\Delta(\xi, \eta) = 0$) the Gram-Charlier series reduces to a 2D Gaussian distribution:

$$p(z'_x, z'_y) = \frac{1}{2\pi\sigma_x\sigma_y}\exp\left[-\frac{1}{2}\left(\frac{z'^2_x}{\sigma^2_x} + \frac{z'^2_y}{\sigma^2_y}\right)\right]. \tag{125}$$

Furthermore, for an isotropic slope distribution $\chi = 0$ so that $z'_x = z_x$ and $z'_y = z_y$, $\sigma_x = \sigma_y$. Therefore $\sigma^2 = \sigma_x^2 + \sigma_y^2 = 2\sigma_x^2 = 2\sigma_x\sigma_y$, and hence

$$\frac{1}{2}\left(\frac{z'^2_x}{\sigma_x^2} + \frac{z'^2_y}{\sigma_y^2}\right)] = \frac{z_x^2 + z_y^2}{2\sigma_x\sigma_y} = \frac{z_x^2 + z_y^2}{\sigma^2}; \quad \text{and} \quad z_x^2 + z_y^2 = \tan^2\theta_n(\sin^2\alpha + \cos^2\alpha) = \tan^2\theta_n$$

since $z_x = \sin\alpha \tan\theta_n$ and $z_y = \cos\alpha \tan\theta_n$. Thus, we obtain a 1D Gaussian:

$$p(\mu_n, \sigma) = p(z_x, z_y) = \frac{1}{\pi\sigma^2}\exp\left(-\frac{z_x^2 + z_y^2}{\sigma^2}\right) = \frac{1}{\pi\sigma^2}\exp\left(-\frac{\tan^2\theta_n}{\sigma^2}\right) = \frac{1}{\pi\sigma^2}\exp\left(-\frac{1 - \mu_n^2}{\sigma^2\mu_n^2}\right). \quad (126)$$

where

$$\mu_n = \cos\theta_n = \frac{\mu + \mu'}{\sqrt{2(1 - \cos\Theta)}} \quad (127)$$

$$\cos\Theta = -\mu\mu' + \sqrt{1 - \mu^2}\sqrt{1 - \mu'^2}\cos(\Delta\phi). \quad (128)$$

A Pseudo Two-Dimensional (Wind-Direction Dependent) Treatment of the BRDF

At the bottom of the atmosphere ($\tau = \tau_{atm}$), the upward reflected radiance $I^+_{refl}(\tau_{atm}, \mu', \phi')$ is connected to the downward incident diffuse radiance $I^-_{inc}(\tau_{atm}, \mu, \phi)$ and the attenuated direct radiance $F_0 e^{-\tau_{atm}/\mu_0}$ through the sea surface reflection that is described by the BRDF $\rho(\mu, \mu', \Delta\phi)$:

$$I^+_{refl}(\tau_{atm}, \mu', \phi') = \mu_0\, \rho(\mu_0, \mu', \phi')\, F_0\, e^{-\tau_{atm}/\mu_0} + \int_0^{2\pi}\int_0^1 \mu\, \rho(\mu, \mu', \Delta\phi) I^-_{inc}(\tau_{atm}, \mu, \phi) d\mu d\phi \quad (129)$$

where $\Delta\phi = \phi' - \phi$, and the solar azimuth angle was set to $\phi_0 = 0°$ so that $\Delta\phi = \phi' - \phi_0 = \phi'$ for the direct beam reflection $\rho(\mu_0, \mu', \phi')$.

The 1D BRDF $\rho(\mu, \mu', \Delta\phi')$ can be written as:

$$\rho(\mu, \mu', \Delta\phi) = \frac{1}{4\mu'\mu(\mu_n)^4} \cdot p(\mu_n, \sigma) \cdot \rho_F \cdot s(\mu, \mu', \sigma) \quad (130)$$

$$\mu_n = \frac{\mu + \mu'}{\sqrt{2(1 - \cos\Theta)}} \quad (131)$$

$$\cos\Theta = -\mu\mu' + \sqrt{1 - \mu^2}\sqrt{1 - \mu'^2}\cos(\Delta\phi). \quad (132)$$

Here $\mu = \cos\theta$, θ being the view zenith angle for the incident light, $\mu' = \cos\theta'$, θ' being the view zenith angle for the reflected light, $\Delta\phi$ is the relative azimuth angle, and Θ is the scattering angle. In Equation (130), ρ_F is the Fresnel reflectance (see Equation (121)), $s(\mu, \mu', \sigma)$ describes the effect of shadowing, and $p(\mu_n, \sigma)$ is the surface slope distribution (Equation (126)).

In a plane-parallel (1D) geometry, the radiance and the BRDF depend only on the difference $\Delta\phi = \phi' - \phi$ in azimuth between the direction of incidence (θ', ϕ') and observation (θ, ϕ). Hence, in a strict plane-parallel geometry, it is impossible to model a wind-direction dependent (azimuthally-asymmetric) BRDF. As a consequence, most treatments of water surface roughness effects are limited to a 1D treatment [76,77].

In the 1D discrete ordinate method, the radiance $I(\tau, \mu, \phi)$ and the BRDF $\rho(\mu, \mu', \Delta\phi)$ are expanded into a Fourier cosine series to isolate the azimuth dependence [14]. The pseudo two-dimensional treatment of the BRDF employs a 2D BRDF to compute the direct beam reflectance, but a 1D (Fourier expanded) BRDF to compute the reflectance due to diffuse, multiply scattered light. A post-processing step, which corrects the direct beam reflectance (1D → 2D), is used for implementation. This post-processing method is similar to the Nakajima-Tanaka (NT) single-scattering correction [78], which retains the multiply scattered radiance, but corrects the singly scattered radiance without considering boundary reflection. In DISORT3, Lin et al. [79] improved the NT procedure by adding a BRDF correction and the same strategy was used to add a 2D BRDF correction [80]. Hence, the correction term for radiance $I_{ss\,corr}(\hat{\tau}, \pm\mu, \phi)$ can be written as:

$$I_{ss\,corr}^{\pm}(\hat{\tau}, \mu, \phi) = I_{ss}^{\pm*}(\hat{\tau}, \mu, \phi) - \tilde{I}_{ss}^{\pm*}(\hat{\tau}, \pm\mu, \phi) + \mu_0 F_0 \left\{ \rho_{2D}(\mu, \phi; -\mu_0, \phi_0) - \rho_{1D}(\mu, \phi; -\mu_0, \phi_0) \right\} e^{-\frac{\hat{\tau}_b}{\mu_0} + \frac{\hat{\tau} - \hat{\tau}_b}{\mu}}. \tag{133}$$

Here $\hat{\tau}$ and $\hat{\tau}_b$ are the scaled optical thicknesses at the height of interest and the lower boundary, respectively [78,79]. On the right hand side of Equation (133), $I_{ss}^{\pm*}(\hat{\tau}, \mu, \phi) - \tilde{I}_{ss}^{\pm*}(\hat{\tau}, \pm\mu, \phi)$ is the original NT correction [78], $\rho_{2D}(\mu, \phi; -\mu_0, \phi_0)$ is the new 2D BRDF used to compute the 2D single-scattering contribution, $\rho_{1D}(\mu, \phi; -\mu_0, \phi_0)$ is the 1D (Fourier expanded) BRDF used to compute the approximate multiple scattering contribution, and $e^{-\frac{\hat{\tau}_b}{\mu_0} + \frac{\hat{\tau} - \hat{\tau}_b}{\mu}}$ is the beam attenuation coefficient.

For multiply scattered light, we use a 1D Gaussian surface slope distribution given by Equation (126), which is widely used in remote sensing applications to represent the slope statistics of water waves with the numerical value of the slope variance parameterized in terms of the wind speed [75]. The 1D BRDF given by Equations (130)–(132) and (126) is suitable for describing "skyglint", that is, the reflectance of downward diffuse light from a rough water surface, because multiple scattering in the atmosphere has made the radiation field approximately 1D, implying that 2D BRDF effects become relatively unimportant for the reflected diffuse skylight [76,77]. Similarly, the slope distribution for a 2D Gaussian surface is given by Equation (125). It will be shown in Section 5.3 that a 2D BRDF treatment is required in RT models to reproduce detailed measurements of the BRDF [80].

4.2. Radiative Transfer Equation—Polarized Radiation

To generalize Equation (112) to apply to polarized radiation, we note that the multiple scattering term $S^{ms}(\tau, u, \phi) = \frac{\omega(\tau)}{4\pi} \int_0^{2\pi} d\phi' \int_{-1}^{1} du' p(\tau, u', \phi'; u, \phi) I(\tau, u', \phi')$ in Equation (113) must be replaced by

$$\mathbf{S}^{ms}(\tau, u, \phi) = \frac{\omega(\tau)}{4\pi} \int_0^{2\pi} d\phi' \int_{-1}^{1} du' \mathbf{P}(\tau, u', \phi'; u, \phi) \mathbf{I}(\tau, u', \phi') \tag{134}$$

where $\mathbf{I}(\tau, u', \phi')$ is the Stokes vector, and $\mathbf{P}(\tau, u', \phi'; u, \phi)$ is the scattering phase matrix (see Section 3.2). The first element of the vector $\mathbf{S}^{ms}(\tau, u, \phi)$ represents the energy per unit solid angle, per unit frequency interval, and per unit time that is scattered by a unit volume in the direction $(u = \cos\theta, \phi)$. Hence, in a plane-parallel (slab) geometry, the integro-differential equation for polarized radiative transfer is expressed in terms of a Stokes vector $\mathbf{I}(\tau, u, \phi)$ as

$$u \frac{d\mathbf{I}(\tau, u, \phi)}{d\tau} = \mathbf{I}(\tau, u, \phi) - \mathbf{S}(\tau, u, \phi) \tag{135}$$

where the source vector is

$$\mathbf{S}(\tau, u, \phi) = \frac{\omega(\tau)}{4\pi} \int_0^{2\pi} d\phi' \int_{-1}^{1} du' \, \mathbf{P}(\tau, u', \phi'; u, \phi) \mathbf{I}(\tau, u', \phi')$$
$$+ \mathbf{Q}(\tau, u, \phi). \tag{136}$$

In the upper slab (slab$_1$, atmosphere), the source term $\mathbf{Q}(\tau, u, \phi)$, due to thermal and beam sources, is given by:

$$\mathbf{Q}_1(\tau, u, \phi) = \frac{\omega(\tau)}{4\pi} \mathbf{P}(\tau, -\mu_0, \phi_0; u, \phi) \mathbf{S}_b e^{-\tau/\mu_0} + [1 - \omega(\tau)] \mathbf{S}_t(\tau)$$
$$+ \frac{\omega(\tau)}{4\pi} \mathbf{P}(\tau, \mu_0, \phi_0; u, \phi) \mathbf{R}_F(-\mu_0, m_{rel}) \mathbf{S}_b \, e^{-\frac{(2\tau_a - \tau)}{\mu_0}}. \tag{137}$$

The first term on the right hand side of Equation (137) describes the incident beam $\mathbf{S}_b$ in direction $(-\mu_0, \phi_0)$, which is attenuated at depth τ by a factor $e^{-\tau/\mu_0}$ and undergoes single scattering into the direction (u, ϕ). For an unpolarized incident beam $\mathbf{S}_b$ has the form

$$\mathbf{S}_b = [I_0/2, I_0/2, 0, 0]^T \quad \text{or} \quad [I_0, 0, 0, 0]^T \tag{138}$$

where the first or second expression corresponds to the choice of Stokes vector representation, $[I_\parallel, I_\perp, U, V]^T$ or $[I, Q, U, V]^T$. The second term on the right hand side of Equation (137) is due to thermal emission, which is unpolarized, and $\mathbf{S}_t(\tau)$ is given by

$$\mathbf{S}_t(\tau) = [B(T(\tau))/2, B(T(\tau))/2, 0, 0]^T \quad \text{or} \quad [B(T(\tau)), 0, 0, 0]^T \tag{139}$$

where B is the Planck function, and where the first or second expression corresponds to the choice of Stokes vector representation. We have set $\mu_0 \equiv |u_0| \equiv |\cos\theta_0|$, where θ_0 is the polar angle of the incident light beam. The third term on the right hand side of Equation (137) describes radiation due to the incident beam $\mathbf{S}_b$ that has been attenuated by the factor $e^{-\tau_a/\mu_0}$ before reaching the air-water interface, undergoing Fresnel reflection given by the reflection matrix $\mathbf{R}_F(-\mu_0, m_{\text{rel}})$, attenuated by the factor $e^{-(\tau_a-\tau)/\mu_0}$ to reach the level τ in the atmosphere, and finally singly scattered from direction (μ_0, ϕ_0) into direction (u, ϕ) described by the factor $\frac{\varpi(\tau)}{4\pi}\mathbf{P}(\tau, \mu_0, \phi_0; u, \phi)$. Thus, the incident beam propagates though the entire atmosphere and a portion of it is reflected upwards by the interface to reach depth τ in the atmosphere, which explains the factor $e^{-(2\tau_a-\tau)/\mu_0}$.

In the lower slab (slab$_2$, water), the source term becomes

$$\begin{aligned}
\mathbf{Q}_2(\tau, u, \phi) =& \; \frac{\varpi(\tau)}{4\pi}\mathbf{P}(\tau, -\mu_0^w, \phi_0; u, \phi)\mathbf{S}_b \, e^{-\tau_a/\mu_0} \\
& \times \mathbf{T}_F(-\mu_0, m_{\text{rel}})\frac{\mu_0}{\mu_0^w}e^{-(\tau-\tau_a)/\mu_0^w} \\
& + [1-\varpi(\tau)]\,\mathbf{S}_t(\tau)
\end{aligned} \tag{140}$$

where $\mathbf{T}_F(-\mu_0, m_{\text{rel}})$ is the Fresnel transmission matrix. The first term in Equation (140) is due to the incident beam $\mathbf{S}_b$ that has been attenuated through the atmosphere by the factor $e^{-\tau_a/\mu_0}$, transmitted into the water by the factor $\mathbf{T}_F(-\mu_0, m_{\text{rel}})\frac{\mu_0}{\mu_0^w}$, further attenuated by the factor $e^{-(\tau-\tau_a)/\mu_0^w}$ to reach depth τ in the water, and singly scattered from the direction $(-\mu_0^w, \phi_0)$ into the direction (u, ϕ) which explains the factor $\frac{\varpi(\tau)}{4\pi}\mathbf{P}(\tau, -\mu_0^w, \phi_0; u, \phi)$. The second term in Equation (140) is due to thermal emission in the water.

Isolation of Azimuth Dependence

We start by expanding the scattering phase matrix in a Fourier series:

$$\mathbf{P}(u', u; \phi' - \phi) = \sum_{m=0}^{M}\{\mathbf{P}_c^m(u', u)\cos m(\phi' - \phi) + \mathbf{P}_s^m(u', u)\sin m(\phi' - \phi)\}. \tag{141}$$

To isolate the azimuth dependence of the radiation field we expand the Stokes vector $\mathbf{I}(\tau, u, \phi)$ in Equation (135) and the source term $\mathbf{Q}_1(\tau, u, \phi)$ in Equation (137) or $\mathbf{Q}_2(\tau, u, \phi)$ in Equation (140) in a Fourier series in a manner similar to the expansion of the scattering phase matrix in Equation (141):

$$\mathbf{I}(\tau, u, \phi) = \sum_{m=0}^{M}\left\{\mathbf{I}_c^m(\tau, u)\cos m(\phi_0 - \phi) + \mathbf{I}_s^m(\tau, u)\sin m(\phi_0 - \phi)\right\} \tag{142}$$

$$\mathbf{Q}_p(\tau, u, \phi) = \sum_{m=0}^{M} \left\{ \mathbf{Q}_{cp}^m(\tau, u) \cos m(\phi_0 - \phi) + \mathbf{Q}_{sp}^m(\tau, u) \sin m(\phi_0 - \phi) \right\} \tag{143}$$

where the subscript s or c denotes sine or cosine mode and the subscript p indicates the slab, p = 1 for slab$_1$, and p = 2 for slab$_2$. Using these expansions it can be shown that we obtain the following equations for the Fourier components (see [3] for details)

$$u \frac{d\mathbf{I}_c^m(\tau, u)}{d\tau} = \mathbf{I}_c^m(\tau, u) - \frac{\omega(\tau)}{4} \int_{-1}^{1} du' \left\{ \mathbf{P}_c^m(\tau, u', u) \, \mathbf{I}_c^m(\tau, u') \, (1 + \delta_{0m}) \right.$$

$$\left. - \mathbf{P}_s^m(\tau, u', u) \, \mathbf{I}_s^m(\tau, u') \right\} - \mathbf{Q}_c^m(\tau, u) \tag{144}$$

$$u \frac{d\mathbf{I}_s^m(\tau, u)}{d\tau} = \mathbf{I}_s^m(\tau, u) - \frac{\omega(\tau)}{4} \int_{-1}^{1} du' \left\{ \mathbf{P}_c^m(\tau, u', u) \, \mathbf{I}_s^m(\tau, u') \right.$$

$$\left. + \mathbf{P}_s^m(\tau, u', u) \, \mathbf{I}_c^m(\tau, u') \right\} - \mathbf{Q}_s^m(\tau, u). \tag{145}$$

The discrete ordinate method consists of replacing the integration over u' by a discrete sum using Gaussian quadrature points u_j (the discrete ordinates) and corresponding weights w_j. One obtains for each Fourier component:

$$u_i \frac{d\mathbf{I}_c^m(\tau, u_i)}{d\tau} = \mathbf{I}_c^m(\tau, u_i)$$

$$- \frac{\omega(\tau)}{4} \sum_{\substack{j=-N \\ j \neq 0}}^{N} w_j \left\{ (1 + \delta_{0m}) \, \mathbf{P}_c^m(\tau, u_j, u_i) \, \mathbf{I}_c^m(\tau, u_j) \right. \tag{146}$$

$$\left. - \mathbf{P}_s^m(\tau, u_j, u_i) \, \mathbf{I}_s^m(\tau, u_j) \right\} - \mathbf{Q}_c^m(\tau, u_i), \quad i = \pm 1, \dots, \pm N$$

$$u_i \frac{d\mathbf{I}_s^m(\tau, u_i)}{d\tau} = \mathbf{I}_s^m(\tau, u_i)$$

$$- \frac{\omega(\tau)}{4} \sum_{\substack{j=-N \\ j \neq 0}}^{N} w_j \left\{ \mathbf{P}_c^m(\tau, u_j, u_i) \, \mathbf{I}_s^m(\tau, u_j) \right. \tag{147}$$

$$\left. + \mathbf{P}_s^m(\tau, u_j, u_i) \, \mathbf{I}_c^m(\tau, u_j) \right\} - \mathbf{Q}_s^m(\tau, u_i), \quad i = \pm 1, \dots, \pm N.$$

The convention for the indices of the quadrature points is such that $u_j < 0$ for $j < 0$, and $u_j > 0$ for $j > 0$. These points are distributed symmetrically about zero, i.e., $u_{-j} = -u_j$. The corresponding weights are equal, i.e., $w_{-j} = w_j$.

Each of the two slabs (atmosphere and water) is divided into several adjacent layers, large enough to resolve vertical changes in the IOPs of each slab. Equations (147) and (148) apply in each layer in the atmosphere or water. As described in some detail elsewhere [3] the solution involves the following steps:

1. the homogeneous version of Equations (147) and (148) with $\mathbf{Q}_c^m = \mathbf{Q}_s^m = 0$ yields a linear combination of exponential solutions (with unknown coefficients) obtained by solving an algebraic eigenvalue problem;
2. analytic particular solutions are found by solving a system of linear algebraic equations;
3. the general solution is obtained by adding the homogeneous and particular solutions;
4. the solution is completed by imposing boundary conditions at the top of the atmosphere and the bottom of the water;

5. the solutions are required to satisfy continuity conditions across layer interfaces in the atmosphere and the water, and last but not least to satisfy Fresnel's equations and Snell's law at the atmosphere-water interface, where there is an abrupt change in the refractive index;

6. the application of boundary, layer interface, and atmosphere-water interface conditions leads to a sparse system of linear algebraic equations, and the numerical solution of this system of equations yields the unknown coefficients in the homogenous solutions.

4.3. Summary of AccuRT

We have described a computational tool, AccuRT, for radiative transfer simulations in a coupled system consisting of two adjacent horizontal slabs with different refractive indices. The computer code accounts for reflection and transmission at the interface between the two slabs, and allows for each slab to be divided into a sufficiently large number of layers to resolve the variation in the IOPs, described in Section 3, with depth in each slab.

The user interface of AccuRT is designed to make it easy to specify the required input including wavelength range, solar forcing, and layer-by-layer IOPs in each of the two slabs as well as the two types of desired output:

*   irradiances and mean radiances (scalar irradiances) at desired vertical positions in the coupled system;
*   total radiances and polarized radiances (including degree of polarization) in desired directions and vertical positions in the coupled system.

5. Examples of Forward-Inverse Modeling

5.1. Introduction

A primary goal in *remote sensing* of the Earth from space is to retrieve information about atmospheric and surface properties from measurements of the radiation emerging at the top-of-the-atmosphere (TOA) at several wavelengths [24,81]. These *retrieval parameters* (RPs), including cloud phase and optical depth, aerosol type and loading, and concentrations of aquatic constituents in an open ocean or coastal water area, depend on the *inherent optical properties* (IOPs) of the atmosphere and the water. If there is a model providing a link between the RPs and the IOPs, a forward radiative transfer (RT) model can be used to compute how the measured TOA radiation field should respond to changes in the RPs, and an inverse RT problem can be formulated and solved to derive information about the RPs [3,82]. A *forward RT model*, employing IOPs that describe how atmospheric and aquatic constituents absorb and scatter light, can be used to compute the *multiply scattered light field* in any particular direction (with specified polar and azimuth angles) at any particular depth level (including the TOA) in a vertically *stratified medium*, such as a coupled atmosphere-water system [83]. In order to solve the *inverse RT problem* it is important to have an accurate and efficient forward RT model. Accuracy is important in order to obtain reliable and robust retrievals, and efficiency is an issue because standard iterative solutions of the *nonlinear inverse RT problem* require executing the forward RT model repeatedly to compute the radiation field and its partial derivatives with respect to the RPs (the *Jacobians*) [82].

In addition to scalar forward RT models, vector RT models that consider polarization are important (see Section 4). Numerous RT models that include polarization effects are available (see Zhai et al. [84] and references therein for a list of papers), and the interest in applications based on vector RT models that apply to coupled atmosphere-water systems is growing. Examples of vector RT modeling pertinent to a coupled atmosphere-water system include applications based on the *doubling-adding method* [85–87], the *successive order of scattering method* [84,88,89], the *matrix operator method* [90,91], and *Monte Carlo methods* [92,93].

The purpose of this section is not provide a comprehensive review of forward-inverse methodology, but rather to provide a few examples of how RT modeling involving coupled atmosphere-water systems described in the previous sections can be used to solve the inverse problem with an emphasis of how machine learning techniques (neural networks) can be used to our advantage.

5.2. Bidirectional Reflectance of Water—Why Is It Important?

The **B**idirectional **R**eflectance **D**istribution **F**unction (BRDF) is defined as the ratio of the reflected radiance to the incident power per unit surface area:

$$\rho(\mu, \phi; -\mu', \phi') = \frac{dI_{\text{refl}}(\tau^*, \mu, \phi)}{I(\tau^*, -\mu', \phi') \, \mu' \, d\mu' d\phi'}. \tag{148}$$

Here $dI_{\text{refl}}(\tau^*, \mu, \phi)$ is the reflected radiance in direction (μ, ϕ), while $I(\tau^*, -\mu', \phi')$ is the incident radiance in direction $(-\mu', \phi')$. Understanding bidirectional effects including sunglint is important for several reasons [63,94]:

1. correct interpretation of ocean color data;
2. comparing consistency of spectral radiance data derived from space observations with a single instrument for a variety of illumination and viewing conditions;
3. merging data collected by different instruments operating simultaneously.

The BRDF defined in Equation (148) has unit per steradian [sr^{-1}]. The remote sensing reflectance defined as $R_{\text{rs}} = I(0^+, \mu, \phi)/F^-(0^+)$, where 0^+ refers the level just above the air-water interface, and $F^-(0^+)$ is the downward irradiance, also has unit [sr^{-1}]. It should be noted that the frequently used bidirectional reflectance factor defined as BRF $= \pi R_{\text{rs}}$ is dimensionless, because π has unit [sr].

A BRDF correction algorithm [63] (denoted as MAG02) was developed for application to open ocean water based on the following expression for the normalized water-leaving radiance nL_w

$$nL_w = L_w \times \frac{\Re_o}{\Re} \times \frac{f_0(\tau_a, W, \text{IOP})}{Q_0(\tau_a, W, \text{IOP})} \times \left[\frac{f(\theta_0, \theta, \Delta\phi, \tau_a, W, \text{IOP})}{Q(\theta_0, \theta, \Delta\phi, \tau_a, W, \text{IOP})} \right]^{-1}$$

where W is the wind speed, and the function $\Re$ accounts for refraction and reflection effects when radiances propagate through the air-water interface. The function f relates the irradiance reflectance ($R = F^+/F^-$) to the IOPs, and the function Q is a bidirectional function, defined as $Q(\theta_0, \theta', \Delta\phi) = F^+(0^-)/I^+(0^-, \theta_0, \theta', \Delta\phi)$. The subscripts "0" on $\Re_o$, f_0 and Q_0 are the values of the three functions evaluated in the nadir direction.

However, the MAG02 algorithm requires knowledge of CHL to derive the f/Q correction factor, and it does not work well in turbid (coastal) water. To remedy these shortcomings, Fan et al. [95] developed a neural network method to correct for bidirectional effects in water-leaving radiances for both clear (open ocean) and turbid (coastal) water. This neural network algorithm directly derives the entire spectral nadir remote sensing reflectances $R_{\text{rs}}(\lambda_i, \theta_0)$ from the angular values $R_{\text{rs}}(\lambda_i, \theta_0, \theta, \Delta\phi)$, without any prior knowledge of the water IOPs. Based on AccuRT simulations, Fan et al. [95] showed that differences in spectral R_{rs} values are significant between clear (open ocean) and turbid (coastal) water, but relatively small between nadir- and slant-viewing directions for a given water type. Consequently, a trained Radial Basis Function Neural Network (RBF-NN) can be used to convert the spectral R_{rs} values from the slant- to the nadir-viewing direction.

To this end, AccuRT was used to simulate R_{rs} values at both nadir- and slant-viewing directions for a 13-layer atmosphere with aerosols added in the bottom 0–2 km layer, by randomly selecting aerosol models based on fraction of small-mode aerosol particles (f_a) and relative humidity (RH). The CCRR bio-optical model, parameterized in terms of CHL, CDOM, and MIN as described above, was used to represent the water IOPs. To obtain the water-leaving radiance, the upward radiance was computed just above the ocean surface twice using AccuRT. Assuming the ocean to be black, i.e., totally absorbing (no scattering), the upward radiance was first computed just above the ocean surface, $I^+_{\text{black}}(0^+, \lambda, \theta_0, \theta, \Delta\phi)$, which includes the radiance due to Fresnel reflection of direct attenuated sunlight and skylight by the air-water interface, but no radiance from the water. The second time the ocean with water and its embedded constituents, was included and the radiance $I^+(0^+, \lambda, \theta_0, \theta, \Delta\phi)$ was computed. Hence, $I^+(0^+, \lambda, \theta_0, \theta, \Delta\phi)$ included the water-leaving radiance as well as the Fresnel reflected direct attenuated sunlight and skylight. Then the water-leaving radiance was obtained from the difference

$$L_w(0^+, \lambda, \theta_0, \theta, \Delta\phi) = I^+(0^+, \lambda, \theta_0, \theta, \Delta\phi) - I^+_{\text{black}}(0^+, \lambda, \theta_0, \theta, \Delta\phi). \tag{149}$$

To work satisfactorily, a neural network must be properly trained. For this purpose AccuRT was used to generate a training dataset containing 30,000 data points of $I^+(0^+, \lambda, \theta_0, \theta, \Delta\phi)$ and $I^+_{\text{black}}(0^+, \lambda, \theta_0, \theta, \Delta\phi)$ at seven wavelengths, 412, 443, 490, 510, 560, 620, and 665 nm, which are similar to the wavelengths used in the MAG02 algorithm. The synthetic dataset was generated by randomly selecting 5000 combinations of the aerosol optical depths at 865 nm ($\tau_a(865)$), the fraction of small aerosol particles (f_a), the relative humidity (RH), and the three ocean parameters: CHL, MIN, and CDOM. To cover a wide range of water and atmospheric IOPs, these six parameters were randomly sampled logarithmically from the following ranges: (i) $\tau_a(865)$: 0.001–0.5, (ii) f_a: 1–95 [%], (iii) RH: 30–95 [%], (iv) CHL: 0.01–100 [mg·m^{-3}], (v) MIN: 0.01–100 [g·m^{-3}], (vi) CDOM: 0.001–10 [m^{-1}]. Then for each case in the 5000 combinations, six combinations of the Sun-sensor geometry were randomly selected in the following ranges: (i) θ_0: 0–80 [°], (ii) θ: 0–70 [°], (iii) $\Delta\phi$: 0–180 [°]. The downward irradiance just above the ocean surface ($F^-(\lambda, 0^+)$) was also computed for each case as well as the remote sensing reflectance:

$$R_{\text{rs}}(\lambda, \theta_0, \theta, \Delta\phi) = \frac{L_w(0^+, \lambda, \theta_0, \theta, \Delta\phi)}{F^-(\lambda, 0^+)} \tag{150}$$

where $L_w(0^+, \lambda, \theta_0, \theta, \Delta\phi)$ is given by Equation (149).

The remote sensing reflectances were arranged into two groups: (i) one consisting of all the angle-dependent remote sensing reflectances $R_{\text{rs}}(\lambda_i, \theta_0, \theta, \Delta\phi)$, the other consisting of the the corresponding nadir remote sensing reflectances, $R_{\text{rs}}(\lambda_i, \theta_0)$. Then a neural network with two hidden layers was created. The first layer used RBFs as neurons, while the second layer used a linear function as neurons.

The input to the neural network training was the three geometry angles $\theta_0, \theta, \Delta\phi$ plus the angle-dependent remote sensing reflectances $R_{\text{rs}}(\lambda_i, \theta_0, \theta, \Delta\phi)$, while the output consisted of the corresponding nadir remote sensing reflectances, $R_{\text{rs}}(\lambda_i, \theta_0)$. After the training, the nadir remote sensing reflectances can be derived from a single equation:

$$R_{\text{rs}}(\lambda_i, \theta_0) = \sum_{j=1}^{N} a_{ij} \exp\left[-b^2 \sum_{k=1}^{N_{\text{in}}} (p_k - c_{jk})^2\right] + d_i \tag{151}$$

where N is the number of neurons, b and c_{jk} are the bias and weight in the first layer, a_{ij} and d_i are the weight and bias in the second layer. These weights and biases are optimized from the training procedure to minimize the error between the neural network derived $R_{\text{rs}}(\lambda_i, \theta_0)$ and the actual $R_{\text{rs}}(\lambda_i, \theta_0)$ values in the training dataset. N_{in} is the number of input parameters, which in our neural network equals 10: three geometry angles plus seven wavelengths. The input parameters are denoted by p_k, which in this case are the three geometry angles and the angle-dependent remote sensing reflectances, $R_{\text{rs}}(\lambda, \theta_0, \theta, \Delta\phi)$, at each of the seven wavelengths.

As discussed by Fan et al. [95], this neural network approach to convert remote sensing reflectances from actual slant-viewing to nadir-viewing directions was tested using synthetic data as well as field measurements. The results can be summarized as follows (see [95] for details):

- The generally anisotropic remote sensing reflectance of oceanic water must be corrected in remote sensing applications that make use of the nadir water-leaving radiance (or remote sensing reflectance) to derive ocean color products.

- The standard MAG02 correction method [63], based on the open ocean assumption, is unsuitable for turbid waters, such as rivers, lakes, and coastal water. The MAG02 method requires the chlorophyll concentration as an input, which is a drawback in remote sensing applications, because the chlorophyll concentration is generally produced from the corrected remote sensing reflectance.
- To meet the need for a correction method that works for water that may be dominated by turbidity or CDOM, Fan et al. [95] developed a neural network method that directly converts the remote sensing reflectance from the slant-viewing to the nadir-viewing direction.
- The neural network was trained using remote sensing reflectances at slant and nadir directions generated by a radiative transfer model (AccuRT), in which scattering phase functions for algal and non-algal particles were adopted. Therefore, the remote sensing reflectance implicitly contains information about the shape of the scattering phase function which affects the BRDF.
- This method uses spectral remote sensing reflectances as input. Hence, it does not require any prior knowledge of the water constituents or their optical properties.
- Tests based on synthetic data show that this method is sound and accurate. Validation using field measurements [96] shows that this neural network method works equally well compared to the standard method [63] for open ocean or chlorophyll-dominated water. For turbid coastal water a significant improvement over the standard method was found, especially for water dominated by sediment particles.

5.3. Sunglint: A Nuisance or Can Can It Be Used to Our Advantage?

For clarity, we should note that "glint" here refers only to Fresnel reflectance from the (calm or "wind-roughened") water surface. In the presence of glint, satellite remote sensing remains a challenging problem [97]. The contribution from glint to the TOA radiance is large enough to dominate the signals received by sensors deployed in space. Algorithms developed for current satellite sensors such as the Sea-viewing Wide Field of view Sensor (SeaWiFS), the MODerate-resolution Imaging Spectroradiometer (MODIS), the MEdium Resolution Imaging Spectrometer (MERIS), the Polarization and Directionality of Earth Reflectances (POLDER) instrument, and the Global Imager (GLI), use different correction algorithms [77,98,99] based on the same principle: estimate the glint contribution inferred from a statistical glint model and a direct beam reflectance and then remove its contribution from the signal received by the sensor.

To analyze remotely sensed radiances obtained by instruments such as SeaWiFS, MODIS, and MERIS, NASA has developed a comprehensive data analysis software package (SeaWiFS Data Analysis System, SeaDAS), which performs several tasks, including cloud screening and calibration, required to convert the raw satellite signals into calibrated TOA radiances. The SeaDAS software package also has tools aimed at quantifying and removing the atmospheric contribution to the TOA radiance (atmospheric correction) as well as contributions from whitecaps and sunglint due to reflections from the ocean surface [81].

In the SeaDAS algorithm (and similar algorithms) a sunglint flag is activated to mask out pixels for which the reflectance or BRDF, exceeds a certain threshold. If the reflectance for a given pixel is above the threshold, the signal is not processed. If the reflectance is below the threshold, a directly transmitted radiance (DTR) approach is used to calculate the TOA sunglint radiance in the SeaDAS algorithm. Thus, it is computed assuming that the direct beam and its reflected portion only experience exponential attenuation through the atmosphere [77], that is

$$I_{glint}^{TOA}(\mu_0, \mu, \Delta\phi) = F_0(\lambda)T_0(\lambda)T(\lambda)I_{GN} \tag{152}$$

$$T_0(\lambda)T(\lambda) = \exp\left\{-[\tau_M(\lambda) + \tau_A(\lambda)]\left(\frac{1}{\mu_0} + \frac{1}{\mu}\right)\right\} \tag{153}$$

where μ_0 and μ are cosines of the solar zenith angle and polar viewing angle, respectively, and the normalized sunglint radiance I_{GN} is the radiance that would result for a transparent atmosphere if the incident solar irradiance were $F_0(\lambda) = 1$. The Rayleigh (molecular) and aerosol optical thicknesses are denoted $\tau_M(\lambda)$ and $\tau_A(\lambda)$, respectively. The downward diffuse incident light (sunlight being multiply scattered by atmospheric molecules and aerosols before hitting the rough sea surface) also contributes to the upward reflectance. In the SeaDAS algorithm, such diffuse light reflectance that accounts for the effect of ocean surface roughness has been included only in the Rayleigh lookup tables [100].

Radiative transfer (RT) simulations may provide a more complete look at the glint problem [87]. An RT model can be used for accurate quantification of contributions not only from direct sunglint, but also from skyglint due to multiply scattered light [101]. Hence, RT simulations can be used to test current correction methods and explore the potential for extending remote sensing into strong glint situations masked out by the current SeaDAS algorithm.

As alluded to in Section 4.1.3, plane parallel RT models assume that the BRDF depends only on the difference in azimuth between the sun-sensor directions. Therefore, they are intrinsically one-dimensional, and cannot be used to simulate the directional dependence of realistic slope distributions that require a wind-direction dependent (hereafter referred to as 2D BRDF) treatment. Also, few studies have focussed on validation of realistic 2D BRDF implementations due to the general lack of a complete set of reflectance measurements that would be suitable for testing purposes.

To enable more realistic simulations of ocean glint reflectance, Lin et al. [80] developed a RT model with a 2D BRDF to mimic the nature of actual sea surface slope distributions (see Section 4.1.3 for details). Reflectance measurements obtained by an instrument deployed on a National Aeronautics and Space Administration (NASA) aircraft were used for validation. The goal was to match simulated reflectances with those measured by the instrument deployed on the NASA airplane, and to use RT simulations as a forward model to invert the measured reflectance in order to retrieve wind direction, sea surface slopes in the crosswind and upwind directions, and aerosol optical thickness. These parameters are important for atmospheric correction, which is the largest source of error and uncertainty in determining water-leaving radiance from space.

5.3.1. BRDF Measurements

NASA CAR Instrument

The measurements used by Lin et al. [80] were obtained under clear sky conditions from the NASA Cloud Absorption Radiometer (CAR) deployed aboard the University of Washington Convair 580 (CV-580) research aircraft [102]. The CAR is an airborne multi-wavelength scanning radiometer that measures scattered light in 14 spectral bands between 0.34 and 2.30 µm. To measure BRDFs, the airplane flew in a circle about 3 km in diameter, taking roughly 2–3 min to complete an orbit about 200 m above the surface. A servo control system is installed to allow the instrument to point at any angle from zenith to nadir, and to compensate for variations in airplane roll angle down to a fraction of a degree. Multiple circular orbits were acquired over a selected surface so that average BRDFs would be smooth. Radiometric calibration was performed at Goddard Space Flight Center prior to and just after the field experiment and a linear change between them is assumed. For more details about the BRDF measurements, see [102] and the official NASA link (http://car.gsfc.nasa.gov/).

CAR measurements provide accurate BRDFs for all geometry angles including zenith viewing angles from the nadir ($\theta = 0°$) to the horizon direction ($\theta = 90°$) over all relative azimuth angles (0–360°). The resolution for both polar and azimuth angles is 1°. The BRDF measurements are also accompanied by concurrent measurements of atmospheric aerosol optical thickness above the airplane and wind speeds from the NOAA Marine Environmental Buoy Database.

5.3.2. Radiative Transfer Simulations

The DISORT code [103] implemented in AccuRT provides accurate computations of singly and multiply scattered radiances in a turbid medium. The DISORT code has been used in a great variety of studies including remote sensing applications. Lin et al. [79] developed an upgraded version called DISORT3 with improved BRDF capabilities. The DISORT3 code can be obtained from the following web site: http://lllab.phy.stevens.edu/disort/. DISORT3 was further modified and optimized [80] to simulate the 2D (wind-direction dependent) nature of the surface reflectance.

Atmospheric Input & Output

To minimize the influence of light backscattered from the water, we used a near infrared wavelength at 1036 nm, with significant water absorption, and adopted two atmospheric layers: one Rayleigh (molecular) layer (2–10 km) and one layer with aerosols and molecules homogeneously mixed (0–2 km). Based on the US standard atmosphere [104], the single-scattering albedo at 1036 nm is $\varpi_{mol} = 0.9610$ in the upper layer (molecular scattering and water vapor absorption, no aerosols) and the upper layer optical thickness is $\tau_{mol} = 0.00645$. In the lower layer, we adopted an aerosol model implemented in SeaDAS [30]. The inherent optical properties (IOPs) of aerosols and molecules were then combined to give a "mixed" single-scattering albedo $\varpi_{mix} = (\beta_{mol} + \beta_{aer})/(\gamma_{mol} + \gamma_{aer}) = 0.9772$, where β_{mol} and β_{aer} are scattering coefficients, and γ_{mol} and γ_{aer} are extinction coefficients for molecules and aerosols, respectively.

The simulated atmospheric output is the radiance $I(\tau, \mu, \phi)$ in arbitrary directions (μ, ϕ). However, for comparison with measurements, we used the bidirectional reflectance factor defined as $BRF(\tau, \mu, \phi) = \pi I(\tau, \mu, \phi)/\mu_0 F_0$, where $\mu = \cos\theta$, θ is the zenith view angle, ϕ is the azimuth angle, τ is the optical thickness (at aircraft altitude), $\mu_0 = \cos\theta_0$, θ_0 being the solar zenith angle, and F_0 is the extraterrestrial solar irradiance. Please note that this BRF is defined such that it would represent the reflected irradiance normalized by the TOA incident irradiance $\mu_0 F_0$ if the radiance were to be isotropic.

5.3.3. Comparison between Measured and Simulated Reflectances

Retrieval Surface Roughness and Aerosol Parameters

Although the variances in the crosswind and upwind directions, σ_c^2 and σ_u^2, can be parameterized in terms of wind speed and direction [75], they may instead be considered to be model input parameters describing the 2D surface slope distribution. For a 1D surface, the slope variance is then automatically given as $\sigma^2 = \sigma_u^2 + \sigma_c^2$. This approach has the advantage that the slope variances do not depend on parameterizations in terms of wind speed, which provides more freedom to reproduce the measured 2D glint pattern by varying σ_c and σ_u.

In addition to reflection from Gaussian surface slope facets, multiple surface reflections, shadowing [see Equation (130)], and polarization effects also influence the glint signal [105,106]. Since multiple reflections and shadowing become important only for very low solar elevations and polarization effects are relatively unimportant at 1035 nm due to weak molecular (and aerosol) scattering, these effects can be ignored [80].

To determine the optimum match between model-simulated and CAR-measured reflectances, we need to invert the measurements to find the best estimate of the model parameters described by the state vector $\mathbf{x} = [\sigma_c^2, \sigma_u^2, \phi_{wind}, \tau_{mix}]^T$ including the four retrieval parameters: (i) the slope variance in the crosswind direction, σ_c^2, (ii) the slope variance in the upwind direction, σ_u^2, (iii) the wind direction, ϕ_{wind}, and (iv) the optical thickness τ_{mix} of the layer with a mixed population of aerosols and molecules. To this end, [80] used a Gauss-Newton/Levenberg-Marquardt non-linear inversion algorithm, in which the residual between reflectances produced by the RT forward model $\mathbf{F}(\mathbf{x})$ and the CAR measurements, stored in the vector $\mathbf{y}$, is minimized. The kth iteration of $\mathbf{x}$ yields:

$$\mathbf{x}_{k+1} = \mathbf{x}_k + \left[\mathbf{J}_k^T \mathbf{J}_k + \gamma_k \mathbf{I} \right] \mathbf{J}_k^T \left(\mathbf{F}(\mathbf{x}_k) - \mathbf{y}_k \right) \tag{154}$$

where the vector $\mathbf{J}$ contains the Jacobians of the forward model, $\mathbf{I}$ is identity matrix, and the parameter γ_k ($0 \leq \gamma_k \leq \infty$) is chosen at each step of the iteration to minimize the residual. If $\gamma_k = 0$ we have a classic Gauss-Newton method, while if γ_k is large we have a steepest descent method.

Figure 3 shows a comparison of model-simulated and measured reflectances at 1036 nm on 20 July 2001. Clearly, model-simulated results agree very well with the measurements, reproducing the main characteristics of the glint pattern. Both the shape of the simulated glint ellipse and its tilt are generally well matched with the measurements.

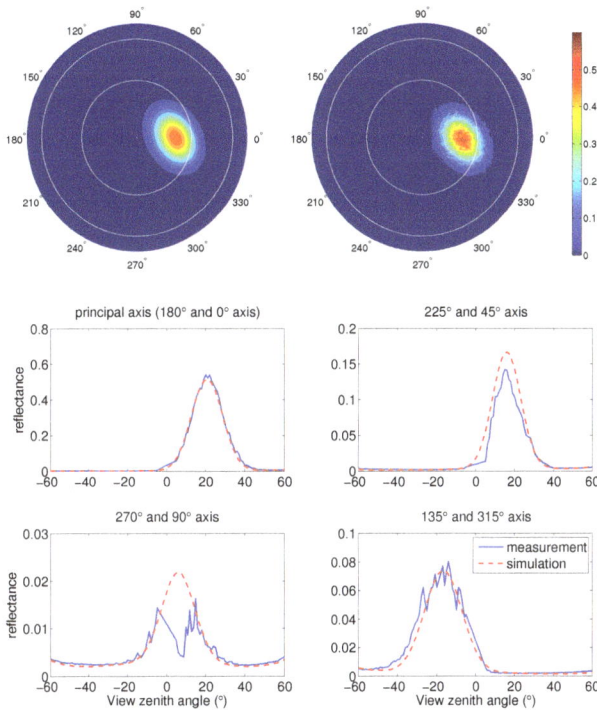

Figure 3. Comparison between model-simulated and measured reflectances. The measurements are in solid blue and the simulations in dashed red lines. The dip in the measured reflectance in the lower left panel is due to aircraft shadowing. [Reproduced from Figures 1a and 2a of [80].]

The shape of the elliptical glint pattern in Figure 3 is determined by the crosswind and upwind slope variances (σ_c^2 and σ_u^2), while the wind direction is determined from the tilt angle. Such a titled ellipse indicates that use of the 2D asymmetric Gaussian BRDF is needed to fit the angular distribution of the reflectance measurements, and that a 1D Gaussian is insufficient because it averages those slopes ($\sigma^2 = \sigma_c^2 + \sigma_u^2$) to give only a circular glint pattern. Figure 4 shows a comparison of 1D and 2D results for the case shown in Figure 3. It is clear that model simulations based on a 1D Gaussian BRDF are unable to match the measured tilted elliptical glint pattern. Hence, employing a 2D BRDF in the data analysis is essential.

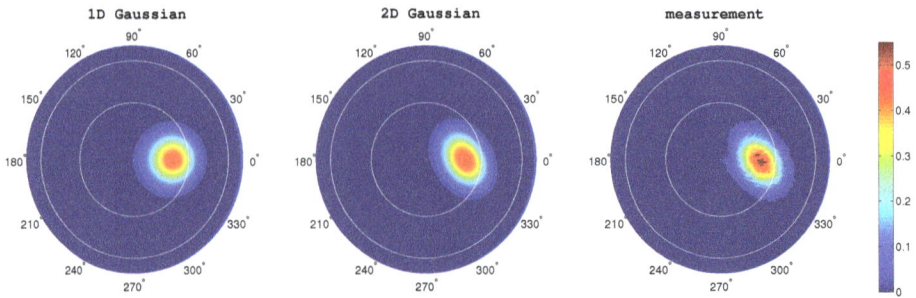

Figure 4. Comparison between model-simulated reflectances assuming a 1D Gaussian BRDF (**left**), a 2D Gaussian BRDF (**middle**), and measurements (**right**) obtained on 10 July 2001.

Since the surface roughness parameters (σ_c^2, σ_u^2, σ^2, and wind direction) inferred from the channel at 1036 nm are independent of wavelength, this information can be applied at shorter wavelengths (472, 672, and 870 nm) to improve the ocean color retrieval in the sun-glint area. The aerosol and ocean parameters can be retrieved from radiances outside the glint region (viewing angles 20–60 degrees and relative azimuth angles 120–240 degrees) using multi-angle reflectances in three CAR channels at 472, 682, and 870 nm. For this purpose, Lin et al. [80] used AccuRT, a RTM for the coupled atmosphere-ocean system (described in Sections 2–4) combined with an optimization technique [see Equation (154)] to retrieve aerosol and water constituent parameters simultaneously [70,107]. The Ahmad et al. aerosol model [30] and the CCRR bio-optical model [58] were used in AccuRT for this retrieval.

In this case, the retrieved aerosol optical depth at 870 nm was 0.086, the Ångstrøm coefficient (472 to 870 nm) was 1.463, the chlorophyll concentration was 0.67 mg·m^{-3}, the colored dissolved organic matter (CDOM) absorption coefficient at 443 nm was 0.07 m^{-1}, and the mineral particle concentration (MIN) was 0.009 g·m^{-3}. When these retrieval results were applied to the BRDF simulation, a very good match to the CAR measurements were obtained, especially in the glint area [80]. Hence, the glint information retrieved from the channel at 1.036 nm can be used in ocean color remote sensing to estimate the glint contribution at visible and NIR ocean color channels.

5.3.4. Summary of Glint Issues

The results discussed above may be summarized as follows:

- A wind-direction dependent Gaussian surface BRDF that uses (1) a 2D slope distribution for singly scattered light, and (2) a 1D slope distribution for multiply scattered light, can be used to successfully simulate BRDF measurements obtained by NASA's Cloud Absorption Radiometer (CAR) at the 1036 nm wavelength.

- Upwind and crosswind slope variances, wind direction, and aerosol optical depth, can be accurately retrieved through forward-inverse modeling.

- The glint parameters (slope variances and wind direction) can be applied to estimate the glint contribution at visible and NIR wavelengths, resulting in a very good match between model-simulated and measured reflectances.

- An advantage of RT simulations of glint reflectance is its inclusion of contributions from the diffuse or multiply scattered light due to scattering by atmospheric molecules and aerosols. The diffuse light reflectance ("skyglint") gives an additional glint signal in addition to "sunglint" resulting from the direct beam reflectance.

- Simulations show that the diffuse glint may contribute more than 4% at 472 nm for a wind speed of 2 m/s, and more than 8% when the wind speed increases to 8 m/s. Hence, the diffuse light reflectance should be considered in the visible bands, especially for large wind speeds.

- A simplified version of the pseudo 2D BRDF and glint reflectance method described above has been implemented in the AccuRT model for the coupled atmosphere-ocean system.

5.4. Retrievals of Atmosphere-Water Parameters from Geostationary Platforms: Challenges and Opportunities

Simultaneous retrieval of aerosol and surface properties by means of inverse techniques based on a coupled atmosphere-surface radiative transfer model and optimal estimation can yield improved retrieval accuracy in complex aquatic environments compared with traditional methods. At high latitudes low solar elevations is a problem, and if one desires to do satellite remote sensing from a geostationary platform in order to study diurnal variations, then large solar zenith and viewing angles become important issues to be resolved.

Satellite remote sensing for such complex situations/environments represents specific challenges due to:

- (i) the complexity of the atmosphere and water inherent optical properties,
- (ii) the unique bidirectional dependence of the water-leaving radiance, and
- (iii) the desire to do retrievals for large solar zenith and viewing angles.

Hence, one needs to consider challenges related to how

1. atmospheric gaseous absorption, absorbing aerosols, and turbid waters can be addressed by using a *coupled* forward model in the retrieval process,
2. corrections for *bidirectional effects* will be accomplished,
3. the *curvature* of the atmosphere will be taken into account, and
4. *uncertainty assessments* and error budgets will be dealt with.

The generic problem is illustrated in Figures 5 and 6, which show that

- there is *a significant change* in sub-surface color with increasing chlorophyll concentration, while at the same time
- there is *only a slight change* in color at the TOA, where *the spectra are dominated by light from atmospheric scattering.*

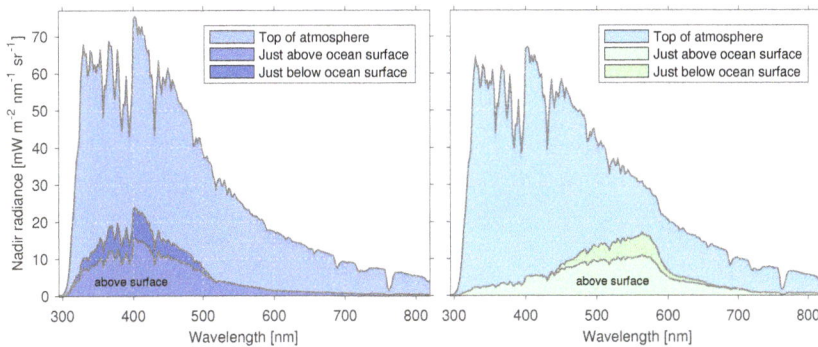

Figure 5. Simulated upward radiance in the nadir direction at the top of the atmosphere and close to the ocean surface. Solar zenith angle = 45°, US Standard atmosphere with aerosol optical depth = 0.23 at 500 nm. (**Left**) Clear water with chlorophyll concentration = 0.1 mg·m^{-3}, MIN = 0.003 g·m^{-3}, CDOM443 = 0.003 m^{-1} (CCRR bio-optical model). (**Right**) Turbid water with chlorophyll concentration = 10 mg·m^{-3}, MIN = 0.1 g·m^{-3}, CDOM443 = 0.1 m^{-1}.

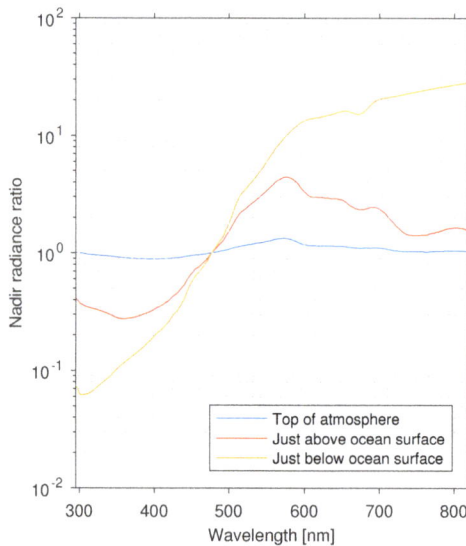

Figure 6. The ratio of the values for turbid water to those for clear water in Figure 5.

Most ocean color algorithms consist of two steps:

1. First, one does an "atmospheric correction" (assuming the water to be black at NIR wavelengths) to determined the water-leaving radiance.
2. Second, one retrieves the desired aquatic parameters from the water-leaving radiance.

In the visible, up to 90% of the radiance measured by a satellite sensor typically comes from the atmosphere implying that:

*   Atmospheric correction becomes a very challenging task unless the near-infrared (NIR) *black-pixel approximation (BPA)* is valid.
*   Estimation of diffuse transmittance is also important, but difficult because it depends on the angular distribution of the radiance just beneath the water surface.

Also, accurate characterization of the atmosphere is important because:

*   a small uncertainty in the atmospheric correction may lead to a big error in the inferred aquatic parameters, and
*   aerosol optical properties vary considerably in space and time.

5.4.1. The OC-SMART Optimal Estimation Approach

To address this situation, the OC-SMART (**O**cean **C**olor—**S**imultaneous **M**arine and **A**erosol **R**etrieval **T**ool) approach was developed. The goal was to improve retrieval accuracy by use of AccuRT forward modeling and Optimal Estimation/Levenberg-Marquardt (OE/LM) inversion [108]:

*   AccuRT: accurate discrete-ordinates radiative transfer model for the *coupled* atmosphere-ocean system; delivers a complete set of simulated radiances and Jacobians (weighting functions).
*   OE/LM inversion: is an iterative, nonlinear least squares cost function minimization with *a priori* and Levenberg-Marquardt regularization.
*   For retrievals of aerosol and aquatic parameters from Ocean Color data, we define a 5-element state vector:

$$\mathbf{x} = \{ \tau_{865}, f_a, \mathbf{CHL}, \mathbf{CDM}, \mathbf{BBP} \}$$

consisting of

- *2 aerosol parameters* (τ_{865} = optical depth at 865 nm and f_a = bimodal fraction of particles),
- *3 marine parameters* (chlorophyll concentration, **CHL**, combined absorption by detrital and dissolved material at 443 nm, **CDM**, and backscattering coefficient at 443 nm, **BBP**).

At each iteration step, the next estimate of the state vector is given by the OE/LM inversion [70]

$$\mathbf{x}_{n+1} = \mathbf{x}_n + [(1+\gamma_n)\mathbf{S}_a^{-1} + \mathbf{J}_n^T\mathbf{S}_m^{-1}\mathbf{J}_n]^{-1}\{\mathbf{J}_n^T\mathbf{S}_m^{-1}(\mathbf{y_m} - \mathbf{y}_n) - \mathbf{S}_a^{-1}(\mathbf{x}_n - \mathbf{x}_a)\}. \tag{155}$$

In Equation (155) $\mathbf{y}_m$ is the vector of measured TOA radiances; $\mathbf{y}_n = F(\mathbf{x}_n, \mathbf{b})$ is the vector of simulated TOA radiances generated by the AccuRT forward model; $\mathbf{y}_n$ is a (non-linear) function of the state vector $\mathbf{x}_n$ of retrieval elements, and $\mathbf{b}$ represents model parameters; $\mathbf{J}_n$ is a matrix of simulated radiance partial derivatives with respect to the state vector elements $\mathbf{x}_n$ (the Jacobians); $\mathbf{x}_a$ and $\mathbf{S}_a$ are the *a priori* state vector and covariance matrix, respectively, and $\mathbf{S}_m$ is the measurement error covariance matrix. γ_n is the Levenberg-Marquardt (LM) regularization parameter, When $\gamma_n \to 0$ Equation (155) becomes the standard Gauss-Newton Optimal Estimation (OE), and when $\gamma_n \to \infty$ it tends to the steepest descent method. AccuRT returns simulated radiances ($\mathbf{y}_n$) and Jacobians ($\mathbf{J}_n$) required to update the state vector estimate ($\mathbf{x}_n$) according to Equation (155) above.

One issue with the OC-SMART approach is that it is relatively slow due to the need to call the forward AccuRT model repeatedly in the iterative inversion to compute radiances and Jacobians. To deal with that problem, one could use AccuRT instead to create a training ensemble in order to construct a Radial Basis Function Neural Network. This approach typically leads to an increase in computational speed by a factor of about 1000. In this way one replaces the AccuRT forward model (thousands of lines of code) with the following single equation (similar to Equation (151)):

$$I_i = \sum_{j=1}^{N} a_{ij} \exp[-b\sum_{k=1}^{K}(P_k - c_{jk})^2] + d_i$$

where I_i it the TOA radiance in channel $i = 1, \ldots, 8$, K = # of input parameters, and $a_{i,j}, b, c_{j,k}, d_i$ are the coefficients to be optimized in the network training.

The Jacobians (partial derivatives) $\mathbf{J}$ are also required for the non-linear optimal estimation using Equation (155). These Jacobians can be calculated by taking partial derivatives of I_i respect to each input parameter:

$$J_{ik} = \frac{\partial I_i}{\partial P_k} = -2\sum_{j=1}^{N} a_{ij}b^2(P_k - c_{jk})\exp\{\sum_{k=1}^{K}(P_k - c_{jk})^2\}. \tag{156}$$

Fan et al. [107] applied the OC-SMART algorithm to a MODIS image obtained on 18 April 2014 over a Norwegian coastal area and compared retrieval results with the standard SeaDAS retrievals as shown in Figure 7 with OC-SMART results at the top and SeaDAS results at the bottom. From left to right, the columns are τ_{869}, f_a, CHL, CDOM and b_{bp}, respectively. The first two columns are retrieved aerosol parameters. Our results are very similar to the SeaDAS retrievals, which makes sense because the two algorithms share the same aerosol model. However, the marine parameters show some differences and we should point out that in OC-SMART, we used the GSM bio-optical model when we retrieved the ocean parameters. The SeaDAS CHL retrieval shown in Figure 7 is from the OC4v6 algorithm. SeaDAS failed when using the GSM model due to negative water-leaving radiances resulting from an incorrect atmospheric correction, which is a common issue in traditional atmospheric correction algorithms. The aerosol optical depths was determined at a near infrared (NIR) wavelength (τ_{869}) and then extrapolated to shorter wavelengths. Since the water-leaving radiance is only a small fraction of the TOA radiance measured by the satellite instrument, a small error in the aerosol retrieval may cause the water-leaving radiance to become negative. We also found the OC-SMART results to be close to the field measurements. The CHL value was 0.56 [mg/m^3] while the field measurements showed an average of 0.86 ± 0.34 [mg/m^3] in the spring season. The CDOM absorption coefficient was 0.15 [m^{-1}] and the field measurements show an average of 0.14 ± 0.06 [m^{-1}].

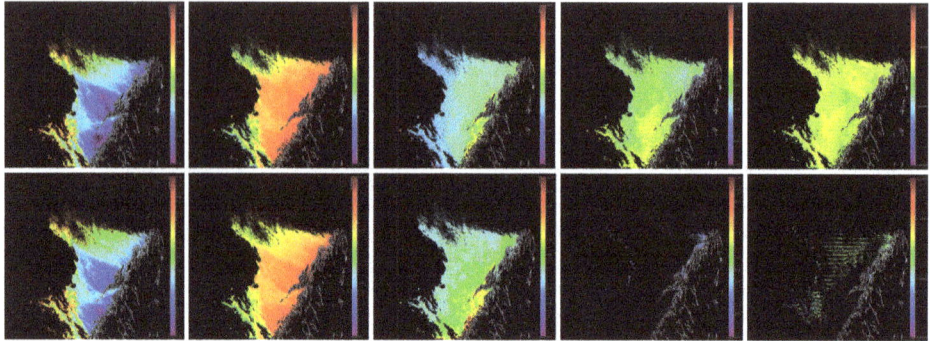

Figure 7. MODIS image comparison between OC-SMART (**top**) and standard SeaDAS (**bottom**) retrievals on 18 April 2014 over a Norwegian coastal area [107]. From left to right: τ_{869}, f, CHL, CDOM and b_{bp}, respectively.

5.4.2. The OC-SMART Multilayer Neural Network Approach

In open ocean areas where the water IOPs are correlated with pigmented particles, standard atmospheric correction (AC) algorithms seem to work reasonably well. However, in turbid coastal water areas the IOPs of suspended inorganic particles, and colored dissolved organic matter (CDOM) may vary independently of pigmented particles. Therefore, in turbid coastal waters standard AC algorithms often exhibit large inaccuracies that may lead to highly uncertain and frequently negative water-leaving radiances (L_w) or remote sensing reflectances (R_{rs} values). To address this problem, Fan et al. [109] introduced a new atmospheric correction algorithm for coastal water areas based on a multilayer neural network (MLNN) method. A coupled atmosphere-ocean radiative transfer model (AccuRT) was used to simulate the Rayleigh-corrected radiance (L_{rc}) at the TOA and the R_{rs} just above the surface simultaneously, and to train a MLNN to derive the aerosol optical depth (AOD) and R_{rs} values directly from the TOA L_{rc}. The method was validated using both a synthetic dataset and Aerosol Robotic Network–Ocean Color (AERONET–OC) measurements.

Extensive testing has shown that this MLNN approach has several advantages [109]:

1. It significantly improved the quality of retrieved remote sensing reflectances (compared to the SeaDAS NIR algorithm) by reducing the average percentage difference (APD) between MODIS retrievals and ground-truth (AERONET-OC) validation data.
2. In highly absorbing coastal water, such as the Baltic Sea, it provides reduction of the APD by more than 60%, and in highly scattering water, such as the Black Sea, it provides reduction of the APD by more than 25%.
3. It is robust and resilient to contamination due to sunglint and adjacency effects of land and cloud edges.
4. It is applicable in extreme conditions such as those encountered for heavily polluted continental aerosols, extreme turbid water, and dust storms.
5. It does not require shortwave infrared (SWIR) bands, and is therefore suitable for all ocean color sensors.
6. It is very fast and suitable for operational use.

Recent addition of training data to make it representative for a variety of water types has shown that this approach can produce a seamless (smooth) transition between turbid coastal water areas and clean open ocean areas.

5.4.3. Issues Specific to Geostationary Platforms

Low Solar Elevations

With geostationary platforms one has the opportunity to investigate diurnal variations in water properties. Hence, there is a desire to obtain useful observations throughout the day from sunrise to sunset. However, for solar zenith angles larger than about $75°$ (and large viewing angles) the *plane-parallel approximation* (PPA) becomes invalid, and we need to take Earth curvature into account, as discussed in a recent paper [110]. An approximate way to deal with this problem is the so-called *pseudo-spherical approximation* (PSA), in which the direct beam single scattering (solar pseudo-source) term is treated in spherical geometry: $e^{-\tau/\mu_0} \to e^{-\tau\, Ch(\mu_0)}$, where the Chapman function $Ch(\mu_0)$ takes curvature into account [14,110], while the multiple scattering term is treated using the PPA. Hence, in the PSA the RTE becomes:

$$u\frac{dI(\tau,u,\phi)}{d\tau} = I(\tau,u,\phi) - \overbrace{\frac{\omega(\tau)}{4\pi}\int_0^{2\pi} d\phi' \int_{-1}^1 du'\, p(\tau,u',\phi';u,\phi)I(\tau,u',\phi')}^{\text{multiple scattering}}$$

$$\underbrace{-\frac{\omega(\tau)}{4\pi}p(\tau,-\mu_0,\phi_0;u,\phi)F_0 e^{-\tau\, Ch(\mu_0)}}_{\text{single scattering}}. \tag{157}$$

It should be pointed out that low solar elevations is generally a problem at high latitudes also for polar-orbiting platforms.

Surface Roughness Considerations

As alluded to above, it would be worthwhile considering the advantage gained by using a 2D rather than a 1D slope distribution to deal with a wind-roughened surface (see Figure 4). Hence, one should consider using

1. a 2D Gaussian surface slope distribution for singly scattered light, and
2. a 1D Gaussian surface slope distribution for multiply scattered light.

Such an approach may be quite successful, as demonstrated in Figure 3, because a 2D BRDF simulates sunglint very well, while a 1D BRDF is sufficient to simulate the smoother (more directionally uniform) skylight.

Use of Vector (Polarized) RT Simulations

Analysis of polarization measurements require vector RT modeling [111], and retrieval results from radiance-only measurements are likely to be improved by employing a vector (polarized) forward RT model to compute the radiances and Jacobians used in the inversion step, particularly for modeling the radiance due to scattering by small aerosol particles at short wavelengths. Also, the increased information from polarization measurements can lead to significant improvements in atmospheric correction due to aerosols [112]. Hence, for ocean color retrievals from geostationary platforms, we should explore the advantage of using the pseudo-spherical approximation combined with

- polarized (vector) radiative transfer simulations,
- a 2D Gaussian distribution of surface slopes,

and neural networks and optimal estimation for

- simultaneous retrieval of atmospheric and marine parameters from multi-spectral as well as hyperspectral measurements of total and polarized (if available) radiances, and
- assessments of retrieval accuracy and error budgets.

6. Remaining Problems

6.1. 3-D Radiative Transfer: The LiDAR Problem

Although we have mostly considered *plane-parallel* systems so far with an emphasis on the coupling between the atmosphere and the underlying surface consisting of a water body or a snow/ice surface, there are many applications that require a three-dimensional (3-D) RT treatment. For a clear (cloud- and aerosol-free) atmosphere, 3-D effects are related to the impact of the Earth's curvature on the radiation field as discussed briefly in Section 5.4.3. To include Earth curvature effects it may be sufficient to employ a *"pseudo-spherical"* treatment, in which the direct solar beam illumination is treated in spherical geometry [110,113], whereas *multiple scattering* is treated in plane-parallel geometry. In fact, this pseudo-spherical approach has been implemented in many RT codes [114,115]. Also, there is a large body of literature on 3-D RT modeling with applications to broken clouds, and readers interested in this topic may want to consult [116] or visit the web-site http://i3rc.gsfc.nasa.gov/.

Finally, 3-D RT modeling may also be important for analysis and interpretation of LiDAR data. The classical "searchlight problem" [117], which considers the propagation of a laser beam through a turbid medium, is relevant in this context. Long-range propagation of a LiDAR beam has been studied both theoretically and experimentally [118]. Monte Carlo simulations are well suited for such studies [119], and use of deterministic models such the *discrete-ordinate method*, briefly discussed in Section 4, have also been pursued to investigate this problem [120,121].

6.2. Time-Dependent Radiative Transfer

Most studies of radiative transfer in the ocean have been concerned with understanding the propagation of sunlight through natural water bodies [3,4,14,122]. For such applications, the transient or *time-dependent* term in the RTE can be ignored, because changes in the incident illumination are much slower than the changes imposed by light propagation through the coupled atmosphere-water system. While this assumption is satisfied for solar illumination, LiDAR systems can use pulses that are shorter than the attenuation distance of seawater divided by the speed of light in water. Also, it has been pointed out [123] that due to multiple light scattering, understanding the LiDAR signal requires a solution of the *time-dependent* RTE. The transient RT problem can be reduced to solving a series of *time-independent* RT problems [124].

6.3. Other Issues

Although inelastic scattering processes (Raman and Brillouin) certainly can be very important and indeed essential in some atmospheric [125–127] and aquatic [128,129] applications, we have limited our discussion to elastic scattering. In fact, inelastic scattering effects can become important for very clear water and sky conditions.

Forward radiative transfer models need accurate IOPs as input. Therefore, in situ measurements are needed of the scattering phase function of hydrosols in clear as well as turbid (coastal) water, including measurements at a scattering angle close to 180°. The 180°-backscatter angle is of particular interest for LiDAR applications. In view of the growing interest in applications based on RT models for coupled systems [84,86,87,90–93], there is also a need for systematic and sustained measurements of the scattering phase (or Muller) matrix, since not much work has been done after the 1984 Voss and Fry publication [130].

To process hyperspectral data one needs algorithms that can handle large data volumes and deal with measurement uncertainties in a proper manner. In general, better bio-optical models are also needed. For example, the approximation that the spectral slope $v = 0$ for CHL > 2.0 in Equation (97) is problematic. Hyperspectral (and potentially polarimetric) measurements are needed to construct more generally applicable bio-optical models.

In Section 5.4.2 we discussed how a forward-inverse modeling approach, based on AccuRT simulations for the coupled atmosphere-water system combined with a multilayer neural network

(MLNN), can be used to infer accurate remote sensing reflectances (R_{rs} values) from clear as well as turbid water [109]. The logical next step is to use these R_{rs} values to infer water IOPs. In fact, several semi-analytic algorithms have been developed for this purpose including the generalized IOP (GIOP) algorithm described by Werdell and colleagues [131].

This GIOP approach is currently the default algorithm used by NASA for processing of MODIS ocean color data. First, the R_{rs} values are obtained after an atmospheric correction step using NIR channels. Ocean IOP parameters are then retrieved from the approximate R_{rs} values (based on many assumptions) by performing a non-linear inversion. The QAA algorithm [132] is another semi(quasi)-analytic algorithm, similar to the GIOP, that performs better than the GIOP in some situations. The GIOP and QAA algorithms generally perform well over the open ocean, but can have significant issues in coastal water because the atmospheric correction method produces inaccurate R_{rs} values that lead to large errors in derived IOPs. Furthermore, the performance of the atmospheric correction is worse the farther the extrapolation is taken from the NIR to shorter wavelengths, which is a problem for ocean color sensors with UV channels. Also, particle size distribution and scattering phase function constitute *a priori* information in the semi-analytic algorithms that cannot be retrieved. These limitations make them unsuitable for application to new sensors with increased information content, such as hyperspectral ocean color sensors. Instruments that can measure the polarization of light will also be extremely useful for improving aerosol retrievals and thus atmospheric correction capabilities. The next generation of sensors composed of hyperspectral polarimeters and ocean-capable LiDAR instruments will enable the use of more advanced retrieval algorithms to more accurately quantify and monitor the atmosphere-ocean system, particularly in coastal areas. Nevertheless, use of accurate R_{rs} values, obtained from new and improved AC methods, such as the MLNN algorithm [109], in existing semi-analytic algorithms (like GIOP and QAA) is expected to significantly enhance their performance when applied to current ocean color sensor data obtained over turbid water.

7. Summary

A review has been provided of forward and inverse radiative transfer modeling in coupled systems consisting of two adjacent, horizontal slabs with different indices of refraction and a rough interface characterized by a Gaussian distribution of surface slopes in one or two dimensions. Such a configuration can be used to simulate radiative transfer in coupled atmosphere-snow/ice-ocean systems. Input and output parameters including boundary conditions for the forward radiative transfer problem were introduced in Section 2, while in Section 3 a review was provided of inherent optical properties (IOPs) of the atmosphere, snow/ice and ocean water. These IOPs are required inputs to the radiative transfer equation introduced in Section 4 both for unpolarized (scalar) and polarized (vector) radiative transfer. Examples of how to solve inverse problems occurring in remote sensing of the environment employing optimization techniques as well as the unique power of machine learning (neural networks) for convenience and efficiency were provided in Section 5. Finally, in Section 6 some remaining radiative transfer problems were discussed.

Author Contributions: K.S. wrote the first draft of the paper. The other authors (B.H., S.S., N.C., Y.F., W.L., Z.L., and J.S.) contributed by reviewing, editing, and making additions to various parts of the paper.

Funding: This work was partially supported by funding from the National Aeronautics and Space Administration (NASA).

Acknowledgments: We thank two anonymous reviewers for constructive comments that led to a significant improvement of this paper.

References

1.   Ricchiazzi, P.; Yang, S.; Gautier, C.; Sowle, D. SBDART: A research and teaching software tool for plane-parallel radiative transfer in the Earth's atmosphere. *Bull. Am. Meteorol. Soc.* **1998**, *79*, 2101–2114. [CrossRef]

2. Key, J.R.; Schweiger, A.J. Tools for atmospheric radiative transfer: Streamer and FluxNet. *Comput. Geosci.* **1998**, *24*, 443–451. [CrossRef]

3. Stamnes, K.; Stamnes, J.J. *Radiative Transfer in Coupled Environmental Systems*; Wiley-VCH: Weinheim, Germany, 2015.

4. Mobley, C.D. *Light and Water*; Academic Press: Cambridge, MA, USA, 1994.

5. Hovenier, J.W.; der Mee, C.D.V.; Domke, H. *Transfer of Polarized Light in Planetary Atmospheres*; Kluwer Academic Publsihers: Dordrecht, The Netherlands, 2004.

6. Rayleigh, L. A re-examination of the light scattered by gases in respect of polarization. II. Experiments on helium and argon. *Proc. R. Soc.* **1920**, *98*, 57–64. [CrossRef]

7. Morel, A. Optical properties of pure water and pure seawater. In *Optical Aspects of Oceanography*; Jerlov, N.G., Nielsen, E.S., Eds.; Academic Press: Cambridge, MA, USA, 1974; pp. 1–24.

8. Morel, A.; Gentili, B. Diffuse reflectance of oceanic waters: its dependence on sun angle as influenced by the molecular scattering contribution. *Appl. Opt.* **1991**, *30*, 4427–4437. [CrossRef] [PubMed]

9. Rayleigh, L. On the light from the sky, its polarization and colour. *Philos. Mag.* **1871**, *41*, 107–120, 274–279, 447–454.

10. Bodhaine, B.; Wood, N.; Dutton, E.; Slusser, J. On Rayleigh optical depth calculations. *J. Atmos. Ocean. Technol.* **1999**, *16*, 1854–1861. [CrossRef]

11. Farinato, R.S.; Rowell, R.L. New values of the light scattering depolarization and anisotropy of water. *J. Chem. Phys.* **1976**, *65*, 593–595. [CrossRef]

12. Zhang, X.; Hu, L. Estimating scattering of pure water from density fluctuation of the refractive index. *Opt. Express* **2009**, *17*, 1671–1678. [CrossRef] [PubMed]

13. Henyey, L.C.; Greenstein, J.L. Diffuse radiation in the galaxy. *Astrophys. J.* **1941**, *93*, 70–83. [CrossRef]

14. Stamnes, K.; Thomas, G.E.; Stamnes, J.J. *Radiative Transfer in the Atmosphere and Ocean*, 2 ed.; Cambridge University Press: Cambridge, UK, 2017.

15. Hovenier, J.W.; van der Mee, C.V.M. Fundamental relationships relevant to the transfer of polarized light in a scattering atmosphere. *Astron. Astrophys.* **1983**, *128*, 1–16.

16. Chandrasekhar, S. *Radiative Transfer*; Dover Publications: Mineola, NY, USA, 1960.

17. De Haan, J.; Bosma, P.; Hovenier, J. The adding method for multiple scattering calculations of polarized light. *Astron. Astrophys.* **1987**, *183*, 371–391.

18. Siewert, C. A discrete-ordinates solution for radiative-transfer models that include polarization effects. *J. Quant. Spectrosc. Radiat. Transf.* **2000**, *64*, 227–254. [CrossRef]

19. Cohen, D.; Stamnes, S.; Tanikawa, T.; Sommersten, E.R.; Stamnes, J.J.; Lotsberg, J.K.; Stamnes, K. Comparison of Discrete Ordinate and Monte Carlo Simulations of Polarized Radiative Transfer in two Coupled Slabs with Different Refractive Indices. *Opt. Express* **2013**, *21*, 9592–9614. [CrossRef] [PubMed]

20. Siewert, C.E. On the equation of transfer relevant to the scattering of polarized light. *Astrophys. J.* **1981**, *245*, 1080–1086. [CrossRef]

21. Siewert, C.E. On the phase matrix basic to the scattering of polarized light. *Astron. Astrophys.* **1982**, *109*, 195–200.

22. Sommersten, E.R.; Lotsberg, J.K.; Stamnes, K.; Stamnes, J.J. Discrete ordinate and Monte Carlo simulations for polarized radiative transfer in a coupled system consisting of two media with different refractive indices. *J. Quant. Spectrosc. Radiat. Transf.* **2010**, *111*, 616–633. [CrossRef]

23. Mishchenko, M.I. Light scattering by randomly oriented rotationally symmetric particles. *J. Opt. Soc. Am. A* **1991**, *8*, 871–882. [CrossRef]

24. Mishchenko, M.I.; Travis, L.D. Satellite retrieval of aerosol properties over the ocean using polarization as well as intensity of reflected sunlight. *J. Geophys. Res.* **1997**, *102*, 16989–17013. [CrossRef]

25. Anderson, G.P.; Clough, S.A.; Kneizys, F.X.; Chetwynd, J.H.; Shettle., E.P. *AFGL Atmospheric Constituent Profiles (0–120 km), AFGL-TR-86-0110 (OPI)*; Optical Physics Division, Air Force Geophysics Laboratory Hanscom AFB: Bedford, MA, USA, 1986.

26. Berk, A.; Anderson, G.P.; Acharya, P.K.; Bernstein, L.S.; Muratov, L.; Lee, J.; Fox, M.; Adler-Golden, S.M.; Chetwynd, J.H.; Hoke, M.L.; et al. *MODTRAN 5: A Reformulated Atmospheric Band Model with Auxiliary Species and Practical Multiple Scattering Options: Update*; Defense and Security, International Society for Optics and Photonics: Orlando, FL, USA, 2005; pp. 662–667.

27. Berk, A.; Conforti, P.; Kennett, R.; Perkins, T.; Hawes, F.; van den Bosch, J. *MODTRAN6: A Major Upgrade of the MODTRAN Radiative Transfer Code*; SPIE Defense+ Security, International Society for Optics and Photonics: Orlando, FL, USA, 2014; p. 90880H.

28. Kneizys, F.X.; Abreu, L.W.; Anderson, G.; Chetwynd, J.; Shettle, E.; Berk, A.; Bernstein, L.; Roberson, D.; Acharya, P.; Rothman, L.; et al. *MODTRAN2/3 Report and LOWTRAN 7 Model*; Technical Report; Phillips Laboratory, Hanscom AFB: Bedford, MA, USA, 1996.

29. Chen, N.; Li, W.; Tanikawa, T.; Hori, M.; Shimada, R.; Aoki, T.; Stamnes, K. Fast yet accurate computation of radiances in shortwave infrared satellite remote sensing channels. *Opt. Express* **2017**, *25*, A649–A664. [CrossRef] [PubMed]

30. Ahmad, Z.; Franz, B.A.; McClain, C.R.; Kwiatkowska, E.J.; Werdell, J.; Shettle, E.P.; Holben, B.N. New aerosol models for the retrieval of aerosol optical thickness and normalized water-leaving radiances from the SeaWiFS and MODIS sensors over coastal regions and open oceans. *Appl. Opt.* **2010**, *49*, 5545–5560. [CrossRef] [PubMed]

31. Hess, M.; Koepke, P.; Schult, I. Optical properties of aerosols and clouds: The software package OPAC. *Bull. Am. Met. Soc.* **1998**, *79*, 831–844. [CrossRef]

32. Davies, C.N. Size distribution of atmospheric particles. *J. Aerosol Sci.* **1974**, *5*, 293–300. [CrossRef]

33. Holben, B.N.; Eck, T.F.; Slutsker, I.; Tanre, D.; Buis, J.P.; Setzer, A.; Vermote, E.; Reagan, J.A.; Kaufman, Y.; Nakajima, T.; et al. AERONET—A federated instrument network and data archive for aerosol characterization. *Remote Sens. Environ.* **1998**, *66*, 1–16. [CrossRef]

34. Holben, B.N.; Tanre, D.; Smirnov, A.; Eck, T.F.; Slutsker, I.; Abuhassan, N.; Newcomb, W.W.; Schafer, J.; Chatenet, B.; Lavenue, F.; et al. An emerging ground-based aerosol climatology: aerosol optical depth from AERONET. *J. Geophys. Res.* **2001**, *106*, 12067–12097. [CrossRef]

35. Hansen, J.E.; Travis, L.D. Light scattering in planetary atmospheres. *Space Sci. Rev.* **1974**, *16*, 527–610. [CrossRef]

36. Hänel, G. The properties of atmospheric aerosol particles as functions of the relative humidity at thermodynamic equilibrium with the surrounding moist air. In *Advances in Geophysics*; Landsberg, H.E., Miehem, J.V., Eds.; Elsevier: New York, NY, USA, 1976; Volume 19.

37. Shettle, E.P.; Fenn, R.W. *Models for the Aerosols of the Lower Atmosphere and the Effects of Humidity Variations on their Optical Properties*; Air Force Geophysics Laboratory, Hanscomb AFB: Bedford, MA, USA, 1979.

38. Yan, B.; Stamnes, K.; Li, W.; Chen, B.; Stamnes, J.J.; Tsay, S.C. Pitfalls in atmospheric correction of ocean color imagery: How should aerosol optical properties be computed? *Appl. Opt.* **2002**, *41*, 412–423. [CrossRef] [PubMed]

39. Du, H. Mie-scattering calculation. *Appl. Opt.* **2004**, *43*, 1951–1956. [CrossRef] [PubMed]

40. Wendisch, M.; Yang, P. *Theory of Atmospheric Radiative Transfer*; John Wiley & Sons: Hoboken, NJ, USA, 2012.

41. Segelstein, D.J. The Complex Refractive Index of Water. Mater's Thesis, Department of Physics, University of Missouri, Kansas City, MO, USA, 1981.

42. Smith, R.C.; Baker, K.S. Optical properties of the clearest natural waters (200–800 nm). *Appl. Opt.* **1981**, *36*, 177–184. [CrossRef] [PubMed]

43. Sogandares, F.M.; Fry, E.S. Absorption spectrum (340–640 nm) off pure water. I. Photothermal measurements. *Appl. Opt.* **1997**, *36*, 8699–8709. [CrossRef] [PubMed]

44. Pope, R.M.; Fry, E.S. Absorption spectrum (380–700 nm) of pure water, II Integrating cavity measurements. *Appl. Opt.* **1997**, *36*, 8710–8723. [CrossRef] [PubMed]

45. Kou, L.; Labrie, D.; Chylek, P. Refractive indices of water and ice in the 0.65 µm to 2.5 µm spectral range. *Appl. Opt.* **1993**, *32*, 3531–3540. [CrossRef] [PubMed]

46. Wiscombe, W.J.; Warren, S.G. A Model for the Spectral Albedo of Snow. I: Pure Snow. *J. Atmos. Sci.* **1980**, *37*, 2712–2733. [CrossRef]

47. Warren, S.G.; Wiscombe, W.J. A Model for the Spectral Albedo of Snow. II: Snow Containing Atmospheric Aerosols. *J. Atmos. Sci.* **1980**, *37*, 2734–2745. [CrossRef]

48. Grenfell, T.S.; Warren, S.G.; Mullen, P.C. Reflection of solar radiation by the Antarctic snow surface at ultraviolet, visible, and near?infrared wavelengths. *J. Geophys. Res.* **1994**, *99*, 18669–18684. [CrossRef]

49. Warren, S.G.; Brandt, R.E. Optical constants of ice from the ultraviolet to the microwave: A revised compilation. *J. Geophys. Res. Atmos.* **2008**, *113*. [CrossRef]

50. Jin, Z.; Stamnes, K.; Weeks, W.F.; Tsay, S.C. The effect of sea ice on the solar energy budget in the atmosphere-sea ice-ocean system: A model study. *J. Geophys. Res.* **1994**, *99*, 25281–25294. [CrossRef]
51. Hamre, B.; Winther, J.G.; Gerland, S.; Stamnes, J.J.; Stamnes, K. Modeled and measured optical transmittance of snow-covered first-year sea ice in Kongsfjorden, Svalbard. *J. Geophys. Res. Oceans* **2004**, *109*. [CrossRef]
52. Jiang, S.; Stamnes, K.; Li, W.; Hamre, B. Enhanced solar irradiance across the atmosphere–sea ice interface: A quantitative numerical study. *Appl. Opt.* **2005**, *44*, 2613–2625. [CrossRef] [PubMed]
53. Stamnes, K.; Hamre, B.; Stamnes, J.J.; Ryzhikov, G.; Birylina, M.; Mahoney, R.; Hauss, B.; Sei, A. Modeling of radiation transport in coupled atmosphere-snow-ice-ocean systems. *J. Quant. Spectrosc. Radiat. Transf.* **2011**, *112*, 714–726. [CrossRef]
54. Ackermann, M.; Ahrens, J.; Bai, X.; Bartelt, M.; Barwick, S.W.; Bay, R.C.; Becka, T.; Becker, J.K.; Becker, K.H.; Berghaus, P.; et al. Optical properties of deep glacial ice at the South Pole. *J. Geophys. Res.* **2006**, *111*. [CrossRef]
55. Fialho, P.; Hansen, A.D.A.; Honrath, R.E. Absorption coefficients by aerosols in remote areas: A new approach to decouple dust and black carbon absorption coefficients using seven-wavelength Aethalometer data. *J. Aerosol Sci.* **2005**, *36*, 267–282. [CrossRef]
56. Twardowski, M.S.; Boss, E.; Sullivan, J.M.; Donaghay, P.L. Modeling the spectral shape of absorption by chromophoric dissolved organic matter. *Mar. Chem.* **2004**, *89*, 69–88. [CrossRef]
57. Uusikivi, J.; Vähätalo, A.V.; Granskog, M.A.; Sommaruga, R. Contribution of mycosporine-like amino acids and colored dissolved and particulate matter to sea ice optical properties and ultraviolet attenuation. *Limnol. Oceanogr.* **2010**, *55*, 703–713. [CrossRef] [PubMed]
58. Ruddick, K.; Bouchra, N.; Collaborators. *Coastcolour Round Robin—Final Report*; 2013. Available online: ftp://ccrropen@ftp.coastcolour.org/RoundRobin/CCRR_report_OCSMART.pdf (accessed on April 12, 2018).
59. Babin, M.; Stramski, A.D.; Ferrari, G.M.; Claustre, H.; Bricaud, A.; Obelesky, G.; Hoepffner, N. Variations in the light absorption coefficients of phytoplankton, nonalgal particles and dissolved organic matter in coastal waters around Europe. *J. Geophys. Res.* **2003**, *108*. [CrossRef]
60. Babin, M.; Morel, A.; Fournier-Sicre, V.; Fell, F.; Stramski, D. Light scattering properties of marine particles in coastal and open ocean waters as related to the particle mass concentration. *Limnol. Oceanogr.* **2003**, *28*, 843–859. [CrossRef]
61. Bricaud, A.; Morel, A.; Babin, M.; Allali, K.; Claustre, H. Mie-Scattering Calculation. *J. Geophys. Res.* **1998**, *103*, 31033–31044. [CrossRef]
62. Loisel, H.; Morel, A. Light scattering and chlorophyll concentration in case 1 waters: A re-examination. *Limnol. Oceanogr.* **1998**, *43*, 847–857. [CrossRef]
63. Morel, A.; Antoine, D.; Gentili, B. Bidirectional reflectance of oceanic waters: Accounting for Raman emission and varying particle scattering phase function. *Appl. Opt.* **2002**, *41*, 6289–6306. [CrossRef] [PubMed]
64. Mobley, C.P.; Sundman, L.K.; Boss, E. Phase function effects on oceanic light fields. *Appl. Opt.* **2002**, *41*, 1035–1050. [CrossRef] [PubMed]
65. Diehl, P.; Haardt, H. Measurement of the spectral attenuation to support biological research in a "plankton tube" experiment. *Oceanol. Acta* **1980**, *3*, 89–96.
66. McCave, I.N. Particulate size spectra, behavior, and origin of nephloid layers over the Nova Scotia continental rise. *J. Geophys. Res.* **1983**, *88*, 7647–7660. [CrossRef]
67. Fournier, G.R.; Forand, J.L. Analytic phase function for ocean water. *Proc. SPIE Ocean Opt. XII* **1994**, *2558*, 194–202.
68. Hu, Y.X.; Wielicki, B.; Lin, B.; Gibson, G.; Tsay, S.C.; Stamnes, K.; Wong, T. Delta-fit: A fast and accurate treatment of particle scattering phase functions with weighted singular-value decomposition least squares fitting. *J. Quant. Spectrosc. Radiat. Transf.* **2000**, *65*, 681–690. [CrossRef]
69. Werdell, P.; Bailey, S. An improved in-situ bio-optical data set for ocean color algorithm development and satellite data product validation. *Remote Sens. Environ.* **2005**, *98*, 122–140. [CrossRef]
70. Li, W.; Stamnes, K.; Spurr, R.; Stamnes, J.J. Simultaneous Retrieval of Aerosols and Ocean Properties: A Classic Inverse Modeling Approach. II. SeaWiFS Case Study for the Santa Barbara Channel. *Int. J. Rem. Sens.* **2008**, *29*, 5689–5698. [CrossRef]
71. Garver, S.A.; Siegel, D. Inherent optical property inversion of ocean color spectra and its biogeochemical interpretation 1. Time series from the Sargasso Sea. *J. Geophys. Res.* **1997**, *102*, 18607–18625. [CrossRef]

72. Garver, S.A.; Siegel, D.; Peterson, A.R. Optimization of a semi-analytical ocean color model for global-scale applications. *Appl. Opt.* **2002**, *41*, 2705–2714.
73. Bohren, C.F.; Huffman, D.R. *Absorption and Scattering of Light by Small Particles*; John Wiley: New York, NY, USA, 1998.
74. Born, M.; Wolf, E. *Principles of Optics*; Cambridge University Press: Cambridge, UK, 1980.
75. Cox, C.; Munk, W. Measurement of the roughness of the sea surface from photographs of the sun's glitter. *J. Opt. Soc. Am.* **1954**, *44*, 838–850. [CrossRef]
76. Masuda, K. Effects of the speed and direction of surface winds on the radiation in the atmosphere—Ocean system. *Remote Sens. Environ.* **1998**, *64*, 53–63. [CrossRef]
77. Wang, M.; Bailey, S.W. Correction of sun glint contamination on the SeaWiFS ocean and atmosphere products. *Appl. Opt.* **2001**, *40*, 4790–4798. [CrossRef]
78. Nakajima, T.; Tanaka, M. Algorithms for radiative intensity calculations in moderately thick atmospheres using a truncation approximation. *J. Quant. Spectrosc. Radiat. Transf.* **1988**, *40*, 51–69. [CrossRef]
79. Lin, Z.; Stamnes, S.; Jin, Z.; Laszlo, I.; Tsay, S.C.; Wiscombe, W.; Stamnes, K. Improved discrete ordinate solutions in the presence of an anisotropically reflecting lower boundary: Upgrades of the DISORT computational tool. *J. Quant. Spectrosc. Radiat. Transf.* **2015**, *157*, 119–134. [CrossRef]
80. Lin, Z.; Li, W.; Gatebe, C.; Poudyal, R.; Stamnes, K. Radiative transfer simulations of the two-dimensional ocean glint reflectance and determination of the sea surface roughness. *Appl. Opt.* **2016**, *55*, 1206–1215. [CrossRef] [PubMed]
81. Gordon, H.R. Atmospheric correction of ocean color imagery in the Earth Observation System era. *J. Geophys. Res.* **1997**, *102*, 17081–17106. [CrossRef]
82. Rodgers, C.D. *Inverse Methods for Atmospheric Sounding: Theory and Practice*; World Scientific: London, UK, 2000.
83. Jin, Z.; Stamnes, K. Radiative transfer in nonuniformly refracting layered media: Atmosphere-ocean system. *Appl. Opt.* **1994**, *33*, 431–442. [CrossRef] [PubMed]
84. Zhai, P.W.; Hu, Y.; Chowdhary, J.; Trepte, C.R.; Lucker, P.L.; Josset, D.B. A vector radiative transfer model for coupled atmosphere and ocean systems with a rough interface. *J. Quant. Spectrosc. Radiat. Transf.* **2010**, *111*, 1025–1040. [CrossRef]
85. Chowdhary, J.; Cairns, B.; Travis, L.D. Case studies of aerosol retrievals over the ocean from multiangle, multispectral photopolarimetric remote sensing data. *J. Atmos. Sci.* **2002**, *59*, 383–397. [CrossRef]
86. Chowdhary, J.; Cairns, B.; Mishchenko, M.I.; Hobbs, P.V.; Cota, G.F.; Redemann, J.; Rutledge, K.; Holben, B.N.; Russell, E. Retrieval of aerosol scattering and absorption properties from photopolarimetric observations over the ocean during the CLAMS experiment. *J. Atmos. Sci.* **2005**, *62*, 1093–1117. [CrossRef]
87. Chowdhary, J.; Cairns, B.; Waquet, F.; Knobelspiesse, K.; Ottaviani, M.; Redemann, J.; Travis, L.; Mishchenko, M. Sensitivity of multiangle, multispectral polarimetric remote sensing over open oceans to water-leaving radiance: Analyses of RSP data acquired during the MILAGRO campaign. *Remote Sens. Environ.* **2012**, *118*, 284–308. [CrossRef]
88. Chami, M.; Santer, R.; Dilligeard, E. Radiative transfer model for the computation of radiance and polarization in an ocean–atmosphere system: Polarization properties of suspended matter for remote sensing. *Appl. Opt.* **2001**, *40*, 2398–2416. [CrossRef] [PubMed]
89. Min, Q.; Duan, M. A successive order of scattering model for solving vector radiative transfer in the atmosphere. *J. Quant. Spectrosc. Radiat. Transf.* **2004**, *87*, 243–259. [CrossRef]
90. Fischer, J.; Grassl, H. Radiative transfer in an atmosphere-ocean system: An azimuthally dependent matrix-operator approach. *Appl. Opt.* **1984**, *23*, 1032–1039. [CrossRef] [PubMed]
91. Ota, Y.; Higurashi, A.; Nakajima, T.; Yokota, T. Matrix formulations of radiative transfer including the polarization effect in a coupled atmosphere-ocean system. *J. Quant. Spectrosc. Radiat. Transfer* **2010**, *111*, 878–894. [CrossRef]
92. Kattawar, G.; Adams, C. Stokes vector calculations of the submarine light field in an atmosphere-ocean with scattering according to the Rayleigh phase matrix: Effect of interface refractive index on radiance and polarization. *Limnol. Oceanogr.* **1989**, *34*, 1453–1472. [CrossRef]
93. Lotsberg, J.; Stamnes, J. Impact of particulate oceanic composition on the radiance and polarization of underwater and backscattered light. *Opt. Express* **2010**, *18*, 10432–10445. [CrossRef] [PubMed]

94. Morel, A.; Gentili, B. Diffuse reflectance of oceanic waters. II. bidirectional aspect. *Appl. Opt.* **1993**, *32*, 2803–2804. [CrossRef] [PubMed]

95. Fan, Y.; Li, W.; Stamnes, K.; Gatebe, C. A neural network method to correct bidirectional effects in water-leaving radiance. *Appl. Opt.* **2016**, *55*, 10–421. [CrossRef] [PubMed]

96. Voss, K.J.; Chapin, A.L. An Upwelling Radiance Distribution Camera System, NURADS. *Opt. Express* **2005**, *13*, 4250–4262. [CrossRef] [PubMed]

97. Kay, S.; Hedley, J.D.; Lavender, S. Sun Glint Correction of High and Low Spatial Resolution Images of Aquatic Scenes: A Review of Methods for Visible and Near-Infrared Wavelengths. *Remote Sens.* **2009**, *1*, 697–730. [CrossRef]

98. Steinmetz, F.; Deschamps, P.Y.; Ramon, D. Atmospheric correction in presence of sun glint: Application to MERIS. *Opt. Express* **2011**, *19*, 9783–9800. [CrossRef] [PubMed]

99. Fukushima, H.; Suzuki, K.; Li, L.; Suzuki, N.; Murakami, H. Improvement of the ADEOS-II/GLI sun-glint algorithm using concomitant microwave scatterometer-derived wind data. *Adv. Space Res.* **2009**, *43*, 941–947. [CrossRef]

100. Wang, M. The Rayleigh lookup tables for the SeaWiFS data processing: accounting for the effects of ocean surface roughness. *Int. J. Remote Sens.* **2002**, *23*, 2693–2702. [CrossRef]

101. Ottaviani, M.; Spurr, R.; Stamnes, K.; Li, W.; Su, W.; Wiscombe, W. Improving the description of sunglint for accurate prediction of remotely sensed radiances. *J. Quant. Spectrosc. Radiat. Transf.* **2008**, *109*, 2364–2375. [CrossRef]

102. Gatebe, C.K.; King, M.D.; Lyapustin, A.I.; Arnold, G.T.; Redemann, J. Airborne spectral measurements of ocean directional reflectance. *J. Atmos. Sci.* **2005**, *62*, 1072–1092. [CrossRef]

103. Stamnes, K.; Tsay, S.C.; Wiscombe, W.; Jayaweera, K. Numerically stable algorithm for discrete-ordinate-method radiative transfer in multiple scattering and emitting layered media. *Appl. Opt.* **1988**, *27*, 2502–2509. [CrossRef] [PubMed]

104. National Oceanic and Atmospheric Administration (NOAA). *US Standard Atmosphere*; Technical Report; NOAA: Washington, DC, USA, 1976.

105. Jin, Z.; Charlock, T.P.; Rutledge, K.; Stamnes, K.; Wang, Y. Analytical solution of radiative transfer in the coupled atmosphere-ocean system with a rough surface. *Appl. Opt.* **2006**, *45*, 7443–7455. [CrossRef] [PubMed]

106. Mobley, C.D. Polarized reflectance and transmittance properties of windblown sea surfaces. *Appl. Opt.* **2015**, *54*, 4828–4849. [CrossRef] [PubMed]

107. Fan, Y.; Li, W.; Stamnes, K.; Stamnes, J.J.; Sørensen, K. Simultaneous Retrieval of AEROSOL and Marine Parameters in Coastal areas Using a Coupled Atmosphere-Ocean Radiative Transfer Model. In Proceedings of the Sentinel-3 for Science Workshop, Venice-Lido, Italy, 2–5 June 2015.

108. Spurr, R.; Stamnes, K.; Eide, H.; Li, W.; Zheng, K.; Stamnes, J. Simultaneous retrieval of aerosol and ocean color: A classic inverse modeling approach: I. Analytic Jacobians from the linearized CAO-DISORT model. *J. Quant. Spectrosc. Radiat. Transf.* **2007**, *104*, 428–449. [CrossRef]

109. Fan, Y.; Li, W.; Gatebe, C.K.; Jamet, C.; Zibordi, G.; Schroeder, T.; Stamnes, K. Atmospheric correction and aerosol retrieval over coastal waters using multilayer neural networks. *Remote Sens. Environ.* **2017**, *199*, 218–240. [CrossRef]

110. He, X.; Stamnes, K.; Bai, Y.; Li, W.; Wang, D. Effects of Earth curvature on atmospheric correction for ocean color remote sensing. *Remote Sens. Environ.* **2018**, *209*, 118–133. [CrossRef]

111. Stamnes, S.; Hostetler, C.; Ferrari, R.; Burton, S.; Lui, X.; Hair, J.; Hu, Y.; Wasilewski, A.; Martin, W.; Van Diedenhoven, B.; et al. Simultaneous polarimeter retrievals of microphysical aerosol and ocean color parameters from the MAPP algorithm with comparison to high spectral resolution lidar aerosol and ocean products. *Appl. Opt.* **2018**, *57*, 2394–2413. [CrossRef] [PubMed]

112. Stamnes, S.; Fan, Y.; Chen, N.; Li, W.; Tanikawa, T.; Lin, Z.; Liu, X.; Burton, S.; Omar, A.; Stamnes, J.; et al. Advantages of measuring the Q Stokes parameter in addition to the total radiance I in the detection of absorbing aerosols. *Front. Earth Sci.* **2018**, *6*, 1–11. [CrossRef]

113. Dahlback, A.; Stamnes, K. A new spherical model for computing the radiation field available for photolysis and heating at twilight. *Planet. Space Sci.* **1991**, *39*, 671–683. [CrossRef]

Appl. Sci. **2018**, *8*, 2682

114. Spurr, R.J.D. VLIDORT: A linearized pseudo-spherical vector discrete ordinate radiative transfer code for forward model and retrieval studies in multilayer multiple scattering media. *J. Quant. Spectrosc. Radiat. Transf.* **2006**, *102*, 316–342. [CrossRef]
115. Rozanov, V.; Rozanov, A.; Kokhanovsky, A.; Burrows, J. Radiative transfer through terrestrial atmosphere and ocean: Software package SCIATRAN. *J. Quant. Spectrosc. Radiat. Transf.* **2014**, *133*, 13–71. [CrossRef]
116. Davis, A.; Marshak, A. *3D Radiative Transfer in Cloudy Atmospheres*; Springer: Berlin, Germany, 2005.
117. Chandrasekhar, S. On the diffuse reflection of a pencil of radiation by a plane-parallel atmosphere. *Proc. Natl. Acad. Sci. USA* **1958**, *44*, 933–940. [CrossRef] [PubMed]
118. Shiina, T.; Yoshida, K.; Ito, M.; Okamura, Y. Long-range propagation of annular beam for lidar application. *Opt. Commun.* **2007**, *279*, 159–167. [CrossRef]
119. Habel, R.; Christensen, P.H.; Jarosz, W. Photon Beam Diffusion: A Hybrid Monte Carlo Method for Subsurface Scattering. In Proceedings of the 24th Eurographics Symposium on Rendering, Zaragoza, Spain, 19–21 June 2013; Volume 32.
120. Barichello, L.; Siewert, C. The searchlight problem for radiative transfer in a finite slab. *J. Comput. Phys.* **2000**, *157*, 707–726. [CrossRef]
121. Kim, A.D.; Moscoso, M. Radiative transfer computations for optical beams. *J. Comput. Phys.* **2003**, *185*, 50–60. [CrossRef]
122. Mobley, C.D.; Gentili, B.; Gordon, H.R.; Jin, Z.; Kattawar, G.W.; Morel, A.; Reinersman, P.; Stamnes, K.; Stavn, R.H. Comparison of numerical models for computing underwater light fields. *Appl. Opt.* **1993**, *32*, 7484–7504. [CrossRef] [PubMed]
123. Mitra, K.; Churnside, J.H. Transient radiative transfer equation applied to oceanographic lidar. *Appl. Opt.* **1999**, *38*, 889–895. [CrossRef] [PubMed]
124. Stamnes, K.; Lie-Svendsen, Ø.; Rees, M.H. The linear Boltzmann equation in slab geometry: Development and verification of a reliable and efficient solution. *Planet. Space Sci.* **1991**, *39*, 1453–1463. [CrossRef]
125. De Beek, R.; Vountas, M.; Rozanov, V.; Richter, A.; Burrows, J. The Ring effect in the cloudy atmosphere. *Geophys. Res. Lett.* **2001**, *28*, 721–724. [CrossRef]
126. Landgraf, J.; Hasekamp, O.; Van Deelen, R.; Aben, I. Rotational Raman scattering of polarized light in the Earth atmosphere: a vector radiative transfer model using the radiative transfer perturbation theory approach. *J. Quant. Spectrosc. Radiat. Transf.* **2004**, *87*, 399–433. [CrossRef]
127. Spurr, R.; de Haan, J.; van Oss, R.; Vasilkov, A. Discrete-ordinate radiative transfer in a stratified medium with first-order rotational Raman scattering. *J. Quant. Spectrosc. Radiat. Transf.* **2008**, *109*, 404–425. [CrossRef]
128. Ge, Y.; Gordon, H.; Voss, K. Simulation of inelastic scattering contributions to the irradiance field in the oceanic variation in Fraunhofer line depths. *Appl. Opt.* **1993**, *32*, 4028–4036. [CrossRef] [PubMed]
129. Kattawar, G.; Xu, X. Filling-in of Fraunhofer lines in the ocean by Raman scattering. *Appl. Opt.* **1992**, *31*, 1055–1065. [CrossRef] [PubMed]
130. Voss, K.J.; Fry, E.S. Measurement of the Mueller matrix for ocean water. *Appl. Opt.* **1984**, *23*, 4427–4439. [CrossRef] [PubMed]
131. Werdell, J.P.; Franz, B.A.; Bailey, S.W.; Feldman, G.C.; Boss, E.; Brando, V.E.; Dowell, M.; Hirata, T.; Lavender, S.; Lee, Z.P.; et al. Generalized ocean color inversion model for retrieving marine inherent optical properties. *Appl. Opt.* **2013**, *52*, 2019–2037. [CrossRef] [PubMed]
132. Lee, Z.P.; Carder, K.L.; Arnone, R. Deriving inherent optical properties from water color: A multi-band quasi-analytical algorithm for optically deep waters. *Appl. Opt.* **2002**, *41*, 5755–5772. [CrossRef] [PubMed]

![applied sciences logo] *applied sciences*

MDPI

Article

Measuring and Modeling the Polarized Upwelling Radiance Distribution in Clear and Coastal Waters

Arthur C. R. Gleason [1], Kenneth J. Voss [1,*], Howard R. Gordon [1], Michael S. Twardowski [2] and Jean-François Berthon [3]

[1]   Physics Department, University of Miami, Coral Gables, FL 33146, USA;
      agleason@physics.miami.edu (A.C.R.G.); gordon@physics.miami.edu (H.R.G.)
[2]   Harbor Branch Oceanographic Institute, Fort Pierce, FL 34946, USA; mtwardowski@fau.edu
[3]   Joint Research Centre of the European Commission, 21027 Ispra, Italy; Jean-Francois.berthon@ec.europa.eu
*   Correspondence: voss@physics.miami.edu; Tel.: +1-305-284-7110

Received: 15 June 2018; Accepted: 8 October 2018; Published: 19 December 2018

Abstract: The upwelling spectral radiance distribution is polarized, and this polarization varies with the optical properties of the water body. Knowledge of the polarized, upwelling, bidirectional radiance distribution function (BRDF) is important for generating consistent, long-term data records for ocean color because the satellite sensors from which the data are derived are sensitive to polarization. In addition, various studies have indicated that measurement of the polarization of the radiance leaving the ocean can used to determine particle characteristics (Tonizzo et al., 2007; Ibrahim et al., 2016; Chami et al., 2001). Models for the unpolarized BRDF (Morel et al., 2002; Lee et al., 2011) have been validated (Voss et al., 2007; Gleason et al., 2012), but variations in the polarization of the upwelling radiance due to the sun angle, viewing geometry, dissolved material, and suspended particles have not been systematically documented. In this work, we simulated the upwelling radiance distribution using a Monte Carlo-based radiative transfer code and measured it using a set of fish-eye cameras with linear polarizing filters. The results of model-data comparisons from three field experiments in clear and turbid coastal conditions showed that the degree of linear polarization (*DOLP*) of the upwelling light field could be determined by the model with an absolute error of ±0.05 (or 5% when the *DOLP* was expressed in %). This agreement was achieved even with a fixed scattering Mueller matrix, but did require in situ measurements of the other inherent optical properties, e.g., scattering coefficient, absorption coefficient, etc. This underscores the difficulty that is likely to be encountered using the particle scattering Mueller matrix (as indicated through the remote measurement of the polarized radiance) to provide a signature relating to the properties of marine particles beyond the attenuation/absorption coefficient.

Keywords: polarization; ocean optics; upwelling radiance distribution; remote sensing

---

1. Introduction

The in-water and water-leaving radiance in the ocean is partially polarized and this has implications for both biological activity [1] and for viewing the ocean from a satellite [2]. The biological implications of the polarization are a current topic of research [3] but the implications for ocean color remote sensing are just being exploited [2,4]. Various studies have indicated that measurement of the polarization of the radiance leaving the ocean can be used to determine particle characteristics [5–7]. In addition to using the polarization of the water leaving radiance for studies of the water properties, polarization may also be important for processing data from ocean color sensors that have unintended polarization sensitivities [8,9].

Measurements of the scalar (without regard to polarization) spectral upwelling radiance distribution have occurred more frequently since the development of the radiance distribution camera, RADS [10], and then the upwelling radiance distribution camera, NuRADS [11]. These instruments use electro-optic camera systems combined with filter changers and fisheye cameras to image the complete upwelling radiance distribution, for a specific wavelength, in one image. The CCD (charge coupled device) resolution and optics allow measurement of the radiance distribution with a 1° angular resolution. These systems have been used in studies of the in-water light field [12] and studies of the angular radiance distribution variations in ocean color algorithms [13,14]. Recently, two new cameras, polarized radiance camera (PolRADS) [15] and downwelling polarized camera (DPOL) [16], have been used to make measurements of the polarized spectral upwelling radiance distribution, as described below.

Measurements of the upwelling, in situ, polarized radiance have typically been done with variations of a Gershun-tube radiometer, obtaining the angular distribution by changing the viewing direction of the radiometer [17,18]. These instruments have advantages such as the simplicity of calibration and the ability to do hyperspectral measurements. Hyperspectral measurements of the polarized light field have been suggested as a method to measure in situ solar stimulated fluorescence [19]. Unfortunately, to obtain the full radiance distribution requires many measurements during which the illumination conditions can change, and spatially varying angular features resulting from the sea surface may not be captured. By making measurements using fisheye lenses, while limited to multi-spectral measurements rather than hyperspectral measurements, higher angular spatial resolution can be obtained nearly simultaneously (typical exposure times are less than 1 s).

Models for the unpolarized bi-directional reflectance distribution function (BRDF) [20,21] have been validated in Case I [13] and Case II [14] waters, but variations in the polarization of upwelling radiance due to sun angle, viewing geometry, dissolved material, and suspended particles have not been systematically documented.

2. Methods

The light field polarization is easily described by use of the four element Stokes vector, I, as in, for example, Bohren and Huffman [22]. The four components of I are prescribed relative to some reference plane, which here we choose as the plane defined by the viewing direction and the nadir direction. Consider an electromagnetic wave of angular frequency ω. If the unit vector $\hat{\ell}$ is parallel to, and the unit vector $\hat{r}$ is perpendicular to, the reference plane (such that $\hat{\ell} \times \hat{r}$ is in the direction of propagation), the electric field can be written as $\vec{E} = E_\ell \hat{\ell} + E_r \hat{r}$, with:

$$E_\ell = E_{\ell 0} \exp[i(\omega t + \delta_\ell)], \tag{1a}$$

$$E_r = E_{r0} \exp[i(\omega t + \delta_r)] \tag{1b}$$

where $E_{\ell 0}$ and E_{r0} are real. The four components of the Stokes vector are then defined as:

$$I \equiv E_\ell E_\ell^* + E_r E_r^* = E_{l0}^2 + E_{r0}^2 \tag{2a}$$

$$Q \equiv E_\ell E_\ell^* - E_r E_r^* = E_{l0}^2 - E_{r0}^2, \tag{2b}$$

$$U \equiv E_\ell E_r^* + E_r E_\ell^* = 2E_{l0}E_{r0}\cos\delta \tag{2c}$$

$$V \equiv i(E_\ell E_r^* - E_r E_\ell^*) = 2E_{l0}E_{r0}\sin\delta \tag{2d}$$

where $\delta \equiv \delta_r - \delta_\ell$. A derived parameter that we will use in our comparison is the degree of linear polarization, *DOLP*, defined as:

$$DOLP = \sqrt{\frac{Q^2 + U^2}{I^2}} \tag{3}$$

Another useful parameter is the angle of the plane of polarization, χ, which is defined as:

$$\chi = \frac{1}{2}\arctan\left(\frac{U}{Q}\right) \tag{4}$$

We measured the Stokes vector over the upwelling hemisphere using the PolRADS and DPOL cameras and simulated it using a Monte Carlo-based radiative transfer (RT) code. Data were available from three field campaigns: Hawaii, USA, in December 2005; the Ligurian Sea in March 2009; and the New York Bight in May 2009.

2.1. Measurements of the Polarized Upwelling Radiance Distribution

In Hawaii, measurements of the polarized upwelling radiance distribution were accomplished using the PolRADS camera system [15]. PolRADS is based on the NuRADS camera system [12], which is a compact (30 cm diameter, 30 cm length), multispectral camera that images the upwelling radiance distribution in six narrow ($\approx$10 nm full-width at half-maximum, FWHM) spectral bands centered at 412, 436, 486, 526, 548, and 615 nm. In the PolRADS instrument, three synchronized NuRADS cameras are used, each with a linear polarizer, to simultaneously acquire images. Combining the images from three NuRADS cameras, when in PolRADS configuration, allows for retrieval of three elements of the Stokes vector (I, Q, and U, but not V), as well as $DOLP$.

In the Ligurian Sea and New York Bight, measurements of the polarized upwelling radiance distribution were accomplished with the DPOL camera system [16]. DPOL is similar to PolRADS in the sense that simultaneous measurements are made using multiple fisheye lenses with different polarizing filters in the optical path. DPOL differs from PolRADS by using optical-fiber bundles to project images from all lenses through a single spectral filter and onto a single CCD array. Eliminating the redundant filters and cameras makes DPOL much smaller than PolRADS, thereby reducing the instrument shadow. Furthermore, DPOL has four lenses, rather than the three on PolRADS, enabling retrieval of all four Stokes vector elements. The spectral filters used on DPOL were also $\approx$10 nm FWHM and were centered at 442, 488, 520, 550, 589, and 650 nm. For both PolRADS and DPOL, the channels above 600 nm will not be used because of instrument self-shading. In addition, because of its absence on DPOL, the 412 nm channel on PolRads will also not be discussed.

Data acquisition for both the PolRADS and DPOL systems used filter changers to rotate interference filters and thus sequentially acquire images in each wavelength band. Typical exposure times were less than one second. However, acquiring a set of images from all wavelengths took about two minutes due to the time required to read and store the data from the CCD. With both systems, typical deployments lasted from one to several hours, enabling multiple acquisitions over a range of solar zenith angles.

Reduction of the raw images consisted of applying calibration factors [15,16] and averaging images in both space and time to reduce environmental noise. After calibration, but before averaging, every image was inspected to find the anti-solar point, to correct the geometry of the image, and to check for obstructions in the field of view such as fish, the power/data cable, the side of the ship, or other instruments. Calibrated images of the Stokes vectors were then averaged in 10-min bins, excluding those that had been flagged as unacceptable in the inspection stage. The symmetry of the images about the principal plane was exploited to further average both halves of each image. In addition, spatial binning of 3×3-pixel windows was performed to produce final average images at a $1° \times 1°$ resolution. Each pixel in the final, reduced image, therefore, could have been an average of up to 90 raw pixels (5 images $\times$ 2 image halves $\times$ 9-pixel window). The mean and coefficient of variation (CV = standard deviation divided by the mean) of the Stokes vector components were computed for each pixel in the reduced image using the up-to-90 raw pixels in the original images. The degree of linear polarization $DOLP$ was then computed for each pixel from the mean I, Q, and U Stokes vector components (Equation (3)). Because V was not available from the PolRADS data, $DOLP$ was calculated rather than the degree of polarization (*DOP*); however, in the upwelling lightfield, they are equivalent because the magnitude of *V* is negligible [23].

For in situ radiometric measurements, instrument self-shading must be considered [24]. In our case, instrument self-shading may also affect the *DOLP* measurements due to camera geometry. The physical displacement of the multiple lenses used means that the images through different polarizers will look into the instrument shadow in slightly different directions. When one lens views a portion of the radiance distribution with less shadow, it will appear brighter in that image. In the algorithm from which the polarization information is derived, for the scene to be unpolarized requires that region of the three camera images to be closely matched in intensity. If the region in one camera is less shaded, it will cause the algorithm to assume that the light field is polarized along the direction of the linear polarizer in that camera. Thus shadowing, which is not symmetric in all the images, will appear as an increase in the *DOLP*, and therefore, a negative model-data *DOLP* difference. This effect should be greater in turbid water and for the PolRADS instrument (because of its larger size) than in clear water or with the DPOL instrument. In our analysis, the area of direct shadow (the anti-solar point) was excluded from the data set. More subtle shadow areas may exist and could cause the measured *DOLP* to be larger than the modelled *DOLP* (as will be seen, this was not generally the case in our data set).

2.2. Modeling the Polarized Upwelling Radiance Distribution

A Monte Carlo model was used to solve the vector radiative transfer equation (Equation (5)) for the propagation of the Stokes vector, *I*, at wavelength λ,

$$\cos(\theta)\frac{dI(\lambda,\tau,\theta,\phi)}{d\tau} = -I(\lambda,\tau,\theta,\phi) + \omega_0 \int_{4\pi} R(\alpha)P(\lambda,\tau,\theta',\phi' \to \theta,\phi)R'(\alpha')I(\lambda,\tau,\theta',\phi')d\Omega' \quad (5)$$

The four components of the column vector $I = [I, Q, U, V]^T$ were defined in Equation (2). The single-scattering albedo, ω_0, the optical depth of the medium, τ, the scattering phase matrix, P, and the rotation matrix, R, are defined below. The right-handed coordinate system in Equation (5) has its origin at the top of the medium, *z*-axis pointed downward, and *x*-axis directed away from the sun. The nadir angle, θ, relative to the *z*-axis, and the azimuth angle, ϕ, in the *x*-*y* plane, define the direction of photon propagation. Photons originate in the solar beam, i.e., with $\theta = \theta_0$, the solar zenith angle, and $\phi_0 = 0$.

The model simulated a horizontally homogenous, two-layer system with a non-absorbing, Rayleigh-scattering atmosphere over an ocean comprised of Rayleigh-scattering water molecules plus scattering and absorbing hydrosols ("particles"). Fresnel reflectance at the air–water interface used the boundary condition defined in Gordon et al. [8]. The single-scattering albedo, ω_0, and the optical depth of the medium, τ, were defined as per usual (e.g., Mobley [25]):

$$\omega_0 = \frac{b_t(\lambda)}{c_t(\lambda)}, \quad (6)$$

$$\text{and } \tau(\lambda) = \int_0^z c_t(\lambda)dz. \quad (7)$$

where the parameters $a_t(\lambda) + b_t(\lambda) = c_t(\lambda)$ are the total absorption, scattering, and attenuation coefficients (in m^{-1}), respectively. For the atmospheric layer, $a_t(\lambda) = 0$, therefore $\omega_0 = 1$, and τ was taken from Teillet [26] assuming standard atmospheric conditions. For the oceanic layer, absorption (a_w) and scattering (b_w) coefficients of seawater were interpolated to PolRADS and DPOL spectral bands from Table 1.1 in Pegau et al. [27]. In the Ligurian Sea and New York Bight experiments, the absorption coefficient of dissolved and particulate constituents (a_{pg}) and the particle scattering (b_p) coefficients were measured in situ, depth-weighted, and interpolated as necessary to the PolRADS and DPOL spectral bands as described by Gleason et al. [14]. No in situ measurements were available from

Hawaii, therefore for that dataset, a_{pg} and b_p were derived from Morel and Gentili [28] (Equations (7) and (8) in reference [28]) using an estimated total chlorophyll concentration of 0.1 g/m^3, the long-term average value at the measurement site. Summing the water and particle contributions gave the total coefficients required for Equations (6) and (7): $a_t = a_w(\lambda) + a_{pg}(\lambda)$ and $b_t = b_w(\lambda) + b_p(\lambda)$. For each image data set, the model was run at the corresponding solar zenith angle.

Scattering events in Equation (5) are represented by a 4 × 4 matrix called the Mueller matrix, M. For example, a photon travelling in direction ζ' with Stokes vector I' would be scattered in a new direction ζ with Stokes vector I by the linear transformation $I(\zeta) = MI'(\zeta')$. Note that M is defined relative to the scattering plane (as is traditional), the plane defined by vectors ζ' and ζ, but we have defined $I(\theta, \phi)$ relative to the viewing direction and the nadir direction. Pre- and post-multiplication by the rotation matrix $R(\alpha)$ is required to account for changes in reference frames: $I(\theta, \phi) = R(\alpha)MR(\alpha')I'(\theta', \phi')$, where

$$R(\alpha) = \begin{pmatrix} 1 & 0 & 0 & 0 \\ 0 & \cos(2\alpha) & \sin(2\alpha) & 0 \\ 0 & -\sin(2\alpha) & \cos(2\alpha) & 0 \\ 0 & 0 & 0 & 1 \end{pmatrix} \tag{8}$$

The rotation angle $\alpha = \cos^{-1}(\hat{l}_l \cdot \hat{l}_r)$ is measured clockwise from the vector $\hat{l}_l$ in the initial reference frame to the vector $\hat{l}_r$ in the rotated reference frame [8].

The scattering phase matrix, P, is a Mueller matrix normalized to the integral of the M_{11} element over all solid angles:

$$P = M/b \tag{9}$$

where

$$b = 2\pi \int_0^\pi M_{11} \sin(\theta) d\theta \tag{10}$$

M_{ij} represents the ith row and jth column of M, and Θ is the scattering angle.

It is advantageous, when investigating the polarization effects of the Muller matrix, to normalize in a different way, specifically to the M_{11} element at each angle Θ (Equation (11)).

$$S_{ij}(\Theta) = M_{ij}(\Theta)/M_{11}(\Theta) \tag{11}$$

Then,

$$P = \tilde{\beta}S \tag{12}$$

where $\tilde{\beta} = M_{11}/b$ is the scattering phase function typically used in scalar radiance transfer models.

For our modeling, P in the atmosphere and P_w, the seawater component of the ocean, were both set to P_r, given by Rayleigh scattering:

$$P_r = \frac{3}{16\pi} \begin{pmatrix} 1 + \cos^2(\Theta) & -\sin^2(\Theta) & 0 & 0 \\ -\sin^2(\Theta) & 1 + \cos^2(\Theta) & 0 & 0 \\ 0 & 0 & 2\cos(\Theta) & 0 \\ 0 & 0 & 0 & 2\cos(\Theta) \end{pmatrix} \tag{13}$$

(We examined a few cases using a code that included the (small) depolarization of Rayleigh scattering in the atmosphere and water by molecular anisotropy and concluded that its omission would not significantly affect the results.) Particle scattering in the ocean used $\tilde{\beta}$ [25,29] and two alternative parameterizations of S. One set of model runs used S from Voss and Fry [30], which was experimentally determined from samples of seawater (referred to here after as V-F). Note that Voss and Fry found all off-diagonal elements of S equal to zero, within experimental error, except for S_{12} and S_{21}. Furthermore, Voss and Fry found $S_{12} \approx S_{21}$. Subsequent laboratory measurements of the Mueller matrix elements for phytoplankton [31–33] have been similar to V-F. Specifically, the $S_{12}(\Theta)$ element of phytoplankton samples tends to have a minimum value of between −0.6 to −0.8 at an angle between 90° and 100°. In contrast, experiments with solutions of suspended marine sediment have revealed smaller absolute values for $S_{12}(\Theta)$ element near $\Theta = 90°$ [32]. Because of this, we decided to carry out a second set of runs with a particle phase matrix having a smaller minimum in S_{12} near 90° (Figure 1). This was accomplished by replacing the V-F $S_{12}(\Theta)$ with that for Rayleigh scattering with a depolarization factor ρ. ρ is the ratio of light scattered with polarization parallel to the plane of incidence to that scattered perpendicular to the plane of incidence, when the incident radiation is polarized perpendicular to the plane of incidence. The element $S_{12}(\Theta)$ as a function of ρ is given by Equation (14) [34]. The other elements of S remained unchanged for this Mueller matrix (referred to as Mod-V-F). Setting $\rho = 0.3$ resulted in a minimum value of $S_{12}(90°) = -0.37$, which is within the range of −0.38 to −0.25 observed by Volten et al. [32].

$$S_{12}(\Theta) = S_{21}(\Theta) = \frac{(1-\rho)\left(\cos(\Theta)^2 - 1\right)}{\left(1 + \cos(\Theta)^2\right) + \left(3 - \cos(\Theta)^2\right)\rho} \tag{14}$$

Kuik et al. [35] presented four inequalities, originally derived by Fry and Kattawar [36] and Hovenier et al. [37], that must be satisfied by the scattering matrix elements of randomly oriented particles having a plane of symmetry. These inequalities are all satisfied by the modified Mueller matrix.

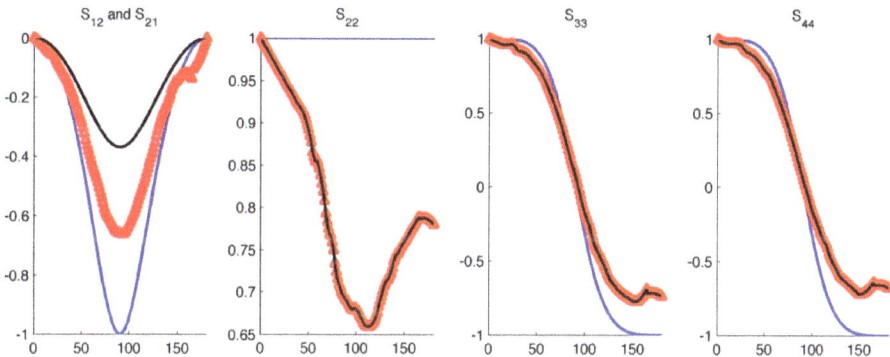

Figure 1. Non-zero, normalized Muller matrix, S, elements for Rayleigh scattering (blue), V-F (red), and Mod-V-F (black).

2.3. Model–Data Comparison

Visual comparisons provided a qualitative check that the Stokes vector components from our RT model were in agreement with the camera data. Quantitative comparisons between the model and data were performed by computing the model–data difference every 10° in nadir from 0° to 80° and every 30° in azimuth from 0° to 180°. Thus, the difference in *DOLP*, $DOLP_{diff}$, between each model output, $DOLP_{model}$, and the corresponding average PolRADS or DPOL image, $DOLP_{data}$, was computed at 63 angles using Equation (15).

$$DOLP_{diff}(\theta, \phi) = DOLP_{model}(\theta, \phi) - DOLP_{data}(\theta, \phi) \tag{15}$$

where $DOLP_{model}(\theta, \phi)$ and $DOLP_{data}(\theta, \phi)$ are the *DOLP* predicted by the RT model and measured by the corresponding average PolRADS or DPOL image, respectively, at the same 63 nadir and azimuth angles.

3. Results

In total, approximately 528 individual images were selected and averaged in 10 min intervals to produce 219 data sets, which were compared with RT model runs (Table 1).

Table 1. Number, N, of reduced images used for model–data comparisons in each of the three datasets in each of four spectral bands. Note, the PolRADS and DPOL cameras used different filters with slightly different band centers; the center of each 10 nm-wide filter is listed in parenthesis.

Experiment	N (λ nm)	N (λ nm)	N (λ nm)	N (λ nm)
Hawaii	27 (436 nm)	31 (486 nm)	28 (526 nm)	24 (548 nm)
Ligurian Sea	9 (442 nm)	9 (488 nm)	13 (520 nm)	14 (550 nm)
New York Bight	7 (442 nm)	8 (488 nm)	9 (520 nm)	7 (550 nm)

3.1. Overall View of the Upwelling Polarization Signal

Figure 2 shows an example image for clear water. In the right column, the data parameters are shown (I, Q, U, $DOLP$, and χ), while in the center column, the results from the RT model are shown, where the model inputs are based on the measured parameters for the data (b_p, a_{pg}, solar zenith angle). As mentioned earlier, during processing of this data, the symmetry about the principal plane was used to average the left and right side of the data (with proper handling of the sign of U/I). Thus, to make these images, this symmetry was used to generate the left side of the data images. The left column is a single scattering calculation, where the Mueller matrix was assumed to be V-F, and the solar zenith angle was the value appropriate for the data. As can be seen, the same broad patterns were visible in the single scattering model, the RT model and data. The I component of the Stokes vector was largest near the horizon in the direction towards the sun. The Q/I and U/I patterns were very similar in all three cases. For the $DOLP$, the single scattering model was significantly different than the data and the RT model. With single scattering, the maximum $DOLP$ was both larger than in the RT model and the data, and occurred for all nadir angles at the scattering angle matching the minimum in the S_{12} and S_{21} elements (90–100°). Interestingly, while it was a subtle effect in the clear water case, the maximum $DOLP$ in the model and data occurred at the same scattering angle, but at an azimuth of 90° relative to the principal plane (the plane containing the anti-solar position and nadir). This can be contrasted with the case of the downwelling sky radiance distribution, where the maximum $DOLP$ occurred in the principal plane, with a decrease in $DOLP$ towards the horizon (for example, Liu and Voss [38]). The same cause was in effect in both cases: for downwelling radiance, the horizon had more multiple scattering than the principal plane at the 90° scattering angle, while for the upwelling radiance, in the water, there was more multiple scattering for the principal plane at a 90° scattering angle than for nadir angles closer to the horizon. The other feature in the data for this clear water case was the separation of the neutral points, areas of zero $DOLP$, on either side of the principal plane, as has been discussed previously [39,40]. Finally, the angle of the polarization plane, χ, was very similar in all three cases, perhaps less well-defined for the data in the area of the low polarization, around the anti-solar position.

An example with more turbid water is shown in Figure 3. Once again, the patterns were generally the same, however in this case the data are noisier. Note that the data shown in this case has not been averaged using the left/right symmetry across the principle plane, thus appears less symmetric than the example in Figure 2. As has been pointed out before [41], the most stable parameter was χ, with the exception of the area with very low $DOLP$, in which case the plane of polarization was not well defined. In general, the images appeared much noisier in the data from the two more turbid sites (New York Bight and Ligurian Sea) than in the clear water off Hawaii. The pattern of Q/I and U/I tended to shift more between each image that goes into these data averages. In the scale of Q/I and U/I ranging from −1 to 1, the standard deviation obtained during averaging the data was approximately 0.05 for the New York Bight and Ligurian Sea data, while it was 0.012 for the clearer Hawaii data. One can also see that in this more turbid case, the single scattering was a poor approximation of all parameters, with the exception of χ.

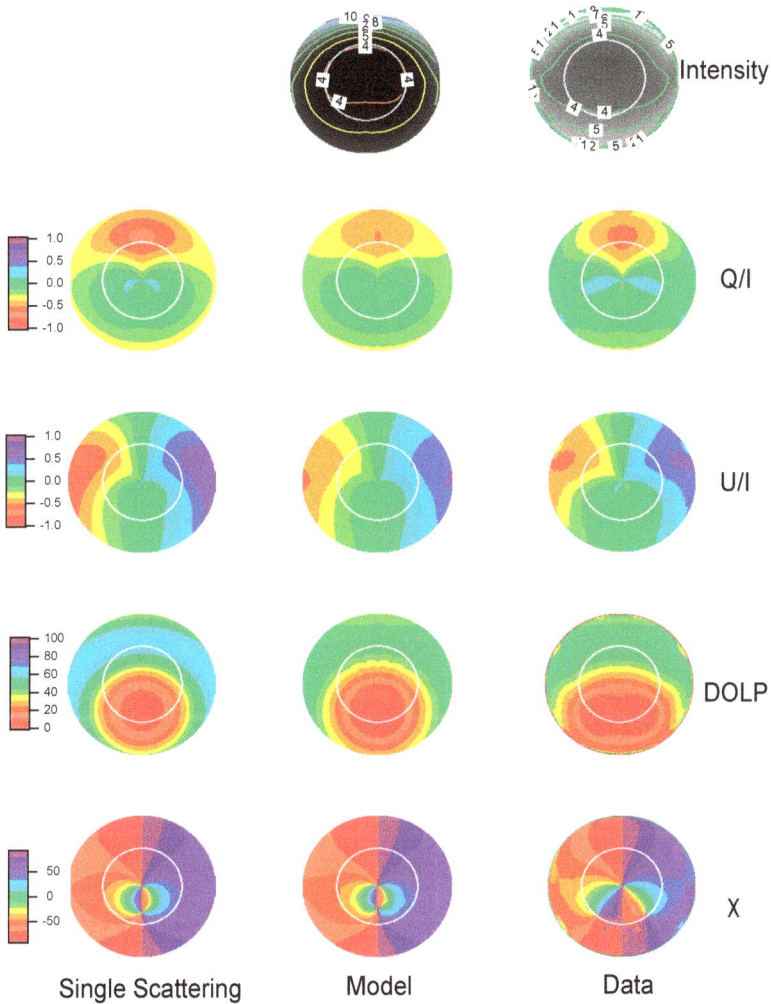

Figure 2. Comparison of the Stokes vector components I, Q/I, U/I, $DOLP$, and χ for clear water. Shown are the calculation for single scattering (left column), RT model (center column), and data (right column). The data were taken on 2 December 2005, at 20:46 UTC off of Oahu, Hawaii. The conditions were: SZA = 48°, 442 nm, Chl = 0.1 mg/m^3, c_t = 0.1 m^{-1} (calculated from Chl as described in text), ω_0 = 0.8, and clear skies. The intensity for the model was adjusted to match the data at nadir. Each image is a fisheye projection. Nadir is in the center of the circle, nadir angle is linearly proportional to radius from center. The angle of the Snell's circle (48° nadir angle) is shown in as the white circle. The principal plane (plane containing the anti-solar point and the nadir direction) is a vertical line through the center of the image. The anti-solar direction is towards the bottom of the image, and the sun direction is towards the top of the image.

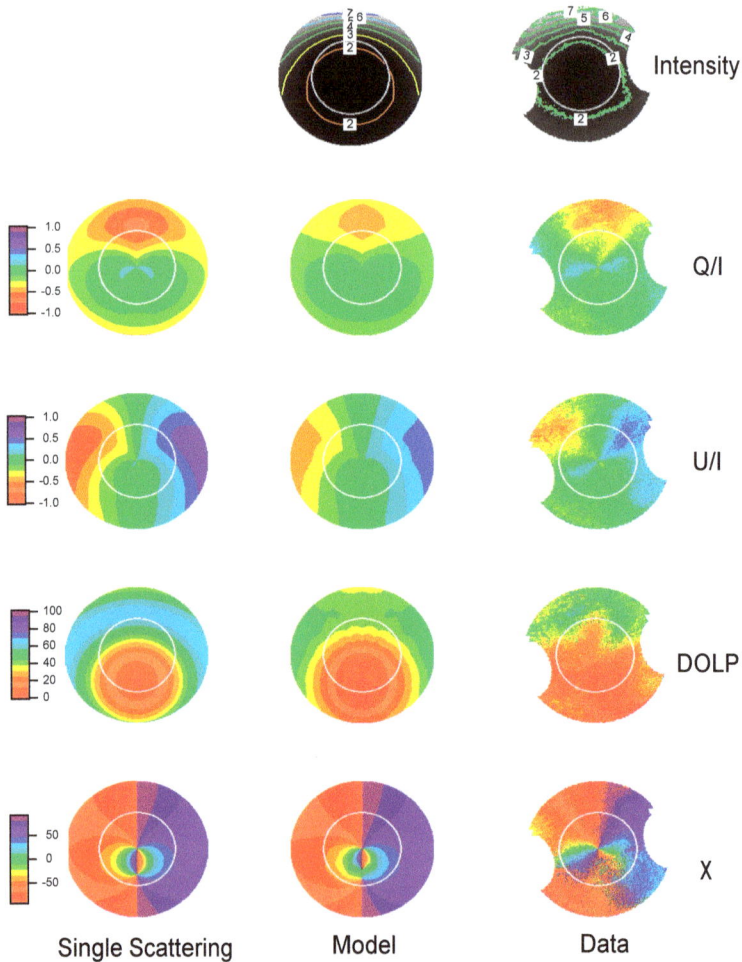

Figure 3. Similar to Figure 2, but for a case with more turbid water. The data were taken on 22 March 2009, at 9:40 UTC in the Ligurian Sea. The conditions were: SZA = 48°, 550 nm, c_t = 0.44 m^{-1}, ω_0 = 0.82, and clear skies. The figure geometry is the same as Figure 2.

3.2. Maximum DOLP and Nadir DOLP

How large can the maximum *DOLP* be in the upwelling radiance? Figure 4 shows this as a function of c_t/a_t and c_t. We used the variation of the optical properties (c_t, b_t, b_{bt}, and a_t) with the wavelength to fill in the data, thus this figure includes all wavelengths with their appropriate optical properties. The quantity c_t/a_t can be shown to equal the mean number of scattering events in the medium taken as a whole [28]. As the water gets more turbid, there is more multiple scattering, which will decrease the *DOLP* [7,42]. The maximum *DOLP* is often outside the Snell's cone, thus not retrievable from above the surface, so we also show the maximum value of the *DOLP* inside the Snell's cone in Figure 4. As can be seen, this was less, sometimes significantly less, than the maximum *DOLP* in the total upwelling field. Another point was that the maximum *DOLP* in the total upwelling field occurred at an azimuth nearly perpendicular to the principal plane, as this position was always at a 90° scattering angle to the sun, and had the largest component of single scattered light due to a combination of light attenuation with depth and the relative smoothness of the scattering function in the backward direction. For the maximum *DOLP* inside the Snell's circle, as can be seen in Figures 2 and 3, this azimuthal angle will move towards the principal plane, and hence got closer to the position where, above the surface, it will be in the region of the solar glitter pattern.

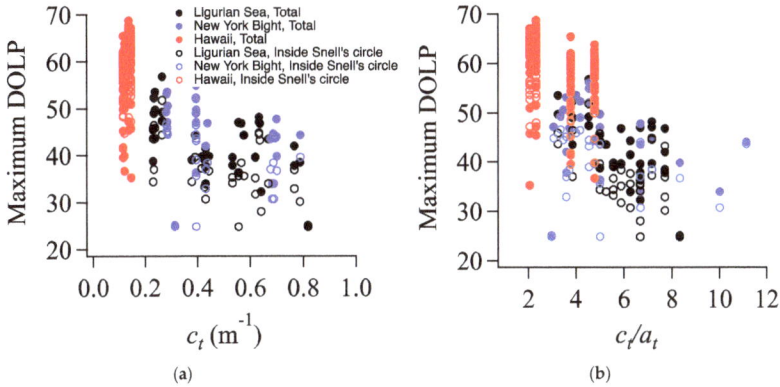

Figure 4. Maximum *DOLP* in upwelling radiance distribution vs c_t (**a**) and c_t/a_t (**b**). The symbol legend is the same for both graphs. Filled circles represent the maximum *DOLP* in the total upwelling field, while the open circles represent the maximum *DOLP* in the portion of the upwelling light field inside the Snell's circle (nadir angle less than 48°).

Another interesting parameter is the *DOLP* at the nadir angle, as this was the direction that many in situ radiometers make their measurement (Figure 5). Here the RT model results are presented rather than the data, as in our definition of the reference frame, the nadir direction is a point of singularity, and the data become excessively noisy at that point.

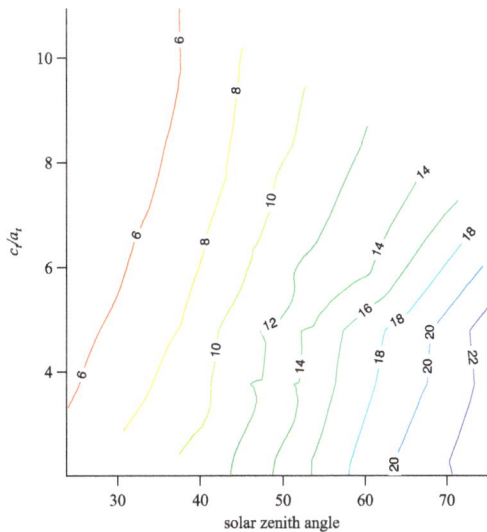

Figure 5. Nadir *DOLP* as a function of solar zenith angle and c_t/a_t, computed from the RT model. Contours are lines of constant *DOLP*.

Note that the nadir light field could be significantly polarized and the *DOLP* could be greater than 20%. Also, as can be seen, the largest values (22%) occurred at large solar zenith angles (when the scattering angle decreased to values closer to 90°), but there was also a trend of decreasing *DOLP* as c_t/a_t increased at a constant solar zenith angle due to increasing multiple scattering. The plane of polarization in all cases was perpendicular to the principal plane.

3.3. Comparison of DOLP Differences

For the first quantitative comparison of the RT model and the data, we looked at scatter plots comparing the measured and modeled *DOLP* (Figure 6). For each cruise, a best linear fit line was calculated, with the y-intercept = 0. The fit was calculated for all of the data, as well as the data inside the Snell's circle; however, the calculated slopes were not significantly different, so we will discuss only the fit for the total image. The slopes were 0.94 (± 0.002, $r^2 = 0.81$), 0.88 (± 0.006, $r^2 = 0.84$), and 0.82 (± 0.008, $r^2 = 0.76$) for Hawaii, New York Bight, and Ligurian Sea, respectively.

Because the *DOLP* was significantly overestimated in the Ligurian Sea data, the alternative normalized Mueller matrix Mod-V-F ($\rho = 0.30$) was tried, with all other parameters the same, resulting in a slope of 1.20 (± 0.009, $r^2 = 0.656$); we concluded that the depolarization of 0.3 was too strong. The true average Mueller matrix for this location could be somewhere between the two we used.

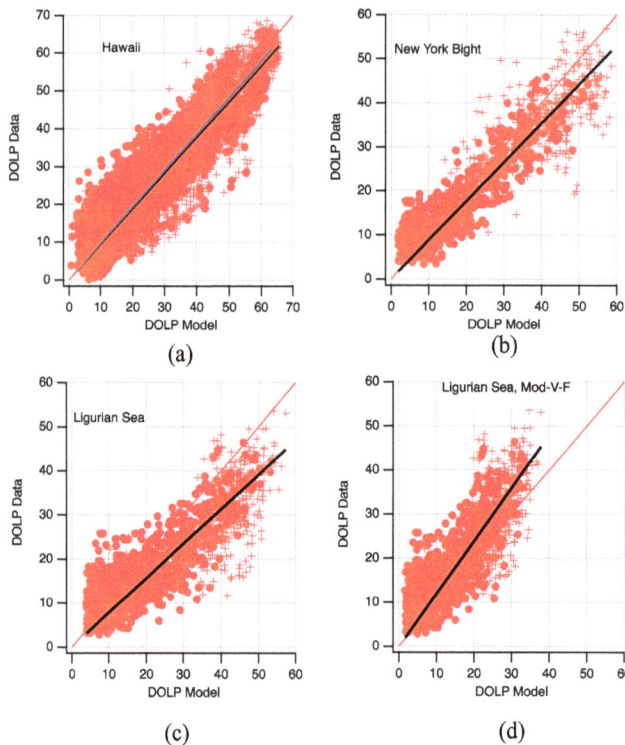

Figure 6. Scatter plots of *DOLP* for model versus data, separated by cruise. Also shown is the best fit line (with y-intercept = 0) (black line) and 1:1 line (red line). The crosses correspond to data outside the Snell's circle (nadir angle larger than 48°), and the dots to points inside the Snell circle. V-F was used in the model for (**a**–**c**), Mod-V-F was used for (**d**).

The correlation between $DOLP_{diff}$ and several environmental parameters was investigated. In Figure 7, we show the $DOLP_{diff}$ variation with solar zenith angle. In this figure, the mean $DOLP_{diff}$ was calculated for each data image, along with the standard deviation (shown as error bars). There was no trend in the data; however, there does appear to have been a slight increase in the magnitude of the deviation at solar zenith angles greater than 70°. In general though, in almost all cases, the mean $DOLP_{diff}$ was within one standard deviation of zero.

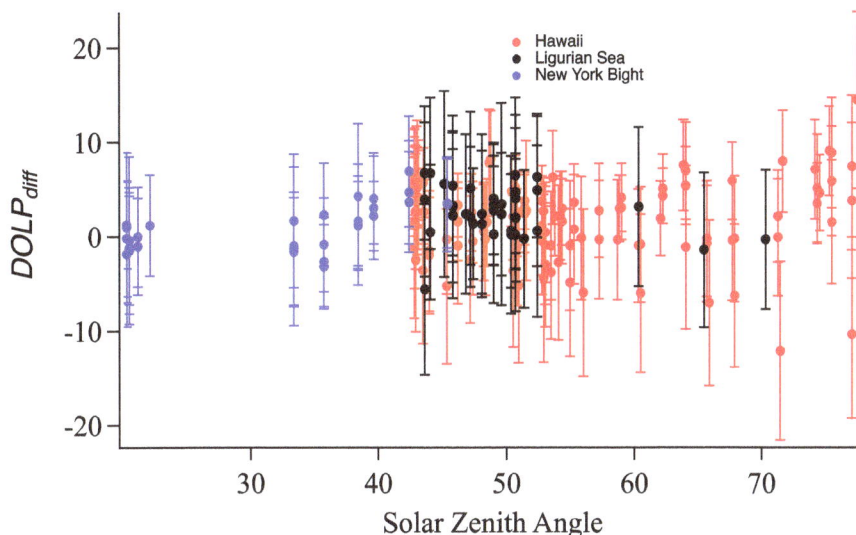

Figure 7. $DOLP_{diff}$ vs solar zenith angle for each data image. The error bars are ± the standard deviation of the mean.

In Figure 8, we look at the $DOLP_{diff}$ dependence on c_t and c_t/a_t. While there was no clear trend in $DOLP_{diff}$ with c_t, a slight trend occurred as a function of c_t/a_t. Low values of c_t/a_t (more absorption, less multiple scattering) had a slightly larger value of $DOLP_{diff}$, while $DOLP_{diff}$ tended towards 0 with larger values of c_t/a_t (more multiple scattering). However, the mean $DOLP_{diff}$ was less than 5% for almost all cases, and this was within the measurement uncertainty of these instruments in the field.

Finally, in Figure 9, we look at the $DOLP_{diff}$ as a function of b_{bt}/b_t (b_{bt}/b_t is the fraction of total scattering which was in the backwards direction, i.e., scattering angles from 90° to 180°). As b_{bt}/b_t increased, $DOLP_{diff}$ appeared slightly higher, indicating less agreement at higher values of b_{bt}/b_t. Low values of b_{bt}/b_t were indicative of a low refractive index, phytoplankton-dominated environment, while higher values of b_{bt}/b_t indicated an environment with other higher refractive index particles, perhaps sediment. V-F was based on measurements in a variety of environments (clear ocean water to more turbid coastal water); however, measurements of the Mueller matrix of phytoplankton cultures also agreed with V-F. It has often been stated that the polarization resulting from higher index particles is less than V-F (i.e., larger *depolarization* at a 90° scattering angle) [32]. It is probably the case in our data set, that as the b_{bt}/b_t ratio increases, the S_{12} and S_{21} elements of the Mueller matrix should also decrease in an absolute value sense, but not as much of a decrease as used in Mod-V-F, where the decrease in this matrix would have caused a significant negative $DOLP_{diff}$.

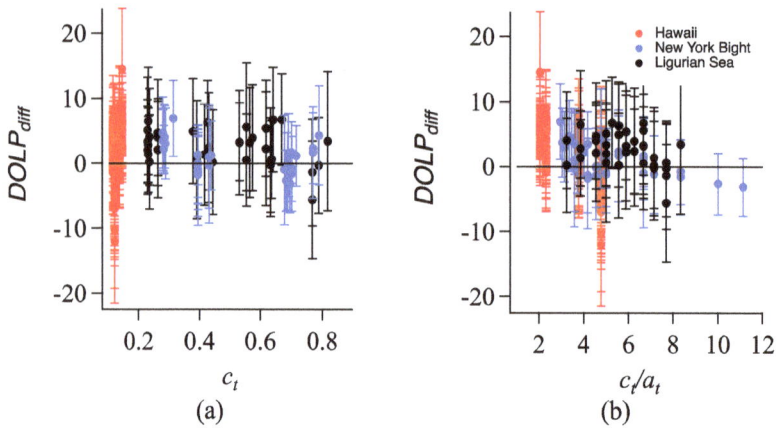

Figure 8. $DOLP_{diff}$ as a function of c_t and c_t/a_t. Color code of data is given in Figure 7. (**a**) shows $DOLP_{diff}$ vs c_t, while (**b**) is $DOLP_{diff}$ vs c_t/a_t.

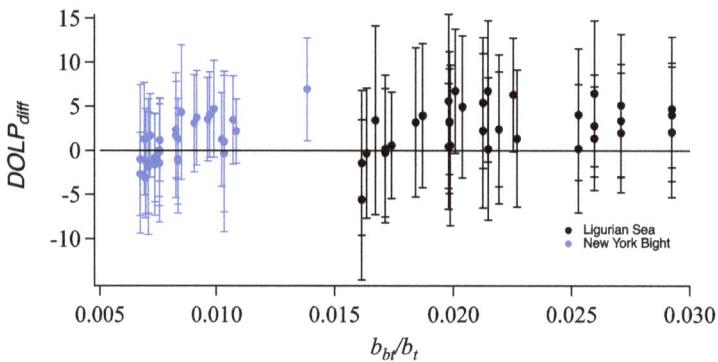

Figure 9. $DOLP_{diff}$ as a function of b_{bt}/b_t. There were no b_{bt} measurements during the Hawaii data collection.

3.4. Comparison of Q/I and U/I Differences

Rather than calculating $DOLP$, we could also show the difference between the model result for Q/I and U/I versus the data, shown in Figure 10 for the Ligurian Sea case. This case is shown because it was neither the best result, which was for Hawaii, nor the worse, which was New York Bight. The advantage of $DOLP$, and the reason most of the results are presented for this parameter, is that it is independent of the frame of reference, thus small errors in locating the solar plane in each image cause smaller differences in the comparison. Q/I and U/I comparisons are with respect to the frame of reference and depend on correctly locating the solar plane in the image, hence the comparison is not quite as good between the model and data result as with the DOLP. The line fits had a slope of 0.66 ($\pm$0.010, $r^2 = 0.70$) for Q/I and 0.71 ($\pm$0.013, $r^2 = 0.67$) for U/I. The fact that the scatter is qualitatively the same in these graphs as in the DOLP figure can be simply understood since Q/I and U/I can be derived from a combination of the $DOLP$ and the plane of polarization, χ. As mentioned earlier, χ is a very stable parameter in the upwelling light field. The spread in the data is slightly larger for U/I versus Q/I because U/I has stronger spatial gradients, and small rotations affect this parameter more than Q/I.

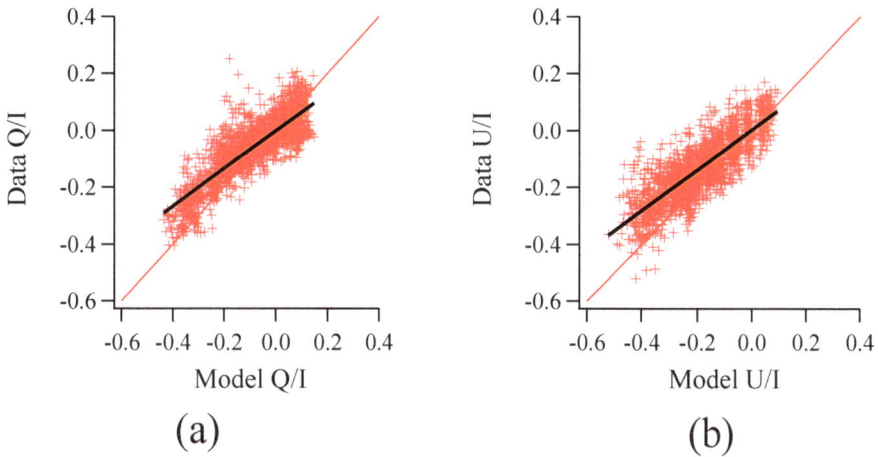

Figure 10. Scatter plots of Q/I (**a**) and U/I (**b**) for model versus data for the Ligurian Sea data set. Also shown is the best fit line between the model and data for each parameter (black line). The 1:1 line is shown in red.

4. Conclusions

We have compared measurements of the polarization properties of the upwelling light field in the marine environment to radiative transfer models based on the measured inherent optical properties (IOPs) and an average normalized Mueller matrix for the scattering. For the Ligurian Sea and New York Bight data sets, the base RT model used the same volume scattering phase function (Petzold) and Mueller matrix (V-F) for the standard model, with the scattering and absorption coefficients as measured coincidently on location. For Hawaii, the IOPs were not measured, but were estimated based on an average *Chl* value of 0.1 mg/m^3 at the measurement location. Even with the fixed scattering matrix parameters, on average, the $DOLP_{diff}$ was less than 5%, and in almost every case was within one standard deviation of zero. Thus, to get more information, or discrimination, from the *DOLP* measurements, the uncertainty of the measured *DOLP* must be smaller than 5%, and the other parameters (scattering phase function, Mueller matrix, and absorption coefficient) need to be known well enough to constrain the *DOLP* model to this accuracy.

Some of the variability in our data was due to instrument noise (on the order of 2%); however, other noise was due to the variability in the environment resulting from the incident light field interacting with the wavy air–sea interface. This was more of a problem for the downwelling light field but was still evident in the upwelling light field. For remote sensing, this radiance (Stokes vector) would have been propagated through the (wavy) air–sea interface, which would introduce more noise.

The 5% agreement of the model with the data shows that the model will provide sufficient accuracy for correcting residual polarization effects in ocean color instruments. However, more work must be done at improving the measurement accuracy of the instrumentation if the goal is to be able to use polarization effects to characterize (and differentiate with respect to) the physical properties of the suspended particles other than c/a.

Author Contributions: Data curation, K.J.V., M.S.T., and J.-F.B.; Formal analysis, A.C.R.G. and H.R.G.; Funding acquisition, K.J.V.; Methodology, A.C.R.G., K.J.V., and H.R.G.; Software, A.C.R.G. and H.R.G.; Supervision, K.J.V.; Writing—original draft, A.C.R.G., K.J.V., and H.R.G.; Writing—review and editing, A.C.R.G., K.J.V., H.R.G., and M.S.T.

Funding: This work was supported by NASA under grants NNX08AH93A and NNX14AP63G (Voss), and the NASA PACE Science Team (Twardowski), and by NOAA under grant NA150AR4320064. The BP'09 cruise (Ligurian Sea) was funded by the NATO Undersea Research Centre.

Acknowledgments: Scott Freeman, Luke Logan, Puru Bhandari, and Nordia Souaidia are thanked for their technical assistance in the field and with the instrumentation and data reduction.

Conflicts of Interest: The authors declare no conflict of interest. The founding sponsors had no role in the design of the study; in the collection, analyses, or interpretation of data; in the writing of the manuscript, or in the decision to publish the results.

References

1. Waterman, T.H. Underwater light and orientation of animals. In *Optical Aspects of Oceanography*; Jerlov, N.G., Nielsen, E.S., Eds.; Academic Press: New York, NY, USA, 1974; pp. 415–443, ISBN 0123849500.
2. Loisel, H.; Duforet, L.; Dessailly, D.; Chami, M.; Dubuisson, P. Investigation of the variations in the water leaving polarized reflectance from the polder satellite data over two biogeochemical contrasted oceanic areas. *Opt. Express* **2008**, *16*, 12905–12918. [CrossRef] [PubMed]
3. Cronin, T.W.; Shashar, N.; Caldwell, R.L.; Marshall, N.; Cheroske, A.G.; Chiou, T.H. Polarization vision and its role in biological signaling. *Int. Comp. Biol.* **2003**, *43*, 549–558. [CrossRef] [PubMed]
4. Mishchenko, M.I.; Cairns, B.; Kopp, G.; Schueler, C.F.; Fafaul, B.A.; Hansen, J.E.; Hooker, R.J.; Itchkawich, T.; Maring, H.B.; Travis, L.D. Accurate monitoring of terrestrial aerosols and total solar irradiance: Introducing the glory mission. *Bull. Am. Meteorol. Soc.* **2007**, *88*, 677–691. [CrossRef]
5. Tonnizio, A.; Gilerson, A.; Harmel, T.; Ibrahhim, A.; Chowdhary, J.; Gross, B.; Moshary, F.; Ahmed, S.; Chami, M. Importance of the polarization in the retrieval of oceanic constituents from the remote sensing reflectance. *J. Geophys. Res.* **2007**, *112*, C05026. [CrossRef]
6. Ibrahim, A.; Gilerson, A.; Chowdhary, J.; Ahmed, S. Retrieval of macro- and micro-physical properties of oceanic hydrosols from polarimetric observations. *Remote Sens. Environ.* **2016**, *186*, 548–566. [CrossRef]
7. Chami, M.; Santer, R.; Dilligeard, E. Radiative transfer model for the computation of radiance and polarization in an ocean-atmosphere system: Polarization properties of suspended matter for remote sensing. *Appl. Opt.* **2001**, *40*, 2398–2416. [CrossRef] [PubMed]
8. Gordon, H.R.; Du, T.; Zhang, T. Atmospheric correction of ocean color sensors: Analysis of the effects of residual instrument polarization sensitivity. *Appl. Opt.* **1997**, *36*, 6938–6948. [CrossRef] [PubMed]
9. Meister, G.; Kwiatkowska, E.J.; Franz, B.A.; Patt, F.S.; Feldman, G.C.; McClain, C.R. Moderate-resolution imaging spectroradiometer ocean color polarization correction. *Appl. Opt.* **2005**, *44*, 5524–5535. [CrossRef] [PubMed]
10. Voss, K.J. Electro-optic camera system for measurement of the underwater radiance distribution. *Opt. Eng.* **1989**, *28*, 241–247. [CrossRef]
11. Voss, K.J.; Chapin, A.L. Upwelling radiance distribution camera system, NURADS. *Opt. Express* **2005**, *13*, 4250–4262. [CrossRef] [PubMed]
12. Voss, K.J. Use of the radiance distribution to measure the optical absorption coefficient in the ocean. *Limnol. Oceanogr.* **1989**, *34*, 1614–1622. [CrossRef]
13. Voss, K.J.; Morel, A.; Antoine, D. Detailed validation of the bidirectional effect in various case 1 waters for application to ocean color imagery. *Biogeosciences* **2007**, *4*, 781–789. [CrossRef]
14. Gleason AC, R.; Voss, K.J.; Gordon, H.R.; Twardowski, M.; Sullivan, J.; Trees, C.; Weidemann, A.; Berthon, J.-F.; Clark, D.; Lee, Z.-P. Detailed validation of the bidirectional effect in various case I and case II waters. *Opt. Express* **2012**, *20*, 7630–7645. [CrossRef] [PubMed]
15. Voss, K.J.; Souaidia, N. Polrads: Polarization radiance distribution measurement system. *Opt. Express* **2010**, *18*, 19672–19680. [CrossRef] [PubMed]
16. Bhandari, P.; Voss, K.J.; Logan, L. An instrument to measure the downwelling polarized radiance distribution in the ocean. *Opt. Express* **2011**, *19*, 17609–17620. [CrossRef] [PubMed]
17. Hoejerslev, N.K.; Aas, E. Spectral irradiance, radiance, and polarization in blue western Mediterranean waters. In Proceedings of the Ocean Optics XIII, Halifax, NS, Canada, 6 February 1997; Volume 2693. [CrossRef]
18. Tonizzo, A.; Zhou, J.; Gilerson, A.; Twardowski, M.S.; Gray, D.J.; Arnone, R.A.; Gross, B.M.; Moshary, F.; Ahmed, S.A. Polarized light in coast waters: Hyperspectral and multiangular analysis. *Opt. Express* **2009**, *17*, 5666–5682. [CrossRef] [PubMed]

19. Gilerson, A.; Zhou, J.; Oo, M.; Chowdhary, J.; Gross, B.M.; Moshary, F.; Ahmed, S. Retrieval of chlorophyll fluorescence from reflectance spectra through polarization discrimination: Modeling and experiments. *Appl. Opt.* **2006**, *45*, 5568–5581. [CrossRef] [PubMed]

20. Morel, A.; Antoine, D.; Gentili, B. Bidirectional reflectance of oceanic waters: Accounting for raman emission and varying particle scattering phase function. *Appl. Opt.* **2002**, *41*, 6289–6306. [CrossRef] [PubMed]

21. Lee, Z.P.; Du, K.; Voss, K.J.; Zibordi, G.; Lubac, B.; Arnone, R.; Weidemann, A. An inherent-optical-property-centered approach to correct the angular effects in water-leaving radiance. *Appl. Opt.* **2011**, *50*, 3155–3167. [CrossRef] [PubMed]

22. Bohren, C.F.; Huffman, D.R. *Absorption and Scattering of Light by Small Particles*; John Wiley and Sons: New York, NY, USA, 1983; ISBN 9783527618156.

23. Ivanoff, A.; Waterman, T.H. Elliptical polarization of submarine illumination. *J. Mar. Res.* **1958**, *16*, 255–282.

24. Gordon, H.R.; Ding, K. Self-shading of in-water optical instruments. *Limnol. Oceanogr.* **1992**, *37*, 491–500. [CrossRef]

25. Mobley, C.D.; Gentili, B.; Gordon, H.R.; Jin, Z.H.; Kattawar, G.W.; Morel, A.; Reinersman, P.; Stamnes, K.; Stavn, R.H. Comparison of numerical-models for computing underwater light fields. *Appl. Opt.* **1993**, *32*, 7484–7504. [CrossRef] [PubMed]

26. Teillet, P.M. Rayleigh optical depth comparisons from various sources. *Appl. Opt.* **1990**, *29*, 1897–1900. [CrossRef] [PubMed]

27. Pegau, S.; Zaneveld JR, V.; Mitchell, B.G.; Mueller, J.L.; Kahru, M.; Wieland, J.; Stramska, M. Inherent optical properties: Instruments, characterizations, field measurements, and data analysis protocols. In *Ocean Optics Protocols for Satellite Ocean Color Sensor Validation, Revision 4, Volume IV*; Mueller, J.L., Fargion, G.S., McClain, C.R., Eds.; NASA/TM-2003-21621/Rev4-Vol IV; Goddard Space Flight Center: Greenbelt, MD, USA, 2002.

28. Morel, A.; Gentili, B. Diffuse reflectance of oceanic waters: Its dependence on sun angle as influenced by the molecular scattering contribution. *Appl. Opt.* **1991**, *30*, 4427–4438. [CrossRef] [PubMed]

29. Petzold, T.J. *Volume Scattering Functions for Selected Ocean Waters*; Scripps Institution of Oceanography Ref. 72–78; University of California: San Diego, CA, USA, 1972; 79p.

30. Voss, K.J.; Fry, E.S. Measurement of the muller matrix for ocean water. *Appl. Opt.* **1984**, *23*, 4427–4439. [CrossRef] [PubMed]

31. Fry, E.S.; Voss, K.J. Measurement of the Mueller matrix for phytoplankton. *Limnol. Oceanogr.* **1985**, *30*, 1322–1326. [CrossRef]

32. Volten, H.; Haan, J.F.; Hovenier, J.W.; Schreurs, R.; Vassen, W.; Dekker, A.G.; Hoogenboom, H.J.; Charlton, F.; Wouts, R. Laboratory measurements of angular distributions of light scattered by phytoplankton and silt. *Limnol. Oceanogr.* **1998**, *43*, 1180–1197. [CrossRef]

33. Svensen, O.; Stamnes, J.J.; Kildemo, M.; Aas LM, S.; Erga, S.R.; Frette, O. Mueller matrix measurements of algae with different shape and size distributions. *Appl. Opt.* **2011**, *50*, 5149–5157. [CrossRef] [PubMed]

34. Adams, J.T.; Aas, E.; Hojerslev, N.K.; BLundgren, B. Comparison of radiance and polarization values observed in the mediterranean sea and simulated in a monte carlo model. *Appl. Opt.* **2002**, *41*, 2724–2733. [CrossRef] [PubMed]

35. Kuik, F.; Stammes, P.; Hovenier, J.W. Experimental determination of scattering matrices of water droplets and quartz particles. *Appl. Opt.* **1991**, *30*, 4872–4881. [CrossRef] [PubMed]

36. Fry, E.S.; Kattawar, G.W. Relationships between elements of the Stokes matrix. *Appl. Opt.* **1981**, *20*, 2811–2814. [CrossRef] [PubMed]

37. Hovenier, J.W.; van de Hulst, H.C.; van der Mee, C.V.M. Conditions for the elements of the scattering matrix. *Astron. Astrophys.* **1986**, *157*, 301–310.

38. Liu, Y.; Voss, K.J. Polarized radiance distribution measurements of skylight: II. Experiment and data. *Appl. Opt.* **1997**, *36*, 8753–8764. [CrossRef] [PubMed]

39. Chowdhary, J.; Cairns, B.; Travis, L.D. Contribution of water-leaving radiances to multiangle, multispectral polarimetric observations over the open ocean: Bio-optical model results for case 1 waters. *Appl. Opt.* **2006**, *45*, 5542–5567. [CrossRef] [PubMed]

40. Voss, K.J.; Gleason, A.C.R.; Gordon, H.R.; Kattawar, G.W.; You, Y. Observation of non-principal plane neutral points in the in-water upwelling polarized light field. *Opt. Express* **2011**, *19*, 5942–5952. [CrossRef] [PubMed]

41. You, Y.; Tonizzo, A.; Gilerson, A.A.; Cummings, M.E.; Brady, P.; Sullivan, J.M.; Twardowski, M.S.; Dierssen, H.M.; Ahmed, S.A.; Kattawar, G. Measurements and simulations of polarization states of understeer light in clear oceanic waters. *Appl. Opt.* **2011**, *50*, 4873–4893. [CrossRef]

42. Timofeyeva, V.A. Degree of polarization of light in turbid media. *Isvestiya Akademii Nauk Sssr Fisika Atmosfery I. Okeana* **1970**, *6*, 513–522.

applied sciences

MDPI

Article

Ocean Color Analytical Model Explicitly Dependent on the Volume Scattering Function

Michael Twardowski [1,2,*] and Alberto Tonizzo [2,*]

[1] Harbor Branch Oceanographic Institute, Florida Atlantic University, Ft. Pierce, FL 34946, USA
[2] Sunstone Scientific LLC, Vero Beach, FL 32963, USA
[*] Correspondence: mtwardowski@fau.edu (M.T.); alberto.tonizzo@gmail.com (A.T.);
Tel.: +1-772-242-2220 (M.T.)

Received: 20 June 2018; Accepted: 1 December 2018; Published: 19 December 2018

Featured Application: Ocean color remote sensing.

Abstract: An analytical radiative transfer (RT) model for remote sensing reflectance that includes the bidirectional reflectance distribution function (BRDF) is described. The model, called ZTT (Zaneveld-Twardowski-Tonizzo), is based on the restatement of the RT equation by Zaneveld (1995) in terms of light field shape factors. Besides remote sensing geometry considerations (solar zenith angle, viewing angle, and relative azimuth), the inputs are Inherent Optical Properties (IOPs) absorption a and backscattering b_b coefficients, the shape of the particulate volume scattering function (VSF) in the backward direction, and the particulate backscattering ratio. Model performance (absolute error) is equivalent to full RT simulations for available high quality validation data sets, indicating almost all residual errors are inherent to the data sets themselves, i.e., from the measurements of IOPs and radiometry used as model input and in match up assessments, respectively. Best performance was observed when a constant backward phase function shape based on the findings of Sullivan and Twardowski (2009) was assumed in the model. Critically, using a constant phase function in the backward direction eliminates a key unknown, providing a path toward inversion to solve for a and b_b. Performance degraded when using other phase function shapes. With available data sets, the model shows stronger performance than current state-of-the-art look-up table (LUT) based BRDF models used to normalize reflectance data, formulated on simpler first order RT approximations between r_{rs} and b_b/a or $b_b/(a + b_b)$ (Morel et al., 2002; Lee et al., 2011). Stronger performance of ZTT relative to LUT-based models is attributed to using a more representative phase function shape, as well as the additional degrees of freedom achieved with several physically meaningful terms in the model. Since the model is fully described with analytical expressions, errors for terms can be individually assessed, and refinements can be readily made without carrying out the gamut of full RT computations required for LUT-based models. The ZTT model is invertible to solve for a and b_b from remote sensing reflectance, and inversion approaches are being pursued in ongoing work. The focus here is with development and testing of the in-water forward model, but current ocean color remote sensing approaches to cope with an air-sea interface and atmospheric effects would appear to be transferable. In summary, this new analytical model shows good potential for future ocean color inversion with low bias, well-constrained uncertainties (including the VSF), and explicit terms that can be readily tuned. Emphasis is put on application to the future NASA Plankton, Aerosol, Cloud, and ocean Ecosystem (PACE) mission.

Keywords: ocean optics; ocean color; remote sensing; radiative transfer approximation; volume scattering function; NASA PACE mission

1. Introduction

Radiative transfer (RT) approximations linking inherent optical properties (IOPs), such as spectral absorption $a(\lambda)$ (m^{-1}) and spectral backscattering $b_b(\lambda)$ (m^{-1}) to ocean color remote sensing reflectance $R_{rs}(\lambda)$ are vital to interpreting R_{rs} because it is not possible to analytically invert the full RT equation [1,2]. Once forward RT approximations are developed, inversions can then be explored to solve for IOPs from R_{rs} and subsequently ocean biogeochemical properties [3]. Ocean color $R_{rs}(\lambda)$ (sr^{-1}) here is defined as $L_w(\lambda)/E_d(\lambda)$, or water leaving radiance (W m^{-2} sr^{-1} nm^{-1}) normalized to above water downwelling irradiance (W m^{-2} nm^{-1}) (see Reference [4] for complete definitions of all optical parameters and Appendix A, Table A1 for notation).

Ocean color expressions to date have almost exclusively relied on first order approximations of RT relating R_{rs} to b_b/a through a proportionality represented as f/Q [5–8], or to $b_b/(a + b_b)$ with multi-term polynomial expressions based originally on Gordon et al. [9] with coefficients represented as l or G ([9–12]; see review by Werdell et al. [13]). The coefficients describing the relationship between R_{rs} and IOPs are detailed in look-up tables (LUTs), or a neural network in the case of [8], with dependencies on geometry (i.e., solar zenith, viewing angle, relative azimuth) and in some cases wavelength, wind speed, atmospheric conditions, and/or chlorophyll concentration [Chl]. The LUTs are generated from full RT computations with so-called synthetic data sets where IOPs and their interrelationships are assumed and referenced to [Chl]. LUT coefficients are thus also implicitly dependent on IOPs.

These ocean color relationships have been tremendously useful to the ocean color community for decades. Morel [14] discusses the first order analytical relationship and the empiricism necessary to invert. Coefficients describing the relationship between R_{rs} and IOPs are dependent on the bidirectional reflectance distribution function (BRDF), which describes the transformation of downwelling irradiance for different solar zenith angles into the distribution of upwelling radiance. Morel et al. [7] describe the current state-of-the-art BRDF model (herein referred to as M02), currently implemented operationally by the ocean color community (NASA Ocean Biology Processing Group (OBPG); [15]) to convert L_w measured at any viewing geometry to a conceptual "exact normalized" L_w, $[L_w]_N^{ex}$, with Sun at zenith and nadir viewing in a non-attenuating atmosphere. This allows any measurement at any Sun-viewing geometry to be directly intercompared. Inversion algorithms to derive IOPs are then typically applied to $[L_w]_N^{ex}$ and are often based on the same type of first order R_{rs} to IOP approximation [13,16]. The current semi-analytical algorithm (SAA) application to water-leaving radiances is thus a two-step process (after applying calibration coefficients, georeferencing, and atmospheric correction): (1) A BRDF correction to carry out the conversion to $[L_w]_N^{ex}$, followed by (2) application of an inversion algorithm to derive IOPs (see [13]).

A key potential source of uncertainty in current SAA approaches is associated with the volume scattering function (VSF; $\beta(\psi)$ m^{-1} sr^{-1}, where ψ is scattering angle), and this uncertainty is ambiguously dispersed in both steps. The VSF dependency is of paramount importance; as Morel and Gentili [5] note, the BRDF " ... is essentially controlled by the shape of the VSF ... ." Historically, very few measurements of the VSF have been available to develop and test SAAs. The M02 BRDF tables were developed from extensive RT modeling using VSF shapes (also known as phase functions $P(\psi)$, defined as the VSF normalized to total scattering, $P(\psi) = \beta(\psi)/b$, with units sr^{-1}) tied to estimated chlorophyll concentrations [Chl]. To obtain $P(\psi)$ for modeling, Morel et al. [7] first mixed two phase functions for populations dominated by large and small particles representing high and low [Chl] extremes, respectively, and then mixed that particulate phase function with the phase function for molecular seawater to a degree that was also linked to estimated [Chl]. The [Chl] estimate was initially derived from R_{rs} with an empirical band ratio algorithm. Uncertainty is thus associated with how close the assumed $P(\psi)$ used in the BRDF modeling matches the actual $P(\psi)$ associated with any given L_w measurement. While the M02 model has been validated in Case 1 waters with atmospheric conditions and IOPs that presumably agree with the underlying atmospheric and bio-optical models [17,18], there have been limited attempts to assess any embedded phase function uncertainties with actual VSF measurements in diverse water types [19,20]. No subsequent BRDF approach has demonstrated enhanced performance for Case 1 waters relative to the M02 approach while being feasible to implement for ocean color. However, Talone et al. [21] have recently shown the Lee et al. [11] LUT-based approach was more accurate for Case 2 waters sampled in the Adriatic, Baltic, and Black Seas.

Uncertainty also arises from exclusively using the b_b term in the inversion algorithm, i.e., in step two of the SAA approach described above. Effort has been devoted to trying to correct this uncertainty by layering additional VSF dependence in the f/Q proportionality [22]. Interestingly, after the current M02 BRDF normalization is applied (i.e., measurement geometry transferred to Sun at zenith and nadir viewing), the scattering parameter that should be most closely linked to $[L_w]_N^{ex}$, and thus presumably should provide the lowest associated uncertainties in IOP retrievals, is $\beta(\pi)$ if single scattering is assumed, which is typically a good approximation [23]. We are not aware however of any algorithm using $\beta(\pi)$ in lieu of b_b. The practical but still arbitrary choice of converting measured L_w to $[L_w]_N^{ex}$ with Sun at zenith and nadir viewing may thus not optimize uncertainties in inversion algorithms. The parameter $\beta(\pi)$ is almost completely unknown in the ocean, as there are virtually no direct measurements of $\beta(\pi)$ in the literature and typical models of particle scattering with simplified particle shapes do not account for possible particle-particle coherent scattering that may cause significant, but poorly understood, enhancement near $\beta(\pi)$ [24]. For example, a recent estimate of $b_b/\beta(\pi)$ made with a combination of airborne lidar and in situ b_b measurements [25] was 50% the value expected from extrapolation of available β measurements [26].

Although the M02 BRDF correction as currently applied can be considered "conceptual" at $L_w(0,\pi)$ without creating a problem from a geometry point of view (i.e., L_w at any viewing geometry can be corrected to a conceptual standard L_w at another geometry as long as it is consistent), it will be problematic if we try to develop algorithms based on $\beta(\pi)$ to reduce uncertainties because the necessary data for algorithm development and validation are lacking. This would also suggest a possible benefit in using the BRDF correction to obtain $[L_w]_N^{ex}$ at another unique geometry (solar zenith, viewing angle, and azimuth), one that was representative of single scattering at an angle that we can measure directly with available instrumentation. To minimize the magnitude of BRDF correction, this angle could be chosen as the centroid angle of the maximum in the frequency distribution of in-water single scattering angles observed for polar orbiting satellites, which is about 150° (Figure 1). Since we can accurately measure β at or near 150° with commercially available instrumentation (e.g., WET Labs ECO, In-situ Marine Optics IMO-SC6, and legacy HOBI Labs Hydroscat sensors), there would appear to be potential to reduce uncertainties with $\beta(150°)$-based algorithms matched to a $\beta(150°)$-based BRDF correction using M02 relative to the current b_b-based algorithms applied to a $\beta(\pi)$-based BRDF correction. In the former approach, uncertainties associated with the phase function are thus mostly restricted to the BRDF correction step.

Herein we explore a different approach, working with a RT expression from Zaneveld [27,28] that explicitly incorporates a dependency on VSF shape, as well as specific viewing geometry. A path to inversion to IOPs is also presented where SAA steps 1 and 2 mentioned above are combined in a single relationship and uncertainties related to VSF shape and other parameters can be directly assessed. This approach has several potential benefits, including (1) a single, fully analytical and invertible expression describing the RT process for all remote sensing geometries, (2) optimal retention of native RT relationships with more degrees of freedom than the first order approximation, directly linked to physically meaningful terms, (3) all parameters including the VSF are explicit in the model with readily characterized uncertainties, (4) model can be readily enhanced by tuning one or more terms rather than developing new LUTs from complete recomputations of full RT, and (5) native viewing geometries produce scattering at angles that can be resolved with available instrumentation. A challenge in an explicitly VSF-dependent approach is amenability to inversion, as some information about the phase function is ostensibly required [1,20]. However, it should be pointed out this is also the case with any model, including the M02 BRDF correction, and subsequent inversion algorithms and their inherent assumptions about VSF shape. In fact, as is assessed herein, the same assumptions of M02 in linking changes in VSF shape with an independent estimate of chlorophyll can be directly applied to this RT model, with the benefit of being able to directly quantify associated errors.

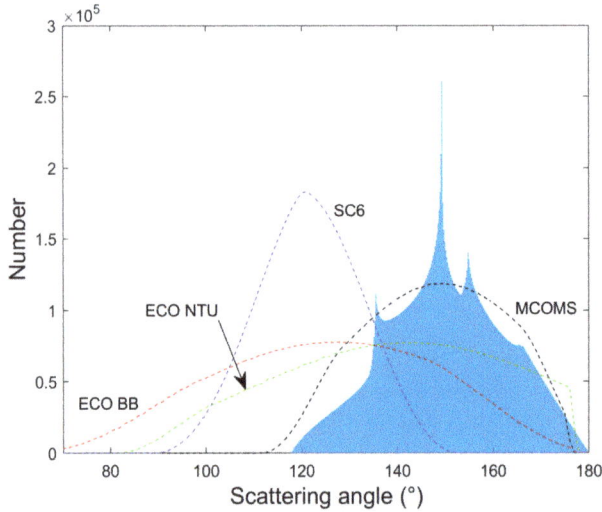

Figure 1. Refracted, in-water scattering angles made between the solar zenith and viewing angle, simulated for the upcoming NASA PACE satellite imager through a complete polar orbit (solid blue). Scattering angle distributions for SeaWiFS were similar. Data courtesy Bryan Franz (NASA GSFC). Angular weighting functions of commercial backscattering sensors WET Labs ECO-BB, ECO-NTU, and MCOMS and IMO-SC6 (see text) are overlaid after scaling by 5×10^6.

In this paper we focus on the performance of a forward implementation of the Zaneveld expression using a modified formulation in terms of IOPs that is now amenable to inversion. Performance of the inversion to IOPs is being fully assessed in ongoing work. A key advance promoting inversion is the finding by Sullivan and Twardowski [26] that the shape of the particulate VSF in the backward direction is relatively constant for a wide range of water types, and so may be represented by a constant function in an algorithm without introducing significant error. Importantly, we have also now overcome a limitation in practically assessing this approach by collecting a database of measured VSFs resolved over a large dynamic range concurrently with other high quality IOPs and radiometry under ideal cloud-free environmental conditions [29].

2. Forward Model Development

Zaneveld [27,28] derived an exact restatement of the RT equation assuming a plane-parallel, optically deep water column in terms of upwelling radiance L_u in viewing direction θ_v:

$$\frac{L_u(\theta_s, \theta_v, \phi)}{E_{od}} = \frac{f_b(\theta_s, \theta_v, \phi)\frac{b_b}{2\pi}}{-\cos(\theta_v)K_{L_u}(\theta_s, \theta_v, \phi) + c - f_L(\theta_s, \theta_v, \phi)b_f}, \tag{1}$$

where E_{od} is the scalar downwelling irradiance, K_{L_u} is the attenuation coefficient for upwelling radiance, b_f is forward scattering, and c is the attenuation coefficient. Parameters f_b and f_L are light field shape factors representing the path radiance term of the RT equation [27,28]. Factor f_b describes the redirection of downwelling radiance into the upwelling viewing angle for a given VSF and is normalized to the redirection that would be observed if the VSF was isotropic. Factor f_b is thus directly linked to the shape of the VSF in the backward direction. The factor f_L is defined similarly but describes the redirection of upwelling radiance into the upwelling viewing angle and is thus directly linked to the shape of the near forward VSF. These shape factors are expected to vary within a relatively narrow range near unity. All terms are a function of depth z and wavelength λ. The Zaneveld expression does not account for any inelastic processes, such as molecular Raman scattering or fluorescence [30–33]. Using physically reasoned approximations for f_b including the assumption of single scattering, the following was obtained by Zaneveld ([28]; his Equation (14)):

$$\frac{L_u(\theta_s, \theta_v, \phi)}{E_{od}} = \frac{\beta(\psi)}{-\cos(\theta_v)K_{Lu}(\theta_s, \theta_v, \phi) + c - f_L(\theta_s, \theta_v, \phi)b_f}, \tag{2}$$

where β is the VSF (including water) and ψ is the in-water scattering angle formed between peak incident sunlight at solar zenith angle θ_s and scattered light traveling in viewing direction (θ_v, Φ), where $\cos(\psi) = \cos(\theta_s)\cos(\theta_v) - \sin(\theta_s)\sin(\theta_v)\cos(\phi)$. Azimuth ϕ is relative to the Sun's direction. Zenith angles θ are in-water, refracted through the air-sea interface and determined from a vertically downward direction. For nadir viewing, $\psi = \pi - \theta_s$ and $-\cos(\theta_v) = 1$ in the denominator. Note approximations of Equation (2) provided in Zaneveld [28] assumed only nadir viewing while we retain the full BRDF functionality here.

In theoretical analyses, Weidemann et al. [34] showed b_b retrievals based on the Zaneveld expression had errors as large as -20% and $+40\%$. However, an extraordinarily wide range of VSF shapes was applied in their simulated data, including VSF shapes for specific particulate subcomponents, such as bacteria, minerals, and phytoplankton. These populations were considered separately as quasi-monodispersions with phase function shapes computed from Lorenz-Mie theory, i.e., assuming homogeneous spheres. These phase function shapes thus had large oscillations with respect to angle, structure that we now know is extremely unrealistic for VSFs representative of bulk in situ particle populations (e.g., [26,35–37]). Weidemann et al. also showed with these phase functions that the Zaneveld approximation of the shape factor at nadir viewing, i.e., $f_b \approx 2\pi\beta(\pi - \theta_s)/b_b$, had an average error of only 5% (their Figure 11) and this included overcast conditions that would not occur in remote sensing. To our knowledge, the only study attempting to test the Zaneveld model with directly measured data was He et al. [20] with a subset of the NOMAD data set, which included no VSF measurements, where performance for the BRDF component of the model was comparable to the current state-of-the-art [7,11].

Equation (2) demonstrates the direct link between $L_u(\theta_s, \theta_v, \phi)$ and $\beta(\psi)$. However, for the Zaneveld expression to be a practical tool for ocean color, the terms K_{Lu} and f_L must be expressed in terms of IOPs. Furthermore, the IOPs in the model should ideally be coefficients that are closely linked to reflectance and directly measurable with good accuracy using existing sensor technology. For example, the term in the denominator $[c - f_L(\theta_s, \theta_v, \phi)b_f]$ has two IOPs that are difficult to determine directly because of acceptance angle issues with standard transmissometer designs [38,39]. However, since f_L is close to unity, this term is also closely related to $a + b_b$, immediately recognizable from commonly used first order approximations. Goals of the next several sections are: 1) To represent Equation (2) entirely in terms of such IOPs, 2) to rework in terms of the commonly used remote sensing reflectance L_u/E_d, since these are the measurements currently available in validation data sets, and 3) to include the inelastic effects associated with water Raman scattering.

2.1. Diffuse Attenuation of Upwelling Radiance K_{Lu}

For the term K_{Lu}, Zaneveld [28] suggested an assumption of equivalency to K_∞, the diffuse attenuation coefficient in the asymptotic regime. Asymptotic theory is based on the principle that the shape of the light field with depth gradually transforms from being dependent on the incident surface light field to being constant, azimuthally symmetric (so L is only a function of θ), and dependent only on IOPs. Attenuation coefficients for all aspects of the light field, i.e., for all radiances and therefore all irradiances, are equivalent in the asymptotic regime and are also IOPs. This is a critical assumption for the purposes of model development since K can then be described only in terms of IOPs. Zaneveld justified the assumption, even though the omnidirectional light field in surface waters is far from asymptotic, due to the decoupling of upwelling radiances to downwelling radiance distributions. Measurements of upwelling radiance fields from the 1960s and 1970s showed a near constant shape and attenuation rate with depth. Recently, Twardowski and Tonizzo [40] confirmed this assumption in RT simulations with no more than 3% error when the sun was at solar zenith and the following relationship held for *above water* solar zeniths θ_s' up to 75°:

$$\frac{K_{Lu} - K_\infty}{K_\infty} = F(\theta_s') = f_1\theta_s'^4 + f_2\theta_s'^3 + f_3\theta_s'^2 + f_4\theta_s' + f_5, \tag{3}$$

which included a full range of possible natural water types. Coefficients f are provided in Appendix A, Table A2. Angle θ_s' is related to θ_s by Snell's Law, i.e., $\theta_s' = \sin^{-1}(1.34\sin(\theta_s))$. Only nadir viewing was considered in the simulations for K_{Lu} in Equation (3), and we make the assumption that K_{Lu} for other viewing angles that define a specific scattering angle ψ can be approximated by Equation (3) with nadir viewing geometry after assigning a θ_s' that provides an equivalent in-water ψ. For example, for $\theta_s' = 60°$, θ_s will be 40.3° and ψ will be 139.7° for nadir viewing; we assume the resulting $F(60°)$ will also be applicable to any off-nadir viewing angle with $\psi = 139.7°$. This assumption has been verified using Hydrolight RT simulations (methodology addressed in Section 3.2) to no worse than 2% in the solar plane and no worse than 5% within the upwelling hemisphere for in-water scattering angles consistent with remote sensing (i.e., Figure 1). This assumption implies a rotational reference frame, where the first order determinant of radiance field shape in the model, i.e., ψ, is preserved. Two potential drawbacks of this assumption are (1) the influence of skylight may be skewed in the rotated reference frame, and (2) the range of viewing angles is restricted since the smallest ψ is ~134° for underwater nadir viewing. The range $\psi > 134°$, however, comprises >95% of the expected scattering angles that will be measured by the PACE imager (Figure 1). Further work with field measurements is needed to verify this assumption. Reformulating Equation (3) in terms of dependency on the in-water scattering angle, we obtain:

$$F(\psi) = f_{A1}\psi^4 + f_{A2}\psi^3 + f_{A3}\psi^2 + f_{A4}\psi + f_{A5}. \tag{4}$$

Coefficients f_A are provided in Appendix A, Table A2.

From Gershun's Law we can set $K_\infty = a/\overline{\mu}_\infty$, where $\overline{\mu}_\infty$ is the average cosine of the asymptotic light field. After inserting Equation (4) into Equation (2) and allowing for $c = a + b$, the following is obtained:

$$\frac{L_u(\theta_s, \theta_v, \phi)}{E_{od}} \cong \frac{\beta(\psi)}{a\left(1 - \cos(\theta_v)\Psi_{K_{Lu}}(\psi)\,\overline{\mu}_\infty^{-1}\right) + b - f_L(\theta_s, \theta_v, \phi)b_f}, \tag{5}$$

$$\text{where } \Psi_{K_{Lu}}(\psi) = K_{Lu}/K_\infty = [1 + F(\psi)]. \tag{6}$$

The parameter $\Psi_{K_{Lu}}$ is assumed spectrally independent with errors over the full range of possible water types estimated at <2% [40]. Dividing numerator and denominator of Equation (5) by b_b, we obtain:

$$\frac{L_u(\theta_s, \theta_v, \phi)}{E_{od}} \cong \frac{\beta(\psi)}{b_b} \bigg/ \left[\frac{a}{b_b}\left(1 - \cos(\theta_v)\Psi_{K_{Lu}}(\psi)\overline{\mu}_\infty^{-1}\right) + f_L(\theta_s, \theta_v, \phi)\left(1 - \widetilde{b}_b^{-1}\right) + \widetilde{b}_b^{-1}\right], \tag{7}$$

where $\widetilde{b}_b = b_b/b$ is the backscattering ratio. For IOPs we now have the backward phase function in the numerator; in the denominator we have the recognizable b_b/a, as well as $\widetilde{b}_b$, an IOP that incorporates information on bulk particle composition (see Section 2.4; [41]), and is not typically associated with ocean color remote sensing vis-à-vis the common first order approximation of r_{rs} assumed proportional to b_b/a.

2.2. Average Cosine of the Asymptotic Light Field $\overline{\mu}_\infty$

The $\overline{\mu}_\infty$ term in Equation (7) must be expressed in terms of IOPs to invert. Using a fit to theoretical calculations of radiance fields by Prieur and Morel [42], Zaneveld [28] recommended $1/\overline{\mu}_\infty$ be modeled empirically with respect to the single scattering albedo ω $(=b/c)$ using a quadratic fit. Additionally, through more detailed RT computations, Berwald et al. [43] found a 4th order dependency of $\overline{\mu}_\infty$ on the albedo.

In the study by Twardowski and Tonizzo [40], $\overline{\mu}_\infty$ was parameterized in terms of b_b/a instead of ω, including an assessment across a full range of environmentally representative phase function shapes using the Fournier-Forand analytical model [44,45]. The data set used in the assessment was not a representative synthetic data set (i.e., [46]), as it included a full range of possible b_b/a values, possible phase functions, and permutations thereof. Representing $\overline{\mu}_\infty$ in terms of b_b/a has two distinct advantages. First, b_b and a can be measured with commercially available in situ instrumentation with accuracies of a few percent [47–49] to enable performance assessment for algorithms. Parameters c and b cannot be measured without significantly larger errors, typically >25–50%, because of the acceptance of near forward scattered light in conventional transmissometer designs [38,39]. Secondly, b and c are not parameters that are closely linked to r_{rs} without additional information, whereas b_b and a are (e.g., Gordon et al. [50]), and a goal here is to represent the entire model in terms of b_b and a to enable inversion.

The Twardowski and Tonizzo [40] parameterization also explicitly depended on η_{bb}, the fraction of b_b attributable to molecular scattering, $\eta_{bb} = b_{bw}/(b_{bp} + b_{bw})$ [6]. The natural range for η_{bb} is from ~0 to ~0.98 [40,51], and the range used for b_b/a was 10^{-4} to 10^{-1}. Since the water components of η_{bb} may be assumed known [52], the effective unknown here is b_{bp}. After extending the analysis from Reference [40] to include near zero b_b/a and increased resolution in η_{bb}, the resulting fit was obtained for $\overline{\mu}_\infty$:

$$\overline{\mu}_\infty\left(\frac{b_b}{a}, \eta_{bb}\right) \approx \left[m_1(\log \eta_{bb})^3 + m_2(\log \eta_{bb})^2 + m_3 \log \eta_{bb} + m_4\right]\left(\log \frac{b_b}{a}\right)^3$$
$$+ \left[m_5(\log \eta_{bb})^3 + m_6(\log \eta_{bb})^2 + m_7 \log \eta_{bb} + m_8\right]\left(\log \frac{b_b}{a}\right)^2$$
$$+ \left[m_9(\log \eta_{bb})^3 + m_{10}(\log \eta_{bb})^2 + m_{11} \log \eta_{bb} + m_{12}\right]\log \frac{b_b}{a}$$
$$+ m_{13}(\log \eta_{bb})^3 + m_{14}(\log \eta_{bb})^2 + m_{15} \log \eta_{bb} + m_{16}. \tag{8}$$

Coefficients m are provided in Appendix A, Table A2. Fits to simulated data were again made with the polyfit function from MATLAB. Absolute errors $\%\delta_{abs}$ for this fit vary from 0.19 to 3.5 for η_{bb} ranging from 0.98 to 0.0098, respectively. Since both b_b/a and η_{bb} are spectrally dependent, $\overline{\mu}_\infty$ will be as well (not shown for clarity).

2.3. Backward Phase Function $\beta(\psi)/b_b$

The scattering parameters in Equation (7) must be expanded into water and particulate components. Expanding $\beta(\psi)/b_b$ gives:

$$\frac{\beta(\psi)}{b_b} = \frac{\beta_p(\psi) + \beta_w(\psi)}{b_{bp} + b_{bw}}, \tag{9}$$

where the p and w subscripts represent particles and molecular seawater, respectively. In Equation (9), the pure seawater terms, which are temperature and salinity specific, can be directly computed with an estimated error of no more than 2% [52]. Introducing a term for the particulate phase function in the backward direction, $P_{bb}(\psi) = \beta_p(\psi)/b_{bp}$, Equation (9) then becomes:

$$\frac{\beta(\psi)}{b_b} = \frac{P_{bb}(\psi)b_{bp} + \beta_w(\psi)}{b_{bp} + b_{bw}}. \tag{10}$$

Note b_{bp} and b_{bw} both have spectral dependencies as does β/b_b, and the unknowns are $b_{bp}(\lambda)$ and $P_{bb}(\psi)$. Equation (10) can be easily rewritten in terms of η_{bb}, with the same unknowns. In coastal waters where η_{bb} is near zero, $\beta_p \gg \beta_w$, $b_{bp} \gg b_{bw}$, and $\beta(\psi)/b_b$ will be approximated by $P_{bb}(\psi)$. For clear ocean waters, phase functions are represented by a mixture of both particles and pure seawater with VSF shapes dependent on η_{bb}.

2.4. Backscattering Ratio $\tilde{b}_b$

Expanding the backscattering ratio $\tilde{b}_b = b_b/b$ in Equation (7) we obtain:

$$\frac{b_b}{b} = \frac{b_{bp} + b_{bw}}{b_p + b_w} = \frac{b_{bp} + b_{bw}}{b_{bp}/\tilde{b}_{bp} + b_w}, \tag{11}$$

where $\tilde{b}_{bp}$ is the particulate backscattering ratio. This $\tilde{b}_{bp}$ is the "true" $\tilde{b}_{bp}$, distinct from the $\tilde{b}_{bp}$ typically derived from measurements that include c data from transmissometers with significant acceptance angle errors (e.g., [38,39]) (we note models linking particle biogeochemical properties and measured $\tilde{b}_{bp}$ should account for the acceptance angle of c measurements [41,53]). Bootstrapping exercises using Equation (7) readily show an impact on reflectance of up to several percent when $\tilde{b}_{bp}$ is varied over the full ~0.003 to ~0.03 dynamic range observed in the oceanic environment [41,53–56]. All terms have spectral dependencies.

2.5. Shape Factor f_L

Zaneveld [28] recommended the term f_L, i.e., the dimensionless upwelling radiance shape factor, could be set to 1.05 with small error. The natural range was estimated at 1 to 1.12 [27]. In the Results, we develop a new model for this term.

2.6. Remote Sensing Reflectance Formulation

Nearly all ocean RT algorithm work over the last several decades has used reflectance defined as L_u/E_d instead of L_u/E_{od}. Zaneveld [28], however, pointed out, as is evident from Equation (1), L_u/E_{od} is most closely aligned with RT theory. The irradiance parameter E_{od} is also less dependent on solar zenith angle than E_d. Furthermore, sensor technology has been available to measure E_{od}, historically from companies Biospherical (www.biospherical.com), Satlantic (www.satlantic.com), and Trios (www.trio.de). Nonetheless, since nearly all field data over the last several decades has focused on E_d, any testing and validation of the RT model described here requires modification in terms of E_d. Substituting $E_d/E_{od} = \overline{\mu}_d$ (the average cosine of downwelling radiance, just below the air-water interface) into Equation (7) gives:

$$r_{rs} \cong \frac{1}{\overline{\mu}_d} \frac{\beta(\psi)}{b_b} \Big/ \left[\frac{a}{b_b}\left(1 - \cos(\theta_v)\Psi_{K_{Lu}}\overline{\mu}_\infty^{-1}\right) + f_L\left(1 - \tilde{b}_b^{-1}\right) + \tilde{b}_b^{-1} \right], \tag{12}$$

where r_{rs} is the classically known remote sensing reflectance just below the water surface. Full dependencies of variables not shown for clarity.

2.7. Average Cosine of the Downwelling Light Field $\overline{\mu}_d$

A model for $\overline{\mu}_d$ is now required in the expression from Equation (12). If the sky is ignored and we assume a negligible fraction of the incident solar beam is scattered in the near-surface, a reasonable first-order approximation of $\overline{\mu}_d$ should be $\mu_w \equiv \cos(\theta_s)$. Adding in a cardioidal radiance distribution for skylight, Morel and Prieur [57] obtained:

$$\frac{1}{\overline{\mu}_d} \approx \frac{0.6}{\mu_w} + \frac{0.4}{0.859}, \tag{13}$$

and noted that for sun angles between $8°$ and $62°$, $\bar{\mu}_d$ varied only from 0.79 to 0.94. Thus, even without knowledge of solar zenith angle, a median value could be used with an accuracy better than 10%.

Numerator values of 0.6 and 0.4 in Equation (13) represent the fractions of direct (E_{dd}/E_d) and diffuse ($H = E_{ds}/E_d$) downwelling light, respectively, which together equal unity. These values primarily depend on θ_s' and horizontal visibility V, the latter of which depends on aerosol optical thickness (AOT). For $20° \leq \theta_s' \leq 60°$ and for H between 0.2 and 0.5, the variability of $\bar{\mu}_d$ is ~7% for a fixed θ_s'.

The term $\bar{\mu}_d$ can be factorized in two parts, one part dependent on the IOPs, the other dependent on the atmospheric conditions and geometry:

$$\bar{\mu}_d\left(\theta_s', V, \frac{b_b}{a}, \eta_{bb}\right) \approx M_d^+\left(\theta_s', V\right) \times M_d^*\left(\frac{b_b}{a}, \eta_{bb}\right). \tag{14}$$

The atmospheric component can be represented as:

$$M_d^+\left(\theta_s', V\right) = \frac{\left[\frac{1-H(\theta_s',V)}{\mu_w} + \frac{H(\theta_s',V)}{0.859}\right]^{-1}}{P^3[\cos(\theta_s')]}. \tag{15}$$

As mentioned, Morel and Prieur [57] assumed $H = 0.4$. Gregg and Carder [58] later provided a relationship for H as a function of θ_s' and V that included skylight. Specifically, the results of ([58]; their Figure 4) can be fit as follows:

$$H(\theta_s', V) = \begin{aligned}&\left[e_1 V^2 + e_2 V + e_3\right]\theta_s'^5 + \left[e_4 V^2 + e_5 V + e_6\right]\theta_s'^4 \\ &+ \left[e_7 V^2 + e_8 V + e_9\right]\theta_s'^3 + \left[e_{10} V^2 + e_{11} V + e_{12}\right]\theta_s'^2 \\ &+ \left[e_{13} V^2 + e_{14} V + e_{15}\right]\theta_s' + e_{16} V^2 + e_{17} V + e_{18}.\end{aligned} \tag{16}$$

Coefficients e are provided in Appendix A, Table A2. Note the typical default value used in Hydrolight is $V = 15$ km.

The P^3 term in the M_d^+ relationship is a 3rd order polynomial in $\cos(\theta_s')$ to correct where Morel and Prieur's original approximation deviates from the Gregg and Carder relationship at large θ_s'. The P^3 term is:

$$P^3\left[\cos\left(\theta_s'\right)\right] = 0.7792 \cos^3\left(\theta_s'\right) - 1.7366 \cos^2\left(\theta_s'\right) + 1.1551 \cos\left(\theta_s'\right) + 0.7842.$$

The IOP-dependent component M_d^* was modeled using the approach in [40] with the extended b_b/a range and η_{bb} resolution discussed in Section 2.2. The final fitted relationship is:

$$M_d^*\left(\frac{b_b}{a}, \eta_{bb}\right) = \begin{aligned}&\left[m_{d,1}^* \log \eta_{bb} + m_{d,2}^*\right]\left(\log \frac{b_b}{a}\right)^3 + \left[m_{d,3}^* \log \eta_{bb} + m_{d,4}^*\right]\left(\log \frac{b_b}{a}\right)^2 \\ &+ \left[m_{d,5}^* \log \eta_{bb} + m_{d,6}^*\right] \log \frac{b_b}{a} + m_{d,7}^* \log \eta_{bb} + m_{d,8}^*.\end{aligned} \tag{17}$$

Coefficients m_d^* are provided in Appendix A, Table A2. Absolute percent error with this relationship relative to computations using full radiative transfer (Hydrolight, see Section 3.2) varies from 0.06% to 0.4% across all θ_s'. Error for the full $\bar{\mu}_d$ expression is <1%. Spectral dependency enters the expression through the V term, which is dependent on spectral AOT.

2.8. Including Inelastic Water Raman Effects

The molecular water Raman scattering contribution to r_{rs} can be included in Equation (12) as an additive term, resulting in the expression:

$$r_{rs}(\theta_s, \theta_v, \phi, V, a, b_b) \cong r_{rs,Raman}\left(\theta_s', a, b_b\right) + \frac{1}{\bar{\mu}_d\left(\theta_s', V, \frac{b_b}{a}, \eta_{bb}\right)} \frac{\beta(\psi)}{b_b} /$$
$$\left[\frac{a}{b_b}\left(1 - \cos(\theta_v)\Psi_{K_{Lu}}(\psi)\bar{\mu}_\infty\left(\frac{b_b}{a}, \eta_{bb}\right)^{-1}\right) + f_L(\theta_s, \theta_v, \phi)\left(1 - \tilde{b}_b^{-1}\right) + \tilde{b}_b^{-1}\right]. \tag{18}$$

Full dependencies for all parameters in the model are shown except for λ; all parameters exhibit dependence on λ in the model except for $\Psi_{K_{Lu}}(\psi)$. We note a similar approach has been taken in adding water Raman effects in other reflectance models (e.g., [59]; reviewed in [13]). The term $r_{rs,Raman}$ can be derived according to Westberry et al. ([60]; see their Equation (7)) with inputs of above water ($z = 0^+$) downwelling irradiance $E_d(0^+,\theta_s')$, a, and b_b. The NASA Generalized IOP (GIOP) inversion model implementation [16] also currently uses this Raman formulation. Terms for other inelastic effects, such as fluorescence from dissolved organic matter and pigments may also be added if representative models are available.

2.9. ZTT Model Summary

Equation (18) is the final model for ocean color reflectance defined as L_u/E_d, called the ZTT model hereafter. In Equation (18) the $\bar{\mu}_d$ term is approximated by Equation (14), the $\beta(\psi)/b_b$ term approximated by Equation (10), the $\Psi_{K_{Lu}}$ term described by Equations (4) and (6), the $\bar{\mu}_\infty$ term described by Equation (8), and $\tilde{b}_b$ represented by Equation (11). The ultimate assignment of the f_L term is addressed in the Results. The geometry variables, $E_d(0^+,\theta_s')$, V, and molecular water scattering parameters in the above can be considered knowns. The model is fully spectral. Key unknowns are b_b and a (or b_b/a and η_{bb}), and since pure seawater absorption a_w in the visible is considered known with good accuracy [61] the effective unknowns are b_{bp} and absorption by non-water constituents a_{pg}. The two additional unknowns are $P_{bb}(\psi)$ and $\tilde{b}_{bp}$. In the forward implementation here, these four parameters must be provided from direct measurements or through some assumptions. In the inversion implementation, these are the parameters that may be solved through techniques to minimize errors in the expression if there are enough spectral bands (i.e., degrees of freedom) in r_{rs}, although *a priori* assumptions may be required for $P_{bb}(\psi)$ and $\tilde{b}_{bp}$.

For ocean color reflectance defined as L_u/E_{od}, the simpler relationship in Equation (7) can be used. Water Raman effects can still be considered by applying an algorithm, such as in Reference [60].

3. Methods

3.1. Synthetic Dataset

A synthetic data set referenced to [Chl] was developed to test the ZTT model and develop an expression for the f_L term (see Section 4.1). Twenty values of [Chl] were assumed, logarithmically spaced between 0.01 and 30 mg m^{-3}. Total absorption coefficient (a) was represented by a sum of four components:

$$a(\lambda) = a_w(\lambda) + a_{ph}(\lambda) + a_d(\lambda) + a_g(\lambda). \tag{19}$$

Pure water absorption a_w was taken from Reference [61]. Phytoplankton absorption a_{ph} was calculated from chlorophyll concentration [Chl] and from spectrally averaged absorption coefficients of micro- and pico-plankton ($\bar{a}_{micro}(\lambda)$ and $\bar{a}_{pico}(\lambda)$, respectively) [62]:

$$a_{ph}(\lambda) = [Chl] * \left(\left[S_f \bar{a}_{pico}(\lambda) \right] + \left[\left(1 - S_f \right) \bar{a}_{micro}(\lambda) \right] \right), \tag{20}$$

where $S_f = [0.25, 0.5, 0.75]$ is the shape mixing factor. Non-algal particulate absorption a_d was given by [63]:

$$a_d(\lambda) = a_d(440) * e^{-0.011(\lambda - 440)}, \tag{21}$$

where $a_d(440)$ is equal to R_d $a_{ph}(440)$, with $R_d = [0.05, 0.1, 0.5, 1, 2]$. Similarly, we used an exponentially decaying expression for the absorption of chromophoric dissolved organic matter, a_g [64]:

$$a_g(\lambda) = a_g(440) * e^{-0.014(\lambda - 440)}, \tag{22}$$

where $a_g(440)$ is equal to R_g $a_{ph}(440)$, with $R_g = [0.1, 0.3, 0.5, 1, 2]$.

Total backscattering b_b was represented by the sum of water and particulate components:

$$b_b(\lambda) = b_{bp}(\lambda) + b_{bw}(\lambda). \tag{23}$$

To derive particulate backscattering b_{bp}, total particulate scattering b_p was first empirically estimated at 550 nm [65]:

$$b_p(550) = 0.416[Chl]^{0.766}, \tag{24}$$

and then extrapolated spectrally [66]:

$$b_p(\lambda) = b_p(550) * \left(\frac{\lambda}{550}\right)^{v([Chl])}, \tag{25}$$

where $v([Chl]) = 0.5(\log[Chl] - 0.3)$ when $0.01 \leq [Chl] \leq 2$ mg m^{-3} and $v([Chl]) = 0$ when $[Chl] > 2$ mg m^{-3}. The empirical relationship from [41] between the particulate backscattering ratio $\tilde{b}_{bp}$ and $[Chl]$ was then used to derive b_{bp}:

$$b_{bp} = b_p \tilde{b}_{bp} = b_p * 0.0096[Chl]^{-0.253}. \tag{26}$$

Pure seawater backscattering b_{bw} was calculated according to Zhang et al. [52].

3.2. Radiative Transfer Simulations

RT simulations of r_{rs} were performed with Hydrolight (Sequoia Scientific, Bellevue, WA), following the procedure in Tonizzo et al. [29]. Fournier-Forand analytical phase functions were derived for each [Chl] iteration following the method of Mobley et al. [67].

Inelastic water Raman scattering was not included in the simulations for the synthetic data set, as this is separately addressed in the model (Equation (18)); fluorescence from any seawater constituents was also not considered. Note water Raman effects were included in full RT simulations for all field data (see [29]). Output wavelengths ranged from 350 to 800 nm at 5 nm resolution. Hydrolight default atmospheric parameters were used for all runs. Computations were run for θ_s' of 10, 30 and 60°. Altogether, 1500 different IOP permutations based on [Chl] were simulated for each θ_s'.

3.3. Field Data Sets

Model performance was assessed using two aggregate data sets. The first is a high quality data set of 23 stations from the Ligurian Sea, waters around the Marine Optics BuoY (MOBY) west of Lanai, the southern California coast, and the New York Bight, collected in 2008 and 2009 as part of NASA Spectral Ocean Radiance Transfer Investigation and Experiment (SORTIE) and NASA Ocean Color Validation (OCVAL) exercises. Radiometric closure was assessed in detail for these data by Tonizzo et al. [29]. [Chl] ranged from 0.24 to 23.85 mg m^{-3}. The VSF was directly measured using the custom Multi-Angle SCattering Optical Tool (MASCOT) [26,36] and radiometric uncertainties were rigorously assessed by Voss et al. [68]. Here we use WET Labs ac-9 absorption measurements corrected for scattering using independent VSF measurements (i.e., the VSF98P correction [29]; also see Stockley et al. [49] for detailed evaluation of this correction). The second data set is the NASA NOMAD database, where 80 data records were identified containing the parameters needed for performance assessments here, i.e., a, b_b, $\tilde{b}_{bp}$, [Chl], and r_{rs}. [Chl] ranged from 0.23 to 10.68 mg m^{-3} in this data set.

3.4. Depth Weighting IOPs

Depth-weighted contributions of IOPs to water-leaving radiance were derived after Zaneveld et al. [69] using a 2-stream, first derivative approximation. The approach finds the IOP for a conceptual homogeneous ocean that reproduces the water-leaving radiance observed in a specific stratified ocean case. The remote sensing depth weighting for generic IOP X is:

$$\langle X \rangle = \int_{z=0}^{z=\infty} Xf(z)dz, \text{ where } f(z) = \frac{d}{dz}exp\left[-2\int_{z=0}^{z=\infty} K(z)dz\right]. \tag{27}$$

The exponential term can also be found in Gordon and Clark [70], where it is mentioned the $2K(z)$ term should actually be $K_u + K_d$, but approximating with K_d alone was not expected to lead to appreciable errors. Several approximations of K_d in terms of IOPs are available from the literature (e.g., [71,72]). The approximation by Lee et al. [73] for averaged K_d within the euphotic zone is used here and conveniently represented in terms of a, b_b and above water solar zenith θ_s':

$$\overline{K_d} = (1+0.005\theta_s')a + 4.18\left(1 - 0.52e^{-10.8a}\right)b_b. \tag{28}$$

For evaluating RT approximations, all IOP terms should ideally be weighted together. For example, Zaneveld et al. [69] considered the simple $R \propto b_b/a$ approximation and demonstrated that $<b_b>/<a>$ was not equivalent to $<b_b/a>$ for an IOP stratified ocean. However, b_b and a are considered independently in the model here and an ultimate objective is to invert the approximation to derive b_b and a independently, so each IOP was individually depth weighted. For all stations sampled in the validation data set, the difference between surface b_b and depth weighted b_b was <1%; with a, this was also the case for most stations, but reached a difference of 5% for one station sampled.

3.5. Metrics for Error Assessment

Mean absolute percent error (MAPE), $\%\delta_{abs}$, is a commonly used metric in assessing performance in r_{rs} match ups:

$$\%\delta_{abs} = 100 * \frac{\delta_{abs}}{\overline{y}}, \quad \delta_{abs} = \frac{\sum_{i=1}^{n}|y_i - \hat{y}_i|}{n}. \tag{29}$$

The MAPE metric takes into account the absolute magnitude of the residuals, giving them equal weight. Other commonly used metrics include root mean square error (RMSE), a measure of accuracy and potential forecasting errors in simulating r_{rs} when the errors may be assumed to be unbiased and normally distributed [74]. RMSE gives greater weight to larger errors than δ_{abs}. Since bias errors are often expected to be more significant than random, normally distributed errors in simulations, $\%\delta_{abs}$ is expected to be the most appropriate metric to assess match ups [29]. Other common metrics (e.g., [75]) are shown in the following plots as appropriate. Also see Werdell et al. [13] for a detailed discussion of performance metrics.

4. Results

4.1. Developing an Expression for the f_L Term

Dynamics of f_L were explored in the synthetic data set. As found by Hoge et al. [76], dependencies on wavelength and θ_s' were most important, and an average spectral shape $f_{L,ave}(\lambda)$ scaled to solar zenith according to $\sin(\theta_s')$ was ultimately found to be most representative (Figure 2):

$$f_L(\theta_s', \lambda) = f_{L,ave}(\lambda)[0.05959\sin(\theta_s') + 0.9728]. \tag{30}$$

Function $f_{L,ave}(\lambda)$ is provided in Appendix A, Table A3. Equation (30) was developed by minimizing residual errors between results from full radiative transfer and the ZTT model for each wavelength and each solar zenith angle for the full synthetic data set. Spectral shapes for f_L were relatively consistent in the results, so the term $f_{L,ave}(\lambda)$ was derived from an average for these results after spectral normalization. The scaling factor in brackets was a suitable function to minimize errors with respect to solar zenith angle. Only nadir viewing was again considered in the simulations for

f_L, and we again make the assumption from Section 2.1 that f_L for other viewing angles that define a specific in-water scattering angle ψ can be approximated by f_L observed at the nadir viewing geometry with equivalent ψ. The potential caveats mentioned in Section 2.1 apply here as well. Restating Equation (30) in terms of ψ, the following is obtained:

$$f_L(\psi, \lambda) = f_{L,ave}(\lambda)[0.07762 \sin(\psi) + 1.0405]. \tag{31}$$

Similarities between $f_{L,ave}(\lambda)$ and a typical ocean absorption spectrum are noted (Figure 2), resulting from the effects of multiple scattering, where higher relative scattering (lower relative absorption) promotes a radiance field closer to that which would result from isotropic scattering, i.e., where f_L approaches unity.

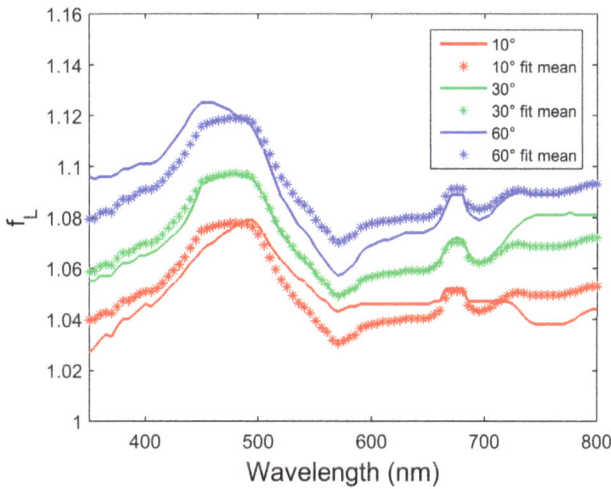

Figure 2. Spectral dependency of the f_L shape function at different θ_s'. Thin lines are solved f_L functions for each θ_s' for the synthetic data set and stars are from the model approximation with constant shape described by $f_{L,ave}(\lambda)$ in Equation (30).

Figure 3A shows results deriving r_{rs} using the ZTT model for the synthetic data set, applying the approximation from Equation (31) for f_L and using all synthetic data IOPs as input. This match up is, thus, effectively assessing combined errors in the ZTT model from the approximations of $f_L, f_b,$ $K_{Lu},$ and $\bar{\mu}_d$. We note f_L was specifically optimized to this data set, so in this respect Figure 3A shows what may be considered a best-case matchup. MAPE $\%\delta_{abs}$ for the full data set was relatively small at 2.68%. The largest errors were observed in three r_{rs} spectra, with the model showing overestimation bias centered around 570 nm. These spectra were associated with the highest [Chl] modeled in the data set and lowest $\tilde{b}_{bp}$.

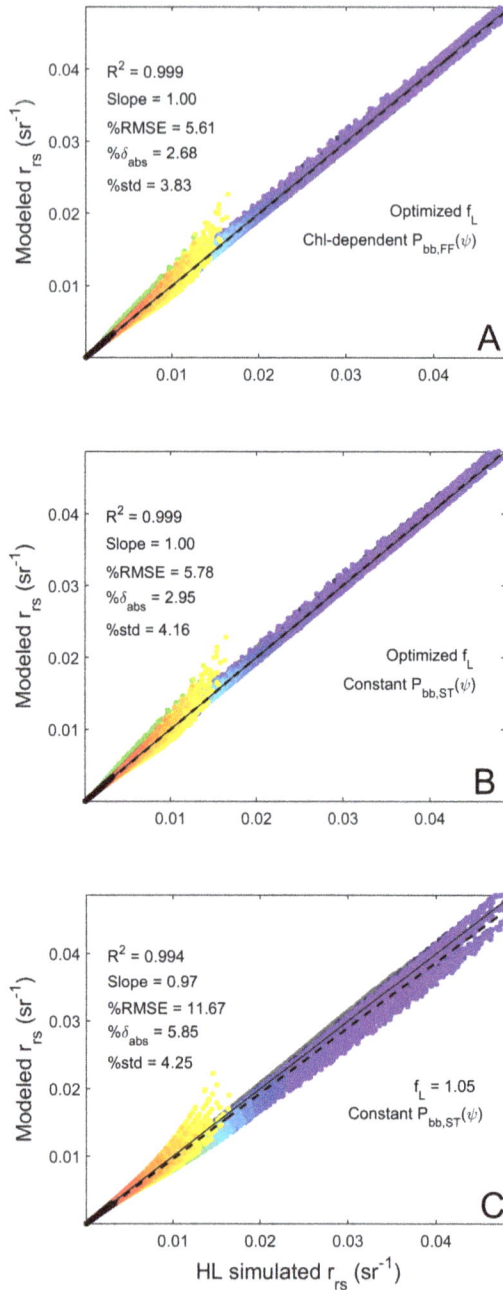

Figure 3. r_{rs} derived from the ZTT model with the synthetic data set compared to r_{rs} simulated using Hydrolight (HL): (**A**) f_L optimized spectrally (i.e., Equation (31) for the synthetic data set with Fournier-Forand phase functions individually determined for each [Chl]; (**B**) f_L optimized spectrally for the synthetic data set with constant $P_{bb,ST}(\psi)$ (see text); and (**C**) f_L set to 1.05 with constant $P_{bb,ST}(\psi)$. Colors represent wavelengths, from 400 to 700 nm with gray used for $350 < \lambda < 400$ nm and $700 < \lambda < 800$ nm.

4.2. Assessing Assumption of Constant $\beta_p(\psi)/b_{bp}$

Sullivan and Twardowski [26] found a high degree of consistency in the shape of the particulate VSF in the backward direction, i.e., $\beta_p(\psi)/b_{bp}$, for a large data set covering a wide dynamic range in bulk particle composition. Deviations from this relationship were no more than 5% through the entire backward angular range. While it is well known the phase function shape over the full angular range varies substantially (e.g., [37,41,45,77]), only the shape in the backward direction that is important for remote sensing and the ZTT model is considered here.

If we assume $\beta_p(\psi)/b_{bp}$ is a constant shape $P_{bb,ST}(\psi)$ after Sullivan and Twardowski [26] in the ZTT model, replacing $P_{bb,FF}(\psi,\tilde{b}_{bp})$ from the synthetic data set, errors increase by only ~0.3% (Figure 3B). We assume $P_{bb,ST}(\psi)$ is constant spectrally after [41] and others. This result is significant, as it demonstrates the potential for eliminating one of the key unknowns in the ZTT model. We note that an averaged Fournier-Forand phase function in the backward direction could be used in the model, which would enhance the analytical character of the model, but the Sullivan and Twardowski [26] averaged phase function, derived from extensive measurements in a wide range of water types, may be more representative of the natural environment. This is assessed further in the next section. Figure 3C shows results of setting f_L to a constant 1.05 after [28] and using a constant $P_{bb,ST}(\psi)$. MAPE increases significantly to 5.85%, showing the importance of using the f_L model in the ZTT. Each of the three distinct data groupings along the 1:1 relationship are associated with the individual solar zeniths that were used, i.e., 10, 30, and 60°, highlighting the influence of the f_L model in removing the effects of solar zenith. All ZTT model runs hereafter use the f_L model from Equation (31). The constant backward phase function $P_{bb,ST}(\psi)$ is used unless another phase function is specified.

4.3. Assessment with High Quality Validation Data

Figure 4A shows results for the ZTT model, with inputs of constant $P_{bb,ST}(\psi)$ and direct measurements of a_{pg}, b_{bp}, and $\tilde{b}_{bp}$, compared to measured r_{rs} for the validation data set [29]. MAPE %δ_{abs} was 16% for all λ (summarized in Table 1; spectral %δ_{abs} provided in Table 2). Interestingly, this result is slightly more favorable than r_{rs} computed with the full radiative transfer (i.e., Hydrolight) using measured phase functions, where %δ_{abs} was 17% for the same data (as reported in Tonizzo et al. [29]). Based on closure analyses for these data [29], this 17% error represented the aggregate inherent error from all sources within the data and computations, i.e., from the IOP and radiometric measurements, as well as any errors from the assumptions within the Hydrolight RT code. For the ZTT model evaluated using measured phase functions instead of the constant $P_{bb,ST}(\psi)$, i.e., the same approach followed with the full RT computations, %δ_{abs} was also 17% (Table 1). The different backward phase functions applied in the ZTT model are shown in Figure 5. Measured backward phase functions showed more variability at individual angles than the average $P_{bb,ST}(\psi)$ derived from a large data set of diverse Case 1 and Case 2 water types, likely the result of small scale hydrosol patchiness along the different light paths for individual measurements at each angle. This observation is addressed further in Section 5. Moreover, the strong agreement in absolute error between full radiative transfer simulations and the ZTT analytical approximation for this diverse data set is encouraging.

Excluding the effects of water Raman from the ZTT model increased absolute error by 3%, demonstrating Raman had a significant effect, especially >580 nm (results not shown).

Figure 4B shows ZTT model performance with a further constraint, setting $\tilde{b}_{bp}$ constant at 0.006, so the only IOP inputs were measured a_{pg} and b_{bp}. With respect to inversion, this approach has the same unknowns, i.e., a_{pg} and b_{bp}, as current algorithms based on the simple first order approximation [16]. Resulting MAPE %δ_{abs} was 17%, comparable to full RT computations. ZTT runs with other fixed $\tilde{b}_{bp}$ are shown in Table 1. Some influence from $\tilde{b}_{bp}$ is apparent, which has not been fully appreciated in previous models based on the first order approximation of r_{rs} to b_b/a.

ZTT model runs with Fournier-Forand phase functions for the particulate component, computed from measured $\tilde{b}_{bp}$ following the method of [67], resulted in %δ_{abs} of 19%, 3% higher than results using the constant $P_{bb,ST}(\psi)$ (Table 1).

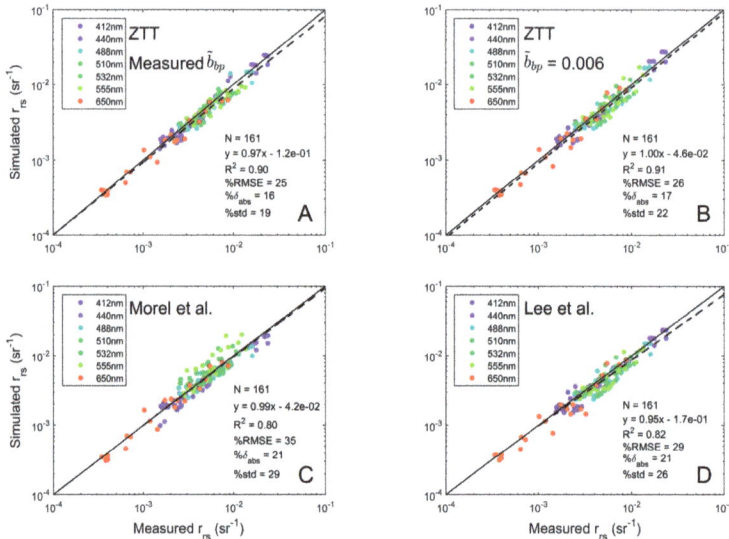

Figure 4. r_{rs} computed from (**A**) ZTT, (**B**) ZTT with $\tilde{b}_{bp}$ fixed at 0.006, (**C**) M02, and (**D**) L11 models, compared to measured r_{rs} in the high quality validation data set from Tonizzo et al. [29]. Note L11 does not account for Raman scattering. Chlorophyll input for M02 was derived from spectral absorption using the Nardelli and Twardowski [78] line height algorithm. MAPE %δ_{abs} was 16%, 17%, 21%, and 21%, in A through D, respectively.

Table 1. Summary MAPE %δ_{abs} results for simulated or calculated r_{rs} vs measured r_{rs} for all λ.

Approach	$\beta_p(\psi)/b_{bp}$ Input	$\tilde{b}_{bp}$ Input	%δ_{abs} SORTIE and OCVAL (23 Stations) [1]	%δ_{abs} NOMAD (80 Stations)
Full RT [2]	directly measured	N/A	17	nd [4]
Full RT [2]	Fournier-Forand [3]	measured	20	nd [4]
ZTT	directly measured	measured	17	nd [4]
ZTT	Fournier-Forand [3]	measured	19	25
ZTT	$P_{bb,ST}(\psi)$	measured	16	20
ZTT	$P_{bb,ST}(\psi)$	0.005	18	22
ZTT	$P_{bb,ST}(\psi)$	0.006	17	23
ZTT	$P_{bb,ST}(\psi)$	0.008	18	25
ZTT	$P_{bb,ST}(\psi)$	0.010	19	27
ZTT	$P_{bb,ST}(\psi)$	0.015	22	29
ZTT	Large and small population phase functions with $\tilde{b}_{bp}$ of 0.19% and 1.4%, blended according to [Chl] [5]	measured	23	26
Morel et al. [7] (M02)	Large and small population phase functions with $\tilde{b}_{bp}$ of 0.19% and 1.4%, blended according to [Chl] [5]	N/A	21	25
Lee et al. [11] (L11)	Blend of Petzold[6] average and 1% $\tilde{b}_{bp}$ Fournier-Forand [3]	N/A	21	26

[1] see [29]; [2] Hydrolight; [3] derived from $\tilde{b}_{bp}$ according to [67]; [4] not determined; [5] algorithm of [7]; [6] [79].

Match ups in r_{rs} were also assessed applying the M02 model, the current standard in BRDF corrections for ocean color remote imagery (Figure 4C). The proportionality between r_{rs} and b_b/a is defined as f/Q, which is a function of viewing geometry, wind speed, and [Chl]. [Chl] can be estimated from an empirical r_{rs} band ratio algorithm. For the assessment here, [Chl] input was either quantified directly from discrete samples or derived from spectral particulate absorption measurements using the line height method with specific absorption $a_p^*(676)$ of 0.0108 m^2 mg^{-1} after [78]. Note that bootstrapping exercises have shown relatively large errors in this estimated [Chl], i.e., up to $+/-50\%$, affect $\%\delta_{abs}$ in match ups by typically no more than 1%. The other inputs were measured a_{pg} and b_{bp}. Match up $\%\delta_{abs}$ for M02 was 21% for the data set from [29] (Table 1). Notably, M02 performed well in the blue spectral region (Table 2).

To derive phase functions, Morel et al. [7] used randomly-oriented spheroidal particles, computing the scattering phase function with the T-matrix method. Because the computations for spheroids were lengthy, a single refractive index n_p of 1.06 was considered with sizes ranging from 0.02 to 14 μm. A Junge-type particle size distribution was assumed, i.e., a power law model, with exponents of 3.1 and 4.2 representing two end-member populations. Particulate scattering phase functions were then calculated from (see Figure 5):

$$P_p(\psi) = \alpha_s([Chl])P_{ps}(\psi) + \alpha_l([Chl])P_{pl}(\psi),$$

with $\alpha_s + \alpha_l = 1$, and $\alpha_s([Chl]) = 0.855[0.5 - 0.25\log_{10}([Chl])]$.

If the Morel et al. [7] approach for deriving the phase function is used as input into the ZTT model, replacing the constant shape $P_{bb,ST}(\psi)$, MAPE $\%\delta_{abs}$ increases significantly from 16% to 23% (Table 1).

Finally, match ups in r_{rs} were assessed in a recently published BRDF model from Lee et al. ([11]; called L11 hereafter) that, like M02, developed a LUT based on RT simulations with a [Chl]-referenced synthetic IOP data set (Figure 4D; Tables 1 and 2). L11 uses a quadratic form of $b_b/(a + b_b)$ split into particulate and molecular seawater components with each term scaled by a G coefficient, with solutions provided in LUTs. In the synthetic data set, Lee et al. [11] used an averaged Petzold phase function and Fournier-Forand phase function assuming $\tilde{b}_{bp} = 1\%$ as particulate phase function endmembers. L11 however does not include water Raman effects. Like M02, match up MAPE $\%\delta_{abs}$ was also 21% for all λ.

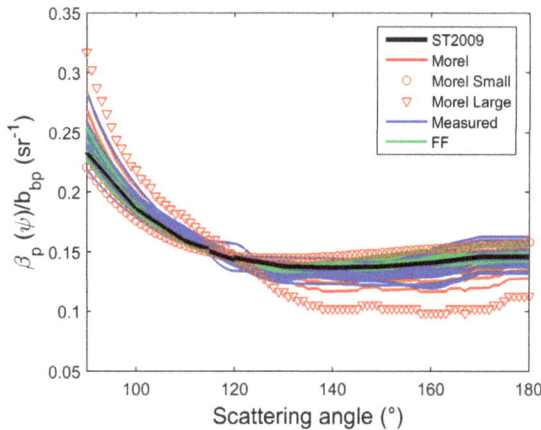

Figure 5. Phase functions in the backward direction used to assess the ZTT model. "ST2009" is $P_{bb,ST}(\psi)$ from [26] with the assumption $P_{bb,ST}(\psi > 170°) = P_{bb,ST}(170°)$; "Morel" is from Morel et al. [7] with endmember populations dominated by small and large particles also shown; "Measured" were directly measured; and "FF" are Fournier-Forand phase functions $P_{bb,FF}(\psi,\tilde{b}_{bp})$ derived from measured $\tilde{b}_{bp}$ following [67].

4.4. Assessment with Global NOMAD Data Set

Figure 6A shows results for the ZTT model applied to a subset of 80 measurements from the NASA NOMAD global data set. IOPs were averaged over the first optical depth. Match up MAPE %δ_{abs} of 20% was observed for this global data set, comparing favorably to the error of 16% observed with the high quality data set from Tonizzo et al. [29] (Table 1; see Table 2 for spectral breakdown). This is especially true considering the different IOP and r_{rs} processing approaches (e.g., there are several scattering error correction options for in-water absorption measurements) used for these data sets. If $\tilde{b}_{bp}$ is constrained to 0.006, MAPE %δ_{abs} increases to 23% (Figure 6B). Results from ZTT runs for other fixed $\tilde{b}_{bp}$ are shown in Table 1, where the dependency of performance on $\tilde{b}_{bp}$ is again apparent.

If the Morel et al. [7] approach for deriving the phase function from [Chl] is used as input into the ZTT model, %δ_{abs} increased from 20% to 26% for the NOMAD data (Table 1). If measured $\tilde{b}_{bp}$ are used to derive $P_{bb,FF}(\psi,\tilde{b}_{bp})$ following [67], %δ_{abs} increased to 25% for the ZTT model.

Table 2. Summary MAPE %δ_{abs} results for calculated r_{rs} vs measured r_{rs} for each λ, for SORTIE and OCVAL data sets (see Figure 4) and NOMAD data (see Figure 6).

Approach	$\beta_p(\psi)/b_{bp}$ Input	$\tilde{b}_{bp}$ Input	λ (nm)	%δ_{abs} SORTIE and OCVAL (23 Stations)	NOMAD (80 Stations)
ZTT	$P_{bb,ST}(\psi)$	measured	412 (410)	20	17
			440	16	20
			488 (490)	16	20
			510	19	19
			532	13	-
			555	14	23
			650	13	-
			665	-	63
ZTT	$P_{bb,ST}(\psi)$	0.006	412 (410)	20	20
			440	16	23
			488 (490)	19	23
			510	13	25
			532	14	-
			555	13	26
			650	22	-
			665	-	63
Morel et al. [7] (M02)	Large and small population phase functions with $\tilde{b}_{bp}$ of 0.19% and 1.4%, blended according to [Chl]	N/A	412 (410)	17	22
			440	14	25
			488 (490)	16	25
			510	18	26
			532	22	-
			555	38	21
			650	17	-
			665	-	71
Lee et al. [11] (L11)	Blend of Petzold average and 1% $\tilde{b}_{bp}$ Fournier-Forand	N/A	412 (410)	22	22
			440	21	26
			488 (490)	24	27
			510	21	27
			532	20	-
			555	19	24
			650	26	-
			665	-	66

Match ups are also shown for implementation of the full M02 and L11 models with the NOMAD data set (Figure 6C,D). MAPE statistics are summarized in Tables 1 and 2. There are several possibilities for the systematically lower measured r_{rs} in the red spectral region for these data: (1) Suboptimal surface extrapolation of radiances due to water Raman effects [80,81], (2) residual scattering errors in reported in-water absorption measurements that may promote a bias to higher values of modeled r_{rs} (e.g., [29,49]), and/or (3) suboptimal corrections for sensor self-shading, which can be strong in clear water at longer wavelengths due to water absorption [82,83].

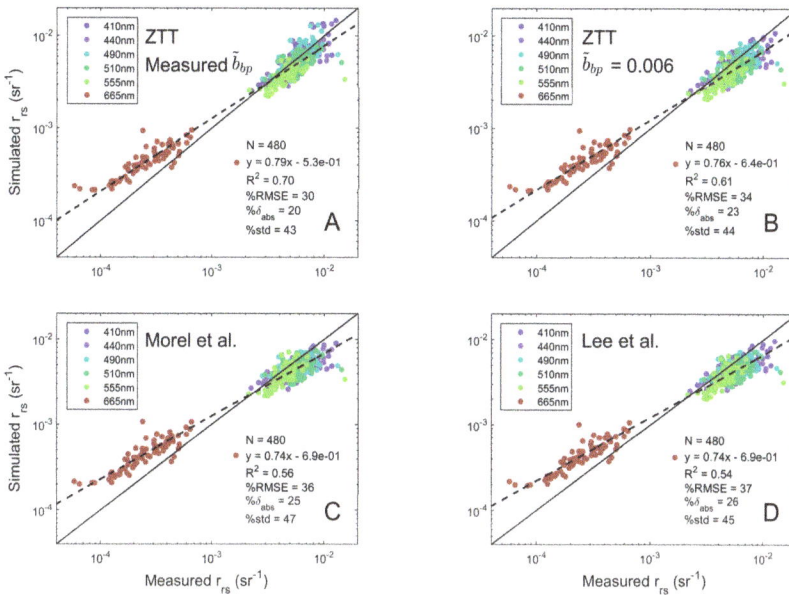

Figure 6. As in Figure 4, but for the NASA NOMAD data set containing a, b_b, r_{rs}, and [Chl]. MAPE $\%\delta_{abs}$ were 20%, 23%, 25%, and 26% in (**A–D**), respectively.

5. Discussion

The ZTT analytical model for remote sensing reflectance is based on the restatement of the RT equation by Zaneveld [27,28] with the following refinements: (1) Full bidirectionality was retained in the final model (no assumption of nadir viewing), (2) the assumption of asymptotic diffuse attenuation for upwelling radiances at the surface was analytically described, (3) the entire model was formulated in terms of b_b and a to enable inversion, (4) the VSF was explicitly included with $\beta_p(\psi)/b_{bp}$ set to a constant shape defined by $P_{bb,ST}(\psi)$ [26], which showed the strongest performance of any phase function, including measured phase functions in disparate, diverse data sets, (5) water Raman effects were added, and (6) for r_{rs}, an analytical model for $\bar{\mu}_d$ was developed that included dependency on atmospheric visibility. IOP inputs to the ZTT model are b_{bp}, a_{pg}, $\beta_p(\psi)/b_{bp}$, and $\tilde{b}_{bp}$. Moreover, the ZTT model enables systematic assessment of the effect of IOPs on the BRDF for the first time.

ZTT showed strong performance in match ups with field data sets spanning a wide range of water types. In terms of MAPE, the model with constant phase function shape $P_{bb,ST}(\psi)$ performed as well as full RT computations. Stronger performance was observed for these available data sets relative to current state-of-the-art LUT-configured models based on the first order approximation between r_{rs} and b_b/a or $b_b/(a + b_b)$, i.e., M02 and L11.

MAPE for the ZTT model was slightly better using $P_{bb,ST}(\psi)$ ($\%\delta_{abs}$ of 16%) when compared to using directly measured backward phase functions for the validation data set (where $\%\delta_{abs}$ was 17%). It may be argued, under certain conditions, the broad average $P_{bb,ST}(\psi)$ may be more accurate than an isolated measurement because specific bias errors in any single VSF measurement at an individual angle may be avoided. This of course assumes such biases may be larger than the natural variability in the shape of the backward phase function. Another explanation is error assessments for the validation data set were made over a relatively few number of stations, 23. To roughly estimate uncertainty in this error metric, single stations were sequentially left out of the error determination for ZTT using $P_{bb,ST}(\psi)$, which gave a total range in $\%\delta_{abs}$ of 14.8 to 16.8% with standard deviation of 0.6%. A ~1% difference in error may thus not be particularly meaningful.

The ZTT model is fully analytical in the sense it is described entirely by equations and "almost fully analytical" with respect to the inclusion of empiricism. Empiricism comes from optimizing the f_L parameter to our synthetic data set based on [Chl]-based bio-optical models, with MAPE improving by a significant ~3% compared to setting f_L equal to a spectrally constant 1.05. Using the constant $P_{bb,ST}(\psi)$ backward phase function shape also introduced empiricism. If sample-specific Fournier-Forand analytical phase functions, based on $\tilde{b}_{bp}$ were used, errors in the validation data set increased by 3–5% (Table 1).

Particulate phase function shapes from Morel et al. [7] were modeled using the T-matrix approach, assuming homogeneous spheroidal shapes with constant n_p of 1.06. Although the endmember phase function shapes were not consistent with $P_{bb,ST}(\psi)$ and $P_{bb,FF}(\psi,\tilde{b}_{bp})$ shapes, many of the blended phase functions were. If $P_{bb,ST}(\psi)$ may be considered widely representative, increasing errors may be expected in applying the M02 model to the lowest and highest Chl conditions where the endmember phase functions would fully manifest. Phase functions in the Chl range 1 to 10 mg m^{-3} were in closest agreement to $P_{bb,ST}(\psi)$. Applying phase functions derived using the Morel et al. [7] approach in the ZTT model increased errors relative to using a constant $P_{bb,ST}(\psi)$ by a significant 7% for the validation data set of [29] (from 16% to 23%) (Table 1). This error was also larger than the 21% error observed for M02 (addressed further below).

Since L11 used a Fournier-Forand phase function where $\tilde{b}_{bp}$ = 1% and the average Petzold phase function as endmembers [shown in 26, their Figure 3], many blended shapes from L11 should be relatively similar in the backward to the $P_{bb,ST}(\psi)$ used in ZTT (although the Petzold shape does exhibit systematic discrepancies). This particular $P_{bb,FF}$ with $\tilde{b}_{bp}$ = 1% is actually a very close match to $P_{bb,ST}$. L11 was developed through least-squares fitting with respect to Hydrolight results, so all the blended phase functions are weighted in the results.

A "typical marine atmosphere" and associated incident sky radiance distribution were used in the Hydrolight RT computations for the synthetic data set, the effects of which enter the ZTT model through spectral optimization of the upwelling radiance shape factor f_L (Section 4.1). The parameters $\Psi_{K_{Lu}}$ and $\bar{\mu}_\infty$ were also solved with Hydrolight RT simulations assuming the same atmosphere. The M02 and L11 models also assumed a single "typical marine atmosphere" in their simulations using Hydrolight based on Reference [58], so this aspect of the models should be consistent. The $\bar{\mu}_d$ term in the ZTT model has added flexibility in accommodating varying atmospheric visibility (closely related to AOT; [58]) inasmuch as V is an explicit term for $\bar{\mu}_d$.

One question arising from this work is, if the phase function shape $P_{bb,ST}$ may indeed be considered highly representative, how would a LUT-based model following the approach of M02 but using $P_{bb,ST}$ perform? To test this, we developed a LUT based on the first order r_{rs} proportionality to b_b/a, using the synthetic data set described in Section 3.1. The particle phase function was constant, consisting of $P_{bb,ST}$ in the backward and a Fournier-Forand phase function with $\tilde{b}_{bp}$ = 0.006 for the forward direction (this had the best performance for the ZTT model in the validation data set [29], with MAPE = 17%). The LUT approach resulted in a MAPE of 22% for the validation data set and 26% for the NOMAD data. This performance is very close to M02 (21% and 25%, respectively) and L11 (21% and 26%, respectively). An insight we can take from this analysis is that a component, such as phase function shape in the different models cannot be considered in a vacuum, i.e., each model has a large set of assumptions that ultimately influence final performance. Furthermore, although not always explicitly stated in the literature, field validation efforts are likely to have had a role in updating assumptions through the course of development and testing of some models to ensure the model was well aligned with available data. For example, in our case, the f_L expression (Section 4.1) was developed from an assumed synthetic data set and the $\tilde{b}_{bp}$ value of 0.006 was recommended as it ultimately provided the best results with the field validation data sets available.

Recent work by Zhang et al. [84] has suggested the possibility of substantial variability in phase function shape in the backward direction, including up to a 40% increase for $\beta_p(\psi)/b_{bp}$ in the near backward relative to $\beta_p(120°)/b_{bp}$. These results are in stark contrast to the comprehensive study of [26] and other published phase functions in the literature (e.g., [35,85]). Other recent works showing increases in the VSF at $\psi > 150°$ [86–88] were made in enclosed cuvettes and did not include any correction for internal reflections (e.g., [89]), which are typically a difficult problem in laboratory VSF devices, since the scattered signal in the backward direction is orders of magnitude smaller than the forward. The Zhang et al. [84] results are also inconsistent with the phase functions used in previous BRDF models, such as M02, which have been shown to have satisfactory performance in validation efforts [17,18], even in many Case II coastal waters [19,90]. The recent theoretical r_{rs} modeling study of Xiong et al. [91] based on the phase function shapes presented in [84], where up to 50% disagreement with L11 was observed, consequently appear unrealistic.

Calibration and possible bias errors in the VSF device MASCOT used in [26] to derive $P_{bb,ST}(\psi)$ have been evaluated in detail [36,48]. The MASCOT is an open path, in situ device with 17 independently calibrated detectors resolving the VSF from $10°$ to $170°$ in $10°$ increments. There are two calibration coefficients for each channel, a dark offset (obtained with the source occluded) and a scaling factor to convert digital counts to VSF units m^{-1} sr^{-1}. Moreover, the observation of a remarkably consistent shape in the backward phase function for VSFs spanning more than four orders of magnitude in dynamic range cannot be explained by any known bias error in these coefficients. In fact this observation argues against any systematic bias, as consistency in phase function shapes during serial particle suspension experiments is a useful method to assess the accuracy of calibrations.

Recently, He et al. [20] developed a BRDF model also based on Zaneveld [28] but it differs from our implementation is several ways: (1) only the BRDF aspect was considered; (2) one wavelength was considered; (3) the formulation was in terms of a LUT (similar to M02 and L11) for aggregated terms; (4) widely varying backward phase function shapes were assumed after [84] described above; (5) the $\bar{\mu}_d$ model depended only on "typical average" sky conditions; and (6) water Raman effects were excluded. Direct comparisons were thus not possible here.

5.1. Assessing Residual Bias in the Model

The assumption of single scattering in the simplification of f_b (Section 2, Equation (2)) is not expected to invoke significant errors at turbidity levels found for most oceanic and coastal waters [5,13]. As shown by Gordon [23], secondary scattering events for light redirected toward the backward direction, i.e., downwelling radiance scattered into the upwelling radiance field, has a negligible effect on the shape of the upwelling radiance distribution because the vast majority of these scattering events are in the near forward direction. The full effects of multiple scattering were included in relationships for other terms in the model as they were determined using full RT modeling.

MASCOT VSF measurements used to determine $P_{bb,ST}(\psi)$ only extended to $170°$, so the application of this function for $\psi > 170°$ in the model is uncertain. There were no $\psi > 170°$ in the data sets tested here, although such scattering angles may be expected for some viewing geometries for the PACE imager (Figure 1). There is a need for a better understanding of phase function shapes for $\psi > 170°$ and this extends to other applications, such as lidar [92].

Only nadir viewing was considered for the analytical relationships for K_{Lu} and f_L and we suggest these relationships may be transferable to other viewing geometries as long as ψ is preserved. This assumption has been verified in Hydrolight simulations to have small errors for representative cases, but cannot be assessed with the field data sets available here, which all have nadir viewing. This is being considered in future work, and, moreover, emphasizes the need for routine measurements of full upwelling radiance distributions in ocean color validation work as validation only with nadir radiance $L_u(\theta_s, \theta_v = \pi)$ neglects the BRDF effects that are essential to any algorithm.

The optimization of f_L was based on the [Chl]-based synthetic data set. A significant deviation from the bio-optical model assumptions of that synthetic data set could lead to a discrepancy in match ups, however it worth pointing out (1) f_L has a relatively weak dependency on IOPs except for the VSF shape in the backward direction (which is effectively accounted for in the solar zenith dependency in Equation (30)), and (2) other state-of-the-art models (i.e., M02 and L11) are constructed entirely around such synthetic data sets. A benefit of the ZTT model is that an analytical relationship for a specific term of the model may be replaced later if a more optimal approach is demonstrated.

As shown in Zaneveld [27,28], L_u normalized to E_{od} is more closely linked to the RT Equation than L_u/E_d, where an additional $\bar{\mu}_d$ term must be added to the model (Section 2.6). Adding this term is expected to increase uncertainty in the RT model and inversion. Scalar irradiance has the additional advantage of being less dependent on solar zenith angle and sky radiance distributions. As discussed in Section 2.6, sensor technology is available to measure E_{od} and inclusion of these measurements in field efforts focused on algorithm development and validation should be encouraged. When the $\bar{\mu}_d$ term must be included in association with E_d measurements, AOT is also a useful validation measurement that may be made in the field [93] and used to derive visibility V [58] for input in Equation (16) or full RT computations.

Lastly, potential uncertainties in the full RT computations must be considered, as this was used to parameterize several ZTT terms. Polarization is not considered in the ZTT model or in Hydrolight, which is likely important [29,33]. There is a need in the research community for commercially available RT code with a focus on the ocean that includes full polarization, along with concomitant measurements of polarized radiance in the field. Neither the Hydrolight simulations or ZTT model included inelastic effects from DOM fluorescence, which could lead to bias errors as high as about 10% in the mid-visible [33]. Another potential source of uncertainty is the water Raman estimation [60], where bias errors as high as 10% can be typical.

5.2. Suitability for Inversion

The ZTT model can be used to directly solve for b_b/a using least-squares minimization of a r_{rs} measurement at any wavelength. This is shown in Figure 7 for the validation data set from Tonizzo et al. [29], assuming a constant $\tilde{b}_{bp}$ of 0.006. The typical value for V of 15 km was assumed. MAPE %δ_{abs} in the inversion was 17%, which matches results from full RT simulations of closure in the forward direction. Residual errors thus appear almost entirely comprised of the inherent errors of the data set arising from IOP and radiometric measurements [29]. With respect to the use of $P_{bb,ST}(\psi)$, another potential approach to estimating VSF shape as it relates to the upcoming PACE mission is using planned multi- and hyper-angular polarimetry data from the Hyper-Angular Rainbow Polarimeter-2 (HARP2; contributed by University of Maryland, Baltimore County) and the Spectrometer for Planetary Exploration—one (SPEXone; contributed by the Netherlands Institute for Space Research) [94]. If information on the backward shape of the VSF could be gleaned with sufficient accuracy from angular polarimetry data (only a few angles would be necessary), this could be used directly in the ZTT model.

The ZTT model described here only addresses in-water RT, although the current approach to normalize water-leaving radiances measured by a satellite imager is expected to directly apply (see [15], Sections 3.1 and 6]). The BRDF is built-in to the Zaneveld model so an additional step to conceptually shift the geometry to a normalized geometry standard would not be required for IOP inversion. A normalized geometry (to nadir viewing or other geometry) could still be applied to intercompare reflectances in images. Translation to water-leaving radiance by accounting for all the reflection and refraction effects through the air-sea interface i.e., $\mathcal{R}(\theta_s,\theta_v)$ [6,15], can still be applied. Similarly, the effects of varying Earth-Sun distance and atmospheric attenuation can be removed following the current approach. Moreover, we would expect a practical implementation of this analytical model to have similar computational requirements as the current M02 LUT-based approach.

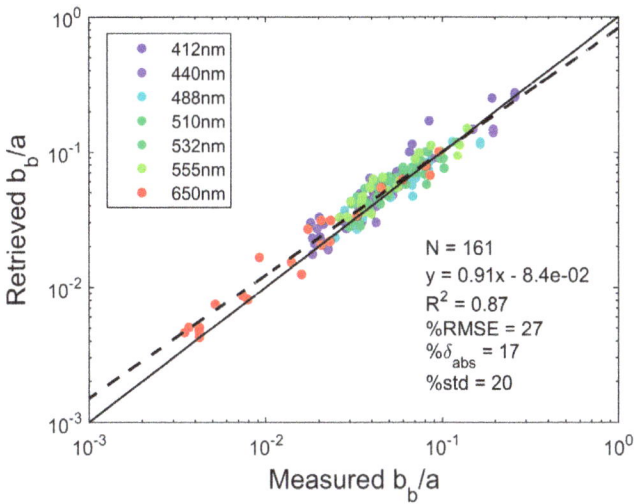

Figure 7. Comparing measured b_b/a to results of inverting ZTT using least-squares minimization individually at each wavelength. Water Raman effects were removed before applying the ZTT model.

The ZTT model shows $\widetilde{b}_{bp}$ exerts some influence on r_{rs} (Table 1) and the potential to invert r_{rs} for $\widetilde{b}_{bp}$ is also being investigated. The parameter $\widetilde{b}_{bp}$ has not historically been associated with r_{rs} and could provide new insights into particle composition in natural waters (e.g., [41]). Even a retrieval of $\widetilde{b}_{bp}$ with relatively large errors could be useful for many research applications.

The ZTT model is fully spectral and can be applied over the full anticipated ocean color hyperspectral wavelength range for the future PACE mission (350–800 nm). A primary goal in model development was parameterization entirely in terms of a and b_b. Inversion approaches, such as the General IOP (GIOP) inversion using a least-squares minimization to solve for IOP subcomponents [16] may consequently be readily applied. Spectral b_{bp} and a_{pg} can be solved using assumptions for subcomponent IOP spectra through least squares minimization of Equation (18). This is the step where empiricism prominently enters the problem as spectral shapes for IOP subcomponents must be assumed. This is being examined in ongoing work. Further investigation is needed to determine IOP subcomponent spectra with a high spectral resolution for error minimization over this full spectral range in inversion, which is currently being assessed by the PACE Science Team [13,95]. Hyperspectral (i.e., 5 nm) resolution for PACE is expected to improve fidelity in inverting for subcomponents (and potentially $\widetilde{b}_{bp}$) as there will be higher degrees of freedom than for multi-spectral imagers [96,97].

Code for the ZTT model in MATLAB is available at ioccg.org.

Author Contributions: Conceptualization, M.T.; Methodology, M.T. and A.T.; Software, A.T.; Validation, A.T. and M.T.; Formal Analysis, M.T. and A.T.; Investigation, A.T. and M.T.; Resources, M.T.; Data Curation, M.T.; Writing—Original Draft Preparation, M.T. and A.T.; Writing—Review and Editing, M.T. and A.T.; Visualization, A.T. and M.T.; Supervision, M.T.; Project Administration, M.T.; Funding Acquisition, M.T.

Funding: Research was funded by the NASA PACE mission under Ocean Biology and Biogeochemistry program grant number [NNX15AN17G] and the HBOI Foundation.

Acknowledgments: Thanks to Tim Moore for kindly providing the data subset from NOMAD, Bryan Franz for providing simulated geometry parameters for the expected orbit of the NASA PACE mission, David Antoine for providing the Morel et al. (2002) endmember phase functions, and to Zhongping Lee for access to the current PACE Science Team synthetic data sets. The authors thank Ron Zaneveld, Tim Moore, Curt Mobley, and Georges Fournier for helpful discussions. We also appreciate very thorough and constructive comments by Dariusz Stramski, Zhongping Lee, and two other anonymous reviewers.

Conflicts of Interest: The authors declare no conflict of interest.

Appendix A

Table A1. Notation.

Symbol	Description	Units
a	total absorption coefficient	m^{-1}
a_x	absorption coefficient, where subscript $x = w, p, d, ph, g, pg, pico,$ and $micro$ specifies water, particulate, non-phytoplankton particulate, phytoplankton, dissolved, particulate plus dissolved, picoplankton, and microplankton	m^{-1}
a^*	specific absorption coefficient	$m^2\,mg^{-1}$
b	total scattering coefficient	m^{-1}
b_p	particulate scattering coefficient	m^{-1}
b_f	forward scattering coefficient	m^{-1}
b_{bx}	backscattering coefficient, where subscript $x = w$ and p specifies water and particulate	m^{-1}
$\tilde{b}_b$	backscattering ratio	-
$\tilde{b}_{bp}$	particulate backscattering ratio	-
β	total volume scattering function (VSF)	$m^{-1}\,sr^{-1}$
β_x	volume scattering function (VSF), where subscript $x = w$ and p specifies seawater and particulate	$m^{-1}\,sr^{-1}$
c	total attenuation coefficient	m^{-1}
[Chl]	chlorophyll concentration	$mg\,m^{-3}$
δ_x	error, where subscript $x = abs$ and rel specifies absolute and relative	-
E_x	planar irradiance, where subscript $x = d, ds,$ and dd specifies downwelling, diffuse downwelling, and direct downwelling	$W\,m^{-2}\,nm^{-1}$
E_{od}	downwelling scalar irradiance	$W\,m^{-2}\,nm^{-1}$
f	model coefficient for relating irradiance reflectance to b_b/a	-
f_b, f_L	radiance shape factors	-
Φ	azimuth angle relative to solar plane	°, rad
G	model coefficient for above-surface remote sensing reflectance	sr^{-1}
ψ	scattering angle	°, rad
H	fraction of diffuse downwelling light (E_{ds}/E_d)	-
η_{bb}	fraction of total backscattering contributed by b_{bw}	-
K_x	diffuse attenuation coefficient, where subscript $x = L_u, u, d,$ and ∞ specifies upwelling radiance, upwelling irradiance, downwelling irradiance, and asymptotic	m^{-1}
L_u	upwelling radiance	$W\,m^{-2}\,nm^{-1}\,sr^{-1}$
L_w	water-leaving radiance	$W\,m^{-2}\,nm^{-1}\,sr^{-1}$
λ	Wavelength	nm
M_d^+	Atmospheric component of $\bar{\mu}_d$	-
M_d^*	IOP component of $\bar{\mu}_d$	-
$\bar{\mu}_\infty$	average cosine of the asymptotic light field	-
$\bar{\mu}_d$	average cosine of the downwelling light field	-
μ_w	cosine of the in-water solar zenith	-
n_p	particulate refractive index, relative to water	-
P	phase function (β/b)	sr^{-1}
P_x	particulate phase function (β_p/b_p), where subscript $x = p, ps,$ and pl specifies particulate, particulate small-dominant, and particulate large-dominant	sr^{-1}
$P_{bb,x}$	backward particulate phase function (β_p/b_{bp}), where subscript $x = ST$ and FF specifies functions from References [26,44]	sr^{-1}
Q	ratio of upwelling irradiance to nadir radiance	sr
r_{rs}	remote sensing reflectance, the ratio of upwelling subsurface radiance to downwelling irradiance	sr^{-1}
R_{rs}	remote sensing reflectance, the ratio of water-leaving radiance to downwelling irradiance	sr^{-1}
R_d	scaling factor for a_d	-
R_g	scaling factor for a_g	-
S_f	mixing factor for a_{ph}	-
θ_x	zenith angle, where $x = s$ and v specifies solar and viewing	°, rad
θ_s'	above water solar zenith angle	°, rad
v	exponent of empirical spectral b_p function	-
V	atmospheric horizontal visibility	km
ω	albedo (b/c)	-
$\Psi_{K_{Lu}}$	ratio of diffuse upwelling attenuation coefficient to asymptotic attenuation coefficient	-
z	depth	m

Table A2. Equation Coefficients. Full 15 decimal scaled fixed point precision in MATLAB is provided to ensure accuracy in calculations.

Equation	Symbol	Value
Equation (3)	f_1	$-5.98948784303628 \times 10^{-8}$
	f_2	$5.95904039870752 \times 10^{-6}$
	f_3	$-6.975283717755 \times 10^{-4}$
	f_4	$2.07111856771792 \times 10^{-3}$
	f_5	$2.69046922858858 \times 10^{-2}$
Equation (4)	f_{A1}	$-3.79435531537314 \times 10^{-7}$
	f_{A2}	$2.42117623125973 \times 10^{-4}$
	f_{A3}	$-5.76056692150838 \times 10^{-2}$
	f_{A4}	6.04944577004764
	f_{A5}	-236.166389774491
Equation (16)	e_1	$-3.37021020153209 \times 10^{-12}$
	e_2	$2.25040435584125 \times 10^{-10}$
	e_3	$-2.25897880448836 \times 10^{-9}$
	e_4	$4.98402568695743 \times 10^{-10}$
	e_5	$-3.67440351688922 \times 10^{-8}$
	e_6	$4.02677827509591 \times 10^{-7}$
	e_7	$-2.52448256032736 \times 10^{-8}$
	e_8	$2.09631870150827 \times 10^{-6}$
	e_9	$-2.43068373614361 \times 10^{-5}$
	e_{10}	$5.98295717192273 \times 10^{-7}$
	e_{11}	$-5.36922068813161 \times 10^{-5}$
	e_{12}	$6.84105803724285 \times 10^{-4}$
	e_{13}	$-5.34168078899319 \times 10^{-6}$
	e_{14}	$4.95201118318049 \times 10^{-4}$
	e_{15}	$-6.09578731164684 \times 10^{-3}$
	e_{16}	$5.32097604773773 \times 10^{-4}$
	e_{17}	$-2.91276619216202 \times 10^{-2}$
	e_{18}	0.589340234481004
Equation (17)	$m_{d,1}^*$	0.00611094400155735
	$m_{d,2}^*$	-0.00104841847722295
	$m_{d,3}^*$	0.0498255758922950
	$m_{d,4}^*$	-0.0117672820980625
	$m_{d,5}^*$	0.128019358635212
	$m_{d,6}^*$	-0.0429896134897322
	$m_{d,7}^*$	0.103528931695373
	$m_{d,8}^*$	0.950921179229178

Table A3. $f_{L,ave}(\lambda)$ function, see Equations (30) and (31).

Wavelength	Value	Wavelength	Value	Wavelength	Value
350	0.990	505	1.018	655	0.992
355	0.990	510	1.013	660	0.993
360	0.992	515	1.009	665	0.998
365	0.992	520	1.005	670	1.000
370	0.992	525	1.002	675	1.001
375	0.995	530	0.999	680	1.000
380	0.997	535	0.996	685	0.995
385	0.997	540	0.995	690	0.994
390	0.998	545	0.992	695	0.993
395	1.000	550	0.989	700	0.994
400	1.000	555	0.987	705	0.994
405	1.000	560	0.985	710	0.996
410	1.002	565	0.982	715	0.997

Table A3. *Cont.*

Wavelength	Value	Wavelength	Value	Wavelength	Value
415	1.003	570	0.981	720	0.999
420	1.006	575	0.982	725	1.000
425	1.008	580	0.983	730	1.000
430	1.010	585	0.984	735	1.000
435	1.013	590	0.986	740	0.999
440	1.016	595	0.987	745	0.999
445	1.020	600	0.988	750	0.999
450	1.023	605	0.988	755	0.999
455	1.024	610	0.989	760	0.999
460	1.025	615	0.989	765	0.999
465	1.025	620	0.989	770	0.999
470	1.026	625	0.990	775	1.000
475	1.026	630	0.990	780	1.000
480	1.026	635	0.990	785	1.001
485	1.026	640	0.990	790	1.002
490	1.026	645	0.990	795	1.002
495	1.024	650	0.990	800	1.002
500	1.022				

References

1. Gordon, H.R. Inverse methods in hydrologic optics. *Oceanologia* **2002**, *44*, 9–58.
2. Zaneveld, J.R.V.; Twardowski, M.S.; Lewis, M.; Barnard, A. Radiative transfer and remote sensing. In *Remote Sensing of Coastal Aquatic Waters*; Miller, R., Del-Castillo, C., McKee, B., Eds.; Springer: Dordrecht, The Netherlands, 2005; pp. 1–20.
3. Twardowski, M.; Lewis, M.; Barnard, A.; Zaneveld, J.R.V. In-water instrumentation and platforms for ocean color remote sensing applications. In *Remote Sensing of Coastal Aquatic Waters*; Miller, R., Del-Castillo, C., McKee, B., Eds.; Springer: Dordrecht, The Netherlands, 2005; pp. 69–100.
4. Mobley, C.D. *Light and Water: Radiative Transfer in Natural Waters*; Academic Press: San Diego, CA, USA, 1994; 592p.
5. Morel, A.; Gentili, B. Diffuse reflectance of oceanic waters. II. Bidirectional aspects. *Appl. Opt.* **1993**, *32*, 6864–6879. [CrossRef] [PubMed]
6. Morel, A.; Gentili, B. Diffuse reflectance of oceanic waters. III. Implication of bidirectionality for the remote-sensing problem. *Appl. Opt.* **1996**, *35*, 4850. [CrossRef] [PubMed]
7. Morel, A.; Antoine, D.; Gentili, B. Bidirectional reflectance of oceanic waters: Accounting for Raman emission and varying particle phase function. *Appl. Opt.* **2002**, *41*, 6289–6306. [CrossRef] [PubMed]
8. Fan, Y.; Li, W.; Gatebe, C.K.; Stamnes, K. Neural network method to correct bidirectional effects in water leaving radiance. *Appl. Opt.* **2016**, *55*, 10–21. [CrossRef] [PubMed]
9. Gordon, H.; Brown, O.B.; Evans, R.H.; Brown, J.W.; Smith, R.C.; Baker, K.S.; Clark, D.K. A semianalytic radiance model of ocean color. *J. Geophys. Res.* **1988**, *93*, 10909–10924. [CrossRef]
10. Park, Y.-J.; Ruddick, K. Model of remote sensing reflectance including bidirectional effects for case 1 and case 2 waters. *Appl. Opt.* **2005**, *44*, 1236–1249. [CrossRef]
11. Lee, Z.-P.; Du, K.-P.; Voss, K.J.; Zibordi, G.; Lubac, B.; Arnone, R.; Weidemann, A. An inherent optical property centered approach to correct the angular effects in water-leaving radiance. *Appl. Opt.* **2011**, *50*, 3155–3167. [CrossRef]
12. Hlaing, S.; Gilerson, A.; Harmel, T.; Tonizzo, A.; Weidemann, A.; Arnone, R.; Ahmed, S. Assessment of a bidirectional reflectance distribution correction of above-water and satellite water-leaving radiance in coastal waters. *Appl. Opt.* **2012**, *51*, 220–237. [CrossRef]
13. Werdell, J.; McKinna, L.I.W.; Boss, E.; Ackleson, S.G.; Craig, S.E.; Gregg, W.W.; Lee, Z.-P.; Maritorena, S.; Roesler, C.S.; Rousseaux, C.S.; et al. An overview of approaches and challenges for retrieving marine inherent optical properties from ocean color remote sensing. *Prog. Oceanogr.* **2018**, *160*, 186–212. [CrossRef]

14. Morel, A. In-water and remote measurements of ocean color. *Bound.-Lay. Meteorl.* **1980**, *18*, 177–201. [CrossRef]

15. Mobley, C.D.; Werdell, J.; Franz, B.; Ahmad, Z.; Bailey, S. *Atmospheric Correction for Satellite Ocean Color Radiometry*; NASA Report NASA/TM–2016-217551; NASA: Washington, DC, USA, 2016; 73p.

16. Werdell, P.J.; Franz, B.A.; Bailey, S.W.; Feldman, G.C.; Boss, E.; Brando, V.E.; Dowell, M.; Hirata, T.; Lavender, S.J.; Lee, Z.-P.; et al. Generalized ocean color inversion model for retrieving marine inherent optical properties. *Appl. Opt.* **2013**, *52*, 2019–2037. [CrossRef] [PubMed]

17. Voss, K.; Morel, A. Bidirectional reflectance function for oceanic waters with varying chlorophyll concentrations: Measurements versus predictions. *Limnol. Oceanogr.* **2005**, *50*, 698–705. [CrossRef]

18. Voss, K.J.; Morel, A.; Antoine, D. Detailed validation of the bidirectional effect in various Case 1 waters for application to ocean color imagery. *Biogeosciences* **2007**, *4*, 781–789. [CrossRef]

19. Gleason, A.; Voss, K.; Gordon, H.R.; Twardowski, M.; Sullivan, J.; Trees, C.; Weidemann, A.; Berthon, J.-F.; Clark, D.; Lee, Z.-P. Detailed validation of ocean color bidirectional effects in various Case I and Case II waters. *Opt. Express* **2012**, *20*, 7630–7645. [CrossRef]

20. He, S.; Zhang, X.; Xiong, Y.; Gray, D. A bidirectional subsurface remote sensing reflectance model explicitly accounting for particle backscattering shapes. *J. Geophys. Res. Oceans* **2017**, *122*, 8614–8626. [CrossRef]

21. Talone, M.; Zibordi, G.; Lee, Z.-P. Correction for the non-nadir viewing geometry of AERONET-OC above water radiometry data: An estimate of uncertainties. *Opt. Express* **2018**, *26*, A541–A561. [CrossRef] [PubMed]

22. Hirata, T.; Hardman-Mountford, N.; Aiken, J.; Fishwick, J. Relationship between the distribution function of ocean nadir radiance and inherent optical properties for oceanic waters. *Appl. Opt.* **2009**, *48*, 3129–3138. [CrossRef] [PubMed]

23. Gordon, H.R. Sensitivity of radiative transfer to small-angle scattering in the ocean: Quantitative assessment. *Appl. Opt.* **1993**, *32*, 7505–7511. [CrossRef] [PubMed]

24. Mishchenko, M.I.; Travis, L.D.; Lacis, A.A. *Scattering, Absorption, and Emission of Light by Small Particles*; Cambridge University Press: Cambridge, UK, 2002; pp. 74–82.

25. Hair, J.; Hostetler, C.; Hu, Y.; Behrenfeld, M.; Butler, C.F.; Harper, D.B.; Hare, R.; Berkoff, R.; Cook, A.; Collins, J.; et al. Combined Atmospheric and Ocean Profiling from an Airborne High Spectral Resolution Lidar. *Eur. Phys. J. Conf.* **2016**, *119*, 22001. [CrossRef]

26. Sullivan, J.M.; Twardowski, M.S. Angular shape of the volume scattering function in the backward direction. *Appl. Opt.* **2009**, *48*, 6811–6819. [CrossRef] [PubMed]

27. Zaneveld, J.R.V. Remotely sensed reflectance and its dependence on vertical structure: A theoretical derivation. *Appl. Opt.* **1982**, *21*, 4146–4150. [PubMed]

28. Zaneveld, J.R.V. A theoretical derivation of the dependence of the remotely sensed reflectance on the inherent optical properties. *J. Geophys. Res.* **1995**, *100*, 13135–13142. [CrossRef]

29. Tonizzo, A.; Twardowski, M.; McLean, S.; Voss, K.; Lewis, M.; Trees, C. Closure and uncertainty assessment for ocean color reflectance using measured volume scattering functions and reflective tube absorption coefficients with novel correction for scattering. *Appl. Opt.* **2017**, *56*, 130–146. [CrossRef]

30. Wolanin, A.; Rozanov, V.; Dinter, T.; Bracher, A. Detecting CDOM Fluorescence Using High Spectrally Resolved Satellite Data: A Model Study. In *Towards an Interdisciplinary Approach in Earth System Science: Advances of a Helmholtz Graduate Research School*; Springer International Publishing: Cham, Switzerland, 2015; pp. 109–121.

31. Zhai, P.W.; Hu, Y.; Winker, D.M.; Franz, B.A.; Boss, E. Contribution of Raman scattering to polarized radiation field in ocean waters. *Opt. Express* **2015**, *23*, 23582–23596. [CrossRef]

32. Rozanov, V.; Dinter, T.; Rozanov, A.; Wolanin, A.; Bracher, A.; Burrows, J. Radiative transfer modeling through terrestrial atmosphere and ocean accounting for inelastic processes: Software package SCIATRAN. *J. Quant. Spectrosc. Radiat. Transf.* **2017**, *194*, 65–85. [CrossRef]

33. Zhai, P.W.; Hu, Y.; Winker, D.M.; Franz, B.A.; Werdell, J.; Boss, E. Vector radiative transfer model for coupled atmosphere and ocean systems including inelastic sources in ocean waters. *Opt. Express* **2017**, *25*, A223–A239. [CrossRef]

34. Weidemann, A.; Stavn, R.; Zaneveld, J.R.V.; Wilcox, M.R. Error in predicting hydrosol backscattering from remotely sensed reflectance. *J. Geophys. Res.* **1995**, *100*, 163–177. [CrossRef]

35. Berthon, J.-F.; Shybanov, E.; Lee, M.; Zibordi, G. Measurements and modeling of the volume scattering function in the coastal northern Adriatic Sea. *Appl. Opt.* **2007**, *46*, 5189–5203. [CrossRef]

36. Twardowski, M.; Zhang, X.; Vagle, S.; Sullivan, J.; Freeman, S.; Czerski, H.; You, Y.; Bi, L.; Kattawar, G. The optical volume scattering function in a surf zone inverted to derive sediment and bubble particle subpopulations. *J. Geophys. Res.* **2012**, *117*, C00H17. [CrossRef]

37. Lee, M.; Korchemkina, E.N. Chapter 4: Volume scattering function of seawater. In *Springer Series in Light Scattering*; Kokhanovsky, A., Ed.; Springer International Publishing: New York, NY, USA, 2017; pp. 151–195.

38. Voss, K.; Austin, R. Beam attenuation measurement error due to small angle scattering acceptance. *J. Atmos. Ocean. Technol.* **1993**, *10*, 113–121. [CrossRef]

39. Boss, E.; Slade, W.H.; Behrenfeld, M.; Dall'Olmo, G. Acceptance angle effects on the beam attenuation in the ocean. *Opt. Express* **2009**, *17*, 1535–1550. [CrossRef] [PubMed]

40. Twardowski, M.; Tonizzo, A. Scattering and absorption effects on asymptotic light fields in seawater. *Opt. Express* **2017**, *25*, 18122–18130. [CrossRef] [PubMed]

41. Twardowski, M.S.; Boss, E.; Macdonald, J.B.; Pegau, W.S.; Barnard, A.H.; Zaneveld, J.R.V. A model for estimating bulk refractive index from the optical backscattering ratio and the implications for understanding particle composition in Case I and Case II waters. *J. Geophys. Res.* **2001**, *106*, 14129–14142. [CrossRef]

42. Prieur, L.; Morel, A. Etude theorique du regime asymptotique: Relations entre characteristiques optiques et coefficient d'extinction relatif a la penetration de la lumibr, e du jour. *Cah. Oceanogr.* **1971**, *23*, 35–48.

43. Berwald, J.; Stramski, D.; Mobley, C.D.; Kiefer, D.A. Influences of absorption and scattering on vertical changes in the average cosine of the underwater light field. *Limnol. Oceanogr.* **1995**, *40*, 1347–1357. [CrossRef]

44. Fournier, G.; Forand, F. Analytic phase function for ocean water. *Ocean Opt. XII* **1994**, *2558*, 194–201.

45. Jonasz, M.; Fournier, G. *Light Scattering by Particles in Water*; Academic Press: Amsterdam, The Netherlands, 2007; 704p.

46. IOCCG. Remote Sensing of Inherent Optical Properties: Fundamentals, Tests of Algorithms, and Applications. In *Reports of the International Ocean-Colour Coordinating Group, No. 5*; Lee, Z.-P., Ed.; IOCCG: Dartmouth, NS, Canada, 2006; 123p.

47. Twardowski, M.S.; Sullivan, J.M.; Donaghay, P.L.; Zaneveld, J.R.V. Microscale quantification of the absorption by dissolved and particulate material in coastal waters with an ac-9. *J. Atmos. Ocean. Technol.* **1999**, *16*, 691–707. [CrossRef]

48. Sullivan, J.; Twardowski, M.; Zaneveld, J.R.V.; Moore, C. Measuring optical backscattering in water. In *Light Scattering Reviews 7: Radiative Transfer and Optical Properties of Atmosphere and Underlying Surface*; Kokhanovsky, A., Ed.; Springer Praxis Books: New York, NY, USA; pp. 189–224. [CrossRef]

49. Stockley, N.D.; Rottgers, R.; McKee, D.; Lefering, I.; Sullivan, J.M.; Twardowski, M.S. Assessing uncertainties in scattering correction algorithms for reflective tube absorption measurements made with a WET Labs ac-9. *Opt. Express* **2017**, *25*, A1139–A1153. [CrossRef]

50. Gordon, H.R.; Brown, O.B.; Jacobs, M.M. Computed relationships between the inherent and apparent optical properties of a flat homogeneous ocean. *Appl. Opt.* **1975**, *14*, 417–427. [CrossRef] [PubMed]

51. Twardowski, M.S.; Claustre, H.; Freeman, S.A.; Stramski, D.; Huot, Y. Optical backscattering properties of the "clearest" natural waters. *Biogeosciences* **2007**, *4*, 1041–1058. [CrossRef]

52. Zhang, X.D.; Hu, L.B.; He, M.-X. Scattering by pure seawater: Effect of salinity. *Opt. Express* **2009**, *17*, 5698–5710. [CrossRef] [PubMed]

53. Twardowski, M.; Jamet, C.; Loisel, H. Analytical Model to Derive Suspended Particulate Matter Concentration in Natural Waters by Inversion of Optical Attenuation and Backscattering. *Proc. SPIE Ocean Sens. Monit. X* **2018**. [CrossRef]

54. Boss, E.; Pegau, W.S.; Lee, M.; Twardowski, M.S.; Shybanov, E.; Korotaev, G.; Baratange, F. The particulate backscattering ratio at LEO 15 and its use to study particles composition and distribution. *J. Geophys. Res.* **2004**, *109*, C01014. [CrossRef]

55. Sullivan, J.M.; Twardowski, M.S.; Donaghay, P.L.; Freeman, S.A. Using scattering characteristics to discriminate particle types in US coastal waters. *Appl. Opt.* **2005**, *44*, 1667–1680. [CrossRef] [PubMed]

56. Stramski, D.; Reynolds, R.A.; Babin, M.; Kaczmarek, S.; Lewis, M.R.; Röttgers, R.; Sciandra, A.; Stramska, M.; Twardowski, M.S.; Franz, B.A.; et al. Relationships between the surface concentration of particulate organic carbon and optical properties in the eastern South Pacific and eastern Atlantic Oceans. *Biogeosciences* **2008**, *5*, 171–201. [CrossRef]

57. Morel, A.; Prieur, L. Analyse spectrale des coefficients d'attenuation diffuse, de retrodiffusion pour diverses regions marines. *Cent. Rech. Oceanogr.* **1975**, *17*, 1–157.

58. Gregg, W.; Carder, K. A simple spectral solar irradiance model for cloudless maritime atmospheres. *Limnol. Oceanogr.* **1990**, *35*, 1657–1675. [CrossRef]
59. Loisel, H.; Stramski, D. Estimation of the inherent optical properties of natural waters from the irradiance attenuation coefficient and reflectance in the presence of Raman scattering. *Appl. Opt.* **2000**, *39*, 3001–3011. [CrossRef]
60. Westberry, T.K.; Boss, W.; Lee, Z.-P. Influence of Raman scattering on ocean color inversion models. *Appl. Opt.* **2013**, *52*, 5552–5561. [CrossRef]
61. Pope, R.M.; Fry, E. Absorption spectrum (380–700 nm) of pure water. II. Integrating cavity measurements. *Appl. Opt.* **1997**, *36*, 8710–8723. [CrossRef]
62. Ciotti, A.M.; Lewis, M.R.; Cullen, J.J. Assessment of the relationships between dominant cell size in natural phytoplankton communities and the spectral shape of the absorption coefficient. *Limnol. Oceanogr.* **2002**, *47*, 404–417. [CrossRef]
63. Roesler, C.S.; Perry, M.J. In situ phytoplankton absorption, fluorescence emission, and particulate backscattering spectra determined from reflectance. *J. Geophys. Res.* **1995**, *100*, 13279–13294. [CrossRef]
64. Twardowski, M.S.; Boss, E.; Sullivan, J.M.; Donaghay, P.L. Modeling the spectral shape of absorbing chromophoric dissolved organic matter. *Mar. Chem.* **2004**, *89*, 69–88. [CrossRef]
65. Loisel, H.; Morel, A. Light scattering and chlorophyll concentration in case 1 waters: A reexamination. *Limnol. Oceanogr.* **1998**, *43*, 847–858. [CrossRef]
66. Morel, A. Are the empirical relationships describing the bio-optical properties of case 1 waters consistent and internally compatible? *J. Geophys. Res.* **2009**, *114*, C01016. [CrossRef]
67. Mobley, C.D.; Sundman, L.K.; Boss, E. Phase function effects on oceanic light fields. *Appl. Opt.* **2002**, *41*, 1035–1050. [CrossRef] [PubMed]
68. Voss, K.J.; McLean, S.; Lewis, M.; Johnson, C.; Flora, S.; Feinholz, M.; Yarbrough, M.; Trees, C.; Twardowski, M.; Clark, D. An example crossover experiment for testing new vicarious calibration techniques for satellite ocean color radiometry. *J. Atmos. Ocean. Technol.* **2010**, *27*, 1747–1759. [CrossRef]
69. Zaneveld, J.R.V.; Barnard, A.H.; Boss, E. A theoretical derivation of the depth average of remotely sensed optical parameters. *Opt. Express* **2005**, *13*, 9052–9061. [CrossRef]
70. Gordon, H.R.; Clark, D.K. Remote sensing optical properties of a stratified ocean: An improved interpretation. *Appl. Opt.* **1980**, *19*, 3428–3430. [CrossRef]
71. Gordon, H.R. Dependence of the diffuse reflectance of natural waters on the sun angle. *Limnol. Oceanogr.* **1989**, *34*, 1484. [CrossRef]
72. Kirk, J.T.O. Estimation of the absorption and scattering coefficients of natural waters by the use of underwater irradiance measurements. *Appl. Opt.* **1994**, *33*, 3276–3278. [CrossRef] [PubMed]
73. Lee, Z.-P.; Du, K.-P.; Arnone, R. A model for the diffuse attenuation coefficient of downwelling irradiance. *J. Geophys. Res.* **2005**, *110*, C02016. [CrossRef]
74. Chai, T.; Draxler, R.R. Root mean square error (RMSE) or mean absolute error (MAE)?—Arguments against avoiding RMSE in the literature. *Geosci. Model. Dev.* **2014**, *7*, 1247–1250. [CrossRef]
75. Brewin, R.J.W. The Ocean Colour Climate Change Initiative: III. A round-robin comparison on in-water bio-optical algorithms. *Remote Sens. Environ.* **2013**. [CrossRef]
76. Hoge, F.; Lyon, P.E.; Mobley, C.D.; Sundman, L.K. Radiative transfer equation inversion: Theory and shape factor models for retrieval of oceanic inherent optical properties. *J. Geophys. Res.* **2003**, *108*, 3386. [CrossRef]
77. Agrawal, Y.; Mikkleson, O. Empirical forward scattering phase functions from 0.08 to 16 deg. for randomly shaped terrigenous 1-21 μm sediment grains. *Opt. Express* **2009**, *17*, 8805–14. [CrossRef] [PubMed]
78. Nardelli, S.; Twardowski, M.S. Improving assessments of chlorophyll concentration from in situ optical measurements. *Opt. Express* **2016**, *24*, A1374–A1389. [CrossRef] [PubMed]
79. Petzold, T.J. *Volume Scattering Functions for Selected Ocean Waters, Report 72–78*; Scripps Institution of Oceanography: La Jolla, CA, USA, 1972.
80. Li, L.; Stramski, D.; Reynolds, R.A. Effects of inelastic radiative processes on the determination of water-leaving spectral radiance from extrapolation of underwater near-surface measurements. *Appl. Opt.* **2016**, *55*, 7050–7067. [CrossRef]
81. Voss, K.; Gordon, H.; Flora, S.; Johnson, B.C.; Yarbrough, M.; Feinholz, M.; Houlihan, T. A method to extrapolate the diffuse upwelling radiance attenuation coefficient to the surface as applied to the Marine Optical Buoy (MOBY). *J. Atmos. Ocean. Technol.* **2017**, *34*, 1423–1432. [CrossRef]

82. Gordon, H.; Ding, K. Self-shading of in-water optical instruments. *Limnol. Oceanogr.* **1992**, *37*, 491–500. [CrossRef]

83. Leathers, R.A.; Downes, T.V.; Mobley, C.D. Self-shading correction for oceanographic upwelling radiometers. *Opt. Express* **2004**, *12*, 4709–4718. [CrossRef] [PubMed]

84. Zhang, X.; Fournier, G.R.; Gray, D.J. Interpretation of scattering by oceanic particles around 120 degrees and its implication in ocean color studies. *Opt. Express* **2017**, *25*, A191–A199. [CrossRef] [PubMed]

85. Lotsberg, J.K.; Marken, E.; Stamnes, J.; Erga, S.R.; Aursland, K.; Olseng, C. Laboratory measurements of light scattering from marine particles. *Limnol. Oceanogr. Methods* **2007**, *5*, 34–40. [CrossRef]

86. Tan, H.; Doerffer, R.; Oishi, T.; Tanaka, A. A new approach to measure the volume scattering function. *Opt. Express* **2013**, *21*, 18697–18711. [CrossRef] [PubMed]

87. Harmel, T.; Hieronymi, M.; Slade, W.; Röttgers, R.; Roullier, F.; Chami, M. Laboratory experiments for intercomparison of three volume scattering meters to measure angular scattering properties of hydrosols. *Opt. Express* **2016**, *24*, A234–A256. [CrossRef] [PubMed]

88. Tan, H.; Oishi, T.; Tanaka, A.; Doerffer, R.; Tan, Y. Chlorophyll-a specific volume scattering function of phytoplankton. *Opt. Express* **2017**, *25*, A564–A573. [CrossRef] [PubMed]

89. Volten, H.; de Haan, J.F.; Hovenier, J.W.; Schreurs, R.; Vassen, W.; Dekker, A.G.; Hoogenboom, H.J.; Charlton, F.; Wouts, R. Laboratory measurements of angular distributions of light scattered by phytoplankton and silt. *Limnol. Oceanogr.* **1998**, *43*, 1180–1197. [CrossRef]

90. Loisel, H.; Morel, A. Non-isotropy of the upward radiance field in typical coastal (Case 2) waters. *Int. J. Remote Sens.* **2001**, *22*, 275–295. [CrossRef]

91. Xiong, Y.; Zhang, X.; He, S.; Gray, D.J. Re-examining the effect of particle phase functions on the remote-sensing reflectance. *Appl. Opt.* **2017**, *56*, 6881–6888. [CrossRef]

92. Hostetler, C.; Behrenfeld, M.J.; Hu, Y.; Hair, J.W.; Schulien, J.A. Spaceborne lidar in the study of marine systems. *Annu. Rev. Mar. Sci.* **2018**, *10*, 121–147. [CrossRef]

93. Dayou, J.; Chang, J.H.W.; Sentian, J. *Ground-Based Aerosol Optical Depth Measurement Using Sunphotometers*; Springer: Singapore, 2014; pp. 9–30.

94. The NASA PACE Mission. Available online: https://pace.oceansciences.org/mission.htm (accessed on 16 June 2018).

95. Boss, E.; Remer, L.A. A novel approach to a satellite mission's science team. *Eos* **2018**, *99*. [CrossRef]

96. Bracher, A.; Bouman, H.A.; Brewin, R.J.W.; Bricaud, A.; Brotas, V.; Ciotti, A.M.; Clementson, L.; Devred, E.; Di Cicco, A.; Dutkiewicz, S.; et al. Obtaining phytoplankton diversity from ocean color: A scientific roadmap for future development. *Front. Mar. Sci.* **2017**, *4*. [CrossRef]

97. Vandermeulen, R.A.; Mannino, A.; Neeley, A.; Werdell, J.; Arnone, R. Determining the optical spectral sampling frequency and uncertainty thresholds for hyperspectral remote sensing of ocean color. *Opt. Express* **2017**, *25*, A785–A797. [CrossRef] [PubMed]

![applied sciences logo] *applied sciences*

MDPI

Article

Assessing Fluorescent Organic Matter in Natural Waters: Towards In Situ Excitation–Emission Matrix Spectroscopy

Oliver Zielinski [1,*], Nick Rüssmeier [1,2], Oliver D. Ferdinand [1], Mario L. Miranda [1,3] and Jochen Wollschläger [1,4]

[1]   Center for Marine Sensors, Institute for Chemistry and Biology of the Marine Environment, University Oldenburg, 26382 Wilhelmshaven, Germany; nick.ruessmeier@uol.de (N.R.); oliver.ferdinand@uol.de (O.D.F.); mario.luis.miranda.montenegro@uol.de (M.L.M.); jochen.wollschlaeger@uol.de (J.W.)
[2]   OFFIS e.V. Institute for Information Technology, 26121 Oldenburg, Germany
[3]   Laboratory of Air and Water Quality, University of Panama, Panamá 4, Panamá
[4]   Helmholtz-Zentrum Geesthacht, Institute of Coastal Research, Max-Planck-Str. 1, 21502 Geesthacht
*   Correspondence: oliver.zielinski@uol.de; Tel.: +49-441-798-3518

Received: 12 July 2018; Accepted: 26 November 2018; Published: 19 December 2018

Featured Application: This proof-of-concept work of in situ EEMS might find applications on different autonomous platforms in biogeochemical observing systems for the marine environment.

Abstract: Natural organic matter (NOM) is a key parameter in aquatic biogeochemical processes. Part of the NOM pool exhibits optical properties, namely absorption and fluorescence. The latter is frequently utilized in laboratory measurements of its dissolved fraction (fluorescent dissolved organic matter, FDOM) through excitation–emission matrix spectroscopy (EEMS). We present the design and field application of a novel EEMS sensor system applicable in situ, the 'Kallemeter'. Observations are based on a field campaign, starting in Norwegian coastal waters entering the Trondheimsfjord. Comparison against the bulk fluorescence of two commercial FDOM sensors exhibited a good correspondence of the different methods and the ability to resolve gradients and dynamics along the transect. Complementary laboratory EEM spectra measurements of surface water samples and their subsequent PARAFAC analysis revealed three dominant components while the 'Kallemeter' EEMS sensor system was able to produce reasonable EEM spectra in high DOM concentrated water bodies, yet high noise levels must be addressed in order to provide comparable PARAFAC components. Achievements and limitations of this proof-of-concept are discussed providing guidance towards full in situ EEMS measurements to resolve rapid changes and processes in natural waters based on the assessment of spectral properties. Their combination with multiwavelength FDOM sensors onboard autonomous platforms will enhance our capacities in observing biogeochemical processes in the marine environment in spatiotemporal and spectral dimensions.

Keywords: natural organic matter; DOM; FDOM; CDOM; Gelbstoff; EEMS; PARAFAC; marine sensors; Kallemeter; FerryBox; Trondheimsfjord; Norway

1. Introduction

The investigation of natural organic matter (NOM) and especially its dissolved fraction (DOM) is of high relevance in aquatic sciences. Studies of NOM and DOM include their sinks and sources as part of the global carbon cycle [1–4], its stability and degradability by abiotic and biotic processes [5,6], as well as natural and anthropogenic ecosystem effects e.g., from hydrocarbons [7]. Complementary to laboratory approaches, the need to measure DOM directly in the water column (in situ) initiated the development of field-applicable optical sensors that can be attached to autonomous vehicles or ship-based platforms [8]. Addressing the optical properties, namely absorption and fluorescence, of certain DOM fractions in situ sensors enable a high-resolution assessment of these ecosystem-relevant parameters or proxies on different observation platforms [9,10]. The absorption of light from fractions of the DOM pool is a long known feature, which was initially investigated by the German chemist Kurt Kalle in the early 20th century [11]. Observing an increased absorption of light towards the blue and ultraviolet (UV) part of the spectrum, he denoted this fraction of DOM as 'Gelbstoff' or 'yellow substance' [12,13], apparently inspired by the yellowish color of the samples taken from the Baltic Sea. Nowadays, this fraction is known as colored or chromophoric dissolved organic matter, short CDOM [14,15]. The measurement principle for CDOM absorption (an inherent optical property, or IOP, of natural water bodies) is based on the reduction of light intensity from a light source to a detector over a certain distance. It typically requires a filtered water sample, since only the dissolved and not the particulate fraction is of interest, and the latter could introduce errors due to light scattering. While in situ filtration is technically possible and practiced e.g., in underway systems [16], it is rather sophisticated and therefore not realized in today's commercially submersible CDOM sensors [17]. For a feasible, but still technically complex, approach to derive CDOM absorption in subsea or underway applications, the scattering loss of photons is suppressed by reflective tubes [18] or a reflecting cavity [19–21]. Yet in the majority of today's observing platforms and ship operated sensor systems, CDOM absorption measurements are not implemented.

The most common sensors within CTD-samplers, FerryBox systems, autonomous underwater vehicles (AUVs), gliders, biogeochemical Argo floats and alike are CDOM fluorometers. Applying a UV light source and detecting an emitted fluorescence signal at a higher wavelength (lower energy), these sensors utilize a principle that Kalle once described as 'skyblue fluorescence' [12,22]. In addition to these single-channel sensors, a few sensor systems commercially exist that enable a selection of several wavelengths pairs [17]. Recently, a matrix fluorescence sensor was presented that can sense a set of 4 detection wavelengths for three different excitation wavelengths, thus spanning a matrix of 12 combinations [23]. While CDOM fluorescence is technically speaking a correct term, it can be mixed up with CDOM (by definition an absorption property) and reports or commercial product descriptions sometimes tend to incorrectly shorten measurement devices as 'CDOM sensors'. To reflect the fact that the fluorescing part of the DOM is a subfraction of the absorbing fraction of DOM, we herein use FDOM as a designator for fluorescent DOM, as common in recent literature [9,24–26]. Laboratory measurement of FDOM is typically performed with UV–VIS spectrofluorometers [27,28]. They enable a free selection of excitation and detection wavelengths and thus different forms of results: (a) emission spectra (with a fixed excitation wavelength); (b) excitation spectra (with a fixed emission wavelength); (c) synchronous scans (excitation and emission changing at the same rate); and (d) excitation–emission matrices (EEM). The latter provides a full scan of a range of excitation wavelengths, performing for each of these excitation wavelengths a full emission scan. The resulting matrix of fluorescence intensities, sometimes referred to as 'optical fingerprint', is reflecting all fluorescing components and its peaks are associated with different components from the DOM pool [14,29]. A classical labeling of the fluorophores in DOM identifies five main peaks A, C, B, T, and M. Those peaks are extensively used to identify the main composition and origin of FDOM. Previous reports indicate the correlation between microbial presence and freshly produced DOM (peaks M, B and T), while allochthonous DOM has been associated to the peaks A and C [29]. EEM spectra are frequently used in recent literature [27,28] and several parameters are calculated from them, including the humification index [30], biological index [31], and recently produced material index [32]. A statistical analysis of EEM spectra is often performed by parallel factor analysis (PARAFAC), a method that identifies the underlying components from a set of spectra [33,34]. EEM spectroscopy (EEMS) and its spectral analysis can be considered the state-of-the-art in FDOM assessment, however, to the best of our knowledge, no submersible sensor exists that enables the high-resolution (in terms of wavelengths intervals) measurement of excitation–emission matrix spectra with in situ equipment.

Herein, we will present (i) design and application of a novel in situ EEMS sensor system (in memory of Kurt Kalle named the 'Kallemeter') that enables full optical fingerprints in a submersible design, and compare these results with (ii) flow-through data from two commercial FDOM sensors, and (iii) spectral components identified in surface water samples from a laboratory UV–VIS spectrofluorometer, based on a field campaign in the Trondheimsfjord, Norway.

2. Materials and Methods

The following section starts with a detailed description of the design of the novel 'Kallemeter' in situ EEMS sensor system, including initial laboratory measurements to illustrate its performance. To enable a comparison of bulk fluorescence detected, two commercial FDOM sensors will briefly be introduced, both operated in a commercially available flow-through sensor system (FerryBox, -4H-JENA engineering, Jena, Germany). This will be followed by a description of the laboratory methods applied on discrete water samples, and the subsequently used statistical method to derive the components of the EEM spectra observed. Finally, we will provide information about the study site itself.

2.1. Design of an In Situ Capable EEMS Sensor System

The newly-developed submersible EEMS sensor system provides comprehensive excitation–emission matrices for in situ fluorescence measurements of organic matter in natural waters. This subchapter presents its design, beginning with a representation of the operation mode in Section 2.1.1. Followed by this, a description of the technical implementation and the system design is given in Section 2.1.2. Section 2.1.3 presents the evaluation of the system performance.

2.1.1. Submersible EEMS Sensor System and Modes of Operation

Investigations on fluorescent organic matter can benefit from spatiotemporal information that enable a comprehensive interpretation of the dataset in context of environmental conditions. The 'Kallemeter' EEMS sensor system supports in situ explorations with three common modes of operation for field studies: vertical profiling, underway sensing (e.g., installed in a moonpool of a vessel), and moored operation (e.g., integrated to a subsea platform). Figure 1 shows the deployment of the EEMS sensor system in profiling mode (Figure 1a) and the deployment in a moonpool of a research vessel for in situ underway investigations (Figure 1b,c). The submersible design offers automated data acquisition from water samples to a maximum water depth of 200 m in a freely selectable time interval, without the support of an operator. Once installed, autonomous long-term investigations from days to weeks are possible, depending on biofouling [35] of the optical sensor unit and sediment load of the water filtration unit.

2.1.2. EEMS Sensor System Design and Technical Implementation

The entire EEMS sensor system is designed for a stand-alone operation as an underwater unit that only needs to be supplied with electrical power from an external power source. In this way, a moored operation with an external 12 V DC battery pack can also be realized. In case of shipborne installations like profiling or underway measurements, the system can be connected to an on-board unit via a 200 m seaworthy cable, as illustrated in Figure 2a. The on-board unit provides 130 V DC power supply and data transmission to the 'Kallemeter' via Ethernet. This gives the opportunity to connect an external PC for remote control of the underwater unit, automatic data backup, or real-time visualization of the in situ measurements. In addition to the in situ fluorescence spectrometer unit, the 'Kallemeter' was equipped with a turbidity meter (STM/bh, Seapoint Sensors Inc., Exeter, NH, USA) and a CTD (Conductivity, Temperature, Depth) (miniCTD probe, ASD Sensortechnik GmbH, Trappenkamp, Germany). The 'Kallemeter' housing is made of a solid PEEK (polyether ether ketone) tube with steel plates at its ends, which also support the dissipation of thermal energy to the aquatic environment through internally forced convection, as this is important to provide long-term stability of the measurement system (light source and optical measurement devices). The EEMS sensor system is controlled through an integrated Microsoft Windows 7 PC, digital I/O converters, and LabVIEW software (version 2013, National Instruments, Austin, TX, USA), using a custom-made control program (named 'EEMsea'). The latter allows for comprehensive adjustments of the measurement procedures, like EEM wavelengths or control of the sampling interval. All EEM datasets are stored in a comma-separated value (CSV) file format and include system- and sampling-relevant metadata in the header (e.g., sampling-identifier, internal temperature, and coefficients), which are assigned to the standardized file structure from PerkinElmer Luminescence Spectrometers ([Info], [Setup], [SampleInfo], [SampleData]; PerkinElmer, Waltham, MA, USA). Therefore, postprocessing and statistical analysis of EEM spectra can easily be applied by standard third party software or Matlab Toolboxes (The MathWorks, Natick, MA, USA) e.g., for PARAFAC [27].

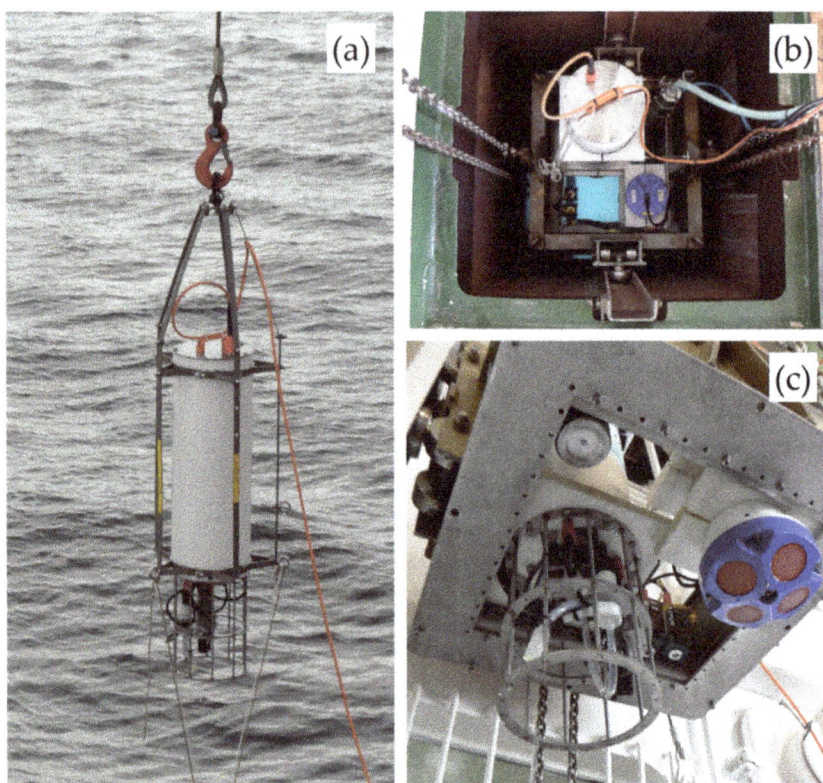

Figure 1. (**a**) Excitation–emission matrix spectroscopy (EEMS) sensor system in profiling mode on a winch of a research vessel. (**b**) EEMS sensor system, installation for in situ underway observations in the moonpool of the research vessel, top-view. Close to the EEMS sensor system, an acoustic Doppler current profiler (ADCP) as well as the water intake pump for the FerryBox system were installed in the moonpool-frame. (**c**) View from below.

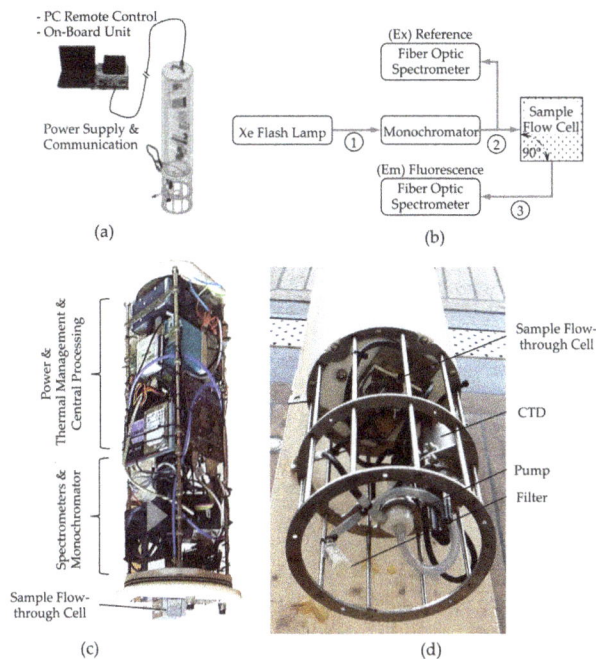

Figure 2. System components of the submersible EEMS sensor system. (**a**) System connected to an on-board unit and remote PC-control (optional). (**b**) Schematic structure of excitation–emission matrices (EEM) fluorescence spectrometer hardware and optical components (connected by optical fibers 1, 2, 3). (**c**) Internal close-up with electronics and optical components. (**d**) Close-up of the outer components of the EEMS sensor system.

The entire EEM fluorescence spectrometer design, which is presented in Figure 2b, consists of the following fluorescence measurement components: In the fluorescence excitation strand, a Xenon flash lamp (PX-2, Ocean Optics, Largo, FL, USA) is used as a broadband light source, which is coupled to a monochromator (CM 110, Spectral Products, Putnam, CT, USA) via an optical fiber (1). The output of the monochromator is coupled via an optical Y-fiber (2) to the reference spectrometer (USB2000+, Ocean Optics, Largo, FL, USA). Through this, a reference measurement of the excitation spectrum is realized. In parallel, the optical Y-fiber (2) connects the monochromator output to a flow-through sample cell in which the sample medium is located. The fluorescence emission strand consists of the detection spectrometer (QE-Pro, Ocean Optics, Largo, FL, USA) coupled via an optical fiber (3) to the sample flow-through cell. In the following, details of the fluorescence measuring system and its components are presented.

Fluorescence Excitation Light Source

Some in situ sensor designs use Xenon flash (or pulsed) lamps due to their high energy output in the ultraviolet (UV) and visible regions of the light spectrum as well as their long lifespan [36]. For these reasons, the 'Kallemeter' also utilizes a Xenon flash lamp (PX-2, Ocean Optics, Largo, FL, USA), which emits pulsed light with a continuous spectrum in a fluorescence relevant excitation wavelength range from 220 nm to 750 nm. A pulse power of maximum 45 microjoules/pulse (with 5 μs pulse duration at 1/3 height of pulse) and a lifetime of 10^9 pulses are expected for this 'expendable part' (estimated 230 days continuous operation @ 50 Hz pulse rate). A stable output from pulse to pulse is guaranteed for up to 100 Hz repetition rate, so the flash delay results in 10 milliseconds. Since the total intensity typically decreases slowly over the operating period [37], a measurement of the excitation spectrum is implemented to provide metadata of the EEMS sensor system status (presented in the subsection "Quality status of the excitation wavelengths").

Control of the Fluorescence Excitation Band

A monochromator (Type CM 110, Spectral Products, Putnam, CT, USA) is used to separate a narrow excitation band from the spectrum of the Xenon flash lamp. A full width at half maximum (FWHM) effective bandwidth of 10 nm is realized (therefore no slits are used) and the excitation wavelengths are selectable from 200 nm to 750 nm in 1 nm steps (with ruled grating AG1200-00300-303). The monochromator provides a coupling and decoupling angle of 14.8° and results to 0.13 numerical aperture (NA).

Quality Status of the Excitation Wavelengths

A wavelength-calibrated reference spectrometer (USB2000+, Ocean Optics, Largo, FL, USA) controls the quality status of the excitation. The internally used grid provides a spectral sensitivity range from 200 nm to 800 nm with a fixed slit width of 100 μm, and a possible integration time from 1 millisecond to 65 s. Reference data are recorded in parallel to the fluorescence measurements but are not included in the calculation of the EEM spectra. Instead, they only serve as metadata for monitoring the quality of the excitation light source state over the operation time.

Water Sampling Flow-Through Cell

Various conceptual instrument designs and principles for fluorescence detection hardware are available for different applications: right angle detection, intersecting cones, flow-through, and fiber optic designs [36]. The 'Kallemeter' sample flow-through cell design combines right angle detection (the commonly used 90° emission–excitation setup in luminescence spectrometry) with fiber optics and intersecting cones in a closed flow-through design, which is illustrated in Figure 3a. In addition, the measuring cell makes use of well-directed multireflective intersections of the optical fiber aperture to achieve multiple excitations and detection cone overlaps. Thus, the optical path within the water sample is enhanced and an effective increase of factor 2.5 of the fluorescence signal for the highest signal-to-noise ratio can be achieved (Appendix A, Table A1). Two mirror elements are used for this purpose, which are arranged opposite to the optical fiber terminations in a way that no direct fluorescence excitation light can enter in the optical fiber of the detector strand. A concave mirror (CM127-025-F01 - Ø1/2" UV Enhanced Al-Coated, f = 25.0, Thorlabs, Newton, NJ, USA) is used in (opposite) combination to the excitation cone. With a focal length of 25 mm, the focus point of the selected concave mirrors is almost at the end of the optical fibers distance of 22 mm. A planar mirror (PF05-03-F01 - Ø1/2" UV Enhanced Al-Coated, Thorlabs, Newton, NJ, USA) is used in (opposite) combination to the emission cone of the optical fibers.

As the sample cell forms the contact point between the optical parts and the water sample, it is installed in a separate housing at the bottom of the EEMS sensor system (see Figure 2c,d). Thus, a decoupling of the fluid-carrying and pressurized parts from the sensitive internal structures of the measurement unit is achieved and a simple sealing and pressure tightness can be realized. The optical interfaces of the sample cell, which are in contact with a seawater sample, are covered with superior transparent quartz glass windows (#45-310. TECHSPEC, Ø 15 mm Uncoated, 1λ Fused Silica Window, Edmund Optics, Barrington, NJ, USA). In order to ensure biogeochemical inertness of the sample cell, a specific adhesive (LS2-6140, NuSil Technology, Carpinteria, CA, USA) was selected through in situ tests as a suitable material for connecting the sample cell components and quartz windows. Even after several days in aqueous solution, this adhesive did not emit any fluorescent substances.

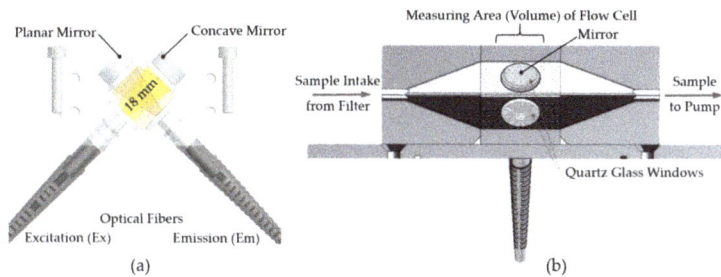

Figure 3. Details of the EEMS sensor system sample flow-through cell. (**a**) Measuring area-section with right angle arrangement of the excitation and emission optical fibers including the concave and planar mirror, respectively. (**b**) Cut-section through the flow-through cell with media intake and quartz glass windows.

The flow-through geometry type results in a greater efficiency in signal output compared to an open-faced design. Thus it is a good choice for waters with a wide range of NOM concentrations and low numbers of particles, as the sensor geometry (sample volume) is well defined [36]. A closed flow-through cell design, as illustrated by the cut-section in Figure 3b, provides shading of the optical inner parts from ambient light. This is a further benefit for an in situ application as it reduces the effect of biofouling [38]. During the sampling process, the closed sample cell is flushed with the water sample. The inlet side of a mini submersible pump for profiling applications (SBE 5M-1, Sea-Bird Scientific, Bellevue, WA, USA) is connected to the water outlet of the sample cell, while a reusable 50-μm nylon screen filtration element (type series F20 filter housing, Wolftechnik Filtersysteme, Germany) is connected to the sample cell water inlet via opaque hoses. This prevents an uncontrolled exchange between the ambient water and the water sample volume during an EEM spectrum measurement. This is particularly important for in situ underway operations when the 'Kallemeter' moves through different water bodies during an ongoing EEM measurement. Note that the setup presented herein was not equipped with a 0.2 μm or 0.45 μm filter, as recommended to separate the dissolved from the particulate fraction of organic matter. This is also the case for common bulk fluorescence sensors, as used in this study for comparison. However, as a pumped flow-through system, the 'Kallemeter' is generally capable of such a filtration unit as used e.g., for in situ spectral absorption measurements of CDOM [17].

Main Fluorescence Detector

The main fluorescence detector unit of the 'Kallemeter' EEMS sensor system is a wavelength-calibrated CCD digital spectrometer with a high dynamic range of ~85.000:1 (QE-Pro, Ocean Optics, Largo, FL, USA). The fluorescence emitted by the water sample is transmitted via an optical fiber, which is installed at the sample flow-through cell. The spectrometer built-in grating enables the detection in a full range of emission wavelengths from 200 nm to 950 nm, which corresponds to the detection range of commonly used laboratory luminescence spectrometers (e.g., the LS55, PerkinElmer, Waltham, MA, USA). The slit of the detector spectrometer has a width of 100 μm. Long-term stability and noise reduction is achieved by internal Peltier cooling of the detector to −10°C, as all fluorescence measurements signals are subject to alterations as a function of temperature. Temperature changes can arise either from self-heating of the instrument or from changes in ambient temperature of the deployment environment. Integration times of 8 milliseconds up to 60 min are selectable to enable stable measurements for environments where sample signal intensity is small. The spectrometer includes further a second order wavelength filter, preventing detection of Rayleigh and Raman scattering of higher orders.

The chosen detector spectrometer offers automatic dark measurements without external shutter units, which is used to obtain related metadata to correct for instrument response variables. As a routine by the 'EEMsea' software, a dark count measurement is carried out once initially per EEM spectrum measuring cycle and is automatically subtracted from the fluorescence data. Investigations of wavelength assigned intensity dark count measurements are presented in Appendix A, Figure A1 through a long-term measurement test (116 h) exhibiting stable operating conditions with no drift or significant invariance over time.

Optical Fibers

Optical fibers in the system setup are solarization-resistant, as ultraviolet radiation below 260 nm degrades transmission in silica fibers, also known as UV degradation. The custom-made designs, as illustrated in Figure 4a, provide maximum fiber diameters (80 single fibers 200 μm core diameter, Ø 2.6 mm bunch, LOPTEK, Berlin, Germany) and achieve a high transmission (96%) in all relevant excitation wavelengths (labeled (1) in Figure 2b). The numerical aperture is NA = 0.22. Figure 4b shows the optical Y-fiber and the single 80 μm fiber to the reference spectrometer that measures the reference excitation spectrum (labeled (2) in Figure 2b). This constellation prevents super-saturation or even damage of the reference spectrometer. For the fluorescence emission (detection) strand, a conventional optical fiber is used (QP1000-2-UV–Vis, Ocean Optics, Largo, FL, USA), connecting the sample flow-through cell to the main fiber optic spectrometer (labeled (3) in Figure 2b).

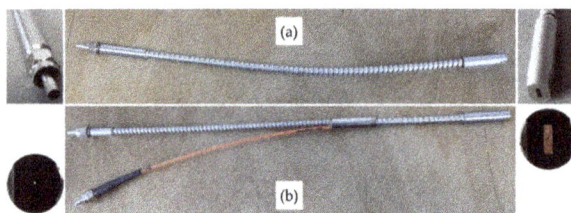

Figure 4. (**a**) Optical fiber coupling of the Xenon flash lamp with the monochromator (labeled (1) in Figure 2b). (**b**) Optical Y-fiber coupling of the monochromator with the reference spectrometer and the water sample flow-through cell (labeled (2) in Figure 2b).

2.1.3. EEMS Sensor System Fluorescence Performance Evaluation

As the complete 'Kallemeter' EEMS sensor system is based on a new design, the system performance for wavelength calibration, fluorescence intensity and signal-to-noise ratio was examined prior to field application. Considering the main focus of this article being the field application, the most important facts and a compilation of regularly practiced calibration methods for the in situ EEMS sensor system will be presented in the following section. All settings for the following measurements are listed in detail in Table A2 of the Appendix A.

The general procedure to calibrate an in situ fluorometer consists of (i) precalibration with tests of pressure, mechanical, and electronic stability as well as precision, (ii) signal output calibration to measure the dark and maximum counts, (iii) internal temperature calibration, (iv) determination and record of offset values with purified water at different temperatures, and (v) manufacturer calibration to obtain the scale factor of the fluorometer [36]. Calibration should be repeated in regular intervals, because environmental conditions can vary and lamp as well as spectrometer performance can degrade over time. Table 1 summarizes the EEMS sensor system specifications.

Signal-To-Noise Ratio

The signal-to-noise ratio (S/N) of the Raman band of tap water is easy to realize and a widely used method to quantify the sensitivity of a spectrofluorometer [39,40]. The S/N (where S denotes signal intensity and N noise intensity) of the Raman band is calculated with the following equation.

$$S/N = (S - N)/\sqrt{N}$$

For the 'Kallemeter' a S/N of 34 was determined for the Raman band measurement of tap water as presented in Figure A2.

Table 1. Overview of the 'Kallemeter' EEMS sensor system, fluorescence hardware specifications.

Parameter	Specification
Light source	Pulsed Xenon flash, 9.9 Watts
Excitation range	220–750 nm
Excitation bandpass	10 nm (FWHM)
Excitation monochromator	Czerny–Turner design, 110 mm focal length
Excitation wavelength precision / accuracy	0.2 nm/±0.6 nm
Emission range	200–950 nm
Emission bandpass	None (full set of wavelengths)
Emission detection spectrograph	(QE-Pro), Fixed, 101 mm focal length
Emission detector	TE-cooled CCD fiber optic spectrometer
Emission integration time	8 ms to 60 min
Excitation reference measurement	(USB2000+), CCD fiber optic spectrometer
Sensitivity	Water-Raman S/N 34

Wavelength Calibration

The wavelength calibration of the EEMS sensor system needs no special adjustments by the user, as both spectrometers used are calibrated by the manufacturers and therefore are supplied with a set of calibration coefficients. For verification we performed a laboratory check with a reference standard for molecular fluorescence spectrophotometry (Compound 610; excitation: 440 nm; emission: 480 nm; nominal concentration of 1×10^{-6} M, Starna Scientific Ltd., Ilford, UK), data not shown. Such a measurement allows the evaluation of the EEMS optical setup over time and corresponding correction methods for the emission wavelengths.

Intensity Calibration with Suwannee River Standard

Since 1982, representatives of the International Humic Substance Society (IHSS) have isolated reference fulvic acids, humic acids, and natural organic matter (NOM) from the Suwannee River in southeastern Georgia, USA. Suwannee River has a remarkable low variance in its elemental composition and is therefore a commonly accepted standard. It has a long history as reference material for multidisciplinary research areas, including the marine community [41]. For the fluorescence measurement of the 'Kallemeter', a sulfuric acid dilution of Suwannee River (sample charge number 2S101H) was used with concentrations of 2.5 ppm and 25 ppm. The EEM spectra of the two concentrations of Suwannee River standard are shown in Figure 5. Receiving reasonable signal intensity in both samples indicated sufficient signal intensity for nominal concentrations at marine study sites.

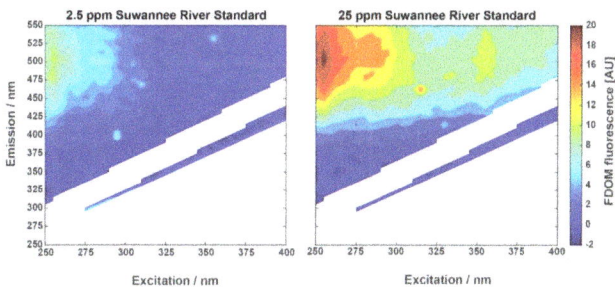

Figure 5. EEM spectra for two concentrations of Suwannee River standard. Excitation wavelengths were used for the x-axis, while emission wavelengths were applied for the y-axis [42]. Color bar (z-axis) denotes relative fluorescence intensity.

2.2. FerryBox-Based Organic Matter Fluorescence Measurements

An important step in the evaluation of the 'Kallemeter' is the comparison with other existing sensors and approaches, here with two commercially available sensors, the Cyclops-7-U (Turner Designs, San Jose, CA, USA) and the MatrixFlu-UV (TriOS, Rastede, Germany) mounted in a FerryBox system (4H-JENA engineering, Jena, Germany) installed on the ship during the research cruise (see Section 2.3). Furthermore a LS-55 spectrofluorometer (PerkinElmer, Waltham, MA, USA) was used as a common standard for laboratory measurements (see Section 2.4), providing complete EEM spectra directly comparable with the 'Kallemeter'. The FDOM sensors either provide only certain parts of the full EEM (MatrixFlu-UV) or only bulk fluorescence data (Cyclops-7-U).

As described previously [43], there are two versions of the MatrixFlu sensor, one for the visible wavelengths, with a focus on phytoplankton, and one for the ultraviolet, dedicated to FDOM components. The latter is used here and referred to as MatrixFlu-UV. The fundamental idea is to measure the typical peaks of organic matter fluorescence, as described by Coble [29], instead of assessing the whole excitation and emission spectrum, thus enabling a rapid detection and a high temporal resolution. This section describes only the sensor's main features, as its technical design has been published elsewhere [23,44].

The MatrixFlu-UV sensor consists of an excitation light source with distinct wavelengths combined with a set of detectors collecting a near field emission of a fluorescence signal at certain wavelengths directly in the water column (open sensor interface). Each combination of an excitation and emission wavelength results in a different fluorescence channel. The optical part contains a single semiconductor device emitting light at three different wavelengths (254 nm, 280 nm, and 320 nm) on the same optical axes. Four single semiconductor detectors are arranged around it collecting the fluorescent signal in an angle of approx. 170° to the optical axes. Different wavebands are selected via customizable narrow bandwidth filters (here 280 nm, 360 nm, 460 nm, and 540 nm). By doing this, all possible combinations result in twelve different channels.

FDOM components referred to in this work are mainly associated to (a) terrestrial or old marine humic-like substances (peak C), (b) "freshly produced" marine humic-like substances (peak M), and protein-like Tryptophan substances (peak T) [45,46]. Humic components show a further related excitation maxima for UV light of higher energy (lower wavelength) but same emission position or range (peak A) [45]. Data from all MatrixFlu-UV channels are collected simultaneously. For the comparison with the 'Kallemeter' and the other bulk fluorescence sensor (described below), Channel 11 was extracted and evaluated, corresponding to peak C (EX/EM 320 nm/460 nm). Data was averaged on every minute values and smoothed by a moving mean filter (window size 13). Factory calibration and further information about the sensor performance are available in [23].

The Cyclops-7-U sensor estimates bulk FDOM concentration by measuring the fluorescence emission at 470 nm (60 nm peak width) for an excitation at 325 nm (120 nm bandwidth filter [47]). Thus, for a comparison with the data from this sensor, the 'Kallemeter' in situ EEM signal in the area corresponding to an excitation of 265 to 385 nm and an emission of 440 to 500 nm have been integrated. This provides a 'Kallemeter' derived bulk fluorescence signal that corresponds to the excitation and emission ranges given in the specifications of the Cyclops-7-U sensor.

2.3. FerryBox Installation and Other Oceanographic Parameters Measured

The water intake of the FerryBox was realized in the moonpool of the research vessel (in approx. 4 m depth) close to the measuring head of the submersed 'Kallemeter' to ensure a comparability of the measurements between the three types of instruments. Besides the FDOM data, readings of standard oceanographic parameters, such as temperature, salinity, and chlorophyll *a* fluorescence were also obtained. All FerryBox data were averaged on minutely values and smoothed by a moving mean filter (window size 11).

2.4. FDOM Laboratory Water Samples

Sampling from the near surface water, here defined as the samples collected between 0.1 and 2.0 m, was obtained with a submersible pump system and stored in 5 L white plastic bottles, at all stations. Samples were filtered in the laboratory immediately after sampling, and stored at 4 °C in previously rinsed glass bottles. CDOM absorption and EEM spectra measurements were completed within 24 h after sampling.

Samples were filtered with a glass syringe device through GF/F Nucleopore filters (0.2 μm, GE Healthcare Whatman, Piscataway Township, NJ, USA). CDOM absorbance spectra were obtained between 220 and 650 nm at 5 nm intervals on a Shimadzu UV-2550PC UV–VIS spectrophotometer (Shimadzu, Kyoto, Japan) with 10 cm quartz cuvettes. All measurements were performed at room temperature using freshly produced ultrapurified water (UPW; Milli-Q 185plus water purification system, Merck Millipore, Burlington, MA, USA) as reference. CDOM absorption was calculated by multiplying absorbance by 2.303/r, where r is the cuvette length (in meter). FDOM EEM spectra measurements were conducted at room temperature on a LS55 spectrofluorometer (PerkinElmer, Waltham, MA, USA) equipped with a Xenon flash lamp, pulsed at line frequency of 60 Hz. The excitation (EX) and emission (EM) scanning ranges were set to 200 to 400 nm (5 nm interval) and 220 to 550 nm (0.5 nm interval), respectively. The scanning speed was 1200 nm min^{-1}, the EX and EM slit was set to 10 nm, and the detector gain was at medium range (775 V). Daily measurements of UPW were used as references. Water Raman scatter peaks were eliminated by subtracting the EEM spectra of UPW blanks. Inner filter effect and Rayleigh scatter peaks were corrected using the drEEM toolbox [33]. Lamp variations were normalized to the Raman peak integral (EX: 350 nm) using freshly produced UPW, therefore producing normalized intensities in Raman units (RU) [39]. EEM spectra were resized due to high S/N in certain regions. Therefore, excitation wavelengths below 250 nm and emission wavelengths below 300 and above 550 nm were excluded from the subsequent PARAFAC modeling process.

PARAFAC modeling was performed with the drEEM toolbox (ver.0.2.0) in MATLAB R2015b (The MathWorks, Natick, MA, USA) according to a method described previously [33], employing the N-way toolbox as the engine of the PARAFAC algorithm [34]. A three-component model was validated by split-half analysis, random initialization analysis, and residuals analysis. The model explained up to 99.95% of the variability within the lab-based LS55 spectrofluorometer EEM dataset.

A similar PARAFAC analysis was applied on the Kallemeter-based EEM spectra along cruise track between station 11 and 15. Correction for inner filter effects was realized by utilizing the measured absorption spectra from the nearest station and using Kallemeter blank measurements from an initial laboratory test with purified water. Four out of 48 spectra were excluded from the analysis, due to high residual error and leverage load, validating a two-component model that explained up to 61% of the variability within the dataset through split-half analysis, random initialization analysis, and residuals analysis. A third component could not be detected.

2.5. Study Site

In order to assess the performance of the 'Kallemeter' in comparison with the MatrixFlu-UV and the FerryBox-mounted Cyclops-7-U fluorescence sensors in the field, the data obtained during an expedition (HE491) with research vessel R/V Heincke from the coastal sea into the Trondheimsfjord (Norway) were analyzed. Figure 6 shows a map of the study site and the cruise track investigated. The instruments were in parallel operation from 19th to 25th of July 2017. The Trondheimsfjord is the third-largest fjord in Norway with a length of 140 km, a volume of 235 km^3, and a surface area of 1420 km^2. It is divided in three basins and has a seaward sill depth of 195 m [48]. It is characterized by comparably high CDOM concentrations as well as vertical and horizontal gradients [49], making it an ideal test site.

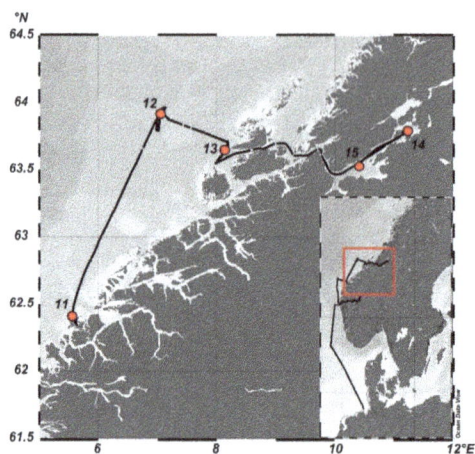

Figure 6. Overview of the study area (insert). Map shows Trondheimsfjord and its adjacent coastal sea with cruise track (solid black line) and stations (red dots) of R/V Heincke expedition HE491.

3. Results

The FerryBox data is well suited to provide a first overview of the overall fjord hydrography. The FerryBox dataset started on 19 July in the Norwegian coastal sea (station 11) heading first northwards (station 12) and then into the Trondheimsfjord (station 13), through the mid-fjord section and towards the innermost part of the fjord (station 14). From there, the transect went back to the mid-section and ended at station 15 on 25 July. There is a considerable difference in all parameters measured between the area outside and within the fjord (Figure 7), indicating the presence of two water bodies with different properties. With the exception of salinity, values of all parameters are higher within the fjord. Salinity is highest towards the ocean side, with values of more than 32, rapidly decreasing inside the fjord towards the mid-section exhibiting values of around 24. Temperature increases slightly from around 14 °C in the outer section towards 15 to 17 °C in the mid- and inner-part of the Trondheimsfjord. Chlorophyll-a fluorescence (here provided in AU, as the sensor has not been specifically calibrated for chl-a before the cruise) shows a gradual increase from the seaward part of the fjord to a maximum at the end of the fjord. FDOM from the Cyclops-7-U fluorometer (provided in AU) shows, inverse to salinity, a four-fold increase and steep gradient from the seaward side of the fjord towards the mid- and inner-part.

Comparing the 'Kallemeter' bulk fluorescence signal extracted from the EEM spectra against the bulk fluorescence signal of the MatrixFlu-UV channel 11 and the FerryBox inbuilt Cyclops-7-U FDOM sensor data (Figure 8), it can be seen that all three signals showed a steep increase on 23 July, when entering Trondheimsfjord. MatrixFlu-UV exhibited a maximum at the end of that day, followed by a decrease on 24 July towards a steady level around half of the signal range. FDOM bulk fluorescence from the FerryBox inbuilt Cyclops-7-U sensor increased in the same steep way towards a higher level on 23 July. Signal intensity stayed on this level with only a slight decrease in the morning of 24 July. 'Kallemeter' bulk fluorescence signal, as indicated by red stars, exhibited the same dynamical behavior as the Cyclops-7-U. A linear regression analysis was performed (Figure 9) of the extracted 'Kallemeter' bulk fluorescence against the Cyclops-7-U sensor (coefficient of determination $R^2 = 0.96$) as well as the corresponding channel of the MatrixFlu-UV ($R^2 = 0.88$), both attached to the flow through system.

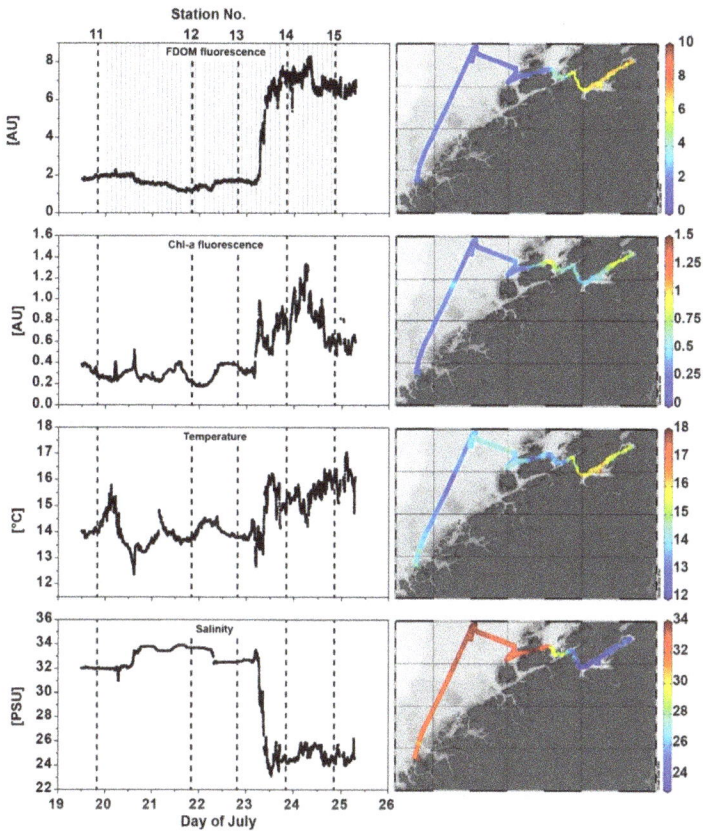

Figure 7. Underway oceanographic data measured by the FerryBox. From top to bottom: Fluorescent dissolved organic matter (FDOM) from the Cyclops-7-U bulk fluorometer, chlorophyll a fluorescence, temperature, and salinity. The left panels display the data as time series, the corresponding right ones as a map plot. Dashed lines indicate the stations where water samples were collected. Gray lines in the top-left panel indicate availability of the in situ Kallemeter-based EEM spectra (N = 48).

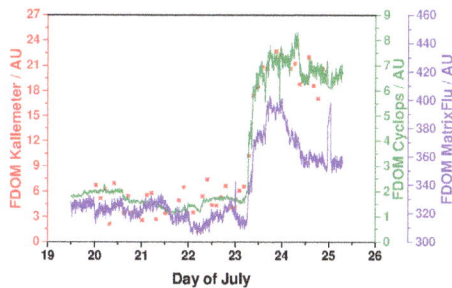

Figure 8. Comparative view on the extracted underway measurements from the MatrixFlu-UV (violet), the FerryBox FDOM Cyclops-7-U sensor (green) and the extracted bulk fluorescence signal from the EEM spectra (red stars), obtained along a transect from the Norwegian coastal sea into the Trondheimsfjord.

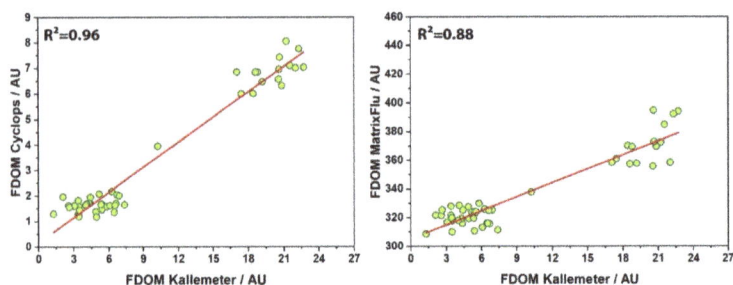

Figure 9. Comparison of extracted 'Kallemeter' bulk fluorescence signal against FerryBox FDOM Cyclops-7-U sensor (**left panel**) as well as the corresponding channel of the MatrixFlu-UV (**right panel**). Red lines represent the results of a linear regression.

In situ underway EEM measurements, measured by the 'Kallemeter' in the moonpool of R/V Heincke, provided a total of 48 spectra in 5 days (20 to 25 July). All fluorescence information and metadata were continuously recorded from open water into the Trondheimsfjord. A comparison of normalized EEM spectra of the lab-based LS55 fluorometer with the corresponding nearest normalized Kallemeter-based EEM spectra at stations 12–15 (Figure 10; station 11 had no close match-up within a time frame of ± 1 h) shows good similarity for stations 14 and 15, where signal intensities are 5-times higher as compared to stations 12 and 13 (compare Figure 8). These lower concentrations of fluorescent organic matter at stations 12 and 13 were only resolved from the apparently more sensitive laboratory LS55 spectrofluorometer, thus illustrating the current limits of the 'Kallemeter' with respect to low concentrations and the effects of noise. Please note that we applied normalization to the EEM spectra to enable a comparison of spectra signatures measured, independent from the signal strength. A non-normalized set of 'Kallemeter' EEM spectra is provided in the appendix (Figure A3) showing for every fourth dataset (a) an EEM spectra with EX wavelengths 230–550 nm, $\Delta\lambda = 10$ nm, integration time of 60 s and two averaged scans, slit = 10 nm, EM wavelengths 189–990 nm, and (b) an emission scan with EM wavelengths 189–990 nm for 250 nm EX wavelength. These Kallemeter-based EEM spectra show an increase in fluorescence intensities inside Trondheimsfjord as compared to the coastal water conditions in the EM wavelength range from 400 to 550 nm.

PARAFAC analysis of the lab-based LS55 spectrofluorometer EEM spectra from water samples gathered at different depths demonstrated the presence of three main components (C1, C2, and C3) in the water samples (Figure 11, top panel). These components were characterized by their excitation and emission pairs (EX/EM) as follows. Component (C1) corresponding to terrestrial humic-like peaks A_C and C, component (C2) corresponding to marine humic-like peaks A_M and M, and component (C3) corresponding to protein-like Tryptophan peak T, as suggested by Coble [29]. The spectral composition of the recorded lab-based LS55 spectrofluorometer EEM spectra (provided in Table 2) shows that the fluorescent intensity of the peaks C1, C2, and C3 remains constant between the open ocean stations (ST11 and ST12) and the outer section of the fjord (ST13). Once the mid- and inner-part of the fjord was reached, we recorded a sharp increment (by a factor of 5) in the fluorescent intensity for the components C1 and C2 (ST14 and ST15), while C3 stays nearly constant.

Performing PARAFAC analysis of the Kallemeter-based EEM spectra along the transect resulted in a model with two components validated (Figure 11, bottom panel). These components, even though influenced by the noisy EEM spectra that suited as model dataset, can be related according to their excitation/emission pairs to the presence of terrestrial humic-like peaks A_C and C, and marine humic-like peaks A_M and M. Although reasonable EEMS where produced by the 'Kallemeter', the presence of noise and suspended materials in the samples must be addressed in the future, in order to minimize the observed displacement on the peak's position.

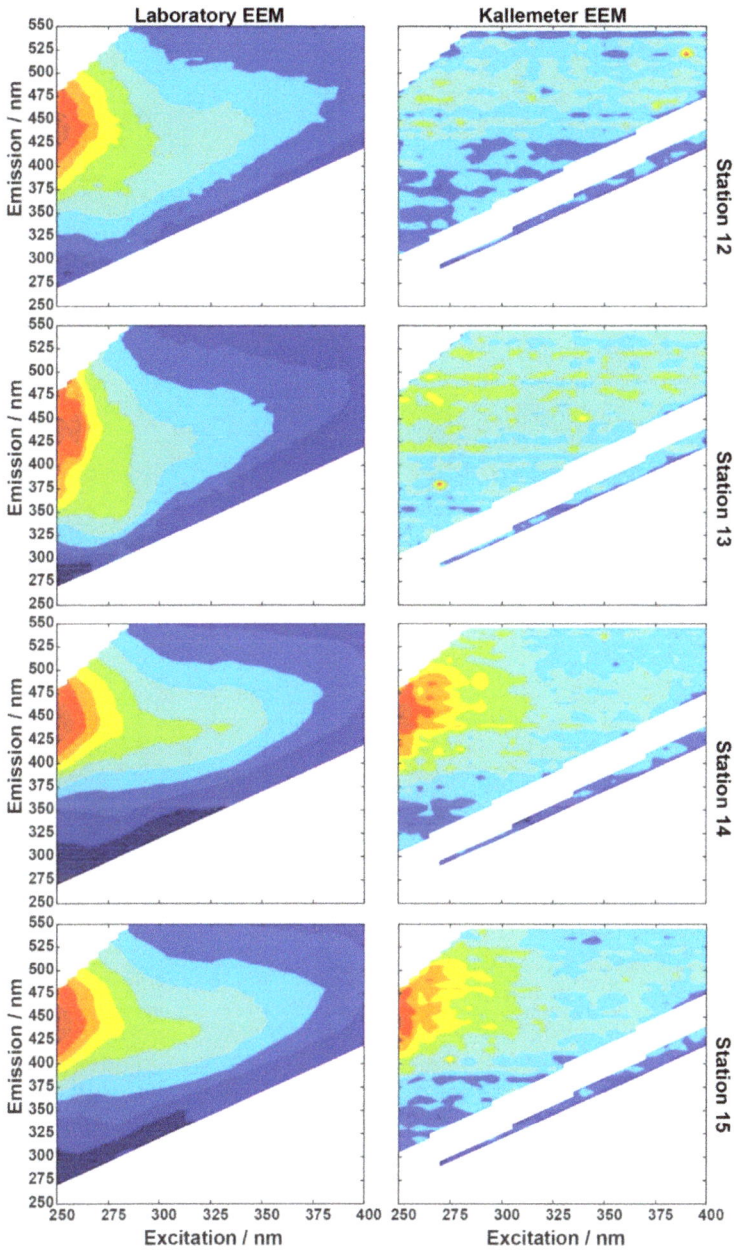

Figure 10. Normalized EEM spectra from lab-based water sample analysis with the LS55 laboratory spectrofluorometer (**left**) and nearby (match-up of ±1 h) in situ measurements with the Kallemeter (**right**) for stations 12–15 of R/V Heincke expedition HE491 along a transect from the coastal ocean into the Trondheimsfjord. Notice that FDOM intensity in stations 14 and 15 are 5-times higher compared to stations 12 and 13.

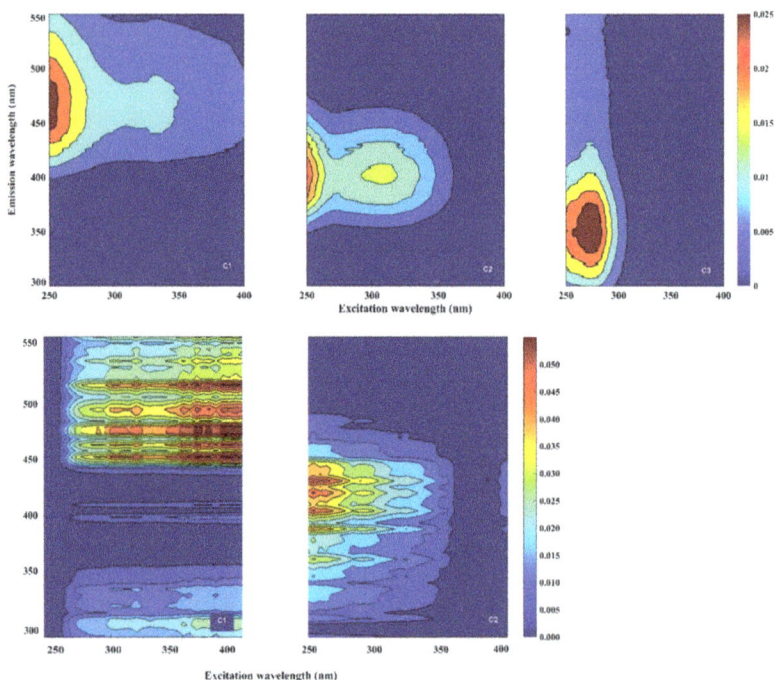

Figure 11. Top panel: PARAFAC-derived principal fluorescent components from lab-based EEM spectra from water samples at stations 11–15. From left to right: (C1) Terrestrial humic-like peaks A_C and C; (C2) marine humic-like peaks A_M and M; and (C3) protein-like Tryptophan peak T, all peaks denoted after Coble [29]. Color bar shows Raman units (RU). Bottom panel: PARAFAC derived principal fluorescent components from Kallemeter-based EEM spectra along the transect.

Table 2. Spectroscopic composition of the lab-based LS55 spectrofluorometer EEM spectra analyzed water samples in Trondheimsfjord. C1, C2, and C3 represent the intensity of PARAFAC components in Raman units.

Sample Number	Station	Depth (m)	C1 (RU)	C2 (RU)	C3 (RU)	Date
1	ST11	0.10	0.0364	0.0302	0.0238	
2	ST11	0.50	0.0349	0.0298	0.0243	20 July 2017
3	ST11	1.50	0.0354	0.0297	0.0283	
4	ST11	2.00	0.0331	0.0289	0.0304	
5	ST12	0.10	0.0252	0.0194	0.0162	
6	ST12	0.50	0.0231	0.0177	0.0106	22 July 2017
7	ST12	1.50	0.0226	0.0174	0.0127	
8	ST12	2.00	0.0227	0.0177	0.0130	
9	ST13	0.10	0.0355	0.0278	0.0214	
10	ST13	0.50	0.0355	0.0295	0.0287	23 July 2017
11	ST13	1.50	0.0323	0.0255	0.0173	
12	ST13	2.00	0.0322	0.0263	0.0191	
13	ST14	0.10	0.1772	0.1054	0.0363	
14	ST14	0.50	0.1745	0.1017	0.0247	24 July 2017
15	ST14	1.50	0.1763	0.1035	0.0248	
16	ST14	2.00	0.1749	0.1028	0.0263	
17	ST15	0.10	0.1471	0.0904	0.0378	
18	ST15	0.50	0.1493	0.0899	0.0233	25 July 2017
19	ST15	1.50	0.1543	0.0928	0.0241	
20	ST15	2.00	0.1530	0.0919	0.0232	

4. Discussion, Limitations, and Conclusions

The main objective of this study was to present a novel in situ EEMS sensor system—the 'Kallemeter'—enabling full optical fingerprints in a submersible design. We presented the design of this system within the methods Section 2.1 and illustrated its application based on laboratory samples of diluted Suwannee River standards with 2.5 ppm and 25 ppm, respectively. Fluorescence exhibited around EX: 240 nm and EM: 450 nm was corresponding to terrestrial humic-like peak C substances, as expected for Suwannee River standards, with increasing intensity and additional visible contributions extending to higher excitation wavelengths [50,51]. The assessment of the 'Kallemeter' EEMS sensor systems' signal-to-noise ratio (S/N) based on the Raman peak of water in a tap-water sample, provided an S/N of 34. Compared to commercially available laboratory spectrofluorometers that exhibit S/N of 500 or higher the 'Kallemeter' is much more limited in resolving low signal intensities [39]. Part of this is based on the different optical components implemented, e.g., a fiber optic spectrometer compared to a monochromator plus photomultiplier setup for the LS55 spectrofluorometer, still considered as the gold-standard for high sensitivity applications. Also the choice of the excitation light was influenced by power- and heat-considerations and therefore restricted to <10 Watts, where other laboratory instruments like the Aqualog (Horiba, Kyoto, Japan) are equipped with up to 150 Watt Xenon flash lamps [52]. The strongest influence can be assumed from the application of fiber optics within the sensor, resulting in losses while coupling light in and out as well as transmitting it in the fiber. The use of fiber optics is a result of the compact design that a submersible housing requires—a potential redesign of the 'Kallemeter' will take this into account and try to partially avoid the use of optical fibers through direct light paths.

The field application of the 'Kallemeter' was realized in the moonpool of R/V Heincke on transect from Norwegian coastal waters into the Trondheimsfjord in July 2017. The system was successfully operated for five days in a fully submerged installation. EEM spectra derived from this application showed very low fluorescence intensities superimposed by noise components for all coastal sea measurements and the outer fjord area, while mid- and inner-fjord measurements exhibited clear broad signals of approximately EX: 250–300 nm and EM: 400–500 nm. These observations are consistent with the spectral component intensities derived from laboratory UV–VIS spectrofluorometer (LS55, PerkinElmer, Waltham, MA, USA) measurements of surface water samples. With respect to components C1 and C2 derived from PARAFACS analysis these were ranging for the outer fjord between 0.02 and 0.04 Raman Units, while mid- and inner-fjord values ranged between 0.09 and 0.18 RU. The direct comparison of normalized EEM spectra from the lab-based spectrofluorometer and the 'Kallemeter' (Figure 10) illustrated both, the capability of the 'Kallemeter' to provide full EEM spectra if fluorescent organic matter concentrations are sufficiently high (stations 14 and 15), and the limitation to resolve low concentrations against the background noise as encountered at stations 12 and 13 in coastal waters. The latter is a result of the lower S/N and sensitivity, discussed above. The application of the 'Kallemeter' EEMS sensor system in its current form therefore seems to be limited towards areas with an increased amount of fluorescent organic matter, typically encountered in near coastal areas, estuaries and inland waters. Furthermore a study of the impact of noise or signal degradation applied on EEM spectra that serve as PARAFAC model input would be of interest to assess general limitations, however that is beyond the scope of this study.

The PARAFAC analysis of the Kallemeter-based EEM spectra enabled us to model two components, corresponding to terrestrial humic-like peaks A_C and C, and marine humic-like peaks A_M and M. This was compared to a PARAFAC analysis of the LS55 spectrofluorometer EEM spectra from near surface water samples at the distinct stations that successfully validated three components. While lab-based components 1 and 2 resemble the in situ Kallemeter components (even though limited by high noise influence), the third component, namely the protein-like Tryptophan peak T, could not be reproduced from the 'Kallemeter' data. Since peak T could be quantified in the laboratory measurements to an average intensity of 0.023 Raman units without any signal increase in the Trondheimsfjord, we consider this as a current limitation for the 'Kallemeter'.

The ability to perform a PARAFAC analysis on automatically measured in situ EEM spectra is a major achievement and, to the best of our knowledge, the first of its kind. We encountered a couple of challenges to that end, that we provide here as current limitations but also avenues for future improvements. (A) Since we did not filter the water entering the measurement cell with 0.2 μm (only applying 50 μm to keep coarse material out) the signal observed is originating from dissolved, colloidal and particulate fractions of organic matter. For low amounts of particulate material this might not be a problem and indeed all in situ FDOM bulk fluorescence sensors (including the two used in this study for comparison) are also measuring unfiltered samples. However, for direct comparison with laboratory measured (filtered) FDOM EEM spectra this can be considered an obstacle. To overcome this, the installation of a filter unit at the inlet, containing an appropriate 0.2 μm filter can be easily realized in future applications of the 'Kallemeter'. It can be speculated that part of the noisiness of the spectra and PARAFAC components can be associated to the influence of particulate material. (B) All EEM spectra need to be corrected for the inner filter effect (IFE). This typically requires a parallel measurement of the absorption spectra of the filtered water sample over the same wavelength range. The 'Kallemeter' itself is not suited to perform this absorption measurement, therefore we applied absorption spectra from water samples analyzed with a laboratory spectrophotometer. These were only available at the five stations investigated and not at the 48 measurement locations of the 'Kallemeter'. Assigning absorption spectra from samples that were gathered up to 24 h earlier or later is introducing uncertainties, even though we took care to respect the different oceanographic gradients encountered in the matching procedure. A much better way to account for this need of parallel absorption measurements will be to use automated in situ capable methods [53]. (C) Applying mirrors in the sample flow through cell, we increased the fluorescence signal intensity by a factor of 2.5, however IFE corrections applied to correct EEM spectra do not consider these and are therefore overcorrecting. Luciani et al. [54] showed how mirrors change the fluorescence signal and can be at the same time utilized to correct for IFE, if measurements with and without mirrors are performed on the same sample. This could be another methodological improvement to overcome constraints mentioned above.

Comparing the results from the bulk fluorescence signal of the 'Kallemeter' (calculated from the EEM spectra) with the flow through data from the commercially available FDOM sensors (a) inside the FerryBox (Cyclops-7-U, Turner Design, San Jose, CA, USA) and (b) the channel 11 from the recently developed matrix fluorescence sensor (MatrixFlu-UV, TriOS, Rastede, Germany), a good correspondence could be observed. The maximum exhibited on 23 July and the following decrease in the 'Kallemeter' fluorescence intensity (compare Figure 8) seems to be better reproduced by the Cyclops-7-U FDOM sensor. The corresponding MatrixFlu-UV channel is able to reproduce the same steep increase, however shows a signal reduction afterwards, not represented by the other methods applied in this study. As this study reports one of the first field datasets achieved with the MatrixFlu-UV sensor we cannot judge on this readings but see a clear need to investigate the sensors responses.

The MatrixFlu-UV sensor was investigated herein only with respect to one of its 12 channels representing a main spectral component in the investigated environment. It is therefore herein utilized as single-channel sensor and its further value in rapid quasi-EEM spectra measurements will be investigated in future works. Other multichannel approaches exist, with two [55] or three [56] combinations technically built up from single EX/EM combinations. A recent study of Nordic seas FDOM intensities [57] utilized a novel three-channel fluorometer (WETStar, Sea-Bird Scientific, Bellevue, WA, USA). This system excites the water sample in a flow-through cell with two LEDs at 280 nm and 310 nm. Two detectors measure the emission at 350 nm and 450 nm, allowing for a combination of channels in specific peak areas: peak C (with EX/EM 310 nm/450 nm), peak A_C (with EX/EM 280 nm/450 nm), and peak T (with EX/EM 280 nm/350 nm). All applications mentioned above underline the potential of multichannel FDOM measurements for specific components dominant in natural water bodies. The technological step from multichannel sensing to real excitation–emission matrix spectroscopy with high spectral resolution in both, the excitation and the emission sides, is huge, since it requires complete different technologies. An interim step with six (up to 12) distinct LED

Appl. Sci. **2018**, *8*, 2685

excitation wavelengths and emission spectra with 7.5-nm spacing (named 'LEDIF') was presented as part of a field-deployable optical system [58].

The 'Kallemeter' represents, to the best of our knowledge, the first submersible EEMS sensor system that performs full optical fingerprints. While the detection side is comparable to the LEDIF approach, using a fiber optical spectrometer, excitation was realized in our system by a Xenon flash lamp and an embedded controlled monochromator. This enabled us to achieve 10-nm steps in the excitation wavelengths selection (and finer steps are technically feasible); however, it also slows down the measurement process to approx. 30 min per individual sample. This is in the same order of magnitude as the LS55 (PerkinElmer, Waltham, MA, USA) UV–VIS laboratory spectrofluorometer used in this study. Faster laboratory systems exist nowadays (e.g., Aqualog, Horiba, Kyoto, Japan) that enable EEM spectra measurements below 5 min that require, however, further technical improvements (e.g., a TE cooled back-illuminated CCD fluorescence detector, and a high power Xenon flash lamp). Modern ocean observing strategies combine sensors with high spatiotemporal resolution but limited information depth with more sophisticated sensors, that provide a high depth of information but are limited in their spatial and temporal resolution [8]. The 'Kallemeter' is one of those highly sophisticated sensors and will likely, as others in that field [59,60], improve in its operational specifications, i.e., size, weight, and power. Multichannel fluorescence sensors for FDOM are already smaller, lightweight, and less power consuming. Their application on different autonomous platforms has been demonstrated [17,55,58]. Thus, they are the linkage towards resolving rapid changes and processes. An assessment of fluorescent organic matter in high spatiotemporal and spectral resolution is therefore technically feasible and likely to become standard in future biogeochemical ocean observatories.

Author Contributions: Conceptualization, O.Z.; Methodology, O.Z., N.R., O.D.F., M.L.M., and J.W.; Validation and analysis, O.Z., N.R., O.D.F., M.L.M., and J.W.; Field investigations, N.R, O.D.F., M.L.M., and J.W.; Project Administration, O.Z. and N.R.; Funding Acquisition and Resources, O.Z.; All authors contributed to the writing and revision process.

Funding: This research was partly funded by the Ministry of Science and Culture of Lower Saxony, Germany, as part of the 'Coastal Ocean Darkening' project, grant number [VWZN3175] and the 'EcoMol' graduate school. Development of the MatrixFlu-UV was partly funded by project NeXOS [grant agreement No. 614102] under the call FP7-OCEAN-2013.2 from the EU Commission as part of the 7th Framework Program, 'The Ocean of Tomorrow'. M.L.M. acknowledges SENACYT-IFARHU for the research scholarship (for Doctoral and Postdoctoral studies 2017/18, Program BIPD-2016). Additional funding was provided by EFRE-project 'MultiFlu', funded by NBank, Hannover, Germany [proposal number 85003576].

Acknowledgments: The authors are thankful to the master and crew onboard the R/V Heincke HE491 campaign. The cruise was supported by the Alfred Wegener Institute for Polar and Marine Research (AWI) and the German Federal Ministry of Education and Research (BMBF). Sincere thanks to Anna Friedrichs and Kathrin Dietrich for assistance in data collection and laboratory analysis of water samples. Additional thanks to Beke Tietjen for supporting the preparation of Figures 8 and 9 in their early form, Ursel Gerken for checking spelling and grammar, and Julian Beck for constructing the flow-through cuvette of the 'Kallemeter'. MatrixFlu-UV was built and provided by TriOS Mess- und Datentechnik GmbH, Rastede, Germany. FerryBox was kindly provided by Wilhelm Petersen, Institute of Coastal Research at the Helmholtz-Zentrum Geesthacht, Germany. Several components of the 'Kallemeter' were kindly provided by Axel Bochert and Jan Boelmann, University of Applied Science, Bremerhaven, Germany. The feedback from the participants of the WOMS 2018 workshop in Toulon (France), two anonymous reviewers, and Dariusz Stramski as editor is gratefully acknowledged.

Conflicts of Interest: The authors declare no conflict of interest. The funders had no role in the design of the study; in the collection, analyses, or interpretation of data; in the writing of the manuscript, and in the decision to publish the results.

Appendix A

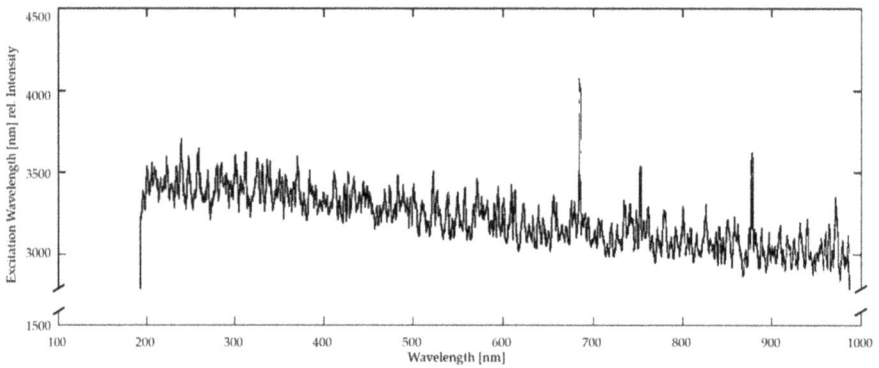

Figure A1. Dark counts of the 'Kallemeter' fluorescence detection spectrometer, measured by a long-term in situ test of 50 full EEM spectra (presented as an overlay) over a period of 116 h exhibiting a constant slight intensity gradient against the ordinate (wavelength).

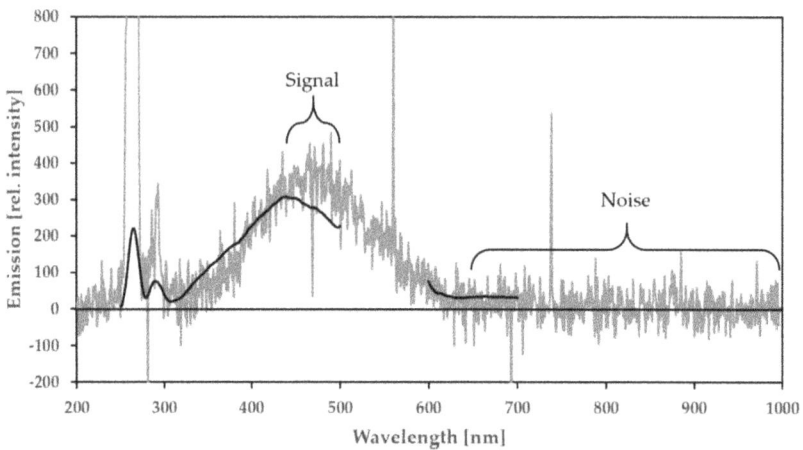

Figure A2. Signal to noise measurement of tap water with the 'Kallemeter' fluorescence detection spectrometer (grey line, integration time 300 s), and the LS55 laboratory spectrofluorometer (black line, 2nd maxima masked), both excitation wavelengths are 260 nm.

Table A1. Comparison of different mirror combinations within the measuring cell. Shown: Mirror on (opposite) excitation and emission side, maximum value of the fluorescence signal, signal-to-noise ratio (S/N). Measured: Compound 610 from Starna Scientific Ltd. (Ilford, UK), flash lamp frequency: 100 Hz, excitation wavelength 440 nm, integration time: 5 s.

Excitation Side	Emission Side	Max [counts]	S/N
Concave	Concave	678	156
Concave	Planar	571	157
Planar	Concave	650	187
Planar	Planar	570	145
Absent	Absent	346	94

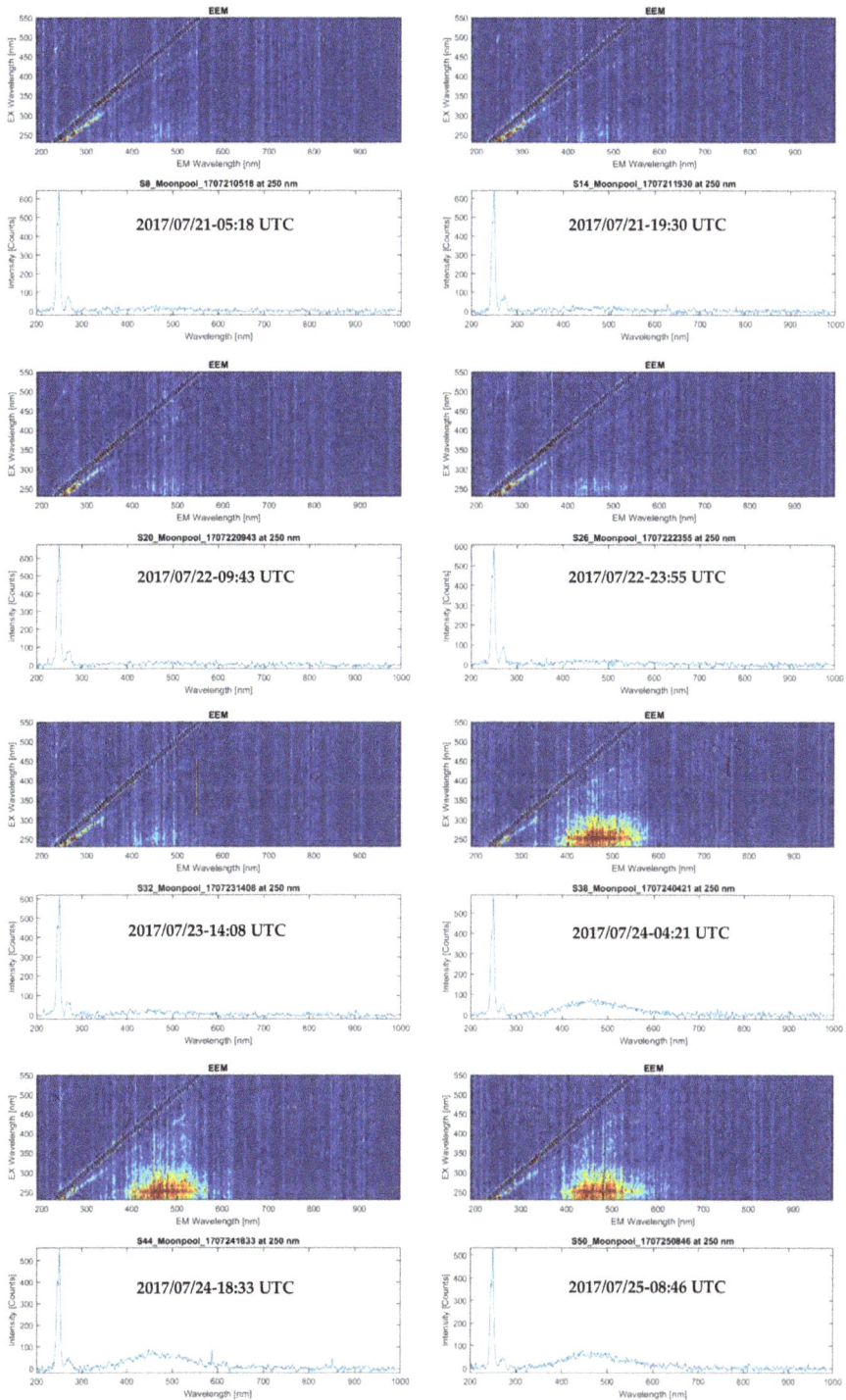

Figure A3. In situ underway measurements of EEM spectra, obtained from the moonpool of R/V Heincke (HE491) along transect from the coastal ocean into the Trondheimsfjord.

Table A2. EEMS measurement parameters for intensity calibration procedures.

Parameter	Settings
Excitation start	200 nm
Scan steps	5 nm
Excitation end	450 nm
Dark correction	True
Dark noise multi scan	20
CM110 offset correction	None
CM110 slits	None
Integration time	6 s
Scans to average	10
Box car	0
Trigger mode	0
Detector temperature	−10 °C
PX-2 connect	Reference
Multiple strobe	On
Flash delay	10 milliseconds
Warm-up	None

References

1. Mostofa, K.; Vione, D.; Mottaleb, M.; Yoshioka, T. (Eds.) *Photobiogeochemistry of Organic Matter: Principles and Practices in Water Environments*, 1st ed.; Springer: Heidelberg, Germany, 2013.
2. Dittmar, T. Reasons behind the long-term stability of marine organic matter. In *The Biogeochemistry of Marine Dissolved Organic Matter*, 2nd ed.; Hansell, D.A., Carlson, C.A., Eds.; Academic Press: Boston, MA, USA, 2015; pp. 369–388.
3. Ridgwell, A.; Arndt, S. Why Dissolved Organics Matter: DOC in Ancient Oceans and Past Climate Change. In *The Biogeochemistry of Marine Dissolved Organic Matter*, 2nd ed.; Hansell, D.A., Carlson, C.A., Eds.; Academic Press: Boston, MA, USA, 2015; pp. 1–20.
4. Watson, J.; Zielinski, O. *Subsea Optics and Imaging*; Woodhead Publishing Limited: Cambridge, UK, 2013.
5. Miranda, M.L.; Mustaffa, N.I.H.; Robinson, T.B.; Stolle, C.; Ribas-Ribas, M.; Wurl, O.; Zielinski, O. Influence of solar radiation on biogeochemical parameters and fluorescent dissolved organic matter (FDOM) in the sea surface microlayer of the southern coastal North Sea. *Elem. Sci. Anth.* **2018**, *6*, 15. [CrossRef]
6. Parlanti, E.; Wörz, K.; Geoffroy, L.; Lamotte, M. Dissolved organic matter fluorescence spectroscopy as a tool to estimate biological activity in a coastal zone submitted to anthropogenic inputs. *Org. Geochem.* **2000**, *31*, 1765–1781. [CrossRef]
7. Drozdowska, V.; Freda, W.; Baszanowska, E.; Rudz, K.; Darecki, M.; Heldt, J.T.; Toczek, H. Spectral properties of natural and oil polluted Baltic seawater—Results of measurements and modeling. *Eur. Phys. J. Spec. Top.* **2013**, *222*, 1–14. [CrossRef]
8. Zielinski, O.; Busch, J.A.; Cembella, A.D.; Daly, K.L.; Engelbrektsson, J.; Hannides, A.K.; Schmidt, H. Detecting marine hazardous substances and organisms: Sensors for pollutants, toxins, and pathogens. *Ocean Sci.* **2009**, *5*, 329–349. [CrossRef]
9. Friedrichs, A.; Ferdinand, O.; Miranda, M.L.; Zielinski, O. SmartFluo goes FDOM: Advancement of the DIY fluorometer approach towards UV excitation. In Proceedings of the OCEANS 2017, Aberdeen, UK, 19–22 June 2017.
10. Wymore, A.S.; Potter, J.; Rodríguez-Cardona, B.; McDowell, W.H. Using In-Situ Optical Sensors to Understand the Biogeochemistry of Dissolved Organic Matter Across a Stream Network. *Water Resour. Res.* **2018**, *54*, 2949–2958. [CrossRef]
11. Kalle, K. Zum Probleme der Meereswasserfarbe. *Ann. Hydrol. Mar. Mitt.* **1938**, *66*, 1–13.
12. Kirk, J.T.O. *Light and Photosynthesis in Aquatic Ecosystems*, 3rd ed.; Cambridge University Press: New York, NY, USA, 2011; p. 638.
13. Kowalczuk, P.; Ston-Egiert, J.; Cooper, W.J.; Whitehead, R.F.; Durako, M.J. Characterization of chromophoric dissolved organic matter (CDOM) in the Baltic Sea by excitation emission matrix fluorescence spectroscopy. *Mar. Chem.* **2005**, *96*, 273–292. [CrossRef]

14. Coble, P. Marine optical biogeochemistry: The chemistry of ocean color. *Chem. Rev.* **2007**, *107*, 402–418. [CrossRef]
15. Para, J.; Coble, P.G.; Charrière, B.; Tedetti, M.; Fontana, C.; Sempéré, R. Fluorescense and absorption properties of chromophoric dissolved organic matter (CDOM) in coastal surface waters of the northwestern Mediterranean Sea, influence of the Rhône river. *Biogeosciences* **2010**, *7*, 4083–4103. [CrossRef]
16. Schüßler, U.; Kremling, K. A pumping system for underway sampling of dissolved and particulate trace elements in near-surface waters. *Deep. Sea Res. Part I Oceanogr. Res. Pap.* **1993**, *40*, 257–266. [CrossRef]
17. Moore, C.; Barnand, A.; Fietzek, P.; Lewis, M.; Sosik, H.; White, S.; Zielinski, O. Optical tools for ocean monitoring and research. *Ocean Sci.* **2009**, *5*, 661–684. [CrossRef]
18. Zaneveld, J.R.V.; Kitchen, J.C.; Moore, C.C. Scattering error correction of reflection-tube absorption meters. In Proceedings of the Ocean Optics XII, Bergen, Norway, 26 October 1994.
19. Tassan, S.; Ferrari, G.M. Variability of light absorption by aquatic particles in the near-infrared spectral region. *Appl. Opt.* **2003**, *42*, 4802–4810. [CrossRef] [PubMed]
20. Fry, E.S.; Kattawar, G.W.; Pope, R.M. Integrating cavity absorption meter. *Appl. Opt.* **1992**, *31*, 2055–2065. [CrossRef] [PubMed]
21. Röttgers, R.; Doerffer, R. Measurements of optical absorption by chromophoric dissolved organic matter using a point-source integrating-cavity absorption meter. *Limnol. Oceanogr. Methods* **2007**, *5*, 126–135. [CrossRef]
22. Determann, S.; Lobbes, J.M.; Reuter, R.; Rullkötter, J. Ultraviolet fluorescence excitation and emission spectroscopy of marine algae and bacteria. *Mar. Chem.* **1998**, *62*, 137–156. [CrossRef]
23. Ferdinand, O.; Friedrichs, A.; Miranda, M.L.; Voß, D.; Zielinski, O. Next generation fluorescence sensor with multiple excitation and emission wavelengths—NeXOS MatrixFlu-UV. In Proceedings of the OCEANS 2017, Aberdeen, UK, 19–22 June 2017.
24. Fukuzaki, K.; Imai, I.; Fukushima, K.; Ishii, K.-I.; Sawayama, S.; Yoshioka, T. Fluorescent characteristics of dissolved organic matter produced by bloom-forming coastal phytoplankton. *J. Plankton Res.* **2014**, *36*, 685–694. [CrossRef]
25. Lee, E.; Yoo, G.; Jeong, Y.; Kim, K.; Park, J.; Oh, N. Comparison of UV–VIS and FDOM sensors for in situ monitoring of stream DOC concentrations. *Biogeosciences* **2015**, *12*, 3109–3118. [CrossRef]
26. Mustaffa, N.I.H.; Ribas-Ribas, M.; Wurl, O. High-resolution variability of the enrichment of fluorescence dissolved organic carbon in the sea surface microlayer of an upwelling region. *Elem. Sci. Anth.* **2017**, *5*, 52. [CrossRef]
27. Kowalczuk, P.; Tilstone, G.H.; Zabłocka, M.; Röttgers, R.; Thomas, R. Composition of dissolved organic matter along an Atlantic Meridional Transect from fluorescence spectroscopy and Parallel Factor Analysis. *Mar. Chem.* **2013**, *157*, 170–184. [CrossRef]
28. Miranda, M.L.; Trozjuck, A.; Voss, D.; Gassmann, S.; Zielinski, O. Spectroscopic evidence of anthropogenic compounds extraction from polymers by fluorescent dissolved organic matter in natural water. *J. Eur. Opt. Soc. Rapid Publ.* **2016**, *11*, 16014. [CrossRef]
29. Coble, P. Characterization of marine and terrestrial DOM in seawater using excitation-emission matrix spectroscopy. *Mar. Chem.* **1996**, *51*, 325–346. [CrossRef]
30. Zsolnay, A.; Baigar, E.; Jimenez, M.; Steinweg, B.; Saccomandi, F. Differentiating with fluorescence spectroscopy the sources of dissolved organic matter in soils subjected to drying. *Chemosphere* **1999**, *38*, 45–50. [CrossRef]
31. Huguet, A.; Vacher, L.; Relexans, S.; Saubusse, S.; Froidefond, J.M.; Parlanti, E. Properties of fluorescent dissolved organic matter in the Gironde Estuary. *Org. Geochem.* **2009**, *40*, 706–719. [CrossRef]
32. Drozdowska, V.; Kowalczuk, P.; Jozefowicz, M. Spectrofluorometric characteristics of fluorescent dissolved organic matter in a surface microlayer in the Southern Baltic coastal waters. *J. Eur. Opt. Soc. Rapid Publ.* **2015**, *10*, 15050. [CrossRef]
33. Murphy, K.R.; Stedmon, C.A.; Graeber, D.; Bro, R. Fluorescense spectroscopy and multi-way techniques. Parafac. *Anal. Methods* **2013**, *5*, 6557–6566. [CrossRef]
34. Bro, R. PARAFAC. Tutorial and applications. *Chemom. Intell. Lab. Syst.* **1997**, *38*, 149–171. [CrossRef]
35. Delauney, L.; Compere, C.; Lehaitre, M. Biofouling protection for marine environmental sensors. *Ocean Sci.* **2010**, *6*, 503–511. [CrossRef]

36. Coble, P.; Spencer, G.; Baker, A.; Darren, M. (Eds.) *Aquatic Organic Matter Fluorescence*; Cambridge University Press: Cambridge, UK, 2014.
37. Ocean Optics. *PX-2 Pulsed Xenon Lamp Stability*; Engineering Note 110-00000-000-04-0704; Ocean Optics: Largo, FL, USA, 2002.
38. Manov, D.V.; Chang, G.C.; Dickey, T.D. Methods for reducing biofouling of morred optical sensors. *J. Atmos. Ocean. Technol.* **2004**, *21*, 958–968. [CrossRef]
39. Lawaetz, A.J.; Stedmon, C.A. Fluorescence Intensity Calibration Using the Raman Scatter Peak of Water. *Appl. Spectrosc.* **2009**, *63*, 936–940. [CrossRef] [PubMed]
40. Heibati, M.; Stedmon, C.A.; Stenroth, K.; Rauch, S.; Toljander, J.; Säve-Söderbergh, M.; Murphy, K.R. Assessment of drinking water quality at the tap using fluorescence spectroscopy. *Water Res.* **2017**, *125*, 1–10. [CrossRef] [PubMed]
41. Perdue, E.M. Standard and Reference Samples of Humic Acids, Fulvic Acids, and Natural Organic Matter from the Suwannee River, Georgia: Thirty Years of Isolation and Characterization. In *Functions of Natural Organic Matter in Changing Environment*; Springer: Heidelberg, Germany, 2013.
42. Stedmon, C.A.; Markager, S. Resolving the variability in dissolved organic matter fluorescence in a temperate estuary and its catchment using PARAFAC analysis. *Limnol. Oceanogr.* **2005**, *50*, 686–697. [CrossRef]
43. Pearlman, J.; Zielinski, O. A new generation of optical systems for ocean monitoring—Matrix fluorescence for multifunctional ocean sensing. *Sea Technol.* **2017**, *2*, 30–33.
44. Delory, E.; Castro, A.; Waldmann, C.; Rolin, J.F.; Woerther, P.; Gille, J.; Del Rio, J.; Zielinski, O.; Golmen, L.; Hareide, N.R.; et al. NeXOS development plans in ocean optics, acoustics and observing systems interoperability. In Proceedings of the 2014 IEEE Sensor Systems for a Changing Ocean (SSCO), Brest, France, 13–17 October 2014.
45. Coble, P. Colored dissolved organic matter in seawater. In *Subsea Optics and Imaging*; Watson, J., Zielinski, O., Eds.; Woodhead: Cambridge, UK, 2013; pp. 98–118.
46. Stedmon, C.A.; Markager, S.; Bro, R. Tracing dissolved organic matter in aquatic environments using a new approach to fluorescence spectroscopy. *Mar. Chem.* **2003**, *82*, 239–254. [CrossRef]
47. Turner Designs. *Optical Specification Guide for Cyclops Submersible Sensors*; Engineering Note 998-2181 Rev. U; Turner Designs: San Jose, CA, USA, 2018.
48. Jacobson, P. Physical Oceanography of the Trondheimsfjord. *Geophys. Astrophys. Fluid Dyn.* **1983**, *26*, 3–26. [CrossRef]
49. Mascarenhas, V.J.; Voß, D.; Wollschlaeger, J.; Zielinski, O. Fjord light regime: Bio-optical variability, absorption budget, and hyperspectral light availability in Sognefjord and Trondheimsfjord, Norway. *J. Geophys. Res. Oceans* **2017**, *122*, 3828–3847. [CrossRef]
50. Carstea, E. Fluorescence Spectroscopy as a Potential Tool for In-Situ Monitoring of Dissolved Organic Matter in Surface Water Systems. In *Water Pollution*; InTech: London, UK, 2012; pp. 47–68.
51. Goldberg, M.; Wiener, E. Fluorescence Measurements of the Volume, Shape, and Fluorophore Composition of Fulvic Acid from the Suwannee River. In *Humic Substances in the Suwannee River, Georgia; Interactions, Properties, and Proposed Structures*; Averett, R.C., Leenheer, A., McKnight, M., Thorn, A., Eds.; United States Geological Survey Water-Supply Paper 2373; USGS: Reston, VA, USA, 1994; pp. 100–113.
52. Gilmore, A.M.; Cohen, S.M. Analysis of chromophoric dissolved organic matter in water by EEMs with HORIBA-Jobin Yvon fluorescence instrument called "Aqualog". *Readout* **2013**, *41*, 19–24.
53. Wollschläger, J.; Voß, D.; Zielinski, O.; Petersen, W. In situ observations of biological and environmental parameters by means of optics-development of next-generation ocean sensors with special focus on an integrating cavity approach. *IEEE J. Ocean. Eng.* **2016**, *41*, 753–762. [CrossRef]
54. Luciani, X.; Redon, R.; Mounier, S. How to correct inner filter effects altering 3D fluorescence spectra by using a mirrored cell. *Chemom. Intell. Lab. Syst.* **2013**, *126*, 91–99. [CrossRef]
55. Cyr, F.; Tedetti, M.; Besson, F.; Beguery, L.; Doglioli, A.M.; Petrenko, A.A.; Goutx, M. A New Glider-Compatible Optical Sensor for Dissolved Organic Matter Measurements: Test Case from the NW Mediterranean Sea. *Front. Mar. Sci.* **2017**, *4*, 89. [CrossRef]
56. Mayerfeld, P. Fluorometers: Integration experiences with unmanned vehicles. In Proceedings of the OCEANS 2017, Anchorage, AK, USA, 18–21 September 2017.

57. Makarewicz, A.; Kowalczuk, P.; Sagan, S.; Granskog, M.A.; Pavlov, A.K.; Zdun, A.; Borzycka, K.; Zabłocka, M. Characteristics of chromophoric and fluorescent dissolved organic matter in the Nordic Seas. *Ocean Sci.* **2018**, *14*, 543–562. [CrossRef]

58. Ng, C.L.; Senft-Grupp, S.; Hemond, H.F. A multi-platform optical sensor for in situ sensing of water chemistry. *Limnol. Oceanogr. Methods* **2012**, *10*, 978–990. [CrossRef]

59. Herfort, L.; Seaton, C.; Wilkin, M.; Roman, B.; Preston, C.M.; Marin, R., III; Seitz, K.; Smith, M.W.; Haynes, V.; Scholin, C.A.; et al. Use of continuous, real-time observations and model simulations to achieve autonomous, adaptive sampling of microbial processes with a robotic sampler. *Limnol. Oceanogr. Methods* **2016**, *14*, 50–67. [CrossRef]

60. Olson, R.; Shalapyonok, A.; Kalb, D.; Graves, S.; Sosik, H. Imaging FlowCytobot modified for high throughput by in-line acoustic focusing of sample particles. *Limnol. Oceanogr. Methods* **2017**, *15*, 867–874. [CrossRef]

![applied sciences logo] *applied sciences*

MDPI

Article

A Brief Review of Mueller Matrix Calculations Associated with Oceanic Particles

Bingqiang Sun [1], George W. Kattawar [2], Ping Yang [1,*] and Xiaodong Zhang [3]

[1] Department of Atmospheric Sciences, Texas A&M University, College Station, TX 77843, USA; sbq1418@gmail.com
[2] Department of Physics and Astronomy and Institute for Quantum Science and Engineering, Texas A&M University, College Station, TX 77843, USA; kattawar@tamu.edu
[3] Department of Earth System Science and Policy, University of North Dakota, Grand Forks, ND 58202, USA; xiaodong.zhang2@und.edu
* Correspondence: pyang@tamu.edu; Tel.: +1-979-845-4923

Received: 19 July 2018; Accepted: 19 August 2018; Published: 19 December 2018

Featured Application: This paper provides guidance for selecting an appropriate method for calculating the Mueller matrix associated with oceanic particles of arbitrary morphologies and refractive indices.

Abstract: The complete Stokes vector contains much more information than the radiance of light for the remote sensing of the ocean. Unlike the conventional radiance-only radiative transfer simulations, a full Mueller matrix-Stokes vector treatment provides a rigorous and correct approach for solving the transfer of radiation in a scattering medium, such as the atmosphere-ocean system. In fact, radiative transfer simulation without considering the polarization state always gives incorrect results and the extent of the errors induced depends on a particular application being considered. However, the rigorous approach that fully takes the polarization state into account requires the knowledge of the complete single-scattering properties of oceanic particles with various sizes, morphologies, and refractive indices. For most oceanic particles, the comparisons between simulations and observations have demonstrated that the "equivalent-spherical" approximation is inadequate. We will therefore briefly summarize the advantages and disadvantages of a number of light scattering methods for non-spherical particles. Furthermore, examples for canonical cases with specifically oriented particles and randomly oriented particles will be illustrated.

Keywords: ocean optics; light scattering; Mueller matrix; volume and surface integral methods

---

1. Introduction

It is well known that the scattering of light by a particle is determined by the detailed characteristics of the scattering particle, particularly its size, chemical composition (thus, the index of refraction), the overall shape, and detailed surface texture (e.g., surface roughness). Oceanic particles vary greatly in size and morphology. While the Lorenz-Mie theory has been used frequently to simulate the optical properties of oceanic particles (e.g., [1–5]), these particles are predominately nonspherical. Significant differences exist in the optical properties simulated by using "equivalent" spheres and non-spherical shapes, such as spheroids (e.g., [6]). In addition, even the simplest biological cell has a membrane and plasma contained within the membrane. Previous studies have shown that accounting for the cell structure can better simulate the optical properties of various phytoplankton species, particularly the scattering at large scattering angles [5,7–11]. Advanced light scattering methods have been developed to deal with complex shape and structure. Here, we briefly summarize light-scattering computational methods for oceanic particles. Light scattering in an absorbing medium has been extensively discussed [12–14]. For generality, however, only a nonabsorbing medium is discussed here. Beginning with Maxwell's equations, in Section 2, we will show exact volume-/surface-integral equations for mapping the near field to the far field. Furthermore, we introduce both the amplitude scattering matrix and the scattering phase matrix. In Section 3, several scattering methods will be introduced. In Section 4, discussions are given that are based on oriented particles and particles in random orientation.

2. Fundamental Concepts for Mueller Matrix Calculations

2.1. Maxwell's Equations and the Volume/Surface-Integral Equations

Since all of the rigorous light-scattering computational methods should obey Maxwell's equations, we will first give a brief introduction to the role that both the volume/surface methods for mapping the near field to far field play in the final solutions. We will only consider time-harmonic electromagnetic waves and dielectric particles. The dielectric particles are assumed to be isotropic and have a linear response to an applied field. In this case, Maxwell's equations in the medium while using SI units are as follows:

$$\nabla \cdot \dot{\mathbf{E}}\,(\vec{r}) = 0, \ \nabla \times \mathbf{E}(\vec{r}) = -\mu \frac{\partial \mathbf{H}(\vec{r})}{\partial t}, \tag{1}$$

$$\nabla \cdot \dot{\mathbf{H}}\,(\vec{r}) = 0, \ \nabla \times \mathbf{H}(\vec{r}) = \varepsilon \frac{\partial \mathbf{E}(\vec{r})}{\partial t}, \tag{2}$$

where **E** and **H** are the electric and the magnetic fields, respectively; ε and μ are the permittivity and permeability of the medium. Using the Fourier transformation, an arbitrary incident field in the time-domain can be transformed into the summation of the fields in the frequency-domain, or the time-harmonic fields. Assuming that the time-harmonic field follows $\exp(-i\omega t)$, where ω is the angular frequency of the electromagnetic wave, Maxwell's equations in a time-independent form become:

$$\nabla \cdot \dot{\mathbf{E}}\,(\vec{r}) = 0, \ \nabla \times \mathbf{E}(\vec{r}) = i\omega\mu\mathbf{H}(\vec{r}), \tag{3}$$

$$\nabla \cdot \dot{\mathbf{H}}\,(\vec{r}) = 0, \ \nabla \times \mathbf{H}(\vec{r}) = -i\omega\varepsilon\mathbf{E}(\vec{r}). \tag{4}$$

Using Equations (3) and (4), the vector Helmholtz equations for the electric and magnetic fields are

$$\left(\nabla^2 + k^2\right)\mathbf{E}(\vec{r}) = 0, \tag{5}$$

$$\left(\nabla^2 + k^2\right)\mathbf{H}(\vec{r}) = 0, \tag{6}$$

where k is the wave number and $k^2 = \omega^2 \mu \varepsilon$. For oceanic particles, the surrounding medium and the scattering particles are assumed to be nonmagnetic, thus $\mu = \mu_0$, where μ_0 is the vacuum permeability. The light speed c in vacuum is equal to $1/\sqrt{\mu_0 \varepsilon_0}$, where ε_0 is the permittivity in vacuum. Consequently, the refractive index m of the medium is $m = c/v = \sqrt{\varepsilon/\varepsilon_0}$, where v is the light speed in the medium. Since the electric and the magnetic fields are dependent on each other, we will use the electric field to describe the electromagnetic field.

The volume integral and surface integral equations of the electric field can be deduced from Maxwell's equations and the vector Green function [15]. In the far-field regime, they can be expressed in the form

$$\mathbf{E}^{sca}\,(\vec{r})\Big|_{r\to\infty} = \frac{\exp(ikr)}{-ikr}\frac{ik^3}{4\pi}\int_V d^3\vec{r}'\left\{(m^2-1)\left[\hat{r}\times\hat{r}\times\mathbf{E}(\vec{r}')\exp(-ik\hat{r}\cdot\vec{r}')\right]\right\}, \tag{7}$$

$$\mathbf{E}^{\text{sca}}\left(\vec{r}\right)\Big|_{r\to\infty} = \frac{\exp(ikr)}{-ikr}\frac{k^2}{4\pi}\hat{r}\times\oint_S d^2\vec{r}'\left\{\left[\hat{n}_s\times\mathbf{E}(\vec{r}')\right] - \frac{\omega\mu_0}{k}\hat{r}\times\left[\hat{n}_s\times\mathbf{H}(\vec{r}')\right]\right\}, \tag{8}$$

where the parameters are given in Figure 1; $\hat{n}_s$ is the outward normal to the surface. It is evident that the scattered far field only depends on the scattered directions with an outgoing spherical wave factor $\exp(ikr)/kr$.

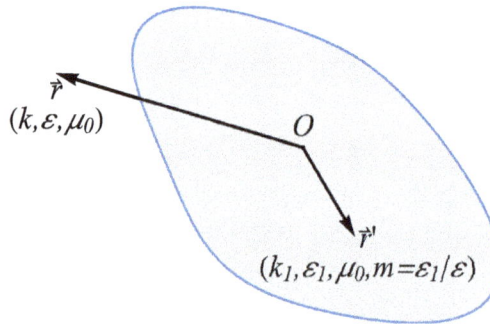

Figure 1. Parameters used for light scattering by a dielectric particle. The field point $\vec{r}$ is outside the scattering particle with wavenumber k, permittivity ε, and permeability μ_0 and the point $\vec{r}'$ is inside the particle with wave number k_1, permittivity ε_1, and permeability μ_0.

2.2. Amplitude Scattering Matrix and Mueller Matrix

Let the incident direction of the incoming wave be along the z-axis of the laboratory frame of reference. The incident direction and the scattered direction define a scattering plane and the incident and scattered fields can be expanded into parallel and perpendicular components with respect to the scattering plane. Consequently, the amplitude scattering matrix $\mathbf{S}$ can be given by [16]

$$\begin{pmatrix} E_\parallel^{\text{sca}} \\ E_\perp^{\text{sca}} \end{pmatrix} = \frac{\exp(ikr - ikz)}{-ikr}\mathbf{S}\begin{pmatrix} E_\parallel^{\text{inc}} \\ E_\perp^{\text{inc}} \end{pmatrix}, \tag{9}$$

where $E_\parallel$ and $E_\perp$ denote the parallel and perpendicular components of the electric field with respect to the scattering plane and $\mathbf{S}$ is a 2×2 complex matrix. The Stokes parameters in a non-absorbing medium are defined based on the measurable quantities, which are normally expressed in terms of a four-element column vector, the Stokes vector $\mathbf{I}$, as follows:

$$\mathbf{I} = \begin{pmatrix} I \\ Q \\ U \\ V \end{pmatrix} = \begin{pmatrix} E_\parallel E_\parallel^* + E_\perp E_\perp^* \\ E_\parallel E_\parallel^* - E_\perp E_\perp^* \\ E_\parallel E_\perp^* + E_\perp E_\parallel^* \\ i\left(E_\parallel E_\perp^* - E_\perp E_\parallel^*\right) \end{pmatrix}. \tag{10}$$

In the above equation, i is the imaginary unit and a constant factor, $\sqrt{\varepsilon/\mu_0}/2$, is neglected since usually relative quantities are measured. The Mueller matrix (also called the scattering phase matrix in the literature) is the transformation matrix from the incident to the scattered Stokes parameters, as follows:

$$\mathbf{I}^{\text{sca}} = \frac{1}{(kr)^2}\mathbf{P}\mathbf{I}^{\text{inc}}, \tag{11}$$

where the 4×4 Mueller matrix $\mathbf{P}$ can be given as quadratic expressions of the amplitude scattering matrix $\mathbf{S}$, as follows [17,18]:

$$\mathbf{P} = \mathbf{A}(\mathbf{S} \otimes \mathbf{S}^*)\mathbf{A}^{-1}, \tag{12}$$

where asterisk denotes the complex conjugate and symbol $\otimes$ denotes the tensor product, and the constant matrix $\mathbf{A}$ is

$$\mathbf{A} = \begin{pmatrix} 1 & 0 & 0 & 1 \\ 1 & 0 & 0 & -1 \\ 0 & 1 & 1 & 0 \\ 0 & i & -i & 0 \end{pmatrix}, \quad \mathbf{A}^{-1} = \frac{1}{2}\mathbf{A}^\dagger, \tag{13}$$

in which the symbol † (sometimes called the dagger) is composed of two operations; namely, complex conjugating (the * symbol), and then transposing the original matrix and the order of these operations is unimportant. Note that the Stokes parameters have the units of irradiance [19], and on the other hand, the corresponding radiance is invariant over distance if no scattering or absorption occurs.

If the incident light is unpolarized, the scattering cross-section can be given in terms of the element $\mathbf{P}_{11}$ by

$$C_{\text{sca}} = \frac{1}{k^2} \int_{4\pi} d\Omega \mathbf{P}_{11}(\theta, \varphi). \tag{14}$$

The phase function is defined as:

$$p = \frac{4\pi}{k^2 C_{\text{sca}}} \mathbf{P}_{11}, \tag{15}$$

and the scattering phase matrix can be defined as:

$$\mathbf{F} = \frac{4\pi}{k^2 C_{\text{sca}}} \mathbf{P}. \tag{16}$$

The symmetry relations of the phase matrix have been extensively discussed in general and also for forward and backward scattering [20–22]. For an arbitrary particle without mirror symmetry in the scattering plane, the scattering phase matrix of a particle in random orientation is in the form

$$\mathbf{F} = \begin{pmatrix} a_1 & b_1 & b_3 & b_5 \\ b_1 & a_2 & b_4 & b_6 \\ -b_3 & -b_4 & a_3 & b_2 \\ b_5 & b_6 & -b_2 & a_4 \end{pmatrix}, \tag{17}$$

where there are only 10 independent parameters. For a particle with mirror symmetry in the scattering plane, the scattering phase matrix of a particle in random orientation is reduced to a block-diagonal matrix, as follows:

$$\mathbf{F} = \begin{pmatrix} a_1 & b_1 & 0 & 0 \\ b_1 & a_2 & 0 & 0 \\ 0 & 0 & a_3 & b_2 \\ 0 & 0 & -b_2 & a_4 \end{pmatrix}, \tag{18}$$

where there are only six independent parameters. Equations (17) and (18) represent the scattering phase matrix of a particle in random orientation. For a collection of particles with a size distribution, the collective scattering phase matrix can also be defined in terms of the distribution and the reader is referred to the book by Mobley [23].

All of the scattering quantities have been presented and we are now faced with the problem of obtaining the near fields or directly the far fields satisfying Maxwell's equations.

3. General Scattering Method for Suspended Particles

The governing principle for light scattering by particles is Maxwell's equations. The scattering solution is called Rayleigh scattering if x « 1 and |mx| « 1 [16,20], where the size parameter x is defined as $2\pi r_v/\lambda$ with r_v being the radius of a sphere or volume-equivalent sphere and λ the incident wavelength in the surrounding medium. The analytical solutions to Maxwell's equations are only effectively available for spheres [16,20]. For a prolate or oblate spheroid, the analytical solution is given in a series of the spheroidal wave functions by Asano and Yamamoto [24] and Asano and Sato [25]. However, the analytical solutions for a spheroid are only computationally effective for small particles due to numerical instability in computing the spheroidal wave functions for large particles. For an infinite circular cylinder, the analytical solution can be easily computed [16]. However, the infinite morphology does not exist in nature. For a particle with spherical symmetry, such as a homogeneous sphere or a multi-layered sphere, the analytical solution can be obtained by using the Lorenz-Mie theory for any size [16,20]. The advantage of using the spherical model is the computational efficiency, while the disadvantage is the appearance of spherical artifacts, such as the rainbow or glory, which have seldom been observed for ocean water (e.g., [26]). For a non-spherical particle, the solution of Maxwell's equations consists of two categories: rigorous and approximate solutions. The rigorous solutions can be further divided into numerically exact solutions and semi-analytical T-matrix solutions.

3.1. Numerically Exact Methods

As the name implies, numerically exact solutions use numerical methods to directly solve Maxwell's equations or the volume or surface integral equations derived from Maxwell's equations. The computational precision depends on the numerical resolution.

The finite difference time-domain method (FDTD) is based on the discretization of Maxwell's equation Equations (1) and (2) both in time and space [27]. The FDTD method uses the Yee grid to discretize the space, which was developed by Yee [28] and reviewed by Taflove [27], and Yang and Liou [29,30]. Since the computational space has to be confined to a finite region, a perfectly matched layer is used to absorb all of the electromagnetic waves in the computational boundary and avoid any artificially reflected electromagnetic waves back into the computational region [31]. The computational region usually with cuboid shape has to encompass the scattering particle so the computational region is always larger than the scattering particle in the FDTD application. The electromagnetic fields on the grids are updated with the advance of time so the FDTD is an initial value problem.

For a time-harmonic field or a field in the frequency domain, Maxwell's equations become the vector Helmholtz equations that are given in Equations (5) and (6). The vector Helmholtz Equations (5) and (6) can be discretized in space while using the finite-element method (FEM) [32]. The boundary condition on the particle surface and the continuity condition on the neighboring grid give a series of linear equations. The FEM is a boundary value problem. A major challenge for applying the FEM to light scattering is choosing the finite region covering the scattering particle so that the field in the computational region satisfies the radiation condition in the far field [33]. Like the FDTD method, the computational region for the FEM is also larger than the region that is occupied by the scattering particle, which can constrain the application regime of the FEM.

Using the vector Green function, the differential equations become the volume-integral or the surface-integral equations given by Equations (7) and (8). Even though the volume-integral and the surface-integral methods are equivalent, the volume-integral method is numerically more stable than the surface-integral method because the volume-integral equation in Equation (7) is a Fredholm integral equation of the second kind whose matrix equation is usually diagonally dominant [34].

The discrete-dipole approximation (DDA) method is a typical volume-integral method. The DDA was first proposed by Purcell and Pennypacker [35] and it was reviewed by Draine [36,37] and by Yurkin and Hoekstra [38]. In the DDA method, the particle volume is discretized into usually cubic cells, as shown in Figure 2. Each cubic cell is represented while using an electric dipole and the excited field at that cell is composed of the original incident field and the field from all other cells but excluding the cell itself. The dipoles generate a series of linear equations and the fields with respect to all dipoles can be obtained by solving the corresponding linear equations. Once the total field with respect to each cell is obtained, the amplitude scattering matrix and Mueller matrix can straightforwardly be computed. It is evident from Equation (7) that the computational region is equal to the volume of a scattering particle. That makes the DDA method computationally efficient when compared to other numerically exact methods.

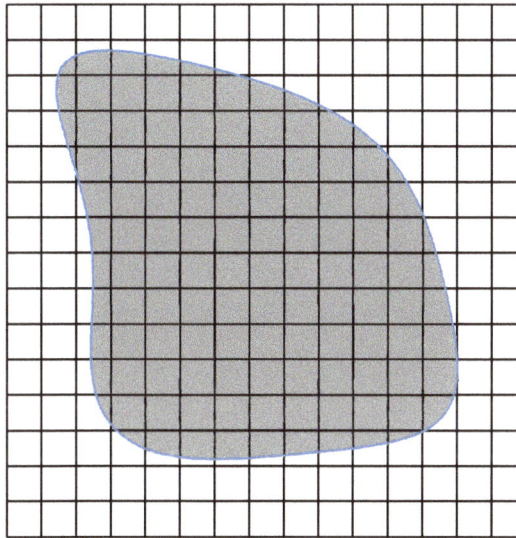

Figure 2. Discretization of a particle volume in the discrete-dipole approximation (DDA) method.

Two numerical implementations of the DDA are a FORTRAN implementation referred to as DDSCAT by Draine and Flatau [39] and a C implementation referred to as ADDA by Yurkin and Hoekstra [40]. The DDSCAT is parallelized for scattering only in different orientations so the memory requirement might restrict the computational capability for large particles. The ADDA is parallelized by distributing grids (dipoles) into different CPUs so the ADDA can handle particles of large sizes. The DDA method has been extensively used to simulate light scattering of oceanic particles. For instance, the light scattering of *Emiliania huxleyi* coccolithophore was simulated while using the DDSCAT by Gordon et al. [41] and using the ADDA by Zhai et al. [42]. Another example of the use of ADDA is the light scattering of dinoflagellates by Liu and Kattawar [43], where the chiral structure of the chromosomes is implemented by using discrete dipoles. This chiral structure leads to optical activity for certain dinoflagellates and another reason for measuring the complete single scattering Mueller matrix, which should be a fruitful area of research in remote sensing of the oceans.

The typical feature of a numerically exact method is that the error asymptotically approaches zero if the corresponding numerical grid that is associated with the method asymptotically reaches zero. Another feature for oceanic particles is that the convergence rate is much faster than the convergence rate of atmospheric particles because the relative refractive indices with respect to oceanic particles are close to unity. Moreover, the composition of a particle using the numerically exact methods can be arbitrary, homogeneous or inhomogeneous, or even different grid by grid.

For all of the numerically exact methods, the Mueller matrix is given in terms of the amplitude scattering matrix and they both depend on the incident direction. Consequently, the light scattering computation of a particle in random orientation while using these methods can usually be given by numerically summing the light scattering for different orientations. The convergence in the random orientation computation becomes significantly more difficult with increasing particle size so the computation will become time-consuming.

3.2. Semi-Analytical T-Matrix Method

The T-matrix method was originally proposed by Waterman [44,45]. The incident and scattered fields are expanded in a series of the vector spherical wave functions. The T-matrix connects the incident and scattered expansion coefficients because of the linearity of Maxwell's equations. The T-matrix of a particle only depends on the intrinsic properties of the particle, such as the refractive index, morphology, and the orientation of the particle frame of reference and its origin location, but not on the incident state. Corresponding to Equations (7) and (8), the T-matrix can be obtained using the surface-integral and volume-integral methods. The computational method of the T-matrix based on the surface integral is called the extended boundary condition method (EBCM) or the null field method, and it was reviewed by Tsang et al. [46], Mishchenko et al. [47,48], Mishchenko and Travis [49], and Doicu et al. [50]. The T-matrix method based on the volume integral equation is called the invariant-imbedding T-matrix method (IITM) and it was originally proposed by Johnson [51]. The IITM was reviewed and developed by Bi et al. [52]. For a particle with axial symmetry, the T-matrix is decoupled into a block-diagonal form, so the computation is significantly simplified. The applications of the EBCM on spheroids, cylinders, and Chebyshev shapes are exceptionally effective [49]. However, when a particle has a large size or an aspect ratio far from unity or one that is asymmetric, the matrices in the T-matrix computation are often ill-conditioned. The T-matrix method that is based on the volume integral is much more stable than the EBCM because the volume integral equation in Equation (7) is a Fredholm integral equation of the second kind, which is often less ill-conditioned [34]. The extreme stability of the IITM has been validated by applying the IITM to particles with large sizes, extreme aspect ratios, or asymmetric particles [52,53]. For instance, the IITM were used to compute the light scattering of oceanic particles, such as *Emiliania huxleyi* coccoliths and coccolithophores by Bi and Yang [54], and diatoms by Sun et al. [11].

When compared to the numerically exact methods, the significant advantage of the T-matrix solution is the analytical realization for a particle with a random orientation. The computational time of T-matrix methods is usually shorter than the numerically exact methods because T-matrix methods use matrix inversion instead of iterations. Moreover, in contrast to the relatively large refractive indices of atmospheric particles, such as ice crystals (m ~1.33) and aerosols (roughly m > 1.5), the relative refractive indices of oceanic particles are usually smaller than 1.2. For this reason, using the T-matrix method for oceanic particles normally yields faster convergence and higher computational efficiency than for atmospheric particles. However, the computational time and memory requirements of the T-matrix are strongly related to the radius of the circumscribed sphere of a particle and its morphology. For instance, for a needle particle with small volume but large circumscribed radius or a complex morphology, such as a porous particle, T-matrix methods are not as efficient as the numerically exact methods.

The numerically exact solutions can provide results for particle size parameters x ~100 or less; for the T-matrix solutions EBCM can reach x ~180, and the IITM can yield accurate results for x ~300. For even larger oceanic particles, approximate solutions must be used.

3.3. Physical-Geometric Optics Method

When the particle size is much larger than the incident wavelength, Maxwell's equations can be approximated by the eikonal equation [55]. The eikonal equation is the theoretical foundation of the geometric optics method. The key process for the geometric optics method is ray-tracing. The ray-tracing process of the geometric optics method consists of two parts: one is the diffracted rays and another is the transmitted rays, including external reflection, refraction without internal reflection, refraction with one internal reflection, and so on. The conventional geometric optics method (CGOM) considers the diffracted and the transmitted rays separately and it assumes equal contributions from the diffracted and transmitted rays under the assumption that the extinction efficiency is 2. Moreover, the CGOM does not consider the ray spreading effect from the near field to the far field, that is, there is no mapping process for the CGOM. The CGOM is applied to compute the light scattering of large particles with a large refractive index, such as ice particles (e.g., [56,57]). The CGOM can be improved by considering the ray spreading effect for a particle in random orientation. The improved geometric optics method (IGOM) can be used to compute light scattering of an intermediate particle or even a small particle [58,59].

For oceanic particles, the diffracted and transmitted rays have strong destructive interference so they cannot be separately handled. The physical-geometric optics method (PGOM) considers not only the interference between the diffracted and transmitted rays, but also the ray spreading effect in the far field [60]. Equations (7) and (8) are fundamental to the PGOM, which substantially extend the applicability of the principles of geometric optics in conjunction with physical optics to from large to moderate particles. For faceted particles, the ray-tracing process can be analytically accomplished since the phase change on a facet is linear [61–63]. The PGOM can be effectively used to compute the light scattering properties of oceanic particles.

4. Computational Results and Discussion

4.1. Dinoflagellate Simulation Using ADDA

Phytoplankton are one of two main categories of oceanic organisms and a significant component of the marine ecosystem that travel along the ocean currents. Many phytoplankton are positioned with preferred orientations due to the ocean flow [64]. Most phytoplankton are single-celled, such as dinoflagellates, diatoms, and coccolithophores. The bloomed phytoplankton can cause huge economic losses and influence environmental health, such as the red tide bloom of dinoflagellates in Florida [65]. Optical properties of an individual or bulked phytoplankton are essential to study phytoplankton populations (e.g., [66]). As mentioned in Section 3, dinoflagellates, diatoms, and coccolithophores have been simulated using the DDA and IITM [11,41–43,54]. Dinoflagellates have a large group of species so we take them as an example to describe the application of a scattering method.

Laboratory observation using transmission electron microscopy showed that the nucleus of dinoflagellates contains cylindrical chromosomes [67–69] and the chromosomes are arranged by ordered helical structures [69,70]. The helical structures are responsible for the strong circularly polarized effect that was observed in dinoflagellates [43,71,72]. The Mueller matrix element P_{14} reflects the circular polarization of a scattering particle and can be used as an index to indicate the strong circularly polarized effect [16,20]. Liu and Kattawar employed the ADDA code to fully simulate a single cell of a dinoflagellate and compute the 16 Mueller matrix elements [43], where the chromosomes are constructed using the plywood model [73]. For computational efficiency, only the nucleus with dozens of randomly positioned chromosomes is simulated. A chromosome is modeled as a cylindrical capsule with many layers, where every layer with fixed diameter contains parallel fibrils and the helical structure is described by making two adjacent layers with a constant rotation angle between them. The height with one period of rotation for the parallel fibrils is called the pitch. The chromosome simulation in the DDA method was performed by constructing fibrils in terms of dipoles and these fibrils were then arranged in layers and each layer was twisted a certain amount to make a helical shaped capsule to represent the chromosome. The diameter, the constant rotation angle, the number of helical periodicities, the pitch, the incident wavelength, and the incident directions can be changed to examine the circularly polarized effect of the helical structure. The important conclusions while using the ADDA code are given by Liu and Kattawar, as follows [43]:

- Strong back scattering signals from Mueller matrix element S_{14} are indeed from the helical structures of the chromosomes.

- Strong S_{14} back scattering signals are observed when the incident wavelength in the ocean is matched with the pitch of the helical structure, even if the chromosomes are under the random orientation condition.
- Strong S_{14} back scattering signals are observed when the incident direction is close to the main axis of the helical structure.
- The helical structure with constant rotation angle has stronger S_{14} back scattering signals than the helical structure with random rotation angle.

These conclusions suggest potential applications on the detection of the dinoflagellate and also the appropriate incident wavelength to match the pitch of helical structure.

4.2. Oceanic Particle Simulation Using ADDA, IITM, and PGOM

Section 4.1 describes an example of dinoflagellates while using the ADDA to compute the 16 Mueller matrix elements that were given by Liu and Kattawar [43], where the chromosome of the dinoflagellate has complex helical structure and is simulated mostly in fixed orientations. Generally, a simple nonspherical shape in random orientation is used to simulate the optical properties of oceanic particles. A hexahedron particle here is used as an example of an oceanic particle to show how Mueller matrix elements can be calculated by three typical methods: the ADDA, the IITM, and the PGOM. The relative refractive index of the particle is set to be 1.12 + i0.0005 and the incident wavelength is 0.658 μm. Only the Mueller matrix of the particle under the random orientation condition is given.

Figure 3 shows the comparisons of the non-zero Mueller matrix elements calculated by the IITM and the ADDA. The volume equivalent radius is 1 μm. The element P_{11} is normalized to give the normalized phase function while other elements are normalized by the element P_{11}. The simulation results calculated by the IITM and the ADDA are perfectly matched since they both are the exact solutions of Maxwell's equations. However, the computation using the IITM is much more efficient than the computation using the ADDA since the random orientation process is realized by ADDA through considering a large number of orientations. On the other hand, the ADDA for a fixed orientation in this case is more efficient than the IITM since the IITM has to compute the T-matrix of the particle, regardless of whether it is in a fixed orientation or under the random orientation condition.

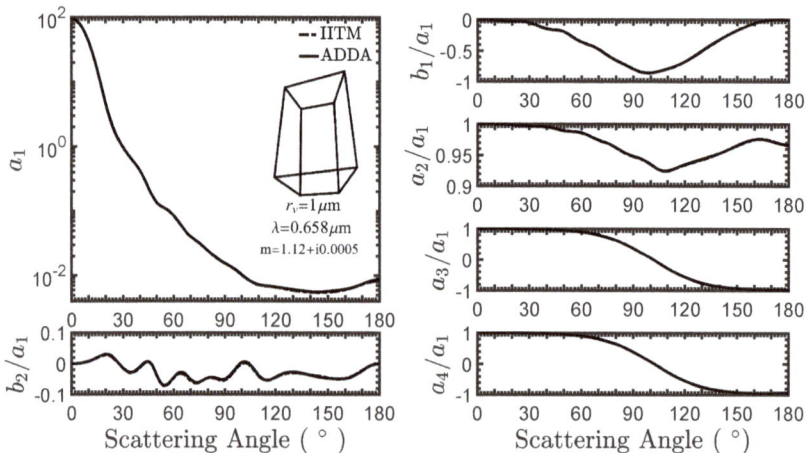

Figure 3. Comparisons of Mueller matrix elements of a hexahedron particle calculated by the invariant-imbedding T-matrix method (IITM) and the ADDA. The volume equivalent sphere radius is 1μm and the incident wavelength is 0.658 μm. The relative refractive index is 1.12 + i0.0005.

Figure 4 shows the comparisons of the Mueller matrix elements calculated by the IITM and the PGOM. The volume equivalent sphere radius is 8 μm. The PGOM results agree quite well with the IITM results, especially for the forward and backward scattering directions. Even though the PGOM is an approximate solution of Maxwell's equation, the process of including the interference between the diffracted and transmitted rays and mapping the near field to the far field significantly enhances its accuracy. The advantage of the PGOM is that it is computationally much more efficient than the IITM.

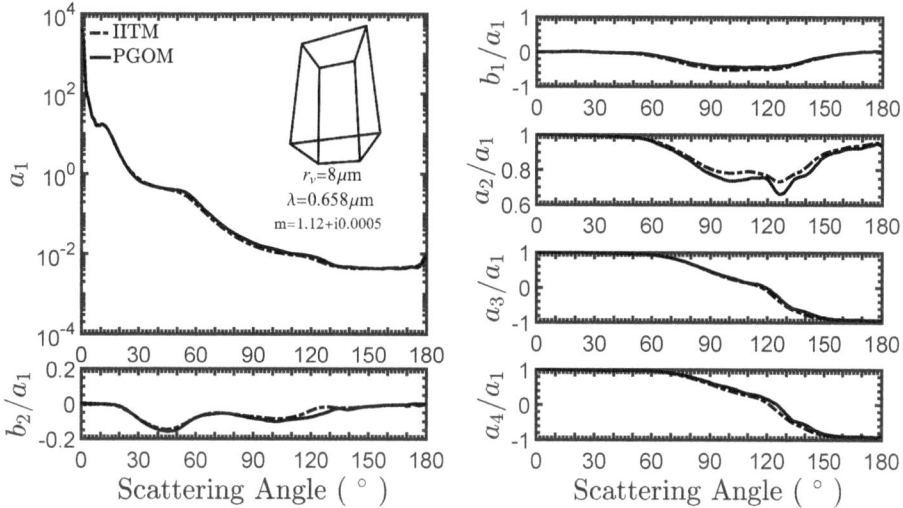

Figure 4. Comparisons of Mueller matrix elements of a hexahedron particle calculated by the IITM and the physical-geometric optics method (PGOM). The volume equivalent sphere radius is 8 μm and the incident wavelength is 0.658 μm. The relative refractive index is 1.12 + i0.0005.

5. Conclusions

A general introduction for calculating the Mueller matrix of suspended particles in the ocean is given. The surface and volume integral equations of the electromagnetic field can be given from Maxwell's equations. Also, the amplitude scattering matrix and the Mueller matrix with respect to light scattering can be defined to describe the polarization state of a suspended particle. To calculate the amplitude scattering matrix and Mueller matrix, the scattering methods are introduced based on the following categories: numerically exact methods, semi-analytical T-matrix methods, and geometric optics methods. For clarity, three typical methods: the DDA method, the IITM, and the PGOM, are briefly presented. Moreover, the Mueller matrix of an arbitrarily generated hexahedron particle under the random orientation condition is computed while using the ADDA and the IITM when the volume equivalent sphere radius is 1 μm and using the PGOM and the IITM for a volume equivalent sphere with a radius of 8 μm, while the incident wavelength is 0.658 μm. Perfect agreement between the ADDA and the IITM are given since both methods are considered to be the exact solutions. The IITM is more computationally efficient than the ADDA when the particle is under the random orientation condition. Excellent agreement between the PGOM and the IITM are obtained especially for the forward and the backward scattering directions. The PGOM is more computationally efficient than the IITM because of the ray-tracing process. Consequently, the Mueller matrix of suspended particles can be computed by using numerically exact methods, T-matrix methods, and the physical-geometric optics method to cover a complete size range.

Author Contributions: Conceptualization, G.W.K.; Methodology, B.S., P.Y., X.Z. and G.W.K.; Software, B.S.; Validation, B.S., P.Y. and X.Z.; Writing-Original Draft Preparation, B.S., P.Y. and X.Z.; Writing-Review & Editing, B.S., P.Y., X.Z. and G.W.K.; Visualization, B.S.; Supervision, P.Y. and G.W.K.; Project Administration, P.Y.; Funding Acquisition, P.Y. and X.Z.

Funding: This research was partly supported by the National Science Foundation (OCE-1459180) and the endowment funds related to the David Bullock Harris Chair in Geosciences at the College of Geosciences, Texas A&M University.

Acknowledgments: This research were conducted with high performance research computing resources provided by Texas A&M University.

Conflicts of Interest: The authors declare no conflict of interest.

References

1. Stramski, D.; Kiefer, D.A. Light scattering by microorganisms in the open ocean. *Prog. Oceanogr.* **1991**, *28*, 343–383. [CrossRef]
2. Morel, A.; Bricaud, A. Theoretical results concerning the optics of phytoplankton, with special references to remote sensing applications. In *Oceanography from Space*; Gower, J.F.R., Ed.; Springer: Berlin, Germany, 1981; pp. 313–327.
3. Morel, A.; Bricaud, A. Inherent optical properties of algal cells, including picoplankton. Theorectical and experimental results. *Can. Bull. Fish. Aquat. Sci.* **1986**, *214*, 521–559.
4. Lerner, A.; Shashar, N.; Haspel, C. Sensitivity study on the effects of hydrosol size and composition on linear polarization in absorbing and nonabsorbing clear and semi-turbid waters. *J. Opt. Soc. Am. A* **2012**, *29*, 2394–2405. [CrossRef] [PubMed]
5. Tzabari, M.; Lerner, A.; Iluz, D.; Haspel, C. Sensitivity study on the effect of the optical and physical properties of coated spherical particles on linear polarization in clear to semi-turbid waters. *Appl. Opt.* **2018**, *57*, 5806–5822. [CrossRef] [PubMed]
6. Clavano, W.R.; Boss, E.; Karp-Boss, L. Inherent Optical Properties of Non-Spherical Marine-Like Particles—From Theory to Observation. In *Oceanography and Marine Biology: An Annual Review*; Gibson, R.N., Atkinson, R.J.A., Gordon, J.D.M., Eds.; Taylor & Francis: Didcot, UK, 2007; pp. 1–38.
7. Quinbyhunt, M.S.; Hunt, A.J.; Lofftus, K.; Shapiro, D.B. Polarized-light scattering studies of marine Chlorella. *Limnol. Oceanogr.* **1989**, *34*, 1587–1600. [CrossRef]
8. Meyer, R.A. Light scattering from biological cells: Dependence of backscattering radiation on membrane thickness and refractive index. *Appl. Opt.* **1979**, *18*, 585–588. [CrossRef] [PubMed]
9. Kitchen, J.C.; Zaneveld, J.R.V. A three-layered sphere model of the optical properties of phytoplankton. *Limnol. Oceanogr.* **1992**, *37*, 1680–1690. [CrossRef]
10. Quirantes, A.; Bernard, S. Light scattering by marine algae: Two-layer spherical and nonspherical models. *J. Quant. Spectrosc. Radiat. Transf.* **2004**, *89*, 311–321. [CrossRef]
11. Sun, B.; Kattawar, W.G.; Yang, P.; Twardowski, S.M.; Sullivan, M.J. Simulation of the scattering properties of a chain-forming triangular prism oceanic diatom. *J. Quant. Spectrosc. Radiat. Transf.* **2016**, *178*, 390–399. [CrossRef]
12. Mundy, W.C.; Roux, J.A.; Smith, A.M. Mie scattering by spheres in an absorbing medium. *J. Opt. Soc. Am.* **1974**, *64*, 1593–1597. [CrossRef]
13. Chylek, P. Light scattering by small particles in an absorbing medium. *J. Opt. Soc. Am.* **1977**, *67*, 561–563. [CrossRef]
14. Mishchenko, M.I.; Yang, P. Far-field Lorez-Mie scattering in an absorbing host medium: Theoretical formalism and FORTRAN program. *J. Quant. Spectrosc. Radiat. Transf.* **2018**, *205*, 241–252. [CrossRef]
15. Morse, P.M.; Feshbach, H. *Methods of Theoretical Physics*; McGraw-Hill: New York, NY, USA, 1953.
16. Bohren, C.F.; Huffman, D.R. *Absorption and Scattering of Light by Small Particles*; John Wiley & Sons: New York, NY, USA, 1983.
17. Parke, N.G., III. Optical Algebra. *J. Math. Phys.* **1949**, *28*, 131–139. [CrossRef]
18. Barakat, R. Bilinear constraints between elements of the 4×4 Mueller-Jones transfer matrix of polarization theory. *Opt. Commun.* **1981**, *38*, 159–161. [CrossRef]

19. Kattawar, G.W.; Yang, P.; You, Y.; Bi, L.; Xie, Y.; Huang, X.; Hioki, S. Polarization of light in the atmosphere and ocean. In *Light Scattering Reviews 10: Light Scattering and Radiative Transfer*; Kokhanovsky, A.A., Ed.; Springer: Berlin, Germany, 2016; pp. 3–39.

20. Van de Hulst, H.C. *Light Scattering by Small Particles*; John Wiley & Sons: New York, NY, USA, 1957.

21. Hu, C.; Kattawar, G.W.; Parkin, M.E.; Herb, P. Symmetry theorems on the forward and backward scattering Mueller matrices for light scattering from a nonspherical dielectric scatterer. *Appl. Opt.* **1987**, *26*, 4159–4173. [CrossRef] [PubMed]

22. Hovenier, J.W.; Mackowski, D.W. Symmetry relations for forward and backward scattering by randomly oriented particles. *J. Quant. Spectrosc. Radiat. Transf.* **1998**, *60*, 483–492. [CrossRef]

23. Mobley, C.D. *Light and Water: Radiative Transfer in Natural Waters*; Academic Press: San Diego, CA, USA, 1994.

24. Asano, S.; Yamamoto, G. Light scattering by a spheroidal particle. *Appl. Opt.* **1975**, *14*, 29–49. [CrossRef] [PubMed]

25. Asano, S.; Sato, M. Light scattering by randomly oriented spheroidal particles. *Appl. Opt.* **1980**, *19*, 962–974. [CrossRef] [PubMed]

26. Voss, K.J.; Fry, E.S. Measurement of the Mueller matrix for ocean water. *Appl. Opt.* **1984**, *23*, 4427–4439. [CrossRef] [PubMed]

27. Taflove, A.; Hagness, S.C. *Computational Electrodynamics: The Finite-Difference Time-Domain Method*; Artech House: Boston, MA, USA, 2000.

28. Yee, S.K. Numerical solution of initial boundary value problems involving Maxwell's equations in isotropic media. *IEEE Trans. Antennas Propag.* **1966**, *14*, 302–307.

29. Yang, P.; Liou, K.N. Finite-difference time domain method for light scattering by small ice crystals in three-dimensional space. *J. Opt. Soc. Am. A* **1996**, *13*, 2072–2085. [CrossRef]

30. Yang, P.; Liou, K.N. Finite difference time domain method for light scattering by nonspherical and inhomogeneous particles. In *Light Scattering by Nonspherical Particles*; Mishchenko, M.I., Hovenier, J.W., Travis, L.D., Eds.; Academic Press: San Diego, CA, USA, 2000; pp. 173–221.

31. Berenger, J.P. A perfectly matched layer for the absorption of electromagnetic waves. *J. Comput. Phys.* **1994**, *114*, 185–200. [CrossRef]

32. Silvester, P.P.; Ferrari, R.L. *Finite Elements for Electrical Engineers*; Cambridge University Press: Cambridge, UK, 1996.

33. Morgan, M.; Mei, K. Finite-element computation of scattering by inhomogeneous penetrable bodies of revolution. *IEEE Trans. Antennas Propag.* **1979**, *27*, 202–214. [CrossRef]

34. Press, W.H.; Flannery, B.P.; Teukolsky, S.A.; Vetterling, W.T. *Numerical Recipes*; Cambridge University Press: Cambridge, UK, 1989.

35. Purcell, E.M.; Pennypacker, C.R. Scattering and absorption of light by nonspherical dielectric grains. *Astrophys. J.* **1973**, *186*, 705–714. [CrossRef]

36. Draine, B.T. The discrete-dipole approximation and its application to interstellar graphite grains. *Astrophys. J.* **1988**, *333*, 848–872. [CrossRef]

37. Draine, B.T. *The Discrete Dipole Approximation for Light Scattering by Irregular Targets*; Academic Press: San Diego, CA, USA, 2000; pp. 131–144.

38. Yurkin, M.A.; Hoekstra, A.G. The discrete-dipole-approximation code ADDA: Capabilities and known limitations. *J. Quant. Spectrosc. Radiat. Transf.* **2011**, *112*, 2234–2247. [CrossRef]

39. Draine, B.T.; Flatau, P.J. User guide for the discrete dipole approximation code DDSCAT 7.3. *arXiv* **2013**, arXiv:1305.6497.

40. Yurkin, M.A.; Hoekstra, A.G. User Manual for the Discrete Dipole Approximation Code ADDA 1.3b4. 2014. Available online: http://a-dda.googlecode.com/svn/tags/rel_1.3b4/doc/manual.pdf (accessed on 6 May 2018).

41. Gordon, R.H.; Smyth, J.T.; Balch, M.W.; Boynton, C.G.; Tarran, A.G. Light scattering by coccoliths detached from *Emiliania huxleyi. Appl. Opt.* **2009**, *48*, 6059–6073. [CrossRef] [PubMed]

42. Zhai, P.W.; Hu, Y.; Trepte, C.R.; Winker, D.M.; Josset, D.B.; Lucker, P.L.; Kattawar, G.W. Inherent optical properties of the coccolithophore: *Emiliania huxleyi. Opt. Express* **2013**, *21*, 17625–17638. [CrossRef] [PubMed]

43. Liu, J.; Kattawar, G.W. Detection of dinoflagellates by the light scattering properties of the chiral structure of their chromosomes. *J. Quant. Spectrosc. Radiat. Transf.* **2013**, *131*, 24–33. [CrossRef]

44. Waterman, P.C. Matrix formulation of electromagnetic scattering. *Proc. IEEE* **1965**, *53*, 805–812. [CrossRef]

45. Waterman, P.C. Symmetry, unitarity, and geometry in electromagnetic scattering. *Phys. Rev. D* **1971**, *3*, 825. [CrossRef]

46. Tsang, L.; Kong, J.A.; Ding, K.H. Scattering of electromagnetic waves. In *Theories and Applications*; Wiley: Hoboken, NJ, USA, 2000; Volume 1.

47. Mishchenko, M.I.; Travis, L.D.; Mackowski, D.W. T-matrix computations of light scattering by nonspherical particles: A review. *J. Quant. Spectrosc. Radiat. Transf.* **1996**, *55*, 535–575. [CrossRef]

48. Mishchenko, M.I.; Travis, L.D.; Lacis, A.A. *Scattering, Absorption, and Emission of Light by Small Particles*; Cambridge University Press: Cambridge, UK, 2002.

49. Mishchenko, M.I.; Travis, L.D. Capabilities and limitations of a current FORTRAN implementation of the T-matrix method for randomly oriented, rotationally symmetric scatterers. *J. Quant. Spectrosc. Radiat. Transf.* **1998**, *60*, 309–324. [CrossRef]

50. Doicu, A.; Wriedt, T.; Eremin, Y.A. *Light Scattering by Systems of Particles: Null-Field Method with Discrete Sources: Theory and Programs*; Springer: Berlin, Germany, 2006; Volume 124.

51. Johnson, B.R. Invariant imbedding T matrix approach to electromagnetic scattering. *Appl. Opt.* **1988**, *27*, 4861–4873. [CrossRef] [PubMed]

52. Bi, L.; Yang, P.; Kattawar, G.W.; Mishchenko, M.I. Efficient implementation of the invariant imbedding T-matrix method and the separation of variables method applied to large nonspherical inhomogeneous particles. *J. Quant. Spectrosc. Radiat. Transf.* **2013**, *116*, 169–183. [CrossRef]

53. Bi, L.; Yang, P. Accurate simulation of the optical properties of atmospheric ice crystals with the invariant imbedding T-matrix method. *J. Quant. Spectrosc. Radiat. Transf.* **2014**, *138*, 17–35. [CrossRef]

54. Bi, L.; Yang, P. Impact of calcification state on the inherent optical properties of *Emiliania huxleyi* coccoliths and coccolithophores. *J. Quant. Spectrosc. Radiat. Transf.* **2015**, *155*, 10–21. [CrossRef]

55. Born, M.; Wolf, E. *Principles of Optics*; Cambridge University Press: Cambridge, UK, 1999.

56. Takano, Y.; Liou, K.N. Solar radiative transfer in cirrus clouds. Part I: Single-scattering and optical properties of hexagonal ice crystals. *J. Atmos. Sci.* **1989**, *46*, 3–19. [CrossRef]

57. Macke, A.; Mueller, J.; Raschke, E. Single scattering properties of atmospheric ice crystals. *J. Atmos. Sci.* **1996**, *53*, 2813–2825. [CrossRef]

58. Yang, P.; Liou, K.N. Geometric-optics—Integral-equation method for light scattering by nonspherical ice crystals. *Appl. Opt.* **1996**, *35*, 6568–6584. [CrossRef] [PubMed]

59. Muinonen, K. Scattering of light by crystals: A modified Kirchhoff approximation. *Appl. Opt.* **1989**, *28*, 3044–3050. [CrossRef] [PubMed]

60. Yang, P.; Liou, K.N. Light scattering by hexagonal ice crystals: Solutions by a ray-by-ray integration algorithm. *JOSA A* **1997**, *14*, 2278–2289. [CrossRef]

61. Bi, L.; Yang, P.; Kattawar, G.W.; Hu, Y.; Baum, B.A. Scattering and absorption of light by ice particles: Solution by a new physical-geometric optics hybrid method. *J. Quant. Spectrosc. Radiat. Transf.* **2011**, *112*, 1492–1508. [CrossRef]

62. Borovoi, A.G.; Grishin, I.A. Scattering matrices for large ice crystal particles. *JOSA A* **2003**, *20*, 2071–2080. [CrossRef] [PubMed]

63. Sun, B.; Yang, P.; Kattawar, G.W.; Zhang, X. Physical-geometric optics method for large size faceted particles. *Opt. Express* **2017**, *25*, 24044–24060. [CrossRef] [PubMed]

64. Nayak, A.R.; Mcfarland, M.N.; Sullivan, J.M.; Twardowski, M.S. Evidence for ubiquitous preferential particle orientation in representative oceanic shear flows. *Limnol. Oceanogr.* **2018**, *63*, 122–143. [CrossRef] [PubMed]

65. Heil, C.A.; Steidinger, K.A. Monitoring, management, and mitigation of Karenia blooms in the eastern Gulf of Mexico. *Harmful Algae* **2009**, *8*, 611–617. [CrossRef]

66. Kiefer, D.A.; Olson, R.J.; Wilson, W.H. Reflectance spectroscopy of marine phytoplankton. part I. optical properties as related to age and growth rate. *Limnol. Oceanogr.* **1979**, *24*, 664–672. [CrossRef]

67. Steidinger, K.A.; Truby, E.W.; Dawes, C.J. Ultrastructure of the red tide dinoflagellate *Gymnodinium breve*. I. General description 2.3. *J. Phycol.* **1978**, *14*, 72–79. [CrossRef]

68. Rizzo, P.J.; Jones, M.; Ray, S.M. Isolation and properties of isolated nuclei from the Florida red tide dinoflagellate *Gymnodinium breve* (Davis). *J. Protozool.* **1982**, *29*, 217–222. [CrossRef] [PubMed]

69. Gautier, A.; Michel-Salamin, L.; Tosi-Couture, E.; McDowall, A.W.; Dubochet, J. Electron microscopy of the chromosomes of dinoflagellates in situ: Confirmation of Bouligand's liquid crystal hypothesis. *J. Ultrastruct. Mol. Struct. Res.* **1986**, *97*, 10–30. [CrossRef]

70. Rill, R.L.; Livolant, F.; Aldrich, H.C.; Davidson, M.W. Electron microscopy of liquid crystalline DNA: Direct evidence for cholesteric-like organization of DNA in dinoflagellate chromosomes. *Chromosoma* **1989**, *98*, 280–286. [CrossRef] [PubMed]
71. Shapiro, D.B.; Quinbyhunt, M.S.; Hunt, A.J. Origin of the Induced circular-polarization in the light-scattering from a dinoflagellate. *Ocean Opt. X* **1990**, *1302*, 281–289.
72. Shapiro, D.B.; Hunt, A.J.; Quinby-Hunt, M.S.; Hull, P.G. Circular-polarization effects in the light-scattering from single and suspensions of dinoflagellates. *Underw. Imaging Photogr. Visibility* **1991**, *1537*, 30–41.
73. Bouligand, Y.; Soyer, M.O.; Puiseux-Dao, S. La structure fibrillaire et l'orientation des chromosomes chez les Dinoflagellés. *Chromosoma* **1968**, *24*, 251–287. [CrossRef] [PubMed]

![applied sciences logo] *applied sciences*

MDPI

Article

Remote Sensing of CDOM, CDOM Spectral Slope, and Dissolved Organic Carbon in the Global Ocean

Dirk Aurin [1,*], Antonio Mannino [2] and David J. Lary [3]

[1]   NASA Goddard Space Flight Center (USRA), Greenbelt, MD 20771, USA
[2]   NASA Goddard Space Flight Center, Greenbelt, MD 20771, USA; antonio.mannino-1@nasa.gov
[3]   University of Texas at Dallas, Richardson, TX 75080, USA; david.lary@utdallas.edu
*   Correspondence: dirk.a.aurin@nasa.gov; Tel.: +1-301-286-8156

Received: 28 July 2018; Accepted: 27 October 2018; Published: 19 December 2018

Featured Application: Methods and algorithms developed in this manuscript may be applied to ocean color satellite or aircraft imagery for the remote sensing of oceanic CDOM spectral absorption, CDOM spectral slope, and DOC.

Abstract: A Global Ocean Carbon Algorithm Database (GOCAD) has been developed from over 500 oceanographic field campaigns conducted worldwide over the past 30 years including in situ reflectances and coincident satellite imagery, multi- and hyperspectral Chromophoric Dissolved Organic Matter (CDOM) absorption coefficients from 245–715 nm, CDOM spectral slopes in eight visible and ultraviolet wavebands, dissolved and particulate organic carbon (DOC and POC, respectively), and inherent optical, physical, and biogeochemical properties. From field optical and radiometric data and satellite measurements, several semi-analytical, empirical, and machine learning algorithms for retrieving global DOC, CDOM, and CDOM slope were developed, optimized for global retrieval, and validated. Global climatologies of satellite-retrieved CDOM absorption coefficient and spectral slope based on the most robust of these algorithms lag seasonal patterns of phytoplankton biomass belying Case 1 assumptions, and track terrestrial runoff on ocean basin scales. Variability in satellite retrievals of CDOM absorption and spectral slope anomalies are tightly coupled to changes in atmospheric and oceanographic conditions associated with El Niño Southern Oscillation (ENSO), strongly covary with the multivariate ENSO index in a large region of the tropical Pacific, and provide insights into the potential evolution and feedbacks related to sea surface dissolved carbon in a warming climate. Further validation of the DOC algorithm developed here is warranted to better characterize its limitations, particularly in mid-ocean gyres and the southern oceans.

Keywords: ocean color database; oceanic carbon; chromophoric dissolved organic matter; dissolved organic carbon; CDOM spectral slope; ocean color remote sensing; algorithm development; ocean color algorithm validation; ocean optics; CDOM climatology; CDOM and ENSO; machine learning

---

1. Introduction

1.1. Background

In 1896, Svante Arrhenius introduced the theory that adding carbon dioxide (CO_2) to the atmosphere enhances the planetary greenhouse effect. Over the intervening century, it became clear that the marine dissolved organic carbon (DOC) pool comprised the vast majority of the organic carbon in the oceans, and was nearly equivalent to the atmospheric pool of CO_2 [1]. In fact, remineralization of just 1% of the DOC in the oceans (e.g., by microbial metabolism and photo-oxidation) would generate a flux of CO_2 into the atmosphere greater than that resulting from all the fossil fuel burned in a year [2]. Recently, Belanger et al. [3] estimated that photoproduction of CO_2 from Chromophoric Dissolved Organic Matter (CDOM) has already increased by ~15% in Arctic waters due to an increase in ultraviolet radiation and the decrease in sea ice associated with global warming. Positive feedbacks such as this have potentially serious consequences for humans and ecosystems alike, and emphasize the urgency to develop robust, global algorithms for retrieving oceanic carbon products remotely and synoptically.

CDOM (refer to Table 1 for terms and abbreviations) is used to describe an often difficult to define fraction of the DOC pool (see Section 1.4) which has historically been called gilvin, *gelbstoff*, or simply "yellow substance". As its name suggests, the presence of CDOM imparts color to the water column through absorption of light by various chromophores, thereby providing an effective means of detecting CDOM remotely from ocean color reflection. Found in all natural waters and generally in highest concentration near shore, CDOM results from the breakdown products of plants and other organic matter into humic materials, and plays a significant role in aquatic photochemistry, photobiology, and as a tracer of the origins of oceanic water masses, e.g., [4,5]. DOC and CDOM can be terrigenous or autochthonous (i.e., deriving from in situ primary and bacterial production in river to ocean waters), with the DOC variously composed of high molecular weight (HMW) humic substances (which tend to be more labile) and low molecular weight (LMW) humics (such as fulvic acids), depending on its origin, labile fraction, age, and whether it has transitioned from fresh waters to marine [6–12]. Most estuarine and nearshore CDOM is terrigenous, and as it mixes in rivers on its transit to marine waters, the amount of HMW material declines from flocculation, photo-oxidation and microbial decomposition leaving marine waters dominated by LMW CDOM (e.g., [6]), a condition imparting a characteristic spectral shape to inherent light absorption by CDOM ($a_g(\lambda)$, where λ is wavelength) [7]. Inherent optical properties (IOPs) of the water column, such as the absorption and backscattering coefficients, depend on the composition and concentration of the dissolved and suspended material present, as well as the size and structure of the particles, and water itself. CDOM concentration—for which $a_g(\lambda)$ is the common proxy following Beer's law—varies widely in the ocean, tending to be highest near river outflows, but may also be high in upwelling regions and other regions of autochthonous, plankton-based production through exudation, excretion, and microbial breakdown of detritus [8]. It is degraded over time both by microbial activity, photooxidation, and other abiotic processes, ultimately resulting in remineralization of the carbon, and release from the ocean as CO and CO_2. In the case of the CDOM fraction of DOC, degradation over time scales of days to millennia can significantly change the magnitude and spectral characteristics of $a_g(\lambda)$.

Table 1. Definition of terms, units, and abbreviations.

	Units	Definition
$a_g(\lambda)$	m^{-1}	CDOM absorption coefficient
$a_d(\lambda)$	m^{-1}	NAP absorption coefficient
$a_{dg}(\lambda)$	m^{-1}	NAP and CDOM absorption coefficient
$a_p(\lambda)$	m^{-1}	Particulate absorption coefficient
$b_{bp}(\lambda)$	m^{-1}	Particle backscattering coefficient
$b_{bt}(\lambda)$	m^{-1}	Total backscattering coefficient
CDOM		Colored Dissolved Organic Matter
Chl	$mg\ m^{-3}$	Chlorophyll concentration
DOC, DOM	$\mu mol\ L^{-1}$	Dissolved Organic Carbon, -Material
$E_s(\lambda)$	$W\ m^{-2}\ nm^{-1}$	Downwelling surface irradiance
$L_w(\lambda)$	$W\ m^{-2}\ nm^{-1}\ sr^{-1}$	Water leaving radiance
$L_{wn}(\lambda)$	$W\ m^{-2}\ nm^{-1}\ sr^{-1}$	Normalized water leaving radiance
POC	$\mu mol\ L^{-1}$	Particulate Organic Carbon
$R_{rs}(\lambda)$	sr^{-1}	Remote sensing reflectance

Table 1. *Cont.*

	Units	Definition
$S_g(\lambda_1\text{-}\lambda_2)$	nm^{-1}	Exponential slope of CDOM and in select spectral range
SPM	$mg\ m^{-3}$	Suspended Particulate Material
TOC	$\mu mol\ L^{-1}$	Total Organic Carbon
AOP		Apparent Optical Properties
GIOP		Generalized IOP Algorithm
GOCAD		Global Ocean Carbon Algorithm Database
HMW		High Molecular Weight
IOCCG		International Ocean-Colour Coordinating Group
IOP		Inherent Optical Properties
LMW		Low Molecular Weight
MLR		Multiple Linear Regression Algorithm
MODIS		Moderate Resolution Imaging Spectroradiometer
NAP		Non-Algal Particulate
NOMAD		NASA bio-Optical Algorithm Dataset
QAA		Quasi-Analytical Algorithm
RFTB		Random Forest Tree Bagger Algorithm
SAA		Semi-Analytical Algorithm
SeaBASS		SeaWiFS Bio-optical Archive and Storage System
SeaWiFS		Sea-viewing Wide Field-of-view Sensor
UV, UVA, UVB		Ultraviolet spectrum, 315–400 nm, 280–315 nm
VIS		Visible spectrum

CDOM absorption is a superposition of the spectral absorption by its varied chromophores, and increases roughly exponentially (or hyperbolically [9]) with decreasing wavelength in the visible (VIS) and ultraviolet (UV) spectral ranges, as described in the next section. CDOM tends to dominate the blue and UV spectrum in many coastal and estuarine environments (e.g., [7,10–12]), and is the most important factor controlling UV and blue light penetration even in the open ocean [13] despite its generally lower concentration and distance from land. Within the visible spectrum, $a_g(\lambda)$ reduces the photosynthetically available radiation supporting phytoplankton and macrophytic growth, and generates heat in the surface layer of the water column, thus affecting mixing [14]. In the UV, CDOM causes surface heating as well, but also acts to protectively shade aquatic organisms, thus reducing the amount of damaging high frequency radiation reaching vulnerable cell structures.

From the passive remote sensing perspective, CDOM reduces the amount of blue light available for reflection out of the water column, and can therefore have a significant impact on ocean color algorithms, for example increasing uncertainty in blue-green band-ratio algorithms designed to estimate chlorophyll-*a* concentration (Chl) from its absorption peak near 443 nm [13,15]. These types of Chl algorithms assume covariance in Chl, CDOM, and other water column constituents (i.e., the "Case 1 waters" assumption [16]). By contrast, semi-analytical algorithms (SAAs) that invert the ocean color signal to retrieve individual component absorption spectra (particles, CDOM, water) are stymied by the presence of non-algal particulates (NAPs), which have a similar spectral shape to CDOM. As a result, these approaches tend to retrieve only the sum of these two elements [17] (and references therein).

1.2. Spectral Shape of CDOM

The CDOM absorption coefficient is generally modeled with an exponentially decaying function with increasing wavelength, λ.

$$a_g(\lambda) = a_g(\lambda_0)\ e^{-Sg\ (\lambda - \lambda_0)} \tag{1}$$

where S_g is the spectral slope parameter and λ_0 is a reference wavelength. S_g in various spectral ranges in the UV and VIS contains information about CDOM's photoreactive state, chemical composition, molecular weight distribution, and origin [4,7,18–21].

While the single exponential model in Equation (1) is accurate within limited wavebands, CDOM spectral slope, S_g, is not constant across the UV and VIS and depends on the wavelength range used, spectral resolution, and reference wavelength λ_0. Furthermore, comparative analysis of CDOM spectral shape as reported in the literature has been confounded by the multitude of methodologies and reference wavebands used historically to calculate S_g [9]. For instance, a linear fit to logarithmically transformed a_g data yields results for S_g biased by higher wavelength absorption, whereas a least-squared difference minimization fitting favors the lower wavelengths where the magnitude of a_g is higher, and is considered more accurate [7,22]. Changes to S_g resulting from photodegradation are wavelength dependent, i.e., increasing below 460 nm and decreasing above 510 nm [23], although when calculated across the VIS from 412–555, slope is expected to increase through the destruction of large humic complexes resulting in lower molecular weight CDOM [24]. This effect appears to reverse over time as more refractory, low-molecular weight compounds are also degraded, thereby reducing CDOM absorption at shorter wavelengths relative to longer, and decreasing spectral slope across the VIS.

All these factors lead to challenges in comparing CDOM spectral slope between studies, and a more standardized approach to CDOM spectral shape measurement still seems warranted [19]. The concept of the spectral slope curve, $S_g(\lambda)$—analogous to the first derivative of S_g with respect to λ—was explored by Loiselle et al. in 2009 [23]. Calculating S_g from natural waters, cultures, and laboratory standards at 20 nm waveband intervals between 200–700 nm, they showed that $S_g(\lambda)$ had complex spectral characteristics including peaks near 390 nm likely indicating a prevalence of autochthonous production of fulvic acid-type CDOM, and near 280 nm possibly due to the release of proteins or phenols by phytoplankton. While the spectral slope curve approach of Loiselle et al. 2009 represents an elegant method for quantifying many subtle characteristics of CDOM spectral shape when compared to, for example, using a single slope parameter across the UV and VIS, it does require relatively high spectral resolution data collection. Historically, this was not always available or reported, and here we focus on a set of eight different spectral ranges commonly seen in the literature and described in detail below.

1.3. Remotely Sensing CDOM and S_g

As interest in CDOM has grown in recent years, numerous empirical ocean color algorithms for retrieving CDOM within limited geographic regions have emerged, e.g., [25–30]. A smaller number of more generally applicable, global empirical algorithms have also been developed, including one for retrieving a unitless index of CDOM prevalence, though it does not retrieve $a_g(\lambda)$ or S_g and depends upon Case 1 assumptions. More recently, Tiwari and Shanmugam published global empirical algorithms for both $a_g(\lambda)$ and S_g [31,32]. These were optimized and tested using field data aggregated in NOMAD (the NASA bio-Optical Marine Algorithm Dataset version 2 [33]) and the synthetic ocean color dataset developed by the International Ocean Colour Coordinating Group (IOCCG) for the purpose of algorithm development and validation [34].

Other approaches to retrieving CDOM remotely depend on the premise that sea-surface reflectance is approximately inversely proportional to the total absorption coefficient [16,35,36], which can be linearly separated into various contributions by particulate and dissolved constituents. This forms the basis to semi-analytical ocean color algorithms (SAAs) for retrieving constituent absorption, e.g., [37–39], but, as already mentioned, owing to the similarity in spectral shape of non-algal particulate (i.e., detrital, microbial, and sedimentary) absorption and $a_g(\lambda)$, SAAs generally retrieve only their sum, a_{dg} [17]. To circumvent this difficulty, empirical methods are sometimes added to SAAs to help distinguish non-algal from dissolved absorption [40–45].

1.4. Remotely Sensing DOC

One of the most challenging aspects of developing robust, global ocean color algorithms for DOC is that the relationship between DOC and CDOM (i.e., the DOC-specific absorption) is highly variable, in some cases negatively correlated (e.g., Southern Ocean, [46]) and often poorly defined, particularly in open ocean areas such as the Sargasso Sea [47,48]. In some cases, the relationship is better constrained within a particular region and season, as shown by measurements made in the Mid-Atlantic Bight on the eastern shelf of North America [28]. Because absorption by CDOM is the only way in which ocean color is impacted by DOC, some other independent knowledge of water type is needed for retrieval of DOC from space.

1.5. Algorithm Development Data

One of the most confounding challenges in the development of both empirical and semi-analytical algorithms is the lack of a large, comprehensive database containing a broad enough dynamic range in optical characteristics to be representative of the majority of the world's oceans, while also having realistic combinations of inherent optical properties, which are not guaranteed in large, synthetic, modeled datasets. NOMAD represents the first (and most recent, as of this writing) major effort to provide the ocean color community with such a dataset. It was aggregated and selected from all of the relevant field data submitted to the NASA SeaBASS archive (http://seabass.gsfc.nasa.gov/), and has been extremely useful to those in the ocean color algorithm community since its original publication in 2005 and update in 2008. However, NOMAD was not focused on CDOM. For example, while it contains about 3700 coincident radiometric and phytoplankton pigment observations, coincident radiometric and CDOM observations number just ~1200. In part, this is because CDOM data collected using in situ instrumentation were excluded for various reasons discussed at greater length below. The remaining CDOM records—those measured from discrete water samples—were modeled spectrally at the preselected NOMAD wavebands after fitting field data to Equation (1), and do not extend into the UV where spectral shape can provide useful insights into the origin and photooxidation state of CDOM. NOMAD does not contain any DOC data observations.

Using the methodology described in the next section, we extend the NOMAD approach to create a global ocean color algorithm development database better suited to DOC and its optical components, CDOM and CDOM spectral slope, ultimately including over 51,000 field observations of surface-averaged inherent optical properties. These are matched to between ~8000 and ~11,000 coincident estimates of sea surface reflectance made from in situ measurements as well as satellite imagery from SeaWiFS and MODIS Aqua and Terra instruments. The global ocean carbon algorithm database (hereafter Global Ocean Carbon Algorithm Database (GOCAD) records are split into independent sets of field stations for training/optimization (i.e., with in situ radiometry) and validation (i.e., with satellite imagery) of algorithms, as described in the Section 2. A basic overview of the most relevant aspects of the global dataset is presented in Section 3.1. In Section 3.2, empirical and SAA approaches to retrieval of DOC, CDOM, and CDOM slope are developed and discussed. Finally, algorithms are applied to global climatological satellite imagery and discussed in Section 3.3.

2. Methodology

2.1. Database Assembly Overview

Field measurements of CDOM, DOC, remote sensing reflectance, $R_{rs}(\lambda)$, and ancillary data and metadata were downloaded from SeaBASS and the Hansell/Carlson collection (https://hansell-lab.rsmas.miami.edu/research/data-collection/index.html) in April 2013. Coincident, Level 2 (L2) SeaWiFS and MODIS Aqua and Terra satellite imagery at all field stations were downloaded from the NASA Ocean Color website (http://oceancolor.gsfc.nasa.gov). Due to the size of aggregated datasets for each of the key constituents (e.g., 117,291 raw CDOM records, 31,474 raw DOC records, 115,773 in situ reflectance records, and ~177,000 matching satellite scenes), extensive automation in the processing, quality control, and merging of the databases was a necessity. A station-by-station analysis (or field experiment-specific analysis, as in [33]) of the data for establishing the customized spatial and temporal thresholds for matching coincident inherent and apparent optical properties and satellite imagery was not feasible. Relatively broad guidelines conducive to automation were established, as described in detail in the following sections. We assume, for example, that geospatial and temporal variability of CDOM and DOC is higher in coastal and shelf waters (defined here as samples collected in waters of 1000 m depth or less) than in the pelagic.

2.2. Field Data

Targeted searches of SeaBASS were conducted for all records containing a_g, DOC, or in situ reflectances (see 2.2.3). Resultant data from the following physical, bio-optical, and biogeochemical fields were also retained where they happened to be present in SeaBASS files: depth, temperature, salinity, a_{nap}, a_p, a_{pg}, a_{dg}, b_{bt}, particulate organic carbon (POC), total organic carbon (TOC), and Chl. Ancillary data including time and date, latitude, longitude, and bottom depth were also retained, as well as complete SeaBASS metadata for each record. Carbon data were downloaded from each of the data repository resources linked in the Hansell/Carlson DOM Data Collection (http://yyy.rsmas. miami.edu/groups/biogeochem/Data.html). These were also queried for all the parameters above and assigned metadata for each cruise. Table 2 provides a complete listing and overview of all the field experiments retained in the final, quality-controlled database.

Table 2. Summary of field data collection campaigns.

Experiment	Principal Investigators	Cruises	Numbers of Stations						Min. Lat	Max. Lat	Min. Lon	Max. Lon	Year(s)
			CDOM	CDOM & IS*	CDOM & SAT**	DOC	DOC & IS*	DOC & SAT**					
MURI	A. Neeley, S. Freeman, J. Chaves, C. McClain	1	0	0	0	6	0	0	19.126	20.692	−157.36	−156.32	2012
EGE3	A. Subramaniam	1	9	0	0	0	0	0	−6.003	3.327	−10.008	7.992	2006
EGE5	A. Subramaniam	1	2	0	0	0	0	0	−5.977	−5.62	5.85	7.997	2007
MANTRA PIRANA	A. Subramaniam	5	41	2	2	0	0	0	3.386	25.002	−158.02	−42.276	2001–2003
MASS BAY	A. Subramaniam	7	39	0	13	0	0	0	41.85	42.619	−70.895	−70.228	2002–2005
IOFFE	A. Khrapko, S. Ershova	1	164	56	51	0	0	0	−66.46	48.59	−67.98	−5.54	2001–2002
B01	A. Mannino	2	15	5	0	16	5	0	36.713	37.786	−76.018	−74.644	2005
B02	A. Mannino	2	28	18	6	30	17	6	36.685	38.918	−76.069	−74.502	2005
B03	A. Mannino	2	26	0	14	29	0	16	36.413	38.87	−76.022	−74.499	2006
B04	A. Mannino	2	30	19	6	46	19	9	36.502	38.908	−76.019	−74.299	2006
B05	A. Mannino	1	15	0	5	15	11	5	36.431	38.586	−76.017	−73.517	2006
BIOD01	A. Mannino	1	15	11	1	15	11	1	42.361	43.574	−70.696	−69.863	2007
BIOD02	A. Mannino	1	17	14	6	17	14	6	42.593	43.708	−70.78	−69.691	2007
BIOD03	A. Mannino	1	13	12	0	12	11	0	41.201	42.812	−76.172	−70.445	2007
CBM01	A. Mannino	1	2	0	0	0	0	0	36.965	37.17	−76.172	−76.029	2004
CBM02	A. Mannino	1	4	0	3	0	0	0	36.987	37.182	−76.163	−76.018	2004
CBM03	A. Mannino	1	4	0	2	0	0	0	36.987	37.182	−76.163	−76.018	2004
CBM04	A. Mannino	1	4	0	0	0	0	0	36.987	37.182	−76.163	−76.018	2004
CBM05	A. Mannino	1	4	0	3	0	0	0	36.987	37.182	−76.163	−76.018	2004
CBM06	A. Mannino	1	4	0	3	0	0	0	36.987	37.182	−76.163	−76.018	2005
CBM07	A. Mannino	1	3	0	2	0	0	0	37.046	37.182	−76.138	−76.018	2005
CBM08	A. Mannino	1	3	0	0	0	0	0	37.047	37.182	−76.137	−76.019	2005
CBM09	A. Mannino	1	5	0	0	0	0	0	36.987	37.181	−76.161	−76.017	2005
CBM10	A. Mannino	1	4	0	3	0	0	0	36.987	37.181	−76.161	−76.019	2005
CBM11	A. Mannino	1	4	0	0	0	0	0	37.046	37.182	−76.136	−76.018	2006
CBM12	A. Mannino	1	4	0	3	0	0	0	36.987	37.182	−76.161	−75.713	2006
CO11	A. Mannino	1	4	2	3	4	2	3	36.969	36.969	−76.017	−75.713	2007
CO12	A. Mannino	1	4	1	3	4	1	3	36.969	36.969	−76.017	−75.71	2007
CO13	A. Mannino	1	4	1	3	3	0	0	36.69	36.969	−76.017	−75.713	2007
CV1	A. Mannino	1	53	0	24	51	0	23	35.745	42.498	−75.706	−65.736	2009
CV2	A. Mannino	1	69	0	35	69	0	35	36.475	43.062	−75.785	−66.088	2009
CV3	A. Mannino	1	43	0	9	43	0	9	37.089	43.112	−75.677	−65.779	2010
CV4	A. Mannino	1	79	0	30	78	0	30	36.073	44.233	−75.911	−65.772	2010
CV5	A. Mannino	1	67	0	24	63	0	24	36.142	44.299	−75.859	−65.775	2010
CV6	A. Mannino	1	92	0	18	92	0	18	36.187	43.928	−75.789	−65.768	2011
D01	A. Mannino	1	4	2	2	4	2	2	36.797	36.966	−76.02	−75.719	2005
D02	A. Mannino	1	6	6	6	6	6	6	36.805	36.973	−76.02	−75.712	2005
D03	A. Mannino	1	5	2	0	5	2	0	36.803	36.968	−76.018	−75.718	2006
D04	A. Mannino	2	6	3	0	6	3	0	36.801	36.966	−76.015	−75.64	2006
OCV1	A. Mannino	1	26	26	18	26	26	18	40.218	40.731	−74.153	−73.479	2007
OCV2	A. Mannino	1	22	18	2	22	18	2	40.208	40.724	−74.151	−73.451	2007–2009
OCV3	A. Mannino	1	8	8	0	8	8	0	40.392	40.739	−74.156	−73.554	2008

Table 2. *Cont.*

Experiment	Principal Investigators	Cruises	CDOM	CDOM & IS*	CDOM & SAT**	DOC	DOC & IS*	DOC & SAT**	Min. Lat	Max. Lat	Min. Lon	Max. Lon	Year(s)
OCV5	A. Mannino	1	11	11	3	11	11	3	39.584	41.028	−73.901	−71.749	2009
PL6	A. Mannino	1	4	0	3	4	0	3	36.802	36.968	−76.017	−75.713	2007
GOMECC-2	A. Mannino, J. Salisbury	1	67	0	1	92	0	2	25.999	43.032	−90.809	−68.01	2012
GEO-CAPE	A. Mannino, M. Mulholland	1	59	53	6	54	51	6	38.098	39.17	−76.491	−76.084	2011
MONTEREY BAY	B. Arnone, R. Gould	1	57	51	11	0	0	0	36.271	36.988	−123.13	−121.81	2003
WOCE P14S P15S	B. Tilbrook	1	0	0	0	107	0	0	−66.99	−0.003	−174.79	173.982	1996
Carbon Transport MS R.	C. Del Castillo	2	9	0	6	0	0	0	28.295	28.925	−89.743	−89.411	2001–2003
GasEx	C. Del Castillo	1	44	10	3	0	0	0	−53.75	−50.14	−38.554	−36.622	2008
Big Bend	C. Hu	10	156	61	36	0	0	0	29.168	29.67	−83.635	−83.201	2010–2011
GEO-CAPE CBODAQ	C. Hu	2	16	12	0	23	11	2	38.098	39.17	−76.487	−76.083	2011
GOM Oil Spill	C. Hu	3	10	0	1	0	0	0	28.584	29.1	−88.42	−87.323	2010
Glider calibration	C. Hu	1	8	0	4	0	0	0	27.452	28.465	−83.992	−83.072	2009–2011
Glider validation	C. Hu	1	13	0	6	0	0	0	27.356	27.48	−83.115	−83.051	2009
SWFL	C. Hu	3	21	3	2	0	0	0	24.827	26.49	−82.318	−81.141	2010–2011
Tampa Bay	C. Hu	5	85	71	5	0	0	0	27.578	27.991	−82.783	−82.408	2008–2012
West Florida Shelf	C. Hu	5	134	67	53	0	0	0	25.057	28.439	−83.785	−81.143	2005–2008
MOCE	C. Trees, D. Clark	2	40	0	17	0	0	0	21.793	36.875	−122	−105.75	1992–1999
BowdoinBuoy	C. Roesler	1	2	0	0	0	0	0	43.762	43.795	−69.988	−69.947	2011
PenBaySurvey	C. Roesler	1	0	0	0	0	0	0	44.26	44.26	−68.983	−68.983	2008
San Diego Coastal Project	D. Stramski, M. Stramska	1	15	0	0	0	0	0	32.558	32.758	−117.26	−117.13	2004–2006
NASA Gulf of Maine	D. Phinney, D. Phinney, J. Brown	4	46	0	6	0	0	0	41	44.245	−70.567	−67.162	1998–1999
NOAA Gulf of Maine	D. Phinney, D. Phinney, J. Brown	4	37	0	4	0	0	0	40.209	44.344	−70.056	−65.549	1996–1998
Panama City Florida	D. Phinney, D. Phinney, J. Brown	1	0	0	0	0	0	0	30.167	30.172	−85.857	−85.852	2001
Plumes and Blooms	D. Siegel	5	88	1	22	0	0	0	34.024	34.464	−120.56	−119.28	2001–2003
CLIVAR A13.5 2010	D. Hansell	1	0	0	0	64	0	4	−54	4.62	−3.002	1.835	2010
CLIVAR I05 2009	D. Hansell	1	0	0	0	1	0	0	−31.19	−31.19	82.564	82.564	2009
CLIVAR I08S 2007	D. Hansell	1	0	0	0	28	0	2	−65.71	−28.32	81.962	95.014	2007
CLIVAR P02 2004	D. Hansell	1	0	0	0	56	0	7	29.991	32.644	−177.99	179.545	2004
CLIVAR P16N 2006	D. Hansell	1	0	0	0	78	0	2	−17	56.28	−153.22	−150	2006
CLIVAR P16S 2005	D. Hansell	1	0	0	0	57	0	1	−71	−16	−150.04	−149.91	2005
CLIVAR P18 2007	D. Hansell	1	0	0	0	72	0	4	−68.91	22.7	−110.04	−102.54	2007
HLY-02-01	D. Hansell	1	0	0	0	21	0	0	64.98	73.431	−169.14	−154.4	2002
HLY-02-03	D. Hansell	1	0	0	0	38	0	1	65.668	73.698	−168.86	−151.94	2002
HLY-0403	D. Hansell	1	0	0	0	36	0	0	65.661	73.827	168.9	−152.02	2004
SR03	D. Hansell	1	0	0	0	24	0	0	−65.57	−44.38	139.658	146.189	2008

Table 2. *Cont.*

Experiment	Principal Investigators	Cruises	CDOM	CDOM & IS*	CDOM & SAT**	DOC	DOC & IS*	DOC & SAT**	Min. Lat	Max. Lat	Min. Lon	Max. Lon	Year(s)
WOCE AR01 A05	D. Hansell	1	0	0	0	45	0	6	24.499	27.622	−79.937	−14.224	1998
ACE-ASIA	G. Mitchell, M. Kahru	1	45	22	10	0	0	0	28.207	38.905	−177	178.05	2001
AMLR	G. Mitchell, M. Kahru	7	47	0	7	0	0	0	−63.01	−57.5	−68.186	−53.296	2000–2007
Aerosols Index	G. Mitchell, M. Kahru	1	21	3	9	0	0	0	−34.53	27.368	−60.615	85.166	1999
CALCOFI	G. Mitchell, M. Kahru	16	179	8	29	0	0	0	29.847	36.057	−124.33	−117.3	1996–2002
Sea_of_Japan	G. Mitchell, M. Kahru	1	17	1	1	0	0	0	34.503	43.302	128.883	139.883	1999
Arc00	G. Cota	1	14	0	0	0	0	0	70.328	72.412	−167.59	−144.63	2000
LAB97	G. Cota	1	10	0	0	0	0	0	44.137	60.38	−58.19	−43.999	1997
Lab2000	G. Cota	1	6	0	2	0	0	0	49.518	60.047	−58.749	−48.899	2000
Lab96	G. Cota	1	10	0	0	0	0	0	52.08	60.999	−58.006	−47.908	1996
ORCA Ches. Light Tower	G. Cota	1	3	0	2	0	0	0	36.9	36.9	−75.71	−75.71	2000
Res95	G. Cota	1	1	0	0	0	0	0	74.645	74.645	−95.91	−95.91	1995
Res96	G. Cota	1	6	0	0	0	0	0	74.644	74.646	−94.915	−94.905	1996
NSF-BWZ	G. Mitchell	2	0	0	0	0	0	0	−62.26	−60.63	−58.375	−55.6	2004–2006
Benthic Ecol. from Space	H. Dierssen, R. Zimmerman	4	8	0	0	0	0	0	24.723	29.847	−85.382	−80.705	2005–2006
Kieber Photochemistry 03	H. Sosik	1	17	0	5	0	0	0	35.278	41.075	−75.218	−71.127	2003
MVCO	H. Sosik	34	129	0	23	0	0	0	41.143	41.342	−70.638	−70.415	2005–2011
GOCAL	J.R.V Zaneveld, W.S. Pegau	6	140	0	48	0	0	0	22.914	31.116	−114.64	−107.75	1996–1999
PREPP	J. Chen	5	47	0	0	0	0	0	22.15	22.555	113.673	114.43	2001
SAB Mapping	J. Nelson, A. Subramaniam	2	18	0	10	0	0	0	30.823	31.993	−81.024	−80.221	2005
GEOTRACES	J. Chaves	1	0	0	0	9	0	0	17.35	36.766	−24.496	−12.825	2010
COOA	J. Salisbury, D. Vandemark, C. Hunt	3	0	0	0	14	0	4	42.861	43.757	−70.66	−69.782	2008
NOAA CSC	J. Brock, A. Subramanian, K. Waters	1	10	0	0	0	0	0	31.335	31.965	−81.128	−80.454	1996
BOA	K. Carder	1	62	62	0	0	0	0	27.579	59.841	−91.768	−15.49	1991–1993
EcoHAB	K. Carder	19	398	208	126	0	0	0	25.3	27.572	−84.394	−81.259	1999–2003
Okeechobee	K. Carder	1	4	4	0	0	0	0	27.149	27.199	−80.794	−80.788	1997
Redtide	K. Carder	2	13	11	7	0	0	0	27.289	28.098	−83.253	−82.866	2005
TOTO	K. Carder	3	86	75	40	0	0	0	24.884	27.5	−82.776	−77.587	1998–2000
ACE-INC	L.W. Harding, Jr.,M. Mallonee, A. Magnuson	6	21	0	0	0	0	0	38.303	38.754	−76.62	−76	2002–2003
BIOCOMPLEXITY	L.W. Harding, Jr.,M. Mallonee, A. Magnuson	11	55	0	15	0	0	0	36.863	39.349	−76.451	−75.878	2001–2004
LMER-TIES	L.W. Harding, Jr.,M. Mallonee, A. Magnuson	17	220	0	22	0	0	0	36.866	39.421	−76.517	−75.749	1996–2000
SGER	L.W. Harding, Jr.,M. Mallonee, A. Magnuson	1	11	0	5	0	0	0	36.95	38.5	−76.481	−75.998	2003

Table 2. *Cont.*

| Experiment | Principal Investigators | Numbers of Stations | | | | | | | Min. Lat | Max. Lat | Min. Lon | Max. Lon | Year(s) |
		Cruises	CDOM	CDOM & IS*	CDOM & SAT**	DOC	DOC & IS*	DOC & SAT**					
ONR-MAB	L.W. Harding, Jr.,M. Mallonee, A. Magnuson	2	31	0	0	0	0	0	36.4	39.134	−75.949	−71.993	1996–1997
Ocean Color Cal Val	M.S. Twardowski, A.H. Barnard, J.R.V. Zaneveld	1	14	13	0	0	0	0	40.208	40.511	−74.054	−73.448	2007
Tokyo Bay	M. Kishino	1	1	0	0	0	0	0	35.223	35.223	139.718	139.718	1984
Global CDOM	N. Nelson, D. Siegel	2	19	0	3	0	0	0	−8.458	7.004	−140.07	−124.35	2005–2006
CLIVAR	N. Nelson, D. Siegel, C. Carlson	9	80	17	4	45	8	4	−68.36	59.5	−150	95.028	2003–2008
TAO 2005	N. Nelson, D. Siegel, C. Carlson	1	61	0	9	0	0	0	−8.89	12	−140.2	−124.35	2005
TAO 2006	N. Nelson, D. Siegel, C. Carlson	2	78	0	7	0	0	0	−8.458	10.012	−140.17	−123.55	2006
BBOP	N. Nelson, D. Siegel	116	78	7	9	40	0	6	31.446	31.815	−64.991	−64.019	1994–2011
Active Fluorescence 2001	R. Morrison, H. Sosik	1	4	0	0	0	0	0	31.919	40.097	−70.528	−69.784	2001
Kieber Photochemistry 02	R. Morrison, H. Sosik	1	69	0	7	0	0	0	38.709	42.511	−75.564	−67.599	2002
GLOBEC	R. Morrison, H. Sosik	5	23	0	9	0	0	0	41.753	43.799	−70.445	−65.685	1997–1999
FRONT	R. Morrison, H. Sosik	3	9	0	4	0	0	0	42.245	40.985	−70.558	−71.75	2000–2002
CLIVAR A16N 2003	R. Freely	1	0	0	0	69	0	8	−6.004	63.295	−29.001	−19.994	2003
CLIVAR A165 2005	R. Freely	1	0	0	0	49	0	3	−60.01	−2.334	−36.21	−24.997	2005
CLIVAR A20 2003	R. Freely	1	0	0	0	27	0	3	7.064	42.637	−53.51	−51.12	2003
CLIVAR A22 2003	R. Freely	1	0	0	0	37	0	6	11.001	39.857	−69.932	−64.161	2003
CLIVAR I09N 2007	R. Freely	1	0	0	0	51	0	2	−28.31	18.004	86.782	95.013	2007
North Carolina 2005	R. Stumpf, P. Tester	5	32	4	20	0	0	0	34.096	35.433	−76.693	−75.756	2005
North Carolina 2006	R. Stumpf, P. Tester	2	12	0	6	0	0	0	34.014	35.228	−76.388	−76.028	2006
Chesapeake Light Tower	R. Zimmerman, G. Cota	3	71	59	32	0	0	0	36.803	36.969	−76.101	−75.551	2005–2007
North Sea	R. Doerffer	1	32	0	0	0	0	0	52.226	55.367	0.591	8.123	1994
ICESCAPE	S.B. Hooker, A. Neeley	3	1609	31	31	85	28	6	56.211	73.828	−168.98	−150.44	2001–2011
B07	S.B. Hooker, M.E. Russ	1	11	0	3	20	0	0	42.65	43.18	−70.868	−70.616	2009
MALINA	S.B. Hooker, V. Wright	1	25	22	0	28	1	0	69.246	72.054	−140.83	−126.5	2009
B08	S.B. Hooker, J. Chaves	1	0	0	0	2	0	1	31.667	31.698	−64.169	−64.164	2009
COASTAL	S.B. Hooker, M.E. Russ	1	13	0	10	2	0	0	42.708	43.434	−70.794	−69.865	2008
USM pCO2	S. Lohrenz	1	1	0	0	0	0	0	28.858	28.858	−89.47	−89.47	2005
Catlin Arctic Survey	V. Hill	1	8	0	0	0	0	0	78.771	78.771	−104.72	−104.72	2011
AMT	W. Balch	5	23211	7703	1918	0	0	0	−47.27	49.716	−55.455	18.611	2005–2011
Gulf of Maine	W. Balch	33	8077	0	3251	0	0	0	42.683	44.058	−70.267	−66.172	2005–2008
Scotia Prince Ferry	W. Balch	47	11521	51	8397	0	0	0	43.604	43.798	−70.026	−66.164	2001–2004
2009oct Chesapeake	W.J. Rhea	1	13	10	4	0	0	0	38.136	39.062	−76.448	−76.229	2009
Totals		535	48574	8857	14568	1957	255	302	−71	78.771	−177.99	179.545	1984–2012

* IS is in situ $R_{rs}(\lambda)$, **SAT is satellite $R_{rs}(\lambda)$

2.2.1. CDOM

CDOM absorption was measured in field experiments using a variety of instruments and protocols. Examples include in-line filtered (generally 0.2 µm) flow-through systems outfitted with ac-9 or ac-S absorption and attenuation meters (Wet Labs) and processed to $a_g(\lambda)$ [49–51], as well as discrete sampling and filtration for bench-top spectrophotometry [52], or in liquid capillary waveguides [53]. Unfortunately, SeaBASS metadata did not historically specify which methods or protocols were used in data collection or processing, but more recently (since approximately 2012), investigators have been required to submit ancillary documentation, such as instrument calibration records, and encouraged to submit documentation retroactively.

CDOM data measured in situ (i.e., with ac-9 or ac-S instruments; 33.5% of the preliminary CDOM dataset) were subject to particulate and bubble contamination, especially in experiments in which an automated in-line flow valve switched between filtered and unfiltered water presenting the opportunity for unfiltered water to reside in the plumbing during CDOM data collection. To identify and eliminate particle contamination, any CDOM records with a notable (i.e., ≥ 0.006 m^{-1}) increase in absorption at 676 nm (a phytoplankton absorption peak) above the absorption curve from 650–715 nm were considered contaminated and removed (109 records).

Nonlinear, least squares minimization was used to fit $a_g(\lambda)$ to Equation (1) for calculating slopes of all hyperspectral a_g into seven spectral ranges: 275–295 nm, 290–600 nm, 300–600 nm, 350–400 nm, 350–600 nm, 380–600 nm, and 412–600 nm. Multispectral ac-9 data were fitted for slope using the six wavebands in the 412–555 nm range. To reduce outliers and noisy data, any CDOM slope data found to be outside of the range 0.005–0.05 nm^{-1} were considered unrealistic and eliminated, together with the $a_g(\lambda)$ data used to calculate them. This accounted for only 125 hyperspectral records in the 300–600 nm range (spectrophotometric), but nearly 6,000 records in the 412–600 nm range (predominantly flow-through). To further reduce outliers and noisy records, S_g and a_g data were eliminated where S_g in any slope range was greater than two standard deviations from the median for the entire database, or where they were outside the 2nd and 98th percentiles. This reduced the database of CDOM by nearly 11,000 records. In addition, 460 records were removed for $a_g(676) > 0.1$ m^{-1}, $a_g(715) > 0.05$ m^{-1}, or an average $a_g(\lambda > 680$ nm$) > 0.05$ m^{-1}, and an additional 1,057 CDOM records with extreme outliers (>4 standard deviations from the median) in the red ($\lambda > 620$) were eliminated.

2.2.2. DOC

While DOC was included in about 850 SeaBASS records, the majority of the carbon data retained after surface and spatial binning (see 2.2.4) were from the Hansell/Carlson datasets. In total, 1957, 625, and 45 stations included DOC, POC, and TOC, respectively. Outliers (1st and 99th percentiles) were eliminated, and stations were merged with the CDOM records after surface and spatial binning. Specifically, Hansell/Carlson data were matched to CDOM field stations if they were within 1 h, 2.5 m depth, and 1 km in continental shelf waters (bottom depth ≤ 1000 m) and within 3 h, 5 m depth, and 5 km off the shelf. Multiple matches within these criteria were averaged and retained if individual measurements were with 1.5 standard deviations of the mean and the coefficient of variability of the ensemble was ≤ 0.25.

2.2.3. In situ Reflectances

SeaBASS searches for field radiometry targeted R_{rs} (or equivalently L_w and E_s, where $R_{rs} = L_w/E_s$, or L_{wn}, where R_{rs} is L_{wn} divided by the top of atmosphere solar irradiance [54]). A total of 135,966 independent field observations of R_{rs} were binned as described in 2.2.4, quality controlled as described in 2.2.5, and then matched to the CDOM database using the same spatial, temporal, and outlier elimination criteria used for DOC (2.2.2).

2.2.4. Bathymetry, Surface Averaging, and Spatial and Spectral Binning

Records with no reported bottom depth (~85% of the database) were matched to the nearest pixel in the UNESCO GEBCO 08 0.05 degree bathymetry grid (http://www.gebco.net/data_and_products/gridded_bathymetry_data/documents/gebco_08.pdf). The purpose of GOCAD is the development of satellite algorithms for surface retrievals, so data collected at depth were discarded as follows: on the continental shelf (defined here as bottom depth 1000 m or less) samples from deeper than 5 m were discarded, as were data from deeper than 10 m off the shelf (~34% of the database combined). 57,127 surface records remained. Samples collected in profile within the surface layer (top 5 m on-shelf, top 10 m off-shelf) were averaged. Samples collected in transect were additionally binned to a 0.5 km grid and averaged.

All absorption related IOPs were matched to the following wavebands with a 2.5 nm tolerance: 245 nm, 1 nm resolution between 250 and 555 nm, 560, 620, 630, 645, 650, 665, 670, 676, 680, 705, and 715 nm. Backscattering data were similarly matched to 1 nm bands from 400 to 700 nm. Hyperspectral in situ $R_{rs}(\lambda)$ were matched to both SeaWiFS bands (412, 443, 490, 510, 555, and 670 nm) and MODIS bands (412, 443 488, 531, 547, 667) by weighting the data to the instrument-specific spectral response functions for SeaWiFS, Aqua, and Terra (https://oceancolor.gsfc.nasa.gov/docs/rsr/rsr_tables/). Multispectral in situ $R_{rs}(\lambda)$ were matched to satellite bands to within a 2.5 nm tolerance.

2.2.5. Additional Quality Controls

In addition to those measures already discussed for CDOM and CDOM slope outliers in Section 2.2.1, $a_d(\lambda)$ records were considered unrealistic and removed at all wavelengths if they exceeded 12 m^{-1} anywhere within the spectral range reported. Similarly, $a_p(\lambda)$ was removed if it exceeded 20 m^{-1}, $b_{bt}(\lambda)$ if it exceeded 0.15 m^{-1}. $R_{rs}(\lambda)$ were eliminated if they exceeded 0.075 sr^{-1} or were less than −0.001 sr^{-1} in any band, or if they were outside the 95th percentile for any given band.

2.3. Satellite Imagery and Matching

Ocean color satellite imagery from SeaWiFS, MODIS-Aqua, and MODIS-Terra that matched the field observations were selected and processed for further analysis. Scripted calls to the NASA GSFC Ocean Color browser (http://oceancolor.gsfc.nasa.gov/cgi/browse.pl) after the 2012.0 MODIS-Aqua reprocessing (http://oceancolor.gsfc.nasa.gov/WIKI/OCReproc20120MA.html) were used to identify and download 1 km nominal nadir resolution L2 SeaWiFS, Aqua, and Terra satellite scenes within 0.05 degrees of field observations on the same day. These were spatially extracted for a 5 × 5 pixel array around the station location. In <1% of stations, high resolution SeaWiFS imagery was not available, and Global Area Coverage (GAC; nominally 4.4 km spacing) scenes were substituted. By default, data were masked based on standard L2 flags using the criteria described in [55]: land, high solar or satellite zenith angle, clouds, sea ice, high light, stray light, glint, low water leaving radiance, and atmospheric correction failure. Extracted satellite data were then evaluated for coincidence with field sampling stations. Criteria were principally based on those of Bailey and Werdell (2006). Specifically, extracted satellite pixel arrays were retained in the database if the overpass occurred within 8 h of field sampling. For each waveband of R_{rs}, negative and outlier pixels within each array (>1.5 standard deviations from the mean) were set to null values. Data were only retained in each waveband if greater than 50% of non-land pixels were still valid, with no fewer than five valid pixels in total. Finally, the mean R_{rs} values for each array were calculated and retained in the database only if those pixels had a coefficient of variation (CV) < 0.25 (rather than <0.15 applied in Bailey and Werdell (2006)). Of the 50,127 field stations with spatially gridded, depth binned and quality controlled CDOM data, 8252 stations had matching quality controlled Aqua imagery, 11,156 matched Terra imagery, and 11,818 matched SeaWiFS imagery.

For the purpose of further quality assurance, several match-up metrics were retained in the final database, including the time difference, CV, the number of matched satellite pixel arrays for each R_{rs}

channel, the areal extent of the matched pixels (nominally ~25 km^2), and the distance between the field sampling location and the central pixel (nominally < ~1 km). Sensor viewing angle, which can significantly increase error in estimates of satellite R_{rs} due to increased uncertainties in the atmospheric correction, is not available on a pixel-by-pixel basis in standard L2 products. On the other hand, ground sample area, which can be approximated from the geographic coordinates of pixel arrays, is a good proxy for sensor viewing angle, with larger areas representing larger viewing angles. Area is also a reasonable metric for accuracy of the geographic collocation, wherein R_{rs} averaged over larger areas of the ocean may not be representative of those measured at the sampling location, depending on the degree of spatial variability of ocean color within the sampled region. These match-up metrics were described in greater detail in [56].

2.4. Statistical Methods

Various metrics and visualization techniques are employed below to gauge the performance of algorithms. Retrieval parameters are compared with the same parameter collected in situ for both the optimization/tuning dataset (i.e., using in situ reflectances) and satellite validation. In addition to common metrics such as the number of samples (N), the standard deviation (*STD*), and the squared correlation coefficient (r^2), we evaluate the adjusted r^2 (r$^{2'}$):

$$r^{2'} = r^2 - \left(1 - r^2\right)\left(\frac{\beta_n}{N - \beta_n - 1}\right),$$
(2)

(where β_n is the number of regressors) which adjusts the r^2 downward to correct for the number of predictive values relative to the number of samples in, for example, multiple linear regression. The root mean square difference (*RMSD*) was also calculated:

$$RMSD = \sqrt{\frac{\Sigma_i^N (mod_i - ref_i)^2}{N}},$$
(3)

where "mod" is the model retrieved parameter and "ref" is the field measurement. The centered-unsigned (or unbiased) *RMSD* (*RMSD*$^{*'}$) was defined as follows:

$$RMSD^{*'} = \sqrt{\frac{\Sigma_i^N (mod_i - mean(mod_i)) - (ref_i - mean(ref_i))^2}{N}},$$
(4)

and the signed *RMSD*$^{*'}$ is simply the *RMSD*$^{*'}$ multiplied by the sign of the difference between the *STD* of the model retrieval and the *STD* of the field data (*RMSD*$^{*'}(\sigma_d)$). The bias and the normalized bias (*Bias**) are also employed:

$$Bias^* = \frac{\Sigma_i^N (mod_i - ref_i)}{N \times STD(ref)}$$
(5)

as well as the percent bias (%*Bias*),

$$\%Bias = 100 \times \frac{mean(mod - ref)}{mean(ref)},$$
(6)

and the mean average percent difference,

$$MAPD = 100 \times mean\left[abs\left(\frac{mod - ref}{ref}\right)\right].$$
(7)

While most of these metrics are fairly straightforward, a few warrant further explanation and context. A powerful graphical tool for assessing the skill of model performance—and comparing one model to another—is the Taylor diagram [57], which combines the *RMSD*$^{*'}$, *STD*, and correlation into a single figure in which proximity to the field data indicates how well the pattern of the modeled data matches the observations. This is made possible in two dimensions because of the relationship between the $RMSD^{*\prime}$, the correlation, and the variances of model and reference. Because the means of the model and reference are removed prior to calculating higher statistics shown in Taylor diagrams, they represent the comparisons between the patterns with any bias removed. For this reason, we have added color here to Taylor diagrams to include *%Bias*. Another graphical assessment used here which accounts for the bias (*Bias**) and adds a sign to the $RMSD^{*\prime}$ is the target diagram [58], in which the *y*-axis represents normalized bias of the model, the *x*-axis is signed-centered $RMSD$, and distance in any direction from the origin to the model is the total $RMSD$. Here, we also include color in our target diagrams to help visualize the $MAPD$.

3. Results and Discussion

3.1. Database Characteristics

GOCAD has over 40 times more CDOM records than NOMAD, and nearly 100 times as many spectra as IOCCG. It contains $a_g(412)$ data that is more normally distributed than either NOMAD or IOCCG, and a Chl distribution similar to NOMAD (Figure 1). IOCCG model data, while covering the same dynamic range as the two field databases, have unrealistically flat distributions of both CDOM and Chl, raising concerns for introducing bias when the dataset is used for algorithm development and optimization. The dynamic range in $a_g(\lambda)$ is larger in GOCAD than NOMAD (e.g., $a_g(412)$ from 0.005 to 2.457 m^{-1}, and from 0.0013 to 1.923m^{-1}, respectively), but the data distribution of GOCAD is narrower than NOMAD and IOCCG ($a_g(412)$ 75th minus 25th percentiles of 0.095 m^{-1}, 0.204 m^{-1}, 0.627 m^{-1}, respectively) around a lower mean absorption level (mean $a_g(412)$ = 0.120 ± 0.133 m^{-1}, 0.194 ± 0.266 m^{-1}, 0.514 ± 0.745 m^{-1}, respectively), reflecting the predominance of low CDOM, offshore data in the database. The large number of field records of CDOM in GOCAD, its range and mean value, all indicate that it is suitable for developing global retrieval algorithms.

The data distributions shown along the bottom row of Figure 1 (with the exception of Chl—see figure caption) show data used to optimize algorithms developed in this study (i.e., from field stations with matching in situ radiometry and IOPs), versus data used for validation—in this case with SeaWiFS wave bands and satellite imagery. For each parameter, distributions of optimization and validation data were compared for similarity to test by analysis of variance (ANOVA) whether the populations share a common mean. Optimization and validation dataset were found to differ ($p \ll 0.01$) for CDOM absorption and spectral slope, but not for DOC and salinity. The difference between the CDOM and DOC match-up datasets results from availability of the data (i.e., stations may not have both CDOM and DOC measurements in addition to in situ radiometry). Based on the distributions shown in Figure 1, as well as the geographic distributions highlighted in Figure 3, differences between the optimization and validation data populations for CDOM absorption and slope appear to derive from slightly fewer near-shore stations being present in the optimization dataset compared to the validation set, although there is clearly some endmember representation in the optimization set for near-shore conditions. We may conclude from this, however, that algorithms for CDOM absorption and spectral slope developed using these optimization data would perform best in oceanic conditions, while regional algorithms may be more accurate in coastal waters, or waters with very high CDOM absorption and low CDOM spectral slope.

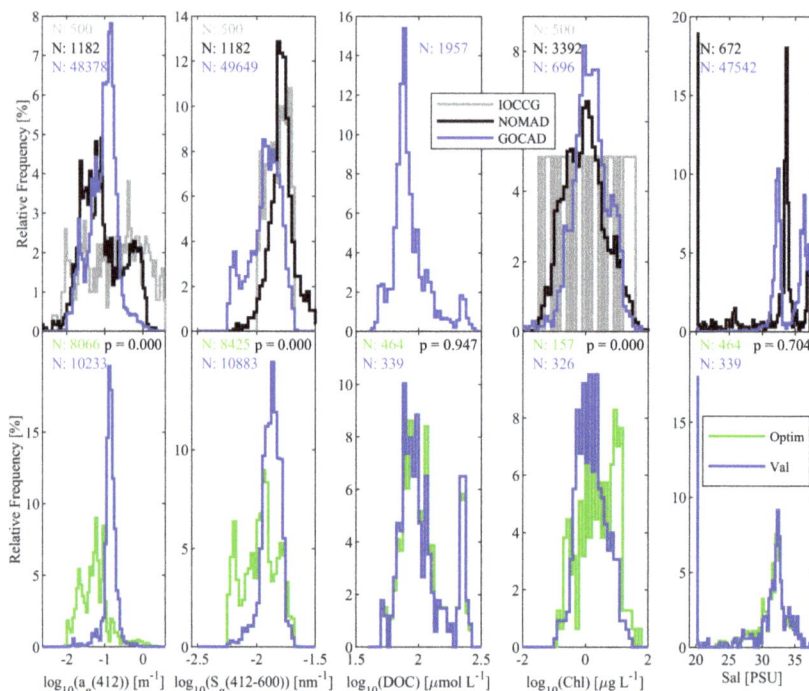

Figure 1. Top row: data distributions and counts (N) of relevant parameters and Chl (for context only) in Global Ocean Carbon Algorithm Database (GOCAD), NASA bio-Optical Marine Algorithm Dataset (NOMAD), and the synthetic ocean color dataset developed by the International Ocean Colour Coordinating Group (IOCCG). Bottom row: comparisons between the subset of GOCAD parameters used in optimization/tuning (Optim) and validation (Val) of algorithms (shown here with SeaWiFS match-ups, but also evaluated for MODIS Terra and Aqua with similar results). Populations of salinity and DOC share a common mean between optimization and validation datasets (ANOVA, $p > 0.01$).

IOCCG and NOMAD contain no UV CDOM data, so direct comparison of spectral slope is only possible in the VIS (Figure 2). The median S_g(412–600) is lower for GOCAD, demonstrating again the predominantly oceanic characteristics (i.e., photodegraded, primarily of marine origin, and presumably refractory) of the CDOM in the database. Slope decreases significantly ($p \ll 0.01$) as the reference wavelength (i.e., the shortest wavelength in the spectral range) increases from 275–412 nm. Overall, the variability in spectral slope for each range is quite low—generally no more than a factor of 2–3. This narrow dynamic range in slope within each waveband presents a challenge for retrieving fine scale differences in CDOM slope by limiting the sensitivity of algorithms built from inherently uncertain ocean color. However, errors in the retrievals should be small relative to the absolute magnitude of the slope even if the algorithm sensitivity (e.g., correlation between retrievals and field measurements) is low.

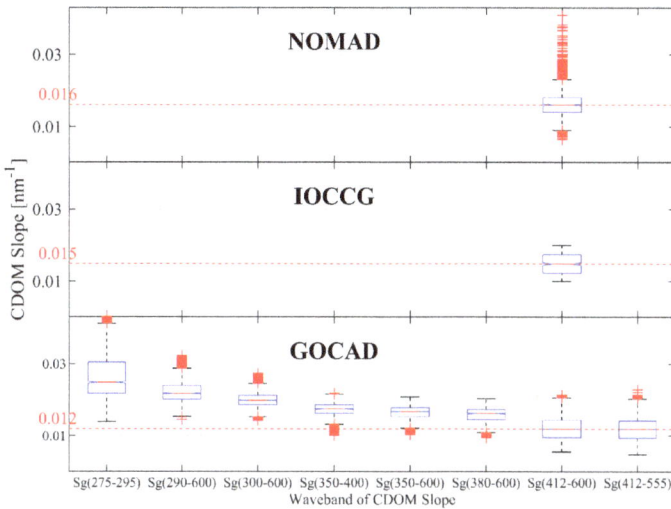

Figure 2. Exponential slope of CDOM in NOMAD, IOCCG, and GOCAD. Median values for $S_{412–600}$ are highlighted in red for comparison. NOMAD and IOCCG lack UV CDOM.

Geographic locations of GOCAD field stations overlap with NOMAD stations (Figure 3). We can see that many of the NOMAD stations were ultimately excluded from GOCAD during the quality assurance analysis described in Section 2.2. Highlighted in the central panel of Figure 3 are those stations with high-quality in situ radiometry, which were set aside for tuning, training, and optimization of ocean color algorithms. The geographic distributions of both the training data and the validation data show a representational combination of stations from both offshore and nearshore waters, which theoretically improves the odds of being able to retrieve a broad dynamic range of bio-optical properties, although as pointed out above the optimization data appears to be slightly dominated by oceanic stations. It is clear from Figure 3 that while there is significant overlap in the CDOM and DOC datasets in certain regions such as northern Alaska and the mid-Atlantic Bight in the Northeastern U.S., globally they follow a somewhat different pattern, and many DOC field stations are not obviously represented in the CDOM dataset.

The dense concentrations of field stations sampled in relatively smaller regions such as the Northeastern U.S. are difficult to resolve at the small scale in Figure 3. Figure 4 shows three-dimensional maps of select sub-regions with $a_g(412)$, $S_g(275–295)$, and $S_g(412–555)$, including the Northeast US and coastal Alaska between the Chukchi Sea and the Beaufort Sea. These are set in broad continental shelves with numerous nearby river outflows. Not surprisingly, CDOM is high throughout the regions shown in Figure 4 with low spectral slope in the UV. CDOM and $S_g(275–295)$ increase and decrease, respectively, in close relation to distance from shore, as expected given the considerations discussed in Section 1.2 and elsewhere. Variability is higher for spectral slope in the VIS ($S_g(412–555)$), but it generally follows the opposite pattern from that in the UV, i.e., decreasing with distance from new sources of CDOM. This is indicative of aging processes as the newly mobilized, near-shore CDOM mixes seaward and photo- and microbial degradation reduce absorption in the UVA relative to the UVB (thus increasing $S_g(275–295)$ and relative to the VIS (thus decreasing $S_g(412–555)$). It may indicate marine sources of CDOM with chromophores that absorb in the UVA and blue rather than terrestrial sources that also absorb in the UVB. These patterns are perhaps clearest at the outflows of the Colville River (~135° W and 70° N) and the Chesapeake Bay (~77° W and 37° N). An interesting exception for $S_g(412–555)$ can be found in the Gulf Stream transect (~70° W and 37°–40° N; GOMECC-2 experiment, Table 2), where slope increases upon entering the productive waters at the edge of the Gulf Stream despite a lack of CDOM increase, and then rapidly declines upon entering the oligotrophic waters south of the Gulf Stream.

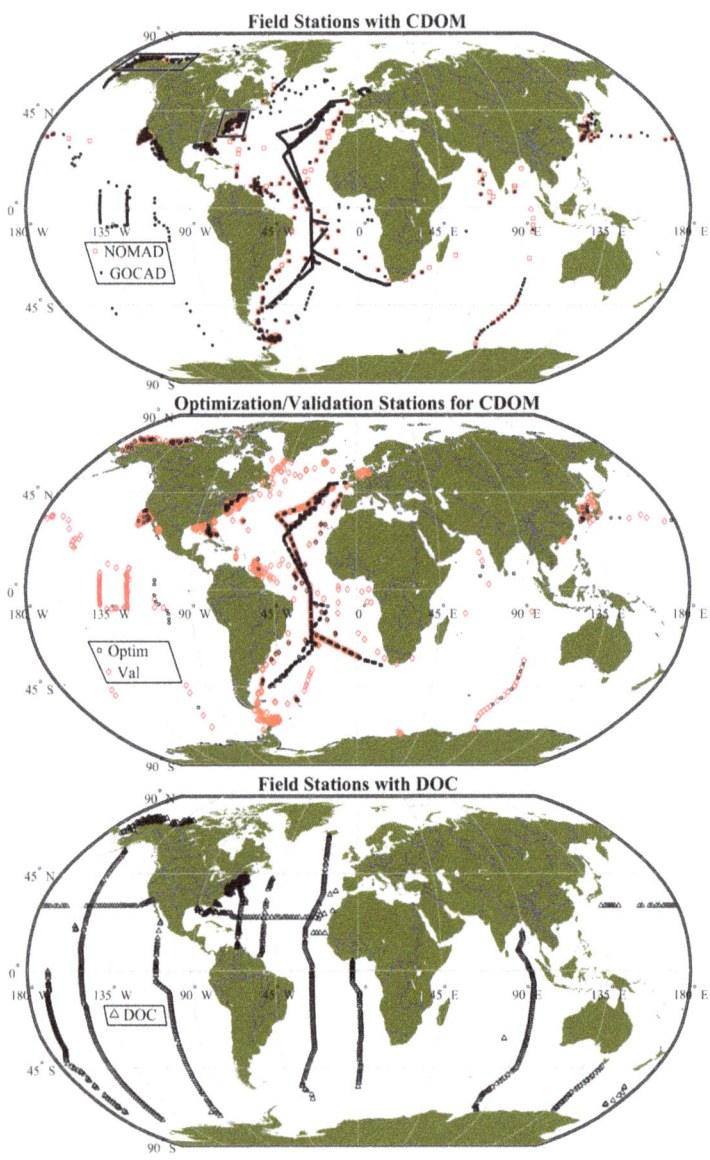

Field Stations with CDOM

Optimization/Validation Stations for CDOM

Field Stations with DOC

Figure 3. Global distribution of GOCAD and NOMAD field stations for CDOM (upper) and DOC (lower). The central panel shows the distributions of data within GOCAD separated into optimization (Optim) and validation (Val) dataset. Stations used in algorithm tuning are shown as red circles, the remainder of stations were available for satellite validation. The boxed subregions in the upper panel are shown in greater detail in Figure 4.

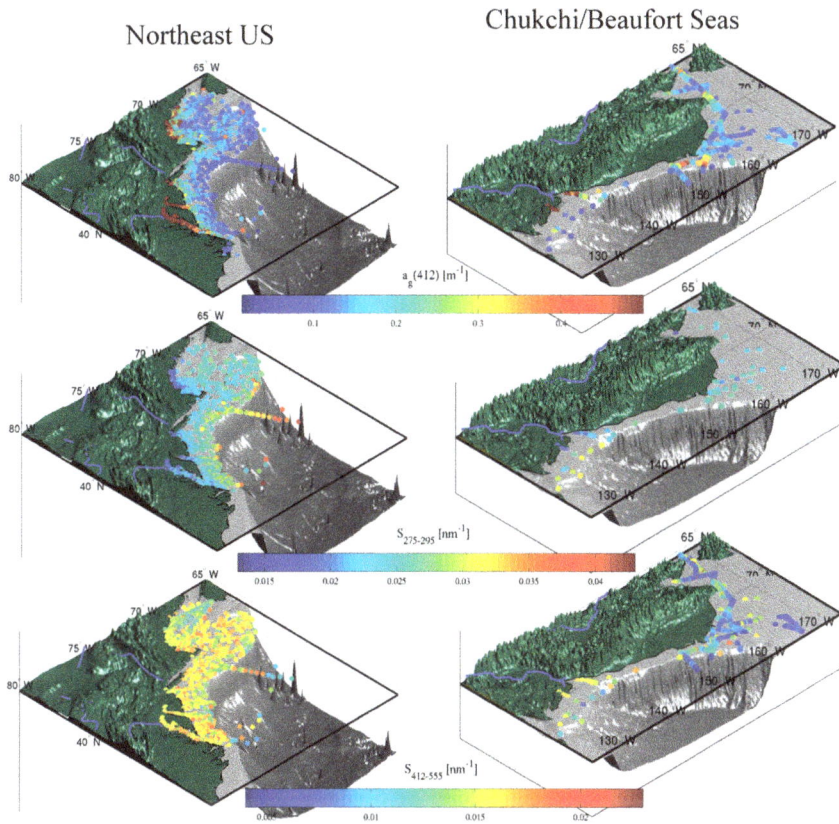

Figure 4. Examples of CDOM absorption at 412 nm (top row), and CDOM spectral slope in the UVB (middle row) and VIS (bottom row) from GOCAD show patterns which reflect the sources and age of CDOM in environments stretching from estuarine, such as the Chesapeake Bay in the eastern U.S., to stations sampled well offshore.

3.2. Algorithm Tuning and Validation

3.2.1. Algorithm Structures and Optimization

Both empirical and semi-analytical approaches to ocean color retrievals of CDOM, S_g, and DOC were explored using the GOCAD dataset. Of the former, a band ratio, single exponential decay model similar to that presented by Mannino (2008) was tested, but found to be better suited for the continental shelf waters for which it was derived rather than for the deep ocean, and will not be presented here. A multiple linear regression (MLR) approach was tested matching the natural logarithm of R_{rs} in four ocean color bands with the logarithm of $a_g(\lambda)$ and S_g at each waveband described in Section 2.2.1, and DOC. The least square difference minimization regression, performed using Matlab's *regstats* function (www.mathworks.com), follows the form:

$$\ln(Y) = \beta_0 + \beta_1 \times \ln(R_{rs}(\lambda_1)) + \beta_2 \times \ln(R_{rs}(\lambda_2)) + \beta_3 \times \ln(R_{rs}(\lambda_3))$$
$$+ \beta_4 \times \ln(R_{rs}(\lambda_4)) \tag{8}$$

where β_0–β_4 are the regression coefficients, Y is the retrieval parameter, and λ_1–λ_4 are the sensor-specific wavelengths (i.e., 443, 488, 531, and 547 nm for MODIS, 443, 490, 510, and 555 nm for

SeaWiFS). Using monthly, binned L3 Aqua imagery for 2010, MLR retrievals were used to establish the 99th percentiles for each retrieval waveband of a_g. Retrievals above these values were considered outside the scope of this global algorithm, and eliminated. Regression coefficients, statistics, and thresholds are presented in Figure 5, and Tables 3–5. Model retrievals plotted against field data are well organized about the 1:1 line with low scatter, particularly in the UVA, which is reflected in high correlation coefficients and low error and bias. *MAPD* is below 30% for all bands below 488 nm; from this band to higher wavelengths, the CDOM signal becomes very weak in most of the global ocean.

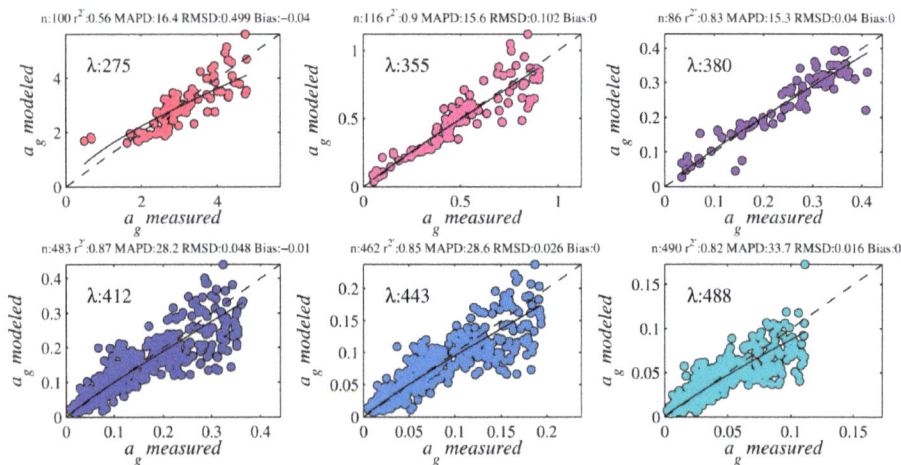

Figure 5. MLR retrievals of CDOM plotted against field data for the tuning dataset (i.e., in situ $R_{rs}(\lambda)$). The solid line shows the fit through the data, and the 1:1 line is dashed.

For the reasons outlined in Section 1.4 (i.e., a large and variable portion of DOC is unpigmented), DOC derived directly from ocean color alone using MLR was not robust (Table 5; MLR1). However, satellite retrievals of sea surface salinity are now available thanks to the Aquarius mission (http://aquarius.nasa.gov/), and for CDOM, salinity was a reasonable choice as an additional proxy for water type considering it will generally reflect proximity to sources of fresh water and CDOM as well as distinguishing water masses (e.g., Gulf Stream). Using GOCAD, a multiple linear regression approach was developed for retrieving DOC from $a_g(355)$ (in place of $R_{rs}(\lambda_1)$ in Equation (8)) and salinity (in place of $R_{rs}(\lambda_2)$), and proved very robust (e.g., $r^2 = 0.91$, %*Bias* = 0). Using CDOM and salinity as predictors significantly improved retrievals of DOC (Table 5; MLR2), with $r^{2'}$ increasing from 0.76 to 0.91, and *MAPD* dropping by about three percentage points. The strength of the correlation between field and retrieved DOC to CDOM and salinity is stronger than expected, considering the many ways in which changes in DOC, CDOM, and salinity may diverge across seasons or from region to region. It is worth pointing out that other factors may be contributing to the stronger statistical performance of MLR2 over MLR1, such as the higher number of coincident predictors and retrievals, as well as the absence of uncertainties associated with reflectance data in the tuning dataset (i.e., DOC is derived directly from CDOM absorption and salinity). Caution is therefore advised when applying this DOC algorithm in regions in which DOC is known to change without commensurate changes in CDOM and/or salinity. For example, the accumulation of DOC in surface subtropical waters including the BATS field station [59,60] appears to be decoupled from CDOM (Norman Nelson and Craig Carlson, personal communication).

Table 3. Coefficients of the MLR algorithm for retrieving CDOM absorption ($ag(\lambda)$) following Equation (8) and metrics of fit for the optimization data set.

	β_0	β_1	β_2	β_3	β_4	N	$r^{2'}$	RMSD [m^{-1}]	MAPD [%]	%Bias [%]	Threshold [m^{-1}]
						MODIS					
[nm]		443	488	531	547						
275	0.089	−0.540	−1.142	3.444	−1.875	100	0.56	0.499	16	−1.4	4.825
355	−2.246	−1.186	−0.558	2.912	−1.336	116	0.90	0.102	16	−0.9	0.9104
380	−2.263	−0.300	−1.882	3.831	−1.787	86	0.83	0.040	15	−1.9	0.4341
412	−2.535	−0.563	−1.294	1.606	0.170	483	0.87	0.048	28	−4.3	0.36419
443	−3.287	−0.727	−0.922	1.278	0.261	462	0.85	0.026	29	−4.3	0.1984
488	−3.722	−0.377	−1.429	1.424	0.300	490	0.82	0.016	34	−5.9	0.1114
						SeaWiFS					
[nm]		443	490	510	555						
275	−2.477	−2.880	2.225	0.480	−0.252	174	0.76	0.659	25	−2.2	4.825
355	−4.199	−2.563	1.214	0.955	−0.040	189	0.87	0.118	26	−2.2	0.9104
380	−4.544	−1.808	0.175	1.181	0.001	150	0.80	0.055	26	−2.4	0.4341
412	−6.004	−0.861	−0.006	−0.346	0.515	8066	0.37	0.035	56	−13.0	0.36419
443	−6.410	−0.743	−0.145	−0.367	0.547	8037	0.33	0.026	58	−13.6	0.1984
490	−7.014	−0.736	0.142	−0.796	0.678	7978	0.28	0.016	65	−15.5	0.1114

Another empirical approach tested here was the machine learning approach known as Random Forests [61,62], which is a method for multivariate, non-linear, non-parametric regression designed to help minimize over-fitting of the training dataset. The method improves on standard decision tree regression performance by using an ensemble of independent decision trees; bootstrapping for the regression is achieved by repeatedly, randomly resampling the original dataset to provide an ensemble of smaller independent datasets, which are each used to grow a decision tree (hence the term random forest tree-bagger, or RFTB). Here, 200 independent decision trees were used, and each tree is trained on approximately 66% of the training dataset. The inputs (i.e., reflectances) and retrievals of the regression (i.e., CDOM, CDOM slope, and DOC) were the same as in the MLR. Model performance and statistics for select bands in the UV and VIS with the training dataset are presented in Figure 6. Comparisons of model retrievals to field data are fairly well correlated, but error and bias are quite high, with *MAPD* reaching several hundred percent.

Table 4. Coefficients of the MLR algorithm for retrieving CDOM spectral slope ($S_g(\lambda)$) following Equation (8) and metrics of fit for the optimization data set.

λ	β_0	β_1	β_2	β_3	β_4	N	$r^{2'}$	RMSD [nm^{-1}]	MAPD [%]	%Bias [%]
						MODIS				
[nm]		443	488	531	547					
275–295	−3.289	0.270	−0.335	1.051	−0.921	322	0.61	0.002	6.4	−0.3
290–600	−3.471	0.127	−0.251	1.025	−0.843	322	0.38	0.002	6.2	−0.3
300–600	−3.607	0.044	−0.153	0.881	−0.722	324	0.30	0.001	5.7	−0.3
350–400	−3.924	−0.242	0.055	0.935	−0.710	331	0.26	0.001	6.8	−0.3
350–600	−3.908	−0.204	0.098	0.609	−0.463	331	0.22	0.001	6.3	−0.3
380–600	−3.912	−0.152	0.127	0.236	−0.173	340	0.14	0.001	6.3	−0.3
412–600	−4.219	−0.180	0.137	0.168	−0.131	782	0.16	0.002	7.4	−0.5
412–555	4.195	−0.162	0.147	0.096	−0.084	760	0.10	0.002	7.6	−0.5
						SeaWiFS				
[nm]		443	490	510	555					
275–295	−3.012	0.427	−0.459	0.357	−0.228	424	0.77	0.002	6.8	−0.4
290–600	−3.425	0.131	−0.085	0.145	−0.130	424	0.46	0.002	6.6	−0.4
300–600	−3.615	0.004	0.014	0.160	−0.129	426	0.29	0.002	6.0	−0.3
350–400	−3.968	−0.298	0.178	0.301	−0.150	433	0.23	0.002	7.4	−0.4
350–600	−4.058	−0.288	0.091	0.356	−0.138	433	0.33	0.001	6.9	−0.4
380–600	−4.072	−0.226	0.088	0.208	−0.051	445	0.32	0.002	7.2	−0.3
412–600	−4.498	−0.466	0.690	−0.202	−0.015	8550	0.06	0.004	28.5	−5.2
412–555	−4.533	−0.455	0.683	−0.214	−0.012	8425	0.05	0.004	28.2	−5.1

Table 5. Coefficients of the MLR algorithm for retrieving DOC following Equation (8) and metrics of fit for the optimization data set.

Algorithm	$ß_0$	$ß_1$	$ß_2$	$ß_3$	$ß_4$	N	$r^{2'}$	RMSD [μmol L^{-1}]	MAPD [%]	%Bias [%]
					MODIS					
[nm]		443	488	531	547					
MLR1	4.923	0.641	−2.424	3.503	1.692	183	0.76	23.9	13.9	−1.5
					SeaWiFS					
[nm]		443	490	510	555					
MLR1	5.272	0.526	−2.982	2.623	0.089	246	0.68	30.3	18.8	−3.0
					$a_g(355)$ and Salinity					
		$a_g(355)$	Sal							
MLR2	192.718	26.790	−3.558	-	-	464	0.91	15.2	10.6	0.0

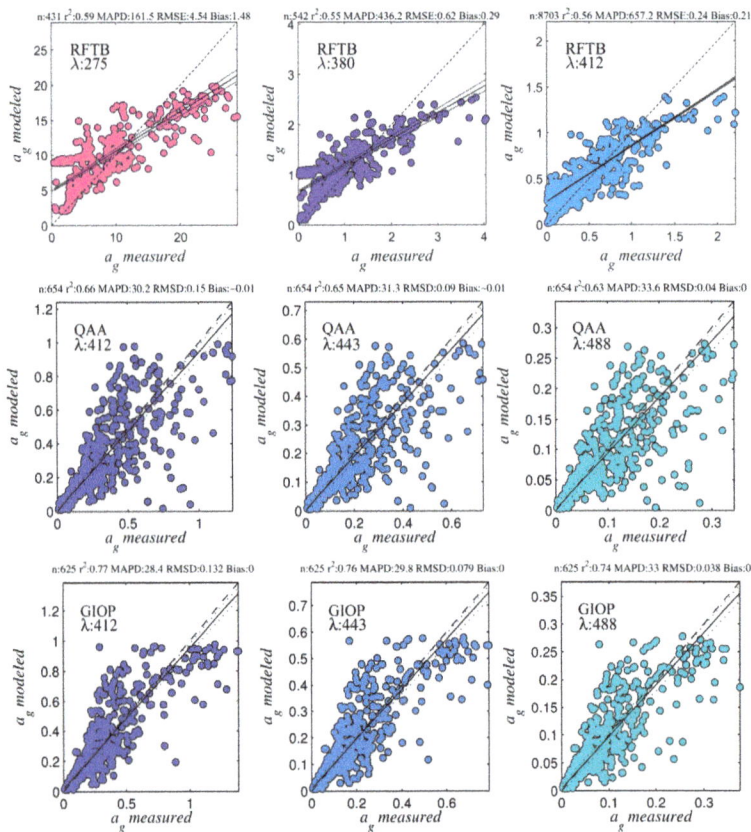

Figure 6. Random forest tree-bagger (RFTB), quasi-analytical algorithm (QAA), and generalized inherent optical property (GIOP) retrievals for tuning datasets.

Semi-analytical approaches included the Quasi-Analytical Algorithm of (QAA) [5,63,64], and the Generalized Inherent Optical Property (GIOP) algorithm [65]. These have the advantage that they are based on theoretical models for how the light field is affected by the inherent properties of the water, but can only retrieve IOPs at those wavebands for which $R_{rs}(\lambda)$ is measured (i.e., they do not extend into the UV for the current and historical suite of satellite sensors). In the future, however, data from GOCAD and elsewhere could be used in the development of linear matrix inversion-type semi-analytical algorithms including basis vector models extending into the UV for CDOM, thereby potentially enabling their retrieval directly using SAAs. In fact, using GOCAD to build more globally representational basis models extending CDOM into the UV may not only provide better retrievals of CDOM, but also of the other concurrently retrieved optical properties from linear matrix inversion. Both the GIOP and QAA invert the $R_{rs}(\lambda)$ to retrieve the water column IOPs following the theory that sea surface reflectance at a given wavelength is proportional to the backscattering coefficient, and inversely proportional to the absorption coefficient [16,66]. Each uses various assumptions, empirical parameterizations, and mathematical inversion techniques to solve for the IOPs and partition them into their constituents. These include the total backscattering coefficient, $b_{bt}(\lambda)$, backscattering by particles, $b_{bp}(\lambda)$, absorption by total particles, by phytoplankton, and by the combination of non-algal particles and CDOM, $a_{dg}(\lambda) = a_d(\lambda) + a_g(\lambda)$, where $a_d(\lambda)$ is non-algal (or detrital) absorption). These latter properties are similar in spectral shape, and therefore difficult to partition, which presents a challenge if we wish to compare the retrievals of SAAs to the other algorithms presented here. Therefore, while we do not re-develop or re-optimize the SAAs here—using them as published—we do utilize GOCAD to facilitate the separation of dissolved and detrital absorption components. Specifically, we solve for $a_g(\lambda)$ by assuming that $a_d(\lambda)$ is a function of the combined backscattering by water, non-algal particles, and the dissolved absorption by CDOM, which we assume does not backscatter, although there is some evidence supporting backscattering by colloids [46]. These SAAs retrieve only the combined $b_{bp}(\lambda)$ from phytoplankton and non-algal particles, but the latter tend to have a higher refractive index, and therefore contribute far more strongly to the backscattering signature, e.g., [67] and references therein. An empirical relationship was developed between $a_d(410)$, $b_{bt}(550)$ and $a_{dg}(410)$, and then $a_g(410)$ was found by subtracting $a_d(410)$ from SAA retrievals of $a_{dg}(410)$:

$$a_d(410) = 0.06822 \times a_{dg}(410) + 1.623 \times b_{bt}(550) + 0.0002123 \tag{9}$$

Due to a paucity of $a_d(\lambda)$ and $b_{bt}(\lambda)$ observations in GOCAD, this relationship was tuned for multiple linear regression using the IOCCG synthetic dataset ($r^{2'} = 0.76$, $RMSD = 0.07$, bias $= -0.004$ m^{-1}, $MAPD = 75\%$, N = 464). $a_g(410)$ was expanded using Equation (1) to other wavebands with the empirical retrieval for $S_g(412$–$555)$ (derived as per Equation (8) and Table 4). Regression statics for the optimization data are shown for the QAA and GIOP in Figure 6, with slightly better results in the GIOP. Although the current version of GOCAD is less well populated with some optical properties than others (i.e., data collection targeted carbon-related properties and only included others if they happened to be in the same SeaBASS file), the digital structures for each property mentioned in this section are included in the database, and future algorithm investigation (particularly using SAAs) would greatly benefit from incorporation of these data into GOCAD or a similar, climate-scale, global database.

3.2.2. Algorithm Validation

This work represents the most rigorously validated set of global CDOM and DOC algorithms to date. Optimization/training of algorithms as described in the previous section was conducted on GOCAD field stations with coincident in situ radiometry. These stations were then set aside from validation, which was performed only on those remaining stations in GOCAD that had coincident satellite imagery (i.e., MODIS Aqua, Terra, and SeaWiFS). In addition to the algorithms already mentioned, two other empirical algorithms based on band ratio approaches were included in validation analysis. The approach of Shanmugam (2011) [31] (hereafter Shan11) used a power-law relationship between the ratio of $R_{rs}(443)/R_{rs}(555)$ and $a_g(350)$ and $a_g(412)$, and performed well using the NOMAD dataset. The ratio of these was then used in another power-law function to solve for $S_g(350$–$412)$. Tiwari and Shanmugam (2011) [32] (hereafter TS11) used linear functions to relate the ratio of $R_{rs}(670)/R_{rs}(490)$ to $a_g(412)$ and $a_g(443)$, and solved for $S_g(412$–$670)$ analytically by inverting Equation (1). As these two algorithms were tuned using SeaWiFS bands, a slight adjustment was made to MODIS input reflectances to obtain the SeaWiFS reflectances required (only MODIS validation is shown here graphically).

The performance of all algorithms in independent validation is weaker than for optimization (Tables 6–8, Figures 7–9). This should not be surprising considering satellite imagery is subject to higher uncertainty associated with atmospheric correction, where the atmosphere comprises ~90% or more of the signal received by the satellite sensor. Furthermore, satellite match-ups exacerbate the issue of temporal and geographic coincidence with in situ measurements. Any regions of moderate to high variability in surface properties will likely not be well captured by the average of a nominally 5 km × 5 km pixel array. Nevertheless, results are encouraging, particularly for the MLR approach and particularly in the UV, where the CDOM signal is strongest (in terms of in situ data) and the SAAs are not currently useful.

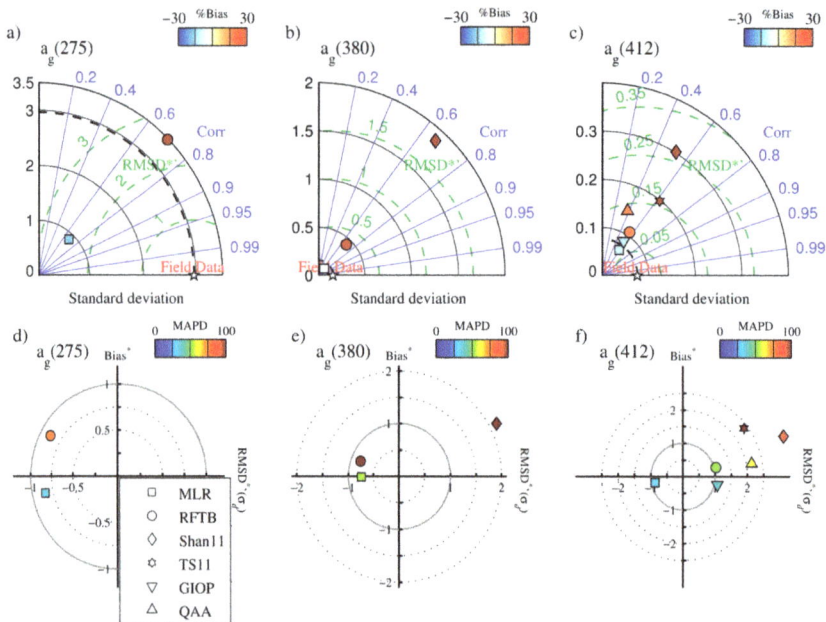

Figure 7. Taylor diagrams (top row) and target plots (bottom row) depicting comparative algorithm performance for retrieving CDOM absorption at 275 nm, 380 nm, and 412 nm from MODIS Aqua.

Figure 7 shows Taylor and target diagrams comparing the CDOM absorption retrieval metrics for various algorithms as described in Section 2.4. In the UVB (275 nm), and UVA (380 nm), only the empirical approaches were feasible, while SAAs (i.e., QAA and GIOP) are also shown at 412 nm. MLR and RFTB perform comparably with respect to correlation between the models and measurements at 275 nm, although MLR does have significantly lower *MAPD* and bias (Table 6; MLR highlighted in bold), and outperforms the RFTB at 380 nm in all but correlation for all sensors. MLR also outperforms all other CDOM absorption algorithms at 412 nm, although GIOP does not appear significantly worse as seen by its proximity to field data in the Taylor plots and the origin in the target diagrams. MLR shows a relatively strong negative bias in most sensors and channels, which is the result of underestimation in high CDOM waters (data not shown).

Table 6. Validation of algorithms for retrieving CDOM absorption (ag(λ)).

Algorithm	λ	N	r^2	RMSE	MAPD	%Bias
	[nm]			[m^{-1}]	[%]	[%]
			MODIS-Aqua			
MLR	275	186	0.47	2.499	33	−17
MLR	380	188	0.45	0.117	54	−1
MLR	412	7626	0.33	0.068	33	−10
RFTB	275	191	0.50	3.100	78	37
RFTB	380	243	0.47	0.400	102	28
RFTB	412	6820	0.30	0.100	52	18
Shan11	350	237	0.45	1.669	134	108
Shan11	412	7748	0.28	0.287	92	72
TS11	412	7299	0.39	0.201	102	86
GIOP	412	6116	0.30	0.077	40	−12
QAA	412	6133	0.14	0.137	59	18
			MODIS-Terra			
MLR	275	291	0.43	2.746	48	−26
MLR	380	326	0.32	0.337	45	−29
MLR	412	10612	0.19	0.081	35	−1
RFTB	275	171	0.35	2.950	72	22
RFTB	380	269	0.35	0.380	125	21
RFTB	412	6962	0.20	0.110	63	34
Shan11	350	345	0.35	1.308	97	75
Shan11	412	10734	0.16	0.311	129	108
TS11	412	9607	0.20	0.202	109	87
GIOP	412	7976	0.12	0.108	51	11
QAA	412	8048	0.05	0.200	91	52
			SeaWiFS			
MLR	275	342	0.25	2.976	49	−57
MLR	380	418	0.32	0.318	58	−53
MLR	412	10233	0.10	0.081	47	23
RFTB	275	199	0.38	2.660	90	40
RFTB	380	423	0.36	0.370	209	41
RFTB	412	11594	0.06	0.110	55	27
Shan11	350	444	0.51	1.425	108	86
Shan11	412	10451	0.26	0.229	100	79
TS11	412	8890	0.20	0.250	152	128
GIOP	412	7863	0.11	0.085	45	11
QAA	412	7904	0.07	0.162	80	53

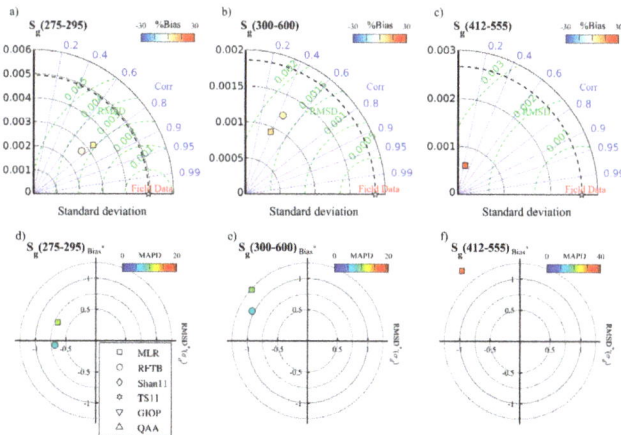

Figure 8. Taylor diagrams (top row) and target plots (bottom row) depicting comparative algorithm performance for retrieving CDOM slope at 275–295 nm, 300–600 nm, and 412–600 nm from MODIS Aqua. Results from Shan11 and TS11 were suppressed to preserve scale.

Table 7. Validation of algorithms for retrieving CDOM spectral slope (S_g).

Algorithm	Waveband	N	r^2	RMSE	MAPD	%Bias
	[nm]			[nm^{-1}]	[%]	[%]
		MODIS-Aqua				
MLR	275–295	187	0.62	0.0034	11	6
MLR	300–600	213	0.15	0.0023	10	8
MLR	412–555	7825	0.06	0.0040	32	23
RFTB	275–295	214	0.58	0.0033	8	−1
RFTB	300–600	244	0.20	0.0019	8	4
Shan11	350–600	188	0.01	0.2750	427	366
TS11	412–555	6223	0.02	0.0290	217	−209
		MODIS-Terra				
MLR	275–295	284	0.41	0.0039	12	3
MLR	300–600	318	0.16	0.0024	10	9
MLR	412–555	10719	0.06	0.0036	27	20
RFTB	275–295	242	0.43	0.0000	10	−3
RFTB	300–600	271	0.11	0.0000	8	5
Shan11	350–600	297	0.03	0.1250	255	176
TS11	412–555	8078	0.00	0.0280	208	−203
		SeaWiFS				
MLR	275–295	350	0.44	0.0047	14	6
MLR	300–600	417	0.11	0.0021	8	4
MLR	412–555	10883	0.03	0.0029	17	−12
RFTB	275–295	350	0.37	0.0053	11	−7
RFTB	300–600	418	0.09	0.0022	9	5
Shan11	350–600	372	0.01	0.1320	202	140
TS11	412–555	4716	0.03	0.0290	203	−198

CDOM spectral slope was only retrievable with empirical approaches. MLR and RFTB performed comparably to each other, although RFTB was not tested at S_g(412–555). In the application of retrieval algorithms for S_g below (Section 3.3), the MLR is used mainly for its simplicity, but we would expect RTFB retrievals to yield nearly equally accurate results. Shan11 and TS11 performed poorly (Table 7; MLR highlighted in bold). Correlations between modeled and measured CDOM slope were weak in the UVA and VIS, but as the dynamic range of the field data is quite low (Figure 2), error and bias were still low in the retrievals (Table 7). In all sensors and bands for the MLR and RFTB, S_g tends to be slightly overestimated in waters with low S_g, and slightly underestimated in waters with high S_g, indicating the weak sensitivity of these empirical approaches also reflected in the low correlation coefficients.

Table 8. Validation of algorithms for retrieving DOC.

Algorithm	N	r^2	RMSE	MAPD	%Bias
			[μmol L^{-1}]	[%]	[%]
		MODIS-Aqua			
MLR1	164	0.23	40.8	41	24.7
MLR2	382	0.89	27.8	16	−13.0
RFTB	161	0.57	27.3	26	13.8
		MODIS-Terra			
MLR1	158	0.23	40.2	32	13.9
MLR2	369	0.90	26.7	15	−12.3
RFTB	114	0.47	29.4	27	12.1
		SeaWiFS			
MLR1	274	0.29	34.2	34	4.5
MLR2	339	0.89	28.9	15	−14.2
RFTB	182	0.30	25.3	23	6.7

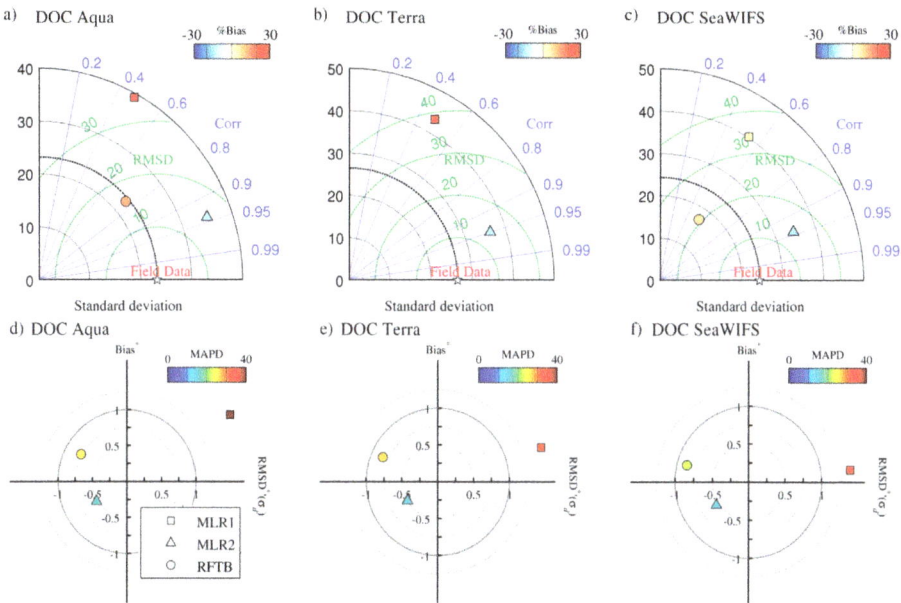

Figure 9. Taylor diagrams (top row) and target plots (bottom row) depicting comparative algorithm performance for retrieving DOC from MODIS Aqua and Terra, and SeaWiFS.

As anticipated, MLR retrievals of DOC using ocean color alone were only weakly correlated with field data (i.e., r^2 < 0.3, Table 8). RFTB performed considerably better, but was unable to match the performance of MLR2 (i.e., regression with retrievals of $a_g(355)$ and known salinity; highlighted in bold in Table 8). Due to the newness of the Aquarius mission, there were too few retrievals available for incorporation in these validation results, and further validation of this approach is encouraged based on these results.

We speculated above (Section 3.1) that small differences between the optimization and validation records may bias algorithm performance to favor oceanic waters. To test this hypothesis, a sensitivity analysis was evaluated for CDOM absorption retrieval by MLR to test correlations between algorithm error (percent error between retrievals and field data) and salinity, water column depth, and $a_g(412)$ measured in the field. We found no sensitivity to these factors (r^2 < 0.04 in each case, n = 29,757 for Aqua, Terra, and SeaWiFS combined), indicating that the algorithm is not optimized in a way that would limit its performance in, for example, high salinity, offshore waters, or fresher waters with high inputs of fresh CDOM. A geographic distribution of algorithm retrieval error (percent error) for $a_g(412)$ and $S_g(412–600)$ is shown in Figure 10.

A similar sensitivity analysis was evaluated for MLR2 (DOC retrieval) performance at validation stations to help identify limitations of the algorithm. We tested the correlation between the DOC retrieval error (percent difference between retrieved and measured DOC) and salinity, water column depth, and DOC concentration, but found no strong trends in the distribution of error (r^2 = 0.17, 0.19, 0.43, respectively, see included figures below), although it could be argued that absolute retrieval error increases somewhat (overestimates) at the extremely high salinity stations, and at extremely low DOC stations. In general, it appears that shallow stations underestimate DOC, and deeper stations tend to overestimate. The geographic distribution of error in algorithm retrievals (Figure 10) revealed no patterns with respect to distance from shore or nearby fluvial sources, but MODIS Aqua retrievals did overestimate DOC in southern oceans (south of 40° S, 41% ± 16%, n = 18) compared to minor underestimates from other sensors and at latitudes north of 40° S (−8% ± 18%, n = 1054).

Care should therefore be taken when evaluating algorithm retrievals in these areas. Only 1,090 stations (all sensors combined) were available for validation of the MLR2 and this analysis of sensitivity, and their distribution is not uniform across the world's oceans, but, as shown in Figure 1, a broad spectrum of water types with a large dynamic range of DOC were represented in both the optimization and validation datasets. Unfortunately, no validation stations for MLR2 were identified for mid-ocean gyres, and therefore the performance of the MLR2 in those waters remains poorly defined, and caution is advised in the interpretation of DOC retrievals in those areas.

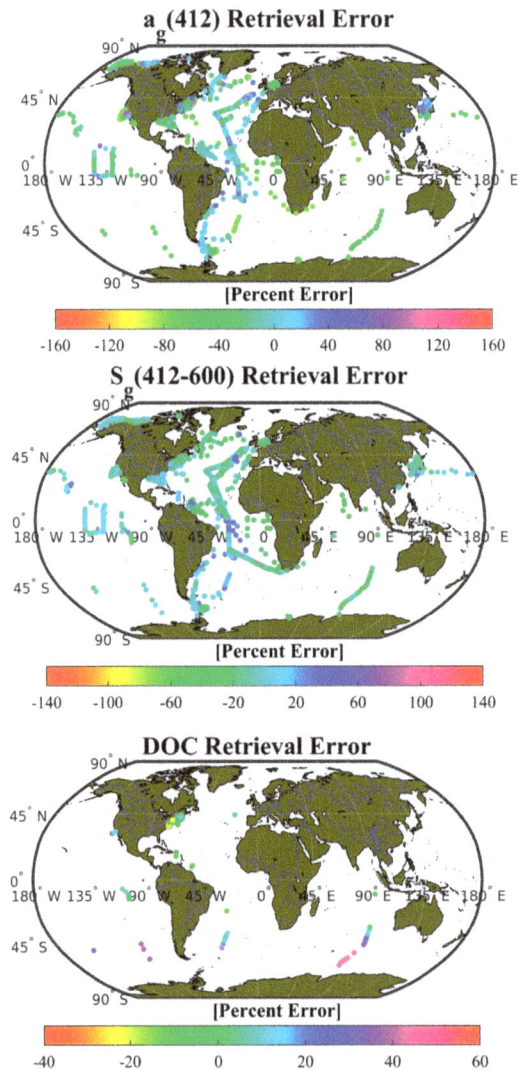

Figure 10. Geographic distribution of error in MLR algorithm retrievals of CDOM absorption and slope in the VIS (top and center), and MLR2 retrievals of DOC (bottom) using validation stations and satellite imagery.

Differences between retrieval statistics across satellite platforms using the MLR approaches were generally small, with Aqua and Terra outperforming SeaWiFS for CDOM absorption (Table 6). All three sensors performed comparably for S_g and DOC (Tables 7 and 8).

3.3. CDOM, S_g, and DOC Climatology

Because the Aquarius mission (providing sea surface salinity) was limited to <4 year data record (~August 2011–June 2015), climatologies for retrieved DOC similar to those presented below for a_g and S_g are not possible. Instead, three years (2011–2013) of coincident MODIS Aqua and Aquarius data were used to generate a three-year mean 9 km global DOC product (Figure 11). An overlay of in situ surface DOC from GOCAD was examined, but not included here because with no temporal coincidence in this comparison, strong biases likely to occur in the field data (e.g., field sampling of high latitudes is proscribed during winter for obvious practical reasons) will not be reflected in the mean DOC satellite product. Nevertheless, the relatively large (±~50%–~100%) disparities apparent in several regions—including high latitudes and the Atlantic subtropical gyre—indicate fundamental weaknesses in the global DOC algorithm. For instance, as mentioned in 3.2.1, the subtropical Atlantic gyre is characterized by an accumulation of DOM not reflected in the CDOM nor apparently traceable with changes in salinity, and is therefore overlooked by the DOC algorithm presented here (MLR2). Based on the tuning statistics, there appears to be merit in the approach, but more study will be required to establish when and where the algorithm works best, and what (if anything) can be done for remotely sensing DOC in regions where no robust optical proxies exist.

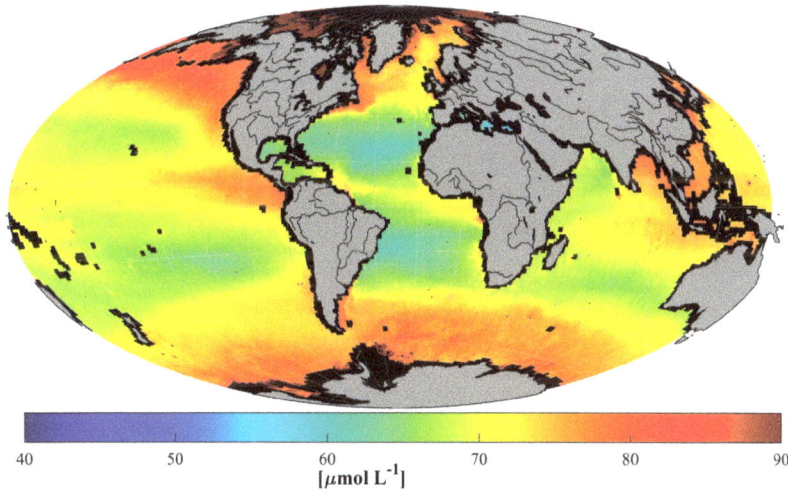

$$[\mu mol\ L^{-1}]$$

Figure 11. Retrieved three-year mean, 9 km nominal resolution DOC from Aquarius and MODIS Aqua using the MLR2 inversion. Validation statistics are reasonably good for the MLR2 (Figure 9, Table 8), but a larger number and wider geographic distribution of validation stations than are currently available is required to thoroughly evaluate the geographic and water-type limitations for MLR2, particularly in the mid-ocean gyres (see text Section 3.2.2). Overestimates of DOC (~41%) retrieved with the MLR2 were found in the southern oceans (S of 40° S)), but only for MODIS Aqua (i.e., not Terra, and no SeaWiFS stations were identified). Elsewhere (i.e., north of 40° S), retrievals tend to slightly underestimate DOC (<10%). Caution is therefore advised in interpreting MLR2 retrievals in mid-ocean gyres, and in the southern oceans using Aqua.

As outlined in the introductory sections, a common assumption made in ocean color remote sensing on a global scale is that CDOM and other water-borne pigmented material covary with Chl.

A valid concern with empirical approaches to retrieve CDOM from ocean color—particularly for those that use some of the same spectral bands as Chl algorithms like OC3M—is that they are essentially tuning themselves to Chl, and not CDOM. While performance metrics for MLR are quite robust (e.g., Figure 5, Table 3), there remains the possibility that this is true in part because these spectral bands are sensitive to Chl, and CDOM is simply covarying with Chl as per Case 1 assumptions. In fact, this was not found to be the case generally from global CDOM and Chl field data in GOCAD ($r^2 = 0.00$, N = 19,446, $\lambda = 412$), nor in the investigation by Siegel et al. (2002), although there exist areas of open ocean outside of the strong influence of terrestrial run-off and upwelling zones where fluctuations in CDOM are clearly driven by local productivity.

To quantify the distinction between retrievals of CDOM using MLR and Chl using OC3M, we calculate retrievals of each over the course of the entire MODIS Aqua mission, and then calculate a residual of the normalized properties, as defined by:

$$\overline{Chl} - \overline{CDOM} = \frac{Chl_i}{median(Chl)} - \frac{a_{g_i}(\lambda)}{median\left(a_g(\lambda)\right)} \tag{10}$$

where the median is taken over an entire composite image to scale each property by its magnitude. For example, MODIS imagery was separated into seasonal composites for the entire Aqua era, then processed to CDOM and Chl, and the residual calculated for each (Figure 12). Bear in mind in this analysis that OC3M-like algorithms for retrieving Chl have been shown to be strongly influenced by not only phytoplankton biomass, but also physiology (particularly in tropical and subtropical regions) as well as the presence of significant absorption by CDOM and non-algal particulates (a_{dg}) [68].

The results show that over much of the world's oceans—particularly at high latitudes, upwelling zones, and regions influenced by large river plumes—normalized CDOM and Chl diverge by as much as a factor of three. Interestingly, the regions shown by Siegel et al. (2013) to be most negatively impacted by a_{dg} in terms of empirical Chl retrievals are the same regions which are shown here to diverge most strongly in terms of the normalized CDOM and Chl residual, indicating that a similar pattern would be expected even when using Chl retrieval algorithms less susceptible to error induced by a_{dg}. The pattern that emerges is that in open ocean regions characterized by strong seasonal blooms such as the North Atlantic and North Pacific, high primary productivity in the presence of lower CDOM (i.e., high residual) is followed after approximately a season by higher CDOM and a collapse in Chl (i.e., low residual). This can be seen in the boreal Spring–Summer transition in the N. Atlantic and Pacific, in the bloom and collapse associated with the reversal of the monsoons in the Arabian Sea between Summer and the following Winter/Spring, and in the Congo and Amazon river plumes over the same period where the residual often shifts by approximately a factor of six between seasons. This observed seasonal lag between peak Chl and peak CDOM helps explain why the two properties rarely covary, as described above for GOCAD, and in [13]. The lag may be explained by the time required for microbial degradation of the bloom's less labile particulate detrital material after the bloom has collapsed.

Application of these algorithms also shows the Spring–Summer transition in the CDOM absorption (left column, Figure 12) as an increase in CDOM from major river outflows such as the Amazon and Congo following peak runoff [69], and in the upwelling region of the Arabian Sea induced by the southwesterly monsoon. In the case of the Amazon River, the CDOM in the distal plume can be seen well into the following season as it drifts slowly eastward across the Atlantic from the retroflection of the North Brazil Current [70], indicating that satellite retrievals of CDOM using the MLR can successfully track surface DOM as it evolves over time scales of weeks to years and over very long distances. The results shown in Figure 12 are broadly similar to those described in [13] for CDM at 440 nm retrieved using the GSM algorithm [71], and to the empirical algorithm of Shanmugam [31] for $a_g(350)$ (their Figure 12), although we show generally higher absorption across the equatorial regions and some parts of the Southern Ocean.

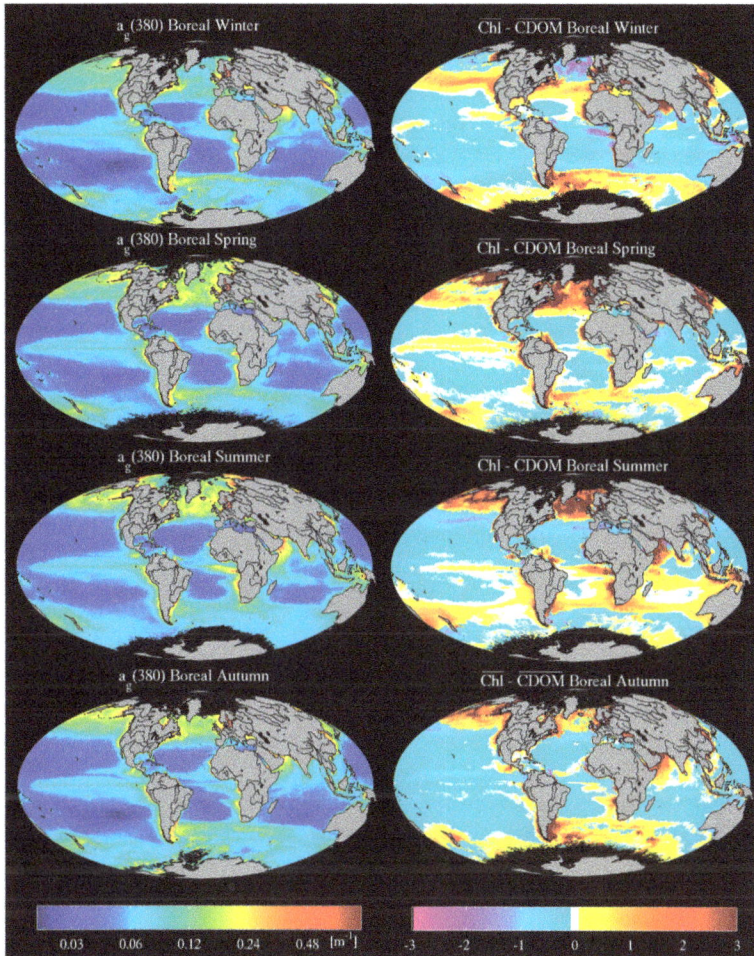

Figure 12. MLR retrievals of a_g(380) by season over the entire MODIS Aqua era (left column), and residuals between Chl and CDOM (right column). Imagery was binned from 4 km resolution monthly composites between August 2002 and January 2014.

Longer time-scale variability in the Aqua-retrieved CDOM was also apparent from the roughly twelve years of monthly, 4 km satellite composites. An examination of the monthly CDOM anomaly ($\Delta a_g(\lambda)$; the monthly $a_g(\lambda)$ divided by the Aqua-era averages for each month) and slope anomaly, $\Delta S_g(\lambda)$, revealed several regions characterized by sharp declines in CDOM during certain years, and elevations in others, as well as the expected inverse proportionality between CDOM and slope in the UV. Figure 13 shows an example of this from seasonal Aqua composites of Δa_g(380) and ΔS_g(275) averaged over periods of El Niño (2002–2005) when surface temperatures are higher, inhibiting vertical nutrient transport and leading to lower primary productivity, and periods of La Niña (2007, 2008, 2010, 2011), which exhibit roughly the opposite dynamics. A feature in the western equatorial Pacific stands out starkly as a crescent stretching from about 10° N to 15° S and spanning nearly the entire 100° longitude range from South America to the Solomon Islands. For brevity, we refer to this as the Western Pacific Crescent (WPC). To test the link between El Niño Southern Oscillation dynamics and the CDOM anomaly in the WPC, an average of monthly $\Delta a_g(380)$ within the WPC is compared with the multivariate ENSO index (MEI [72]) and $\Delta S_g(275\text{–}295)$ (Figure 14). The MEI provides a convenient index for tracking the dominant characteristics associated with ENSO, namely sea-surface pressure, temperature, wind stress, and cloudiness. Positive MEI represent the warmer El Niño cycle associated with lower wind stress, a flattening in the trans-Pacific thermocline, and inhibited productivity in the equatorial Pacific, and negative MEI represents La Niña, which is cooler, and more productive. The coupling between MEI and $\Delta a_g(380)$ and MEI and $\Delta S_g(275)$ is remarkably strong and well correlated (Figure 14; r = −0.77, r = 0.80, respectively, and $p \ll 0.01$ in each case). As expected, CDOM and UVB slope are also very well correlated (r = −0.90, $p \ll 0.01$). The tight correlation between CDOM anomaly, CDOM slope anomaly and MEI may help to predict broad changes in surface CDOM in a future in which warmer sea surface temperatures are expected, particularly in the western Pacific, as the long-term warming trend leads to oceanic conditions favorable to El Niño-like conditions [73]. In fact, sustained deficits in surface CDOM available for photooxidation and microbial remineralization across the WPC, as demonstrated here, is likely to result in a lower partial pressure of CO_2 derived from CDOM, and may therefore increase the flux of CO_2 into the ocean from the atmosphere, although this effect would largely be offset by the decrease in solubility associated with warmer temperatures in the surface ocean. Another consequence of lower CDOM across this region in a warming regime may be decreased surface heating through CDOM absorption, potentially providing some degree of negative feedback to the surface warming trend.

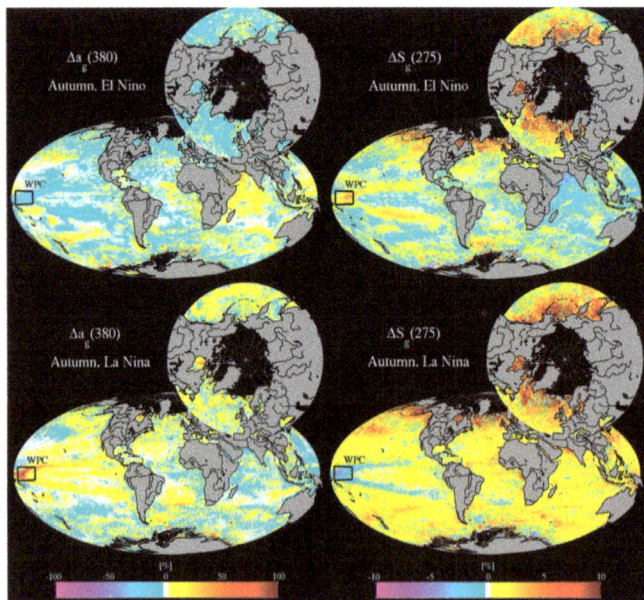

Figure 13. CDOM anomaly (left) and slope anomaly (right) from MLR applied to MODIS Aqua during Autumn in El Niño years (2002–2005; top panel) and La Niña years (2007, 2008, 2010, 2011; bottom panel). The Western Pacific Crescent (WPC) feature is defined here as the broad region exhibiting a notable decline in CDOM during El Niño years, and enhancement during La Niña. UV slope shows the opposite pattern, with lower slopes during La Niña, although the percentage change is roughly an order of magnitude lower. The box shows the portion of the WPC subsampled for comparison with MEI (See Figure 14).

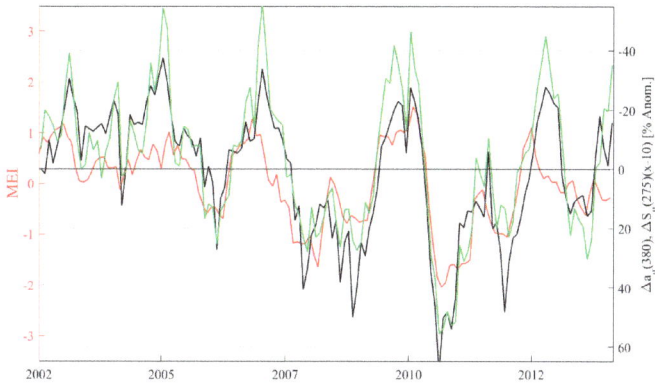

Figure 14. Multivariate ENSO Index (MEI, in red), CDOM anomaly at 380 nm (black), and UVB slope anomaly (green, scaled by a factor of −10 for clarity) over the entire MODIS Aqua era for the region of interest highlighted in Figure 13. Strong negative and positive correlations exist between MEI and CDOM and slope anomalies, respectively (see text).

4. Summary

The importance of characterizing and tracking change in global oceanic dissolved carbon over climatic evolutions is only possible synoptically using earth-observing technology. As methods to measure DOM sources and sinks continue to improve using laboratory and in situ optical techniques, algorithms and orbital sensor technology must keep pace. With technological and methodological improvements, however, come inevitable challenges. The nearly three decades of field data presented here are by necessity compromised in that, for example, no one standard method was employed for the measurement of spectral CDOM absorption. Similarly, there is obviously no standard algorithm for global ocean color retrievals of CDOM, as the algorithms must also continuously evolve with our knowledge of the parameters they retrieve. Many approaches have proven robust in retrieving CDOM absorption and its spectral slope over the years, though most are regionally optimized with little or no provision for what ties them together (e.g., proxies for optical water types). Global algorithms have been hampered by relatively small datasets of coincident radiometry with CDOM and CDOM slope extending into the UV, and DOC retrievals have been especially challenging due to the highly variable and often unpredictable fraction of chromophoric content.

In this study, we aggregate a global dataset approximately forty times the size of previous global, bio-optical databases. Naturally, despite our best efforts to ensure consistency in the data through quality control, the data within are subject to error and uncertainty, largely because methodologies and technology have evolved over thirty years. Quantification of the uncertainty in field estimates of the parameters retrieved here must necessarily precede uncertainty estimates in the algorithms used to derive them. Efforts are currently underway at NASA and elsewhere to do just that. New field data are always being collected and archived all over the world by various academic and public-sector agencies, but only a fraction is broadly distributed through invaluable archives like SeaBASS, in part because submission is only required of those principal investigators funded by NASA. Future algorithm development efforts should facilitate more cooperation and collaboration with other agencies collecting field data around the world encouraging sharing of data within a reasonable time after collection. To date, GOCAD and SeaBASS coverage in regions such as the Mediterranean and the oceans around Australia is astonishingly poor. Efforts must be sustained to continue bringing newly collected and historical data into GOCAD, NOMAD, and similar global, long-term bio-optical databases, and to expand them to include an even more comprehensive suite of inherent optical properties, which help support the development of more robust semi-analytical approaches. GOCAD was designed using nested and comprehensive Matlab structures conducive to expansion for both newly collected datasets as well as more complete suites of physical, optical, radiometric, and biogeochemical data. Based on our experience with SeaBASS, more rigorous quality control and documentation standards should be applied not only to recent and new submissions, but retroactively to historic data as well. While some algorithms performed better than others in this non-exhaustive comparison, it is important that algorithms continue to evolve as the data used to develop them improves, incorporating more than minor adjustments to empirical coefficients, and moving algorithms for oceanic carbon closer to theoretical, analytically-based approaches.

A representational suite of approaches, including empirical, semi-analytical, and machine-learning algorithms was evaluated against GOCAD field data for retrieving a_g in six wavebands between 275 and about 490 nm, S_g in eight wavebands in the UV and VIS, and DOC using a wide variety of metrics. Ultimately, the most versatile and best performing of those tested was a simple, empirical set of relationships based on multiple linear regression between four wavebands of remote sensing reflectance (440–555 nm), with the exception of DOC which also required sea surface salinity (e.g., from Aquarius) to act as a proxy to optical water type. Results varied, with CDOM retrievals showing regression coefficients to field data (r^2) generally over 0.80 for field radiometry to within 16%–34%, depending on the wavelength, and within 33%–54% for MODIS Aqua validation. CDOM slopes retrievals were best in the UVB (e.g., $r^2 = 0.62$, $MAPD = 11\%$ in satellite validation of $S_g(275–295)$), while DOC algorithms only optimized well after the inclusion of salinity ($r^2 = 0.91$, $MAPD = 15\%$), and did not perform well in validation (e.g., RMSE = 27–29 µmol L^{-1}). Our analysis of the sensitivity of the DOC algorithm performance to factors such as salinity, DOC, water column depth, and geographic location ultimately proved inconclusive, exposing only a small anomaly involving overestimates of DOC in the southern oceans using MODIS Aqua imagery. Further validation—particularly in mid–ocean gyres where DOC varies very weakly or not at all with CDOM absorption, and salinity changes are very small—is clearly warranted prior to application of the DOC algorithm in those regions.

Application of CDOM algorithms to monthly and climatological Aqua imagery demonstrated that global retrievals of CDOM do not covary well with similar empirical retrievals of Chl, but rather appear to follow Chl on a seasonal lag depending on the region and source of dissolved material. This helps explain the lack of correlation between CDOM and Chl found in global GOCAD field data and described in previous studies, and further challenges the use of Case 1 assumptions in bio-optical remote sensing. Surface CDOM concentration varies in regions such the western equatorial Pacific by about 150% over the course of long-term climatological shifts associated with ENSO, fluctuating in tight correlation with the MEI and CDOM slope. Algorithms developed here may be applied to tracking ENSO behavior in the future, as well as observing changes in CDOM character and concentration associated with global warming.

Author Contributions: Conceptualization, D.A. and A.M.; Data Curation, D.A.; Formal Analysis, D.A. and D.L.; Funding Acquisition, A.M.; Investigation, D.A.; Methodology, D.A. and A.M.; Project Administration, A.M. and D.A.; Resources, A.M.; Software, D.A. and D.L.; Validation, D.A. and D.L.; Visualization, D.A.; Writing—original draft, D.A.; Writing—review & editing, D.A. and A.M.

Funding: This research was funded by the NASA ROSES Science of Aqua and Terra—MODIS project.

Acknowledgments: We would like to acknowledge all investigators who have contributed to the SeaBASS database and the various Hansell/Carlson datasets (see Table 2). Thanks also to Jeremy Werdell, Chris Proctor, Norman Kuring, Chris Moellers, and Jason Leffler of the NASA Ocean Biology Processing Group for data acquisition and programming support.

Conflicts of Interest: The authors declare no conflict of interest.

References

1. Siegenthaler, U.; Sarmiento, J. Atmospheric carbon dioxide and the ocean. *Nature* **1993**, *365*, 119–125. [CrossRef]
2. Hedges, J.I. Why dissolved organics matter. *Biogeochem. Mar. Dissolved Org. Matter* **2002**, 1–33.

3. Belanger, S.; Babin, M.; Xie, H.; Krotkov, N.; Larouche, P.; Vincent, W.F. Cdom Photooxidation in the Arctic Coastal Waters: New Approach Using Satellite Information and Implications of Climate Change. In Proceedings of the Ocean Optics XVIII, Montreal, QC, Canada, 9–13 October 2006.

4. Carder, K.L.; Steward, R.G.; Harvey, G.R.; Ortner, P.B. Marine humic and fulvic acids: Their effects on remote sensing of chlorophyll. *Limnol. Oceanogr.* **1989**, *34*, 68–81. [CrossRef]

5. Lee, Z.; Arnone, R.; Hu, C.; Werdell, P.J.; Lubac, B. Uncertainties of optical parameters and their propagations in an analytical ocean color inversion algorithm. *Appl. Opt.* **2010**, *49*, 369–381. [CrossRef] [PubMed]

6. Sempéré, R.; Cauwet, G. Occurrence of organic colloids in the stratified estuary of the krka river (Croatia). *Estuar. Coast. Shelf Sci.* **1995**, *40*, 105–114. [CrossRef]

7. Blough, N.V.; Del Vecchio, R. Chromophoric dom in the coastal environment. In *Biogeochemistry of Marine Dissolved Organic Matter*; Hansell, D.A., Carlson, C.A., Eds.; Academic Press: San Diego, CA, USA, 2002; pp. 509–546.

8. Nelson, N.B.; Siegel, D.A. Chromophoric dom in the open ocean. *Biogeochem. Mar. Dissolved Org. Matter* **2002**, 547–578.

9. Twardowski, M.S.; Boss, E.; Sullivan, J.M.; Donaghay, P.L. Modeling the spectral shape of absorption by chromophoric dissolved organic matter. *Mar. Chem.* **2004**, *89*, 69–88. [CrossRef]

10. Aurin, D.A.; Dierssen, H.M.; Twardowski, M.S.; Roesler, C.S. Optical complexity in long island sound and implications for coastal ocean color remote sensing. *J. Geophys. Res. Oceans* **2010**, *115*. [CrossRef]

11. Del Vecchio, R.; Subramaniam, A. Influence of the amazon river on the surface optical properties of the western tropical north atlantic ocean. *J. Geophys. Res. Oceans* **2004**, *109*. [CrossRef]

12. Sauer, M.J.; Roesler, C.S. Seasonal and spatial cdom variability in the gulf of maine. In Proceedings of the Ocean Optics XVIII, Montreal, QC, Canada, 9–13 October 2006.

13. Siegel, D.A.; Maritorena, S.; Nelson, N.B.; Hansell, D.A.; Lorenzi-Kayser, M. Global distribution and dynamics of colored dissolved and detrital organic materials. *J. Geophys. Res. Oceans* **2002**, *107*. [CrossRef]

14. Pegau, W.S. Inherent optical properties of the central arctic surface waters. *J. Geophys. Res. Oceans* **2002**, *107*, 8035. [CrossRef]

15. Carder, K.L.; Chen, R.F.; Lee, Z.; Hawes, S.K.; Cannizzaro, J.P. *MODIS* Ocean Science Team Algorithm Theoretical Basis Document. *ATBD* **2003**, *19*, 7–18.

16. Morel, A.; Prieur, L. Analysis of variations in ocean color. *Limnol. Oceanogr.* **1977**, *22*, 709–722. [CrossRef]

17. Lee, Z. *Remote Sensing of Inherent Optical Properties: Fundamentals, Tests of Algorithms, and Applications*; International Ocean–Colour Coordinating Group: Dartmouth, NS, Canada, 2006.

18. Del Vecchio, R.; Blough, N.V. Photobleaching of chromophoric dissolved matter in natural waters: Kinetics and modeling. *Mar. Chem.* **2002**, *78*, 231–253. [CrossRef]

19. Helms, J.R.; Stubbins, A.; Ritchie, J.D.; Minor, E.C. Absorption spectral slopes and slope ratios as indicators of molecular weight, source, and photobleaching of chromophoric dissolved organic material. *Limnol. Oceanogr.* **2008**, *53*, 955–969. [CrossRef]

20. Stedmon, C.A.; Markager, S. Behavior of the optical properties of coloured dissolved organic matter under conservative mixing. *Estuar. Coast. Shelf Sci.* **2003**, *57*, 973–979. [CrossRef]

21. Del Castillo, C.E.; Miller, R.L. Horizontal and vertical distributions of colored dissolved organic matter during the southern ocean gas exchange experiment. *J. Geophys. Res. Oceans* **2011**, *116*. [CrossRef]

22. Stedmon, C.A.; Markager, S.; Kaas, H. Optical properties and signatures of chromophoric dissolved organic matter (CDOM) in danish coastal waters. *Estuar. Coast. Shelf Sci.* **2000**, *51*, 267–278. [CrossRef]

23. Cartisano, C.M.; Del Vecchio, R.; Bianca, M.R.; Blough, N.V. Investigating the sources and structure of chromophoric dissolved organic matter (CDOM) in the north pacific ocean (NPO) utilizing optical spectroscopy combined with solid phase extraction and borohydride reduction. *Mar. Chem.* **2018**, *204*, 20–35. [CrossRef]

24. Twardowski, M.S.; Donaghay, P.L. Photobleaching of aquatic dissolved materials: Absorption removal, spectral alteration, and their interrelationship. *J. Geophys. Res. Oceans* **2002**, *107*. [CrossRef]

25. D'Sa, E.J.; Hu, C.; Muller–Karger, F.E.; Carder, K.L. Estimation of colored dissolved organic matter and salinity fields in case 2 waters using seawifs: Examples from florida bay and florida shelf. *Proc. Indian Acad. Sci. Earth Planet. Sci.* **2002**, *111*, 197–207. [CrossRef]

26. Johannessen, S.C.; Miller, W.L.; Cullen, J.J. Calculation of uv attenuation and colored dissolved organic matter absorption spectra from measurements of ocean color. *J. Geophys. Res. Oceans* **2003**, *108*, 3301. [CrossRef]

27. Kahru, M.; Mitchell, B.G. Seasonal and nonseasonal variability of satellite–derived chlorophyll and colored dissolved organic matter concentration in the california current. *J. Geophys. Res. Oceans* **2001**, *106*, 2517–2529. [CrossRef]

28. Mannino, A.; Russ, M.E.; Hooker, S.B. Algorithm development and validation for satellite–derived distributions of doc and cdom in the u.S. Middle atlantic bight. *J. Geophys. Res.* **2008**, *113*, C07051. [CrossRef]

29. Matthews, M.W. A current review of empirical procedures of remote sensing in inland and near–coastal transitional waters. *Int. J. Remote Sens.* **2011**, *32*, 6855–6899. [CrossRef]

30. Tehrani, N.; Sa, E.; Osburn, C.; Bianchi, T.; Schaeffer, B. Chromophoric dissolved organic matter and dissolved organic carbon from sea–viewing wide field–of–view sensor (SEAWIFS), moderate resolution imaging spectroradiometer (MODIS) and meris sensors: Case study for the northern gulf of mexico. *Remote Sens.* **2013**, *5*, 1439–1464. [CrossRef]

31. Shanmugam, P. A new bio-optical algorithm for the remote sensing of algal blooms in complex ocean waters. *J. Geophys. Res. Oceans* **2011**, *116*. [CrossRef]

32. Tiwari, S.P.; Shanmugam, P. An optical model for the remote sensing of coloured dissolved organic matter in coastal/ocean waters. *Estuar. Coast. Shelf Sci.* **2011**, *93*, 396–402. [CrossRef]

33. Werdell, P.J.; Bailey, S.W. An improved in situ bio–optical data set for ocean color algorithm development and satellite data product validation. *Remote Sens. Environ.* **2005**, *98*, 122–140. [CrossRef]

34. Maritorena, S.; Lee, Z.; Du, K.P.; Loisel, H.; Doerffer, R.; Roesler, C.; Lyon, P.; Tanaka, A.; Babin, M.; Kopelevich, O.V. Chapter 2: Synthetic and in situ data sets for algorithm testing. In *Remote Sensing of Inherent Optical Properties: Fundamentals, Tests of Algorithms and Applications*; Lee, Z., Ed.; International Ocean Colour Coordinating Group: Dartmouth, NS, Cabada, 2006; pp. 13–18.

35. Gordon, H.R.; Brown, O.B.; Evans, R.H.; Brown, J.W.; Smith, R.C.; Baker, K.S.; Clark, D.K. A semi–analytic radiance model of ocean color. *J. Geophys. Res. Atmos.* **1988**, *93*, 10909–10924. [CrossRef]

36. Gordon, H.R.; Brown, O.B.; Jacobs, M.M. Computed relationships between the inherent and apparent optical properties of a flat homogenous ocean. *Appl. Opt.* **1975**, *14*, 417–427. [CrossRef] [PubMed]

37. Garver, S.A.; Siegel, D.A. Inherent optical property inversion of ocean color spectra and its biogeochemical interpretation. 1. Time series from the sargasso sea. *J. Geophys. Res. Oceans* **1997**, *102*, 18607–18625. [CrossRef]

38. Lee, Z.P.; Carder, K.L. Effect of spectral band numbers on the retrieval of water column and bottom properties from ocean color data. *Applied Optics* **2002**, *41*, 2191–2201. [CrossRef] [PubMed]

39. Roesler, C.S.; Perry, M.J. In–situ phytoplankton absorption, fluorescence emission, and particulate backscattering spectra determined from reflectance. *J. Geophys. Res. Oceans* **1995**, *100*, 13279–13294. [CrossRef]

40. Dong, Q.; Shang, S.; Lee, Z. An algorithm to retrieve absorption coefficient of chromophoric dissolved organic matter from ocean color. *Remote Sens. Environ.* **2013**, *128*, 259–267. [CrossRef]

41. Matsuoka, A.; Babin, M.; Doxaran, D.; Hooker, S.; Mitchell, B.; Bélanger, S.; Bricaud, A. A synthesis of light absorption properties of the pan–arctic ocean: Application to semi–analytical estimates of dissolved organic carbon concentrations from space. *Biogeosci. Discuss.* **2013**, *10*, 17071–17115. [CrossRef]

42. Odermatt, D.; Gitelson, A.; Brando, V.E.; Schaepman, M. Review of constituent retrieval in optically deep and complex waters from satellite imagery. *Remote Sens. Environ.* **2012**, *118*, 116–126. [CrossRef]

43. Swan, C.M.; Nelson, N.B.; Siegel, D.A.; Fields, E.A. A model for remote estimation of ultraviolet absorption by chromophoric dissolved organic matter based on the global distribution of spectral slope. *Remote Sens. Environ.* **2013**, *136*, 277–285. [CrossRef]

44. Tilstone, G.H.; Peters, S.W.; van der Woerd, H.J.; Eleveld, M.A.; Ruddick, K.; Schönfeld, W.; Krasemann, H.; Martinez–Vicente, V.; Blondeau–Patissier, D.; Röttgers, R. Variability in specific–absorption properties and their use in a semi–analytical ocean colour algorithm for meris in north sea and western english channel coastal waters. *Remote Sens. Environ.* **2012**, *118*, 320–338. [CrossRef]

45. Zhu, W.; Yu, Q.; Tian, Y.Q.; Chen, R.F.; Gardner, G.B. Estimation of chromophoric dissolved organic matter in the mississippi and atchafalaya river plume regions using above-surface hyperspectral remote sensing. *J. Geophys. Res. Oceans* **2011**, *116*. [CrossRef]

46. Stramski, D.; Woźniak, S.B. On the role of colloidal particles in light scattering in the ocean. *Limnol. Oceanogr.* **2005**, *50*, 1581–1591. [CrossRef]

47. Nelson, N.B.; Siegel, D.A. The global distribution and dynamics of chromophoric dissolved organic matter. *Annu. Rev. Mar. Sci.* **2013**, *5*, 447–476. [CrossRef] [PubMed]

48. Weishaar, J.L.; Aiken, G.R.; Bergamaschi, B.A.; Fram, M.S.; Fujii, R.; Mopper, K. Evaluation of specific ultraviolet absorbance as an indicator of the chemical composition and reactivity of dissolved organic carbon. *Environ. Sci. Technol.* **2003**, *37*, 4702–4708. [CrossRef] [PubMed]

49. Sullivan, J.M.; Twardowski, M.; Zanefeld, J.R.V.; Moore, C.M.; Barnard, A.H.; Donaghay, P.L.; Rhoades, B. Hyperspectral temperature and salt dependencies of absorption by water and heavy water in the 400–750 nm spectral range. *Appl. Opt.* **2006**, *45*, 5294–5309. [CrossRef] [PubMed]

50. Twardowski, M.S.; Sullivan, J.M.; Donaghay, P.L.; Zaneveld, J.R.V. Microscale quantification of the absorption by dissolved and particulate material in coastal waters with an ac–9. *J. Atmos. Ocean. Technol.* **1999**, *16*, 691–707. [CrossRef]

51. Zaneveld, J.R.V.; Kitchen, J.C.; Moore, C.M. Scattering error correction of reflecting–tube absorption meters. In *Ocean optics XII*; Jaffe, J.S., Ed.; International Society for Optics and Photonics: Bellingham, WA, USA, 1994; pp. 44–55.

52. Mitchell, B.G.; Stramska, M.; Wieland, J. Determination of spectral coefficients of particles, dissolved material and phytoplankton for discrete water samples. In *Ocean Optics Protocols for Satellite Ocean Color Sensor Validation*; National Aeronautics and Space Administration: Greenbelt, MA, USA, 2003.

53. Mueller, J.L. Inherent optical properties: Instruments, characterizations, field measurements and data analysis protocols. In *Ocean Optics Protocols for Satellite Ocean Color Sensor Validation*; National Aeronautics and Space Administration: Greenbelt, MA, USA, 2003.

54. Thuillier, G.; Hersé, M.; Foujols, T.; Peetermans, W.; Gillotay, D.; Simon, P.; Mandel, H. The solar spectral irradiance from 200 to 2400 nm as measured by the solspec spectrometer from the atlas and eureca missions. *Sol. Phys.* **2003**, *214*, 1–22. [CrossRef]

55. Bailey, S.W.; Werdell, P.J. A multi–sensor approach for the on–orbit validation of ocean color satellite data products. *Remote Sens. Environ.* **2006**, *102*, 12–23. [CrossRef]

56. Aurin, D.A.; Mannino, A. A database for developing global ocean color algorithms for colored dissolved organic material, cdom slope, and dissolved organic carbon. In Proceedings of the Ocean Optics XXI, Glasgow, Scotland, 8–12 October 2012.

57. Taylor, K.E. Summarizing multiple aspects of model performance in a single diagram. *J. Geophys. Res. Atmos.* **2001**, *106*, 7183–7192. [CrossRef]

58. Jolliff, J.K.; Kindle, J.C.; Shulman, I.; Penta, B.; Friedrichs, M.A.; Helber, R.; Arnone, R.A. Summary diagrams for coupled hydrodynamic–ecosystem model skill assessment. *J. Mar. Syst.* **2009**, *76*, 64–82. [CrossRef]

59. Carlson, C.A.; Hansell, D.A.; Tamburini, C. Doc persistence and its fate after export within the ocean interior. *Microb. Carbon Pump Ocean* **2011**, 57–59.

60. Hansell, D.A.; Carlson, C.A. Biogeochemistry of total organic carbon and nitrogen in the sargasso sea: Control by convective overturn. *Deep Sea Res. Part II* **2001**, *48*, 1649–1667. [CrossRef]

61. Breiman, L. Random forests. *Mach. Learn.* **2001**, *45*, 5–32. [CrossRef]

62. Lary, D.J.; Zewdie, G.K.; Liu, X.; Wu, D.; Levetin, E.; Allee, R.J.; Malakar, N.; Walker, A.; Mussa, H.; Mannino, A.; Aurin, D. Machine learning applications for earth observation. In *Earth Observation Open Science and Innovation*; Mathieu, P.P., Aubrecht, C., Eds.; Springer: Cham, Switzerland, 2018.

63. Lee, Z.P.; Carder, K.L.; Arnone, R. Chapter 10: The quasi–analytical algorithm. In *Remote Sensing of Inherent Optical Properties: Fundamentals, Tests of Algorithms, and Applications*; Stuart, V., Ed.; IOCCG: Dartmouth, NS, Canada, 2006; pp. 73–79.

64. Lee, Z.P.; Carder, K.L.; Arnone, R.A. Deriving inherent optical properties from water color: A multiband quasi–analytical algorithm for optically deep waters. *Appl. Opt.* **2002**, *41*, 5755–5772. [CrossRef] [PubMed]

65. Werdell, P.J.; Franz, B.A.; Bailey, S.W.; Feldman, G.C.; Boss, E.; Brando, V.E.; Dowell, M.; Hirata, T.; Lavender, S.J.; Lee, Z. Generalized ocean color inversion model for retrieving marine inherent optical properties. *Appl. Opt.* **2013**, *52*, 2019–2037. [CrossRef] [PubMed]

66. Morel, A.; Bricaud, A. Theoretical results concerning light absorption in a discrete medium and application to the specific absorption of phytoplankton. *Deep Sea Res.* **1981**, *28*, 1357–1393. [CrossRef]

67. Twardowski, M.S.; Boss, E.; Macdonald, J.B.; Pegau, W.S.; Barnard, A.H.; Zaneveld, J.R.V. A model for estimating bulk refractive index from the optical backscattering ratio and the implications for understanding particle composition in case i and case ii waters. *J. Geophys. Res. Oceans* **2001**, *106*, 14129–14142. [CrossRef]

68. Siegel, D.; Behrenfeld, M.; Maritorena, S.; McClain, C.; Antoine, D.; Bailey, S.; Bontempi, P.; Boss, E.; Dierssen, H.; Doney, S. Regional to global assessments of phytoplankton dynamics from the seawifs mission. *Remote Sens. Environ.* **2013**, *135*, 77–91. [CrossRef]

69. Dai, A.; Qian, T.; Trenberth, K.E.; Milliman, J.D. Changes in continental freshwater discharge from 1948 to 2004. *J. Clim.* **2009**, *22*, 2773–2792. [CrossRef]

70. Lentz, S.J. The amazon river plume during amassds: Subtidal current variability and the importance of wind forcing. *J. Geophys. Res. Oceans* **1995**, *100*, 2377–2390. [CrossRef]

71. Maritorena, S.; Siegel, D.A.; Peterson, A.R. Optimization of a semianalytical ocean color model for global–scale applications. *Appl. Opt.* **2002**, *41*, 2705–2714. [CrossRef] [PubMed]

72. Wolter, K.; Timlin, M.S. Monitoring enso in coads with a seasonally adjusted principal component index. In Proceedings of the 17th Climate Diagnostics Workshop, Norman, OK, USA, January 1993.

73. Collins, M.; An, S.-I.; Cai, W.; Ganachaud, A.; Guilyardi, E.; Jin, F.-F.; Jochum, M.; Lengaigne, M.; Power, S.; Timmermann, A. The impact of global warming on the tropical pacific ocean and el niño. *Nat. Geosci.* **2010**, *3*, 391–397. [CrossRef]

applied
sciences

MDPI

Article

Influence of Three-Dimensional Coral Structures on Hyperspectral Benthic Reflectance and Water-Leaving Reflectance

John D. Hedley [1,*], Maryam Mirhakak [2], Adam Wentworth [3] and Heidi M. Dierssen [2]

[1] Numerical Optics Ltd., Tiverton EX16 8AA, UK
[2] Department of Marine Sciences, University of Connecticut, Groton, CT 06340, USA;
   ma.mirhakak@gmail.com (M.M.); heidi.dierssen@uconn.edu (H.M.D.)
[3] Materials Science and Engineering Department, University of Connecticut, Storrs, CT 06269, USA;
   awentworth2@bwh.harvard.edu
* Correspondence: j.d.hedley@numopt.com; Tel.: +44-1884-675070

Received: 31 July 2018; Accepted: 30 September 2018; Published: 19 December 2018

Featured Application: Hyperspectral remote sensing of coral reefs. We present simplifying factors and guidance on handling or avoiding the potential variability in spectral reflectances caused by the structural nature of corals and illumination conditions. Results presented here can be implemented in remote sensing algorithms for bottom mapping of coral reefs, and guide data collection practice.

Abstract: Shading and inter-reflections created by the three-dimensional coral canopy structure play an important role on benthic reflectance and its propagation above the water. Here, a plane parallel model was coupled with a three-dimensional radiative transfer canopy model, incorporating measured coral shapes and hyperspectral benthic reflectances, to investigate this question under different illumination and water column conditions. Results indicated that a Lambertian treatment of the bottom reflectance can be a reasonable assumption if a variable shading factor is included. Without flexibility in the shading treatment, nadir view bottom reflectances can vary by as much as $\pm 20\%$ (or $\pm 9\%$ in above-water remote sensing reflectance) under solar zenith angles (SZAs) up to 50°. Spectrally-independent shading factors are developed for benthic coral reflectance measurements based on the rugosity of the coral. In remote sensing applications, where the rugosity is unknown, a shading factor could be incorporated as an endmember for retrieval in the inversion scheme. In dense coral canopies in clear shallow waters, the benthos cannot always be treated as Lambertian, and for large solar-view angles the bi-directional reflectance distribution functions (BRDF) hotspot propagated to above water reflectances can create up to a 50% or more difference in water-leaving reflectances, and discrepancies of 20% even for nadir-view geometries.

Keywords: remote sensing; hyperspectral; shallow water; coral; derivative; radiative transfer; canopy

---

1. Introduction

Remote sensing of coral reefs is an important complementary survey technique for science, monitoring, and management, being able to cover substantially larger areas than in-situ surveys albeit at lower accuracy [1,2]. A well-established paradigm in coral reef remote sensing is that hyperspectral data offers the best results for discrimination of both benthic habitats and specific benthic types, such as corals, algae, and sand [3,4]. Applied studies have shown that benthic habitat classification accuracy increases with the number of available spectral bands [3,5], specifically in the 400–700 nm range since the opacity of water beyond 700 nm limits use of NIR wavelengths to only the shallowest waters

(<2 m). Habitats are broad mapping categories, whereas mapping of specific types, e.g., corals vs. algae, is very challenging since their spectral reflectances can be similar. Modelling and sensitivity analyses based on reflectance spectra measured, either in-situ using diver-operated spectroradiometers, or ex-situ in the lab, imply that spectral separation of benthic types by hyperspectral reflectances is possible [4,6,7], and there are increasing numbers of applied demonstrations of discriminations of benthic types (e.g., coral, sand, and algae) from airborne hyperspectral imagery [8–14].

These studies imply that capability for discrimination of coral reef types is reliant on the discrimination of spectral features of the pigments that are present, such as peridinin in the coral symbionts, and accessory pigments of algae [8]. Pigment-based reflectance features are present at the surface of the corals and other benthic types, but the coral reef environment is structurally complex, with multiple benthic types typically present within even the highest spatial resolution imagery (e.g., pixels 1 m). Vertical structure in the reef canopy and the benthic types themselves gives rise to shading, inter-reflections, and a generally complex interaction with the light environment [15,16]. Most sensitivity analyses to date work with empirically measured reflectances of individual types, at a scale roughly equivalent to their morphology, and use a linear mixing model, which essentially assumes that the individual types can be treated in the same way as a spatially flat Lambertian patchwork [4,7]. However, even in a carefully defined lab experiment, the linear mixing model does not always work well with structural components [17]. Given the complexity of light interaction with a mixed structural canopy, it is not immediately obvious to what extent a pigment based hyperspectral reflectance feature present at the scale of a coral surface translates to an above water measurement at remote sensing scales.

The purpose of the study presented here was to investigate propagation of hyperspectral reflectances from the surfaces of structural benthos (corals and surrounding substrate) to water leaving reflectance. The study combined a three-dimensional canopy model [18], with a plane parallel water-column model (HydroLight [19]) (Figure 1). The model was parameterized by 3D reconstructions of actual coral shapes and surface reflectances derived from hyperspectral images of the same corals. A variety of investigations were conducted (Table 1), to track the main factors that mitigate the transmission of hyperspectral features, and to justify practical simplifications to provide useful results that can be immediately applied in other contexts. Investigations included the effects of shading, and water attenuation and scattering, both in the region of individual corals and within mixtures of corals and surrounding substrates. The three-dimensional model was used to generate fully populated bi-directional reflectance distribution functions (BRDFs) of mixed canopies, and these were input to HydroLight to model above water reflectance for different depths, water inherent optical properties (IOPs), and solar-view geometries.

In summary, specific objectives were:

1. Investigate how the spectral reflectance at the scale of coral macro-morphology (shape) relates to coral surface-scale spectral reflectance.
2. Quantify the error arising from making simplifying modelling assumptions, such as using a single reflectance spectrum to represent a coral under different light environments, or excluding water attenuation and scattering between coral structures within the canopy.
3. Characterize the BRDF of assemblages of corals in different densities.
4. Apply those BRDFs as the bottom boundary in HydroLight, to model BRDF effects in the remote sensing reflectance arising from canopy BRDF properties in different depths and conditions.

The results indicated that whilst structural factors can introduce substantial variability in spectral reflectances, this variability can be avoided or accommodated by simplifying factors. The major variations in spectral reflectance over corals, due to structure and illumination, can be captured by incorporating a spectrally flat shading factor or black shade endmember into linear mixing models. Above-water BRDF effects on reflectance can be 20% or more for dense shallow canopies when viewing close to the solar plane in typical remote sensing solar-view geometries, but are less apparent if the solar-view geometry is at 90° to the solar plane.

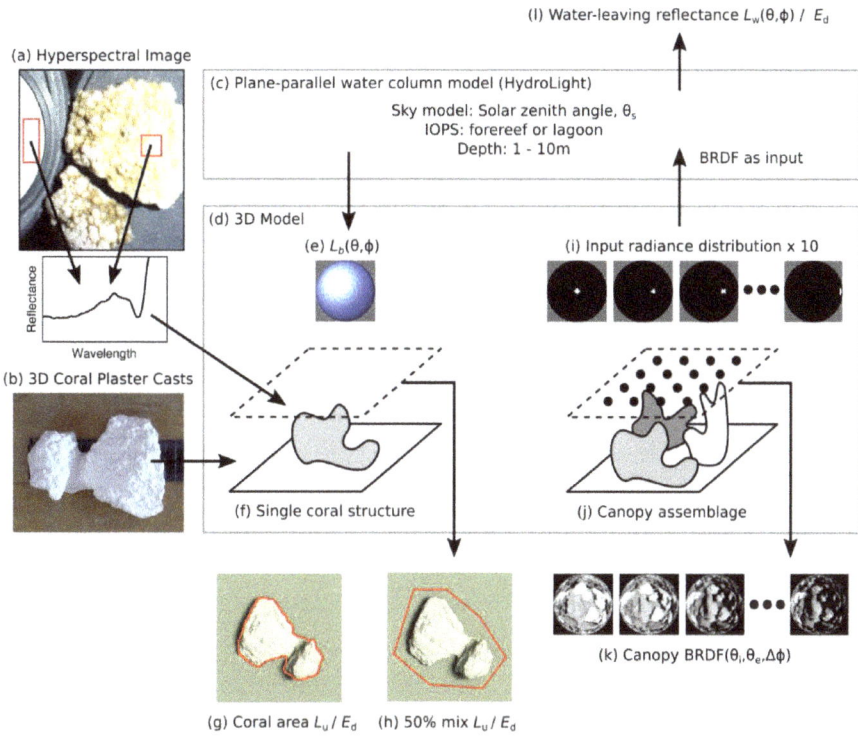

Figure 1. Model setup: (**a**) Coral surface reflectances were derived from hyperspectral image, and (**b**) shape from 3D reconstructions from plaster casts; (**c**) A plane parallel water column model was coupled with (**d**) a 3D canopy model in two ways: (1) (**e**) To model bottom of water column light fields over (**f**) single structures and estimate (**g**) reflectance over the coral and (**h**) of a 50% mix with surrounding substrate; and (2) (**i**) Directional incident radiance at different angles over (**j**) assemblages of structures were used to characterize (**k**) the bi-directional reflectance distribution function (BRDF). The BRDF was then input to the water column model to give (**l**) water-leaving reflectances.

2. Materials and Methods

2.1. Overview

Toward the aims listed above, three-dimensional models of actual coral shapes were input to a 3D canopy radiative transfer model [18] coupled to a plane-parallel water column model, HydroLight [19]. Six different modelling exercises were conducted (Table 1, Figure 1). These were structured to progressively incorporate complexity and justify the design of subsequent modelling steps: from reflectance over a single coral shape only, to then including substrate, then include assemblages of corals, and finally the water column. Intermediate results from these activities also provided results of interest, for example, on the validity of using a linear mixing model for bottom of water column reflectance. The following sections give details on the input data and model set up.

Table 1. Different properties and treatments modelled in this study and the aim of each investigation. "Res." indicates if the high- or low-resolution coral structure models were used. "I." indicates if interstitial scattering and attenuation by the water around the coral shapes was included (N-No, Y-Yes). "Position" indicates the notional location of the estimated reflectance.

	Modelled Property	Res.	I.	Derived Reflectance and Purpose of Activity	Position
1	Reflectance averaged over each of the 16 coral shapes, for the coral area only (substrate masked out). Bottom of water column incident radiance distribution generated by HydroLight with 32 treatments, combining: Solar zenith angles 10° and 50°; depths 1 m and 10 m; forereef and lagoon IOPs; and four rotational positions of the coral shape.	high	N	$\pi L_u/E_d$ Effect of shape on reflectance over coral area only.	above canopy
2		low	N	$\pi L_u/E_d$ Effect of resolution of 3D models.	above canopy
3	Reflectance averaged over a single coral shape, plus in each case, surrounding substrate to give a 50% mix in terms of areal cover. Incident radiance distributions as above.	low	N	$\pi L_u/E_d$ Validity of linear mixing model for reflectance over coral and surrounding substrate.	above canopy
4		low	Y	$\pi L_u/E_d$ Effect of ignoring water scattering and attenuation within canopy.	above canopy
5	Azimuthally averaged BRDF function for six canopy assemblages ranging from 39% to 75% coral cover. BRDF tabulated in HydroLight standard directional discretization.	low	N	$\pi \, BRDF(\theta_i, \theta_e, \Delta\varphi)$ Demonstrate canopy BRDF effects and generate function for input to HydroLight.	above canopy
6	Above-water remote sensing reflectance as a function of solar-view geometry, depth, and IOPs, for six canopy BRDFs generated as above.	low	N	$\pi L_w(\theta_s, \theta_v, \Delta\varphi)/E_d$ Demonstrate above water BRDF effects under typical solar-view geometries.	above water

2.2. Coral Surface Reflectances and Shape

The three-dimensional shape and surface reflectance of 16 corals were used as input to the model. Live corals of various species, including *Porites compressa*, *Porites evermanni*, and *Montipora capitate*, were sampled from Mahukona Beach Park, Potters, and Pauko Bay Boat Ramp in Big Island, Hawaii, in February 2017. For the purposes of this paper, corals were grouped based on sampling depth, referred to as shallow 'S' (<8 m) or deep 'D' (10 to 20 m) (Table 2). For practical reasons the corals were relatively small (<20 cm diameter). Corals were stored in open buckets filled with natural salt water and kept in a cool water bath that was shaded from the direct sun. Coral samples were imaged in the air, immediately after being removed from the water, using a tripod-mounted 710 hyperspectral imager (Surface Optics Corporation). This instrument records an image cube of 520 × 696 pixels with spectral information at 128 spectral bands, and with 5 nm spacing from 380 to 1040 nm. All images were made under natural illumination at noon ±2 h, to minimize any shading effects on the coral surfaces. Images were obtained within 24 h of collection, and before any noticeable degradation. A Labsphere Spectralon™ Diffuse Reflectance Standard with a reflectance value of 20% was placed in the image frame during each measurement (Figure 1a). Using the reference, for each coral, a single surface reflectance, $R(\lambda)$, i.e., the reflectance at the scale of the coral surface, assuming the surface is locally flat, was estimated by taking the mean over one or more small areas of the image where the coral surface was relatively horizontal (Figure 1a). While in reality, the reflectance over a coral surface is typically variable, in the scope of this study, we were interested in the effect of coral macro-morphology and canopy structure on reflectance. Hence, each coral was treated as if it had a uniform surface reflectance. The availability of horizontal areas for extracting surface reflectance possibly biased the measured reflectances slightly, but the sample areas where sufficiently large to mitigate small scale influences, such as low pigmented apical polyps (Figure 1a). For the underlying substrate, a spectral reflectance of dead coral rock was used, this data was from a previous study, collected with a GER 1500 spectroradiometer as described in Reference [17]. All spectra were resampled to 168 bands in 2 nm intervals from 400 nm to 736 nm, as this was the spectral resolution of the modelling software.

Table 2. Coral shapes used in this study. "Group" indicates grouping based on depth of collection, shallow (S, <8 m) or deep (D, 10–20 m). Coral 03 was from a site subjected to terrestrial run-off and was not placed in either group. Polygon counts for the high- and low-resolution models are given. The final column is discussed in the results and is the maximum percentage difference that choice of resolution leads to in reflectance averaged over the coral area in any band.

ID	Species		Polygon Count		Rugosity	Max L_u/E_d Difference
		Group	Low Res.	High Res.		Low vs. High
03	*Montipora capitata*	n/a	2306	152,185	1.98	0.36
08	*Porites evermanni*	D	1712	69,945	2.74	0.51
10	*Porites evermanni*	D	1832	75,524	2.16	0.28
11	*Montipora capitata*	D	1830	103,394	1.77	0.29
12	*Porites evermanni*	D	2150	51,954	2.16	0.24
16	*Porites compressa*	D	1454	89,703	5.43	1.33
17	*Porites evermanni*	S	2079	124,092	2.05	0.27
18	*Porites compressa*	D	1898	103,430	5.82	0.47
19	*Porites evermanni*	S	1804	183,283	1.61	0.22
31	*Porites compressa*	D	1226	56,726	2.96	0.43
32	*Montipora capitata*	S	2461	100,886	2.01	0.31
33	*Porites compressa*	S	2160	98,502	2.90	0.47
35	*Pocillopora* sp.	S	2966	240,353	3.06	0.21
38	*Porites evermanni* or *lutea*	S	1908	95,265	1.72	0.17
39	*Montipora patula*	S	2264	133,594	1.81	0.39
40	*Porites evermanni* or *lutea*	S	1403	56,696	1.45	0.17

To capture the shape of each field-sampled coral, they were impressed into silicone rubber putty after imaging and the moulds were allowed to air dry. These moulds were transported back to the University of Connecticut; each mold was filled with plaster to recreate the coral shape (Figure 1b). Digital 3D models of the plaster cast of natural corals were created using Autodesk ReCapTM photogrammetry software (version 5.0.1.30). The process involved taking between 20 and 40 overlapping photographs of the plaster corals and converting them into 3D digital models consisting of a mesh of vertices and triangular surface polygons. Then, the 3D models were edited in Autodesk® MeshmixerTM (version 3.4.35) to isolate the coral shape from the background. For each coral, two models were produced: High resolution, ranging from 50,000 to 200,000 triangles, and low-resolution, typically ~2000 triangles (Table 1). The low-resolution models are computationally easier to handle, and the first intended test was to determine if low-resolution models would be sufficient. For interpretation of results, a surface rugosity measure [20] was calculated for each coral shape, being the ratio of the coral surface area to its projected area (i.e., nadir view areal extent) in the high-resolution models.

2.3. Radiative Transfer Modelling

The radiative transfer modelling was split into two components: (1) a three-dimensional canopy model designed to evaluate optical properties just above the canopy or above individual coral structures (Figure 1d) [18,21], and (2) a plane-parallel water column model (Figure 1c), HydroLight [19]. All modelling was conducted hyperspectrally in 168 bands of 2 nm step, from 400 to 736 nm.

The 3D model has been described and used in a number of previous publications [18,21–23]. The solution method involves breaking down all surfaces and volumes into discrete elements, triangular surface polygons (as already provided by the 3D digital models), and cubic voxels, for scattering and attenuating media (the water). In this application, the coral shapes were placed on a flat underlying substrate that was decomposed into 100 × 100 squares (each being two triangles) over an area of approximately 20 cm × 20 cm. The model domain can either have light incident from above with horizontally periodic boundaries (i.e., the model set-up repeats in all horizontal directions), or be embedded into a "far-field" radiance distribution characterized from all directions (appropriate for modelling an isolated structure). The model is solved by calculating light transfer between all pairs of elements and propagating the incident light through the system by iteration, for a predetermined number of passes or until convergence is detected. Surfaces are treated as locally Lambertian, but volumetric elements embody full directional scattering, according to a scattering coefficient combined with a phase function [18]. As an update to previous work, this model is now implemented on modern Graphics Processing Units (GPUs), which facilitates improved performance on computationally demanding applications with many bands and elements.

Since the coral structures were small (~20 cm max. height), it would be inefficient to use the 3D model for the full water column, since depths up to 10 m were of interest. More efficiently, the 3D model was coupled to HydroLight in two ways:

1. HydroLight was used to generate a bottom of water column radiance distribution, and this was used to illuminate the individual coral structures, standing on flat dead coral substrate, by being used as the input radiance boundary condition (Figure 1e,f). From these model runs, the reflectance over the coral structure was determined under different illumination conditions to deduce the average reflectance and its variability due to shading and other effects, across a range of illumination conditions (Figure 1g,h). Treatments were applied corresponding to the range limits of interest: Depths of 1 m and 10 m; solar zenith angles of 10° and 50°, azimuth angles of 0°, 90°, 180°, 270°, and two sets of Inherent Optical Properties (IOPs) denoted "forereef" and "lagoon" (described below). Water surface roughness was set to correspond to a wind speed of 5 ms^{-1} but note that only water leaving radiance was used in the results, so surface reflectance was excluded. This method was used for tests 1 to 4, listed in Table 1.

2. Mixtures of 3D living coral structures were assembled on flat dead coral rock substrate, in areal cover densities from 39% to 75%. The 3D model was used with horizontal periodic boundaries to generate a canopy BRDF function, which was then used as a bottom boundary condition in HydroLight (Figure 1i,j). By this method, spectral water-leaving reflectances derived as water-leaving radiance at a given wavelength (λ) normalized to downwelling planar irradiance ($L_w(\theta, \Delta\varphi, \lambda)/E_d(\lambda)$), could be derived above the water for different view directions under different solar zenith angles and water column conditions (Figure 1l). Incorporating the water column this way spatially averages the results, and scope of the results corresponds to pixels larger than the coral structures, i.e., pixels ≥ 1 m, since the coral structures were less than 20 cm across. This modelling covered the same range of depths, solar zenith angles, and the two IOP treatments mentioned above. This aspect of the work corresponds to activities 5 and 6 in Table 1.

One important test was to determine if water attenuation and scattering within the canopy, i.e., at the scale of, and in between, coral shapes was at all important with respect to the derived reflectances. The 3D model can incorporate volumetric absorption and scattering but it is computationally slow, especially for computations in 168 bands. The "interstitial" IOPs were incorporated by embedding the shape in a mesh of $16 \times 16 \times 16$ voxels in activity 4 (Table 1), with IOPs consistent with the corresponding HydroLight modelling (see below). In the other modelling activities, interstitial IOPs were not included. Since the coral shapes were small, within the canopy path lengths of only a few centimeters through the water were achieved. Scattering is predominantly in the forward direction so on short path lengths it has very little effect; the primary issue was absorption on the paths between surface patches. An absorption coefficient of 0.4 (approximately as at 700 nm) gives losses of just 4% on a 10 cm path, so whilst it was not expected that excluding interstitial IOPs would have a substantial effect, a specific test was conducted, and this is evaluated in the results section.

2.4. Inherent Optical Properties

The two IOP treatments "forereef" and "lagoon" (Figure 2) were configured using the HydroLight 5.3 New Case 2 model [19] to produce total absorption, $a(\lambda)$, and attenuation, $c(\lambda)$, values very similar to representative AC-S measurements taken in forereef and lagoon locations at Glovers Reef in Belize and in Palau, Micronesia in 2006 [7]. Both treatments had chlorophyll set at 0.12 mg·m^{-3}, but the lagoon treatment had more scattering and more coloured dissolved organic matter (CDOM). The independent

CDOM $a_g(440)$ (not associated with phytoplankton) was 0.008 m^{-1} for forereef and 0.04 m^{-1} for lagoon, scattering was introduced by a calcareous sand component of 0.01 and 0.4 mg·m^{-3}, respectively. These component concentrations were derived by trial and error to produce bulk IOPs close to the measured values, which represented the limit of the range for high clarity (forereef) and high scattering (lagoon) in the AC-S dataset (see figures in [7]). Under the scattering phase function, Petzold's phase function was used throughout, as described in Reference [24]. These IOPs were within the range of those measured across different reefs of the Pacific [25].

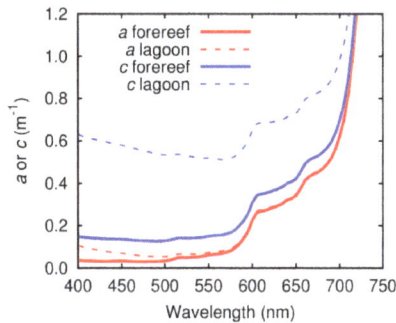

Figure 2. Total absorption, $a(\lambda)$, and attenuation, $c(\lambda)$, as used in the two water column inherent optical property (IOP) treatments, forereef and lagoon (includes the contribution of pure water itself).

2.5. Coral Canopy Assemblages

Incorporating individual corals into a water column model would correspond to a single coral surrounded by an infinite extent of bare substrate, which is not a situation of practical interest for remote sensing. Propagation of reflectance through a plane-parallel water column model, such as HydroLight, necessitates constructing a "pseudo plane-parallel" canopy to characterize a horizontally invariant bottom boundary BRDF. To this aim, six assemblages of corals on flat substrate were constructed by randomly manually placing and rotating the individual low-resolution coral models in an area of 20 cm × 20 cm (periodically repeating). Three assemblages were generated using corals from the shallow group, and three from the deep group. Viewed from above, the areal cover of the corals versus the substrate varied from 39% to 75%. Assemblages were named based on the group and percentage cover, specifically: S46, S58, S75, D39, D46, D74. There was no intention to specifically compare or construct any hypothesis concerning the shallow and deep groups, but it was considered prudent to be consistent and not to mix deep corals with shallow. The assemblages contained between 7 and 15 individual coral shapes, where some shapes appeared twice or three times in specific assemblages, but in different rotational positions. These randomly constructed canopies may not be strictly representative of a real canopy, but they do incorporate a mixture of typical coral morphologies and surface reflectances and enable an estimate of the optical properties of assemblages of such structures. In these models the total number of surface polygons, including substrate, varied from 35,000 to 68,000.

2.6. Coral Model Outputs and Processing

For activities 1 to 4 (Table 1), the main output of interest was reflectance just above the canopy under naturalistic illumination conditions, for a typical range of situations, i.e., the kind of reflectance that could be used as a simple bottom reflectance boundary condition in a remote sensing model (Figure 1g,h). The key questions were if this reflectance was consistent under different illumination conditions (hence robust), and the extent of the effect of shading and other spectral mixing processes due to structure. Toward this aim, for each coral shape and illumination treatment, the model generated a 512 × 512 pixel hyperspectral nadir-view image of the coral in orthographic projection (no perspective) (Figure 1g,h). For activities 1 to 3, the upward radiance directly above the coral shape only (no substrate) was extracted from this image and averaged to give an average nadir view upward radiance, $L_u(\lambda)$, over the coral shape. Reflectance was calculated as $L_u(\lambda)/E_d(\lambda)$ and multiplied by π, for the purpose of comparing to the surface scale diffuse reflectance. For activity 4, which was concerned with spectral mixing with the substrate reflectance (Table 1), the upward radiance was averaged over the shape plus enough surrounding substrate to make a 50% areal cover mix between the coral shape and substrate (Figure 1h). The extent of this area was manually determined as a roughly consistent border around the shape. Activities 1 to 4 illuminated corals with azimuth angles of 0°, 90°, 180°, and 270°, so shadows and other effects would be variable between these repeats, corresponding to the same shape illuminated from different directions.

The evaluation of BRDFs required that the full BRDF function be tabulated in a form suitable to use as input to HydroLight. HydroLight requires azimuthally averaged BRDFs, i.e., as a function of relative azimuth only, so the form of the function is BRDF (θ_i, θ_e, $\Delta\varphi$) (units sr^{-1}) in terms of quad-averaged radiance in the HydroLight directional discretization (directional segments of 10° × 15°, [19]). The incident and exitant zenith angles are denoted by θ_i and θ_e, and only the relative azimuth angle, $\Delta\varphi$, is relevant. The 3D model can work directly with the input and output quad-averaged radiances, where a single run of the model illuminates the canopy with radiance from a single directional quad, and the full hemisphere of exitant quad-averaged radiances is averaged over a 5 × 5 grid of points over the canopy (Figure 1k), see Reference [21] for more details. A minimum of 10 model runs can populate the azimuthally averaged function, one run for each incident zenith angle quad position, θ_i (Figure 1i). Here, three runs per zenith angle were conducted with the canopy rotated by 120° each time. Mirror image symmetry was assumed BRDF (θ_i, θ_e, $\Delta\varphi$) = BRDF (θ_i, θ_e, $-\Delta\varphi$), and the reciprocity condition BRDF (θ_i, θ_e, $\Delta\varphi$) = BRDF (θ_e, θ_i, $-\Delta\varphi$) was checked to ensure there were no major errors and then enforced. Each value in the final function was the average of 12 values in general, although not all values were independent.

3. Results and Discussion

3.1. Effect of Coral Shape on Nadir View Reflectance

Pseudo-color renderings of the 512 × 512 pixel images used to derive the upward radiance over the corals are shown in Figure 3. These model outputs are for high resolution 3D models, for one of the 32 bottom of water column illumination treatments. The RGB images were derived from the 168 band model output by forming the product of the spectrum in each pixel with red, green, and blue visual tristimulus functions [24]. The surface reflectance for some corals was darker than the substrate reflectance, e.g., coral 03 (Figure 3), but for most, the coral surface reflectance was lighter. The necessity of obtaining the reflectance from a horizontal surface area in the hyperspectral image (Figure 1a) may bias the reflectance compared to the true total surface average, but this is of minor consequence in this study, since the overall brightness of the reflectance is of secondary importance to the spectral shape.

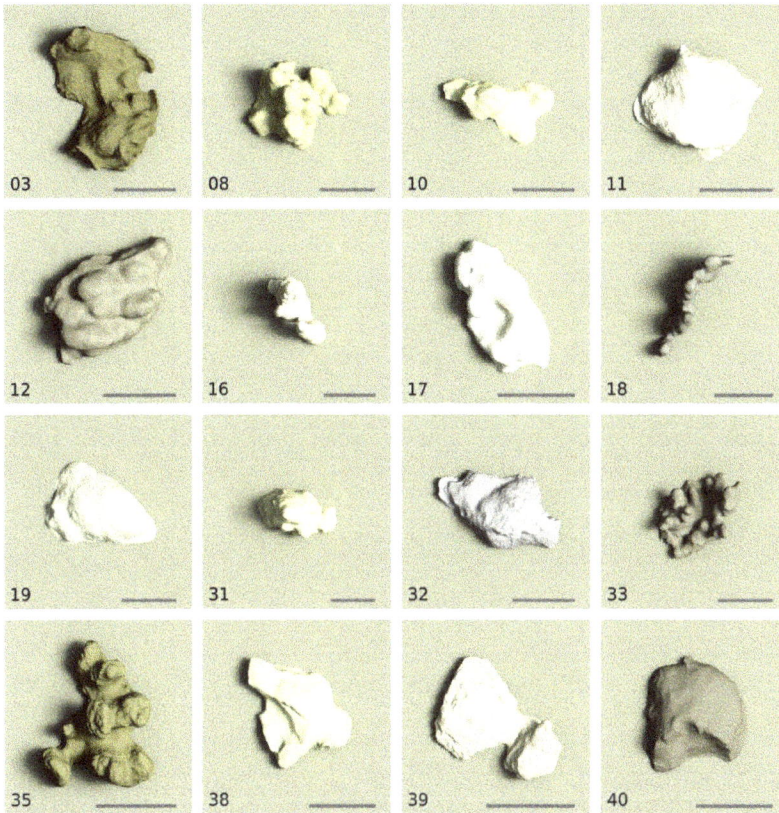

Figure 3. High resolution model outputs, as a nadir-view orthogonal projected rendering above each coral, converted to RGB using the tristimulus functions [24]. Illumination conditions are solar zenith $\theta_s = 50°$, forereef IOPs, depth 1 m. These are the high-resolution models without interstitial IOPs (activity 1, Table 1). Note: Corals are not shown in the same relative scale, bar is 5 cm.

With respect to the modelled reflectance over the coral area (Figure 1g), for all corals, the spectra averaged over the coral area had a very similar shape to the coral surface reflectance but was darker due to shading within the coral structure (Figure 4). The different illumination conditions introduced some variability in the reflectance but in general, this variability was small and certainly smaller than the shading effect. Coral 16 had the highest variability in reflectance under different illuminations, as this coral was a relatively vertical structure with an overall irregular shape when viewed from above; therefore the level of shading was quite dependent on azimuth angle of the incident light. Regarding corals shapes that were flatter (e.g., coral 39), or with numerous protrusions (coral 35), different lighting conditions produced little variation in the average reflectance (Figure 4).The first useful observation from these results was that in general, the average reflectance over the coral shape up to 690 nm can be represented by a scaled version of the surface reflectance, where the scaling factor < 1 effectively introduces shading as a black endmember in a linear mixing model, i.e., the transformation of surface reflectance $R(\lambda)$ is of the form $R'(\lambda) = R(\lambda) \times f$, where $0 \leq f \leq 1$ and for a flat Lambertian surface $f = 1$. Shading endmembers have been used in mineral applications [26]. For each coral, a single scaling factor was deduced as the median value of the fit from 400 to 690 nm to each of the 32 treatments (Table 3, Figure 4). These scaling factors varied from 0.50 (coral 16) to 0.84 (coral 39), and for flatter coral shapes, the scaling factor was higher since less shading was introduced. Corals 18 and 31, which were also relatively vertical structures, had scaling factors of 0.59 and 0.54, respectively. Above 690 nm, the fit to the simple shadow model was not as good, as reflectances over the coral area tended to be higher (Figure 4). It is likely the reason for this is the very high coral surface reflectance above 690 nm (Figure 4). A part of the coral surface that is shaded will "see" in its hemispherical field of view other parts of the coral surface. In wavelengths where the coral surface is highly reflective, those surrounding surfaces will reflect light to the point in question, and so the shading effect will be less overall.

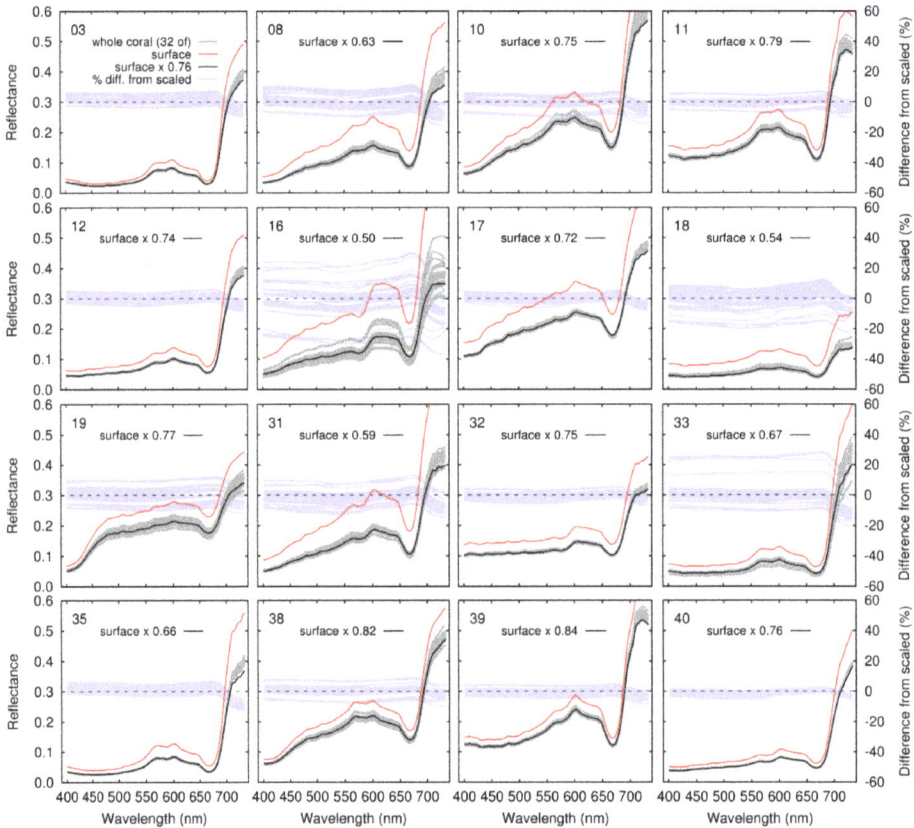

Figure 4. Reflectance over coral shape area only (no substrate) at the bottom of the water column with no interstitial water scattering or attenuation included. Plots show surface reflectance (red), reflectance over coral area under 32 treatments (grey), surface reflectance scaled by shading factor (black), and % difference of each treatment from scaled surface reflectance (light blue, right hand y-axis).

In each coral, most of the reflectances under different illumination conditions were within ±20% of the surface reflectance scaled by the mean shade factor (Figure 4). This shading factor, which in our data ranges from 0.50 to 0.84, could be of immediate use in approaches that use diffuse reflectance as a bottom boundary condition, for example, model inversion techniques for image processing [27]. If basic inputs are surface reflectances, a variable shading factor (effectively a shade endmember) could be included as an additional parameter to be estimated in the inversion, or fixed at a reasonable value, e.g., ~0.7, or certainly less than one. A variable factor would not only accommodate the difference between the coral shapes shown here but would also accommodate the residual error in each coral, due to variation under illumination conditions (Figure 4). However, this also points to the importance of the design of in-situ methodologies used for collection of spectral reflectance data [10,11,28,29]. Protocols which include multiple sampling over benthos will likely include a certain amount of the shading factor already, but this is scale-dependent, i.e., for small branching corals it may be completely included, whereas for larger massive morphologies the measurements may be closer to surface reflectances. Moreover, protocols that use shading or artificial light sources may or may not include the relevant amount of shading for a remote sensing context. Careful consideration of the relative scale of measurements in the context of the application is required. From our results, the maximum error in the brightness of the endmember reflectance, when using a surface scale reflectance instead of macro-morphology scale, could be a factor of as much as 0.5 (or 2). It may be worthwhile to consider if spectral libraries should be standardized or normalized in this respect, to either include or not include the shading factor. With this information, there would be the possibility to apply a post-hoc factor, at least as a rough estimate (between 0.5 and 0.84 in our data, e.g., ~0.67), or as a species-specific value (Table 3). Another consideration is that for very high spatial resolution imagery (pixels $\leq$ 0.5 m), where benthic structure is at a similar scale to the pixels, shading effects may vary from pixel to pixel, over for example large massive corals, where the sides and top may be imaged in different pixels.

Table 3. Shading factors for coral area only and coral with surrounding substrate in 50% mix. Final column shows the maximum difference in the reflectance of any band up to 690 nm, when interstitial attenuation and scattering is included in the canopy model.

ID	Species	Shading Scale Factor Coral only		Shading Scale Factor 50% Coral-Substrate Mix		Max L_u/E_d diff. with Interstitial IOPs ($\leq$690 nm)
		Range	Median	Range	Median	
03	*Montipora capitata*	0.71–0.79	0.76	0.83–0.88	0.85	2.6%
08	*Porites evermanni*	0.57–0.69	0.63	0.68–0.77	0.74	2.8%
10	*Porites evermanni*	0.71–0.82	0.75	0.75–0.84	0.81	1.6%
11	*Montipora capitata*	0.75–0.84	0.79	0.80–0.85	0.84	1.0%
12	*Porites evermanni*	0.71–0.77	0.74	0.79–0.85	0.82	1.7%
16	*Porites compressa*	0.39–0.66	0.50	0.51–0.64	0.59	3.1%
17	*Porites evermanni*	0.69–0.75	0.72	0.75–0.79	0.78	1.0%
18	*Porites compressa*	0.50–0.65	0.54	0.55–0.68	0.65	3.0%
19	*Porites evermanni*	0.69–0.85	0.77	0.78–0.86	0.83	1.9%
31	*Porites compressa*	0.54–0.65	0.59	0.64–0.71	0.70	2.8%
32	*Montipora capitata*	0.72–0.78	0.75	0.79–0.83	0.81	1.2%
33	*Porites compressa*	0.51–0.74	0.67	0.71–0.81	0.78	2.6%
35	*Pocillopora* sp.	0.62–0.68	0.66	0.70–0.81	0.74	2.1%
38	*Porites evermanni* or *lutea*	0.76–0.88	0.82	0.82–0.87	0.86	1.5%
39	*Montipora patula*	0.81–0.88	0.84	0.83–0.88	0.87	0.7%
40	*Porites evermanni* or *lutea*	0.75–0.79	0.76	0.87–0.89	0.88	1.7%

3.2. Effect of 3D Model Resolution

To determine if the low-resolution 3D coral models were sufficient, the modelling of reflectance over coral shape area (Figure 1g, results in Figures 3 and 4, and Table 3) was duplicated using the low-resolution versions of the coral 3D structures (activity 2, Table 1). Across all corals and illumination treatments, the difference in any band when using the low-resolution 3D models was not greater than 0.51%, apart from coral 16 which had a maximum difference of 1.33%. In most corals, the maximum difference was less than a third of a per cent (Table 2, final column). Visually the outputs were very similar (Figure 5). Therefore, in general, using the low-resolution structures made almost no difference to the reflectance averaged over the coral area, and there seemed little need to continue using the high-resolution models, which with up to ~100,000 polygons cause the 3D radiative transfer model to run substantially slower. For this reason, the remaining modelling activities used only the low-resolution models.

(a) high resolution (240353 triangles)   (b) low resolution (2966 triangles)

Figure 5. (**a**) High-resolution model versus; (**b**) low-resolution model of coral 35, scale bar is 5 cm.

3.3. Linear Mixing of Reflectance over Coral Shapes and Surrounding Substrate

As with the reflectance over the coral shape only, reflectance over a 50% areal mix of coral and surrounding substrate was consistently darker than a 50% linear mix of the surface reflectances, but could be well represented in wavelengths less than 690 nm by a uniformly scaled version of that linear mix (Figure 6). The shading factor when substrate was included was a consistently higher value (less shading) than the values over coral shape only, where the median for each coral ranged from 0.59 (coral 16) to 0.88 (coral 40) (Table 3). Individual values were from 4% to 20% higher than the corresponding values over the coral area only. This occurred because the flat substrate does not introduce any shading itself but is shaded by the coral structure (Figure 3). Including surrounding substrate introduces relatively less shading than would occur within the area of the structure. Therefore, both in the experimental set-up and the physical world, the shading factor is likely dependent on the ratio of coral cover to substrate.

The previous suggestion of including a variable black shade endmember in a linear mixing model for bottom reflectance would seem to be further supported by these findings. The shading factor is dependent on the areal cover of flat substrate included, and the difference was variable from 4% to 20%. A-priori inclusion of the shading factor, even by careful collection of in-situ spectra to include shadows, cannot account for the variability due to structural context. However, again the figures presented here assume that the basic input reflectances are surface reflectances, where for flat substrates such as sand or dead coral rock this is likely to be the case, whilst for corals some a-priori inclusion of shading is likely. The possible range of shading factors would have to be modified accordingly.

Figure 7 shows that median shading factors were a clear function of surface rugosity (coral surface area divided by projected top-down area) for the reflectance over the coral area only (Figure 7a), and for coral shape and substrate mixes (Figure 7b). Intuitively, the shading factor should be 1 when rugosity is 1 (flat surface), and very high rugosities should tend to show some positive shading factor greater than zero. To express these constraints, a simple two-factor negative exponential model was fit to the plots in Figure 7, of which $y = (1 - A) \times (\exp[-S \times (x - 1)] + A$, fit for S (slope) and, A, the asymptotic shading factor. This function gives a good fit in both cases but is just indicative of the general shape of the relationship, since the validity of the concept of an asymptotic shading factor is debatable. The slightly reduced shading effect when flat substrate is included, can be seen in the smaller magnitude slope (S) of the best-fit line in Figure 7b. Note the surface rugosity value refers only to the coral shape. If rugosity in Figure 7b were calculated incorporating the flat substrate, the rugosity values would be lower and the relationship would be closer to Figure 7a. Solar zenith angle of 10° vs. 50° only had a small effect on the shading factor (Figure 7b), but as expected the effect increased with higher rugosity. Several factors were relevant to the observed small effect of the solar zenith angle, i.e., due to refraction the corresponding sub-surface zenith angles are less, at approximately 7°–35°, and side-ward illumination reduces the irradiance on some surface facets, but at the same time increases it on others. Considering Figure 2, produced with $\theta_s = 50°$, for most corals, only a small part of the area is actually visibly in shadow. Those with the most shaded area correspond to those where solar zenith angle has the most effect, e.g., corals 08, 33, and 35, corresponding to the group with rugosity ~3 in Figure 7b. Overall Figure 7 indicates that rugosity, or equivalently species morphology, can be a robust indicator of the magnitude of the appropriate shading factor. This is consistent with empirical studies on BRDFs, which have shown that morphology and shadowing between branches are key determiners of spectral reflectance [16]. If a variable shading factor could be reliably estimated as a black endmember contribution in a model inversion image analysis [27], this estimate would carry information on benthic rugosity and benthic type.

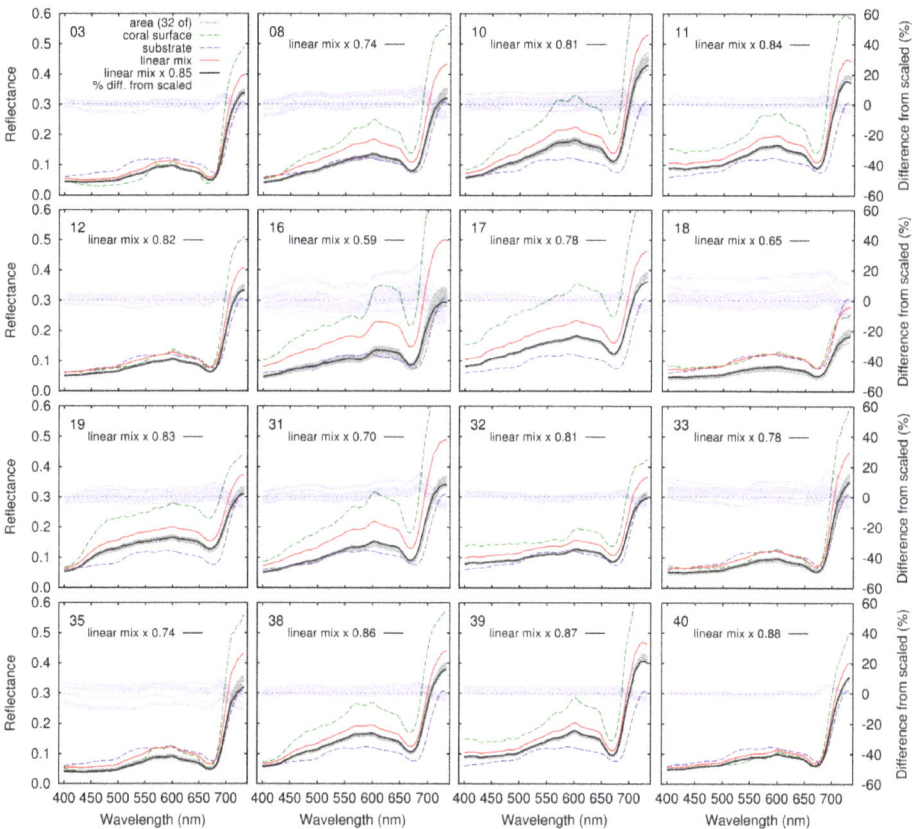

Figure 6. Reflectance over coral structure and surrounding substrate to give a 50% mix in areal cover. Plots show the coral surface and substrate surface reflectances (thin and thick green lines), and a 50% linear mix of those reflectances (red), reflectance over the coral and substrate area under 32 treatments (grey), surface reflectance scaled by shading factor (black), and % difference of each treatment from scaled surface reflectance (light blue, right hand y-axis).

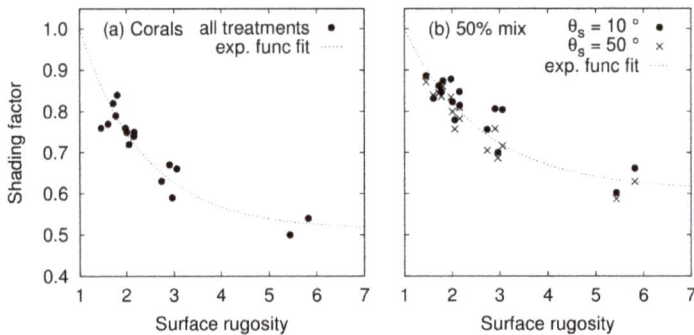

Figure 7. Median shading factor as a function of surface rugosity: (**a**) Over coral area only (Figure 4), and; (**b**) over coral area and surrounding substrate in a 50% areal mix (Figure 6) with data separated into solar zenith angles of 10° and 50°. Lines are of the form $y = (1 - A) \times \exp[-S \times (x - 1)] + A$ and are the least squares best fit, giving (**a**) $A = 0.51$, $S = 0.72$; (**b**) $A = 0.61$, $S = 0.61$.

3.4. Effect of Interstitial Scattering and Absorption

The difference when including interstitial IOPs (scattering and absorption in the water around and within the coral shape) in the modelling of reflectance averaged over a 50% mix of coral and substrate was in general, less than 3% in any band for wavelengths less than 690 nm (Table 3), and for most coral shapes of the order of 2% or less. This showed that the previous back-of-envelope calculation that suggested a 4% error at 700 nm, based on absorption over path lengths of 10 cm, was quite reasonable. Beyond 700 nm, absorption by pure water is high and the discrepancy when omitting interstitial IOPs became larger, up to 5%. However, in a practical application, unless corals almost touch the water surface, the absorption in the water column above the canopy will dominate over the within-canopy effect, so the discrepancies above 690 nm are of minor consequence. The cost of 2% or 3% in accuracy when omitting the interstitial IOPs comes at the advantage, not only of substantially faster computation in the 3D model, but also the BRDF of the canopy can be decoupled from the IOPs of the overlying water column. That is, a BRDF can be computed based on coral canopy structure and surface reflectance, and then the overlying water column can be specified with independently varying IOPs. Therefore, this answers point 4 in Table 1, and is the strategy that was taken in the BRDF and water column modelling activities (5 and 6 in Table 1). However, this result is scale dependent and only holds in this case because the coral structures were small. With larger corals and reef structures, where the vertical distances can be a meter or more, the effect of interstitial water scattering absorption is likely a more significant component of the reflectance.

3.5. Canopy Assemblage BRDF Effects

The discussion so far has concentrated on nadir view bottom of water column reflectance over individual corals under naturalistic light conditions, calculated based on the upward directed radiance and downward irradiance, i.e., $\pi L_u(\lambda)/E_d(\lambda)$. Whilst the previous results were fairly insensitive to illumination conditions, such as solar zenith angle (Figures 4 and 6), view angle effects were not considered. Direct use of these results as a bottom boundary requires the assumption of Lambertian reflectance. As such, they are appropriate for incorporation into models where this assumption is implicit, for example, in model inversion techniques for image analysis [27]. However, to correctly model propagation of light through the water column in a physically exact model, such as HydroLight, the full BRDF of canopy assemblages is required (activities 5 and 6, Table 1).

The BRDF function is calculated using light incident only for a specific direction in an otherwise black radiance field (Figure 1i), so in the model outputs, shading effects are very apparent (Figure 8). Visually, Figure 8 may appear to contradict Figure 7b, since shading as a function of incident zenith angle in Figure 8 appears stronger than implied by Figure 7b. To reconcile this first note that Figure 7b is expressed in terms of above-surface solar zenith angles of 10° and 50°, and so most closely corresponds to Figure 8a,b (sub-surface incident angles of 10° and 30°), Figure 8c is a more extreme example. Furthermore, Figure 8 arises from unidirectional illumination, whereas Figure 7b arises from a more diffuse illumination, being the sub-surface propagated solar and sky illumination. A final point is that the increased shadow from Figure 8a to 8b, although visually very apparent, is probably not more than 10% of the actual image area, so even under unidirectional illumination it likely would not affect the shading factor by more than 10%, which is comparable to the differences seen for rugosities ~3 in Figure 7b.

(a) $\theta_i = 10°$ (b) $\theta_i = 30°$ (c) $\theta_i = 50°$

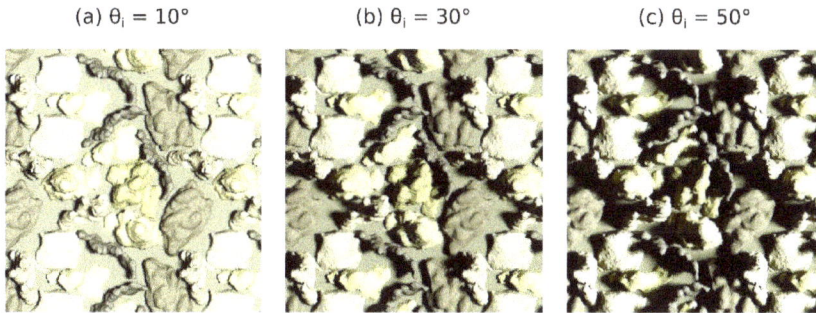

Figure 8. Examples from BRDF generation of canopy using reflectances of deep corals at 74% coral coverage (D74). Shown as nadir-view ($\theta_e = 0°$) orthogonal projected rendering above each canopy, corresponding to incident radiance from: (**a**) $\theta_i = 10°$; (**b**) $\theta_i = 30°$ and; (**c**) $\theta_i = 50°$, respectively, azimuth $\varphi s = 0°$. RGB images are created from hyperspectral data using the tristimulus functions and represent an area of 30 cm × 30 cm.

One of the densest canopy assemblages, D74 (74% coral cover), also tended to have vertically higher coral structures and consequentially showed the strongest BRDF effect, with an almost linear effect of view angle in the incident plane ($\Delta\varphi = 0$), when incident light was at a 50° zenith angle (Figure 9e). The reflected radiance decreased by a factor of seven when varying the view direction from almost horizontal in the same direction as the incident light, to horizontally toward the incident light (Figure 9e). In all canopy assemblages there is a clear hotspot effect [30], where brightness is at a maximum when the incident and view directions are the same. The BRDFs discussed here were spatially averaged, and the hot spot occurs because the viewing geometry determines the fraction of shadowed area detected. When the viewing and incident light directions are the same, shadows are maximally obscured by the illuminated surfaces. Interestingly, for the canopy structures modelled here, the hotspot effect is stretched out to larger view angles (Figure 9). This is likely a geometrical consequence of an increasingly horizontal view onto vertical structures illuminated from the side (Figure 8).

In all canopy assemblages, there were only small BRDF effects at 90° to the incident plane ($\Delta\varphi = 90°$) (Figure 9b,d,f). Typically reflectance decreased slightly as the view angle moved from nadir, although the pattern was stronger for denser canopies, such as D74 (Figure 9), and even included a slight increase in reflectance close to horizontal, being a pattern also seen in models of dense seagrass canopies [21].

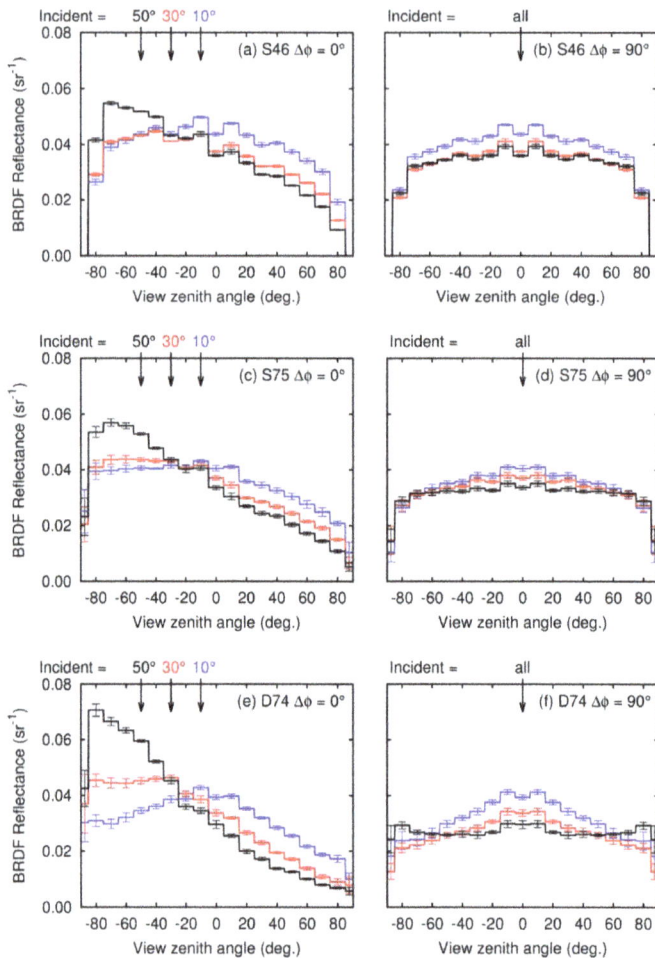

Figure 9. Bidirectional reflectance functions (BRDFs) at 550 nm, of three of the composite coral canopies for shallow (S, 0–10 m) and deep (D, 10–20 m) corals with areal cover: (**a,b**) 46%; (**c,d**) 75%; and (**e,f**) 74%, in the incident plane (a, c, e, $\Delta\varphi = 0°$) and at 90° to the incident plane (**b,d,f**, $\Delta\varphi = 90°$). Arrows show direction of incident light, each plot shows θi of 10°, 30° and 50°. Negative view zenith angle for $\Delta\varphi = 0°$ (**a,c,e**) means backward reflection (source and view point in the same hemisphere). Error bars are ±1 standard error on 12 values from assumed reciprocity, rotational and mirror symmetries.

3.6. Above-Water BRDF Effects

Canopy BRDFs contributed to the directional pattern in water leaving radiance under naturalistic sky illumination primarily for dense canopies in shallow, clear water (Figure 10). For the 46% cover canopy, S46, a slight BRDF effect of increased retro-reflection in the incident plane, leading to effective propagation of the hotspot effect, was evident for depths less than 5 m with a low sun position, $\theta_s = 50°$ (Figure 10a). In deeper waters and at 90° to the incident plane L_w, it decreased with increased view angle (Figure 10b,d,f). In deeper waters over sparser canopies, the water column itself becomes a contributor to the above water BRDF, where at 10 m depth the BRDF response in the solar plane (Figure 10a) resembles the response at 90° to the solar plane (Figure 10b).

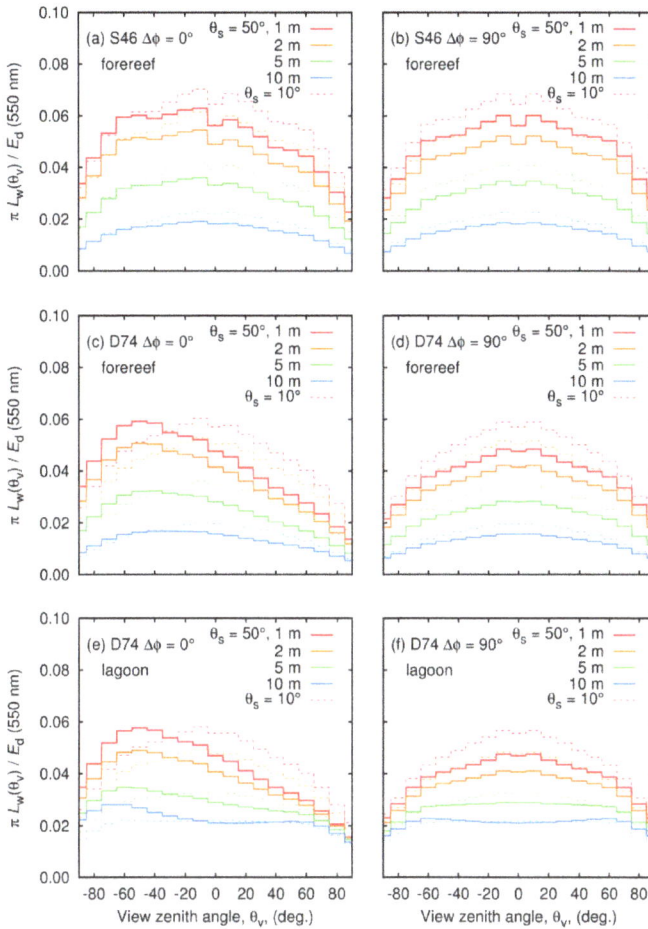

Figure 10. Modelled influence of three-dimensional coral BRDF effects (from Figure 9) on water-leaving reflectance (π Lw/Ed) at 550 nm for: (**a,b**) Composite canopies S46 (46% shallow coral cover) and; (**c,d,e,f**) D74 (74% deep coral cover), in the incident plane (**a,c,e**, $\Delta\varphi = 0°$) and at 90° to the incident plane (**b,d,f**, $\Delta\varphi = 90°$). Each plot shows results of incident solar zenith angles of 10° (dotted lines) and 50° (solid lines), and depths 1, 2, 5, and 10 m. Negative view zenith angle for $\Delta\varphi = 0°$ (**a,c,e**) means backward reflection (source and view point in the same hemisphere).

Regarding the dense 74% coral cover canopy D74, water-leaving BRDF effects in the incident plane were quite prominent for both the forereef IOP treatment and the higher scattering lagoon waters (Figure 10c,e). For example, when the water depth was 1 m, the difference in $L_W(550)$ between a view zenith angle of 50° toward and away from a sun at $\theta_s = 50°$ was about a factor of 1.5 (~0.06 vs. ~0.04, Figure 10c,e); whereas this factor was ~1.2 for view zenith angles at ± 20°. At 90° to the incident plane, even for dense canopies, BRDF effects are less apparent (Figure 10d,f). It is clear that for these modelled canopies, the assumption of a Lambertian bottom is primarily violated close to the retro-reflected (hotspot) direction for large zenith angles (>30°) but could be a more reasonable assumption under other solar-view geometries. The difference in nadir view water-leaving reflectance at 550 nm for canopy D74 at 1 m, for solar zenith angle of 50° vs. 10°, was 18% (Figure 10c,d, ~0.047 vs. ~0.057). This is equivalent to the variation previously shown in reflectance over individual corals under different illumination conditions (Figures 4 and 6), which for some corals was as much as ±20%, but here the variation was reduced by the water column, to effectively ±9%. Mobley et al. [31] estimated ignoring non-Lambertian bottom effects would in general, cause errors of less than 10% in remote sensing reflectance. While not strictly a comparison to a Lambertian bottom assumption, a discrepancy of 18% implies the potential for larger errors.

To avoid across-swath BRDF effects in airborne imagery, the advice is therefore broadly the same as for avoiding surface glint [32]; thus flying toward or away from the sun will ensure the cross track view is at 90° to the solar plane, and a solar zenith angle > 30° both avoids glint and slightly flattens the BRDF response (Figure 10d,f).

4. Conclusions

A number of conclusions from this work can be made that relate (1) to using Lambertian bottom boundary conditions and linear mixing models, and (2) to data collection of reflectances and imagery. In summary:

- If the input reflectances for different benthic types can be considered 'pure' surface reflectances, then structural complexity implies the introduction of a shading scale factor or a black shade endmember in mixing models is required. In our data, that shading scale factor could be anywhere from 0.5 to 0.9, whereas for a genuinely flat substrate it would be 1.0.
- The magnitude of the shading factor is inversely related to surface rugosity; hence the shade factor value could be set based on the expected bottom types, or if derivable in an image analysis, it would give additional information on benthic type.
- Nadir viewing reflectance over the bottom (L_u/E_d) for individual corals varies as much as ~20% (more only in exceptional cases) under the range of light environments found on reefs, under typical remote sensing conditions.
- Regarding view directions beyond a few 10s of degrees from nadir for dense canopies in shallow water shading, hotspot effects become relevant, leading to potentially a 50% difference (factor of 1.5) in above water reflectance. At narrower remote sensing geometries (view zenith angles < 20°), differences can be 20% (factor of 1.2).
- For all modelled canopy assemblages, there were only small BRDF effects at a relative viewing angle of 90° to the incident solar plane. The advice for minimizing cross-track BRDF effects in airborne imagery is therefore consistent with that for minimizing surface glint, i.e., fly in a direction close to the solar plane with a solar zenith angle greater than 30°.
- Collection of in-situ spectra of benthic types requires consideration of what level of shading is already included, or whether the data is close to being a surface reflectance measurement. However, including shading in field measurements cannot fully accommodate the influence of canopy interactions, because there are effects at canopy scale involving different benthic types.
- These modelling results provide useful concepts and parameter ranges, which can assist in the interpretation of empirical data and the development of image processing algorithms.

Author Contributions: Conceptualization J.D.H., H.M.D.; Methodology, J.D.H., M.M., A.W., H.M.D.; Writing-Original draft preparation, J.D.H.; Writing–Review & Editing J.D.H., M.M., A.W., H.M.D.

Funding: This work was funded by award #NNX15A32G of the NASA Biodiversity and Ecological Forecasting Program.

Acknowledgments: The authors extend thanks to Brandon Russell and Jeff Godfrey for assistance in data acquisition and field work. Part of Brandon Russell's time was funded by the NASA Ocean Biology and Biogeochemistry program and the COral Reef Airborne Laboratory (CORAL) project. Coral collection was conducted under special activity permit (SAP) 2018-02 issued on 2 February 2017, by the State of Hawaii Department of Land and Natural Resources.

Conflicts of Interest: The authors declare no conflict of interest. The funding sponsors had no role in the design of the study; in the collection, analyses, or interpretation of data; in the writing of the manuscript and in the decision to publish the results.

References

1. Hedley, J.D.; Roelfsema, C.M.; Chollet-Ordaz, I.; Harborne, A.R.; Heron, S.F.; Weeks, S.; Skirving, W.J.; Strong, A.E.; Eakin, C.M.; Christensen, T.R.L.; et al. Remote sensing of coral reefs for monitoring and management: A review. *Remote Sens.* **2016**, *8*, 118. [CrossRef]

2. Mumby, P.J.; Skirving, W.; Strong, A.E.; Hardy, J.T.; LeDrew, E.F.; Hochberg, E.J.; Stumpf, R.P.; David, L.T. Remote sensing of coral reefs and their physical environment. *Mar. Pollut. Bull.* **2004**, *48*, 219–228. [CrossRef] [PubMed]

3. Mumby, P.J.; Green, E.P.; Edwards, A.J.; Clark, C.D. Coral reef habitat-mapping: How much detail can remote sensing provide? *Mar. Biol.* **1997**, *130*, 193–202. [CrossRef]

4. Hochberg, E.J.; Atkinson, M.J. Capabilities of remote sensors to classify coral, algae, and sand as pure and mixed spectra. *Remote Sens. Environ.* **2003**, *85*, 174–189. [CrossRef]

5. Capolsini, P.; Andréfouët, S.; Rion, C.; Payri, C. A comparison of Landsat ETM+, SPOT HRV, Ikonos, ASTER, and airborne MASTER data for coral reef habitat mapping in South Pacific islands. *Can. J. Remote Sens.* **2007**, *23*, 87–200. [CrossRef]

6. Lubin, D.; Li, W.; Dustan, P.; Mazel, C.H.; Stamnes, K. Spectral Signatures of coral reefs: Features from space. *Remote Sens. Environ.* **2001**, *75*, 127–137. [CrossRef]

7. Hedley, J.D.; Roelfsema, C.; Phinn, S.R.; Mumby, P.J. Environmental and sensor limitations in optical remote sensing of coral reefs: Implications for monitoring and sensor design. *Remote Sens.* **2012**, *4*, 271–302. [CrossRef]

8. Hedley, J.D.; Mumby, P.J. Biological and remote sensing perspectives of pigmentation in coral reef organisms. *Adv. Mar. Biol.* **2002**, *43*, 277–317. [PubMed]

9. Mumby, P.J.; Hedley, J.D.; Chisholm, J.R.M.; Clark, C.D.; Ripley, H.; Jaubert, J. The cover of living and dead corals from airborne remote sensing. *Coral Reefs* **2004**, *23*, 171–183. [CrossRef]

10. Goodman, J.A.; Ustin, S.L. Classification of benthic composition in a coral reef environment using spectral unmixing. *J. Appl. Remote Sens.* **2007**, *1*, 011501.

11. Hochberg, E.J.; Atkinson, M.J. Spectral discrimination of coral reef benthic communities. *Coral Reefs* **2000**, *19*, 164–171. [CrossRef]

12. Lesser, M.P.; Mobley, C.D. Bathymetry, water optical properties, and benthic classification of coral reefs using hyperspectral remote sensing imagery. *Coral Reefs* **2007**, *26*, 819–829. [CrossRef]

13. Leiper, I.A.; Phinn, S.R.; Roelfsema, C.M.; Joyce, K.E.; Dekker, A.G. Mapping Coral Reef Benthos, Substrates, and Bathymetry, Using Compact Airborne Spectrographic Imager (CASI) Data. *Remote Sens.* **2014**, *6*, 6423–6445. [CrossRef]

14. Garcia, R.A.; Lee, Z.; Hochberg, E.J. Hyperspectral Shallow-Water Remote Sensing with an Enhanced Benthic Classifier. *Remote Sens.* **2018**, *10*, 147. [CrossRef]

15. Joyce, K.E.; Phinn, S.R. Bi-directional reflectance of corals. *Int. J. Remote Sens.* **2002**, *23*, 389–394. [CrossRef]

16. Miller, I.; Foster, B.C.; Laffan, S.W.; Brander, R.W. Bidirectional reflectance of coral growth-forms. *Int. J. Remote Sens.* **2016**, *37*, 1553–1567. [CrossRef]

17. Hedley, J.D.; Mumby, P.J.; Joyce, K.E.; Phinn, S.R. Spectral unmixing of coral reef benthos under ideal conditions. *Coral Reefs* **2004**, *23*, 60–73. [CrossRef]

18. Hedley, J.D. A three-dimensional radiative transfer model for shallow water environments. *Optics Express* **2008**, *16*, 21887–21902. [CrossRef] [PubMed]

19. Mobley, C.D.; Sundman, L. Hydrolight 5.2 User's Guide. Sequoia Scientific. Available online: https://www.sequoiasci.com/wp-content/uploads/2013/07/HE52UsersGuide.pdf (accessed on 25 July 2018).

20. Young, G.C.; Dey, S.; Rogers, A.D.; Exton, D. Cost and time-effective method for multi-scale measures of rugosity, fractal dimension, and vector dispersion from coral reef 3D models. *PLoS ONE* **2017**, *12*, e0175341. [CrossRef] [PubMed]

21. Hedley, J.D.; Enríquez, S. Optical properties of canopies of the tropical seagrass *Thalassia testudinum* estimated by a three-dimensional radiative transfer model. *Limnol. Oceanogr.* **2010**, *55*, 1537–1550. [CrossRef]

22. Hedley, J.D.; McMahon, K.; Fearns, P. Seagrass canopy photosynthetic response is a function of canopy density and light environment: A model for Amphibolis griffithi. *PLoS ONE* **2014**, *9*, e111454. [CrossRef] [PubMed]

23. Hedley, J.D.; Russell, B.; Randolph, K.; Dierssen, H. A physics-based method for the remote sensing of seagrasses. *Remote Sens. Environ.* **2015**, *174*, 134–147. [CrossRef]

24. Mobley, C.D. *Light and Water: Radiative Transfer in Natural Waters*; Academic Press: San Diego, CA, USA, 1994; ISBN 0-12-502750-8.

25. Russell, A.B.; Hochberg, E.; Dierssen, H.M. Comparison of water column optical properties across geomorphic zones of Pacific coral reefs. *Limnol. Oceanogr.* under review.

26. Adams, J.B.; Smith, M.O.; Johnson, P.E. Spectral mixture modelling: A new analysis of rock and soil types at the Viking Lander I site. *J. Geophys. Res.* **1986**, *91*, 8098–8112. [CrossRef]

27. Dekker, A.G.; Phinn, S.R.; Anstee, J.; Bissett, P.; Brando, V.E.; Casey, B.; Fearns, P.; Hedley, J.; Klonowski, W.; Lee, Z.P.; et al. Intercomparison of shallow water bathymetry, hydro–optics, and benthos mapping techniques in Australian and Caribbean coastal environments. *Limnol. Oceanogr. Methods* **2011**, *9*, 396–425. [CrossRef]

28. Roelfsema, C.M.; Marshall, J.; Phinn, S.R.; Joyce, K. *Underwater Spectrometer System 2006 (UWSS04)—Manual*; Biophysical Remote Sensing Group, Centre for Spatial Environmental Research, University of Queensland: Queensland, Australia, 2006.

29. Roelfsema, C.M.; Phinn, S.R. Spectral Reflectance Library of Healthy and Bleached Corals in the Keppel Islands, Great Barrier Reef. PANGAEA. Available online: https://doi.org/10.1594/PANGAEA.872507 (accessed on 25 July 2018).

30. Liang, S. *Quantitative Remote Sensing of Land Surfaces*; Wiley: Hoboken, NJ, USA, 2004; ISBN 978-0471281665.

31. Mobley, C.D.; Zhang, H.; Voss, K.J. Effects of optically shallow bottoms on upwelling radiances: Bidirectional reflectance distribution function effects. *Limnol. Oceanogr.* **2003**, *48*, 337–345. [CrossRef]

32. Kay, S.; Hedley, J.D.; Lavender, S. Sun glint correction of high and low spatial resolution images of aquatic scenes: A review of methods for visible and near-infrared wavelengths. *Remote Sens.* **2009**, *1*, 697–730. [CrossRef]

applied
sciences

MDPI

Article

Assessing the Impact of a Two-Layered Spherical Geometry of Phytoplankton Cells on the Bulk Backscattering Ratio of Marine Particulate Matter

Lucile Duforêt-Gaurier [1,*], David Dessailly [1], William Moutier [2] and Hubert Loisel [1]

[1]   Univ. Littoral Cote d'Opale, Univ. Lille, CNRS, UMR 8187, LOG,
     Laboratoire d'Océanologie et de Géosciences, F 62930 Wimereux, France;
     david.dessailly@univ-littoral.fr (D.D.); hubert.loisel@univ-littoral.fr (H.L.)
[2]   Royal Meteorological Institute of Belgium, 1180 Brussels, Belgium; william.moutier@gmail.com
*   Correspondence: lucile.duforet@univ-littoral.fr; Tel.: +33-321-99-64-21

Received: 2 August 2018; Accepted: 21 November 2018; Published: 19 December 2018

Abstract: The bulk backscattering ratio ($\tilde{b}_{bp}$) is commonly used as a descriptor of the bulk real refractive index of the particulate assemblage in natural waters. Based on numerical simulations, we analyze the impact of modeled structural heterogeneity of phytoplankton cells on $\tilde{b}_{bp}$. $\tilde{b}_{bp}$ is modeled considering viruses, heterotrophic bacteria, phytoplankton, organic detritus, and minerals. Three case studies are defined according to the relative abundance of the components. Two case studies represent typical situations in open ocean, oligotrophic waters, and phytoplankton bloom. The third case study is typical of coastal waters with the presence of minerals. Phytoplankton cells are modeled by a two-layered spherical geometry representing a chloroplast surrounding the cytoplasm. The $\tilde{b}_{bp}$ values are higher when structural heterogeneity is considered because the contribution of coated spheres to light backscattering is higher than homogeneous spheres. The impact of heterogeneity is; however, strongly conditioned by the hyperbolic slope ξ of the particle size distribution. Even if the relative abundance of phytoplankton is small (<1%), $\tilde{b}_{bp}$ increases by about 58% (for $\xi = 4$ and for oligotrophic waters), when the heterogeneity is taken into account, in comparison with a particulate population composed only of homogeneous spheres. As expected, heterogeneity has a much smaller impact (about 12% for $\xi = 4$) on $\tilde{b}_{bp}$ in the presence of suspended minerals, whose increased light scattering overwhelms that of phytoplankton.

Keywords: ocean optics; backscattering ratio; phytoplankton; coated-sphere model; bulk refractive index; seawater component

1. Introduction

Seawater constituents (water molecules, suspended particles, dissolved substances, and air bubbles) impact the propagation of light through absorption and scattering processes. In natural waters, suspended particulate matter is mostly composed of phytoplankton, heterotrophic organisms, viruses, biogenic detritus, and mineral particles. Absorbing and scattering characteristics of water constituents are described by the inherent optical properties (IOP) [1] which do not depend on the radiance distribution but solely on the concentration and chemical composition of dissolved organic matter, and the concentration, size distribution and chemical composition of particulate matter. All IOPs can be defined from the absorption coefficient, a, and the volume scattering function, β. For instance, the scattering, b, and backscattering, b_b, coefficients are obtained from the integration of β over all scattering angles, and only backward scattering angles, respectively.

Owing to the availability of commercial optical backscattering sensors and flow-through attenuation and absorption meters, in situ measurements of bulk IOP have now been routinely performed for more than two decades. While these measurements allow a better description of the IOP variability in natural waters, they can also be used as proxies for the estimation of the bulk particulate matter. For instance, the spectral slope of the particulate beam attenuation coefficient, c_p, is tightly linked to the slope of the particle size distribution (PSD), ξ, assuming a Junge-type distribution of PSD [2–4]. The particulate backscattering ratio b_{bp}/b_p is used to obtain information about the particle composition. Indeed, based on the Lorentz-Mie scattering calculations that assume marine particles as homogeneous spheres, an analytical relationship between b_{bp}/b_p, ξ and $\tilde{n}_r$ was generated [5]. This latter equation is used in conjunction with in situ measurements of b_{bp}, b_p, and c_p to describe the variability of the physical nature (i.e., refractive index) of the bulk particulate matter in oceanic and coastal environments [6–10].

In the past, many theoretical and experimental studies, mainly dedicated to phytoplankton, showed that while the absorption, attenuation and total scattering of algal cells are correctly described using the homogeneous sphere model, such model is less appropriate for simulating backscatter. Indeed, the structural heterogeneity and inner complexity of phytoplankton cells (gas vacuoles, chloroplast, silica wall, etc.) explain why the measured backscattering signal is higher than predicted by the Lorentz-Mie theory for homogeneous spheres [11–20]. The underestimation of b_{bp} by homogeneous spheres may explain the fact that in situ observations of backscattering are significantly higher than theoretical simulations [21–23].

In this paper, we examine the impact of particle structural heterogeneity on the bulk backscattering ratio for realistic combinations of optically significant constituents. The purpose of our study is not to provide a new analytical relationship for b_{bp}/b_p as a function of $\tilde{n}_r$ and ξ but rather to assess the sensitivity of b_{bp}/b_p to the modeled structural heterogeneity of phytoplankton cells for some realistic water bodies. Typical phytoplankton bloom and no bloom conditions, as defined in Stramski and Kiefer [24], will first be examined. Then, the last case study will account for the presence of mineral particles, which have a great effect on the scattering properties.

Because the bulk scattering (b) and backscattering (b_b) coefficients of a water body result from additive contributions of all individual constituents that scatter light, we will consider various sub-populations of marine particles, namely organic detritus, minerals, heterotrophic bacteria, viruses, and phytoplankton. Robertson Lain et al. [23] showed that the two-layered sphere model is appropriate for modeling of remote-sensing reflectance and IOPs in high biomass Case 1 waters. The real refractive index of the chloroplast and the relative volume of the chloroplast are key parameters impacting the backscattering efficiency of phytoplankton cells. This was recently confirmed by two studies where measurements of light scattering by phytoplankton cultures were well reproduced by the two-sphere model [15,16]. For these reasons, in this study, phytoplankton optical properties have been simulated considering a two-layered sphere model. The size range of the different considered particles (viruses, bacteria, phytoplankton, and organic detritus), as well as their real and imaginary refractive index values are defined from literature [21,25].

To establish the foundations of the present study, the different theoretical considerations as well as the two different numerical codes used for the calculations are first presented. Then, we describe the different sub-populations of particles and their associated size distribution, refractive index, and internal structure used to simulate their optical properties. The impact of the modeled structural heterogeneity of phytoplankton cells is then discussed for the three realistic water bodies as mentioned previously.

2. Theoretical Considerations

2.1. Backscattering Cross Section for Polydisperse Particle Assemblages

Light scattering is produced by the presence of an object (such as a particle) with a refractive index different from that of the surrounding medium. The refractive index is expressed in complex form as $n(\lambda) = n_r(\lambda) + i\, n_i(\lambda)$, where λ is the wavelength of the radiation in vacuum in units of nm. The real part determines the phase velocity of the propagating wave and the imaginary part accounts for the absorption. Please note that the refractive index is a relative value dependent upon the surrounding medium, i.e., relative to the refractive index of the medium. The single scattering process by a particle is described by the scattering cross section $C_{sca}(D, \lambda)$ (units m^2) and the scattering phase function $\tilde{F}(D, \theta, \lambda)$ (dimensionless) as defined by Mishchenko et al. [26] (Equations (4.51)–(4.53), pp. 100–101):

$$\frac{1}{2} \int_0^\pi \tilde{F}(D, \lambda, \theta) \sin\theta \, d\theta = 1 \tag{1}$$

As particles are here assumed to be spherical, $\tilde{F}$ depends only on the particle diameter D, the scattering angle θ within the arbitrary azimuthal plane of scattering, and the wavelength λ. In the following, λ is omitted for brevity. To account for polydisperse particulate assemblages, the particle size distribution (PSD) is defined. For the present study, we adopt a power-law PSD (also named the Junge-like PSD) which is commonly used to represent the size distribution of marine particles in natural waters [5,24,27,28]. The ensemble-average normalized phase function is:

$$\tilde{F}(\theta) = \int_{D_{min}}^{D_{max}} \tilde{F}(D, \theta) \times A D^{-\xi} dD \tag{2}$$

where D_{min} and D_{max} define the particle diameter range, ξ is the hyperbolic slope of PSD, and $A D^{-\xi}$ (units, µm^{-1}) is the relative differential particle size distribution. As in many theoretical studies, the relative PSD is normalized such that the integral over the entire size range is unity. It follows that $\tilde{F}(\theta)$ represents the average normalized phase function per particle. Equation (2) can be written for the scattering cross section replacing $\tilde{F}(D, \theta)$ with $C_{sca}(D)$ and $\tilde{F}(\theta)$ with C_{sca}. The backscattering cross section of the polydisperse assemblage is defined as:

$$C_{sca}^{bb} = \frac{C_{sca}}{2} \int_{\pi/2}^\pi \tilde{F}(\theta) \sin\theta \, d\theta \tag{3}$$

It can be easily seen from Equations (1)–(3) that the integration of $\tilde{F}(\theta)$ between 0 and π gives C_{sca}, the scattering cross section of the polydisperse population. Many numerical codes (including those described in Section 3) use the normalized phase function $\tilde{F}(\theta)$ to describe the angular distribution of the scattered radiation. However, in hydrologic optics, the volume scattering function (VSF), $\tilde{\beta}(\theta)$ (m^{-1} sr^{-1}), is more commonly used instead of $\tilde{F}(\theta)$ [29]. The relationship between $\tilde{\beta}(\theta)$ and $\tilde{F}(\theta)$ is:

$$\tilde{\beta}(\theta) = \frac{N C_{sca}}{4\pi} \tilde{F}(\theta) \tag{4}$$

with N the number of particles per cubic meter.

2.2. The Bulk Backscattering Ratio

Marine particles are divided into five different categories: viruses (VIR), heterotrophic bacteria (BAC), phytoplankton (PHY), organic detritus (DET), and minerals (MIN). Table 1 displays the size ranges and the refractive indices of the different components as defined by previous studies [21,24,25]. The ensemble-average values of $\tilde{F}_j(\theta)$, $C_{sca,j}$, and $C_{sca,j}^{bb}$ are computed from Equations (1)–(3) for each particulate component j.

Table 1. Summary of the seawater constituents.

Component (*j*)	Sphere Model	D_{min}-D_{max} (µm)	n_r	n_i
Viruses	homogeneous	0.03–0.2	1.05	0
Heterotrophic bacteria	homogeneous	0.2–2	1.05	1.0×10^{-4}
Phytoplankton cells	two or three-layered	0.3–40	1.044 *	1.5×10^{-3} *
Organic detritus	homogeneous	0.05–500	1.04	2.3×10^{-5}
Minerals	homogeneous	0.05–500	1.18	1.0×10^{-4}

* the values represent the equivalent refractive indices (Equation (12)). The refractive indices of the spheres representing the chloroplast and cytoplast are described in Table 2. λ = 532 nm.

The total normalized phase function and total scattering cross section of the water body are obtained as in Mishchenko et al., 2002 (Equations (4.74) p. 102 and (3.13) p. 71):

$$\tilde{F}^{tot}(\theta) = \frac{\sum_{j=1}^{5} N_j \, C_{sca,j} \, \tilde{F}_j(\theta)}{\sum_{j=1}^{5} N_j \, C_{sca,j}} \tag{5}$$

$$C_{sca}^{bb,\,tot} = \sum_{j=1}^{5} N_j \, C_{sca,j}^{bb} \tag{6}$$

where N_j is the relative concentration (i.e., the relative number of particles per unit volume of water) of the considered component. C_{sca}^{tot} is defined by replacing $C_{sca,j}^{bb}$ with $C_{sca,j}$ in Equation (6).

The total (i.e., bulk) backscattering coefficient (b_{bp}) (units m^{-1}) of the water body is the sum of the relevant $b_{bp,j}$ associated with each *j*th group. $b_{bp,j}$ is equal to the polydisperse $C_{sca,j}^{bb}$ weighted by the particle concentration of the *j*th group:

$$b_{bp} = \sum_{j=1}^{5} b_{bp,j} = N_{TOT} \times C_{sca}^{bb,\,tot} \tag{7}$$

with N_{TOT} the total particle concentration (particles per m^3) in the water body. Similarly, b_p is defined from Equation (7) by replacing b_{bp} with b_p and $C_{sca}^{bb,\,tot}$ with C_{sca}^{tot}. The bulk backscattering ratio $\tilde{b}_{bp}$ is the dimensionless ratio:

$$\tilde{b}_{bp} = \frac{b_{bp}}{b_p} \tag{8}$$

In this study, we will use the bulk particulate real refractive index ($\tilde{n}_r$), which reproduces the bulk scattering properties of a water body. It represents the mean refractive index weighted by the scattering cross sections of all the particles:

$$\tilde{n}_r = \frac{\sum_{j=1}^{5} n_{r,j} \times N_j \, C_{sca,j}}{\sum_{j=1}^{5} N_j \, C_{sca,j}} \tag{9}$$

Similarly, the bulk imaginary refractive index ($\tilde{n}_i$) is defined as follows:

$$\tilde{n}_i = \frac{\sum_{j=1}^{5} n_{i,j} \times N_j \, C_{abs,j}}{\sum_{j=1}^{5} N_j \, C_{abs,j}} \tag{10}$$

where $C_{abs,j}$ is the absorption cross section of particles.

Table 2. Refractive index (n_r(chlp) + $i\,n_i$(chlp)) of the sphere representing the chloroplast for two morphological models. The refractive index of the sphere representing the cytoplast is constant ($1.02 + i\,1.336 \times 10^{-4}$). The equivalent refractive index of the cell is $1.044 + i\,1.5 \times 10^{-3}$.

Model * (%cyt-%chlp)	80%–20%	70%–30%	80%–18.5%–1.5%
n_r	1.140	1.100	1.144
n_i	6.966×10^{-3}	4.688×10^{-3}	7.531×10^{-3}

* The percentages represent the relative volume of the model cytoplasm and chloroplast.

2.3. The Scattering Coefficient as Measured by In Situ Transmissometers

In field measurements, b_p is derived from the total absorption and beam attenuation coefficients (a and c, respectively) as measured by instruments such as WETLabs ac9 and its later variants. Any detector has a finite field of view (FOV), therefore beam transmissometers are defined by their acceptance angle $\theta_{acceptance}$, which differs from $0°$. If we want to compare, in a future study, our theoretical results to available in situ measurements, b_p must be derived from the scattering cross section, rebuilt from the normalized phase function integrated between $\theta_{acceptance}$ and π instead of 0 and π [30]. To make a distinction, when C_{sca} is calculated by integrating the scattering function between $\theta_{acceptance}$ and π, the symbols $C_{sca}^{\theta_a}$, $b_p^{\theta_a}$ and $\widetilde{b_{bp}^{\theta_a}}$ ($= b_{bp}/b_p^{\theta_a}$) will be used. As in Twardowski et al. [5], we set the acceptance angle to $1°$, which is consistent with acceptance angles of commercially available beam transmissometers such as the WETLabs C-Star ($1.2°$) or WETLabs ac9 ($0.93°$) ([30] and references therein).

3. Numerical Modeling of the Marine Particle Scattering

The Meerhoff Mie program version 3.0 [31] and the ScattnLay code [32,33] are used to simulate the scattering and absorbing properties of homogeneous and multilayered spheres, respectively. Radiative transfer computations are carried out given the wavelength of the incident radiation equal to 532 nm and the refractive index of sea water equal to 1.34. The Meerhoff Mie program allows simulations of a polydisperse ensemble of spheres with a large choice of PSD. The outputs are the ensemble-average quantities per particle $\tilde{F}(\theta)$, C_{sca} and C_{sca}^{bb} (Equations (2) and (3)). The ScattnLay code performs computations only for monodisperse particles. To obtain the normalized phase function and cross sections for a polydisperse population, a numerical integration over the size range must be done separately (Figure 1, N^{o}S2). Particular attention must be paid to the integration step to guarantee the accuracy of the numerical integration.

The Meerhoff Mie program is used to generate a first dataset named DS1 based on computations for homogeneous spheres for the same case studies as in Twardowski et al. [5]. n_r ranges from 1.02 to 1.2 (with a 0.2 increment), n_i is set to 0.005, $D_{min} = 0.012$ μm, $D_{max} = 152$ μm, and ξ is between 2.5 and 5. Please note that Twardowski et al. [5] did not mix different particle components with different refractive indices, as they studied $\widetilde{b_{bp}^{\theta_a}}$ for a polydisperse population of particles having the same refractive index. In this case, Equations (5)–(7) are not useful as $\widetilde{b_{bp}^{\theta_a}}$ is directly related to $C_{sca}^{bb}/C_{sca}^{\theta_a}$.

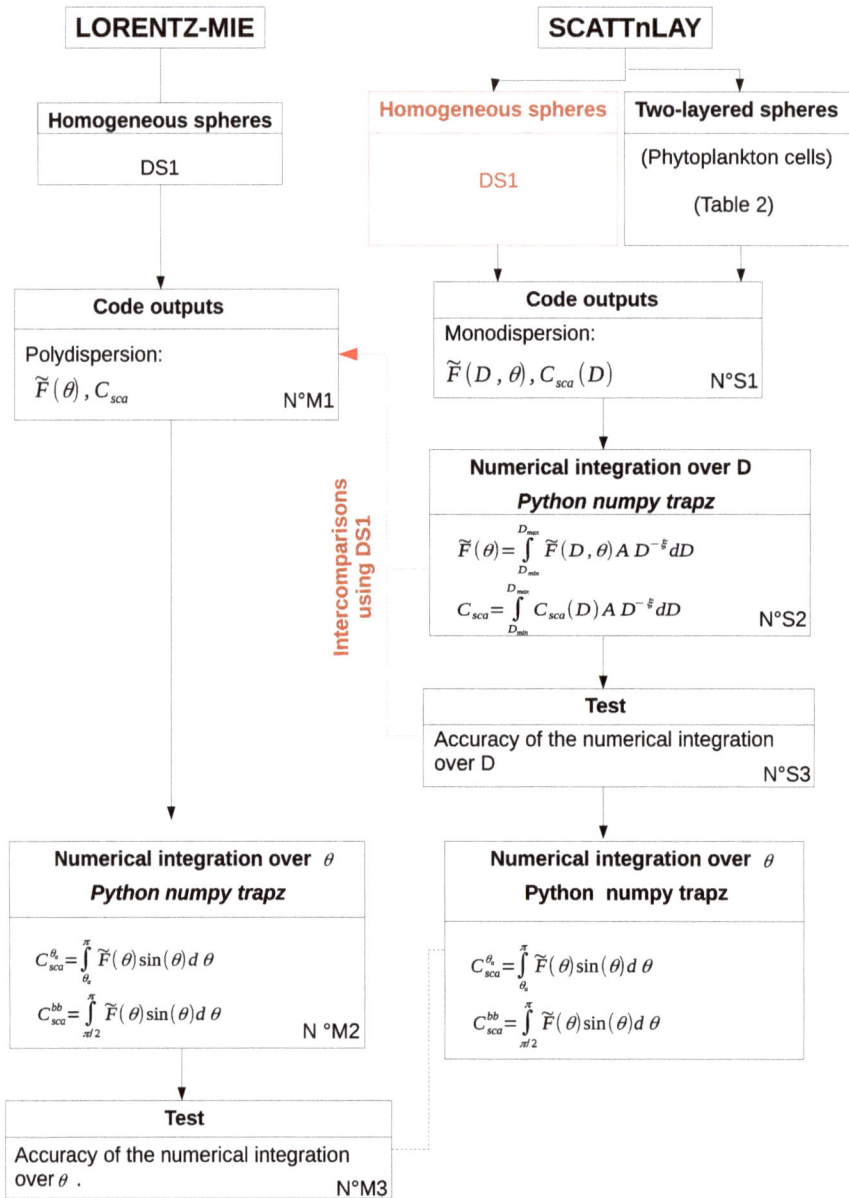

Figure 1. Flow chart of the integration procedure applied to the MIE and ScattnLay outputs.

In the second dataset (named DS2), a distinction is made between VIR, BAC, PHY, DET, and MIN in terms of internal structure, refractive index, and size range. The scattering properties of phytoplankton cells are modeled using the two-layered sphere model as in Robertson Lain et al. [23]. These investigators showed that a chloroplast layer (chlp) surrounding the cytoplasm (cyt) is an optimal morphology to simulate optical properties of algal cells. Based on their study, the value of the real part of the refractive index of the sphere representing the cytoplasm is fixed to 1.02, and the value of the imaginary part at 532 nm is 1.336×10^{-4} [23]:

$$n_i(\text{cyt, 532 nm}) = n_i(\text{cyt, 400 nm}) \times \exp[-0.01 \times (532 - 400)] \tag{11}$$

with $n_i(\text{cyt, 400 nm}) = 0.0005$. Concerning the sphere representing the chloroplast, $n(\text{chlp})$ is calculated according to the Gladstone and Dale formula [34]:

$$\sum_k n_k \times \vartheta_k = n_{equ}, \tag{12}$$

where n_k and ϑ_k are the complex refractive index and the relative volume of the k-th layer, and n_{equ} is the complex equivalent refractive index of the whole particle. The knowledge of the complex equivalent refractive index is useful to compare the simulations of heterogeneous spheres among themselves, regardless of the number of layers and the relative proportion of each layer. The complex equivalent refractive index is kept constant ($n_{equ} = 1.044 + i\,1.5 \times 10^{-3}$). The refractive index of the sphere representing the chloroplast is described in Table 2 according to the relative volume of the modeled chloroplast (20% or 30%). We also tested a three-layered sphere model. The outer layer represents the cell membrane. We assumed that the cell membrane is non-absorbing and have a $n_r = 1.09$ [13]. The second layer represents the chloroplast and the third layer the cytoplasm. The values of n_r and n_i for the cytoplasm are identical to n_r and n_i for the two-layered sphere. The values of n_r and n_i for the chloroplast are adjusted according to Equation (12) to keep the complex equivalent refractive index of the cell constant (Table 2). The relative volumes are 1.5%, 18.5% and 80% for the modeled cell membrane, chloroplast, and cytoplasm, respectively.

In DS2, multilayered sphere models are not implemented for viruses, heterotrophic bacteria, organic detritus, and minerals because of the paucity of relevant information about their optical and morphometrical properties. As we cannot gather enough accurate information about the internal structure of such particles, the homogeneous sphere model is used. The suitable n_r and n_i values for viruses, heterotrophic bacteria, organic detritus, and minerals are obtained from [25] (Table 1).

4. Abundance of the Various Particulate Components

The relative concentrations N_j associated with each particle group are chosen to realistically represent the mix of marine components and to ensure that the overall size distribution matches the Junge power law (Tables 3–5 and Figure 2).

In situ laser diffraction measurements of the PSD in different oceanic regions showed that the size distribution of marine particles can be approximated by the Junge-like power law [35–37]. As discussed by Reynolds et al. [36], the power law with a single slope is a convenient empirical descriptor of the PSD, but we have to keep in mind that, in some cases (e.g., in nearshore waters and in the presence of specific populations of phytoplankton) the particle size distribution deviates from the Junge-like power law [36–38]. Relatively steep hyperbolic slopes (around 4) are encountered in open ocean waters, whereas less steep slopes (around 3.3) are characteristic of phytoplankton bloom and/or production of particle aggregates. In the present study, results are discussed for $\xi = 3$, 3.5, and 4 as the vast majority of hyperbolic slopes are in this range (Figure 11 in [36,37]). Results are shown also for $\xi = 2.5$ and 5 but hyperbolic slopes greater than 4 are much less likely to occur. Likewise, $\xi < 3$ are rare excepted when there is biological growth in the relatively large size classes and/or aggregation. To compare with typical particulate abundances estimated in natural waters, a total abundance (N_{TOT}) of 1.1262×10^{14} particles per m^3 is considered to be in Stramski et al. [25]. Three case studies are defined. The first one represents oligotrophic-like waters with no phytoplankton bloom and no-mineral particles: the phytoplankton abundance (N_{PHY}) spans from 1.1×10^9 (for $\xi = 4.9$) to 4.6×10^{11} (for $\xi = 2.5$) particles per m^3 (0.001%–0.41% of N_{TOT}). The second one represents waters with a phytoplankton bloom and no minerals, where N_{PHY} is higher as compared to the oligotrophic-like case: N_{PHY} ranges between 8.3×10^9 (for $\xi = 4.9$) and 2.3×10^{12} (for $\xi = 2.5$) particles per m^3 (0.007%–2% of N_{TOT}). The third one represents coastal-like waters with minerals and no bloom conditions: minerals are added proportionally to obtain a bulk real refractive index $\tilde{n}_r$ around 1.1. The mineral abundance (N_{MIN}) spans from 4.8×10^{12} (for $\xi = 4.9$) to 1.3×10^{13} (for $\xi = 2.5$) particles per m^3 (4.2%–11.7% of N_{TOT}).

The abundances of the different particle components can be directly compared to the abundances provided in Stramski et al. [25] as N_{TOT} is identical. In Stramski et al. [25], $\xi = 4$, so comparisons are possible only for this value (Table 6). We note that N_{PHY} is of the same order of magnitude. Stramski et al. [25] have higher concentrations of DET and MIN and lower concentrations of VIR and BAC. In their paper, the authors explained that the concentrations of DET and MIN were chosen to obtain realistic contributions of detrital and mineral absorption. However, they cautioned against attaching particular significance to their selected DET and MIN concentrations in the context of how well these values can represent realistic concentrations in specific oceanic water bodies. The abundances of viruses and bacteria (N_{VIR} and N_{BAC}), used in this study, agree with the Stramski and Kiefer values [24]. Stramski and Kiefer [24] (Table 1 in their paper) used N_{VIR} between 3.0×10^9 and 4.6×10^{14} particles per m^3, N_{BAC} between 3.0×10^{11} and 1.5×10^{12} particles per m^3. Middleboe and Brussard, 2017 [39] confirmed that viral abundance can reach up to 10^{14} particles per m^3. For phytoplankton, Stramski and Kiefer made a distinction between prochlorophytes, cyanobacteria, ultrananoplankton, larger nanoplankton, and microplankton. Over these different phytoplankton groups, N_{PHY} ranges between 1.0×10^{11} for picoplankton to 3.0×10^5 for microplankton. They used $N_{PHY} \geq 5 \times 10^{11}$ particles per m^3 when there is a bloom of phototrophic picoplankton.

Table 3. Relative abundance of viruses (VIR), bacteria (BAC), phytoplankton (PHY), and organic detritus (DET) with the corresponding bulk refractive index (Equations (9) and (10)) for the water body with no bloom conditions and no minerals (oligotrophic-like).

ξ	$\tilde{n}_r$	$\tilde{n}_i$	Relative Abundance N_j (%)			
			VIR	BAC	PHY	DET
2.5	1.040	4.280×10^{-4}	78.85	5.349	0.4059	15.39
3	1.042	7.570×10^{-4}	84.74	2.120	0.1002	13.04
3.5	1.043	1.034×10^{-3}	88.50	0.8244	0.0281	10.64
4	1.045	9.931×10^{-4}	91.15	0.3178	0.0084	8.528
4.9	1.047	6.718×10^{-4}	94.35	5.651×10^{-2}	0.0010	5.588

Table 4. Relative abundance of viruses (VIR), bacteria (BAC), phytoplankton (PHY), and organic detritus (DET) with the corresponding bulk refractive index (Equations (9) and (10)) for the water body with phytoplankton bloom conditions and no minerals (phytoplankton bloom).

ξ	$\tilde{n}_r$	$\tilde{n}_i$	Relative Abundance N_j (%)			
			VIR	BAC	PHY	DET
2.5	1.041	6.195×10^{-4}	51.96	3.760	1.995	42.29
3	1.041	1.048×10^{-3}	61.91	1.599	0.6165	35.88
3.5	1.042	1.313×10^{-3}	69.84	0.6575	0.1922	29.31
4	1.043	1.362×10^{-3}	76.18	0.2650	0.0600	23.49
4.9	1.044	1.194×10^{-3}	84.55	0.0499	7.367×10^{-3}	15.40

Table 5. Relative abundance of viruses (VIR), bacteria (BAC), phytoplankton (PHY), and organic detritus (DET) with the corresponding bulk refractive index (Equations (9) and (10)) for waters with minerals and no bloom conditions (coastal-like).

			Relative Abundance N_j (%)				
ζ	$\tilde{n}_r$	$\tilde{n}_i$	VIR	BAC	PHY	DET	MIN
2.5	1.103	7.322×10^{-4}	70.96	5.311	3.650×10^{-1}	11.68	11.68
3	1.108	9.361×10^{-4}	78.04	2.105	8.801×10^{-2}	9.882	9.882
3.5	1.119	6.253×10^{-4}	83.03	0.819	2.391×10^{-2}	8.066	8.066
4	1.131	1.376×10^{-4}	86.75	0.3155	6.902×10^{-3}	6.462	6.462
4.9	1.145	9.794×10^{-6}	91.47	5.607×10^{-2}	7.782×10^{-4}	4.23	4.23

Table 6. Comparisons between abundances defined in the present study and abundances defined by Stramski et al. [25]. The hyperbolic slope ζ is 4 and N_{TOT} is 1.1262×10^{14} particles per m^3.

	Abundance (Particles per m^3)				
Case Study	VIR	BAC	PHY	DET	MIN
Oligotrophic-like	1.0265×10^{14}	3.5796×10^{11}	9.4680×10^{9}	9.6046×10^{12}	0
Phytoplankton bloom	8.5799×10^{13}	2.9846×10^{11}	6.7587×10^{10}	2.6455×10^{13}	0
Coastal-like	9.7702×10^{13}	3.5536×10^{11}	7.7733×10^{9}	7.2774×10^{12}	7.2774×10^{12}
Stramski et al. [25]	2.5000×10^{12}	1.0000×10^{11}	2.4759×10^{10}	8.2500×10^{13}	2.7500×10^{13}

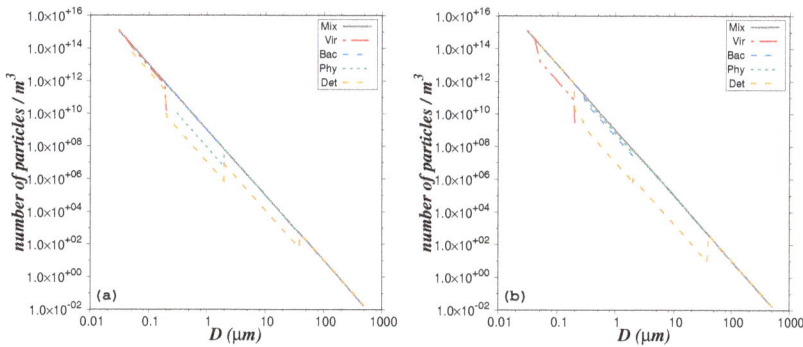

Figure 2. Composite PSD as derived from individual PSDs of the five considered particle groups for (**a**) the oligotrophic-like water body and (**b**) the phytoplankton bloom water body. $N_{TOT} = 1.1262 \times 10^{14}$ particles per m^3 and $\zeta = 4$.

An indication of the total chlorophyll-a concentration is given for the oligotrophic-like, phytoplankton bloom, and coastal-like case studies. For that purpose, we considered the median intracellular chlorophyll-a values given in Brotas et al. [40]. These authors used Brewin et al. model [41] to calculate the fractional contributions of pico, nano, and microplankton to total phytoplankton chlorophyll biomass. Then, they derived the intracellular chlorophyll-a per cell for each size class from the results of cell enumeration (microscope counts and flow cytometry) and the chlorophyll-a concentration for that size class given by the Brewin et al. model. The computed median intracellular chlorophyll-a values were 0.004, 0.224, and 26.78 pg Chla cell^{-1} for pico, nano, and microplankton, respectively. In our study, we multiplied the corresponding intracellular chlorophyll-a content by the numerical abundance of pico-, nano-, and micro-plankton as derived from the PSD and we summed the chlorophyll-a concentration per class to obtain the total chlorophyll-a concentration ([Chla], units mg m^{-3}) (Table 7).

Table 7. Total chlorophyll-a concentration for the three case studies.

	Oligotrophic-Like	Phytoplankton Bloom	Coastal-Like
ξ	[Chla]	[Chla]	[Chla]
3	8.35	11.51	7.497
3.5	0.773	1.580	0.6889
4	0.102	0.341	0.0884

The chlorophyll-a concentration (mg m^{-3}) is estimated, as an indication, using the relative abundance of phytoplankton cells described in Tables 3–5 and considering N_{TOT} is 1.1262×10^{14} particles per m^3.

We emphasize that total chlorophyll-a concentrations are given as an indication as they depend on the abundance of phytoplankton, which in turn depends on N_{TOT} and ξ. For the oligotrophic-like case study, [Chla] ranges from 0.10 for $\xi = 4$ to 8.4 mg m^{-3} for $\xi = 3$. However, in oligotrophic waters, in situ measurements of PSD showed that ξ values are around 4. For ξ between 3.5 and 4, [Chla] is less than 1 mg m^{-3}, which is typical [Chla] in oligotrophic waters. In bloom conditions, the hyperbolic slope can be less than 4. For example, Buonassissi and Dierssen [35] found ξ around 3.3 in bloom conditions. For $\xi = 3.3$, we found [Chla] of 1.92 mg m^{-3}. For the coastal case study, [Chla] is low as compared to in situ [Chla] values in coastal areas. This is because we considered a high load of minerals as compared to phytoplankton abundance.

5. Results

5.1. Accuracy of Numerical Computations

A numerical integration over θ is required to derive $b_p^{\theta_a}$ and b_{bp} from the normalized phase function (Section 2). Due to the sharp increase of the normalized phase function in the forward scattering directions (Figure 3), the selection of the relevant angular step for the numerical integration is crucial. For that purpose, the impact of angular step ($\Delta\theta$) on the calculation of $\widetilde{b_p^{\theta_a}}$ is studied using Lorentz-Mie simulations in DS1 (Figure 1, N^{o}M2, M3). The normalized phase function of polydisperse particles $\tilde{F}(\theta)$ exhibits a maximum around $\theta = 0^o$ [26]. For small ξ value, that is when the proportion of large-sized particles compared to smaller particles increases, the forward peak is sharper. Indeed, for particles with a large diameter as compared to the wavelength, $\tilde{F}(D,\theta)$ displays a sharp forward peak [26] due the concentration of light near $\theta = 0^o$ caused by diffraction. The presence of the peak in $\tilde{F}(\theta)$ requires several integration points large enough to provide the desired numerical accuracy. The numerical integration over θ (Figure 1, N^{o}M2) is performed using the "Trapz" function from the Numpy package with Python. The "Trapz" function performs an integration along the given axis using the composite trapezoidal rule. To test the accuracy of the integration and to find the correct integration step, $\Delta\theta$, we compare the result of the numerical integration of $\tilde{F}(\theta)$ between 0 and π to its theoretical value (=2) (Figure 1, N^{o}M3). When $\Delta\theta = 0.05^o$, corresponding to a total number of integration steps (N$_\theta$) of 3600, the numerical integration value of $\tilde{F}(\theta)$ is in the range [1.999–2.000] for small ξ. For larger ξ, it is in the range [1.800–1.999]. When the value of the numerical integration is in the range [1.800–1.999], a renormalization factor is applied to $\tilde{F}(\theta)$ to ensure that the result of the numerical integration is exactly 2. We could also increase the number of integration points, but it would increase the computation time. Using a renormalization factor for large ξ is a good compromise to guarantee the accuracy and save computation time.

For two-layered spheres (i.e., phytoplankton cells), the ScattnLay code provides only normalized phase functions for monodisperse particles (Figure 1, N^{o}S1), so the numerical integration over the particle diameter range (Equation (2)) is realized as a separate calculation with the Python "Trapz" function (Figure 1, N^{o}S2). For monodisperse particles, the normalized phase function displays a forward peak as explained above but can also display a sequence of maxima and minima due to interference and resonance features [26,42]. The frequency of the maxima and minima over the range of θ increases with both increasing n_r and size parameter ($=\pi D/\lambda$). To test the accuracy of the numerical integration over the particle diameter range (Figure 1, N^{o}S3), we ran the ScattnLay code for DS1 case studies and compared $\tilde{F}(\theta)$ and C_{sca} rebuilt from Equation (2) with Lorentz-Mie computations as the Lorentz-Mie code provides the polydisperse phase functions and cross section as outputs (Figure 1, N^{o}M1). Note that even a narrow polydispersion washes out the interference and resonance features, which explains why most natural particulate assemblages do not exhibit such patterns [26,42] (Figure 3). A perfect match is obtained between the ScattnLay-rebuilt-polydisperse and Lorentz-Mie-polydisperse $\tilde{F}(\theta)$ and C_{sca} values when the integration step (ΔD) is set to 0.01 μm for D in the range [0.03, 2 μm]; 0.1 μm for D in the range [2, 20 μm]; 2.0 μm for D in the range [20, 200 μm]; and 10.0 μm for D in the range [200, 500 μm].

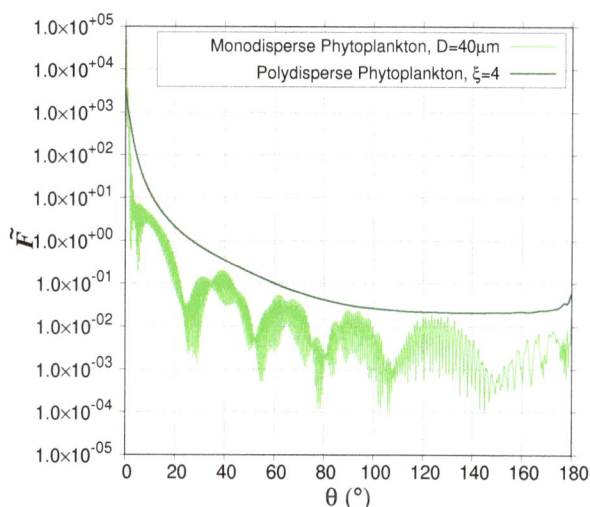

Figure 3. Interference and resonance features observed for the scattering phase function of monodisperse particles (light green). The major low-frequency maxima and minima are called the "interference structure". The high-frequency ripples are resonance features. The interference and resonance feature are washed out for a polydisperse assemblage of particles (dark green).

Using the DS1 data set, the impact of the angular integration on the backscattering ratio $\widetilde{b}_{bp}^{\theta_a}$ is examined as a function of the hyperbolic slope ξ for different values of the real refractive index and two values of total angular steps (i.e., $N_\theta = 750$ and 3600) (Figure 4). The impact of the integration is noticeable only for ξ values lower than about 3 and relatively high n_r values. When the number of angular steps increases, the curves become flatter at low ξ values. Differences in the curve shape are reduced if we increase the angular step. For $\Delta\theta = 0.24^o$ ($N_\theta = 750$), the present results of the Lorentz-Mie calculations (solid lines in Figure 4) perfectly match those previously obtained by Twardowski et al. [5] (not shown). However, in this case ($N_\theta = 750$), the numerical integration is not accurate enough as the integration of Equation (1) gives values between 1.999 ($\xi = 4.9$) and 1.04 ($\xi = 2.5$). In the following, $\Delta\theta$ is set to 0.05^o ($N_\theta = 3600$) and Figure 4 (dashed lines) will be the reference figure for homogeneous spheres.

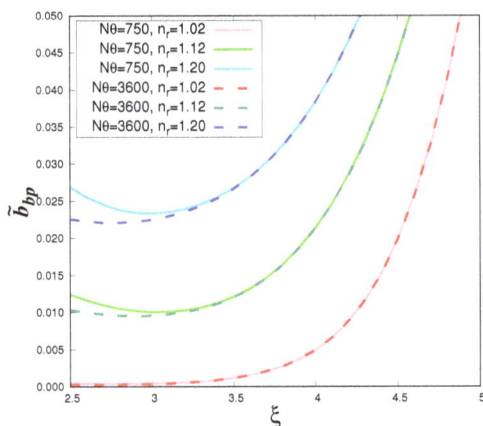

Figure 4. Results of Lorentz-Mie calculations (DS1) of the particulate backscattering ratio $\widetilde{b_{bp}^{\theta_a}}$ as a function of the hyperbolic slope, ξ, and different values of n_r and N_θ. The imaginary part of the refractive index = 0.005 as in Twardowski et al. [5]. This figure can be compared to Figure 1 in Twardowski et al. [5].

5.2. Impact of the Structural Heterogeneity of Phytoplankton Cells on the Bulk Particulate Backscattering Ratio

The impact of phytoplankton cell structural heterogeneity on $\widetilde{b_{bp}^{\theta_a}}$ is examined as a function of ξ for the three previously described water bodies (oligotrophic-like, phytoplankton bloom, coastal-like) considering the 80%–20% phytoplankton morphological model (Figure 5a). The real and imaginary bulk refractive indices for oligotrophic-like, phytoplankton bloom, and coastal-like case studies, vary with ξ as the relative proportions of the different particle components, having different n_r and n_i, vary with ξ (Tables 1–5 and Figure 5b,c). For the no-mineral water bodies (oligotrophic-like and phytoplankton bloom), $\tilde{n}_r$ stays around 1.04 ± 0.007 (Figure 5b). In contrast, $\tilde{n}_i$ shows large variation with ξ for both oligotrophic-like and phytoplankton bloom water bodies (Figure 5c). In bloom conditions, $\tilde{n}_i$ increases as the relative proportion of phytoplankton increases as compared to the no bloom conditions. In agreement with typical values of the oceanic bulk imaginary refractive index [43], the $\tilde{n}_i$ values for the particulate populations considered here are always lower than 0.002. In the presence of mineral particles (coastal-like), $\tilde{n}_r$ increases as MIN have a higher n_r than VIR, BAC, PHY and DET. Its values are between 1.103 ($\xi = 2.5$) and 1.145 ($\xi = 4.9$). Values of $\tilde{n}_i$ vary between 9.79×10^{-6} and 9.44×10^{-4}.

The impact of the structural heterogeneity of phytoplankton cells is evaluated by comparison with Lorentz-Mie calculations (particulate assemblages composed of homogeneous spheres only, regardless of the particle group) performed for low and high bulk refractive index. These case studies with homogeneous spheres only are named "Homogeneous reference cases". The real and imaginary values of the bulk refractive indices are 1.045 and 9.93×10^{-4} for the "Homogeneous reference case 1", 1.043 and 1.36×10^{-3} for the "Homogeneous reference case 2", and 1.131 and 1.37×10^{-4} for the "Homogeneous reference case 3", respectively (Figure 5b,c). These values of $\tilde{n}_r$ and $\tilde{n}_i$ were chosen to be equal to values of $\tilde{n}_r$ and $\tilde{n}_i$ obtained for the oligotrophic-like, phytoplankton bloom, and the coastal-like case study when $\xi = 4$. "Homogeneous reference cases 1 and 2" with low $\tilde{n}_r$ represent phytoplankton-dominated Case 1 waters and are compared with the oligotrophic-like and phytoplankton bloom water body, respectively. "Homogeneous reference case 3" with high $\tilde{n}_r$ represents mineral-dominated Case 2 waters and is compared with the coastal-like water body. The variation of $\widetilde{b_{bp}^{\theta_a}}$ due to structural heterogeneity of phytoplankton cells is evaluated using the relative absolute difference calculated between the homogeneous reference cases (named

x in Equation (13)) and oligotrophic, phytoplankton bloom or coastal-like water bodies (named y in Equation (13)):

$$\Delta \epsilon = \frac{|x - y|}{(x + y)} \times 200 \ (\%) \tag{13}$$

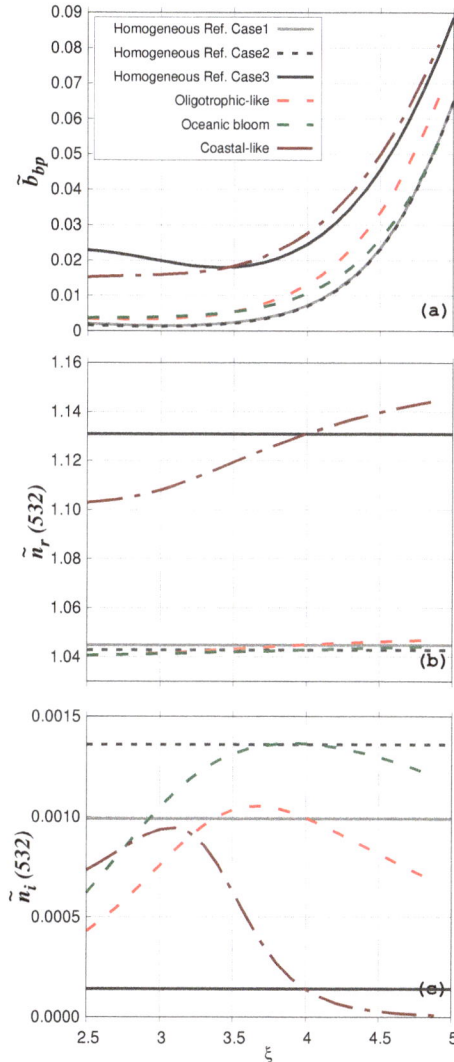

Figure 5. (a) Particulate backscattering ratio $\widetilde{b_{bp}^{\theta_a}}$ as a function of the hyperbolic slope for the oligotrophic-like (red dashed line), phytoplankton bloom (green dashed line), and coastal-like (brown dashed line) water bodies as described in Section 4. Black and gray lines are for homogeneous reference cases. The gray solid line corresponds to $n_r = 1.045$, $n_i = 9.93 \times 10^{-4}$, the black dashed line to $n_r = 1.1043$, $n_i = 1.36 \times 10^{-3}$, and the black solid line to $n_r = 1.131$, $n_i = 1.37 \times 10^{-4}$, respectively. Phytoplankton cells are modeled as two-layered spheres with a relative volume of the cytoplasm of 20% (%cyt-%chl = 80–20). (b) as in panel (a) but for the real refractive index. (c) as in panel (a) but for the imaginary part of the refractive index.

Even if the numerical relative abundance of phytoplankton is very small for the oligotrophic-like water body (=8.4 × 10^{-3}%), the structural heterogeneity increases the $\widetilde{b_{bp}^{\theta_a}}$ value by 58% compared to the homogeneous case ("Homogeneous reference case 1"). This is consistent with previous studies showing the large contribution of coated spheres to the backscattering signal [15,16,18,19,22,44]. The value of $\Delta \epsilon$ calculated between the oligotrophic-like and phytoplankton bloom water bodies is smaller (=22% at ζ = 4) even if $\tilde{n}_i$ is different (9.93 × 10^{-4} for oligotrophic-like against 1.36 × 10^{-3} for phytoplankton bloom). This latter pattern provides evidence that the structural heterogeneity (coated-sphere model) has a greater impact on the particulate backscattering ratio than the tested increase in the bulk imaginary refractive index. The relative absolute differences between the phytoplankton bloom and "Homogeneous reference case 2" is 41%. When mineral particles are taken into account, $\Delta \epsilon$ is 12% between the "Homogeneous reference case 3" and the coastal-like water body. This smaller difference is because phytoplankton have a smaller impact on the bulk scattering when highly scattering particles such as minerals are added.

The impact of the relative volume of the cytoplasm on $\widetilde{b_{bp}^{\theta_a}}$ is now evaluated by comparing the change of $\widetilde{b_{bp}^{\theta_a}}$ as a function of ζ for the 80%–20% and 70%–30% models for the oligotrophic-like and phytoplankton bloom water bodies (Figure 6). The mean relative difference in $\widetilde{b_{bp}^{\theta_a}}$ is about 5.41% with a maximum value of 11.5 % (ζ = 3) for oligotrophic-like case study. In bloom conditions, the mean relative difference reaches 13.0% with a maximum value of 23.5% (ζ = 3.2). Figure 7 compares simulated $\widetilde{b_{bp}^{\theta_a}}$ when phytoplankton cells are modeled as two-layered spheres (80%–20%) or three-layered spheres (80%–18.5%–1.5%). For the oligotrophic-like waters, relative absolute differences are small. They range between 0.0174% and 1.81% with a mean value of 0.444%. For phytoplankton bloom case study, they are between 9.84 × 10^{-3} and 2.86% with a mean value of 0.894%.

Regardless of the morphological model used to optically simulate phytoplankton cells, the $\widetilde{b_{bp}^{\theta_a}}$ reaches an asymptote when ζ decreases for phytoplankton bloom water bodies ($\widetilde{b_{bp}^{\theta_a}}$ = 0.005 for ζ = 3.5 and 0.004 for ζ = 2.5). The value of the asymptote is consistent with previous observations [5], which showed the lowest backscattering ratio (about 0.005) in waters with high chlorophyll-a concentration.

The contribution of the different particle groups to the backscattering ratio is presented in Figure 8 for the 80%–20% model and ζ = 4. For coastal-like waters, the minerals contribute more than 80% of the total $\widetilde{b_{bp}^{\theta_a}}$, whereas they contribute only 6.5% to the total particulate abundance. This percentage agrees with the results of Stramski et al. [25] (Figure 12 in their paper). Such high contribution to backscattering is due to the high real refractive index of minerals. As in Stramski et al. [25], these results show the important role of minerals even when they are less abundant than organic living and non-living particles. In oligotrophic-like waters, the contribution of heterotrophic bacteria ranges from a few to about 30% with a maximum for ζ between 3.5 and 4, which agrees with Stramski and Kiefer [24]. The contribution of viruses is quite high, about 40–60% for ζ between 4 and 5. This high contribution is explained by the extreme value of viral abundance (around 1 × 10^{14} particles per cubic meter) used in this study [24]. As for bacteria, the contribution of phytoplankton ranges between a few and 30% with a maximum around ζ = 3. For a ζ value of 4, typical of oligotrophic waters, the contribution is around 10%. For the phytoplankton bloom study case, the contribution of phytoplankton cells is between 10% and 60%; maximum values are reached for a PSD slope between 3 and 3.5. Such high percentages are due to the higher backscattering cross section of phytoplankton as compared to the other particles (Figure 9). The low phytoplankton abundance is offset by the high $C_{sca,PHY}^{bb}$ so that the backscattering coefficient of phytoplankton represents a significant contribution to the total backscattering.

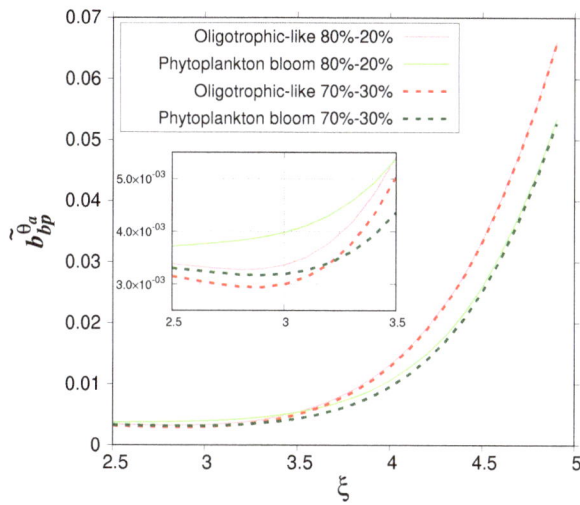

Figure 6. Particulate backscattering ratio as a function of the hyperbolic slope for oligotrophic-like and phytoplankton bloom water bodies. Phytoplankton cells are modeled as two-layered spheres with a relative volume of the chloroplast of 20 % and 30 %, as indicated.

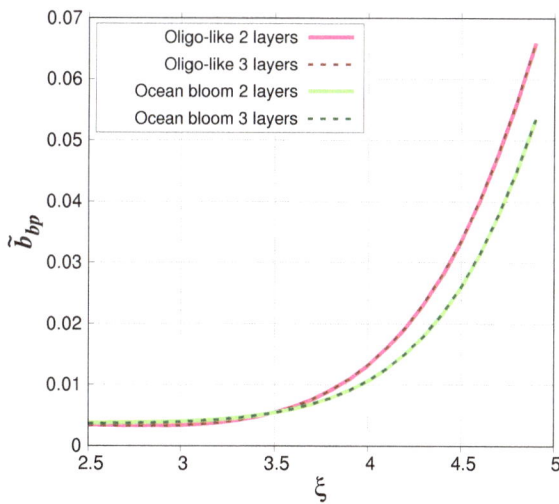

Figure 7. Particulate backscattering ratio as a function of the hyperbolic slope for oligotrophic-like and phytoplankton bloom water bodies. Phytoplankton cells are modeled as two-layered spheres (80%–20%) or three-layered spheres (80%–18.5%–1.5%), as indicated.

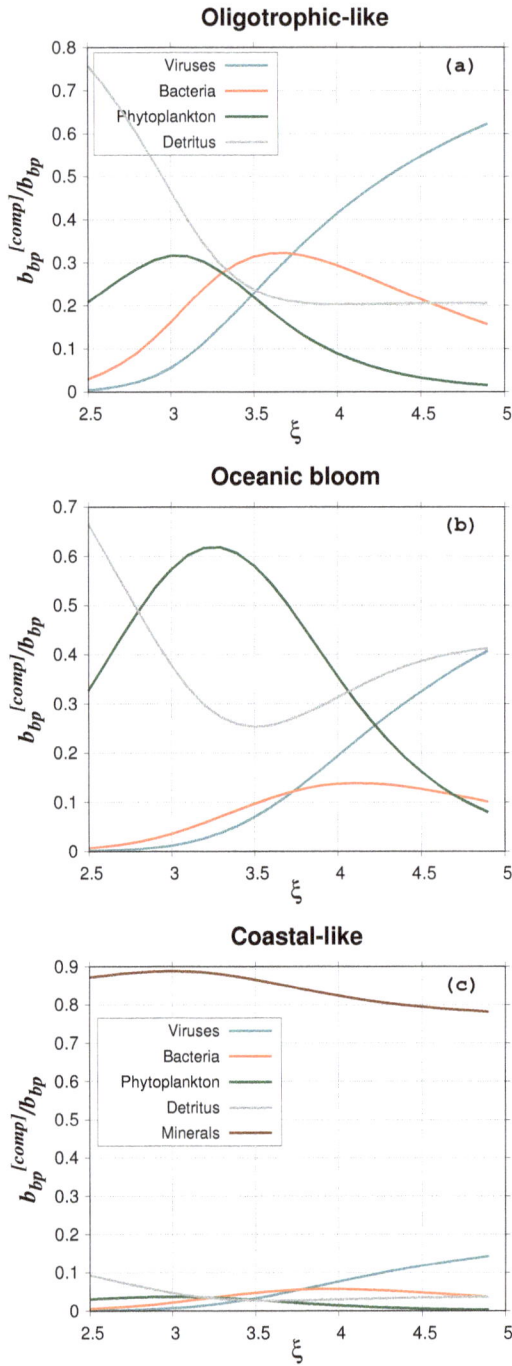

Figure 8. Contribution of the different particle groups the total bulk backscattering ratio for (**a**) oligotrophic-like, (**b**) phytoplankton bloom, and (**c**) coastal-like water bodies. The phytoplankton cells are modeled as a two-layered sphere (80%–20%).

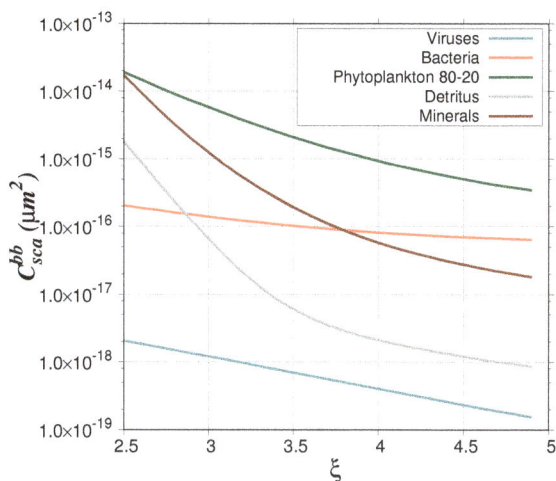

Figure 9. Backscattering cross sections, C_{sca}^{bb}, of the different particle groups. The phytoplankton cells are modeled as a two-layered sphere (80%–20%).

6. Concluding Remarks

Modeling phytoplankton cells as two-layered spheres tends to increase the bulk backscattering ratio because heterogeneous spheres are more efficient backscatterers. Even if the phytoplankton component has the smallest relative abundance, its impact on $\widetilde{b}_{bp}^{\theta_a}$ can be important depending on the hyperbolic slope of the Junge distribution. This is because phytoplankton cells have the highest backscattering cross section. For $\xi = 4$, the relative difference (in absolute value), $(\Delta\epsilon)$, between $\widetilde{b}_{bp}^{\theta_a}$ for the oligotrophic-like and the reference case, having the same bulk refractive index but composed exclusively of homogeneous spheres, can reach about 58%. When minerals are added, the impact of phytoplankton decreases as the scattering by minerals dominates.

Considering different sub-populations of particles with different refractive index implies that the bulk refractive index varies with the value of the hyperbolic slope of the PSD, as the contribution of each scattering component varies. Consequently, the development of models, similar to the one proposed by Twardowski et al. [5] to derive $\tilde{n}_r$ from $\widetilde{b}_{bp}^{\theta_a}$ and ξ, but accounting for phytoplankton heterogeneity, is not straightforward. Other aspects of the problem and other parameters such as the adoption of a 2D or 3D model, the proportion between cytoplasm and chloroplast, or the variation of phytoplankton refractive index according to the considered species, would have to be taken into account. One option would be to develop a look-up table approach based on the main parameters driving the $\widetilde{b}_{bp}^{\theta_a}$ variability. For that purpose, other simulations would have to be performed to be able to identify the pertinent parameters.

In this study, a large set of numerical simulations, as well as a proper methodology have been developed to simulate the particulate scattering properties of a water body in its complexity. We show that a special care should be taken in the integration step size when the particulate scattering coefficients are calculated from the particulate scattering function, especially for relatively small values of the PSD slope. We show that an integration angular step of $0.05°$ ($N_\theta = 3600$) is required to obtain the required accuracy considering the inputs (refractive indices and size range) used in this study.

The method is adapted to be used repeatedly to model a very large variety of particulate assemblages. While the present study has been limited to three case studies, additional calculations can be conducted to better represent the variability encountered in oceanic environments in terms of particulate community and its complexity in terms of mixing, morphology, size, and chemical nature.

For example, in waters with significant presence of specific populations of phytoplankton or under bloom conditions, the Junge-like approximation of PSD is usually unsatisfactory [36–38,45]. We also showed that the relative volume of the model chloroplast to the cytoplasm or the presence of third layer to model the cell wall can all affect the backscattering ratio (about 5–25%). It means that changes in the phytoplankton structural heterogeneity can explain partly the natural variability of the backscattering ratio, particularly in bloom conditions.

Further work is required, mainly experimental studies, to better characterize the internal structure and chemical composition of viruses, heterotrophic bacteria, biogenic detritus, and particle aggregates. This will allow, as Robertson Lain et al. [23] did for phytoplankton, suitable models to be developed to describe properly, in numerical code, the morphological properties of various types of particles to provide more realistic simulations of their optical properties.

Author Contributions: The research is completed through collaboration of all authors. L.D.-G. and H.L. were the team leaders of this work who were responsible for coordination. D.D. was in charge of program coding and assisted in data analysis. W.M. was involved in discussions. L.D.-G. analyzed the data and wrote the paper.

Funding: The authors gratefully acknowledge the support from the French Space Agency (CNES) through the COYOTE project (CNES/TOSCA program).

Acknowledgments: The authors thank Michael Twardowski, (Harbor Branch Oceanographic Institute) for the valuable advice. We appreciate thoughtful comments from the anonymous reviewers and Dariusz Stramski, who provided time and effort to improve this manuscript. Experiments presented in this paper were carried out using the CALCULCO computing platform, supported by SCoSI/ULCO (Service COmmun du Système d'Information de l'Université du Littoral Côte d'Opale).

Conflicts of Interest: The authors declare no conflict of interest.

References

1. Preisendorfer, R.W. *Hydrologic Optics, Volume 1: Introduction*; Springfield National Technical Information Service; Office of Naval Research: Arlington, VA, USA, 1976.
2. Morel, A. *The Scattering of Light by Seawater: Experimental Results and Theoretical Approach*; Translation by George Halikas of the paper published in French in AGARD Lecture Series; N°61; North Atlantic Treaty Organization: Neuilly-sur-Seine, France, 1973.
3. Boss, E.; Pegau, W.S.; Gardner, W.D.; Zaneveld, J.R.V.; Barnard, A.H.; Twardowski, M.S.; Chang, G.C.; Dickey, T.D. The spectral particulate attenuation and particle size distribution in the bottom boundary layer of a continental shelf. *J. Geophys. Res.* **2001**, *106*, 9509–9516. [CrossRef]
4. Boss, E.; Twardowski, M.S.; Herring, S. Shape of the particulate beam attenuation spectrum and its relation to the size distribution of oceanic particles. *Appl. Opt.* **2001**, *40*, 4885–4893. [CrossRef] [PubMed]
5. Twardowski, M.; Boss, E.; Macdonald, J.; Pegau, W.; Barnard, A.; Zaneveld, J. A model for estimating bulk refractive index from optical backscattering ratio and the implications for understanding particle composition in case I and case II waters. *J. Geophys. Res.* **2001**, *106*, 14129–14142. [CrossRef]
6. Boss, E.; Pegau, W.S.; Lee, M.; Twardowski, M.; Shybanov, E.; Korotaev, G.; Baratange, F. Particulate backscattering ratio at LEO 15 and its use to study particle composition and distribution. *J. Geophys. Res.* **2004**, *109*, C01014. [CrossRef]
7. Loisel, H.; Mériaux, X.; Berthon, J.F.; Poteau, A. Investigation of the optical backscattering to scattering ratio of marine particles in relation to their biogeochemical composition in the eastern English Channel and southern North Sea. *Limnol. Oceanogr.* **2007**, *52*, 739–752. [CrossRef]
8. Nasiha, H.J.; Shanmugam, P. Estimating the Bulk Refractive Index and Related Particulate Properties of Natural Waters from Remote-Sensing Data. *IEEE J. Sel. Top. Appl. Earth Obs. Remote Sens.* **2015**, *8*, 5324–5335. [CrossRef]
9. Snyder, W.A.; Arnone, R.A.; Davis, C.O.; Goode, W.; Gould, R.W.; Ladner, S.; Lamela, G.; Rhea, W.J.; Stavn, R.; Sydor, M.; et al. Optical scattering and backscattering by organic and inorganic particulates in U.S. coastal waters. *Appl. Opt.* **2008**, *47*, 666–677. [CrossRef]
10. Sullivan, J.; Twardowski, M.; Donaghay, P.; Freeman, S. Use of optical scattering to discriminate particle types in coastal waters. *Appl. Opt.* **2005**, *44*, 1667–1680. [CrossRef]

11. Meyer, R.A. Light scattering from biological cells: Dependence of backscatter radiation on membrane thickness and refractive index. *Appl. Opt.* **1979**, *18*, 585–588. [CrossRef]

12. Bricaud, A.; Zaneveld, J.R.V.; Kitchen, J.C. Backscattering efficiency of coccolithophorids: Use of a three-layered sphere model. *Proc. SPIE* **1992**, *1750*, 27–33.

13. Kitchen, J.C.; Zaneveld, J.R.V. A three-layered sphere model of the optical properties of phytoplankton. *Limnol. Oceanogr.* **1992**, *37*, 1680–1690. [CrossRef]

14. Stramski, D.; Piskozub, J. Estimation of scattering error in spectrophotometric measurements of light absorption by aquatic particles from three-dimensional radiative transfer simulations. *Appl. Opt.* **2003**, *42*, 3634–3646. [CrossRef]

15. Moutier, W.; Duforêt-Gaurier, L.; Thyssen, M.; Loisel, H.; Mériaux, X.; Courcot, L.; Dessailly, D.; Rêve, A.H.; Grégori, G.; Alvain, S.; et al. Evolution of the scattering properties of phytoplankton cells from flow cytometry measurements. *PLoS ONE* **2017**, *12*. [CrossRef] [PubMed]

16. Poulin, C.; Zhang, X.; Yang, P.; Huot, Y. Diel variations of the attenuation, backscattering and absorption coefficients of four phytoplankton species and comparison with spherical, coated spherical and hexahedral particle optical models. *J. Quant. Spectrosc. Radiat. Transf.* **2018**, *217*, 288–304. [CrossRef]

17. Quirantes, A.; Bernard, S. Light scattering by marine algae: Two-layer spherical and nonspherical models. *J. Quant. Spectrosc. Radiat. Transf.* **2004**, *89*, 311–321. [CrossRef]

18. Vaillancourt, R.D.; Brown, C.W.; Guillard, R.L.; Balch, W.M. Light backscattering properties of marine phytoplankton: Relationships to cell size, chemical composition and taxonomy. *J. Plankton Res.* **2004**, *26*, 191–212. [CrossRef]

19. Volten, H.; Haan, J.F.; Hovenier, J.W.; Schreurs, R.; Vassen, W.; Dekker, A.G.; Hoogenboom, H.J.; Charlton, F.; Wouts, R. Laboratory measurements of angular distributions of light scattered by phytoplankton and silt. *Limnol. Oceanogr.* **1998**, *43*, 1180–1197. [CrossRef]

20. Witkowski, K.; Król, T.; Zielinski, A.; Kuten, E. A light-scattering matrix for unicellular marine phytoplankton. *Limnol. Oceanogr.* **1998**, *43*, 859–869. [CrossRef]

21. Stramski, D.; Boss, E.; Bogucki, D.; Voss, K.J. The role of seawater constituents in light backscattering in the ocean. *Prog. Oceanogr.* **2004**, *61*, 27–56. [CrossRef]

22. Whitmire, A.L.; Pegau, W.S.; Karp-Boss, L.; Boss, E.; Cowles, T.J. Spectral backscattering properties of marine phytoplakton cultures. *Opt. Express* **2010**, *18*, 15073–15093. [CrossRef]

23. Robertson Lain, L.; Bernard, S.; Evers-King, H. Biophysical modelling of phytoplankton communities from first principles using two-layered spheres: Equivalent Algal Populations (EAP) model. *Opt. Express* **2014**, *22*, 16745–16758. [CrossRef] [PubMed]

24. Stramski, D.; Kiefer, D.A. Light scattering by microorganisms in the open ocean. *Prog. Oceanogr.* **1991**, *28*, 343–383. [CrossRef]

25. Stramski, D.; Bricaud, A.; Morel, A. Modeling the inherent optical properties of the ocean based on the detailed composition of the planktonic community. *Appl. Opt.* **2001**, *40*, 2929–2945. [CrossRef] [PubMed]

26. Mishchenko, M.I.; Travis, L.D.; Lacis, A.A. *Scattering, Absorption and Emission of Light of Small Particles*; Cambridge University Press: Cambridge, UK, 2002; ISBN 9780521782524.

27. Jonasz, M. Particle size distribution in the Baltic. *Tellus* **1983**, *B35*, 346–358. [CrossRef]

28. Loisel, H.; Nicolas, J.M.; Sciandra, A.; Stramski, D.; Poteau, A. Spectral dependency of optical backscattering by marine particles from satellite remote sensing of the global ocean. *J. Geophys. Res.* **2006**, *111*, C09024. [CrossRef]

29. Morel, A.; Bricaud, A. Inherent optical properties of algal cells, including picoplankton. Theoretical and experimental results. *Can. Bull. Fish. Aquat. Sci.* **1986**, *214*, 521–559.

30. Boss, E.; Slade, W.H.; Behrenfeld, M.; Dall'Olmo, G. Acceptance angle effects on the beam attenuation in the ocean. *Opt. Express* **2009**, *17*. [CrossRef]

31. Dolman, V.L. *Meerhoff Mie Program User Guide*; Internal Report Astronomy Department, Free University: Amsterdam, The Netherlands, 1989.

32. Peña, O.; Pal, U. Scattering of electromagnetic radiation by a multilayered sphere. *Comput. Phys. Commun.* **2009**, *180*, 2348–2354. [CrossRef]

33. Yang, W. Improved recursive algorithm for light scattering by a multilayered sphere. *Appl. Opt.* **2003**, *42*, 1710–1720. [CrossRef]

34. Aas, E. Refractive index of phytoplankton derived from its metabolite composition. *J. Plankton Res.* **1996**, *18*, 2223–2249. [CrossRef]

35. Buonassissi, C.J.; Dierssen, H.M. A regional comparison of particle size distributions and the power law approximation in oceanic and estuarine surface waters. *J. Geophys. Res.* **2010**, *115*, C10028. [CrossRef]

36. Reynolds, R.A.; Stramski, D.; Wright, V.M.; Woźniak, S.B. Measurements and characterization of particle size distributions in coastal waters. *J. Geophys. Res.* **2010**, *115*, C08024. [CrossRef]

37. Reynolds, R.A.; Stramski, D.; Neukermans, G. Optical backscattering by particles in Arctic seawater and relationships to particle mass concentration, size distribution, and bulk composition. *Limnol. Oceanogr.* **2016**, *61*, 1869–1890. [CrossRef]

38. Woźniak, S.B.; Stramski, D.; Stramska, M.; Reynolds, R.A.; Wright, V.M.; Miksic, E.Y.; Cichocka, M.; Cieplak, A.M. Optical variability of seawater in relation to particle concentration, composition, and size distribution in the nearshore marine environment at Imperial Beach, California. *J. Geophys. Res.* **2010**, *115*, C08027. [CrossRef]

39. Middleboe, M.; Brussaard, C.P.D. Marine Viruses: Key Players in Marine Ecosystems. *Viruses* **2017**, *9*, 302. [CrossRef] [PubMed]

40. Brotas, V.; Brewin, R.; Sá, C.; Brito, A.C.; Silva, A.; Mendes, C.R; Diniz, T.; Kaufmann, M.; Tarran, G.; Groom, S.B.; et al. Deriving phytoplankton size classes from satellite data: Validation along a trophic gradient in the eastern Atlantic Ocean. *Remote Sens. Environ.* **2013**, *134*, 66–77. [CrossRef]

41. Brewin, R.J.W.; Sathyendranath, S.; Hirata, T.; Lavender, S.; Barciela, R.M.; Hardman-Mountford, N.J. A three-component model of phytoplankton size class for the Atlantic Ocean. *Ecol. Model.* **2010**, *221*, 1472–1483. [CrossRef]

42. Mishchenko, M.; Lacis, A. Manifestations of morphology-dependent resonances in Mie scattering matrices. *Appl. Math. Comput.* **2000**, *116*, 167–179. [CrossRef]

43. Bricaud, A.; Roesler, C.; Zaneveld, J.R.V. In situ methods for measuring the inherent optical properties of ocean waters. *Limnol. Oceanogr.* **1995**, *40*, 393–410. [CrossRef]

44. Zaneveld, J.R.V.; Kitchen, J.C. The variation in the inherent optical properties of phytoplankton near an absorption peak as determined by various models of cell structure. *J. Geophys. Res.* **1995**, *100*, 309–313. [CrossRef]

45. Reynolds, R.A.; Stramski, D.; Mitchell, B.G. A chlorophyll-dependent semianalytical reflectance model derived from field measurements of absorption and backscattering coefficients within the Southern Ocean. *J. Geophys. Res.* **2001**, *106*, 7125–7138. [CrossRef]

applied
sciences

MDPI

Article

Measurements of the Volume Scattering Function and the Degree of Linear Polarization of Light Scattered by Contrasting Natural Assemblages of Marine Particles

Daniel Koestner *, Dariusz Stramski and Rick A. Reynolds

Marine Physical Laboratory, Scripps Institution of Oceanography, University of California San Diego, La Jolla, CA 92093-0238, USA; dstramski@ucsd.edu (D.S.); rreynolds@ucsd.edu (R.A.R.)
* Correspondence: dkoestne@ucsd.edu; Tel: +1-631-357-9656

Received: 13 August 2018; Accepted: 13 November 2018; Published: 19 December 2018

Abstract: The light scattering properties of seawater play important roles in radiative transfer in the ocean and optically-based methods for characterizing marine suspended particles from in situ and remote sensing measurements. The recently commercialized LISST-VSF instrument is capable of providing in situ or laboratory measurements of the volume scattering function, $\beta_p(\psi)$, and the degree of linear polarization, $DoLP_p(\psi)$, associated with particle scattering. These optical quantities of natural particle assemblages have not been measured routinely in past studies. To fully realize the potential of LISST-VSF measurements, we evaluated instrument performance, and developed calibration correction functions from laboratory measurements and Mie scattering calculations for standard polystyrene beads suspended in water. The correction functions were validated with independent measurements. The improved LISST-VSF protocol was applied to measurements of $\beta_p(\psi)$ and $DoLP_p(\psi)$ taken on 17 natural seawater samples from coastal and offshore marine environments characterized by contrasting assemblages of suspended particles. Both $\beta_p(\psi)$ and $DoLP_p(\psi)$ exhibited significant variations related to a broad range of composition and size distribution of particulate assemblages. For example, negative relational trends were observed between the particulate backscattering ratio derived from $\beta_p(\psi)$ and increasing proportions of organic particles or phytoplankton in the particulate assemblage. Our results also suggest a potential trend between the maximum values of $DoLP_p(\psi)$ and particle size metrics, such that a decrease in the maximum $DoLP_p(\psi)$ tends to be associated with particulate assemblages exhibiting a higher proportion of large-sized particles. Such results have the potential to advance optically-based applications that rely on an understanding of relationships between light scattering and particle properties of natural particulate assemblages.

Keywords: marine optics; inherent optical properties; volume scattering function; degree of linear polarization; marine particles; light scattering measurements; LISST-VSF instrument

1. Introduction

It has long been recognized that inherent light-scattering properties of natural waters are of crucial importance and have strong potential for wide-ranging applications in aquatic sciences, including oceanography. These properties are essential inputs to the radiative transfer models used to compute the ambient light fields in natural water bodies [1–4]. The variability in the light scattering properties of seawater is driven primarily by the concentration of suspended particles, particle size distribution, and composition through particle refractive index, internal structure, and shape. Hence, scattering measurements carry potentially useful information about characteristics of natural particle assemblages. For example, the scattering and backscattering coefficients of suspended particles have been shown to provide useful proxies of mass concentration of total suspended particulate matter (SPM), particulate inorganic carbon (PIC), and particulate organic carbon (POC) in the ocean [5–7]. Multi-angle light scattering measurements provide a means to estimate the particle size distribution [8–11], including the submicrometer size range [12–14]. The angular pattern of light scattering can also contain useful information about the composition of particulate assemblages, including the bulk refractive index of particles [15–19]. In addition, measurements of the scattering matrix that provide information about polarization effects of light scattering [15,20–23] have the potential for identifying and discriminating different types of particles, such as phytoplankton species or minerals, which are present in complex natural assemblages [24–34]. Despite the potential usefulness of information provided by light scattering measurements, the complexity and variability in composition of natural particulate assemblages impose significant challenges in achieving an understanding of bulk light-scattering properties of seawater in terms of detailed compositional characteristics of particulate matter [35].

The volume scattering function, $\beta_p(\psi, \lambda)$, and the degree of linear polarization, $DoLP_p(\psi, \lambda)$ of light scattered by marine particles are of primary interest in this study. Here, ψ denotes the scattering angle, λ the light wavelength in vacuum, and the subscript p indicates that the quantity is associated with particles. When the subscript p is omitted, the quantity describes the scattering by the entire suspension with additive contributions from both water molecules and suspended particles. The volume scattering function, $\beta(\psi, \lambda)$ [in units of $m^{-1} \, sr^{-1}$], is one of the fundamental inherent optical properties (IOPs) of seawater, which describes the scattered intensity as a function of scattering angle per unit incident irradiance per unit volume of small sample of water [2]. Several light-scattering related IOPs can be derived from $\beta(\psi, \lambda)$. For example, integrating $\beta(\psi, \lambda)$ over all scattering directions gives the total spectral scattering coefficient, $b(\lambda)$ [m^{-1}]. In this integration, it is commonly assumed that light scattering by an assemblage of randomly-oriented scatterers (molecules and particles) in natural waters is azimuthally symmetric about the incident direction of light beam. When $\beta(\psi, \lambda)$ is normalized by $b(\lambda)$, the resulting scattering phase function $\tilde{\beta}(\psi, \lambda)$ [sr^{-1}] provides a useful indicator of the angular shape of the volume scattering function. In optical remote sensing applications based on measurements with above-water sensors (e.g., from satellites or aircraft), the spectral backscattering coefficient, $b_b(\lambda)$ [m^{-1}], is particularly useful. This coefficient can be obtained by integrating $\beta(\psi, \lambda)$ over the range of backward scattering angles [2].

The volume scattering function provides incomplete information, in the sense that it does not contain information about polarization effects associated with light scattering. A complete characterization of elastic incoherent interactions of light at arbitrary wavelength λ with a sample volume of seawater is provided by a 4×4 scattering matrix, often referred to as the phase matrix or Mueller matrix [20–23]. This matrix describes a linear transformation of irradiance and polarization of an incident beam described by a 4-component Stokes vector into the intensity and polarization of the scattered beam that is also described by its corresponding Stokes vector. $\beta(\psi, \lambda)$ is related to the first element of the scattering matrix, $p_{11}(\psi, \lambda)$, and can be obtained from a measurement using unpolarized light for illumination of sample and measuring the total scattered intensity. The degree of linear polarization of scattered light, $DoLP(\psi, \lambda)$, describes the proportion of linearly polarized light relative to total intensity of the scattered light beam. As described in greater detail below, for various assemblages of particles including suspended marine particles and when the incident light beam is unpolarized, this quantity can be derived from the first two elements of the scattering matrix, which requires measurements involving linear polarization [29,36,37].

Despite the relative importance of $\beta(\psi, \lambda)$ and $DoLP(\psi, \lambda)$ of seawater and the associated particulate components $\beta_p(\psi, \lambda)$ and $DoLP_p(\psi, \lambda)$, the ocean optics community has historically relied mostly on simplified theoretical models (such as Mie scattering theory for homogenous spheres) and a limited dataset of measurements made with custom-built light scattering instruments. For example, over the past several decades, a limited dataset of $\beta(\psi, \lambda)$ measurements made by Petzold [38] was widely used as a standard input for the particulate scattering phase function for radiative transfer modeling in the ocean. Comprehensive determinations of the scattering matrix for natural seawater have been very scarce [39–42]. These determinations showed that the off-diagonal matrix elements for seawater are very small or negligible, indicating very small effects associated with optical activity or orientational anisotropy of seawater scatterers [37,42]. More recently, several light scattering sensors have been developed for in situ deployments or laboratory use [43–46], but to our knowledge, none of these sensors are commercially available. While measurements with these new sensors have already significantly contributed to the increase of available datasets of $\beta(\psi, \lambda)$ (or $\beta_p(\psi, \lambda)$ which can usually be satisfactorily estimated by subtracting the contribution associated with water molecules) in various oceanic environments [44,47,48], the determinations of $DoLP_p(\psi, \lambda)$ for natural assemblages of marine particles remain very scarce, as indicated by the rarity of scattering matrix measurements of seawater.

Recently, a new light scattering instrument, the LISST-VSF (Sequoia Scientific, Inc., Bellevue, WA, USA), has become commercially available, and provides the capability of determining both the volume scattering function and the degree of linear polarization of scattered light at a single light wavelength (532 nm) with high angular resolution over the range ~0.1° to 155° [49]. It is capable of both in situ and benchtop measurements on water samples. This commercial instrument is expected to enable routine measurements by different groups of investigators, so it has the potential to enhance our understanding of light scattering properties of seawater and marine particles and advance the related applications. In this study, we report on LISST-VSF measurements of $\beta_p(\psi)$ and $DoLP_p(\psi)$ and size and compositional characteristics for contrasting natural particulate assemblages from marine coastal and offshore environments. The particulate scattering (b_p) and backscattering (b_{bp}) coefficients have also been determined from measured $\beta_p(\psi)$.

To fully realize the potential of such quantitative determinations for seawater samples from this new instrument, we also conducted an evaluation of the LISST-VSF performance through a series of laboratory experiments using samples of National Institute of Standards and Technology (NIST) certified standard polystyrene beads ranging in diameter between 100 nm and 2 μm. These measurements were compared with theoretical simulations of light scattering by bead suspensions using Mie scattering computations. With this approach, we developed corrections to the determinations of $\beta_p(\psi)$ and $DoLP_p(\psi)$ from LISST-VSF measurements. A validation of the corrected measurements was performed using independent measurements of multi-angle light scattering with another instrument, the DAWN-EOS (Wyatt Technology Corporation, Santa Barbara, CA, USA).

2. Methods

The description of methods includes two main parts: first, a description of laboratory experiments and Mie scattering calculations for standard polystyrene beads which were carried out to evaluate the performance of the LISST-VSF instrument and develop a calibration correction; second, a description of measurements on natural assemblages of marine particles from coastal and offshore oceanic environments.

2.1. Laboratory Experiments and Mie Scattering Calculations to Evaluate LISST-VSF

In order to evaluate the LISST-VSF instrument, light scattering and beam attenuation measurements were made in the laboratory on samples of nearly monodisperse standard polystyrene spherical beads with mean nominal diameters of 100, 200, 400, 500, 700, and 2000 nm, which were suspended in water (Table 1). In addition to LISST-VSF, two other instruments were used in these experiments, a DAWN-EOS for measuring multi-angle light scattering and a dual beam UV/VIS spectrophotometer Lambda 18 (Perkin-Elmer, Inc., Waltham, MA, USA) equipped with a 15-cm integrating sphere (Labsphere, Inc., North Sutton, NH, USA) for measuring the beam attenuation coefficient of particles in suspension. The use of standard beads ensures that Mie scattering calculations for homogeneous spherical particles can be used to calculate the Mueller matrix elements for these particles to determine reference (expected) values of the volume scattering function and the degree of linear polarization. The comparison of measurements with such reference values allows for evaluation of performance of LISST-VSF instrument and formulation of calibration correction functions for improved determinations of the volume scattering function and the degree of linear polarization from this instrument. This type of approach, which combines measurements on standard well-characterized particles with accurate scattering calculations, has been previously used for the evaluation, calibration, and characterization of light scattering instruments [29,43,44,50]. Although the evaluation results presented in this study are relevant to the specific version of the LISST-VSF instrument used in our laboratory, most methodological aspects are generally applicable to evaluation of other light scattering instruments.

Table 1. Information on the polystyrene bead size standards used to create laboratory sample suspensions for experiments. The nominal bead diameter (D), catalog number, and actual mean diameter $\overline{D}$ ($\pm$ standard error of estimate) and standard deviation of the mean (SD) provided by the manufacturer (Thermo Fisher Scientific, Inc.) is listed. The particulate beam attenuation coefficient at light wavelength 532 nm of the master sample as determined with a spectrophotometer, c_p^{SPEC}, is listed in addition to specific dilution names and factors (e.g., DF1, DF2, etc.) of the master suspension used for LISST-VSF measurements at different PMT gain settings. The dilution factors in italic font denote the experimental data used for generation of the final correction functions CF_f and BF_f, and those in boldface font denote the six examples used for statistical evaluation in Table 2.

Nominal D [nm]	Catalog No.	$\overline{D}$ [nm]	SD [nm]	c_p^{SPEC} [m^{-1}]	Dilution Factor (PMT 500)	Dilution Factor (PMT 550)
100	3100A	100 ± 3	7.8	58.63	DF1: 96, DF2: 48.5, DF3: 32.7	DF1: 96, *DF2: 48.5,* DF3: 32.67
200	3200A	203 ± 5	5.3	46.26	DF1: 96, *DF2: 48.5,* DF3: 32.7	DF1: 96, *DF2: 48.5,* **DF3: 32.7**
400	3400A	400 ± 9	7.3	51.44	DF1: 87.4, **DF2: 44.2,** DF3: 29.8	DF1: 87.4
500	3500A	508 ± 8	8.5	20.64	**DF2: 20**	
700	3700A	707 ± 9	8.3	50.93	DF1: 96, **DF2: 48.5**	
2000	4202A	2020 ± 15	21	18.21	**DF2: 20**	

2.1.1. Instrumentation

A LISST-VSF instrument (S/N 1475) was equipped with a custom designed 2 L sample chamber for benchtop laboratory use. This chamber effectively rejects ambient light and promotes good mixing conditions to maintain particles in suspension. For sample illumination the LISST-VSF uses a frequency-doubled YAG laser to produce a beam of light at a wavelength of 532 nm with a Gaussian beam profile of 3 mm in diameter. A single measurement takes approximately 4 s and consists of two scans of a 15-cm path within the sample, each with a different linear polarization state of the incident beam, i.e., parallel and perpendicular to the scattering plane. Scattered intensity is measured at multiple scattering angles ψ from 0.09° to 15.17° with 32 logarithmically-spaced ring detectors and from 14° to 155° with 1° interval using a fixed axis Roving Eyeball sensor equipped with photomultiplier tubes (PMTs). For the Roving Eyeball, scattered light is split between two PMTs with a polarizing prism allowing for only parallel or perpendicularly polarized light to be detected by each PMT. To enable measurements of large dynamic range of scattered intensity with a single PMT, the laser power is dimmed by a factor of 8 for the angular range 14–63° and returned to full power for 64–155°. The beam attenuation coefficient, c, is also measured at light wavelength of 532 nm for the 15-cm path length of the sample.

For incoherent elastic scattering of light at a given wavelength λ by a collection of particles suspended in water, the Stokes vector of incident light beam, $S_i = [I_i\ Q_i\ U_i\ V_i]^T$, where T represents the transpose operation, is transformed into the Stokes vector of scattered beam, $S_s(\psi)$, by a scattering matrix, $P(\psi)$. For an ensemble of randomly-oriented particles exhibiting certain symmetry properties and no optical activity, the scattering matrix simplifies to 6 independent non-zero elements [20,36,51]

$$S_s(\psi) = \begin{bmatrix} I_s(\psi) \\ Q_s(\psi) \\ U_s(\psi) \\ V_s(\psi) \end{bmatrix} = P(\psi)S_i = C \begin{bmatrix} p_{11}(\psi) & p_{12}(\psi) & 0 & 0 \\ p_{12}(\psi) & p_{22}(\psi) & 0 & 0 \\ 0 & 0 & p_{33}(\psi) & p_{34}(\psi) \\ 0 & 0 & -p_{34}(\psi) & p_{44}(\psi) \end{bmatrix} \begin{bmatrix} I_i \\ Q_i \\ U_i \\ V_i \end{bmatrix}, \tag{1}$$

where λ has been omitted for brevity, C is a constant factor (for a given sample, light wavelength, and measurement geometry), $p_{11}(\psi)$ represents the scattering phase function, and the reference plane is the scattering plane containing the incident and scattered directions [20,51,52]. This form provides a reasonable description of the measured scattering matrix by suspensions of randomly-oriented marine particles, including various specific types of particles present in seawater [24,29,33,36,37,42]. In the case of unpolarized incident light (i.e., Q_i, U_i, and V_i are all zero), the volume scattering function $\beta(\psi)$ equals (to within a constant factor) $p_{11}(\psi)$, and the degree of linear polarization $DoLP(\psi)$ can be determined from [29,36,37,53]

$$DoLP(\psi) = \frac{-p_{12}(\psi)}{p_{11}(\psi)} = \frac{-Q_s(\psi)}{I_s(\psi)}. \tag{2}$$

Positive values of $DoLP(\psi)$ are for dominantly perpendicular polarization and negative values for dominantly parallel polarization. We note that this definition of $DoLP(\psi)$ has been widely used for characterizing the inherent scattering properties of various types of particles beyond aquatic particles, such as aerosol particles and cosmic dust [30,54–58].

The LISST-VSF measurements of forward scattering within the angular range 0.09–15.17° are made with two linear polarization states of the incident beam, but with no polarization analyzers of the ring detectors. For the ring detectors, the calibrated $\beta(\psi)$ in absolute units is a standard output of the manufacturer's processing software. The absolute calibration is based on the manufacturer-provided conversion from ring detector counts to physical units using radiant sensitivity of ring detectors [59,60]. Detection of scattered light within the angular range 14–155° using the Roving Eyeball sensor employs measurements made with two linear polarization states of the incident beam and the corresponding two linear polarization states of the scattered light. The four measurement configurations allow for the determination of relative values of $p_{11}(\psi)$, $p_{12}(\psi)$, and $p_{22}(\psi)$. The calibrated $\beta(\psi)$ values within the Roving Eyeball angular range are obtained by scaling the $p_{11}(\psi)$ data from the Roving Eyeball sensor. Specifically, the scattering measurements from the first angles of the Roving Eyeball sensor are forced to match the calibrated $\beta(\psi)$ values from the overlapping last ring detectors. The $DoLP(\psi)$ values are obtained from Equation (2) using $p_{11}(\psi)$ and $p_{12}(\psi)$, and are also included in the standard output of the manufacturer's processing code.

We also used a DAWN-EOS multi-angle light scattering instrument which provided independent measurements of $\beta(\psi)$ and $DoLP(\psi)$ of polystyrene beads suspended in water. These measurements were made with a sample placed in a 20 mL cylindrical glass vial. The DAWN-EOS instrument used in this study has been previously characterized and calibrated for such measurement configuration [61]. This instrument uses a diode-pumped frequency-doubled Nd-YAG laser at light wavelength 532 nm with a Gaussian beam profile of 62 μm in diameter. The interrogated sample volume is on the order of 10 nL. The incident beam can be linearly polarized both parallel and perpendicular to the scattering plane. The intensity of scattered light is measured simultaneously with eighteen photodiode detectors and no polarization analyzers, enabling measurements within a range of scattering angles from 22.5°

to 147°. To encompass the large dynamic range of scattered intensity, three selectable gain settings are available for each detector (gain factors of 1, 21, or 101).

As the DAWN-EOS detectors have no polarization analyzers, they only measure the first parameter of Stokes vector of the scattered light, $I_s(\psi)$. Here we define $I_{s\|}(\psi)$ for the parallel polarization of the incident beam and $I_{s\perp}(\psi)$ for the perpendicular polarization of the incident beam. The matrix elements $p_{11}(\psi)$ and $p_{12}(\psi)$ can be obtained (to within a constant factor) from DAWN-EOS measurements as

$$p_{11}(\psi) = \frac{I_{s\|}(\psi) + I_{s\perp}(\psi)}{2} \tag{3}$$

$$p_{12}(\psi) = \frac{I_{s\|}(\psi) - I_{s\perp}(\psi)}{2}, \tag{4}$$

which allows for determination of $DoLP(\psi)$ from Equation (2). The calibration procedure described in Babin et al. [61] allows for determination of $\beta(\psi)$ in absolute units. Importantly, the calibration procedure of DAWN-EOS is fundamentally different from the calibration procedure of LISST-VSF. The manufacturer's calibration of LISST-VSF is based on a nominal radiant sensitivity of ring detectors (amperes of photoelectric current per watt of optical power) traceable to the National Institute of Standards and Technology [59,60]. In contrast, the calibration of DAWN-EOS is based on measurements of light scattered at 90° by pure toluene with the incident beam having a linear perpendicular polarization [61]. This calibration relies on the known magnitude of molecular scattering by toluene. The two different methods employed in calibration of LISST-VSF and DAWN-EOS allow for comparisons of independent estimates of $\beta(\psi)$ obtained by these instruments. We also recall that the $DoLP(\psi)$ estimates obtained with the two instruments within the common range of scattering angles are based on different polarization measurement configurations used by these instruments.

A Lambda 18 spectrophotometer was used to collect independent measurements of the spectral beam attenuation coefficient, $c(\lambda)$, of polystyrene beads suspended in water. These measurements were made for comparisons with the beam attenuation data obtained with LISST-VSF, and also to aid in the preparation of samples with appropriate concentrations of polystyrene beads to ensure that measurements with LISST-VSF and DAWN-EOS were made within the single scattering regime. The spectrophotometric measurements were made in the spectral range from 290 nm to 860 nm with 1 nm interval, but only data at 532 nm are used in this study. The general applicability of laboratory spectrophotometers with proper modifications to enable measurements of beam attenuation of particle suspensions, including colloidal samples, has long been recognized [62,63]. In our study, a sample of particle suspension was measured in a 1-cm quartz cuvette placed at a significant distance from the detector (~25 cm from the entrance of the integrating sphere), and field stops were aligned within the light path to reduce the size of the beam and acceptance angle of the detector to less than 1°. This measurement geometry has been used in our previous studies of spectral beam attenuation by various particle assemblages [64,65].

2.1.2. Experimental Procedure

Baseline measurements of 0.2 μm filtered water were collected with all three instruments used in the experiments; LISST-VSF, DAWN-EOS, and Lambda 18 spectrophotometer. These baseline measurements were subtracted from subsequent measurements taken on particle suspensions to determine the optical properties associated with suspended particles only, i.e., the particulate volume scattering function, $\beta_p(\psi)$, the particulate degree of linear polarization, $DoLP_p(\psi)$, and the particulate beam attenuation coefficient, c_p.

Original manufacturer's stock samples of standard polystyrene beads (100, 200, 400, 500, 700, and 2000 nm in diameter) were used to generate master samples using 0.2 μm filtered, deionized, and degassed water as a medium (with the exception of 2000 nm beads which used 0.2 μm filtered seawater). In the process of preparation of master samples, the particle concentration was optimized to ensure that spectrophotometric measurements of beam attenuation coefficient can be performed either directly or with small dilution factor (~3) on these samples over 1-cm path length with sufficiently high signal but negligible multiple scattering effects. The c_p values for master samples ranged from about 18 m^{-1} to 58 m^{-1} (Table 1).

The master sample was diluted for measurements with the LISST-VSF to avoid oversaturation of PMT detectors and multiple scattering over the longer path length (15 cm). For baseline measurements, the LISST-VSF sample chamber was filled with 1900 mL of 0.2 µm filtered water. The final samples of particle suspensions were created by addition of 20 to 100 mL of master sample to the LISST-VSF chamber. For most beads examined in our experiments, more than one particle suspension differing in terms of particle concentration was measured with LISST-VSF (Table 1). The different particle concentrations were achieved by different dilution of master sample within LISST-VSF chamber. Owing to different dilution factors ranging from 20 to 96 (labeled as DF1, DF2, and DF3 in Table 1), the c_p values of LISST-VSF samples ranged from about 0.5 m^{-1} to 1.8 m^{-1}. For a single bead size, concentration, and PMT gain, a series of LISST-VSF measurements was composed of 200 measurements taken in rapid succession (recall that a measurement refers to two scans, each with a different polarization of incident beam). This measurement series was divided into five sets of 20 measurements and one set of 100 measurements to enable manual gentle mixing of sample before each set of measurements. In addition, for the 2000 nm bead suspensions a magnetic stir bar which operated on low speed and changed direction of rotation every 30 s was used to prevent particle settling during the measurement.

Several LISST-VSF baseline measurements of 0.2 µm filtered water were collected for each experiment, i.e., for each examined bead size. However, for reasons of consistency and out of the desire to use an optimal baseline representative of the least contaminated 0.2 µm filtered water, a single baseline was used for processing of all experimental data collected for various bead sizes and concentrations except for 2000 nm sized beads which used 0.2 µm filtered seawater. This baseline was determined on the basis of finding a measurement which exhibited minimal scattering signal detected by Roving Eyeball and ring detectors and maximum directly transmitted light detected by the laser transmission sensor. We note, however, that for each PMT gain setting of the Roving Eyeball sensor a separate baseline was determined.

Measurements using the DAWN-EOS instrument were collected for 100, 200, 400, and 700 nm beads. Dilution factors of master samples for DAWN-EOS measurements were between 300 and 3000, depending on bead size. The gain settings for each detector were adjusted to the highest setting that would avoid saturation of signal with incident perpendicular polarization of light. For 400 and 700 nm bead suspensions, two different dilutions were measured. For each polarization state (i.e., perpendicular and parallel) of incident light, we acquired 1440 measurements with a sampling frequency of 8 Hz over 3 min. For a given sample, this data acquisition protocol was repeated three times. Each of these three replications was made with a different randomly-chosen orientation of sample cylindrical vial within the instrument. The sample was gently mixed between these replicate measurements. The baseline measurements of 0.2 µm filtered water were acquired using the same protocol.

As mentioned above, the optical measurements were made on sufficiently-diluted samples to ensure negligible effects of multiple scattering over a pathlength used by a given instrument. A criterion for a single scattering regime is generally defined in terms of small optical thickness of the sample, $\tau \ll 1$, where τ is a product of the beam attenuation coefficient, c, and pathlength, r [51,66]. Also, a simple practical test for ensuring that multiple scattering effects are negligible is to verify a direct proportionality between the measured optical signal and the concentration of particles in suspension by conducting a series of measurements on the same sample with different dilutions [51]. Our measurements on bead samples with different dilutions showed an excellent 1:1 relationship between the LISST-VSF measurement and the bead concentration over the range of beam attenuation coefficient up to at least 2 m^{-1}. The single scattering regime can also be determined by the condition $\tau(1-g) \ll 1$, where g is the average cosine of the scattering angle of the volume scattering function [23,66]. For the 100 nm polystyrene beads, the g value is 0.115, which yields the most restrictive condition in our study, $\tau \ll 1.13$. For all bead samples measured with LISST-VSF, including all bead sizes and sample dilutions, τ was always less than about 0.3. This condition was also satisfied for samples measured with a spectrophotometer. For the measurements with DAWN-EOS, the τ values were even smaller. For the natural seawater samples examined in our study (which is described below in Section 2.2), the g values (for the total volume scattering function including the contribution by pure seawater) were about 0.9 or somewhat higher, which yields less restrictive criterion $\tau \ll 10$. Our measurements of natural samples clearly satisfied this single scattering condition, as the highest value of c for the natural samples measured with LISST-VSF was about 2.6 m^{-1}, so τ was always less than about 0.45, given that the maximum pathlength for LISST-VSF is 17.5 cm for the scattering angle of 150°.

2.1.3. Data Processing

Processing of LISST-VSF data was done with a standard processing code provided by manufacturer (version of 2013) to determine $\beta_p(\psi)$, $DoLP_p(\psi)$, and c_p, denoted hereafter as $\beta_p^{LISST*}(\psi)$, $DoLP_p^{LISST*}(\psi)$, and c_p^{LISST*} respectively (the asterisk indicates that the variable is derived from the standard processing code without additional corrections developed in this study). Some details specific to the processing and quality control of our experimental data are provided below.

As a first step in data processing, the baseline values in raw counts were subtracted from each LISST-VSF measurement of raw counts acquired on samples of bead suspensions. To account for light attenuation along the path between the scattering volume and the detector, an attenuation correction factor was calculated using the average c_p^{LISST*} from the series of measurements and the length of the path for each scattering angle. Further, to account for the difference in sensitivity of the two Roving Eyeball PMT detectors, a factor α is used to adjust the measured counts of one PMT detector relative to the other [67]. The value of $\alpha = 0.9335$ was determined by averaging all median values of α derived from each series of measurements for each bead size, particle concentration, and PMT gain. The α parameter was observed to be nearly constant over the period of experiments (~18 months, the coefficient of variation < 5%). For each series of measurements a specific scaling factor was determined to convert $p_{11}(\psi)$ in PMT counts to $\beta_p^{LISST*}(\psi)$ in absolute units [m^{-1} sr^{-1}] for scattering angles 14–155° measured by the Roving Eyeball sensor. First, for each measurement from a given series of measurements, a scaling factor was determined by matching the PMT counts measured with Roving Eyeball sensor between 15° and 16° with $\beta_p^{LISST*}(\psi)$ in absolute units obtained from measurements with the last two ring detectors at 13.01° and 15.17°. Then, using these determinations, the average scaling factor for a given series of measurements was calculated and used for further data processing. Note that this scaling was not needed for the determination of $DoLP_p^{LISST*}(\psi)$ for the Roving Eyeball angular range, which is calculated from $p_{11}(\psi)$ and $p_{12}(\psi)$ determined in PMT counts following Equation (2).

Quality control of data was performed by removing the first set of 20 measurements (the remaining four sets with 20 measurements each were retained) and the first 20 measurements from the set of 100 measurements. We observed that this was necessary to ensure reasonable stability in the measured scattering signal. The mean and standard deviation values for each angle based on all of the 160 remaining measurements in the series were determined, and the outlying single measurements within the series were identified and rejected from subsequent analysis. Typically, 120 to 130 measurements from a given series of 200 measurements passed the quality criteria.

Example data of uncorrected $\beta_p^{LISST*}(\psi)$ for 200 nm and 2000 nm bead suspensions are shown in Figure 1. The series of measurements that remained after quality control and the median values of $\beta_p^{LISST*}(\psi)$ derived from the series of measurements are shown. We also note that the median values were very close to mean values for our data (<1% difference for most scattering angles). The results for 2000 nm beads show a distinct pattern with several scattering maxima and minima due to constructive and destructive interference of the scattered light from a nearly monodisperse population of beads that are large relative to the wavelength of light. The 200 nm beads are smaller than the wavelength of light leading to a more featureless shape of $\beta_p^{LISST*}(\psi)$. The variability between the individual measurements is largest at very small scattering angles, i.e., approximately <4°, where the scattering signal for submicron particles is low relative to our baseline measurements. Apart from small scattering angles, the coefficient of variation (CV) for each scattering angle calculated from a series of measurements on 200 nm beads is generally very small, ranging from ~3% to <1%, with the smallest values at angles greater than 64° where full laser power is used. The measurements of 2000 nm beads exhibit somewhat higher CV, i.e., between about 3% and 6%. The higher values of CV are observed mostly near the angles where minima of $\beta_p^{LISST*}(\psi)$ occur.

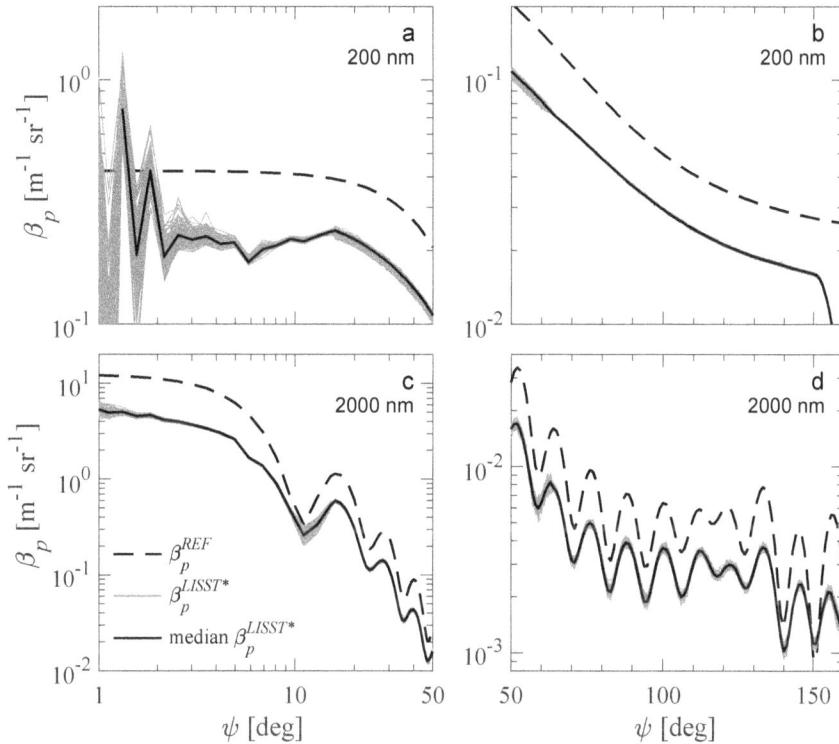

Figure 1. Measurements of the particulate volume scattering function, $\beta_p(\psi)$, at light wavelength of 532 nm for 200 nm (**a,b**) and 2000 nm (**c,d**) diameter polystyrene beads suspended in water. The left panels depict the angular range of 1–50° with logarithmic scaling, and the right panels depict the range 50–160° with linear scaling. The expected reference value, $\beta_p^{REF}(\psi)$, obtained from Mie scattering calculations is indicated as a dashed line. Quality-controlled but uncorrected measurements obtained with the LISST-VSF (gray lines, number of measurements $N = 128$) and the median value (solid black line) are shown.

The DAWN-EOS measurements for four bead sizes were also used to calculate $\beta_p(\psi)$ and $DoLP_p(\psi)$, denoted as $\beta_p^{DAWN}(\psi)$ and $DoLP_p^{DAWN}(\psi)$. First, for each time series of 1440 measurements with DAWN-EOS, the highest 2% of data was rejected, as these data are assumed to result from sample contamination with rare, larger particles. Each set of measurements then consists of 1411 measurements of both $I_{s\parallel}(\psi)$ and $I_{s\perp}(\psi)$ for a specific orientation of sample vial. These measurements were averaged to represent that orientation. Such results were then averaged for three vial orientations. This protocol was applied to both the sample and baseline measurements, with the exception that baseline values were calculated by averaging the lowest 5% of data. The final $I_{s\parallel}(\psi)$ and $I_{s\perp}(\psi)$ for the beads were calculated by subtracting the average baseline from the average sample data. These particulate $I_{s\parallel}(\psi)$ and $I_{s\perp}(\psi)$ were then used to determine $p_{11}(\psi)$ and $p_{12}(\psi)$ according to Equations (3) and (4), from which $\beta_p^{DAWN}(\psi)$ [61] and $DoLP_p^{DAWN}(\psi)$ (Equation (2)) were determined. Note that two dilutions of the master suspension for 400 and 700 nm beads were measured with DAWN-EOS and the average of the two was used to represent these bead sizes. As a final step, the determined $\beta_p^{DAWN}(\psi)$ values were rescaled using relevant dilution factors to obtain final results representing the particle concentration in LISST-VSF samples and enable direct comparisons with LISST-VSF measurements. Note that such rescaling is not necessary for $DoLP_p^{DAWN}(\psi)$.

With regard to processing of data acquired with a Lambda 18 spectrophotometer, the spectral data of measured optical density $OD(\lambda)$ (i.e., measurements made in the absorbance mode of the spectrophotometer) were converted (after subtraction of baseline measurement) into the particulate beam attenuation coefficient [m^{-1}] using the relationship $c_p(\lambda) = \ln(10)\ OD(\lambda)/0.01$, where ln is the natural logarithm and 0.01 is the path length in meters. The final particulate beam attenuation coefficient obtained from spectrophotometric measurements is denoted as c_p^{SPEC}. The estimates of c_p^{LISST*} from LISST-VSF measurements were calculated with the standard manufacturer's processing code. Because each LISST-VSF measurement consists of two linear polarization states of the incident beam, the average of these two is used as the final estimate of c_p^{LISST*}. As a final step, the determined c_p^{SPEC} values were multiplied by relevant dilution factors to obtain final results representing particle concentration in LISST-VSF samples and enable direct comparisons with LISST-VSF measurements.

2.1.4. Determination of Correction Functions

In addition to $\beta_p^{LISST*}(\psi)$, Figure 1 shows results for the 200 nm and 2000 nm polystyrene beads based on Mie scattering calculations (more details about these calculations are provided below). These results are significantly higher (nearly a factor of 2) than the measured values of $\beta_p^{LISST*}(\psi)$. We assume that the Mie scattering calculations for samples of spherical polystyrene beads are sufficiently accurate to provide reference values for such samples.

In order to correct for the mismatch between the measured and reference values, a calibration correction function $CF(\psi)$ is defined as

$$CF(\psi) = \frac{\beta_p^{REF}(\psi)}{\beta_p^{LISST*}(\psi)}, \tag{5}$$

where $\beta_p^{REF}(\psi)$ is a reference volume scattering function determined according to

$$\beta_p^{REF}(\psi) = \widetilde{\beta}_p^{Mie}(\psi) b_p^{REF}, \tag{6}$$

where $\widetilde{\beta}_p^{Mie}(\psi)$ is the scattering phase function [sr^{-1}] obtained from Mie scattering computations and b_p^{REF} is the reference particulate scattering coefficient [m^{-1}]. Note that all quantities in Equations (5) and (6) are for the LISST-VSF light wavelength of 532 nm.

For each examined suspension of standard polystyrene beads, $\widetilde{\beta}_p^{Mie}(\psi)$ was determined from Mie scattering computations for homogeneous spherical particles. We used the Mie scattering code for homogeneous spheres of Bohren and Huffman [20], which included our modifications to account for polydispersity of the sample, i.e., to use particle size distribution as input to the code rather than just a single particle diameter as in the original code. The computations were performed assuming a relative particle size distribution (PSD) of Gaussian shape, with 300 evenly spaced size bins about the nominal mean diameter ± 3 standard deviations, as provided by the manufacturer for each bead size (Table 1). The use of such PSDs allows us to account for the realistic, small degree of polydispersity of each sample. The Mie computations also require input of the refractive index of particles. Based on the study of Ma et al. [68] we assumed that the complex refractive index of polystyrene relative to water at

532 nm is $m = 1.193 + 0.0003i$, where the first component is the real part and the second component is the imaginary part of refractive index. Note that the imaginary part is very small because polystyrene is a weakly absorbing material in the examined spectral region.

Equation (6) also requires b_p^{REF}, which was determined from the combination of beam attenuation measurements and Mie scattering calculations as

$$b_p^{REF} = c_p^{LISST*} \frac{Q_b^{Mie}}{Q_c^{Mie}}, \tag{7}$$

where Q_b^{Mie} and Q_c^{Mie} are the single-particle scattering and attenuation efficiency factors, respectively, obtained from Mie computations. Because the populations of examined beads exhibit a slight degree of polydispersity, the calculated Q_b^{Mie} and Q_c^{Mie} represent the average values of efficiency factors for a given particle population [69]. Given very weak light absorption of polystyrene beads at 532 nm, the ratio $\frac{Q_b^{Mie}}{Q_c^{Mie}}$ was found to be >95%. We also note that in addition to c_p^{LISST*}, we have another potential measurement of beam attenuation coefficient from the spectrophotometer (c_p^{SPEC}). Figure 2 shows that the measurements of c_p^{LISST*} and c_p^{SPEC} are consistent, and generally agree very well.

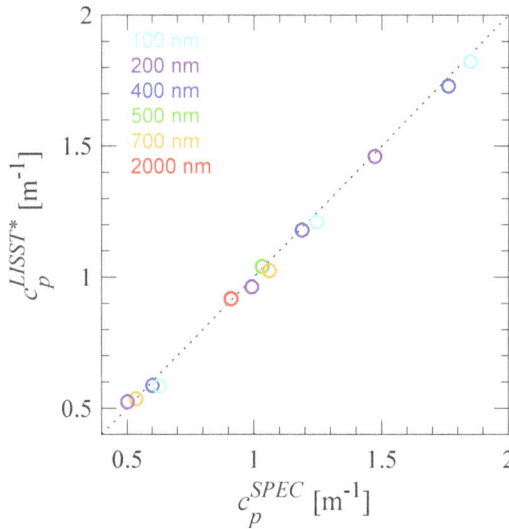

Figure 2. Comparison of measurements of the particulate beam attenuation coefficient, c_p, at 532 nm obtained with a spectrophotometer with measurements from the LISST-VSF. The comparison is depicted for suspensions of polystyrene beads of six different diameters as indicated in the legend, and the 1:1 line is plotted for reference (dotted black line). Appropriate dilution factors have been applied to account for the different particle concentrations used in measurements with each instrument. The presented values correspond to samples measured with the LISST-VSF.

By combining Equations (6) and (7), $\beta_p^{REF}(\psi)$ can be determined for each LISST-VSF measurement as

$$\beta_p^{REF}(\psi) = \tilde{\beta}_p^{Mie}(\psi) \, c_p^{LISST*} \frac{Q_b^{Mie}}{Q_c^{Mie}}. \tag{8}$$

Note that the estimates of $\beta_p^{REF}(\psi)$ can vary between individual measurements because of variations in c_p^{LISST*}. Finally, by combining Equations (5) and (8), $CF(\psi)$ can be determined for each LISST-VSF measurement as

$$CF(\psi) = \frac{\widetilde{\beta}_p^{Mie}(\psi)\, c_p^{LISST*}\, \frac{Q_b^{Mie}}{Q_c^{Mie}}}{\beta_p^{LISST*}(\psi)}. \tag{9}$$

The application of this protocol to every individual measurement of $\beta_p^{LISST*}(\psi)$ helps to better capture the variability between individual measurements during a given series of LISST-VSF measurements on a given sample, for example due to imperfect mixing in the 2 L sample chamber or potential electronic fluctuations in the instrument.

The smaller-sized particle standards (<500 nm in diameter) appear as the best candidates for determination of $CF(\psi)$ because they produce a relatively featureless pattern of angular scattering (see the results for 200 nm beads in Figure 1). The measurements with larger beads (500 nm to 2000 nm) were not used in these determinations because the angular scattering pattern includes multiple maxima and minima (see the results for 2000 nm beads in Figure 1), which render the comparison of $\beta_p^{LISST*}(\psi)$ and $\beta_p^{REF}(\psi)$ particularly sensitive to even small uncertainties in measurements or theoretical calculations. The results obtained with beads of 100, 200, and 400 nm in diameter were considered in the determinations of final correction function $CF_f(\psi)$ within four angular ranges, as described below:

$$CF_f(\psi) = (CF_{100}(\psi) + CF_{200}(\psi) + CF_{400}(\psi))/3 \text{ for } \psi = 0.09\text{--}60° \tag{10a}$$

$$CF_f(\psi) = CF_{200}(\psi) \text{ for } \psi = 61\text{--}128° \tag{10b}$$

$$CF_f(\psi) = (CF_{200}(\psi) + CF_{400}(\psi))/2 \text{ for } \psi = 129\text{--}150° \tag{10c}$$

$$CF_f(\psi) = (CF_{100}(\psi) + CF_{200}(\psi) + CF_{400}(\psi))/3 \text{ for } \psi = 151\text{--}155°. \tag{10d}$$

The $CF_{200}(\psi)$ data obtained with 200 nm beads provide the main contribution to the determination of $CF_f(\psi)$. The $CF_{100}(\psi)$ data obtained with 100 nm beads are used partially because of increased uncertainty in the PSD of these beads (*CV* of nominal mean diameter is 7.8%, see Table 1). The $CF_{400}(\psi)$ data obtained with 400 nm beads are also used partially and cover the backscattering angles, where these particular data are useful for correction of an apparent artifact near 130–140°, which is rather minor but has been consistently observed with our LISST-VSF instrument for various natural particle assemblages. The calculations of $CF_{100}(\psi)$, $CF_{200}(\psi)$, and $CF_{400}(\psi)$ were made using data for particle concentrations and PMT gains which ensured sufficient signal for the ring detectors while avoiding PMT saturation of the Roving Eyeball sensor (see dilution factors in italic font in Table 1). For a given bead standard, the final values of correction function at different angles were determined as the median values of all the relevant determinations.

The final $CF_f(\psi)$ was smoothed in the angular range 2.56–155° with a 3-point and then a 5-point moving average. In addition, $CF_f(\psi)$ within the near-forward angular range 0.09–4.96° was set to a constant value of $CF_f(\psi_{32})$, where $\psi_{32} = 15.17°$ corresponds to the last ring detector. The rationale for this assumption is that the scattering signal produced by the examined beads for the first 25 rings ($\psi = 0.09\text{--}4.96°$) is comparable to the baseline, while there is good signal relative to the baseline for the last ring detector.

The final correction simply involves the multiplication of uncorrected $\beta_p^{LISST*}(\psi)$ by the correction function $CF_f(\psi)$,

$$\beta_p^{LISST}(\psi) = \beta_p^{LISST*}(\psi)\, CF_f(\psi), \tag{11}$$

where $\beta_p^{LISST}(\psi)$ is the corrected LISST-VSF measurement of volume scattering function (note that the superscript * is removed from this symbol).

We also determined a correction function for $DoLP_p^{LISST*}(\psi)$,

$$BF(\psi) = DoLP_p^{LISST*}(\psi) - DoLP_p^{REF}(\psi), \tag{12}$$

where $BF(\psi)$ quantifies a correction for potential bias in $DoLP_p^{LISST*}(\psi)$ obtained from the standard processing code applied to LISST-VSF measurements and $DoLP_p^{REF}(\psi)$ is a reference degree of linear polarization determined from Mie scattering calculations of the two scattering matrix elements, $p_{11}^{Mie}(\psi)$ and $p_{12}^{Mie}(\psi)$, for a given sample of standard beads. The results for $BF_{100}(\psi)$, $BF_{200}(\psi)$, $BF_{400}(\psi)$, and the final correction function $BF_f(\psi)$ were obtained using a procedure similar to that for $CF_{100}(\psi)$, $CF_{200}(\psi)$, $CF_{400}(\psi)$, and $CF_f(\psi)$. The correction of $DoLP_p^{LISST*}(\psi)$ simply requires a subtraction of $BF_f(\psi)$,

$$DoLP_p^{LISST}(\psi) = DoLP_p^{LISST*}(\psi) - BF_f(\psi), \tag{13}$$

where $DoLP_p^{LISST}(\psi)$ is the corrected degree of linear polarization within the range of scattering angles from $16°$ to $150°$. Because the $DoLP_p^{LISST*}(\psi)$ data output from standard processing of LISST-VSF measurements begins at $\psi = 16°$, no correction for the forward scattering angles of the ring detectors ($\psi < 16°$) was determined.

2.2. Measurements and Analysis of Natural Seawater Samples

Optical measurements with the LISST-VSF and ancillary analyses of natural particle assemblages were performed on seawater samples collected between summer 2016 and spring 2017 in contrasting marine environments, namely, in open ocean waters off the coast of Southern California, nearshore ocean waters at the pier of the Scripps Institution of Oceanography (SIO Pier) in La Jolla, and the tidal estuary of the San Diego River. Overall 17 samples representing a broad range of natural particle assemblages were analyzed. Most samples (number of samples $N = 11$) were collected at the SIO Pier. These samples were collected during typical dry weather conditions, phytoplankton bloom events, and after heavy rain. The tidal estuary samples ($N = 3$) include three tidal states between low and high tide. The offshore samples ($N = 3$) were collected in the Santa Barbara Channel, about 8 km off San Diego Bay, and about 2 km off SIO Pier. Seawater samples were collected just beneath the sea surface using either Niskin bottles or a bucket, except for one offshore sample (off San Diego Bay) that was collected at the subsurface chlorophyll-*a* maximum at a depth of 18 m. All samples were analyzed in the laboratory within 24 h of sampling.

To characterize the concentration and composition of particulate matter for each sample, we determined the dry mass concentration of total suspended particulate matter, SPM [g m^{-3}], mass concentration of particulate organic carbon, POC [mg m^{-3}], and mass concentration of the pigment chlorophyll-*a*, Chla [mg m^{-3}]. For these determinations, the particles were collected on glass-fiber filters (GF/F Whatman) by filtration of appropriate volumes of seawater (150–2100 mL depending on the sample). SPM was determined following a gravimetric method using pre-washed and pre-weighted filters [7,70]. The determinations of POC were made on precombusted filters with a standard CHN analysis involving high temperature combustion of sample filters [7,71,72]. Chla was determined spectrophotometrically using a Lambda 18 spectrophotometer and placing 1-cm cuvettes containing acetone extracts of the samples inside the integrating sphere. The measured absorbance values at 630, 647, 665, and 691 nm (after subtraction of acetone baseline values) were used in the calculation of Chla [73]. For each seawater sample, replicate determinations of SPM and POC were made on separate sample filters. The final SPM and POC are average values of replicate determinations. The replicates for SPM and POC agreed generally to within 15% and 10%, respectively. No replicates were taken for Chla. In addition to information about particle concentration, SPM, POC, and Chla provide useful proxies of bulk composition of particulate matter. The organic and inorganic fractions of SPM can be characterized using the ratio POC/SPM, and the contribution of phytoplankton to SPM using Chla/SPM [65]. These ratios are expressed on a [g/g] basis.

The measurements of particle size distribution (PSD) were made with a Coulter Multisizer 3 (Beckman Coulter, Brea, CA, USA) equipped with a 100 μm aperture, which allows particle counting and sizing in the range of volume-equivalent spherical diameter from 2 μm to 60 μm. Within this size range we used 300 log-spaced size bins to provide high resolution PSDs. For each experiment, 0.2 μm filtered seawater was used as a blank that was subtracted from sample measurements.

Approximately 10 to 15 replicate measurements of 2 mL subsamples of each seawater sample were collected. After removing outliers, the remaining measurements were summed and divided by the total analyzed volume to produce an average density function of PSD in particle number per unit volume per width of size bin. For each sample the power function fit with a slope parameter, ζ, was determined using these PSD data over the size range 2–50 µm. In these determinations, the linear regression analysis was applied to log-transformed data, and the last size bins with very low particle counts were ignored. Although the measured PSDs often showed significant deviations from the power function fits, we use the slope parameter ζ as a particle size metric, because this is the most common parameterization of size distribution of marine particles [23,74]. Additionally, assuming spherical particles, the particle volume distributions were determined from particle number distributions for each sample. From particle volume distributions, we calculated the percentile-based particle diameters such as the median diameter, D_V^{50}, and the 90th percentile diameter, D_V^{90}. These parameters have been shown to provide potentially useful metrics in the analysis of relationships between the optical and particle size properties in seawater [65].

Measurements and processing of data collected with LISST-VSF for natural seawater samples were made following a protocol similar to that described above for standard polystyrene bead samples. For each experiment, baseline measurements were taken on 0.2 µm filtered seawater obtained from a given seawater sample. However, a single baseline selected from the lowest measured baselines was used for data processing of all seawater samples to ensure a consistent baseline unaffected by possible variations associated with the imperfect purity of 0.2 µm filtered seawater prepared during different experiments. To ensure scattering measurements were acquired in a single-scattering regime, samples with an average c_p over 3.0 m^{-1} were diluted using 0.2 µm filtered seawater. Dilution was necessary only for the two most turbid samples collected in the San Diego River Estuary. Between four and eight sets of 50 measurements were collected for each seawater sample with gentle hand mixing between the measurement sets, while a magnetic stir bar was on very low speed changing direction of rotation every 30 s. All results from LISST-VSF measurements for natural seawater samples shown in this paper represent the $CF_f(\psi)$-corrected volume scattering function of particles, $\beta_p^{LISST}(\psi)$, and $BF_f(\psi)$-corrected degree of linear polarization of particles, $DoLP_p^{LISST}(\psi)$. For a given sample the final values of $\beta_p^{LISST}(\psi)$ and $DoLP_p^{LISST}(\psi)$ correspond to the median values of the series of measurements that passed the quality control criteria.

To determine the particulate scattering, b_p^{LISST}, and particulate backscattering, b_{bp}^{LISST}, coefficients, the corrected measured $\beta_p^{LISST}(\psi)$ was first extrapolated in the angular range 150–180°. The extrapolated portion of $\beta_p^{LISST}(\psi)$ was obtained by fitting a specific function to the data of $\beta_p^{LISST}(\psi)$ in the angular range 90–150°. We used two methods for fitting and extrapolating $\beta_p^{LISST}(\psi)$. The first method is based on a non-linear least squares best fit of the analytical function proposed by Beardsley and Zaneveld [75]. The second method is based on a linear mixing model that finds a non-negative least squares best fit for combined contributions of four end members representing shapes of volume scattering functions associated with scattering by small and large particles, as described in Zhang et al. [76].

A backscattering factor, κ, was determined for the fitted volume scattering function as

$$\kappa = \frac{b_{bp}^{fit}}{b_{bp,150}^{fit}}, \tag{14}$$

where b_{bp}^{fit} is the particulate backscattering coefficient determined by the integration of the fitted function in the angular range 90–180° and $b_{bp,150}^{fit}$ is the coefficient determined by the integration of the fitted function in the range 90–150°. The final estimate of backscattering coefficient, b_{bp}^{LISST}, was calculated as

$$b_{bp}^{LISST} = \kappa\, b_{bp,150}^{LISST}, \tag{15}$$

where $b_{bp,150}^{LISST}$ is obtained by the integration of $\beta_p^{LISST}(\psi)$ in the angular range 90–150°. The final estimate of scattering coefficient, b_p^{LISST}, was calculated as the sum of b_{bp}^{LISST} and the forward scattering coefficient obtained from the integration of $\beta_p^{LISST}(\psi)$ in the angular range 0.09–90°.

The calculations of b_p^{LISST} and b_{bp}^{LISST} were made for each seawater sample using the two methods for fitting and extrapolation. The particulate backscattering ratio, $\tilde{b}_{bp}^{LISST} = b_{bp}^{LISST}/b_p^{LISST}$ was also calculated. We note that the κ values for all examined seawater samples were found to range between 1.125 and 1.138 and 1.118–1.120 for the Beardsley and Zaneveld [75] and Zhang et al. [76] methods, respectively. An example illustration of fitting and extrapolation methods for one sample collected during high tide at the San Diego River estuary is depicted in Figure 3. As seen, both the Beardsley and Zaneveld [75] and Zhang et al. [76] fitted functions are in good agreement with the measured data of $\beta_p^{LISST}(\psi)$ in the angular range 90–150°. However, the extrapolated portion of the Beardsley and Zaneveld [75] function in the angular range 150–180° has somewhat higher values compared with the Zhang et al. [76] function. Nevertheless, the estimates of b_{bp}^{LISST} for this sample obtained from the two extrapolation methods differ only by 0.5%. For all other seawater samples the difference was also small, not exceeding 1.5%. The final results of b_p^{LISST} and b_{bp}^{LISST} for seawater samples presented in this study are based on the Zhang et al. [76] method.

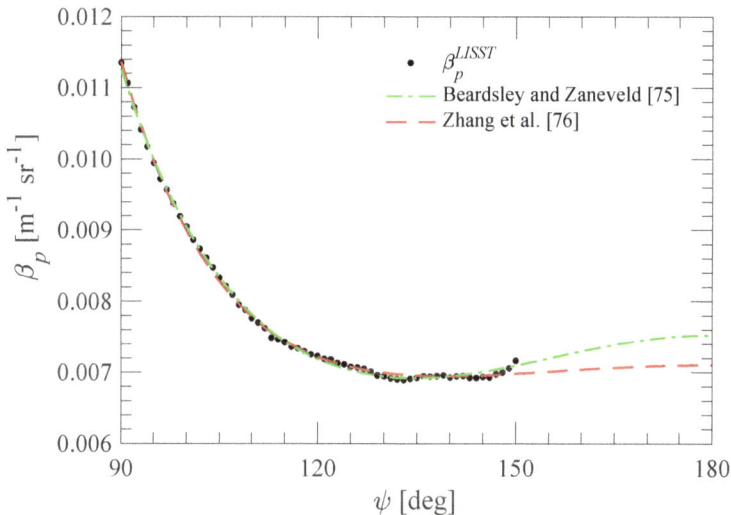

Figure 3. Measured values of the particulate volume scattering function $\beta_p(\psi)$ obtained with the LISST-VSF after correction (circles) for scattering angles 90–150° and illustration of the results of two model relationships (Beardsley and Zaneveld [75], Zhang et al. [76]) fitted to the data. The illustrated example measurement was made on a natural sample collected from the San Diego River estuary.

3. Results and Discussion

3.1. Correction Functions for LISST-VSF

The results for $CF_{100}(\psi)$, $CF_{200}(\psi)$, $CF_{400}(\psi)$, and $CF_f(\psi)$ are plotted in Figure 4. The final correction function $CF_f(\psi)$ indicates that $\beta_p^{LISST*}(\psi)$ is lower than $\beta_p^{REF}(\psi)$ by a factor of about 2, and also exhibits some angular variability. One consistent feature in the forward scattering region, which is independent of the bead size, is a sharp increase in $CF_f(\psi)$ with a peak at ring 26 ($\psi = 5.84°$). We observed a similar but inverse feature consistently in natural seawater samples, which suggests that the behavior of the correction function at these angles is credible. Within the angular range of data from the Roving Eyeball sensor (16–150°), the $CF_f(\psi)$ values remain generally in the range between

1.7 and 1.9. For angles larger than 150°, we did not obtain consistent results of the correction function for different bead sizes (not shown), so this angular range is omitted from our analysis of LISST-VSF measurements. Note also that $CF_{400}(\psi)$ differs greatly from $CF_{100}(\psi)$ and $CF_{200}(\psi)$ within the angular range between about 65° and 120°. This can be attributed to the uncertainty in the determinations of $CF_{400}(\psi)$ associated with a well-pronounced minimum in the volume scattering function for the 400 nm beads in this angular range. Therefore, the $CF_{400}(\psi)$ data in this angular range were not used in the determination of final $CF_f(\psi)$.

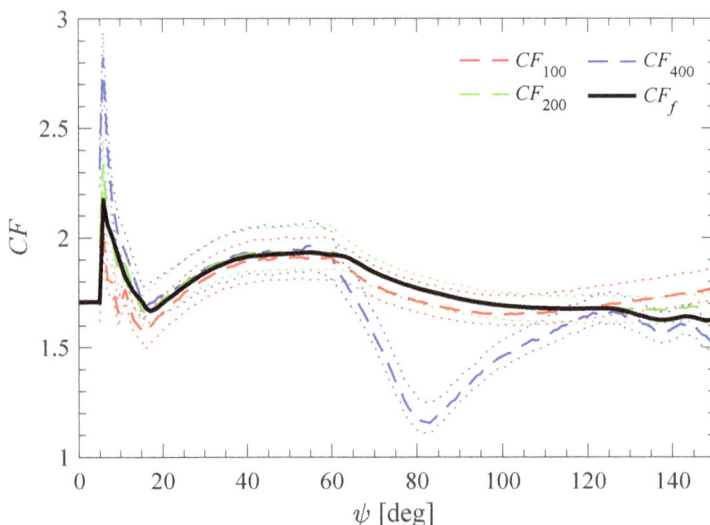

Figure 4. Correction functions, $CF(\psi)$, for the LISST-VSF measurements of particulate volume scattering function $\beta_p^{LISST*}(\psi)$ over the angular range 4.96–150° determined for 100, 200, and 400 nm polystyrene bead suspensions. For each individual bead size, dashed lines represent the median values and the dotted lines indicate the 25th and 75th percentiles determined from the series of measurements. The final computed correction function $CF_f(\psi)$ is shown in black, and includes the constant value used for the near-forward angular range from 0.09° to 4.96°.

The results for $BF_{100}(\psi)$, $BF_{200}(\psi)$, $BF_{400}(\psi)$, and $BF_f(\psi)$, are shown in Figure 5. As seen, $BF_f(\psi)$ is negative within the examined angular range and varies within a relatively narrow range of values between about -0.02 and -0.04. Similar to the results for $CF_{400}(\psi)$, the distinct feature of positive bias observed in the $BF_{400}(\psi)$ data around the scattering angle of 80° can be attributed to the uncertainty associated with a minimum in the volume scattering function for the 400 nm beads in this angular range. This portion of $BF_{400}(\psi)$ data was not used in the determination of final $BF_f(\psi)$.

The performance of the final correction function $CF_f(\psi)$ within the range of scattering angles from 0.09° to 150° was evaluated by comparing the corrected LISST-VSF measurements of volume scattering function, $\beta_p^{LISST}(\psi)$, with reference values of $\beta_p^{REF}(\psi)$ for six samples of polystyrene beads (100, 200, 400, 500, 700, and 2000 nm in diameter; see the dilution factors for these samples indicated in boldface in Table 1). The beads with diameters of 500, 700, and 2000 nm were not used in the generation of the final correction function, so they provide completely independent data for evaluating the performance of $CF_f(\psi)$. The evaluation with the data for 100, 200, and 400 nm beads is also useful because the final $CF_f(\psi)$ was determined by averaging the results obtained with multiple bead sizes and concentrations of these samples, and not from a single bead size and concentration. Results of independent measurements obtained with the DAWN-EOS on four bead suspensions (100, 200, 400, and 700 nm) are also included in the evaluation analysis for additional comparisons.

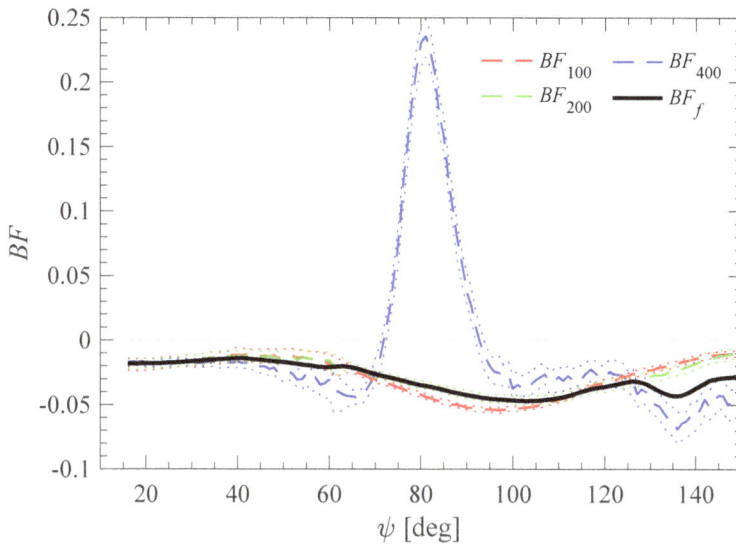

Figure 5. Correction functions, $BF(\psi)$, for LISST-VSF measurements of the degree of linear polarization of light scattered by particles, $DoLP_p^{LISST*}(\psi)$ over the angular range 16–150° determined for 100, 200, and 400 nm polystyrene bead suspensions. For each individual bead size, dashed lines represent the median values and the dotted lines indicate the 25th and 75th percentiles determined from the series of measurements. The final computed correction function $BF_f(\psi)$ is shown in black.

The comparisons of $\beta_p^{LISST}(\psi)$ and $\beta_p^{REF}(\psi)$ are shown in Figure 6 for the six polystyrene bead samples. The presented values of $\beta_p^{LISST}(\psi)$ are the median values for each angle from each measurement series. The measured values of $\beta_p^{DAWN}(\psi)$ are additionally depicted for the 100, 200, 400, and 700 nm diameter beads. In general, the magnitude and angular dependence of $\beta_p^{LISST}(\psi)$ exhibits good agreement with reference values for all bead diameters. Notable differences occur within the minima of volume scattering function, for example near the angle of 80° for the 400 nm beads (Figure 6c). This issue has been mentioned above in the context of determinations of $CF_f(\psi)$ and $BF_f(\psi)$. The agreement observed between $\beta_p^{LISST}(\psi)$ and $\beta_p^{DAWN}(\psi)$ lends additional credence to the determined correction function $CF_f(\psi)$ and its application to LISST-VSF measurements.

Figure 7a illustrates the relationship between $\beta_p^{LISST}(\psi)$ measured at all angles between 3.02° and 150° and $\beta_p^{REF}(\psi)$ for the corresponding angles for the six bead samples. The overall agreement is quite good over a range spanning nearly 4 orders of magnitude. The regions of largest disagreement correspond to angles measured with the ring detectors, as well as angles corresponding to sharp minima or maxima in volume scattering function which are observed for the larger beads. Although the measured minima and maxima occur essentially at the same angles as predicted by Mie scattering calculations, the measured magnitude of minima or maxima can differ by a few tens of percent from the reference values. This is illustrated by plots of percent differences between the measured and reference values (Figure 7b). The oscillations and peaks (both positive and negative) in percent differences correspond to the minima and maxima in the angular patterns of volume scattering function.

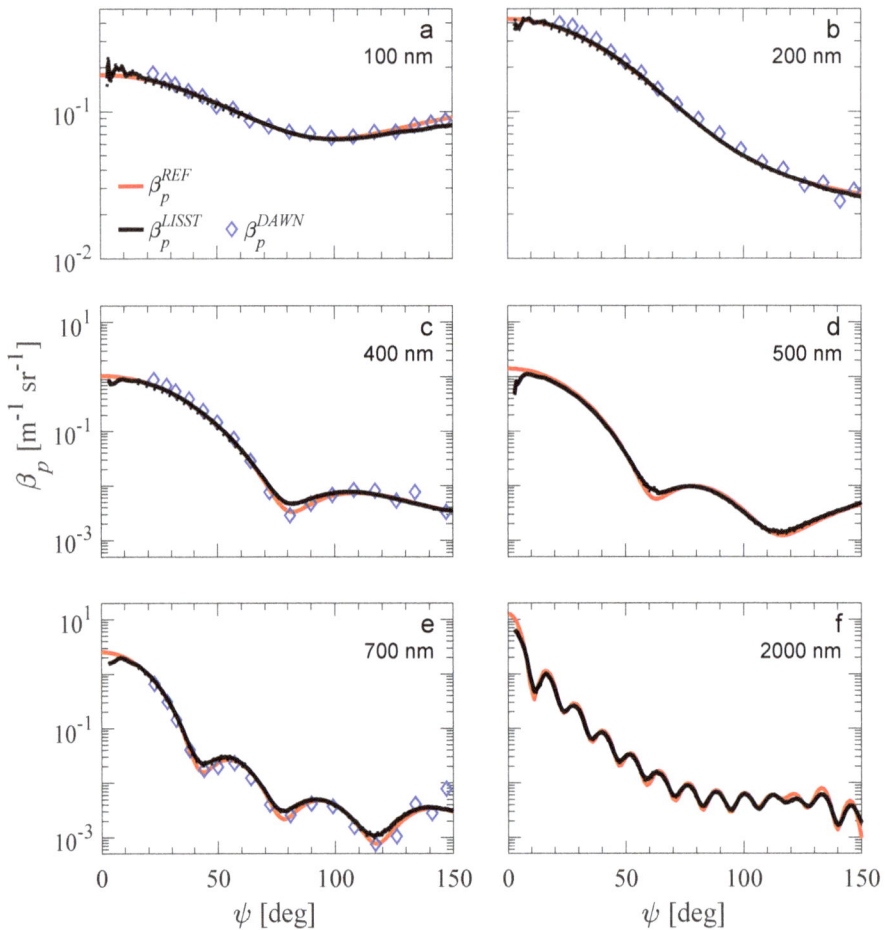

Figure 6. Comparison of $\beta_p(\psi)$ measurements on suspensions of polystyrene beads of varying diameter with reference values, $\beta_p^{REF}(\psi)$. The $\beta_p^{LISST}(\psi)$ data represent CF_f-corrected median values obtained from a series of measurements with the LISST-VSF. Independent measurements of $\beta_p(\psi)$ obtained with the DAWN-EOS instrument are also shown as diamonds in panels a, b, c and e. The bead diameters are indicated in the legend.

Table 2 includes several statistical parameters that quantify the agreement between the data of $\beta_p^{LISST}(\psi)$ and $\beta_p^{REF}(\psi)$ illustrated in Figure 7a. In this analysis we ignore $\psi < 3.02°$ due to generally low scattering signal relative to baseline for these ring detectors. The values of statistical parameters support the overall good agreement; for example, the median ratio (*MR*) of $\beta_p^{LISST}(\psi)$ to $\beta_p^{REF}(\psi)$ is very close to 1, and the median absolute percent difference (*MAPD*) between $\beta_p^{LISST}(\psi)$ and $\beta_p^{REF}(\psi)$ is only ~4%. These median values indicate no overall bias in the corrected measurements of $\beta_p^{LISST}(\psi)$ relative to the reference values of $\beta_p^{REF}(\psi)$ and small statistical differences between $\beta_p^{LISST}(\psi)$ and $\beta_p^{REF}(\psi)$. Table 2 also includes the statistical parameters for a subset of data presented in Figure 7a. In this subset, the forward scattering measurements with ring detectors were excluded, so the angular range is 16–150°. The statistical parameters for this subset are generally improved compared with the dataset covering the angular range 3.02–150°. For example, the root mean square difference (*RMSD*) is smaller (0.015 m^{-1} sr^{-1} vs. 0.21 m^{-1} sr^{-1}) and the slope of linear regression is closer to 1 (0.958 vs.

0.723). The improvements in the statistical parameters after removing the ring detector data are related primarily to much larger values of volume scattering function at forward scattering angles compared with larger angles, and a tendency to negative bias in $\beta_p^{LISST}(\psi)$ relative to $\beta_p^{REF}(\psi)$ at forward angles.

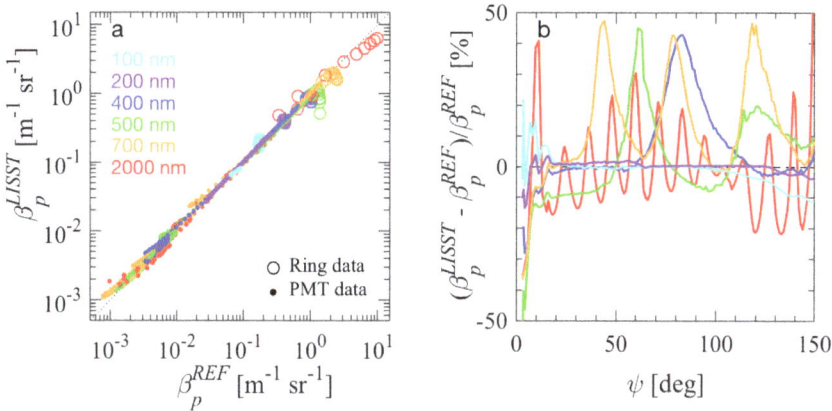

Figure 7. (a) Scatter plot of β_p^{LISST} vs. β_p^{REF} for polystyrene beads of varying diameters as indicated. Data obtained with the ring detectors and Roving Eyeball sensor are plotted separately, and the 1:1 line is plotted for reference (dotted black line). (b) Residuals expressed as percentages between β_p^{LISST} and β_p^{REF} for each bead size as a function of scattering angle.

To further validate the correction of LISST-VSF measurements with the $CF_f(\psi)$ function, we performed comparisons for approximate scattering and backscattering coefficients, $b_{p,150}$ and $b_{bp,150}$, respectively. The approximate scattering coefficient $b_{p,150}$ was obtained by integrating the volume scattering function within the angular range from 0.09° to 150°. The approximate backscattering coefficient $b_{bp,150}$ was obtained by the integration from 90° to 150°. These calculations were made for the uncorrected measured $\beta_p^{LISST*}(\psi)$, CF_f-corrected measured $\beta_p^{LISST}(\psi)$, and reference $\beta_p^{REF}(\psi)$. We also used Mie scattering calculations to estimate the underestimation of the scattering and backscattering coefficients for the examined polystyrene beads caused by the integration of $\beta_p^{REF}(\psi)$ up to 150° as opposed to 180°. We found that the approximate scattering coefficient, $b_{p,150}^{REF}$, can be lower by as much as 7% compared with the "true" scattering coefficient b_p^{REF}. This result was observed for 100 nm beads. For backscattering the approximate coefficient $b_{bp,150}^{REF}$ was found to be lower by as much as 24% for the 500 nm beads. Although the LISST-VSF measurements extend to 150° rather than 180°, the approximate coefficients are still useful for our validation exercise because most of the angular range and magnitude of total scattering and backscattering coefficients are included in the integration up to 150°. In addition, this validation analysis includes all 20 experiments conducted in this study, and not just the six example experiments presented in Figures 6 and 7.

Figure 8 compares the reference values of $b_{p,150}^{REF}$ and $b_{bp,150}^{REF}$ with LISST-VSF values determined from uncorrected $\beta_p^{LISST*}(\psi)$ and CF_f-corrected $\beta_p^{LISST}(\psi)$. In these calculations we used the median values of $\beta_p^{LISST*}(\psi)$ and $\beta_p^{LISST}(\psi)$ for each measurement series from all 20 experimental combinations of bead sizes, concentrations, and PMT gains listed in Table 1. For all experiments, the approximate coefficients, $b_{p,150}^{LISST*}$ and $b_{bp,150}^{LISST*}$, derived from uncorrected $\beta_p^{LISST*}(\psi)$ are nearly half of the reference values of $b_{p,150}^{REF}$ and $b_{bp,150}^{REF}$. After $CF_f(\psi)$ correction the approximate coefficients $b_{p,150}^{LISST}$ and $b_{bp,150}^{LISST}$ exhibit a nearly 1:1 relationship with $b_{p,150}^{REF}$ and $b_{bp,150}^{REF}$. The statistical parameters that quantify the overall good agreement between $b_{p,150}^{LISST}$ and $b_{p,150}^{REF}$ and between $b_{bp,150}^{LISST}$ and $b_{bp,150}^{REF}$ are listed in Table 2.

Table 2. Statistical results evaluating the comparison of corrected data from the LISST-VSF measurements with reference values obtained from Mie scattering calculations. For β_p^{LISST}, the results are shown for the angular range 3.02–150° which includes the ring data and for the range 16–150° without the ring data. R is the Pearson correlation coefficient and the coefficients A and B are the slope and y-intercept, respectively, determined from a type II linear regression between individual pairs of X_i and Y_i values where Y_i represents measured values and X_i reference values. The mean bias (MB) was calculated as $1/N \times \sum_{i=1}^{N} (Y_i - X_i)$ and MR represents the median ratio of Y_i / X_i. The root mean squared deviation, $RMSD$, was calculated as $\sqrt{\frac{1}{N} \sum_{i=1}^{N} (Y_i - X_i)^2}$, and the median absolute percent difference, $MAPD$, was calculated as the median value of $\left| \frac{Y_i - X_i}{X_i} \right| \times 100$. N is the number of data points used in the analysis.

Data	R	A	B	MB	MR	RMSD	MAPD	N
β_p^{LISST} (w/ rings)	0.987	0.72	0.031 m^{-1} sr^{-1}	-0.028 m^{-1} sr^{-1}	1.00	0.210 m^{-1} sr^{-1}	3.94%	876
β_p^{LISST} (w/o rings)	0.998	0.96	0.002 m^{-1} sr^{-1}	-0.002 m^{-1} sr^{-1}	1.00	0.015 m^{-1} sr^{-1}	3.39%	810
$b_{p,150}^{LISST}$	0.995	1.04	-0.048 m^{-1}	-0.007 m^{-1}	1.00	0.043 m^{-1}	2.30%	20
$b_{bp,150}^{LISST}$	0.999	0.99	0.0001 m^{-1}	-0.001 m^{-1}	1.00	0.006 m^{-1}	3.70%	20
$DoLP_p^{LISST}$	0.989	0.91	0.046	0.016	0.99	0.065	5.00%	810

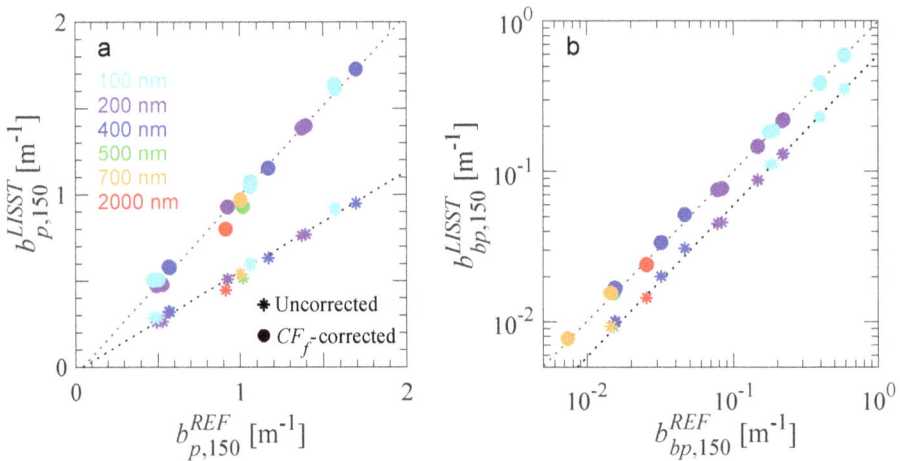

Figure 8. (a) Scatter plot comparing reference values of the particulate scattering coefficient computed over the angular range 0.09–150°, $b_{p,150}^{REF}$, with values determined from the LISST-VSF, $b_{p,150}^{LISST}$, before (asterisks) and after (circles) correction with CF_f. A type II linear regression model fit to the data is indicated by the dotted lines. (b) Similar to (a), but for the particulate backscattering coefficient computed over the range 90–150°.

Similarly to the validation analysis of $CF_f(\psi)$, the performance of the correction function $BF_f(\psi)$ was evaluated by comparing the corrected LISST-VSF measurements of the degree of linear polarization, $DoLP_p^{LISST}(\psi)$, with reference values of $DoLP_p^{REF}(\psi)$ for six samples of polystyrene beads (100, 200, 400, 500, 700, 2000 nm in diameter). Figure 9 depicts these comparisons. The values of $DoLP_p^{DAWN}(\psi)$ measured with DAWN-EOS are also depicted for the 100, 200, 400, and 700 nm beads. For all bead sizes, the magnitude and angular dependence of $DoLP_p^{LISST}(\psi)$ exhibits generally a very good agreement with both the reference values and DAWN-EOS measurements. For larger beads, notable differences occur within the minima of the degree of linear polarization, for example near the angle of 80° for the 400 nm beads (Figure 9c).

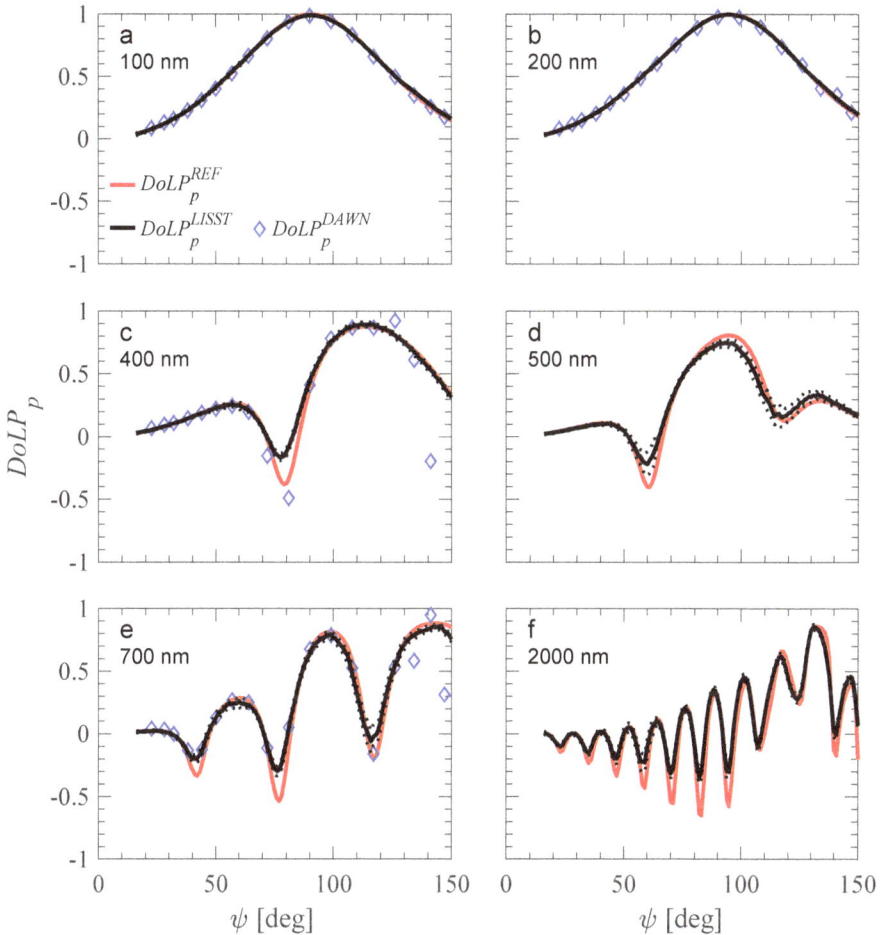

Figure 9. Similar to Figure 6, but for measured and reference values of particulate degree of linear polarization $DoLP_p$. Measurements obtained with the LISST-VSF were corrected with BF_f.

Figure 10a is a scatter plot of $DoLP_p^{LISST}(\psi)$ vs. $DoLP_p^{REF}(\psi)$ which includes all data for the six bead samples presented in Figure 9. In the region of negative values which correspond to the minima in the angular pattern of the degree of linear polarization, the $DoLP_p^{LISST}(\psi)$ exhibits a positive bias relative to $DoLP_p^{REF}(\psi)$. This bias is seen in the form of peaks in the angular pattern of the difference between $DoLP_p^{LISST}(\psi)$ and $DoLP_p^{REF}(\psi)$ for larger bead sizes (Figure 10b). The peak amplitudes generally range from 0.05 to 0.3. Importantly, however, aside from these features the data of $DoLP_p^{LISST}(\psi)$ vs. $DoLP_p^{REF}(\psi)$ are distributed close to the 1:1 line within the major part of the region of positive values (Figure 10a). This includes the region of maximum values of the degree of linear polarization of scattered light from natural seawater samples, which are observed at scattering angles near 90° or greater. The overall good agreement between BF_f-corrected measured $DoLP_p^{LISST}(\psi)$ and $DoLP_p^{REF}(\psi)$ is supported by the statistical parameters shown in Table 2 which are calculated on the basis of the entire dataset presented in Figure 10. For example, the *RMSD* and *MAPD* values are small, 0.065 and 5%, respectively. Also, despite some negative bias for data with negative values of the degree of linear polarization, the *MR* for the ratio of $DoLP_p^{LISST}(\psi)$ to $DoLP_p^{REF}(\psi)$ for the entire dataset is 0.993, indicating essentially no bias. These statistics would improve if the data within the minima in the angular pattern of the degree of linear polarization were removed from the analysis.

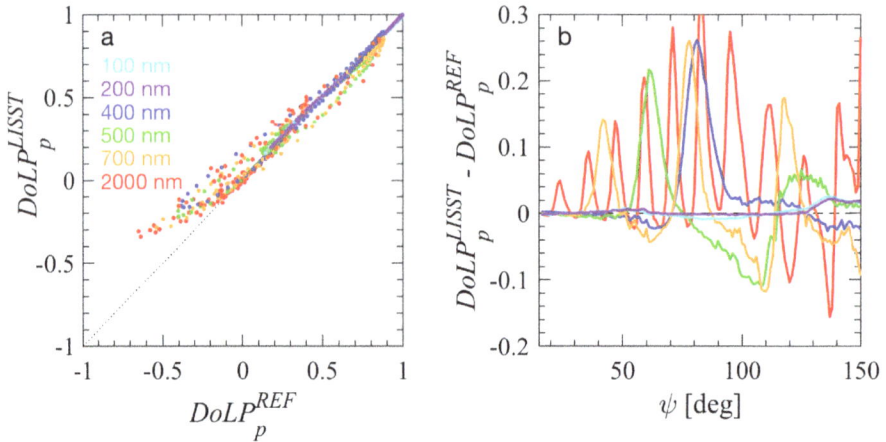

Figure 10. Similar to Figure 7, but for *DoLP*$_p$. All data are obtained with the Roving Eyeball sensor, and the residuals between *DoLP*$_p^{LISST}$ and *DoLP*$_p^{REF}$ in (**b**) are expressed as absolute differences.

3.2. Measured Light Scattering Properties of Natural Particulate Assemblages

Figure 11 depicts the corrected measured volume scattering function, $\beta_p^{LISST}(\psi)$, and the degree of linear polarization, *DoLP*$_p^{LISST}(\psi)$, for three contrasting natural assemblages of particles. The selected parameters describing the particulate and optical properties of these samples are provided in Table 3. Sample A was obtained ~8 km offshore from the subsurface chlorophyll-*a* maximum at a depth of 18 m, sample B was collected just beneath the sea surface at SIO Pier during a calm sunny summer day, and sample C was collected ~2 km inland at the San Diego River Estuary during low tide. The SPM range covers about one order of magnitude from 0.36 g m^{-3} for sample A to 3.18 g m^{-3} for sample C, which is reflected in significant range of the particulate scattering coefficient, b_p^{LISST}, from 0.36 m^{-1} to 2.23 m^{-1}. Chla was also lowest for sample A (0.75 mg m^{-3}) but highest for sample B (2.5 mg m^{-3}). Thus, whereas the offshore sample A represents a particle concentration that is within the range of observations in relatively clear open ocean waters, sample C is representative of more turbid coastal or nearshore waters [5,7]. Samples A and B have similarly high values of the ratio POC/SPM (0.43 and 0.47, respectively) and relatively high values of Chla/SPM (2.1 × 10^{-3} and 2.2 × 10^{-3}, respectively), indicating organic-dominated particulate assemblages with significant contribution of phytoplankton. In contrast, sample C has much lower values of POC/SPM (0.14) and Chla/SPM (3.8 × 10^{-4}), indicating inorganic-dominated particulate assemblage and relatively small role of phytoplankton, despite significant chlorophyll-*a* concentration (1.21 mg m^{-3}).

These differences in particle properties between the three samples are responsible for the differences in the magnitude and angular shape of $\beta_p^{LISST}(\psi)$ and *DoLP*$_p^{LISST}(\psi)$ presented in Figure 11 and the optical parameters listed in Table 3. These optical parameters include the particulate backscattering ratio, $\tilde{b}_{bp}^{LISST}$, the ratio of $\beta_p^{LISST}(45°)$ to $\beta_p^{LISST}(135°)$, and the maximum value of *DoLP*$_p^{LISST}(\psi)$ denoted as *DoLP*$_{p,max}^{LISST}$. This maximum value occurs at a scattering angle ψ_{max} which is also provided in Table 3. The offshore sample A has an intermediate value of $\tilde{b}_{bp}^{LISST}$ and the highest *DoLP*$_{p,max}^{LISST}$ of about 0.77 associated with the smallest ψ_{max} of 92°. Sample B from the SIO Pier has the lowest $\tilde{b}_{bp}^{LISST}$ of 0.008 among the three samples, suggesting a relatively steep slope of particle size distribution, relatively low bulk particle refractive index, or both [17]. This sample also shows the least steep near-forward scattering pattern (Figure 11c), which suggests a higher proportion of small particles relative to larger particles compared with the two other samples. Finally, the most turbid and least organic sample, sample C, exhibits an enhanced proportion of backscattering with the highest $\tilde{b}_{bp}^{LISST}$ of 0.027. While this result may suggest a relatively high bulk particle refractive index [17] consistent with the lowest POC/SPM ratio among the three samples, the additional influence of particle size distribution cannot be ruled out. Sample C shows steep near-forward scattering pattern (Figure 11c), which is typically indicative of an increased proportion of large particles relative to small particles. Note that sample C also has the lowest $\beta_p^{LISST}(45°)/\beta_p^{LISST}(135°)$ ratio of 12, which indicates a higher degree of symmetry in the angular pattern of scattering about $90°$, which is consistent with the relatively high value of $\tilde{b}_{bp}^{LISST}$ for this sample. In addition, sample C has the lowest $DoLP_{p,max}^{LISST}$ of 0.58.

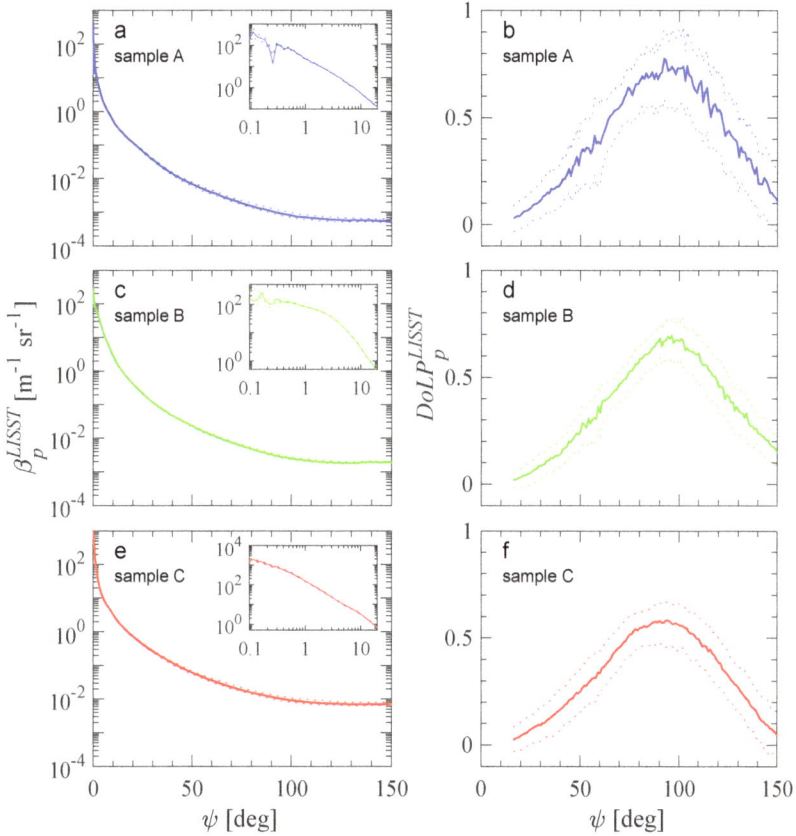

Figure 11. Measurements of β_p^{LISST} and $DoLP_p^{LISST}$ obtained with the LISST-VSF on natural seawater samples from the San Diego region representing (**a,b**) subsurface offshore waters, (**c,d**) SIO Pier, and (**e,f**) San Diego River Estuary. Solid lines represent median values while dotted lines indicate the 10th and 90th percentiles obtained from a series of measurements on each sample. Insets in (**a,c,d**) display greater detail on the near-forward scattering range.

Table 3. General information on particle characteristics and median values of optical quantities derived from LISST-VSF measurements for the three example natural seawater samples depicted in Figure 11. Values of selected optical quantities estimated from the measurements of Petzold [38] are shown for comparison.

Sample ID	Chla [mg m^{-3}]	SPM [g m^{-3}]	POC/SPM [g/g]	b_p [m-1]	$\tilde{b}_{bp}$ [dim]	$\dfrac{\beta_p(45°)}{\beta_p(135°)}$	$DoLP_{p,max}$ [dim]	ψ_{max} [deg]
A	0.75	0.36	0.43	0.36	0.012	16.5	0.77	92
B	2.49	1.13	0.47	1.75	0.008	17.1	0.69	96
C	1.21	3.18	0.14	2.23	0.022	12.0	0.58	94
Petzold Measurements								
Clear				0.03	0.015	18.0		
Coastal				0.19	0.009	17.8		
Turbid				1.74	0.020	12.2		

For comparison, Table 3 also includes the values for the selected optical parameters estimated from measurements reported by Petzold [38] for clear ocean waters (off Bahamas), coastal waters (San Diego coastal region), and turbid waters (San Diego Harbor). These measurements span a generally similar range of scattering angles as the LISST-VSF (10–180° in 5° increments), but are based on a spectrally broader incident beam (75 nm full width half maximum) centered at 514 nm. The estimates of particulate volume scattering function β_p from Petzold's measurements were obtained by subtracting pure seawater contribution β_w from the measured total β. The β_w value was calculated assuming a water temperature of 15 °C and salinity of 33 PSU [77]. Although Petzold's data include measurements made in clearer waters compared with our samples, the range of values for the dimensionless parameters associated with the shape of angular scattering pattern, $\tilde{b}_{bp}^{LISST}$ and $\beta_p^{LISST}(45°)/\beta_p^{LISST}(135°)$, is very similar. Specifically, our data for the offshore sample A are similar to Petzold's data from clear waters, sample B from the SIO Pier aligns with Petzold's data from coastal San Diego waters, and sample C from San Diego River Estuary with Petzold's data from the San Diego Harbor.

We note that the dotted lines in Figure 11 reflect some variations in $\beta_p^{LISST}(\psi)$ and $DoLP_p^{LISST}(\psi)$ between the individual measurements within a given set of measurements for each sample. For example, on the basis of the collection of 200 to 400 measurements for each of the two linear polarization states of the incident beam and the scattered light, the coefficient of variation at $\psi = 90°$ was 14%, 8%, and 13% for β_p^{LISST} and 20%, 13%, and 15% for $DoLP_p^{LISST}$ for samples A, B, and C, respectively. These variations between the individual measurements that have been taken in rapid succession do not necessarily reflect the measurement precision, as they can also be associated with actual variations in the sample, for example the fluctuations in the presence of relatively rare large particles within the interrogated sample volume. Another important point is that the small negative values of $DoLP_p^{LISST}$ observed for some individual measurements at forward scattering angles <30° (see the 10th percentile dotted lines in Figure 11b,d,f) are not necessarily an indication of measurement uncertainty because the negative values, especially in this angular range, are physically possible for certain types of particles [30,54,56,58].

Figure 12 depicts scatter plots of the relationships between the dimensionless optical parameters, $\tilde{b}_{bp}^{LISST}$ and $DoLP_{p,max}^{LISST}$, and the dimensionless particulate compositional properties, POC/SPM and Chla/SPM, for all 17 samples examined in this study. The overall range of POC/SPM in our dataset is 0.04 to 0.6. The presented data have been divided into three groups according to the values of POC/SPM as follows: the least organic-dominated (or the most mineral-dominated) data with POC/SPM < 0.15; the most organic-dominated data with POC/SPM > 0.3, and the intermediate data with $0.15 \leq$ POC/SPM ≤ 0.3. The selected boundary values of POC/SPM for discriminating between the organic-dominated and mineral-dominated groups of data differ from those used in our previous studies [65,78], but appear to adequately reflect the patterns in the present data. In particular, the most mineral-dominated samples with POC/SPM < 0.15 form a clear cluster of data points with the highest $\tilde{b}_{bp}^{LISST}$ (Figure 12a) and the lowest Chla/SPM (Figure 12b,d). We also note that no data were collected for POC/SPM between 0.15 and 0.2, so we will refer to all data with POC/SPM > 0.2 as highly organic because they all represent highly significant or dominant role of organic particles.

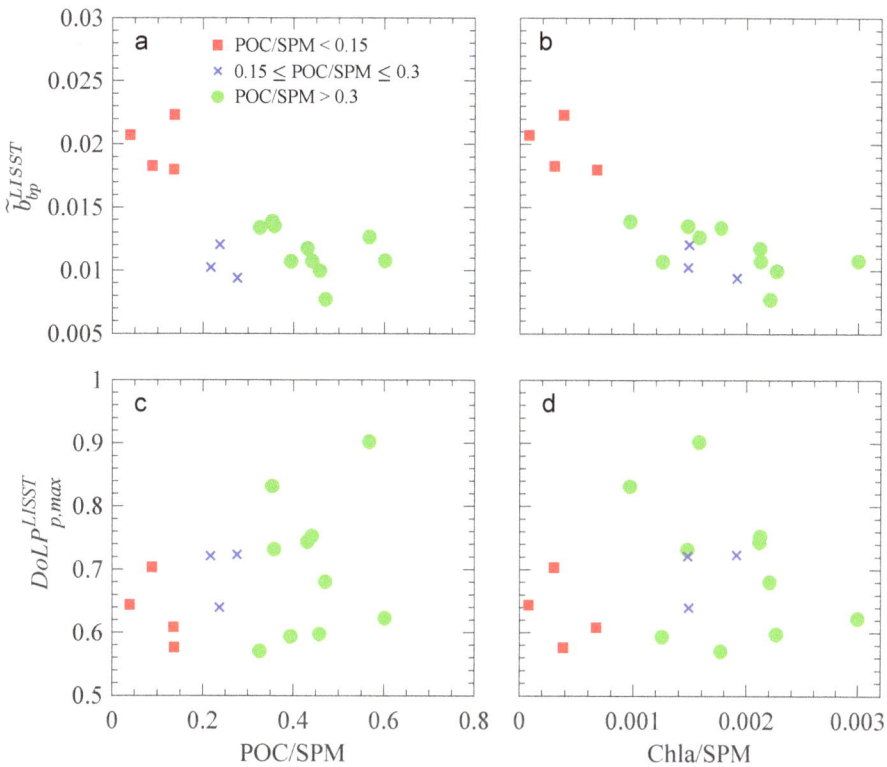

Figure 12. LISST-VSF measurements of (**a,b**) the particulate backscattering ratio, $\tilde{b}_{bp}^{LISST}$, and (**c,d**) the maximum value of the degree of linear polarization of scattered light, $DoLP_{p,max}^{LISST}$, as a function of the POC/SPM or Chla/SPM ratio. The data are divided into three groups defined by the range of POC/SPM as indicated in the legend.

The scatter plot for the data of $\tilde{b}_{bp}^{LISST}$ vs. POC/SPM suggests the presence of a relational trend with significant negative correlation between the variables (the correlation coefficient $R = -0.73$). While the $\tilde{b}_{bp}^{LISST}$ values are clearly highest for POC/SPM < 0.15, the organic-dominated samples have consistently lower $\tilde{b}_{bp}^{LISST}$. This result is consistent with the notion that mineral-dominated particulate assemblages with relatively high bulk refractive index of particles tend to have higher backscattering ratio $\tilde{b}_{bp}$ compared with organic-dominated assemblages with lower refractive index [17]. However, we also note that highly organic samples with POC/SPM > 0.2 show no clear relationship and essentially no correlation between $\tilde{b}_{bp}^{LISST}$ and POC/SPM ($R = -0.11$). This result may be attributable to the effect of other particle characteristics on $\tilde{b}_{bp}^{LISST}$, such as variations in refractive index of particles associated with changes in the composition of particulate organic matter, variations in particle size distribution, or both. The scatter plot of $\tilde{b}_{bp}^{LISST}$ vs. Chla/SPM (Figure 12b) provides interesting insight into this question, as this relationship is significantly better compared with $\tilde{b}_{bp}^{LISST}$ vs. POC/SPM. Whereas the correlation between $\tilde{b}_{bp}^{LISST}$ and Chla/SPM for all data is strong ($R = -0.85$), the subset of data for highly organic samples (POC/SPM > 0.2) has also a relatively high correlation coefficient of -0.51. This is an important result, suggesting that for particulate assemblages with high organic content, the backscattering ratio $\tilde{b}_{bp}$ tends to decrease with increasing proportion of phytoplankton in the particulate assemblage. It is likely that the relationship in Figure 12b is largely driven by a decrease in the bulk particle refractive index with increasing proportion of phytoplankton in the particulate assemblage. Because this trend also holds for the subset of highly organic samples, it may indicate that live phytoplankton cells have generally lower refractive index than non-living organic particles.

In contrast to $\tilde{b}_{bp}^{LISST}$, the $DoLP_{p,max}^{LISST}$ data show no clear relational trend and very weak correlation with POC/SPM ($R = 0.31$), indicating that the maximum degree of linear polarization does not provide a useful optical signature for the organic vs. inorganic content of particulate assemblages in our dataset (Figure 12c). A similar result with no correlation ($R = 0.07$) is observed for $DoLP_{p,max}^{LISST}$ vs. Chla/SPM, indicating that varying proportion of phytoplankton in total particulate assemblage has no discernible systematic effect on the maximum degree of linear polarization (Figure 12d). We also determined that there is no significant correlation between $DoLP_{p,max}^{LISST}$ and $\tilde{b}_{bp}^{LISST}$ in our dataset ($R = -0.22$), as well as between ψ_{max} and POC/SPM or Chla/SPM ($R = -0.09$ and 0.06, respectively). It is also of interest to note that the range of our $DoLP_{p,max}^{LISST}$ data is generally consistent with the range of values reported in literature for natural seawater samples, although the reported range in some earlier studies extends to somewhat lower values, as low as about 0.4 [15,39–42,79,80].

The assessment of potential presence of systematic effects of particle size distribution (PSD) on $\tilde{b}_{bp}^{LISST}$ and $DoLP_{p,max}^{LISST}$ is presented in Figure 13. In this assessment, we use two PSD metrics: the 90th percentile diameter, D_V^{90}, derived from the particle volume distribution, and the power function slope, ζ, derived from the particle number distribution. We also tested other percentile-based diameters such as the median D_V^{50} but no improvements in the examined relationships were observed. Figure 13a,b shows no trend in the data of $\tilde{b}_{bp}^{LISST}$ associated with variations in the particle size metrics, even though these metrics vary over a significant dynamic range. This is the case for the entire dataset as well as a subset of highly organic samples with POC/SPM > 0.2, which supports the interpretation of results presented in Figure 12b in terms of the role of refractive index. The data of $DoLP_{p,max}^{LISST}$ vs. D_V^{90} show the potential for the presence of a relational trend (Figure 13c). Although the scatter in these data points is significant and correlation is weak ($R = -0.47$), the lowest values of $DoLP_{p,max}^{LISST}$ tend to occur along with the highest values of D_V^{90}. This result indicates that the decrease in the maximum degree of linear polarization tends to be associated with particulate assemblages exhibiting a higher proportion of large-sized particles. The potential usefulness of the relationship between the degree of linear polarization and particle size has been proposed for the first time in 1930 [81], and the trend observed in our data is consistent with those early results.

4. Concluding Remarks

Our laboratory measurements combined with Mie scattering calculations for samples of standard polystyrene beads illustrate the value of such an approach for evaluating the calibration and performance of light scattering instruments. For the specific version of LISST-VSF instrument and data processing code used in our study, we determined the calibration correction functions for improved determinations of the particulate volume scattering function $\beta_p(\psi)$ and the degree of linear polarization $DoLP_p(\psi)$. The required correction was found to be particularly significant for $\beta_p(\psi)$ (a correction factor of ~1.7 to 1.9). The improved determinations of $\beta_p(\psi)$ and $DoLP_p(\psi)$ were validated with measurements on independent samples, and also using another independently-calibrated light scattering instrument, DAWN-EOS. Although the correction functions developed in this study are applicable only to the specific version of LISST-VSF instrument and the data processing code used in this study, our results emphasize a general need for evaluating the performance of light scattering instruments and minimizing the associated uncertainties in quantitative determinations from measurements.

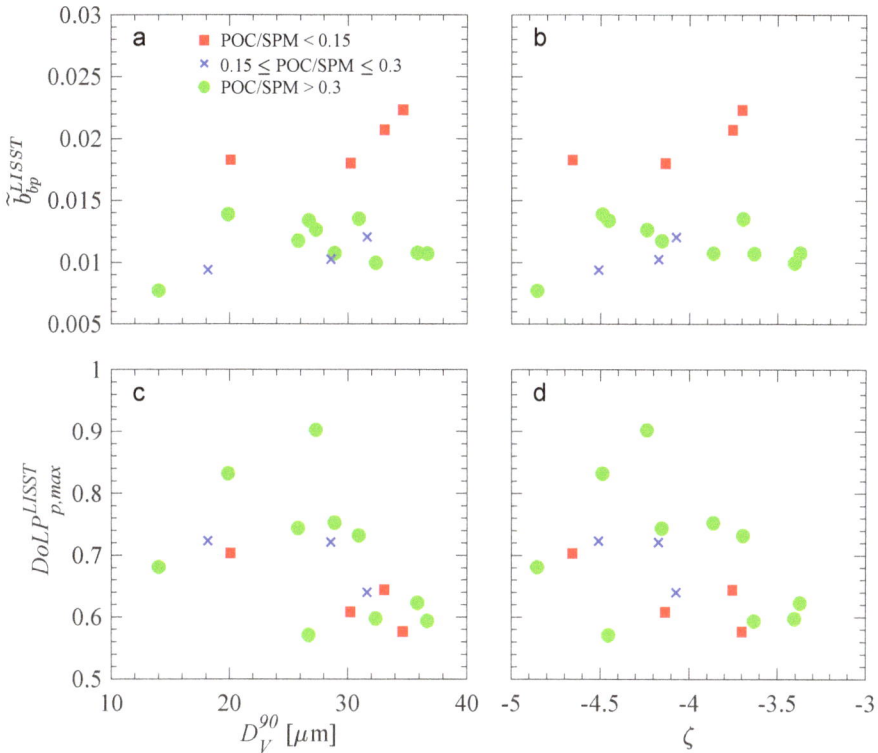

Figure 13. Similar to Figure 12, but with optical quantities shown as a function of the particle size metrics (**a,c**) D_V^{90}, representing the diameter corresponding to the 90th percentile of the particle volume distribution, and (**b,d**) ζ, the power law slope of the particle number distribution.

The improved protocol for measurements of light scattering with our LISST-VSF instrument was applied to measurements taken on 17 natural seawater samples from coastal and offshore marine environments characterized by contrasting assemblages of suspended particles. The particulate volume scattering function, degree of linear polarization, and backscattering ratio were determined from LISST-VSF measurements. For our dataset, these light scattering properties exhibit significant variations related to a broad range of measured particle properties characterizing the organic vs. inorganic composition and size distribution of particulate assemblages. For example, we observed negative relational trends between the particulate backscattering ratio and the increasing proportions of organic particles or phytoplankton in the total particulate assemblage. These proportions were parameterized in terms of the measured ratio of particulate organic carbon (POC) or chlorophyll-*a* (Chla) concentration to the total dry mass concentration of suspended particulate matter (SPM). The observed trends can be useful in the development of optical approaches for characterizing the composition of particulate assemblages. Our results also suggest a potential trend between the maximum degree of linear polarization of light scattered by particles and particle size metrics. Specifically, the decrease in the maximum degree of linear polarization observed at scattering angles close to $90°$ tends to be associated with particulate assemblages exhibiting a higher proportion of large-sized particles.

Earlier theoretical studies have shown that changes in the angular shape and the maximum value of the degree of linear polarization depend on particle refractive index and size distribution [82,83]; however, the experimental data of the degree of linear polarization of scattered light for natural marine particle assemblages are very scarce. Our results provide a contribution to filling this gap. This type of data can also be useful for improving an understanding of the polarization properties of marine light fields including polarization of water-leaving radiance and advancing related applications, including remote sensing applications [84–90]. The various potential applications of angular light scattering measurements, including the polarization effects associated with light scattering by marine particles, call for further efforts in this research area.

Author Contributions: All authors made intellectual contributions to the study design and participated in the collection of experimental data. D.K. and R.A.R. processed the data and conducted Mie scattering calculations, and D.K. analyzed the results. D.K. and D.S. wrote the paper, and all authors reviewed and provided final edits in the manuscript.

Acknowledgments: This work was supported by NASA Terrestrial Hydrology and Ocean Biology and Biogeochemistry Programs (Grant #NNX13AN72G) and the NASA Earth and Space Science Fellowship Program (Grant #NNX14AK93H). We acknowledge Edward R. Blocker, Eric Chen, and Linhai Li for assistance during measurements and processing of data. We also thank Xiaodong Zhang for providing computer code for backscattering extrapolation method and related discussions, and Wayne Slade for discussions about LISST-VSF instrument and processing code. We thank Hubert Loisel and three anonymous reviewers for valuable comments on the manuscript.

Conflicts of Interest: The authors declare no conflict of interest. The funding sponsors had no role in the design of the study; in the collection, analyses, or interpretation of data; in the writing of the manuscript, and in the decision to publish the results.

References

1. Kattawar, G.W.; Adams, C.N. Stokes vector calculations of the submarine light field in an atmosphere–ocean with scattering according to a Rayleigh phase matrix: Effect of interface refractive index on radiance and polarization. *Limnol. Oceanogr.* **1989**, *34*, 1453–1472. [CrossRef]
2. Mobley, C.D. *Light and Water: Radiative Transfer in Natural Waters*; Academic Press: San Diego, CA, USA, 1994.
3. Mobley, C.D.; Sundman, L.K.; Boss, E. Phase function effects on oceanic light fields. *Appl. Opt.* **2002**, *41*, 1035–1050. [CrossRef] [PubMed]
4. Kattawar, G.W.; Yang, P.; You, Y.; Bi, L.; Xie, Y.; Huang, X.; Hioki, S. Polarization of light in the atmosphere and ocean. In *Light Scattering Reviews 10*; Kokhanovsky, A.A., Ed.; Springer: Berlin, Germany, 2016; pp. 3–39.
5. Babin, M.; Morel, A.; Fournier-Sicre, V.; Fell, F.; Stramski, D. Light scattering properties of marine particles in coastal and open ocean waters as related to the particle mass concentration. *Limnol. Oceanogr.* **2003**, *48*, 843–859. [CrossRef]
6. Balch, W.M.; Gordon, R.G.; Bowler, B.C.; Drapeau, D.T.; Booth, E.S. Calcium carbonate measurements in the surface global ocean based on moderate-resolution imaging spectroradiometer data. *J. Geophys. Res.* **2005**, *110*, C07001. [CrossRef]
7. Stramski, D.; Reynolds, R.A.; Babin, M.; Kaczmarek, S.; Lewis, M.R.; Röttgers, R.; Sciandra, A.; Stramska, M.; Twardowski, M.S.; Franz, B.A.; et al. Relationships between the surface concentration of particulate organic carbon and optical properties in the eastern South Pacific and eastern Atlantic Oceans. *Biogeosciences* **2008**, *5*, 171–201. [CrossRef]
8. Bale, A.J.; Morris, A.W. In situ measurement of particle size in estuarine waters. *Estuar. Coast. Shelf Sci.* **1987**, *24*, 253–263. [CrossRef]
9. Agrawal, Y.C.; Pottsmith, H.C. Instruments for particle size and settling velocity observations in sediment transport. *Mar. Geol.* **2000**, *168*, 89–114. [CrossRef]
10. Agrawal, Y.C.; Whitmire, A.; Mikkelsen, O.A.; Pottsmith, H.C. Light scattering by random shaped particles and consequences on measuring suspended sediments by Laser Diffraction. *J. Geophys. Res.* **2008**, *113*, C04023. [CrossRef]
11. Reynolds, R.A.; Stramski, D.; Wright, V.M.; Woźniak, S.B. Measurements and characterization of particle size distributions in coastal waters. *J. Geophys. Res.* **2010**, *115*, C08024. [CrossRef]
12. Wyatt, P.J.; Villalpando, D.N. High-precision measurement of submicrometer particle size distributions. *Langmuir* **1997**, *13*, 3913–3914. [CrossRef]
13. Wyatt, P.J. Submicrometer particle sizing by multiangle light scattering following fractionation. *J. Colloid Interface Sci.* **1998**, *197*, 9–20. [CrossRef] [PubMed]

14. Uitz, J.; Stramski, D.; Baudoux, A.C.; Reynolds, R.A.; Wright, V.M.; Dubranna, J.; Azam, F. Variations in the optical properties of a particle suspension associated with viral infection of marine bacteria. *Limnol. Oceanogr.* **2010**, *55*, 2317–2330. [CrossRef]

15. Morel, A. Diffusion de la lumière par les eaux de mer: Resultats expérimentaux et approche théorique. In *Optics of the Sea*; North Atlantic Treaty Organization AGARD Lecture Series, No. 61; Technical Editing and Reproduction Ltd: London, UK, 1973; pp. 3.1.1–3.1.76.

16. Ackleson, S.G.; Spinrad, R.W. Size and refractive index of individual marine particulates: A flow cytometric approach. *Appl. Opt.* **1988**, *27*, 1270–1277. [CrossRef] [PubMed]

17. Twardowski, M.S.; Boss, E.; Macdonald, J.B.; Pegau, W.S.; Barnard, A.H.; Zaneveld, J.R.V. A model for estimating bulk refractive index from the optical backscattering ratio and the implications for understanding particle composition in case I and case II waters. *J. Geophys. Res.* **2001**, *106*, 14129–14142. [CrossRef]

18. Sullivan, J.M.; Twardowski, M.S.; Donaghay, P.L.; Freeman, S.A. Use of optical scattering to discriminate particle types in coastal waters. *Appl. Opt.* **2005**, *44*, 1667–1680. [CrossRef] [PubMed]

19. Zhang, X.; Huot, Y.; Gray, D.J.; Weidemann, A.; Rhea, W.J. Biogeochemical origins of particles obtained from the inversion of the volume scattering function and spectral absorption in coastal waters. *Biogeosciences* **2013**, *10*, 6029–6043. [CrossRef]

20. Bohren, C.F.; Huffman, D.R. *Absorption and Scattering of Light by Small Particles*; Wiley: New York, NY, USA, 1983.

21. Bickel, W.S.; Bailey, W.M. Stokes vectors, Mueller matrices, and polarized scattered light. *Am. J. Phys.* **1985**, *53*, 468–478. [CrossRef]

22. Kattawar, G.W. Polarization of light in the ocean. In *Ocean Optics*; Spinrad, R.W., Carder, K.L., Perry, M.J., Eds.; Oxford University Press: New York, NY, USA, 1994; p. 202.

23. Jonasz, M.; Fournier, G.R. *Light Scattering by Particles in Water: Theoretical and Experimental Foundations*; Academic Press: San Diego, CA, USA, 2007.

24. Fry, E.S.; Voss, K.J. Measurement of the Mueller matrix for phytoplankton. *Limnol. Oceanogr.* **1985**, *30*, 1322–1326. [CrossRef]

25. Quinby-Hunt, M.S.; Hunt, A.J.; Lofftus, K.; Shapiro, D. Polarized-light scattering studies of marine Chlorella. *Limnol. Oceanogr.* **1989**, *34*, 1587–1600. [CrossRef]

26. Wyatt, P.J.; Jackson, C. Discrimination of phytoplankton via light-scattering properties. *Limnol. Oceanogr.* **1989**, *34*, 96–112. [CrossRef]

27. Shapiro, D.B.; Quinby-Hunt, M.S.; Hunt, A.J. Origin of the induced circular-polarization in the light scattered from a dinoflagellate. In *Ocean Optics X*; Spinrad, R.W., Ed.; SPIE: Bellingham, WA, USA, 1990; Volume 1302, pp. 281–289.

28. Witkowski, K.; Wolinski, L.; Turzynski, Z.; Gedziorowska, D.; Zielinski, A. The investigation of kinetic growth of Chlorella vulgaris cells by the method of integral and dynamic light-scattering. *Limnol. Oceanogr.* **1993**, *38*, 1365–1372. [CrossRef]

29. Volten, H.; De Haan, J.F.; Hovenier, J.W.; Schreurs, R.; Vassen, W.; Dekker, A.G.; Hoogenboom, H.J.; Charlton, F.; Wouts, R. Laboratory measurements of angular distributions of light scattered by phytoplankton and silt. *Limnol. Oceanogr.* **1998**, *43*, 1180–1197. [CrossRef]

30. Volten, H.; Muñoz, O.; Rol, E.; Haan, J.D.; Vassen, W.; Hovenier, J.W.; Nousiainen, T. Scattering matrices of mineral aerosol particles at 441.6 nm and 632.8 nm. *J. Geophys. Res.* **2001**, *106*, 17375–17401. [CrossRef]

31. Volten, H.; Muñoz, O.; Hovenier, J.W.; Waters, L.B.F.M. An update of the Amsterdam light scattering database. *J. Quant. Spectrosc. Radiat. Transf.* **2006**, *100*, 437–443. [CrossRef]

32. Svensen, Ø.; Stamnes, J.J.; Kildemo, M.; Aas, L.M.S.; Erga, S.R.; Frette, Ø. Mueller matrix measurements of algae with different shape and size distributions. *Appl. Opt.* **2011**, *50*, 5149–5157. [CrossRef] [PubMed]

33. Muñoz, O.; Moreno, F.; Guirado, D.; Dabrowska, D.D.; Volten, H.; Hovenier, J.W. The Amsterdam–Granada light scattering database. *J. Quant. Spectrosc. Radiat. Transf.* **2012**, *113*, 565–574. [CrossRef]

34. Liu, J.P.; Kattawar, G.W. Detection of dinoflagellates by the light scattering properties of the chiral structure of their chromosomes. *J. Quant. Spectrosc. Radiat. Transf.* **2013**, *131*, 24–33. [CrossRef]

35. Stramski, D.; Boss, E.; Bogucki, D.; Voss, K.J. The role of seawater constituents in light backscattering in the ocean. *Prog. Oceanogr.* **2004**, *61*, 27–56. [CrossRef]

36. Hovenier, J.W.; Volten, H.; Muñoz, O.; Van der Zande, W.J.; Waters, L.B.F.M. Laboratory studies of scattering matrices for randomly oriented particles: Potentials, problems, and perspectives. *J. Quant. Spectrosc. Radiat. Transf.* **2002**, *79*, 741–755. [CrossRef]

37. Kokhanovsky, A.A. Parameterization of the Mueller matrix of oceanic waters. *J. Geophys. Res.* **2003**, *108*, 3175. [CrossRef]

38. Petzold, T.J. *Volume Scattering Functions for Selected Ocean Waters*; SIO Ref. 72–78, Scripps Institution of Oceanography Visibility Lab; University of California: San Diego, CA, USA, 1972.

39. Beardsley, G.F. Mueller scattering matrix of sea water. *J. Opt. Soc. Am.* **1968**, *58*, 52–57. [CrossRef]

40. Kadyshevich, Y.A.; Lyubovtseva, Y.S.; Rozenberg, G.V. Light-scattering matrices of Pacific and Atlantic ocean waters. *Izv. Acad. Sci. USSR Atmos. Ocean. Phys.* **1976**, *12*, 106–111.

41. Kadyshevich, Y.A. Light-scattering matrices of inshore waters of the Baltic Sea. *Izv. Acad. Sci. USSR Atmos. Ocean. Phys.* **1977**, *13*, 77–78.

42. Voss, K.J.; Fry, E.S. Measurement of the Mueller matrix for ocean water. *Appl. Opt.* **1984**, *23*, 4427–4439. [CrossRef] [PubMed]

43. Lee, M.; Lewis, M. A new method for the measurement of the optical volume scattering function in the upper ocean. *J. Atmos. Ocean. Technol.* **2003**, *20*, 563–572. [CrossRef]

44. Sullivan, J.M.; Twardowski, M.S. Angular shape of the oceanic particulate volume scattering function in the backward direction. *Appl. Opt.* **2009**, *48*, 6811–6819. [CrossRef] [PubMed]

45. Tan, H.; Doerffer, R.; Oishi, T.; Tanaka, A. A new approach to measure the volume scattering function. *Opt. Express* **2013**, *21*, 18697–18711. [CrossRef] [PubMed]

46. Chami, M.; Thirouard, A.; Harmel, T. POLVSM (Polarized Volume Scattering Meter) instrument: An innovative device to measure the directional and polarized scattering properties of hydrosols. *Opt. Express* **2014**, *22*, 26403–26428. [CrossRef] [PubMed]

47. Zhang, X.; Lewis, M.; Lee, M.; Johnson, B.; Korotaev, G. The volume scattering function of natural bubble populations. *Limnol. Oceanogr.* **2002**, *47*, 1273–1282. [CrossRef]

48. Twardowski, M.S.; Zhang, X.; Vagle, S.; Sullivan, J.; Freeman, S.; Czerski, H.; You, Y.; Bi, L.; Kattawar, G. The optical volume scattering function in a surf zone inverted to derive sediment and bubble particle subpopulations. *J. Geophys. Res.* **2012**, *117*, C00H17. [CrossRef]

49. Slade, W.H.; Agrawal, Y.C.; Mikkelsen, O.A. Comparison of measured and theoretical scattering and polarization properties of narrow size range irregular sediment particles. In *Oceans San Diego*; IEEE: San Diego, CA, USA, 2013; pp. 1–6.

50. Slade, W.H.; Boss, E.S. Calibrated near-forward volume scattering function obtained from the LISST particle sizer. *Opt. Express* **2006**, *14*, 3602–3615. [CrossRef] [PubMed]

51. Van de Hulst, H.C. *Light Scattering by Small Particles*; Dover Publications: New York, NY, USA, 1981.

52. McCartney, E.J. *Optics of the Atmosphere: Scattering by Molecules and Particles*; Wiley: New York, NY, USA, 1976.

53. Mishchenko, M.I.; Travis, L.D. Light scattering by polydisperse, rotationally symmetric nonspherical particles: Linear polarization. *J. Quant. Spectrosc. Radiat. Transf.* **1994**, *51*, 759–778. [CrossRef]

54. Yanamandra-Fisher, P.A.; Hanner, M.S. Optical properties of nonspherical particles of size comparable to the wavelength of light: Application to comet dust. *Icarus* **1999**, *138*, 107–128. [CrossRef]

55. Petrova, E.V.; Jockers, K.; Kiselev, N.N. Light scattering by aggregates with sizes comparable to the wavelength: An application to cometary dust. *Icarus* **2000**, *148*, 526–536. [CrossRef]

56. Muñoz, O.; Volten, H.; Hovenier, J.W.; Min, M.; Shkuratov, Y.G.; Jalava, J.P.; van der Zande, W.J.; Waters, L.B.F.M. Experimental and computational study of light scattering by irregular particles with extreme refractive indices: Hematite and rutile. *Astron. Astrophys.* **2006**, *446*, 525–535. [CrossRef]

57. Muinonen, K.; Zubko, E.; Tyynelä, J.; Shkuratov, Y.G.; Videen, G. Light scattering by Gaussian random particles with discrete-dipole approximation. *J. Quant. Spectrosc. Radiat. Transf.* **2007**, *106*, 360–377. [CrossRef]

58. Zubko, E. Light scattering by irregularly shaped particles with sizes comparable to the wavelength. In *Light Scattering Reviews 6*; Kokhanovsky, A.A., Ed.; Springer: Berlin, Germany, 2012; pp. 39–74.

59. Agrawal, Y.C. The optical volume scattering function: Temporal and vertical variability in the water column off the New Jersey coast. *Limnol. Oceanogr.* **2005**, *50*, 1787–1794. [CrossRef]

60. Agrawal, Y.C.; Mikkelsen, O.A. Empirical forward scattering phase functions from 0.08 to 16 deg. for randomly shaped terrigenous 1–21 µm sediment grains. *Opt. Express* **2009**, *17*, 8805–8814. [CrossRef] [PubMed]

61. Babin, M.; Stramski, D.; Reynolds, R.A.; Wright, V.M.; Leymarie, E. Determination of the volume scattering function of aqueous particle suspensions with a laboratory multi-angle light scattering instrument. *Appl. Opt.* **2012**, *51*, 3853–3873. [CrossRef] [PubMed]

62. Heller, W.; Tabibian, R.M. Experimental investigations on the light scattering of colloidal spheres. II. Sources of error in turbidity measurements. *J. Colloid Sci.* **1957**, *12*, 25–39. [CrossRef]

63. Bateman, J.B.; Weneck, E.J.; Eshler, D.C. Determination of particle size and concentration from spectrophotometric transmission. *J. Colloid Sci.* **1959**, *14*, 308–329. [CrossRef]

64. Stramski, D.; Babin, M.; Woźniak, S.B. Variations in the optical properties of terrigenous mineral-rich particulate matter suspended in seawater. *Limnol. Oceanogr.* **2007**, *52*, 2418–2433. [CrossRef]

65. Woźniak, S.B.; Stramski, D.; Stramska, M.; Reynolds, R.A.; Wright, V.M.; Miksic, E.Y.; Cieplak, A.M. Optical variability of seawater in relation to particle concentration, composition, and size distribution in the nearshore marine environment at Imperial Beach, California. *J. Geophys. Res.* **2010**, *115*, C08027. [CrossRef]

66. Bohren, C.F. Multiple scattering of light and some of its observable consequences. *Am. J. Phys.* **1987**, *55*, 524–533. [CrossRef]

67. *LISST-VSF Multi-Angle Polarized Light Scattering Meter: User'S Manual Revision A*; Sequoia Scientific: Bellevue, WA, USA.

68. Ma, X.; Lu, J.Q.; Brock, R.S.; Jacobs, K.M.; Yang, P.; Hu, X.H. Determination of complex refractive index of polystyrene microspheres from 370 to 1610 nm. *Phys. Med. Biol.* **2003**, *48*, 4165–4172. [CrossRef] [PubMed]

69. Morel, A.; Bricaud, A. Inherent optical properties of algal cells including picoplankton: Theoretical and experimental results. *Can. Bull. Fish. Aquat. Sci.* **1986**, *214*, 521–559.

70. Van der Linde, D.W. Protocol for determination of total suspended matter in oceans and coastal zones. *JRC Tech. Note I* **1998**, *98*, 182.

71. Parsons, T.R.; Maita, Y.; Lalli, C.M. *A Manual of Chemical and Biological Methods for Seawater Analysis*; Elsevier: New York, NY, USA, 1984.

72. Knap, A.; Michaels, A.; Close, A.; Ducklow, H.; Dickson, A. *Protocols for the Joint Global Ocean Flux Study (JGOFS) Core Measurements*; UNESCO: Paris, France, 1994.

73. Ritchie, R.J. Universal chlorophyll equations for estimating chlorophylls a, b, c, and d and total chlorophylls in natural assemblages of photosynthetic organisms using acetone, methanol, or ethanol solvents. *Photosynthetica* **2008**, *46*, 115–126. [CrossRef]

74. Bader, H. The hyperbolic distribution of particle sizes. *J. Geophys. Res.* **1970**, *75*, 2822–2830. [CrossRef]

75. Beardsley, G.F., Jr.; Zaneveld, J.R.V. Theoretical dependence of the near-asymptotic apparent optical properties on the inherent optical properties of sea water. *J. Opt. Soc. Am.* **1969**, *58*, 373–377. [CrossRef]

76. Zhang, X.; Fournier, G.R.; Gray, D.J. Interpretation of scattering by oceanic particles around 120 degrees and its implication in ocean color studies. *Opt. Express* **2017**, *25*, A191–A199. [CrossRef] [PubMed]

77. Zhang, X.; Hu, L.; He, M. Scattering by pure seawater: Effect of salinity. *Opt. Express* **2009**, *17*, 5698–5710. [CrossRef] [PubMed]

78. Reynolds, R.A.; Stramski, D.; Neukermans, G. Optical backscattering by particles in Arctic seawater and relationships to particle mass concentration, size distribution, and bulk composition. *Limnol. Oceanogr.* **2016**, *61*, 1869–1890. [CrossRef]

79. Ivanoff, A. Optical method of investigation of the oceans: The p-ß diagram. *J. Opt. Soc. Am.* **1959**, *49*, 103–104. [CrossRef]

80. Ivanoff, A.; Jerlov, N.; Waterman, T.H. A comparative study of irradiance, beam transmittance and scattering in the sea near Bermuda. *Limnol. Oceanogr.* **1961**, *6*, 129–148. [CrossRef]

81. Hatch, T.; Choate, S.P. Measurement of polarization of the Tyndall beam of aqueous suspension as an aid in determining particle size. *J. Franklin Inst.* **1930**, *210*, 793–804. [CrossRef]

82. Chami, M.; Santer, R.; Dilligeard, E. Radiative transfer model for the computation of radiance and polarization in an ocean–atmosphere system: Polarization properties of suspended matter for remote sensing. *Appl. Opt.* **2001**, *40*, 2398–2416. [CrossRef] [PubMed]

83. Lotsberg, J.K.; Stamnes, J.J. Impact of particulate oceanic composition on the radiance and polarization of underwater and backscattered light. *Opt. Express* **2010**, *18*, 10432–10445. [CrossRef] [PubMed]

84. Waterman, T.H. Polarization patterns in submarine illumination. *Science* **1954**, *120*, 927–932. [CrossRef] [PubMed]

85. Ivanoff, A. Polarization measurements in the sea. In *Optical Aspects of Oceanography*; Jerlov, N.G., Steeman-Nielsen, E., Eds.; Academic Press: London, UK; New York, NY, USA, 1974; pp. 151–175.

86. Chami, M. Importance of the polarization in the retrieval of oceanic constituents from the remote sensing reflectance. *J. Geophys. Res.* **2007**, *112*, C05026. [CrossRef]

87. Loisel, H.; Duforet, L.; Dessailly, D.; Chami, M.; Dubuisson, P. Investigation of the variations in the water leaving polarized reflectance from the POLDER satellite data over two biogeochemical contrasted oceanic areas. *Opt. Express* **2008**, *16*, 12905–12918. [CrossRef] [PubMed]

88. Tonizzo, A.; Gilerson, A.; Harmel, T.; Ibrahim, A.; Chowdhary, J.; Gross, B.; Ahmed, S. Estimating particle composition and size distribution from polarized water-leaving radiance. *Appl. Opt.* **2011**, *50*, 5047–5058. [CrossRef]

89. Ibrahim, A.; Gilerson, A.; Chowdhary, J.; Ahmed, S. Retrieval of macro- and micro-physical properties of oceanic hydrosols from polarimetric observations. *Rem. Sens. Environ.* **2016**, *186*, 548–566. [CrossRef]

90. Zhai, P.W.; Knobelspiesse, K.; Ibrahim, A.; Franz, B.A.; Hu, Y.; Gao, M.; Frouin, R. Water-leaving contribution to polarized radiation field over ocean. *Opt. Express* **2017**, *25*, A689–A708. [CrossRef] [PubMed]

applied
sciences

MDPI

Article

Remote Sensing of Coral Reefs: Uncertainty in the Detection of Benthic Cover, Depth, and Water Constituents Imposed by Sensor Noise

Steven G. Ackleson *, Wesley J. Moses and Marcos J. Montes

Naval Research Laboratory, Washington, DC 20375, USA; wesley.moses@nrl.navy.mil (W.J.M.);
marcos.montes@nrl.navy.mil (M.J.M.)
* Correspondence: steve.ackleson@nrl.navy.mil

Received: 20 August 2018; Accepted: 20 November 2018; Published: 19 December 2018

Featured Application: Remote sensing of shallow coral reefs. We present an analysis of sensor noise impacts on the detection of key coral reef ecological parameters: Bottom depth, benthic cover, and water constituent concentration related to water quality. The results will help guide the requirements for future satellite-based remote sensors for application to shallow water coastal environments.

Abstract: Coral reefs are biologically diverse and economically important ecosystems that are on the decline worldwide in response to direct human impacts and climate change. Ocean color remote sensing has proven to be an important tool in coral reef research and monitoring. Remote sensing data quality is driven by factors related to sensor design and environmental variability. This work explored the impact of sensor noise, defined as the signal to noise ratio (SNR), on the detection uncertainty of key coral reef ecological properties (bottom depth, benthic cover, and water quality) in the absence of environmental uncertainties. A radiative transfer model for a shallow reef environment was developed and Monte Carlo methods were employed to identify the range in environmental conditions that are spectrally indistinguishable from true conditions as a function of SNR. The spectrally averaged difference between remotely sensed radiance relative to sensor noise, ε, was used to quantify uncertainty in bottom depth, the fraction of benthic cover by coral, algae, and uncolonized sand, and the concentration of water constituents defined as chlorophyll, dissolved organic matter, and suspended calcite particles. Parameter uncertainty was found to increase with sensor noise (decreasing SNR) but the impact was non-linear. The rate of change in uncertainty per incremental change in SNR was greatest for $SNR < 500$ and increasing SNR further to 1000 resulted in only modest improvements. Parameter uncertainty was complicated by the bottom depth and benthic cover. Benthic cover uncertainty increased with bottom depth, but water constituent uncertainty changed inversely with bottom depth. Furthermore, water constituent uncertainty was impacted by the type of constituent material in relation to the type of benthic cover. Uncertainty associated with chlorophyll concentration and dissolved organic matter increased when the benthic cover was coral and/or benthic algae while uncertainty in the concentration of suspended calcite increased when the benthic cover was uncolonized sand. While the definition of an optimal SNR is subject to user needs, we propose that SNR of approximately 500 (relative to 5% Earth surface reflectance and a clear maritime atmosphere) is a reasonable engineering goal for a future satellite sensor to support research and management activities directed at coral reef ecology and, more generally, shallow aquatic ecosystems.

Keywords: remote sensing; coral reef; sensor noise; retrieval uncertainty

---

1. Introduction

Coral reefs are among the most biologically diverse and productive ecosystems [1] and provide a variety of goods and services to many tropical and sub-tropical coastal nations [2,3]. Coral reef health and economic value are on the decline worldwide in response to direct human impacts and global changes in climate [4] and this trend is expected to continue [5–7]. Within human-dominated environments, rapid and potentially irrecoverable changes in the state of coral reef ecosystems have been described as regime shifts [8,9], the most notable cause of which is the systematic removal of herbivores that prevent algae from overgrowing live coral [10]. Thus, the benthic cover offers a key metric to describe changes in coral reef health and is typically characterized optically as three primary endmembers; healthy corals and associated organisms, minimal coral cover dominated by fleshy macroalgae overgrowth, and uncolonized calcareous sand and dead coral rubble [11,12].

Coral reef benthic components, both fauna and flora, are highly diverse in appearance. Hochberg et al. [13] analyzed over 13,000 reflectance spectra of benthic coral reef components and identified 12 characteristic spectra representing algae, soft and hard coral, and sediments. Subtle differences in benthic reflectance can be detected in the above-water light field. Airborne imaging spectrometers providing continuous, high resolution sampling across the visible and near-infrared spectrum have been shown to yield more accurate discrimination between shallow water benthic types compared to multispectral systems that provide data at a small number of broad spectral bands [14–18]. Hedley and Mumby [19] examined the detailed spectral reflectance of various coral species and reported that spectral unmixing approaches, such as derivative analysis, are potentially effective discrimination techniques. Botha et al. [20] used radiative transfer simulations to investigate the effect of spectral sampling to discriminate between coral endmembers and concluded that increasing spectral sampling and resolution increased the depth from 2 m to 6 m, at which primary reef features can be identified within clear coastal water. As a result of these and other similar findings, remote sensing methods operating within the visible and near infrared portions of the light spectrum are rapidly becoming incorporated into coral reef monitoring efforts [21]. In addition to high spectral fidelity, coral reef scientists desire high spatial resolution, expressed as ground sampling distance (*GSD*), of less than a few tens of meters in order to limit sub-pixel variability and provide meaningful ecological information [22]. Building upon the lessons of modeling and in situ spectrometry and the positive applications of airborne imaging spectrometers with high spatial resolution, future satellite sensors are under development that will strive to provide similar data on a global scale. The NASA Surface Biology and Geology sensor, formerly the Hyperspectral Infrared Imager, for example, is envisioned as an imaging spectrometer that will provide continuous spectra between 380 and 2500 nm with 10 nm channels and a spatial resolution of 30 m [23].

Remote sensing of coral reefs is a challenging problem from environmental and engineering perspectives. In order to retrieve meaningful benthic and water column signals the confounding effects associated with environmental variability, e.g., atmospheric conditions and glint from the water surface must either be removed from the data or avoided. Uncertainty associated with such corrections is often referred to as environmental noise. From an engineering perspective, the sensor must be designed to collect as much light per pixel as possible in order to minimize signal variability associated with system electronics and random variations in light intensity. This variability is commonly referred to as sensor noise and expressed as the ratio of signal to noise (*SNR*). The total amount of noise embedded within a remotely sensed signal is the combination of noise from both environmental and sensor sources.

Light energy received by a remote sensor is expressed as photon flux, ϕ_q, and defined as the average number of photons received by a sensor per unit time per unit area,

$$\phi_q = \frac{\lambda}{hc} X_{sys} L_{sat}, \tag{1}$$

where h is Planck's constant, c is the speed of light, X_{sys} accounts for system design attributes, including field of view, aperture and integration time, and L_{sat} is radiance [24]. Variations in ϕ_q attributed to the sensor include readout noise associated with errors in reading detector signals, digitization noise due to rounding errors in the conversion of analog signals to digitized values, dark noise resulting from electric current in the system even when no photons are incident on the detector, and random variations in the number of photons received, referred to as shot noise. Recent advances in sensor design have greatly reduced readout, digitization, and dark noise across the designed signal range so that shot noise forms the primary source of measurement uncertainty. Thus, our investigation assumes that all non-environmental signal uncertainty is in the form of shot noise.

Shot noise forms a Poisson distribution in ϕ_q with a standard deviation σ_q equal to the square root of the signal. It follows, therefore, that

$$SNR = \phi_q / \sigma_q = \frac{\lambda}{hc} X_{sys} L_{sat} / \sqrt{\frac{\lambda}{hc} X_{sys} L_{sat}}. \tag{2}$$

Since *SNR* is a function of the radiance received, sensor uncertainty is expressed as a reference signal to noise ratio, SNR^*, reported for a typical at-sensor radiance, L_{typ}, and representing a specified set of environmental conditions, including solar zenith angle and mean Earth-Sun distance. For aquatic applications, the reference condition typically includes a clear maritime atmosphere overlying a surface reflectance of 5%. With knowledge of the reference noise level *SNR* for any scene radiance, L_{sat}, can be computed as

$$SNR = SNR^* \sqrt{\frac{L_{sat}}{L_{typ}}}. \tag{3}$$

It can be shown that while *SNR* is formally defined in terms of photon flux, it applies equally to radiance [25];

$$SNR = \frac{\phi_q}{\sigma_q} = \frac{L_{sat}}{\sigma_L}, \tag{4}$$

where σ_L is the standard deviation of radiance in units of radiance, e.g., W m^{-2} nm^{-1} sr^{-1}. Thus, the uncertainty in radiance resulting from sensor shot noise alone may be computed as $\sigma_L = L_{sat}/SNR$.

A recent model analysis of remotely sensed coral reef benthic cover concluded that under most conditions environmental variability dominates the total noise and that sensor design efforts to maximize *SNR* in order to retrieve more subtle benthic signatures are overemphasized [26]. This is an important point because relaxing *SNR* requirements within the engineering trade space can result in potential increases in spectral sampling and spatial resolution. The purpose of this work is to reexamine the problem posed by this earlier work and to extend the analysis to include uncertainty in the retrieval of bottom depth and water column properties that impact light attenuation. The analysis omits all noise imposed by the water surface and the atmosphere and focuses solely on the impacts of *SNR*. Uncertainty is defined as the envelope of environmental parameter values imposed by sensor noise that bracket the true (reference) values. We consider continuous sampling at a resolution of 1 nm within the visible and near-infrared portions of the electromagnetic spectrum (400 nm $\leq \lambda \leq$ 750 nm). This spectral sampling is not meant to represent a specific future satellite sensor, but rather an experimental design to avoid uncertainty imposed by under sampling the spectral domain. While *GSD* is not explicitly addressed, the implications of sub-pixel variability are implicit in the results, since benthic cover heterogeneity was considered in the analysis. The problem is addressed using an analytical radiative transfer model representing a shallow reef system with an overlying clear, stable atmosphere.

2. Approach

The approach compares scene radiances at the top of the atmosphere representing a reference condition, i.e., benthic cover, bottom depth, and water clarity, with test conditions that deviate from the reference condition. Parameters that define a reference condition are highlighted with a prime symbol (′) while the symbol is omitted from parameters representing test conditions. For example, scene radiance at the top of the atmosphere is written as L'_{sat} for a reference condition and L_{sat} for a test condition. The prime symbol is also omitted in instances of general reference. In all conditions, the atmosphere was defined using a low optical depth maritime aerosol (Table 1) and was held constant along with solar zenith angle throughout the analysis. The water surface was flat and downwelling sun and sky irradiance reflected from the water surface was not included in the water signal.

Table 1. Atmospheric parameter definitions and quantities representing a clear maritime atmosphere used in all model computations.

Parameter	Quantity
Aerosol model	Maritime
Aerosol relative humidity	90%
Aerosol optical depth (550 nm)	0.06
Atmospheric model	Tropical
Precipitable water vapor	5.0 cm
Ozone	0.34 atm-cm
Solar zenith angle	30°
Sun-earth distance	149.6×10^9 km

Reference conditions were compared with radiances representing other possible coral reef conditions, generated by randomly and independently adjusting water and benthic properties within realistic ranges. The advantages of this Monte Carlo approach over systematic adjustments of individual parameters are that (1) fewer computations were required to estimate parameter uncertainty, (2) all uncertainties were obtained simultaneously, and (3) all parameter interactions were included in the results. In each case, the randomly generated test spectra were compared with the reference spectrum and determined to be either indistinguishably similar or sufficiently unique based on a spectral difference criterion, ε, defined by the condition-dependent *SNRs* of the two signals. A flow chart of the modeling procedure is shown in Figure 1. If the reference and test spectra were indistinguishable, then the test condition was recorded and a new random test condition was generated. Random tests continued until enough indistinguishable conditions were encountered to reasonably resolve the range in parameter values. This typically required between 10^5 and 10^7 test conditions. The ensemble of indistinguishable test conditions was then used to identify the range in parameter values as a measure of detection uncertainty.

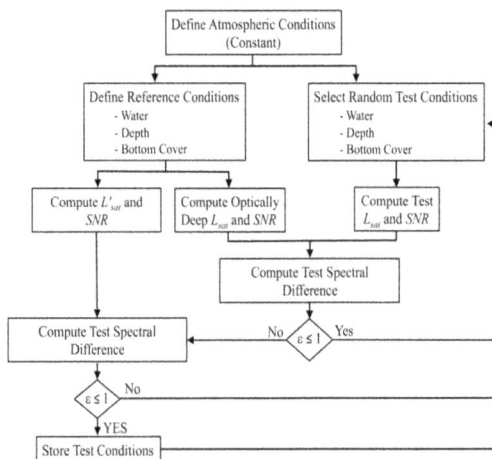

Figure 1. Computation flow chart. SNR, signal to noise ratio.

2.1. Shallow Water Light Model

Light propagation within a shallow reef environment was modeled following the original two-flow irradiance propagation concepts of Schuster [27] and adapted for a shallow ocean [28–30]. While all of the terms are spectrally dependent, wavelength notation λ is generally omitted for brevity and inserted as necessary for clarity.

Water remote sensing reflectance measured just above the water surface is expressed as

$$R_{rs} = \frac{0.52\, r_{rs}}{1 - 1.7\, r_{rs}}, \tag{5}$$

$$r_{rs} = r_{rs,\infty} + \left[\frac{\rho_b}{\pi} - r_{rs,\infty}\right] e^{-(K_d + K_u)\, D}, \tag{5a}$$

$$r_{rs,\infty} = \sum_{i=1}^{2} g_i \left(\frac{b_b}{a + b_b}\right)^i, \tag{5b}$$

$$K = \frac{a}{\mu}\left[1 + (0.425\, \mu - 0.19)\frac{b}{a}\right]^{\frac{1}{2}}, \tag{5c}$$

where r_{rs} sr^{-1} is the radiance reflectance measured just below the water surface, ρ_b is the irradiance reflectance of the shallow bottom, K m^{-1} is the attenuation coefficient for diffuse light, the subscripts d and u refer to the downwelling and upwelling irradiance streams, respectively, and D m is bottom depth. Equation (5) propagates underwater reflectance through the air/water interface assuming a flat surface [31]. The reflectance from infinitely deep water (Equation (5b)) is approximated as a polynomial [32], where b_b m^{-1} is the backscatter coefficient of the water mixture (pure water plus suspended particles) and a m^{-1} is the absorption coefficient of the water mixture (pure water plus all absorbing dissolved and particulate matter). The factors $g_1 = 0.084$ and $g_2 = 0.125$ were derived by Gordon et al. [32] through extensive Monte Carlo simulations. Diffuse attenuation (Equation (5c)) was expressed as a function of a, the total scatter coefficient of the water mixture (b m^{-1}), and the average cosine of the irradiance stream (μ), according to Kirk [33]. For K_d, the attenuation of downwelling irradiance, $\mu = cos(\theta_w)$, where θ_w is the in-water solar zenith angle. For upwelling irradiance attenuation, K_u, $\mu = 0.7$.

The inherent optical properties of the water column (a, b, and b_b) are defined as linear combinations of pure water (subscript w) and three primary constituents: Phytoplankton (subscript p), colored dissolved organic matter (subscript g), and suspended non-algal particulate matter (subscript d), assumed to be non-absorbing calcite;

$$a = a_w + a_p + a_g, \tag{6a}$$

$$b = b_w + b_p + b_d, \text{ and} \tag{6b}$$

$$b_b = b_{bw} + b_{bp} + b_{bd}. \tag{6c}$$

Pure water optical properties [34] were considered constant. Phytoplankton optical properties were spectrally dependent and defined as the product of chlorophyll concentration (C_{chl} mg m^{-3}) and the chlorophyll-specific coefficients for absorption ($a^*{}_{chl}$ m^2 mg^{-1}) and scatter ($b^*{}_{chl}$ m^2 mg^{-1}). The spectral shapes for $a^*{}_{chl}$ and $b^*{}_{chl}$ were defined as the average of spectra reported by Stramski et al. [35] for laboratory cultures of phytoplankton. Absorption due to dissolved organic matter was computed as $a_{g,\lambda} = a_{g,450}\, e^{0.014\,(450-\lambda)}$, where $a_{g,450}$ is the absorption at the reference wavelength $\lambda = 450$ nm [36]. Light scatter from suspended calcite particles was computed as $b_d = C_{cal} b^*_{cal} \left[\frac{660}{\lambda}\right]^{-1.45}$, where C_{cal} g m^{-3} is calcite concentration and b^*_{cal} m^2 g^{-1} is the calcite-specific scattering coefficient. For this study, $b^*_{cal} = 1.034$ and is based on light scatter measurements reported for calcite-dominated water within intense coccolithophore blooms [37].

Backscatter from pure water, b_{bw} m^{-1}, was computed according to Zhang and Hu [38] as $b_{bw} = 0.002\left[\frac{450}{\lambda}\right]^{4.3}$. Backscatter from phytoplankton was estimated from data reported by Stramski et al. [35] and computed as $b_{bp} = 0.001\,b_p$. Backscatter from suspended calcite particles was computed as $b_{bd} = 0.043\,b_d$ [37].

Benthic reflectance was computed as

$$\rho_b = \sum_{i=1}^{N} B_i \rho_{bi}, \tag{7}$$

where i refers to the bottom type and N is the number of types considered. The fractional benthic cover of each bottom type B_i is bounded by 0 and 1 and $\sum_{i=1}^{N} B_i = 1$.

Radiance received at the satellite was computed with the Tafkaa atmospheric model [39,40];

$$L_{sat} = \frac{\tau_g}{\pi E_o cos(\theta_s)}\left[\rho_{atm} + \frac{\pi R_{rs}\tau_d\tau_u}{1 - \bar{s}\pi R_{rs}}\right], \tag{8}$$

where E_o is the extraterrestrial solar irradiance, θ_s (= 30°) is the solar zenith angle, τ_d, τ_u, and τ_g are atmospheric transmittances for downwelling and upwelling irradiance, and absorption by atmospheric gases, respectively, ρ_{atm} is the atmospheric contribution to the upwelling radiance at the satellite, and $\bar{s}$ accounts for atmospheric backscattering of water reflectance.

2.2. Spectral Difference

Reference sensor noise, SNR^*, was defined for an Earth surface reflectance of 5%, i.e., $\pi R_{rs} = 0.05$, regardless of wavelength. Equation (8) was then used to compute L_{typ}. Note that while SNR^* was adjusted throughout the analyses for various reference conditions, L_{typ} remained constant.

The spectral difference criterion, ε, was defined as the difference in the top of the atmosphere radiance representing a reference condition and a test condition relative to sensor noise. Since SNR changes with scene radiance, the joint radiance variability σ when comparing the reference radiance, L'_{sat}, and a test radiance, L_{sat}, was computed as

$$\sigma = \sqrt{\left[\frac{L'_{sat}}{SNR'_L}\right]^2 + \left[\frac{L_{sat}}{SNR_L}\right]^2}. \tag{9}$$

The between-spectra difference averaged across the spectral range of interest was computed as

$$\varepsilon = \frac{1}{N}\sum_{i=1}^{N}\left|L'_{sat} - L_{sat}\right|_i/\sigma_i, \tag{10}$$

where N is the number of discrete spectral bands. If $\varepsilon < 1$, i.e., the average radiance difference was smaller than σ, then the two spectra were regarded as indistinguishable from system noise. Equation (10) is consistent with the z-statistic for comparing the similarity of two normally distributed populations [41].

Equation (10) does not offer an absolute measure of spectral discrimination. For example, wavelengths in the red and near-IR portions of the spectrum are absorbed at higher rates than those in the mid- and short-range portions of the visible spectrum, due to absorption by pure water. Hochberg et al. [13] noted this effect and suggested that remote sensing approaches to coral reef environments are best achieved at wavelengths shorter than 580 nm. Likewise, coral reefs ecosystems are often sources of colored dissolved organic matter [42] that increases water absorption in the blue portion of the spectrum. As ε approaches the threshold value, $|L_{sat} - L'_{sat}|/\sigma$ will generally be greater than the threshold in some portions of the spectrum, likely in the mid-visible range where light transmission is a maximum, and less than the threshold in the blue and red regions of the spectrum.

2.3. Model Scenarios

Several scenarios were considered to investigate the effects of environmental conditions on the detection uncertainty for benthic and water column properties as a function of sensor *SNR* (Table 2). In each scenario three reference *SNR* values were considered; 100, 500, and 1000. The benthic cover was defined as fractional contributions from the three primary coral reef endmembers; average brown hermatypic coral (B_c), green fleshy algae (B_a), and uncolonized calcareous sand (B_s). In addition, the average coral spectrum was compared with that of a specific coral species, *Porites astreoides* ($B_{c,p}$), in order to test the impact of *SNR* on coral species discrimination. Total benthic reflectance was computed as linear combinations of the reflectance spectra representing the specified bottom types (Figure 2) using data reported by Myers et al. [43] and collected using methods reported in Mazel [44].

Table 2. Summary of Scenario Conditions.

	Global Scenario Reference Conditions		
	SNR = 100, 500, 1000		
	1 m $\leq D' \leq$ 10 m; $\Delta D' = 1$ m		
	C'_{chl}=0.1 mg m^{-3}; $a'_{g,450}$=0.2 m^{-1}; C'_{cal}=0.3 g m^{-3}		
Scenario	**Benthic Cover**	**Bottom Depth**	**Water Constituents**
S1	$B'_c = 1$; $B'_a = B'_s = 0$ $0 \leq B_c,\ B_a,\ B_s \leq 1$	$0.1\ m \leq D \leq OD^*$	$0 \leq C_{chl} \leq 1$ mg m^{-3} $0 \leq a_{g,450} \leq 1$ m^{-1} $0 \leq C_{cal} \leq 1$ g m^{-3}
S2		*OD = Optically Deep	
S3			$C_{chl} = C'_{chl}$ mg m^{-3} $a_{g,450} = a'_{g,450}$ m^{-1} $C_{cal} = C'_{cal}$ g m^{-3}
S4	$B'_c = 1$; $B'_{c,p} = B'_s = 0$ $0 \leq B_c,\ B_{c,p},\ B_s \leq 1$	$D = D'$	
S5	S5a $B_c = B'_c = 1$ $B_a = B'_a = 0$ $B_s = B'_s = 0$	S5b $B_c = B'_c = 0$ $B_a = B'_a = 0$ $B_s = B'_s = 1$	$0 \leq C_{chl} \leq 1$ mg m^{-3} $0 \leq a_{g,450} \leq 1$ m^{-1} $0 \leq C_{cal} \leq 1$ g m^{-3}

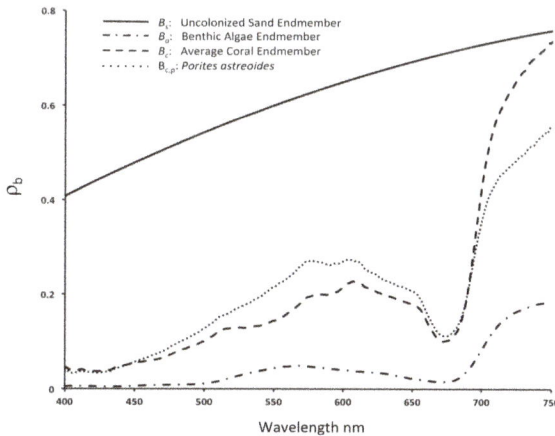

Figure 2. Benthic reflectance spectra used in model computations.

Water constituent concentrations for all reference conditions were set low to represent a relatively clear water column; C'_{chl} = 0.1 mg m^{-3}, $a'_{g,450}$ = 0.2 m^{-1}, and C'_{cal} = 0.3 g m^{-3}. The conditions are similar to the least clear waters examined by Hedley and others [26]. Diffuse attenuation for this condition, expressed as the spectrally weighted K_d,

$$\overline{K}_d = \frac{\sum_{i=400}^{750} E_{si} K_{di}}{\sum_{i=400}^{750} E_{si}}, \tag{11}$$

where E_s is the solar and sky irradiance at the water surface, was 0.409 m^{-1}. Test constituent concentrations were varied randomly between zero and maximum values judged to be reasonable representations of most coral reef environments; $C_{chl} = 1.0$ mg m^{-3}, $a_{g,450} = 1.0$ m^{-1}, and $C_{cal} = 1.0$ g m^{-3}. These conditions span the clearest waters to be encountered within coral reef environments and episodic turbid conditions that might be encountered in close proximity to population centers. The reference bottom depth for each scenario ranged from 1 to 10 m in increments of 1 m. For test conditions D was varied randomly between 0.1 m and the maximum depth at which the test condition could be distinguished from optically deep water.

In the first scenario, *S1*, the objective was to compute the uncertainty in the signal representing 100% coral imposed by variability in water constituents, bottom depth, and benthic cover type. Since all parameters were allowed to vary without constraints, this scenario might in practice represent the application of remote sensing to an environment where the user has no a priori knowledge other than assumptions regarding parameter value ranges. The resulting uncertainties, therefore, represented the combined variability in all parameters. The reference benthos was defined as 100% coral cover having a reflectance equal to the average coral endmember, $B'_c = 1.0$, and test conditions included random mixtures of B_c, B_a, and B_s.

In scenario *S2*, uncertainties associated with endmember benthic cover and bottom depth retrieval were investigated with the assumption that the water constituent concentrations were known. This scenario is perhaps analogous to a situation where the coral reef is rather remote with typically clear water conditions and no direct influence of degraded water quality imposed by a nearby population center. The reference conditions were identical to *S1*, test water conditions were set equal to the reference condition, and bottom depth and benthic cover were varied randomly across the prescribed ranges.

In scenario *S3*, uncertainty in benthic cover was investigated with the assumption that water clarity and bottom depth are known. This scenario is analogous to a coral reef monitoring program where water quality is measured on a frequent basis, detailed bathymetric data are available, and the objective is to detect changes in benthic cover type. The reference conditions were identical to *S1*, water conditions and depth were set equal to the reference conditions, and only the benthic cover was allowed to vary randomly.

In scenario *S4*, the impact of *SNR* on coral species discrimination was investigated. The reference benthic cover was assumed to be 100% coral and represented by the average coral reflectance spectra, B'_c. The test benthic conditions consisted of random mixes of the coral species *Porites astreoides* ($B_{c,p}$) and the benthic algae endmember (B_a). Water constituent concentration and depth were constant and equivalent to the reference conditions specified in *S1*.

Finally, in scenario *S5*, we investigated the impact of *SNR* on water property retrieval with knowledge of endmember benthic cover and bottom depth. As in *S3*, this scenario is analogous to the situation in a coral reef monitoring program where benthic cover and bottom depth have been previously mapped and the objective is to quantify changes in water quality. In addition to changes in bottom depth, this scenario was devised to test the potential impact of benthic cover on water constituent uncertainty. Two separate reference conditions were considered: 100% coral cover and 100% uncolonized sand. In both cases, benthic cover and bottom depth were held constant and equal to the reference condition and only water constituent concentrations were allowed to change independently.

3. Results

3.1. Bottom Depth

The maximum depth at which benthic cover could be detected, i.e., where $\varepsilon > 1$ when comparing the reference condition with the signal representing optically deep water, was affected by a combination of the concentration of water constituents (as the determinants of K), the benthic composition (that controls ρ_b), and the noise envelope surrounding the reference condition (defined by SNR). For example, considering $SNR = 500$, the extinction depths for benthic cover representing the three endmember bottom types were 18.4, 11.4, and 21.6 m for B'_c, B'_a, and B'_s, respectively (Figure 3a). The difference results from the relative contrast between the optically shallow and deep-water reflectance. The extinction depth for benthic algae was nearly half that of uncolonized sand because the spectrally averaged ρ_b for B'_a was more similar to $r_{rs,\infty}$ and, therefore, became indistinguishable at a shallower depth compared with B'_s, which was the least similar to optically deep water. For a given bottom type (results for 100% coral cover are presented as an example), the extinction depth decreased with SNR (Figure 3b), because the envelope of uncertainty increased with sensor noise (inversely with SNR). For example, given a bottom composed of 100% coral, the extinction depth was 20.9 m for $SNR = 1000$ and was more than 1.5 times the extinction depth (12.6 m) under the same environmental conditions when $SNR = 100$.

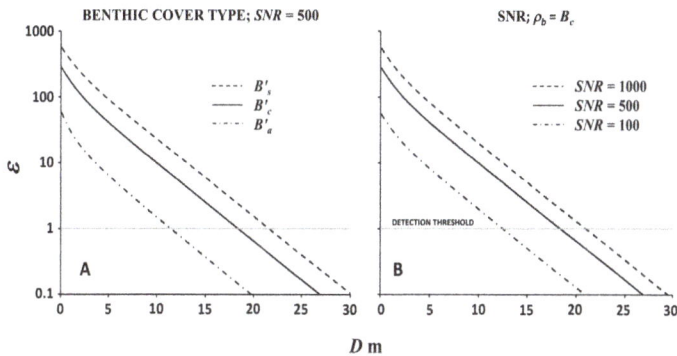

Figure 3. Average spectral difference parameter, ε, as a function of depth, where L'_{sat} representing shallow reef conditions is compared with optically deep water. The water column is relatively clear ($C'_{chl} = 0.1$ mg m^{-3}, $a'_{g,450} = 0.2$ m^{-1}, and $C'_{cal} = 0.3$ g m^{-3}) and the threshold of spectral separation is $\varepsilon = 1.0$ (gray line). Impacts of cover type (**A**) are shown for B'_c (solid), B'_a (dot-dash), and B'_s (dash) and the response to SNR (**B**) is shown for B'_c, where $SNR = 100$ (dot-dash), 500 (solid), and 1000 (dash).

Uncertainty in the detection of bottom depth increased with increasing reference depth, expressed as the dimensionless optical depth D_o ($= D' \overline{K}_d$) and sensor noise, expressed as the difference between the upper and lower bounds of spectral similarity and the reference condition, ΔD (Figure 4). Positive values indicate an overestimation of D' and negative values indicate underestimation. The rate of increase in uncertainty with respect to the reference bottom depth was greatest for $SNR = 100$ and progressively diminished for $SNR = 500$ and 1000. Retrieval uncertainty was greatest when no assumptions were made about the environment, i.e., scenario *S1*, where water constituent concentration was allowed to vary across the prescribed ranges. However, when water constituent concentration was constant and set equal to the reference condition (*S2*), uncertainty in bottom depth retrieval decreased for each of the SNR values (bottom panel in Figure 4), although uncertainty associated with $SNR = 100$ remained noticeably greater than for $SNR = 500$ and 1000.

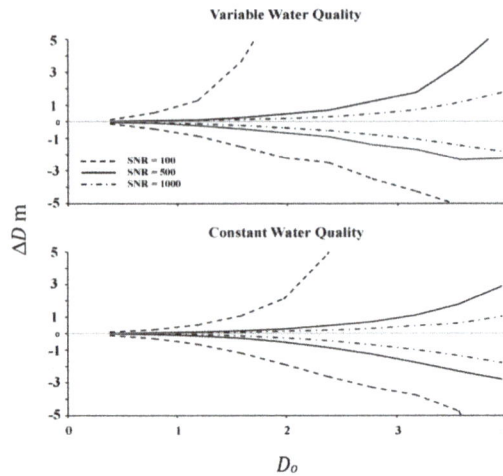

Figure 4. Uncertainty in water depth (ΔD), expressed as the difference between the upper and lower bounds of the test conditions relative to the reference condition, plotted against optical depth (D_o) for variable water constituent concentration (S1, upper figure) and constant constituent concentration (S2, lower figure).

3.2. Benthic Cover

Uncertainty in the retrieval of endmember benthic cover increased with sensor noise (decreasing *SNR*) and the reference bottom depth (Figure 5). As the bottom depth increased, the lower confidence bound for B'_c decreased from the reference condition of 100% coral cover while B'_a and B'_s increased from the reference condition of 0% cover. Allowing water constituent concentrations and bottom depth to vary across the allowable range (S1), coral and benthic algae were indistinguishable for *SNR* = 100 and $D_o > 1.1$; $-1 \leq \Delta B_c \leq 0$ and $0.79 \geq \Delta B_a \geq 0$, where Δ indicates the difference between the upper and lower bounds of the test conditions and the reference condition. Negative values indicate under estimation of reference benthic cover while positive values indicate overestimation. Increasing *SNR* to 500 decreased the underestimation of B'_c to -0.28 and the overestimation of B'_a to 0.23. This trend continued as *SNR* increased to 1000, but the rate of change as a function of *SNR* was considerably less. Uncertainty in distinguishing between B'_c (or B'_a) and B'_s was generally lower, due to a greater contrast between the vegetated and uncolonized substrates. The increase in uncertainty in benthic cover with increasing bottom depth is well understood and reported. The impact of *SNR* on benthic cover uncertainty was also reported by Hedley et al. [26] and expressed as the statistical overlap between specific conditions with noise applied, but their analysis did not suggest a large impact of *SNR* compared with environmental noise.

Constraining water constituent concentration to the reference condition (S2) resulted in a decrease in uncertainty in benthic cover detection regardless of *SNR*. For example, when $D_o = 1.1$, underestimation of B'_c decreased from the S1 values to -0.78 for *SNR* = 100, -0.18 for *SNR* = 500, and -0.09 for *SNR* = 1000. Detection uncertainty between coral and benthic algae and between vegetated and uncolonized substrates decreased similarly. Incorporating knowledge of bottom depth (S3) decreased uncertainty further; for $D_o = 1.1$, the underestimation of B'_c decreased to -0.58, -0.12, and -0.06 for *SNR* = 100, 500, and 1000, respectively.

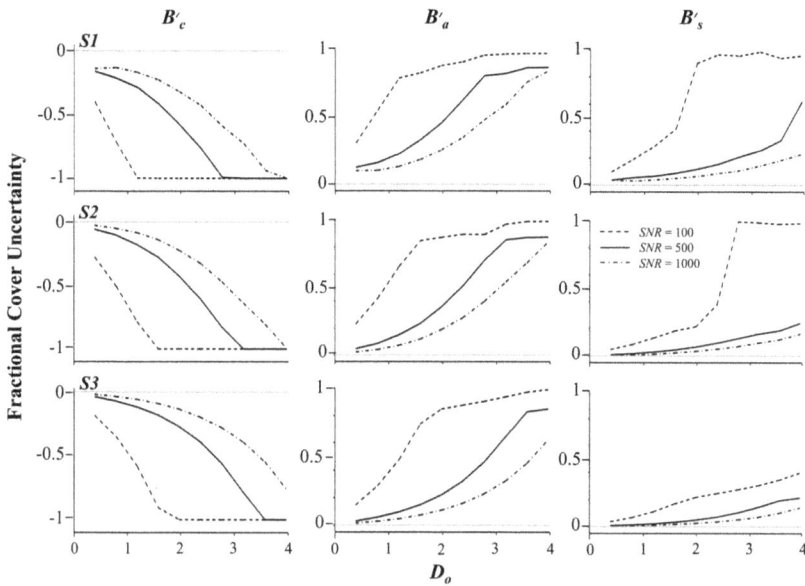

Figure 5. Uncertainty in the detection of endmember benthic cover relative to the reference condition, indicated as zero on the ordinate in each graph. *S1* (top row of graphs) represent variable water constituent concentration and depth across the allowable ranges. *S2* (middle row of graphs) represent variable depth while setting water constituent concentration constant and equal to the reference condition. *S3* (bottom row of graphs) represent depth and water constituent concentration constant and equal to the reference condition.

Uncertainty in distinguishing between the average coral endmember and the coral species *Porites astreoides* ($B'_{c,p}$), scenario *S4*, was consistently greater than the uncertainty in distinguishing between coral and benthic algae because the coral reflectance spectra were more similar to each other than the average coral and algae spectra (Figure 6). At the shallowest depths considered, $D' \leq 3$ m ($D_o \leq 1.3$), uncertainty in $B'_{c,p}$ was between two and three times greater than the uncertainty in B'_a. Uncertainty increased with bottom depth and the difference in uncertainty between the coral covers and that of algae gradually diminished as the signatures from each benthic cover gradually became more similar, due to water attenuation. At the same time, detection uncertainty was significantly impacted by sensor noise. At $D' = 1$ m ($D_o = 0.41$), uncertainty in coral cover, $\Delta B'_{c,p}$, for $SNR = 100$ was 0.29 relative to the reference condition $B'_c = 1$ (100% average coral cover). In other words, a fractional cover of 29% *Porites a.* was confused with 100 % average coral cover. At $D' = 2$ m, the uncertainty increased to 0.52 (52% cover). However, increasing *SNR* to 500 significantly reduced this uncertainty to 0.06 (6%) at 1 m and 0.1 (10%) at 2 m. For $SNR = 1000$, uncertainty decreased slightly further to ≤ 0.052 (5.2%).

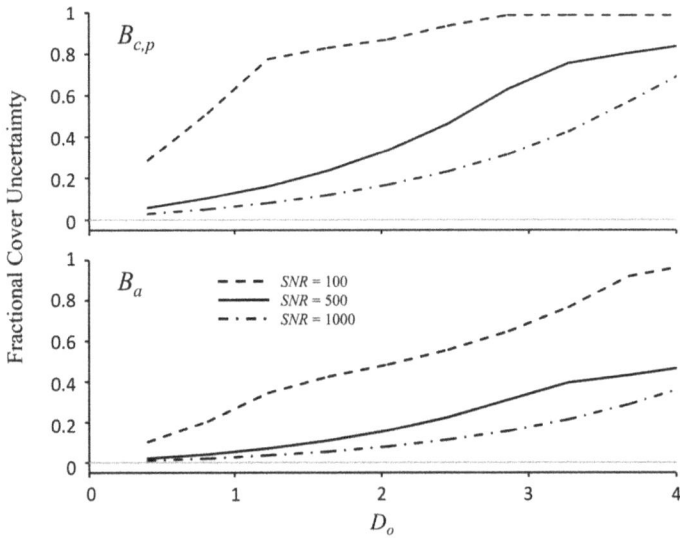

Figure 6. Detection uncertainty in the fractional cover of Porites ($B'_{c,p}$) and benthic algae (B'_a) relative to the reference condition (scenario S4, $B'_c = 1$). The uncertainty in the fractional cover is expressed as the difference between the upper bound of the test condition and the reference condition (solid gray line). Note that uncertainty in this scenario can only be positive (overestimated), since in this scenario $B'_{c,p}$ and B'_a are zero.

3.3. Water Constituents

Uncertainty associated with the detection of water constituent concentration was affected by *SNR*, bottom depth, and benthic cover (Figure 7). Higher *SNR* consistently reduced detection uncertainty. However, while benthic cover uncertainty decreased with D', uncertainty in water constituent concentration increased in shallow water, due to the diminished water signal. Both $a'_{g,450}$ and C'_{cal} were generally detectable with low uncertainty for $SNR \geq 500$ and $D_o > 1$; $\Delta a_{g,450} < \pm 0.05$ m^{-1} and $\Delta C'_{cal} < \pm 0.04$ g m^{-3} respectively. At $D' = 1$ m ($D_o = 0.4$), retrieval uncertainty for these parameters increased significantly, especially for $SNR = 100$. If the benthic cover was spectrally similar to the material suspended or dissolved within the overlying water column, uncertainty increased. For example, uncertainty in the detection of chlorophyll concentration was relatively large for scenarios where the reference benthos was composed of healthy coral (in S1 and S5a), although increasing *SNR* did result in a modest reduction in uncertainty. This is because absorption by phytoplankton is similar to absorption by zooxanthellae, the chlorophyll-containing coral symbiont (in S5a), and the combination of coral and benthic algae (in S1). For $SNR = 100$, the reference concentration was indistinguishable from the entire range of test concentration, $0 \leq C_{chl} \leq 1$ mg m^{-3}. However, uncertainty in C_{chl} detection improved when the benthic cover was composed of uncolonized sand (in S5b). In this case, increasing *SNR* from 100 to 500 resulted in a significant decrease in detection uncertainty. At the same time, uncertainty in detecting C_{cal} was smaller when the benthic cover was coral or algae and increased when the substrate was uncolonized sand.

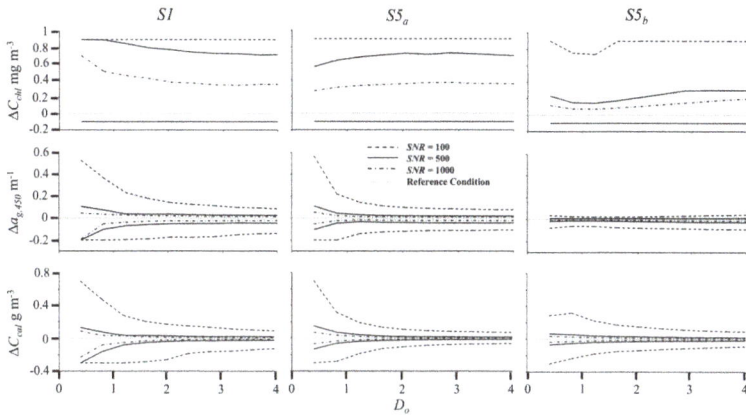

Figure 7. Water constituent uncertainty (*y*-axis) for variable depth and benthic cover (*S1*) and water depth and benthic cover equal to the reference condition (*S5$_a$* and *S5$_b$*). Scenario *S5$_a$* represents benthic cover of 100% coral and *S5$_b$* represents 100% uncolonized sand. The ordinate scale represents the difference between the upper and lower boundaries of parameter uncertainty, within the ranges specified, and the reference condition, indicated as 0 in all graphs.

4. Discussion

It is shown that *SNR* can impose significant limitations on the ability to detect benthic cover, bottom depth, and water constituents for all but the shallowest reef conditions considered (*D* = 1 m). The threshold criteria for signal detection, ε = 1, essentially quantifies the average difference between a reference condition and many test conditions relative to sensor noise imposed on the two signals. If ε > ($\leq$) 1, then the two signals are (are not) detectable as two distinct conditions. Since ε expresses the spectrally averaged difference, one should expect that ε > 1 in some parts of the spectrum while in other parts ε < 1. The criteria might, therefore, be viewed as too constraining since there will likely be unique information remaining where ε > 1. However, the amount of retrievable information will diminish as portions of the spectrum become progressively immersed within the envelope of sensor noise. Thus, ε may be viewed as a reasonable proxy for the difference between two similar spectra.

Hedley et al. [26] conducted a sensitivity analysis of sensor and environmental sources of retrieval uncertainty applied to a hypothetical airborne sensor and concluded that *SNR* had only a limited influence on the overall uncertainty relative to environmental sources. While many environmental sources, such as atmospheric constituent concentration and glint from a wind-roughened water surface, can easily exceed those associated with *SNR*, such uncertainty is often spectrally correlated in ways that are well understood and, therefore, can be mitigated. *SNR*, on the other hand, forms the foundation of irreconcilable uncertainty to which environmental sources of uncertainty are added. It is, therefore, prudent to make sensor development decisions with knowledge of how *SNR* impacts the overall uncertainty in the retrieval of key environmental parameters.

From an optical remote sensing perspective, one of the strongest indicators of coral health is bleaching; a brightening of the coral host, due to the absence of the microalgal symbiont, zooxanthellae [44]. When present in healthy coral, absorption by photosynthetic pigments contribute largely to the overall reduction in coral reflectance relative to the calcium carbonate foundation secreted by the coral host. During a bleaching event, the zooxanthellae vacate the coral host and the overall reflectance of the coral increases and takes on an appearance more similar to uncolonized calcareous sediment. Remote detection of bleaching is similar to the problem of estimating healthy coral cover when surrounded by uncolonized sand, illustrated in scenarios *S1*, *S2*, and *S3*. For *SNR* = 100, the difference between the maximum and minimum confidence bounds in the detection of uncolonized sand at the shallowest depths considered, $D' = 1$ m, ranged from 0.1 (*S1*) with unconstrained water constituent concentration and bottom depth to 0.04 with a priori knowledge water constituent concentration and bottom depth. Increasing *SNR* to 1000 decreased detection uncertainty by an order of magnitude. The probability of detecting a bleaching event, assuming that the reflectance of bleached coral is similar to uncolonized sand, and the resulting change in the fraction of healthy coral cover would be reasonably high given these uncertainty limits. However, uncertainty increases with bottom depth for all *SNR*s considered, particularly for *SNR* = 100. In addition to *SNR* the likelihood of detecting a bleaching event depends on the contrast between healthy and bleached coral. Bleached coral is not always as bright as the surrounding uncolonized sand [13] and the fraction of symbiont loss does not always scale with reflectance [45,46]. Therefore, the uncertainty estimates with regard to bleaching are likely optimistic.

The joint retrieval of both benthic cover and water constituent information is problematic, regardless of *SNR*. The least uncertainty in water constituent concentration occurred in optically deep water where the effects of benthic cover were minimal, but the benthic cover was highly uncertain. Conversely, the most accurate detection of benthic cover occurred in optically shallow water, but uncertainty in water constituent concentration was greatest. There is perhaps a glimmer of hope for the shallow water conundrum in that water constituent uncertainty is lower when the water constituent and benthic cover result in dissimilar effects on the total water reflectance. If the water constituent material looks like the benthic cover material, uncertainty in the retrieval of either the water constituent or benthic cover is large. The converse is also true. Within shallow water, uncertainty associated with water column chlorophyll and dissolved organic matter (both results in darker water) is minimal when the bottom is composed of bright, uncolonized sand whereas uncertainty in suspended calcite sediment (which increases water reflectance) is smallest when the bottom is composed of dark coral or algae. This raises the possibility of constructing water constituent retrieval algorithms that are benthic cover specific. For example, one algorithm could potentially retrieve C_{chl} and $a_{g,450}$ over bright, shallow features while an algorithm for retrieving C_{cal} could apply only to areas of dark benthic cover. Such an approach, if successful, would likely resolve spatial variability in water quality within a heterogeneous reef environment and would be an improvement over the practice of retrieving water constituent concentration from adjacent deep water signals that are then extended to the entire adjacent shallow reef. Regardless, it is important to note that *SNR* will have a significant impact on the detection uncertainty, especially when *SNR* < 500 and D_0 < 1.

This study suggests that *SNR* for shallow coral reef applications could be optimized around a value of 500. For *SNR* < 500, uncertainties associated with detecting changes in benthic cover type, bottom depth and water constituent concentration increase rapidly and limit remote sensing utility to the shallowest portions of a reef environment. As *SNR* increases beyond 500 the parameter uncertainty decreases, but at a much slower rate. Detection of coral reef benthic features, e.g., coral reef species, seagrasses, benthic algae, and uncolonized sand, has been demonstrated utilizing airborne scanning spectrometers with *SNR* approaching 500, computed as the top of atmosphere radiance from a clear maritime atmosphere and a 5% reflectance surface [14–18]. In comparison, data from the Landsat-8 Operational Land Imager (OLI) with *SNR* between 237 and 367 and four relatively broad visible bands have been applied to surveys of shallow reef flats [47,48] with far less effectiveness. Roelfsema et al. [49], for example, combined empirically derived geomorphic-ecological rules with OLI data to map benthic cover type to a maximum depth of 20 m at selected locations within the Great Barrier Reef system. However, the OLI imagery was only used to characterize benthic cover directly where $D < 0.75$ m. In the deeper reef areas, only single-band determinations were used to characterize the bottom as either uncolonized sand (light pixels) or combinations of coral, rock, and algae (dark pixels). Combined, these reports are in agreement with the range in uncertainty in benthic cover detection as a function of depth and *SNR* indicated in the *S1*, *S2*, and *S3* scenarios.

The engineering constraints on *SNR* for a sensor deployed on an aircraft are very different compared with an identical sensor placed in low Earth orbit (LEO). The number of photons collected per pixel is a product of the amount of time that the pixel is viewed (integration time), the solid angle subtended by the pixel (instantaneous field of view), and the spectral sampling. The most advanced aircraft sensors to date include the Airborne Visible/Infrared Imaging Spectrometer—AVIRIS [50], and the Portable Remote Imaging Spectrometer—PRISM [51]. These sensors collect data with high spatial resolution (GSD < 10 m), fine spectral sampling and resolution (generally < 5 nm), and peak $SNR \geq 500$ within the mid-visible portion of the spectrum. This is possible because the sensors can be flown at low altitude resulting in a wider field of view for a specified GSD and with slow speed resulting in longer integration time. Together, these attributes result in high photon flux per pixel. A LEO sensor orbits at an altitude of approximately 2000 km and travels with a ground speed about 28,000 km h^{-1}. Therefore, for a specific GSD, an identical LEO sensor will collect only a fraction of the photons that would be received by an aircraft sensor for a specified GSD resulting in a lower SNR. The SNR of a LEO sensor could be increased by incorporating a larger aperture in order to collect more light, but this would also increase the manufacturing and launch costs. The design GSD, i.e., the instantaneous field of view per pixel, could be increased with minimal impact on size and weight, but this will result in coarser spatial resolution and potentially significant loss of fine scale information due to spectral mixing. Finally, the sensor could operate with lower spectral resolution, thus increasing the number of photons per spectral band, but this would lead to loss of information necessary to discriminate between spectrally dissimilar properties. Thus, there exists a well-defined SNR trade space between GSD, spectral sampling, and the cost of manufacturing and launching a LEO sensor, all of which impact the amount of retrievable environmental information.

The next generation of Earth imaging systems, motivated largely by reasoning articulated in the latest NASA Decadal Survey of Earth remote sensing needs [52], will attempt to optimize the sensor design tradeoffs in order to most effectively address concerns about coastal ecosystem responses to population and climate change [53]. Coral reefs are a key environment globally due to their high biodiversity and societal services. This study indicates that SNR is a key sensor design attribute for remote sensing in support of coral reef ecological research and management and, more generally, investigations of shallow coastal environments and that data value is diminished considerably when $SNR < 500$.

5. Conclusions

There are three primary conclusions of this research.

(1) SNR was found to have significant and non-linear impacts on the uncertainty in detecting change in key environmental parameters associated with shallow coral reef systems (bottom depth, benthic cover, and the concentration of particulate and dissolved materials within the water column). In general, uncertainty decreased with increasing SNR and the rate of change in uncertainty per incremental change in SNR was greatest for $SNR < 500$. Increasing SNR from 100 to 500 resulted in significant decreases in parameter uncertainty, but increasing SNR further to 1000 resulted in only modest improvements in uncertainty.

(2) Regardless of SNR, benthic cover uncertainty increased with bottom depth as reported by previous researchers. However, uncertainty in water constituent concentration increased with decreasing water depth, particularly when the geometric depth was less than 1 optical depth. For $D_o > 1$, uncertainty associated with $a_{g,450}$ and C_{cal} was relatively small for $SNR \geq 500$. Uncertainty in the C_{chl} was high regardless of SNR for all scenarios except when the benthic cover was 100% uncolonized sand and $SNR \geq 500$. Regardless, retrieval of water constituent concentration in shallow water is problematic.

The sensitivity of uncertainties in water constituent concentrations to benthic cover is unrelated to SNR, but nonetheless is impacted by sensor noise. The results suggest that it may be possible to retrieve water constituent concentration in shallow water environments by parsing the imagery based on benthic brightness, where algorithms developed for absorbing matter, such as chlorophyll and dissolved organic matter, might be better retrieved over areas of high benthic reflectance, such as uncolonized sand, while the concentration of highly scattering suspended particulate matter, such as calcite sediment, may be better retrieved in areas of dark benthic cover, such as healthy coral and algae.

(3) Lastly, given the change in parameter uncertainty with sensor noise and the likely trade-offs that will be required for the development of future spaceborne imaging systems with application to coastal ecosystems, we conclude that $SNR = 500$ is a reasonable sensor design goal in order to optimize the utility of such a sensor for application to shallow coral reef environments. Uncertainty in submerged feature detection and retrieval increases significantly as SNR approaches a value of 100. While this may result in significant immediate savings in sensor development and launch costs, the potential loss of environmental information may have larger societal costs in the long run. On the other hand, increasing SNR above 500 will likely result in much higher sensor development costs that will be hard to justify based on expected scientific benefits.

Author Contributions: All authors contributed significantly to the writing and/or final editing of the manuscript. S.G.A. was the lead author and was responsible for model development and analysis and writing the manuscript. W.J.M. was responsible for developing concepts of SNR. M.J.M. was responsible for defining a clear maritime atmosphere, running the Tafkaa atmospheric model.

Funding: Financial support for this project is provided by the National Aeronautic and Space Administration HyspIRI Program through grants NNH15AB471 and NNX16AB05G and the research funds provided by the U.S. Naval Research Laboratory.

Conflicts of Interest: There are no conflicts of interest on the part of any of the authors.

Mathematical Symbols, Units, and Definitions

Symbol	Unites	Definition
a	m^{-1}	Absorption coefficient for water and all impurities
b	m^{-1}	Scattering coefficient for water and all impurities
b_b	m^{-1}	Backscattering coefficient for water and all impurities
B		Fraction of benthic cover
c	$m\,s^{-1}$	Speed of light
C	$mass\,m^{-3}$	Water constituent concentration
D	m	Bottom depth
D_o		Optical depth
E_o	$W\,m^{-2}\,nm^{-1}$	Extraterrestrial solar irradiance
h	$J\,s$	Planck's constant
K	m^{-1}	Diffuse light attenuation coefficient
$\overline{K}$	m^{-1}	Spectrally weighted diffuse attenuation coefficient
L_{sat}	$W\,m^{-2}\,nm^{-1}\,sr^{-1}$	Earth radiance received by a satellite sensor
L_{typ}	$W\,m^{-2}\,nm^{-1}\,sr^{-1}$	Typical top-of-atmosphere Earth radiance
r_{rs}	sr^{-1}	In-water remote sensing reflectance
R_{rs}	sr^{-1}	Above-water remote sensing reflectance
$\overline{s}$		Atmospheric scatter of Earth surface reflectance
SNR		Signal to noise ratio
X_{sys}		Collective remote sensor attributes
ε		Spectral-averaged radiance difference criterion
θ_s	radians	Solar zenith angle
λ	nm	Wavelength of light
μ		Average cosine of irradiance
ρ_{atm}		Atmospheric reflectance due to path radiance
ρ_b		Benthic reflectance
σ	$W\,m^{-2}\,nm^{-1}\,sr^{-1}$	Radiance standard deviation
σ_q	$photons\,s^{-1}$	Photon flux standard deviation
τ		Atmospheric transmittance
ϕ_q	$photons\,s^{-1}$	Photon flux

References

1. Odum, H.T.; Odum, E.P. Trophic structure and productivity of a windward coral reef community on Eniwetok Atoll. *Ecol. Monogr.* **1955**, *25*, 291–320. [CrossRef]
2. Spurgeon, J.P.G. The economic value of coral reefs. *Mar. Pol. Bull.* **1992**, *24*, 529–536. [CrossRef]
3. Moberg, F.; Folke, C. Ecological goods and services of coral reef ecosystems. *Ecol. Econ.* **1999**, *29*, 215–233. [CrossRef]
4. Hughes, T.P.; Baird, A.H.; Bellwood, D.R.; Card, M.; Connolly, S.R.; Folke, C.; Grosberg, R.; Hoegh-Guldberg, O.; Jackson, J.B.C.; Kleypas, J.; et al. Climate Change, Human Impacts, and the Resilience of Coral Reefs. *Science* **2003**, *301*, 929–933. [CrossRef] [PubMed]
5. Kleypas, J.A.; McManus, J.W.; Menez, L.A.B. Environmental limits to coral reef development: Where do we draw the line? *Am. Zool.* **1999**, *39*, 146–159. [CrossRef]
6. Anthony, K.R.N.; Kline, D.I.; Diaz-Pulido, G.; Dove, S.; Hoegh-Guldberg, O. Ocean acidification causes bleaching and productivity loss in coral reef builders. *Proc. Natl. Acad. Sci. USA* **2008**, *105*, 17442–17446. [CrossRef] [PubMed]
7. Hoegh-Guldberg, O. The impact of climate change on coral reef ecosystems. In *Coral Reefs: An Ecosystem in Transition*; Dubinsky, Z., Stambler, N., Eds.; Springer: Dordrecht, The Netherlands, 2011; pp. 391–403. ISBN 978-94-007-0113-7.
8. Nyström, M.; Folke, C.; Moberg, F. Coral reef disturbance and resilience in a human-dominated environment. *Trends Ecol. Evol.* **2000**, *15*, 413–417. [CrossRef]
9. Knowlton, N.; Jackson, J.B.C. Shifting baselines, local impacts, and global change on coral reefs. *PLoS Biol.* **2008**, *6*, e54. [CrossRef]
10. Bellwood, D.R.; Hughes, T.P.; Folke, C.; Nyström, M. Confronting the coral reef crisis. *Nature* **2004**, *6*, 827–833. [CrossRef]
11. Done, T.J. Ecological criteria for evaluating coral reefs and their implications for managers and researchers. *Coral Reefs* **1995**, *14*, 183–192. [CrossRef]
12. Connell, J.H.; Hughes, T.P.; Wallace, C.C. A 30-year study of coral abundance, recruitment, and disturbance at several scales in space and time. *Ecol. Monogr.* **1997**, *67*, 461–488. [CrossRef]
13. Hochberg, E.J.; Atkinson, N.J.; Andréfouët, S. Spectral reflectance of coral reef bottom-types worldwide and implications for coral reef remote sensing. *Remote Sens. Environ.* **2003**, *85*, 159–173. [CrossRef]
14. Louchard, E.M.; Reid, R.P.; Stephens, F.C. Optical remote sensing of benthic habitats and bathymetry in coastal environments at Lee Stocking Island, Bahamas: A comparative spectral classification approach. *Limnol. Oceanogr.* **2003**, *48*, 511–521. [CrossRef]
15. Karpouzli, E.; Malthus, T.J.; Place, C.J. Hyperspectral discrimination of coral reef benthic communities in the western Caribbean. *Coral Reefs* **2004**, *23*, 141–151. [CrossRef]
16. Goodman, J.A.; Ustin, S.L. Classification of benthic composition in a coral reef environment using spectral unmixing. *J. Appl. Remote Sens.* **2007**, *1*, 1–17. [CrossRef]
17. Lesser, M.P.; Mobley, C.D. Bathymetry, water optical properties, and benthic classification of coral reefs using hyperspectral remote sensing imagery. *Coral Reefs* **2007**, *25*, 819–829. [CrossRef]
18. Dekker, A.G.; Phinn, S.R.; Anstee, J.; Bissett, P.; Brando, V.E.; Casey, B.; Fearns, P.; Hedley, J.; Klonowski, W.; Lee, Z.P.; et al. Intercomparison of shallow water bathymetry, hydro-optics, and benthos mapping techniques in Australian and Caribbean coastal environments. *Limnol. Oceanogr. Methods* **2011**, *9*, 396–425. [CrossRef]
19. Hedley, J.D.; Mumby, P.J. A remote sensing method for resolving depth and subpixel composition of aquatic benthos. *Limnol. Oceanogr.* **2003**, *48*, 480–488. [CrossRef]
20. Botha, E.J.; Brando, V.E.; Anstee, J.M.; Dekker, A.G.; Sagar, S. Increased spectral resolution enhances coral detection under varying water conditions. *Remote Sens. Environ.* **2013**, *131*, 247–261. [CrossRef]
21. Hedley, J.D.; Roelfsema, C.M.; Chollett, I.; Harborne, A.R.; Heron, S.F.; Weeks, S.J.; Skirving, W.J.; Strong, A.E.; Eakin, C.M.; Christensen, T.R.L.; et al. Remote sensing of coral reeffs for monitoring and management: A review. *Remote Sens.* **2016**, *8*, 118. [CrossRef]
22. Andréfouët, S.; Berkelmans, R.; Odriozola, L.; Done, T.; Oliver, J.; Muller-Karger, F. Choosing the appropriate spatial resolution for monitoring coral bleaching events using remote sensing. *Coral Reefs* **2002**, *21*, 147–154. [CrossRef]

23. Lee, C.M.; Cable, M.L.; Hook, S.J.; Green, R.O.; Ustin, S.L.; Mandl, D.J.; Middleton, E.M. An introduction to the NASA Hyperspectral Infrared Imager (HyspIRI) mission and preparatory activities. *Remote Sens. Environ.* **2015**, *167*, 6–19. [CrossRef]

24. Moses, W.J.; Bowles, J.H.; Lucke, R.L.; Corson, M.R. Impact of signal-to-noise ratio in a hyperspectral sensor on the accuracy of biophysical parameter estimation in case II waters. *Opt. Express* **2012**, *2*, 4310–4330. [CrossRef] [PubMed]

25. Gillis, D.B.; Bowles, J.H.; Montes, M.J.; Moses, W.J. Propagation of sensor noise in oceanographic hyperspectral remote sensing. *Opt. Exp.* **2018**, *26*, A818–A831. [CrossRef] [PubMed]

26. Hedley, J.D.; Roelfsema, C.M.; Phinn, S.R.; Mumby, P.J. Environmental and sensor limitations in optical remote sensing of coral reefs: Implications for monitoring and sensor design. *Remote Sens.* **2012**, *4*, 271–302. [CrossRef]

27. Schuster, A. Radiation through a foggy atmosphere. *Astrol. J.* **1905**, *21*, 1–22. [CrossRef]

28. Philpot, W.D.; Ackleson, S.G. *Remote Sensing of Optically Shallow, Vertically Inhomogeneous Waters: A Mathematical Model*; DEL-SG-12-81; Delaware Sea Grant Collage Program, University of Delaware: Newark, DE, USA, 1981; pp. 283–299.

29. Maritorena, S.; Morel, A.; Gentili, B. Diffuse reflectance of oceanic shallow waters: Influence of water depth and bottom albedo. *Limnol. Oceanogr.* **1994**, *39*, 1689–1703. [CrossRef]

30. Ackleson, S.G.; Smith, J.P.; Rodriguez, L.M.; Moses, W.J.; Russell, B.J. Autonomous coral reef survey in support of remote sensing. *Front. Mar. Sci.* **2017**, *4*, 325. [CrossRef]

31. Lee, Z.P.; Carder, K.L.; Mobley, C.D.; Steward, R.G.; Patch, J.S. Hyperspectral remote sensing for shallow waters: 2. Deriving bottom depths and water properties by optimization. *Appl. Opt.* **1999**, *38*, 3831–3843. [CrossRef]

32. Gordon, H.R.; Brown, O.B.; Evans, R.H.; Brown, J.W.; Smith, R.C.; Baker, K.S.; Clark, D.K. A semianalytic radiance model of ocean color. *J. Geophys. Res.* **1988**, *93*, 10909–910924. [CrossRef]

33. Kirk, J.T.O. Dependence of relationship between inherent and apparent optical properties of water on solar altitude. *Limnol. Oceanogr.* **1984**, *29*, 350–356. [CrossRef]

34. Pope, R.M.; Fry, E.S. Absorption spectrum (380–700 nm) of pure water. II. Integrating cavity measurements. *Appl. Opt.* **1997**, *36*, 8710–8723. [CrossRef] [PubMed]

35. Stramski, D.; Bricaud, A.; Morel, A. Modeling the inherent optical properties of the ocean based on the detailed composition of the planktonic community. *Appl. Opt.* **2001**, *40*, 2929–2945. [CrossRef] [PubMed]

36. Bricaud, A.; Morel, A.; Prieur, L. Absorption by dissolved organic matter of the sea (yellow substance) in the UV and visible domains. *Limnol. Oceanogr.* **1981**, *26*, 43–53. [CrossRef]

37. Ackleson, S.G.; Balch, W.M.; Holligan, P.M. Response of water-leaving radiance to particulate calcite and chlorophyll-a concentrations: A model for Gulf of Maine coccolithophore blooms. *J. Geophys. Res.* **1994**, *99*, 7483–7499. [CrossRef]

38. Zhang, X.; Hu, L. Estimating scattering of pure water from density fluctuation of the refractive index. *Opt. Express* **2009**, *17*, 1671–1678. [CrossRef] [PubMed]

39. Gao, B.; Montes, M.J.; Ahmad, Z.; Davis, C.O. Atmospheric correction algorithm for hyperspectral remote sensing of ocean color from space. *Appl. Opt.* **2000**, *39*, 887–896. [CrossRef]

40. Montes, M.J.; Gao, B.-C.; Davis, C.O. *NRL Atmospheric Correction Algorithms for Oceans: Tafkaa User's Guide*; NRL/MR/7230-04-8760; The U.S. Naval Research Laboratory: Washington, DC, USA, 2004.

41. Devore, J.L. *Probability and Statistics for Engineering and the Sciences*, 9th ed.; Cengage Learning: Boston, MA, USA, 2016; pp. 1–225. ISBN 978-1-305-25180-9.

42. Boss, E.; Zaneveld, J.R.V. The effect of bottom substrate on inherent optical properties: Evidence of biogeochemical processes. *Limnol. Oceanogr.* **2003**, *48*, 346–354. [CrossRef]

43. Myers, M.R.; Hardy, J.T.; Mazel, C.H.; Dustan, P. Optical spectra and pigmentation of Caribbean reef corals and macroalgae. *Coral Reefs* **1999**, *18*, 179–186. [CrossRef]

44. Baker, A.C.; Glynn, P.W.; Riegl, B. Climate change and coral reef bleaching: An ecological assessment of long-term impacts, recovery trends and future outlook. *Estuar. Coast. Shelf Sci.* **2008**, *80*, 435–471. [CrossRef]

45. Mazel, C.H. Measurement of spectral fluorescence and reflectance of benthic marine organisms and substrates. *Opt. Eng.* **1997**, *36*, 2612–2617. [CrossRef]

46. Enríquez, S.; Méndez, E.R.; Iglesias-Prieto, R. Multiple scattering on coral skeletons enhances light absorption by symbiotic algae. *Limnol. Oceanogr.* **2005**, *50*, 1025–1032. [CrossRef]

47. Morfitt, R.; Barsi, J.; Levy, R.; Markham, B.; Micijevic, E.; Ong, L.; Scaramuzza, P.; Vanderwerff, K. Landsat-8 Operational Land Imager (OLI) radiometric performance on-orbit. *Remote Sens.* **2015**, *7*, 2208–2237. [CrossRef]

48. El-Askary, H.; Abd El-Mawla, S.H.; El-Hattab, M.M.; El-Raey, M. Change detection of coral reef habitats using Landsat-5 TM, Landsat 7 ETM, and Landsat 8 OLI data in the Red Sea (Hurghada, Egypt). *Int. J. Remote Sens.* **2014**, *35*, 2327–2346.

49. Roelfsema, C.; Kovacs, E.; Ortiz, J.C.; Wolff, N.H.; Callaghan, D.; Wettle, M.; Ronan, M.; Hamylton, S.M.; Mumby, P.J.; Phinn, S. Coral reef habitat mapping: A combination of object-based image analysis and ecological modeling. *Remote Sens. Environ.* **2008**, 27–41. [CrossRef]

50. Green, R.O. Lessons and key results from 30 years of imaging spectroscopy. In *Imaging Spectrometry XIX, Proceedings of SPIE 9222, San Diego, CA, USA, 17–21 August 2014*; SPIE: Bellingham, WA, USA, 2014. [CrossRef]

51. Mouroulis, P.; Van Gorp, B.; Green, R.O.; Dierssen, H.; Wilson, D.W.; Eastwood, M.; Boardman, J.; Gao, B.; Cohen, D.; Fraklin, B.; et al. Portable remote imaging spectrometer coastal ocean sensor: Design, characteristics, and first flight results. *Appl. Opt.* **2014**, *53*, 1364–1380. [CrossRef] [PubMed]

52. National Academies of Sciences, Engineering, and Medicine. *Thriving on Our Changing Planet: A Decadal Strategy for Earth Observation from Space*; The National Academies Press: Washington, DC, USA, 2018. [CrossRef]

53. Muller-Karger, F.E.; Hestir, E.; Ade, C.; Turpie, K.; Roberts, D.A.; Siegel, D.; Miller, R.J.; Humm, D.; Izenberg, N.; Keller, M.; et al. Satellite sensor requirements for monitoring essential biodiversity variables of coastal ecosystems. *Ecol. Appl.* **2018**, *28*, 749–760. [CrossRef] [PubMed]

applied
sciences

MDPI

Article

Advantages and Limitations to the Use of Optical Measurements to Study Sediment Properties

Emmanuel Boss [1,*], **Christopher R. Sherwood** [2], **Paul Hill** [3] and **Tim Milligan** [3]

[1] University of Maine, Orono, ME 04469, USA
[2] U.S. Geological Survey, Woods Hole, MA 02543, USA; csherwood@usgs.gov
[3] Dalhousie University, 1355 Oxford Street, Halifax, NS B3H 4R2, Canada; paul.hill@dal.ca (P.H.);
    milligantg@gmail.com (T.M.)
*   Correspondence: emmanuel.boss@maine.edu; Tel.: +1-207-745-3061

Received: 6 September 2018; Accepted: 12 December 2018; Published: 19 December 2018

Featured Application: Sediment characteristics and dynamics are studied using in situ optical measurements.

Abstract: Measurements of optical properties have been used for decades to study particle distributions in the ocean. They are useful for estimating suspended mass concentration as well as particle-related properties such as size, composition, packing (particle porosity or density), and settling velocity. Measurements of optical properties are, however, biased, as certain particles, because of their size, composition, shape, or packing, contribute to a specific property more than others. Here, we study this issue both theoretically and practically, and we examine different optical properties collected simultaneously in a bottom boundary layer to highlight the utility of such measurements. We show that the biases we are likely to encounter using different optical properties can aid our studies of suspended sediment. In particular, we investigate inferences of settling velocity from vertical profiles of optical measurements, finding that the effects of aggregation dynamics can seldom be ignored.

Keywords: particle dynamics; optical properties; suspended sediment

---

1. Introduction

Optical properties have long been used to study suspended particles and their dynamics (e.g., reviews by [1–3]). The most commonly measured optical properties are attenuation and scattering at different angles (both forward and back). Other optical devices, including ambient radiation sensors, cameras, and holographic instruments, also produce valuable data, but this paper will focus primarily on measurements of attenuation and scattering. Measurement volumes are typically small (from a few mL to tens of mL) and temporal averaging can increase the likelihood that rare large particles are sampled. Optical measurements can provide relatively direct estimates of mass or volume concentrations and particle size, and they also can be used to infer information about particle density, composition, and settling velocity. The primary advantages of using optical properties to study suspended particles are that they can be obtained at high frequency over long periods, and they are relatively non-invasive. Interpretation of optical measurements, however, are complicated by the fact that measurements are affected by all the particles in the suspension, but as we explain below, they do not respond to all particles equally. Other known disadvantages of optical instruments are that they saturate at high particle concentrations (e.g., [2]); they are intrusive and can produce turbulent wakes; they can be affected by ambient light; they can have large power demand; and they are susceptible to bio-fouling.

1.1. Optical Proxies of Properties of Sediment Particles

1.1.1. Volume and Mass Concentration

Most of the variation in optical signals measured in the field is due to changes in suspended particle mass concentrations (SPM; e.g., [4]). Optical estimates of SPM typically are made with measurements of attenuation or scattering at long visible wavelengths (e.g., 660 nm) or at infrared wavelengths (e.g., 850 nm), minimizing the impacts from varying dissolved materials and particulate absorption [5]. Transmitted light, which is reduced by scattering and absorption by particles, is measured in the near forward direction by so-called transmissometers, and scattered light can be measured at an angle near 90° by nephelometers, or at an angle greater than 90° by optical backscatter sensors (OBS). A multi-site comparison of the application of backscattering, side-scattering, and attenuation as proxies for SPM demonstrated their ability to predict SPM within 36%, 51%, and 54% respectively, for 95% of all cases [4]. The differences likely are due to variable sensitivity of each property to particle size, packing, and composition [1,6–8]. For example, the acceptance angle of transmissometers acts to filter out responses from larger particles [9]. SPM has also been estimated from space-based measurements of radiance (e.g., [10]). Remotely-sensed reflectance is most sensitive to the particulate backscattering coefficient in red and NIR wavelengths [11].

1.1.2. Size

Suspended sediment ranges in size from sub-μm-sized clay platelets to mm-sized sand and even larger flocs. It also encompasses plankton, non-algal organic particles, and aggregates that can be mixture of both organic and inorganic particles. We ignore in this paper particles capable of sinking at speeds >10 mm s^{-1}. These are rarely in suspension and, when they are in suspension, concentrations may saturate optical instruments. Optical proxies for size information include size distributions inverted from measurements of near forward scattering at several angles [12], the exponents of power-law fits of the particulate attenuation or backscattering spectrum [13–15] (but see [16]), and the fluctuation in optical signals, which can be used to obtain the average size of suspended particles [17]. Images of particles have also been used to derive size distributions, particularly of larger flocs and aggregates (e.g., [1,3]).

Theoretically, the maximal response of attenuation or scattering per volume (equivalent to mass if density is constant) occurs for single-grain sediment near $(D/\lambda)(n-1) \sim 1$ where λ is the wavelength in water (= $0.75\lambda_{air}$), D is the particle diameter, and n is the index of refraction of the particle relative to water ([18], Figure 1). For solid inorganic particles with $n = 1.15$ at $\lambda_{air} = 660$ nm, the maximal attenuation per mass occurs for small particles with diameters between 0.8–3.2 μm. This dependency decreases as $1/D$ as D increases, and it increases for larger indices of refraction (Figure 1). Increases in the index of refraction are typically associated with increases in the inorganic fraction in the particle suspension [19].

The dependence of attenuation or scattering per unit of suspended mass on size should limit their use for estimation of SPM in suspensions with varying particle size, yet they are reasonably precise proxies for SPM across a range of environments, as discussed previously. This paradox is resolved if particles in suspension are not primarily single solid particles, but rather are agglomerations of particles separated by relatively large volumes of interstitial fluid and transparent organic material [1,20]. Terminology used to describe particle agglomerations in suspension can be ambiguous. Here we refer to a "floc" as an agglomeration of material that forms in suspension at relatively short times scales (e.g., tides) that is susceptible to breakup under increasing shear stress. We use the term "aggregate" to refer to an agglomeration that has undergone multiple cycles of resuspension and deposition, during which it has become more compacted and more strongly bound. While the component particles in flocs and aggregates may be similar, the density and settling velocity of aggregates is greater [21]. Flocs and aggregates typically sink faster than their component particles [22,23]. Flocs are broken by shear, including that generated by turbulence and by sinking, limiting their maximal size to about

1 cm [20,22,24–26]. Aggregates, because they are more compact and stronger, are not broken easily by shear. In terms of their optical properties, flocs, if sufficiently porous, can maintain the efficiency of scattering of the particles of which they are made [27], resulting in mass-specific optical properties that are independent of floc size. These theoretical results were validated in laboratory and field experiments that found that attenuation and backscattering to SPM ratios remained relatively constant despite large changes in particle size [1,28]. More densely packed aggregates theoretically should have mass-specific optical attenuation, scattering and backscattering coefficients that decrease with increasing aggregate size, but at a rate that is less than the $1/D$ dependence of solid particles. This theoretical relationship has not been demonstrated with measurements.

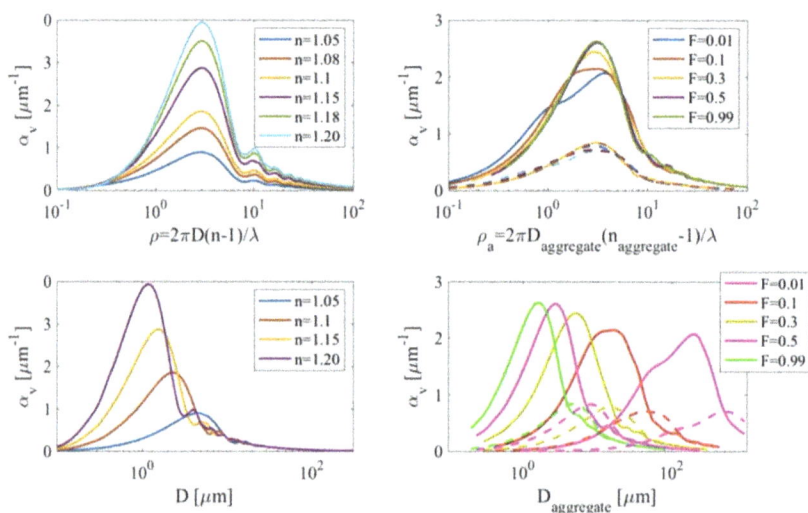

Figure 1. Top panels: volume-specific beam attenuation (α_v for solid particles (left) and aggregates (right) as function of $(2\pi D)/\lambda_{water}$ ($n-1$) where D is diameter, λ_{water} the wavelength in water and n the index of refraction (which increases between organic and inorganic particles). For aggregates, $n_{aggregate} = 1 + F(n-1)$, where F is the solid fraction and n the index of refraction of the particles that comprise the aggregate. In all cases, the volume used to compute α_v is that of the solid fraction. Bottom panels: same as on top but plotted as function of particle diameter. In all the computations $\lambda_{air} = 660$ nm and $n' = 0.0001$, where n' is the imaginary part of the index of refraction, representing absorption. For aggregates we use $n = 1.15$ (solid) and $n = 1.05$ (dashed) typical of inorganic and organic materials, respectively.

1.1.3. Composition

Composition of suspended particles ranges from inorganic clays and silts of varied mineralogy to organic particles including both pigmented phytoplankton and non-algal particles. An optical proxy for composition (separating dominance by organic and inorganic particles) is the ratio of backscattering to total scattering [29,30]. Fluid-filled organic particles such as plankton have lower indexes of refraction compared to inorganic particles, resulting in a lower backscattering relative to total scattering. The ratio of chlorophyll-containing particles to total particles, estimated from the ratio of chlorophyll absorption to the particulate beam attenuation, has also been used as a compositional proxy in locations where the particle assembly includes a significant inorganic component [31]. In addition, the ratio of particulate organic carbon to SPM was found to correlate well with the ratio of particulate absorption at 675 nm to that at 570 nm [32]. It should be noted that optically derived knowledge of composition and/or size can lead to improved estimates of particulate concentration from optical measurements and associated composition (or size)-specific algorithms relative to those that do not account for variations in particle composition and/or size [1,16,32,33].

1.1.4. Packing

A proxy for bulk particle density has been derived from the ratio of beam attenuation (a mass concentration proxy) to the total particulate volume (obtained by inverting the forward scattering measurements with the Laser In Situ Scattering and Transmission (LISST) sensor [28,34,35]). If particle volume increases due to aggregation (with a significant fluid fraction), particle density will decrease. This will happen up to the point where the particle has a fractal dimension that is reduced below two, at which point most light passes through the particle and attenuation is caused only by component particles. Because the beam attenuation is also closely linked to the summed cross-sectional area of all the particles, it follows that the above proxy should behave similarly to the suspension's volume-to-cross-sectional-area ratio, or Sauter diameter [34].

1.2. Particle Dynamics

In this section, we focus on the bottom boundary layer (BBL; for a recent review of the fluid dynamics of the BBL see [36]). The particle assembly C is a sum of many types of particles $C = \sum_i C_i$, each with a specific settling speed, composition, size and any other property that is relevant to either hydrodynamic or optical properties.

The general conservation equation for particles can be written as:

$$\frac{\partial C_i}{\partial t} + \nabla(C_i(U - w_i)) = \nabla(K\nabla C_i) + \sum_j f(C_{i-j}, C_j) + S_i \tag{1}$$

where C_i is the mass concentration of particles of type i, t is time, U is the 3-D velocity field, w_i is the settling velocity of these particles (we use the convention that it is positive downward), K is a diffusion coefficient, $f(C_{i-j}, C_j)$ represents aggregation and disaggregation dynamics creating and destroying C_i-type particles, and S_i represents other sources and sinks (e.g., resuspension/deposition or biological production/consumption of particles).

To solve Equation (1) for a given flow field, one needs boundary conditions (e.g., flux or concentration of particles at the boundaries of the domain) and initial conditions (e.g., the state of the particles at time $t = 0$). Even then, Equation (1) cannot be solved analytically except for very simple cases such as we will address next.

Equation (1) can be simplified to represent only the vertical dimension (z, positive upwards), assuming that horizontal advection of horizontal gradients is negligible relative to vertical processes. After time averaging (or assuming steady-state), the steady balance between downward settling, upward diffusion, exchange among particle types (aggregation/disaggregation), and sources/sink is:

$$-w_i \frac{dC_i}{dz} = \frac{d(K_{eddy}dC_i)}{dz^2} + \sum_j f(C_{i-j}, C_i) + S_i \tag{2}$$

where C_i is the time- (and horizontally) averaged concentration of particles of type i, K_{eddy} is an eddy diffusion coefficient (representing the mixing by the small-scale turbulent field), and w_i is a constant settling velocity for each particle type i.

Large aggregation rates are associated with large particle concentrations, e.g., following a major resuspension event or during an algal bloom. Large disaggregation rates are associated with large fluid shears, e.g., in the wave boundary layer, which is a layer a few centimeters thick next to the bottom. If particle concentrations and fluid shears do not vary greatly throughout a boundary layer, then an equilibrium size distribution develops, and the aggregation and disaggregation terms cancel one another. An assumed equilibrium under certain circumstances is unlikely to emerge in bottom boundary layers, for example when near-bed wave-generated shears are much greater than current-generated shears higher in the boundary layer.

The law-of-the-wall is often invoked in the BBL, which is consistent with a linear eddy diffusivity profile where $K_{eddy} = \kappa u_* z$, and κ is von Kármán's constant (~ 0.4), and u_* is the friction velocity (a function of BBL turbulence, e.g., due to wave and current shear). Neglecting aggregation dynamics and sources/sinks, and integrating Equation (2) in the vertical and solving the resulting differential equation results in the Rouse equation [37,38] for a homogenous population of particles of type i:

$$C_i(z) = C_i(z_a) \left(\frac{z}{z_a} \right)^{-\frac{w_i}{\kappa u_*}} \tag{3}$$

where z_a is a reference elevation where a particle concentration is assigned. This is one of several different analytical solutions for the balance of settling and turbulent mixing, which vary depending on assumptions about the eddy diffusivity profile [39,40]. An alternative derivation of Equation (3) using probability theory has also been proposed [41]. Equation (3) predicts a profile that is linear in $\log(C_i)$ versus $\log(z)$, and all of the other forms predict similar decreases in concentration with elevation above the bottom. Because mass concentrations can be summed, the bulk concentration profile is:

$$C_b(z) = \sum_i C_i(z) = \sum_i \left(C_i(z_a) \left(\frac{z}{z_a} \right)^{-\frac{w_i}{\kappa u_*}} \right) \tag{4}$$

It follows that the bulk particle concentration will also decrease with height above the bottom. Note, however, that if a suspension comprises sub-populations of particle types with differing settling velocities, each will have a different vertical profile (Equation (3)), and the profile of the summed concentration will follow Equation (4), which does not, in general, follow the linear shape in log-log space. In fact, large deviations in the bulk profile from Equation (3), under steady conditions in a turbulent BBL, are likely indications that a suspension contains particles with diverse settling velocities.

1.3. An Equation for the Vertical Distribution of an Optical Property

The mass concentration of a population of particle type i is $C_i = N_i V_i \rho_I$, where ρ_i is individual particle density, N_i is the number concentration of particles of type i, and V_i is individual particle volume. By the Beer–Lambert law the bulk optical response (e.g., backscattering or beam attenuation) of a sub-population $b_{x,i}$, is the simple sum of the individual contributions so that:

$$b_{x,i} = \alpha_{v,i} N_i V_i \tag{5}$$

and the bulk response to the combined sub-populations is:

$$b_x = \sum_i b_{x,i} \tag{6}$$

where $\alpha_{v,i}$, is a volume-specific optical property. Volume- or mass-specific optical properties typically are calculated by using Mie theory (which assumes homogeneous spherical particles) for solid particles, or by using other models for flocs (e.g., [31]). The calculations of $\alpha_{v,i}$ require as input particle diameter, index of refraction (a function of composition), wavelength of light, and, for flocs, the fractal dimension that relates volume concentration to mass concentration. The results are resonance-like functions of size (Figure 1). Although the application of Mie solutions to backscattering has been challenged based on observations (e.g., [42]), it is reasonable to relate constant values of $\alpha_{v,i}$ to specific particle types. Both Equations (5) and (6) rely on the Beer–Lambert law, with underlying assumptions that the light is monochromatic with parallel rays and, most importantly, that the particles do not scatter the light multiple times. This is clearly not the case as particle concentrations rise but, at low concentrations, we may be able to assume that Equations (5) and (6) are valid. Substituting optical response for concentration in Equation (3) yields an equation for the optical-response profile for a homogenous sub-population i with identical $\alpha_{v,i}$, V_i, and w_i:

$$b_{x,i}(z) = b_{x,i}(z_a) \left(\frac{z}{z_a} \right)^{-\frac{w_i}{\kappa u_*}} \tag{7}$$

Equations (6) and (7) can be combined to yield an equation for the bulk response of a heterogeneous population:

$$b_x(z) = \sum_i b_{x,i}(z_a) \left(\frac{z}{z_a} \right)^{-\frac{w_i}{\kappa u_*}} = \sum_i \frac{\alpha_{v,i} C_i(z_a)}{\rho_i} \left(\frac{z}{z_a} \right)^{-\frac{w_i}{\kappa u_*}} \tag{8}$$

As with particle concentrations, we find that the information provided by a profile of optical properties will depend on how heterogeneous the suspension is. Unlike the concentration profile, the averaging done by the optical properties depends on how the optics respond to different particles (Equation (8)). Hence, the more biased an optical property is towards a specific particle type, the better it can provide information on its specific settling speed.

2. Observations

Field data were obtained from a profiling instrument platform deployed at the Martha's Vineyard Coastal Observatory (MVCO) south of Martha's Vineyard, MA, USA, at the 12-m isobath in the summer of 2011, as part of the Office of Naval Research (ONR)-funded Optics and Acoustics and Stress In Situ (OASIS) experiment [43]. The platform was mounted on a pivoting arm that profiled every 20 min from 10 cm to 2 m above bottom. The platform was equipped with variety of optical and acoustical sensors. Here we discuss data from a Sequoia LISST 100-X (Sequoia Scientific, Bellevue, WA, USA), a EcoBB2f (WETLabs, Philomath, OR, USA) triplet measuring dissolved organic fluorescence and backscattering at two wavelengths (532 and 650 nm), and a WETLabs AC-9 spectral absorption and beam attenuation meter. Water was pumped from an intake at the tip of the arm into the 10-cm pathlength sampling volume of the AC-9. An automatic valve periodically routed the water sample through a 0.2-μm filter to remove particulates, leaving the dissolved fraction, to obtain calibration-independent particulate properties [44]. Shear velocity u_{*c} [cm/s] associated with mean flow in the bottom boundary layer (BBL) was inferred from a pair of acoustic Doppler velocimeter (Sontek, San-Diego, CA, USA) measurements (cf. [45]), and model estimates of the wave–current combined maximum shear velocity in the wave boundary layer (relevant to sediment resuspension) were determined using a version of the Grant–Madsen model [46,47]. We have plotted sign(u) u_{*c} in Figure 2, where u is the east–west component of the current velocity, to differentiate flood (east, positive) from ebb (west, negative). The location is dominated by the east–west semi-diurnal tidal currents, northward swell, and periodic storms (Figure 2). The measurements discussed here were taken in the bottom boundary layer, which is mostly well mixed.

Waves and currents varied during the experiment, and we have selected four periods with distinctly different forcing and optical responses (Figure 2). Two periods (Maria and Ophelia) were associated with offshore passage of hurricanes and arrival of swell waves at the study site. Another period (Spring tides) was associated with moderate wave conditions and strong spring tidal currents. The fourth period (Calm) was characterized by low waves and weaker neap tidal currents. For each of the four periods identified in Figure 2, we show the distribution of the following optical properties (or properties inferred from optical measurements):

1. Beam attenuation $c_p(650)$ [m^{-1}] measured by the AC-9 and particulate backscattering coefficient ($b_{bp}(650)$ [m^{-1}] measured by the EcoBB2F provide proxies of particulate concentration (e.g., [5], where higher values associated with higher particle concentrations.
2. Exponent of the power-law fits of the particulate beam attenuation γ_{cp} [dimensionless] and backscattering γ_{bbp} [dimensionless] ($c_p = c_p(\lambda_0)(\lambda/\lambda_0)^{-\gamma_{cp}}$, with an analogous formula for γ_{bbp})

provide proxies for size distribution in the finer sizes (e.g., [13]). Lower values are associated with larger size averaged particles. γ_{cp} is biased towards the smaller (0.5 to 10 μm) particles in the population [9], and γ_{bbp} may be more sensitive to larger particles [15].

3. Sauter diameter D_s [μm] is determined from the ratio of LISST measurements of volume and area concentrations, summed over size classes i as $D_s = 1.5 \sum V_i / \sum A_i$ and reciprocal of particle density $\rho_a^{-1} = \sum V_i / c_p$ [m ppm^{-1} = μm], using the LISST-based c_p. Both are proxies for packing: larger values of D_s indicate larger, less-dense particle populations, and larger values of ρ_a^{-1} also indicate less-dense particle populations.

4. Particulate backscattering ratio $b_{bp}(532)/b_p(532)$ measured by the EcoBB2F ($b_{bp}(532)$) and by differencing of particulate attenuation and particulate absorption from the AC-9 ($b_p(532)$) was a proxy of composition. Increasing values of this ratio are associated with inorganic particles [29,30]. For very small particles, this ratio is also sensitive to size, increasing for smaller particles.

5. Chlorophyll to attenuation ratio $Chl/c_p(650)$ is another proxy of composition where higher values are associated with higher phytoplankton-based organic content [31].

6. LISST-based size distribution spanning from 2–250 μm at 32 size bins and using a spherical kernel.

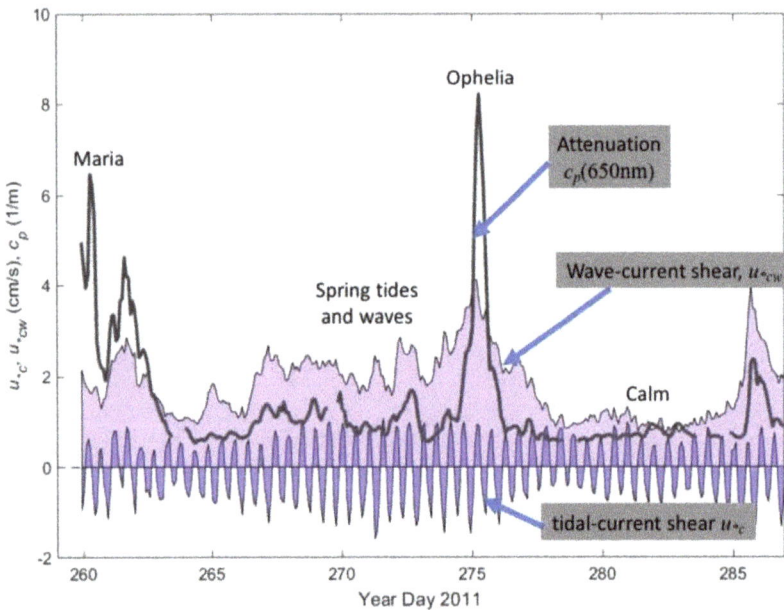

Figure 2. Time series of conditions at the 12-m Martha's Vineyard Coastal Observatory (MVCO) site during the Optics and Acoustics and Stress In Situ (OASIS) deployment in 2011: beam attenuation at 650 nm measured at 1 m above the bottom (gray), tidal current shear velocity sign(u)u_{*c} (blue), and combined wave-current shear velocity u_{*cw} (purple). Notice labels describing specific periods.

We selected 1-h intervals during each of the different periods identified in Figure 2 when the concentration profiles estimated from attenuation decreased monotonically with elevation above the seabed, and the profiles were relatively well approximated by a linear profile in log-log coordinates (Equation (5)). These profiles were consistent with the steady-state Rouse balance discussed above, suggesting that we might be able to neglect the effects of horizontal gradients and temporal transients. The data represent the average of three consecutive profiles, each of which took 20 min to complete. All properties are displayed as a function of elevation above the bottom in Figures 3–6. Trends in vertical profiles and representative values (mean and standard deviation) for each time period are summarized in Tables 1 and 2.

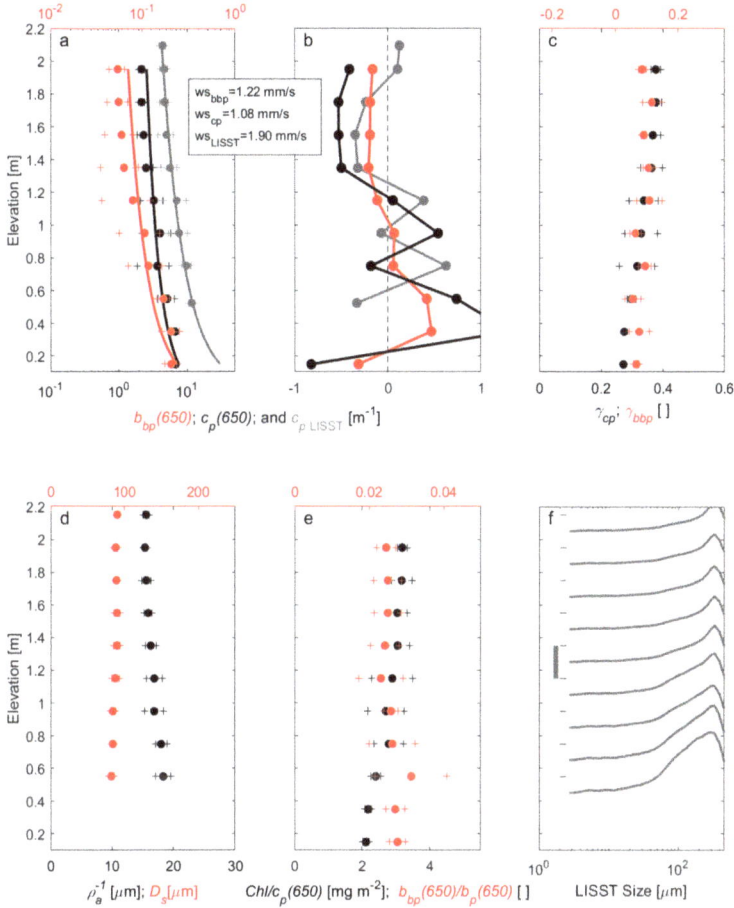

Figure 3. Profiles of optical parameters at the MVCO site on day 261.1 when waves from Hurricane Maria were coming to shore (see Figure 2). (**a**) Beam attenuation (c_p(650), black), particulate backscattering (b_{bp}(650), red), and LISST attenuation (gray). (**b**) Deviation from the Rouse-profile fits for the same parameters (i.e., observed minus fit). (**c**) Power-law exponent of c_p(650) (black) and b_{bp}(650) (red). (**d**) Sauter diameter (red) and inverse particle density (black). (**e**) Ratio of chlorophyll divided by c_p(650) (black) and backscattering ratio (red). (**f**) Spectra of LISST volume concentration as a function of size at nine elevations. In panels (**a**, **c**, **d**, and **e**), standard deviation about the mean values for three consecutive profiles (60 min) are shown with crosses. Dashed lines in panel (**a**) are log-log (Rouse) fits to the data. In panel (**f**), the elevations for each spectrum are indicated by black lines, and the gray vertical scale indicates 10 μL/L. Numbers in the box denote settling velocities based on Rouse fits to backscattering, AC-9 particulate attenuation at 660 nm, and LISST attenuation at 670 nm.

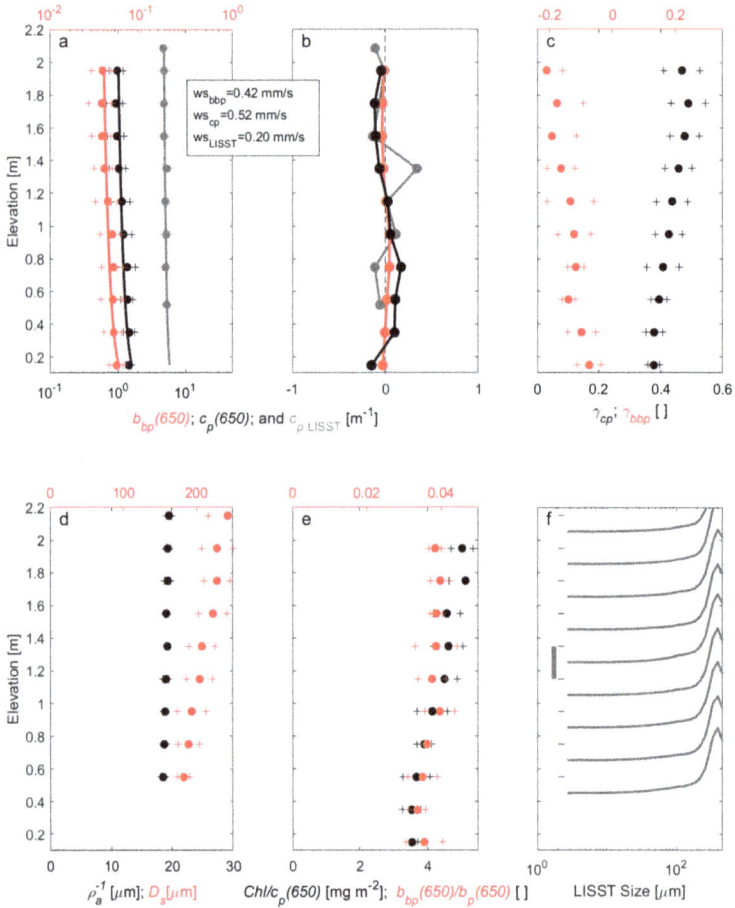

Figure 4. Profiles of optical parameters at MVCO on day 268.62 during spring tides and moderate waves. Panels are as described in Figure 3.

Table 1. Trends in particle parameters as function of depth based on the average of three profiles in the BBL in each of the four periods denoted in Figure 2. ↘ and ↗ denote profiles that are decreasing or increasing (respectively) with elevation above the bottom, ~ denotes that the trend is weak, and | denotes that there is no trend with elevation.

Parameter	Maria	Spring Tide	Ophelia	Calm
$b_{bp}(650)$		↘	↘	~↘
$c_p(650)$	↘	↘	↘	↘
γ_{bbp}	↗	↘	↗	~↗
γ_{cp}	↗	↗	↗	|
D_s	|	↗	~↗	~↘
ρ_a^{-1}	~↘	↗	~↗	~↘
$b_{bp}(650)/b_p(650)$	~↗	↗	|	|
$Chl/c_p(650)$	↗	↗	~↗	↗

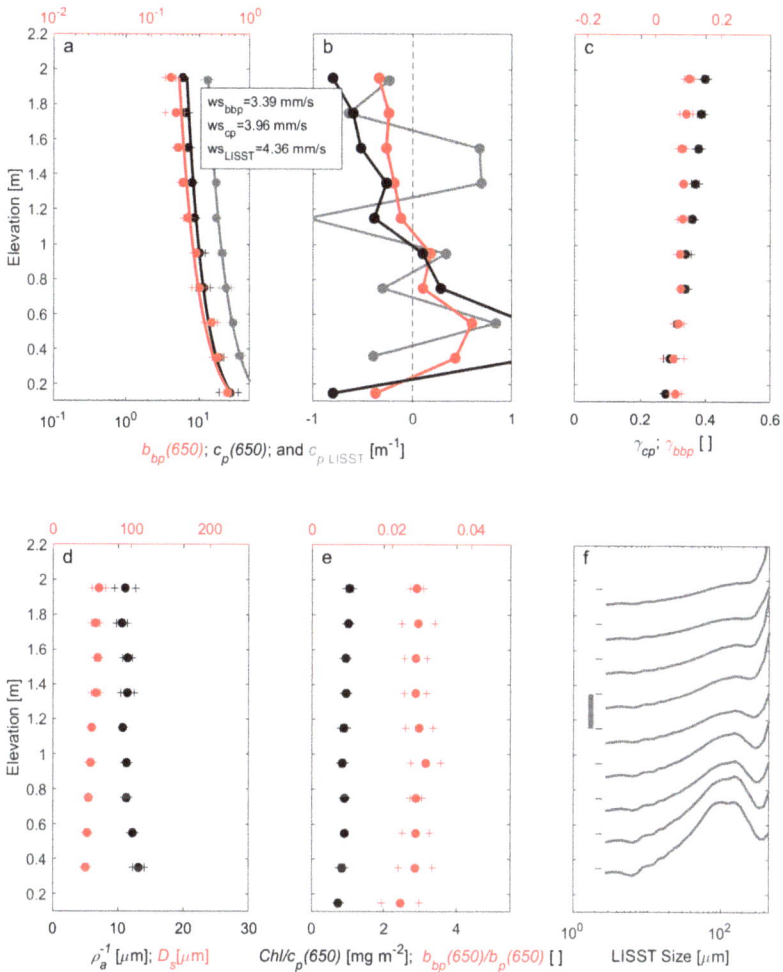

Figure 5. Profiles of optical parameters at MVCO on day 275.20 during the passage of Hurricane Ophelia. Panels are as described in Figure 3.

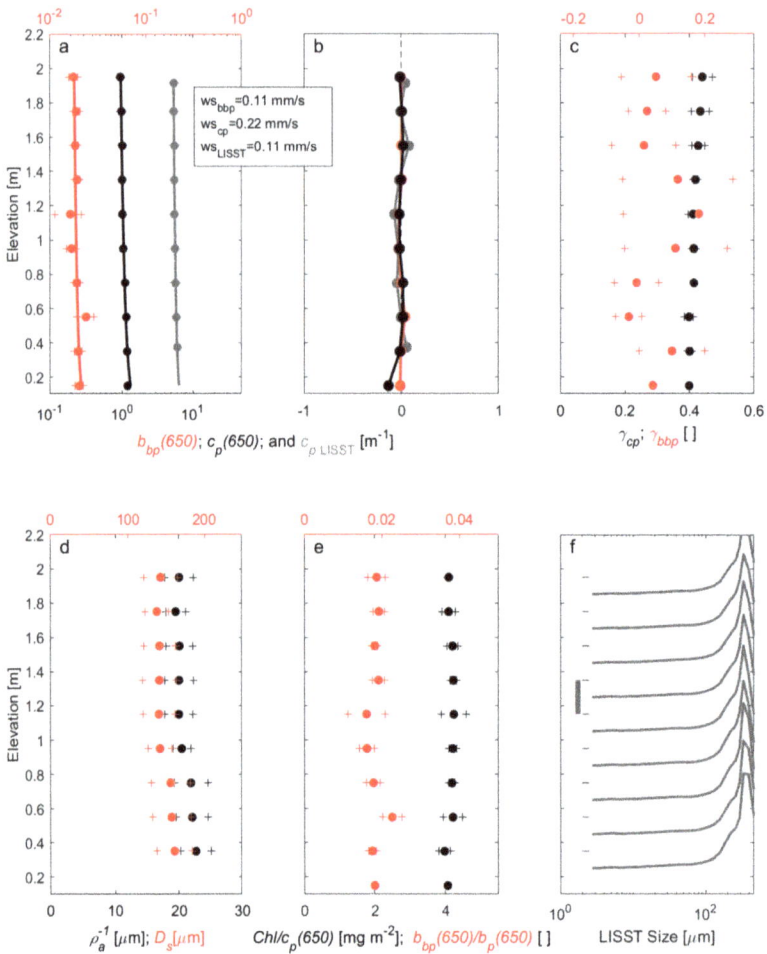

Figure 6. Profiles of optical parameters at MVCO on day 281.93 during calm conditions. Panels are as described in Figure 3.

3. Results

3.1. Suspended Particulate Material (SPM)

The overall concentration of SPM fluctuated through the experiment (Figure 2), as indicated by variations $c_p(650)$ (Figure 2) and $b_{bp}(650)$ (not shown). The time series of $c_p(650)$ at 1 m above the bottom indicates increased SPM during Maria and Ophelia, somewhat reduced SPM during the spring tides, and low SPM during the calm period. Mean values of $c_p(650)$ in the 60-min profiles ranged from about 1 m^{-1} during the calm period to 12 m^{-1} during Ophelia, when wave-induced resuspension increased SPM in the BBL. Mean values of $b_{bp}(650)$ generally covaried with $c_p(650)$, ranging from 0.02 m^{-1} to 0.30 m^{-1} for the same periods. The $c_p(650)$ and $b_{bp}(650)$ concentrations decreased with elevation, and the profiles examined here were nearly linear in log-log space (Rouse-like). However, there was typically more scatter in the profiles of $b_{bp}(650)$, as evidenced by their lower r^2 (Table 2), especially during periods of lower concentrations.

Table 2. Mean and standard deviation (in brackets) of optical particle parameters computed based on the average of three profiles in the BBL in each of the four periods denoted in Figure 2. The wave-current and current shear velocities and the settling velocities inferred from fitting the Rouse profile to c_p and b_{bp} are also listed, with regression coefficient r^2 in brackets.

Parameter	Maria	Spring Tide	Ophelia	Calm
$b_{bp}(650)$ [m^{-1}]	0.11(0.06)	0.04(0.01)	**0.30**(0.14)	0.02(0.002)
$c_p(650)$ [m^{-1}]	3.87(1.77)	1.19(0.22)	**11.81**(6.51)	1.06(0.09)
γ_{bbp}	**0.09**(0.02)	−0.14(0.06)	0.0(0.01)	0.06(0.07)
γ_{cp}	0.33(0.04)	**0.43**(0.04)	0.34(0.04)	0.42(0.01)
D_s [µm]	87.3(3.0)	**210.7**(20.5)	50.4(6.0)	147.0(8.7)
ρ_a^{-1} [m ppm^{-1}]	16.6(1.1)	19.1(0.3)	11.5(0.8)	**20.8**(1.1)
$b_{bp}(650)/b_p(650)$	0.03(0.002)	**0.04**(0.002)	0.03(0.002)	0.02(0.002)
$Chl/c_p(650)$ [µgm^{-2}]	2.8(0.4)	**4.3**(0.6)	0.9(0.1)	4.2(0.1)
u_{*wc} [cm/s]	2.3	2.4	**4.1**	0.8
u_{*c} [cm/s]	0.6	0.7	**1.8**	0.4
ws_{bbp} [cm/s]	1.22(0.72)	0.42(0.72)	**3.39**(0.94)	0.11(~0.0) *
ws_{cp} [cm/s]	1.08(0.80)	0.52(0.66)	**3.96**(0.99)	0.22(0.95)

* fit was not significant. Bold is used to highlight the largest values in each row.

3.2. Settling Velocities Assuming Rouse Profiles

Mean apparent settling velocities inferred from $c_p(650)$, $c_{p,\text{LISST}}$ and $b_{bp}(650)$ profiles ranged from about 0.1 to 4 mm/s. Although the settling velocities estimated from each instrument during a given period were similar, and showed similar trends over the four periods, $w_{s,cp(650)}$ was usually larger than $w_{s,bbp(650)}$. Particulate beam attenuation measured by the AC-9 is less sensitive to large, low-density particles, so settling-velocity estimates from $c_p(650)$ were likely to favor smaller, denser, faster-settling aggregates and single-grained particles. Inferred settling speeds were largest during passage of Maria and Ophelia. They were intermediate during spring tides, and they were lowest during the calm period. During passage of the two storms the residuals of the Rouse fits showed vertical structure, with negative residuals higher in the boundary layer and positive residuals lower in the boundary layer. This pattern indicates that, nearer the bed, the profiles during storms were steeper than the Rouse balance.

3.3. Size

The spectral exponents of attenuation and backscatter increased slightly with height above the bed during the calm periods and during the passage of Maria and Ophelia. During the spring tides, the spectral exponent of attenuation increased with height above the bed, but the spectral exponent of backscatter decreased with height above the bed. Values of γ_{cp} were smaller during passage of the two storms than during the calm period and during spring tides. Values of γ_{bbp} were smallest during spring tides, even taking on negative values. Values were higher during the storms, and they were scattered during the calm period.

Sauter diameter and the reciprocal of floc density generally were well correlated because they are based on similar measurements. Sauter diameter was smaller during passage of Ophelia, when it was ~50 µm, and during passage of Maria, when it was ~100 µm. During the spring tides, Sauter diameter was largest, with diameters ~200 µm. During the calm period, Sauter diameter was ~150 µm. During Maria, the spring tides, and the calm period, inverse densities were similar and equal to ~20 µm. Values were ~10 µm during the passage of Ophelia. During the passage of the two storms, inverse density and Sauter diameter values diverged near the bed, and during the spring tides, they diverged higher in the boundary layer.

Size distributions from the LISST provide a more complete understanding of the vertical and temporal changes in size distribution. During the calm period and during spring tides, the size distributions had well-defined modes at the upper end of the LISST range. The widths of the modes were relatively constant with height above the bed during the calm period, but they were wider nearer the bed during the spring tide period. During passage of Maria, distributions were unimodal, but they were more skewed to smaller sizes than during the calmer periods. During passage of Ophelia, curious bi-modal distributions emerged. One mode was centered on diameters of ~100 μm, and the other was located in the largest diameter class. This coarser mode was dominant farther away from the bed, and the finer mode grew larger nearer the bed.

3.4. Composition

The backscattering ratio was relatively high (>0.02) during the full deployment, indicating that the suspensions were dominated by inorganic particles. The ratio was smaller during the passage of the two storms than it was during the spring tides. The backscatter ratio was not correlated to the Chl/c_p ratio, which increased away from the bed in each period, consistent with chlorophyll-containing particles being at lower concentration relative to other particles closer to the bed. Chlorophyll profiles typically showed no trend with elevation (not shown), consistent with the hypothesis that the increase in the Chl/c_p ratio away from the bed was due to enrichment in faster-settling non-algal particles near the bottom.

4. Discussion

4.1. Inferences from Optical Properties

The vertical profiles of the attenuation and backscattering coefficients were fit reasonably well by Rouse profiles, and the estimated settling velocities increased with increasing shear stresses, as expected. The magnitudes of the estimated settling velocities are within ranges expected for flocs and for the fine sands typical of the seabed at the MVCO site. The ~0.1 mm s^{-1} estimated settling velocities during the calm period were small, and they indicate material that behaved as slowly sinking washload that is evenly distributed throughout the boundary layer. The ~0.4 mm s^{-1} estimated settling velocities during the spring tides are in the range of typical floc-settling velocities, and the ~1 mm s^{-1} estimated settling velocities during the passage of Maria are typical of flocs (e.g., [48]). The ~4 mm s^{-1} observed during the passage of Ophelia are larger than typical floc-settling velocities, but they are representative of very fine sand (D = ~80 μm; [49]) or densely packed aggregates [21].

Interestingly, boundary shear stresses during passage of Maria were similar to boundary shear stresses during the spring-tide period, yet the estimated settling velocities were more than two times larger during Maria. Residual plots of the Rouse fits indicate that the profiles had smaller gradients (lower settling velocities) during spring tides than the profiles near the bed during passage of Maria. In addition, Sauter diameters were smaller during passage of Maria, as were the spectral exponents of attenuation. Particle size was inverted during the spring tides, with larger particles farther above bottom. The spectral exponent of backscattering was lower during the spring tides than during passage of Maria, and it decreased with height above bottom.

Despite the similar stresses during these two periods, these observations suggest different particle dynamics. The larger Sauter diameters, inverted particle size, and smaller values of γ_{bbp} are consistent with the hypothesis that flocs dominated the suspension during the spring tides. The flocs were fragile, and they were disrupted near the bed, but they were able to reform higher in the boundary layer. The breakage of flocs near the bed decreased bulk estimated settling velocities from the profiles by transferring mass near the bed from faster-sinking large flocs (~1 mm s^{-1}) into slower-sinking smaller flocs or single grains. Overall, however, the extensive packaging of small particles into large flocs during spring tides was associated with small values of γ_{bbp} [15]. During the passage of Maria, we hypothesize that higher bottom shear stresses associated with the larger swell caused resuspension of fine sand typical of the site (D = ~125 μm; [50]) and caused a greater degree of floc breakup. Resuspension of sand caused larger gradients in optical properties near the bed. Resuspended sand also can account for the broadened peak in the particle size distributions. Greater floc breakup can account for smaller values of γ_{cp} during passage of Maria, because particle mass was transferred out of large flocs that were not sensed by the AC-9 into smaller flocs or single grains that were [15]. It can also account for large values of γ_{bbp} [15].

During passage of Ophelia, when stress was highest, particle sizes were smallest. This result suggests that floc breakup was more extensive during Ophelia (e.g., [26]). The finer mode in the LISST size distributions was similar to the fine sands at the site (D = ~125 μm), and the presence of this sand, with typical settling velocities of 8 mm s^{-1} [51], steepened the near-bed profiles of the optical properties. The appearance of the coarser mode at the upper limit of the LISST size range, however, is not consistent with greater floc breakup. We hypothesize that these particles were resuspended bed aggregates that were tougher than flocs and not prone to breakup (cf. [21]). The fact that they were relatively more abundant higher above the bottom suggests that they had settling velocities smaller than that of the fine sands (<8 mm s^{-1}), consistent with the estimated settling velocities in the outer part of the profile of 4 mm s^{-1}.

The hypothesis of sand resuspension during passage of the two storms can be reconciled with observed variations in the backscatter ratio, which was lower during passage of the two storms than it was during the period of spring tides. Permeable sand beds can store organic matter in the interstitial pore spaces [52]. We propose that during resuspension events, this organic material was resuspended, which caused backscatter ratios to decrease. The observation that the Chl/c_p ratio did not change indicates that the organic matter in the bed interstices was degraded and did not contain chlorophyllous material.

The relative magnitudes of Sauter diameter and inverse density in the different periods may also argue for different particle dynamics during passage of the storms. During these periods, the ratios of Sauter diameter to inverse density were smaller than during the periods of spring tides and calm. This ratio reduces to the ratio of the beam attenuation coefficient measured by the LISST to the particle area estimated by the LISST. The inversion of near-forward scattered light measured by the LISST to particle area assumes that a single refractive index applies to all particles. We hypothesize that small organic particles liberated from the bed during resuspension had attenuation to particle area ratios that were smaller than the ratios for particles during the spring tides and during the calm period.

4.2. Broader Advantages and Disadvantages of Optical Measurements

Measurements of optical properties often behave in ways consistent with our expectations. When they do, their advantages are that they provide robust, relatively non-intrusive, well-understood first-order information about the suspended particle populations. Their main disadvantages are their sensitivity to ambient light and limitations in turbid conditions, tendency to foul, and, in some cases, their large size, power requirements, or data storage limits. The more subtle disadvantages are the biases that each instrument has, often based on wavelength, angles of illumination, or acceptance angles. These disadvantages can be turned to advantage by using combinations of different measurements, each with a different bias, as we have shown here. Combinations of optical measurements (attenuation and backscatter at various wavelengths, measurements of chlorophyll, size data from the LISST) provide more information about the size and composition of the particle population.

Simple, time-tested optical measurements such as attenuation and backscatter remain valuable, even in complicated settings like MVCO. Their main advantage is they yield simple interpretations in terms of SPM that capture the first-order variations that often can be linked to wave- and current-induced bottom stress. The main disadvantages are that they can be biased by size and composition, and the conversion from attenuation or backscatter to SPM can vary, depending on instruments, particle size, composition [1,2].

The advantages of direct measurements of size from the LISST are clear: they provide information that cannot be measured directly with simpler optical sensors, and the size information is very useful for interpreting the dynamic behavior of the particles. One important disadvantage is that LISST instruments are large and intrusive. In addition, they have upper limits on sizes and, when larger particles are present, they can bias results by elevating reported volumes in smaller size classes [53]. Finally, when flocs are present, the LISST may be sensitive to smaller component particles, leading to over-representation of microflocs when macroflocs are present [54].

Near-bottom profiles of optical properties allow the observations of particle populations to be interpreted in terms of sediment dynamics. In particular, near-bottom profiles of size from the LISST have not been previously reported, and offer the advantage of direct observations of size changes in a region with strong gradients in concentration and turbulent shear.

The nearly-ubiquitous appearance of Rouse-like profiles in optical measurements provides estimates of settling velocity that can be linked to particle size. The disadvantage of these interpretations is that the bulk profile can be confounded by changes in particle density (and therefore settling velocity) or skewed by overlapping profiles of diverse particle populations with varying vertical distributions. In all the cases illustrated here, relatively good Rouse fits provided settling-velocity estimates, but they could vary by an order of magnitude between profiles. Additional information was required to evaluate the underlying particle dynamics, and both floc dynamics and resuspension were found to be important.

Addition of acoustical measurements [55] could further constrain the particle dynamics and distribution as their sensitivity to composition, size, and packing differs from optical measurements. For example, for acoustic backscattering in the MHz frequency range, aggregation decreases the signal [56] while being sensitive to large single-grain particles (whereas the optical response is not).

Models of resuspension and settling have been used for years to examine the dynamics giving rise to Rouse-like vertical profiles in the BBL [35,36,57–60]. However, these models have not been coupled to an optical model so that their output could be validated with optical (and/or acoustical) observations. Models that now include resuspension as well as aggregation/disaggregation dynamics (e.g., [61]) can be coupled with optical models (for single grains as well as flocs), making it possible to express modeled variables (e.g., spatial and temporal concentrations of different types of particles) as optical measurements and allow for direct comparison with field observations. This also opens the possibility for assimilating such observational data into the models. Thus, the development of optical models of sediment dynamics and their evaluation on real-world datasets of is critically important.

5. Conclusions

This paper has described a suite of optical measurements designed to provide information about suspended particle populations. We have discussed the biases of these measurements and demonstrated their value in analyses of field observations. In conclusion:

- Near-bottom profiles of optical properties are valuable because they sample particle populations in a region with strong gradients in turbulence and concentrations.
- Profiles with combinations of instruments can be used to make inferences about sediment dynamics in the bottom boundary layer. Resuspension of bottom material and dynamics of aggregation and disaggregation are especially important at the MVCO study site.
- Aggregation/disaggregation dynamics cannot be neglected when interpreting profiles of properties sensitive to the small particles (e.g., beam attenuation) as the flocs are both a sink and source for fine particles.
- Combinations of optical instruments provide information about suspended particle population that individual instruments cannot, because of their individual design and biases. Many of the disadvantages associated with individual optical sensors can be turned to advantages when multiple sensors are used.

Author Contributions: P.H., T.M., E.B., and C.S. conceived, designed, and performed the experiments. C.S. and E.B. processed the data; E.B. conceived the paper and wrote the first draft; C.S., P.H. and T.M. contributed to the writing.

Acknowledgments: This work was supported by the Office of Naval Research and the United States Geological Survey Coastal and Marine Geology Program. The unique instrument platform and data acquisition system was designed and built by technical staff lead by Marinna Martini at the United States Geological Survey Woods Hole Coastal and Marine Science Center. This team was also responsible for deployment and recovery of the instrumentation. We thank the Woods Hole Oceanographic Institution (WHOI) MVCO staff for support during this experiment, and we thank the captains and crews of the R/V Connecticut and the R/V Tioga. Any use of trade, product, or firm names is for descriptive purposes only and does not imply endorsement by the United States Government. This paper has benefited significantly from insightful comments from D. Stramski, A. Aretxabaleta and two anonymous reviewers.

Data Set: Sherwood: C.R., Dickhudt, P.J., Martini, M.A., Montgomery, E.T., and Boss, E.S., 2012, Profile measurements and data from the 2011 Optics, Acoustics, and Stress In Situ (OASIS) project at the Martha's Vineyard Coastal Observatory: U.S. Geological Survey Open-File Report 2012–1178, at http://pubs.usgs.gov/of/2012/1178/.

Data Set License: The data presented in this paper are in the public domain in the United States because they come from the United States Geological Survey, an agency of the United States Department of Interior.

Conflicts of Interest: The authors declare no conflict of interest. The founding sponsors had no role in the design of the study; in the collection, analyses, or interpretation of data; in the writing of the manuscript, and in the decision to publish the results.

Abbreviations

The following abbreviations are used in this manuscript:

BBL	bottom boundary layer
LISST	Laser In Situ Scattering and Transmission
MVCO	Martha's Vineyard Coastal Observatory
OASIS	Optics and Acoustics and Stress In Situ
ONR	Office of Naval Research
SPM	suspended particulate mass
USGS	USA Geological Survey
WHOI	Woods Hole Oceanographic Institution

References

1. Hill, P.S.; Bos, E.; Newgard, J.P.; Law, B.A.; Milligan, T.G. Observations of the sensitivity of beam attenuation to particle size in a coastal bottom boundary layer. *J. Geophys. Res.* **2011**, *116*, C02023. [CrossRef]
2. Downing, J. Twenty-five years with OBS sensors: The good, the bad, and the ugly. *Cont. Shelf Res.* **2006**, *26*, 2299–2318. [CrossRef]
3. Mikkelsen, O.A.; Hill, P.S.; Milligan, T.G.; Chant, R.G. In situ particle size distributions and volume concentrations from a LISST100 laser particle sizer and a digital floc camera. *Cont. Shelf Res.* **2005**, *25*, 1959–1978. [CrossRef]
4. Boss, E.; Taylor, L.; Gilbert, S.; Gundersen, K.; Hawley, N.; Janzen, C.; Johengen, T.; Purcell, H.; Robertson, C.; Schar, D.W.; et al. Comparison of inherent optical properties as a surrogate for particulate matter concentration in coastal waters. *Limnol. Ocean. Meth.* **2009**, *7*, 803–810. [CrossRef]
5. Stramski, D.; Babin, M.; Wozniak, S. Variations in the optical properties of terrigenous mineral-rich particulate matter suspended in seawater. *Limnol. Oceanogr.* **2007**, *52*, 2418–2433. [CrossRef]
6. Stemmann, L.; Boss, E. Plankton and particle size and packaging: From determining optical properties to driving the biological pump. *Annu. Rev. Mar. Sci.* **2012**, *4*, 263–290. [CrossRef] [PubMed]
7. Stramski, D.; Boss, E.; Bogucki, D.; Voss, K.J. The role of seawater constituents in light backscattering in the ocean. *Prog. Ocean.* **2004**, *61*, 27–55. [CrossRef]
8. Hill, P.S.; Bowers, D.G.; Braithwaite, K.M. The effect of suspended particle composition on particle 389 area-to-mass ratios in coastal waters. *Meth.Oceanogr.* **2013**, *7*, 95–109. [CrossRef]
9. Boss, E.; Slade, W.H.; Behrenfeld, M.; Dall'Olmo, G. Acceptance angle effects on the beam attenuation in the ocean. *Opt. Exp.* **2009**, *17*, 1535–1550. [CrossRef]
10. Stumpf, R.P. Sediment transport in chesapeake bay during floods: Analysis using satellite and surface observations. *J. Coast. Res.* **1988**, *4*, 1–15.

11. Nechad, B.; Ruddick, K.G.; Park, Y. Calibration and validation of a generic multisensor algorithm for mapping of total suspended matter in turbid waters. *Remote. Sens. Envir.* **2010**, *114*, 854–866. [CrossRef]

12. Agrawal, Y.; Pottsmith, H. Instruments for particle size and settling velocity observations in sediment transport. *Mar. Geol.* **2000**, *168*, 89–114. [CrossRef]

13. Slade, W.H.; Boss, E. Spectral attenuation and backscattering as indicators of average particle size. *Appl. Opt.* **2015**, *54*, 7264–7277. [CrossRef] [PubMed]

14. Boss, E.; Twardowski, M.S.; Herring, S. Shape of the particulate beam attenuation spectrum and its relation to the size distribution of oceanic particles. *Appl. Opt.* **2001**, *40*, 4885–4893. [CrossRef] [PubMed]

15. Tao, J.; Hill, P.S.; Boss, E.S.; Milligan, T.G. Variability of suspended particle properties using optical measurements within the Columbia River Estuary. *J. Geophys. Res. Oceans* **2018**, *123*. [CrossRef]

16. Reynolds, R.A.; Stramski, D.; Neukermans, G. Optical backscattering of particles in Arctic seawater and relationships to particle mass concentration, size distribution, and bulk composition. *Limnol. Oceanogr.* **2016**, *61*, 1869–1890. [CrossRef]

17. Briggs, N.T.; Slade, W.H.; Boss, E.; Perry, M.J. Method for estimating mean particle size from high-frequency fluctuations in beam attenuation or scattering measurements. *Appl. Opt.* **2013**, *52*, 6710–6725. [CrossRef]

18. van de Hulst, H.C. *Light Scattering by Small Particles*; John Wiley and Sons: Dover, UK, 1981.

19. Carder, K.L.; Betzer, P.R.; Eggimann, D.W. Physical, chemical and optical measures of suspended-particle concentrations: Their intercomparison and application to the West African Shelf. In *Suspended Solids in Water*; Gibbs, J., Ed.; Plenum: New York, NY, USA, 1974; pp. 173–193.

20. Winterwerp, J.C.; Van Kesteren., W.G.M. *Introduction to the Physics of Cohesive Sediment Dynamics in the Marine Environment*; Elsevier: Amsterdam, The Netherlands, 2004.

21. Milligan, T.G.; Kineke, G.C.; Blake, A.C.; Alexander, C.R.; Hill, P.S. Flocculation and sedimentation in the ACE Basin, South Carolina. *Estuaries* **2001**, *24*, 734–744. [CrossRef]

22. Hill, P.S.; Syvitiski, J.P.; Cowan, E.A.; Powell, R.D. In situ observations of floc settling velocities in Glacier Bay, Alaska. *Mar. Geol.* **1998**, *145*, 85–94. [CrossRef]

23. Johnson, C.; Li, X.; Logan, B. Settling velocities of fractal aggregates. *Environ. Sci. Technol.* **1996**, *30*, 1911–1918. [CrossRef]

24. Winterwerp, J.C. A simple model for turbulence induced flocculation of cohesive sediment. *J. Hydraulic Res.* **1998**, *36*, 309–326. [CrossRef]

25. Winterwerp, J.C.; Manning, A.J.; Martens, C.; de Mulder, T.; Vanlede, J. A heuristic formula for turbulence-induced flocculation of cohesive sediment. *Estuar. Coast. Shelf Sci.* **2006**, *68*, 195–207. [CrossRef]

26. Hill, P.S.; Voulgaris, G.; Trowbridge, J.H. Controls on floc size in a continental shelf bottom boundary layer. *J. Geophys. Res.* **2001**, *106*, 9543–9549. [CrossRef]

27. Boss, E.; Slade, W.H.; Hill, P. Effect of particulate aggregation in aquatic environments on the beam attenuation and its utility as a proxy for particulate mass. *Opt. Exp.* **2009**, *17*, 9408–9420. [CrossRef]

28. Slade, W.H.; Boss, E.; Russo, C. Effects of particle aggregation and disaggregation on their inherent optical properties. *Opt. Exp.* **2011**, *19*, 7945–7959. [CrossRef] [PubMed]

29. Twardowski, M.; Boss, E.; MacDonald, J.B.; Pegau, W.S.; Barnard, A.H.; Zaneveld, J.R.V. A model for estimating bulk refractive index from the optical backscattering ratio and the implications for understanding particle composition in case I and case II waters. *J. Geophys. Res.* **2001**, *106*, 129–142. [CrossRef]

30. Loisel, H.; Meriaux, X.; Berthon, J.-F.; Poteau, A. Investigation of the optical backscattering to scattering ratio of marine particles in relation to their biogeochemical composition in the eastern English Channel and southern North Sea. *Limnol. Oceanogr.* **2007**, *52*, 739–752. [CrossRef]

31. Boss, E.; Pegau, W.S.; Lee, M.; Twardowski, M.; Shybanov, E.; Korotaev, G.; Baratange, F. Particulate backscattering ratio at LEO 15 and its use to study particle composition and distribution. *J. Geophys. Res.* **2004**, *109*. [CrossRef]

32. Wozniak, S.B.; Stramski, D.; Stramska, M.; Reynolds, R.A.; Wright, V.M.; Miksic, E.Y.; Cichocka, M.; Cieplak, A.M. Optical variability of seawater in relation to particle concentration, composition, and size distribution in the nearshore marine environment at Imperial Beach, California. *J. Geophys. Res.* **2010**, *115*, C08027. [CrossRef]

33. Neukermans, G.; Reynolds, R.A.; Stramski, D. Optical classification and characterization of marine particle assemblages within the western Arctic Ocean. *Limnol. Oceanogr.* **2015**, *61*, 1472–1494. [CrossRef]

34. Neukermans, G.; Loisel, H.; Meriaux, X.; Astoreca, R.; McKee, D. In situ variability of mass-specific beam attenuation and backscattering of marine particles with respect to particle size, density, and composition. *Limnol. Oceanog.* **2012**, *57*, 124–144. [CrossRef]

35. Hurley, A.J.; Hill, P.S.; Milligan, T.G.; Law, B.A. Optical methods for estimating apparent density of sediment in suspension. *Meth. Oceanog.* **2016**, *17*, 153–168. [CrossRef]

36. Trowbridge, J.H.; Lentz, S.J. The bottom boundary layer. *Ann. Rev. Mar. Sci.* **2018**, *10*, 397–420. [CrossRef] [PubMed]

37. Rouse, H. Modern concepts of the mechanics of turbulence. *ASCE Trans.* **1937**, *102*, 463–543.

38. Rouse, H. *An Analysis of Sediment Transportation in the Light of Fluid Turbulence*; United States Department of Agriculture: Washington, DC, USA, 1939.

39. Dyer, K.R. *Coastal and Estuarine Sediment Dynamics*; John Wiley: Chichester, UK, 1986.

40. Orton, P.M.; Kineke, G.C. Comparing calculated and observed vertical suspended-sediment distributions from a Hudson River Estuary turbidity maximum. *Estuarine Coastal Shelf Sci.* **2001**, *52*, 401–410. [CrossRef]

41. Kumbhakar, M.; Ghoshal, K.; Singh, V.P. Derivation of Rouse equation for sediment concentration using Shannon entropy. *Phys. A* **2017**, *465*, 494–499. [CrossRef]

42. Dall'Olmo, G.; Westberry, T.K.; Behrenfeld, M.J.; Boss, E.; Slade, W.H. Significant contribution of large particles to optical backscattering in the open ocean. *Biogeosciences* **2009**, *6*, 947–967. [CrossRef]

43. Sherwood, C.R.; Dickhudt, P.J.; Martini, M.A.; Montgomery, E.T.; Boss, E.S. *Profile Measurements and Data from the 2011 Optics, Acoustics, and Stress In Situ (OASIS) Project at the Martha's Vineyard Coastal Observatory*; United States Geological Survey: Reston, VA, USA, 2012.

44. Slade, W.H.; Boss, E.; Dall'Olmo, G.; Langner, M.R.; Loftin, J.; Behrenfeld, M.J.; Roesler, C.; Westberry, T.K. Underway and moored methods for improving accuracy in measurement of spectral particulate absorption and attenuation. *J. Atmos. Ocean. Tech.* **2010**, *27*, 1733–1746. [CrossRef]

45. Trowbridge, J.H. On a technique for measurement of turbulent shear stress in the presence of surface waves. *J. Atmos. Oceanic Technol.* **1998**, *15*, 290–298. [CrossRef]

46. Grant, W.D.; Madsen, O.S. Combined wave and current interaction with a rough bottom. *J. Geophys. Res. Oceans* **1979**, *84*, 1797–1808. [CrossRef]

47. Madsen, O.S. Spectral Wave-Current Bottom Boundary Layer Flows. In Proceedings of the 24th International Conference Coastal Engineering Research Council, Kobe, Japan, 23–28 October 1994; ACSE: Reston, VA, USA, 1995.

48. Fox, J.M.; Hill, P.S.; Milligan, T.G.; Ogston, A.S.; Boldrin, A. Floc fraction in the waters of the Po River prodelta. *Cont. Shelf Res.* **2004**, *24*, 1699–1715. [CrossRef]

49. Wiberg, P.L.; Smith, J.D. Calculations of the critical shear stress for motion of uniform and heterogeneous sediments. *Water Resour. Res.* **1987**, *23*, 1471–1480.

50. Traykovski, P.; Richardson, M.D.; Mayer, L.A.; Irish, J.D. Mine burial experiments at the Martha's Vineyard Coastal Observatory. *IEEE J. Oceanic Eng.* **2007**, *32*, 150–166. [CrossRef]

51. Dietrich, W.E. Settling velocity of natural particles. *Water Resour. Res.* **1982**, *18*, 1615–1626. [CrossRef]

52. Law, B.A.; Hill, P.S.; Milligan, T.G.; Zions, V. Erodibility of aquaculture waste from different bottom substrates. *Aquacult. Environ. Interact.* **2016**, *8*, 575–584. [CrossRef]

53. Davies, E.J.; Nimmo-Smith, W.A.M.; Agrawal, Y.C.; Souza, A.J. LISST-100 response to large particles. *Mar. Geol.* **2012**, *307*, 117–122. [CrossRef]

54. Graham, G.W.; Davies, E.J.; Nimmo-Smith, W.A.M.; Bowers, D.G.; Braithwaite, K.M. Interpreting LISST-100X measurements of particles with complex shape using digital in-line holography. *J. Geophys. Res. Oceans* **2012**, *117*. [CrossRef]

55. Lynch, J.F.; Irish, J.D.; Sherwood, C.R.; Agrawal, Y.C. Determining suspended sediment particle size information from acoustical and optical backscatter measurements. *Cont. Shelf Res.* **1994**, *14*, 1139–1165. [CrossRef]

56. Russo, C. An Acoustical Approach to the Study of Marine Particles Dynamics Near the Bottom Boundary Layer. Ph.D. Thesis, University of Maine, Orono, ME, USA, 2011.

57. Dyer, K.R.; Soulsby, R.L. Sand transport on the continental shelf. *Annu. Rev. Fluid Mech.* **1988**, *20*, 295–324. [CrossRef]

58. McLean, S.R. On the calculation of suspended load for noncohesive sediments. *J. Geophys. Res. Oceans* **1992**, *97*, 5759–5770. [CrossRef]

59. Gelfenbaum, G.; Smith, J.D. *Experimental Evaluation of a Generalized Suspended-Sediment Transport Theory;* AAPG: Tulsa, OK, USA, 1986; pp. 133–144.
60. Pal, D.; Ghoshal, K. Vertical distribution of fluid velocity and suspended sediment in open channel turbulent flow. *Fluid Dyn. Res.* **2016**, *48*, 035501. [CrossRef]
61. Sherwood, C.R.; Aretxabaleta, A.L.; Harris, C.K.; Rinehimer, J.P.; Verney, R.; Ferré, B. Cohesive and mixed sediment in the Regional Ocean Modeling System (ROMS v3.6) implemented in the Coupled Ocean-Atmosphere-Wave-Sediment Transport Modeling System (COAWST r1234). *Geosci. Model Dev.* **2018**, *11*, 1849–1871. [CrossRef]

*applied
sciences*

MDPI

Article

Estimating Underwater Light Regime under Spatially Heterogeneous Sea Ice in the Arctic

Philippe Massicotte [1,*], Guislain Bécu [1], Simon Lambert-Girard [1], Edouard Leymarie [2] and Marcel Babin [1]

[1] Takuvik Joint International Laboratory (UMI 3376) Université Laval (Canada) Centre National de la Recherche Scientifique (France), Québec, QC G1V 0A6, Canada; guislain.becu@takuvik.ulaval.ca (G.B.); Simon.Lambert-Girard@takuvik.ulaval.ca (S.L.-G.); marcel.babin@takuvik.ulaval.ca (M.B.)
[2] Laboratoire d'Océanographie de Villefranche, Sorbonne Université, CNRS, LOV, F-06230 Villefranche-sur-Mer, France; leymarie@obs-vlfr.fr
* Correspondence: philippe.massicotte@takuvik.ulaval.ca or pmassicotte@hotmail.com

Received: 8 September 2018; Accepted: 23 November 2018; Published: 19 December 2018

Abstract: The vertical diffuse attenuation coefficient for downward plane irradiance (K_d) is an apparent optical property commonly used in primary production models to propagate incident solar radiation in the water column. In open water, estimating K_d is relatively straightforward when a vertical profile of measurements of downward irradiance, E_d, is available. In the Arctic, the ice pack is characterized by a complex mosaic composed of sea ice with snow, ridges, melt ponds, and leads. Due to the resulting spatially heterogeneous light field in the top meters of the water column, it is difficult to measure at single-point locations meaningful K_d values that allow predicting average irradiance at any depth. The main objective of this work is to propose a new method to estimate average irradiance over large spatially heterogeneous area as it would be seen by drifting phytoplankton. Using both in situ data and 3D Monte Carlo numerical simulations of radiative transfer, we show that (1) the large-area average vertical profile of downward irradiance, $\overline{E_d}(z)$, under heterogeneous sea ice cover can be represented by a single-term exponential function and (2) the vertical attenuation coefficient for upward radiance (K_{Lu}), which is up to two times less influenced by a heterogeneous incident light field than K_d in the vicinity of a melt pond, can be used as a proxy to estimate $\overline{E_d}(z)$ in the water column.

Keywords: apparent optical properties; 3D Monte Carlo numerical simulations; downward irradiance; upward radiance; sea ice heterogeneity; vertical attenuation coefficient; melt ponds

1. Introduction

The vertical distribution of underwater light is an important driver of many aquatic processes such as primary production by phytoplankton, and photochemical reactions such as the photodegradation of organic matter. Hence, an adequate description of the underwater light regime is mandatory to understand energy fluxes in aquatic ecosystems. In open water, when assuming an optically homogeneous water column, downward irradiance at any given wavelength follows, as a first approximation, quite well a monotonically exponential decrease with depth, which can be modelled as follows [1] (Equation (1)):

$$E_d(z) = E_d(0^-) \, e^{-K_d(z) \, z} \tag{1}$$

where $E_d(z)$ is the downward plane irradiance (W m^{-2}) at depth z (m), $E_d(0^-)$ is the downward plane irradiance (W m^{-2}) just below the surface and $K_d(z)$ is the diffuse vertical attenuation coefficient (m^{-1}) describing the rate at which downward irradiance decreases with increasing depth. K_d is one of the most commonly used apparent optical properties (AOP) of seawater, and a good estimation of this parameter is important for measuring or modelling primary production. K_d may vary with depth because of changes in seawater inherent optical properties (IOPs), the angular structure of the light field, and the effects of inelastic radiative processes such as Raman scattering by water molecules and fluorescence by phytoplankton pigments or dissolved organic matter. As Kirk [1] pointed out, for practical considerations in oceanography and limnology, the K_d value, even when averaged within the euphotic zone, provides a useful proxy to represent the downward irradiance attenuation in the upper water column. For example, to determine primary production based on simulated on-deck incubations or photosynthetic parameters derived from photosynthesis–irradiance curves (P vs. E curves) requires measured or estimated values of K_d (e.g., Morel [2]). Nowadays, K_d is relatively easy to estimate using commercially available radiometers.

The ice-infested regions of the Arctic ocean are characterized by a complex mosaic made of sea ice with snow, melt ponds, ridges, and leads [3–5]. Phytoplankton are exposed to a highly variable light regime while drifting under these heterogeneous features (e.g., Lange et al. [6]). Estimating primary production of phytoplankton under sea ice requires an approach that is adequate to capture this large-area variability in the light field. In situ incubations at single locations of seawater samples inoculated with ^{14}C or ^{13}C are not appropriate because they reflect primary production under local light conditions, which is not representative of the range of irradiance experienced by drifting phytoplankton over a large area. One classical approach that is more adequate consists in conducting on-deck simulated 24-h incubations of seawater samples inoculated with ^{14}C or ^{13}C and applying the light attenuation at the depths of sample collections, using natural illumination and neutral filters. An alternative approach consists in calculating primary production using modelled or measured daily time series of incident irradiance, sea ice transmittance and in-water vertical attenuation coefficients, combined with photosynthetic parameters determined from P–E curves measured with short (under two hours) incubations of seawater samples inoculated with ^{14}C. The latter two methods require that the vertical profile of the irradiance experienced by drifting phytoplankton be appropriately determined, which is challenging due to surface heterogeneity. Traditionally, one or very few $E_d(z)$ profiles are measured at discrete locations under sea ice (e.g., Mundy et al. [7]). Such measurements, however, do not capture the variability induced by sea ice features. In recent studies, to better document the spatial variability of $E_d(z)$, radiometers were attached to either remotely operated vehicles (ROV) [4] or a surface and under-ice trawl (SUIT), a net developed for deployment in ice-covered waters, typically behind an icebreaker [6]. Both a ROV and a SUIT allow a better description of the light field right under sea ice, which is more appropriate for determining average irradiance experienced by drifting phytoplankton. Such under-ice measurements can then be combined with averaged K_d values to propagate light at depth.

Estimating irradiance at depth for primary production measurement or calculation using K_d values derived from only a few discrete vertical profiles of $E_d(z)$ under heterogeneous sea ice is problematic whatever the platform for radiometer deployment. Let us consider that phytoplankton, by continuously drifting horizontally relative to sea ice, are exposed to fluctuations in irradiance due to surface heterogeneity, and that the relevant light metrics for primary production in such conditions is irradiance at any depth averaged over some horizontal area. When measuring an irradiance profile at one given location under sea ice, as the depth of the upward-looking detector increases, light from a larger area on the underside of the ice enters the detector field of view. In other words, the detector "sees" different things at different depths. One consequence is that $E_d(z)$ measured that way may not follow the usual monotonically exponential decrease with increasing depth (Equation (1)). For example, irradiance profiles measured beneath low-transmission sea ice (e.g., white ice) relative to surrounding areas showing melt ponds, show subsurface light maxima. The literature reports subsurface maxima varying between 5 m and 15 m in depth [5,8,9]. Conversely, it is also important to note that K_d estimations are biased when profiles are measured beneath an area of high transmission (e.g., a melt pond) relative to surrounding areas [5]. Indeed, with depth, light decreases more quickly than what would be expected from the IOPs of the water column. In the field, this situation is more difficult to identify compared to profiles showing subsurface maxima because the former measurements may appear to follow a single exponential decrease but would not produce a diffuse attenuation coefficient that adequately describes the water mass. So, two vertical light profiles measured a few meters apart under sea ice are often very different. More importantly, local measurements of light under heterogeneous sea ice do not provide an adequate description of the average light field as it would be seen by drifting phytoplankton cells at different depths. This makes estimations of primary production and the interpretation of biogeochemical data challenging in the presence of sea ice.

To fit vertical profiles of $E_d(z)$ under bare ice that do not follow an exponential decay under sea ice covered with melt ponds, Frey et al. [8] proposes a simple geometric model (Equation (2)).

$$E_d(z) = \pi E_d(0^-)(1 + P(N-1)\cos^2\phi)e^{-K_d(z)\,z} \tag{2}$$

where $E_d(0^-)$ is the irradiance directly below the ice/snow, P the areal fraction of the ice cover, N the ratio between ice and melt ponds transmittance and ϕ a fitting parameter defined as $\arctan(R/z)$ with R the radius of the ice patch and z the depth. A major drawback of this method is that additional field observations of N and P are required to adequately parametrise the model, which makes its use more difficult. To address this concern (among others), Laney et al. [9] proposed a semi-empirical parametrisation that includes a second exponential coefficient in Equation (1) to model light decrease at the interface between the ice and ocean water at the bottom of the ice layer (Equation (3)):

$$E_d(z) = E_d(0^-)e^{-K_d(z)\,z} - (E_d(0^-) - E_d(NS))\,e^{-K_{NS}(z)\,z} \tag{3}$$

where $E_d(0^-)$ is the irradiance that would be observed under homogeneous snow or ice cover, $E_d(NS)$ is the irradiance under-ice, and $K_{NS}(z)$ describes the decrease of $E_d(0^-)$ just under the ice layer. Both the methods by Frey et al. [8] and Laney et al. [9] make it possible to propagate local $E_d(z)$ vertically under low transmission ice. However, these methods cannot identify and correct for inflated K_d when profiles are measured beneath an area of high transmission relative to surrounding areas. Additionally, when trying to determine primary production by phytoplankton that drift under sea ice and therefore are not static under sea ice features, what matters is the average shape of the vertical $E_d(z)$ profile, which may possibly be predictable using a large-area $\overline{K_d}$ as under a wavy open ocean surface [10].

In this study, using both in situ data and 3D Monte Carlo numerical simulations of radiative transfer, we show that the vertical propagation of average $E_d(z)$, $\overline{E_d}(z)$, is reasonably well approximated by a single exponential decay with a so-called large area K_d, $\overline{K_d}$, under sea ice covered in melt ponds. We further demonstrate that $\overline{K_d}$ can be estimated from the vertical attenuation coefficient for upward radiance (K_{Lu}) because the latter is apparently less affected by local surface features of the ice cover. We implicitly assume that primary production can be adequately modelled using $\overline{E_d}(z)$, and we conclude that K_{Lu} is an appropriate AOP for predicting the vertical variations in $\overline{E_d}(z)$ under sea ice.

2. Material and Methods

2.1. Study Site and Field Campaign

The field campaign was part of the GreenEdge project (www.greenedgeproject.info) which was conducted on landfast ice southeast of the Qikiqtarjuaq Island in the Baffin Bay (67.4797 N, 63.7895 W). The field operations took place at an ice camp where the water depth was 360 m, from 20 April to 27 July 2016 (Figure A1 included in Appendix A). During the sampling period, the study site experienced changes in the snow cover and landfast ice thickness of 0–49 cm and 106–149 cm, respectively.

2.2. In Situ Underwater Light Measurements

During the campaign, a total of 83 vertical light profiles were acquired using a factory-calibrated ICE-Pro (an ice floe version of the C-OPS, or Compact-Optical Profiling System, from Biospherical Instruments Inc., San Diego, CA, USA) equipped with both downward plane irradiance $E_d(z)$ (W m^{-2}) and upward radiance $L_u(z)$ (W m^{-2} sr^{-1}) radiometers. The ICE-Pro system is a negatively buoyant instrument with a cylindrical shape 10 inches in diameter and is not designed for free-fall casts (as opposed to its open-water version). To perform the profiles, the frame was manually lowered into an auger hole that had been cleaned of ice chunks. Once it was underneath the ice layer, fresh clean snow was shovelled back in the hole to prevent the creation of a bright spot right on top of the sensors. Great care was taken not to pollute the hole surroundings (footsteps, water and slush spillage from the auger drilling, etc.). The operator then stepped back 50 m, while keeping the sensors right under the ice, to avoid any human shadow on top of the profile. The frame was then lowered manually at a constant descent rate of approximately 0.3 m s^{-1}. The above-surface atmospheric reference sensor was fixed on a steady tripod standing on the floe approximately 2 m above the surface and above all neighbouring ice camp features. Data processing and validation were performed using a protocol inspired by the one proposed by Smith and Baker [11]. Measurements were made at 19 wavelengths: 380, 395, 412, 443, 465, 490, 510, 532, 555, 560, 589, 625, 665, 683, 694, 710, 765, 780 and 875 nm. For this study, E_d and L_u spectra were interpolated linearly between 400 and 700 nm every 10 nm. In situ diffuse attenuation coefficients (K) for both E_d (K_d) and L_u (K_{Lu}) were calculated on a 5 m sliding window (10–15 m, 15–20 m, . . ., 70–75 m, 75–80 m) starting at 10 m depth to reduce the effects of surface heterogeneity. A total of 72,044 non-linear models were calculated to estimate both K coefficients from Equation (1) (83 profiles × 14 depths × 31 wavelengths × 2 radiometric quantities (E_d, L_u)). A conservative R^2 of 0.99 was used essentially to filter out noisy profiles. 42,407 models were kept for subsequent analysis.

2.3. 3D Monte Carlo Numerical Simulations of Radiative Transfer

2.3.1. Theory and Geometry

3D numerical Monte Carlo simulation is a convenient approach for modelling the light field under spatially heterogeneous sea surfaces [5,12–14]. They are simple to understand and versatile, and incident light, IOPs and geometry can be easily changed. In this study, we used SimulO, a 3D Monte Carlo software program that simulates the propagation of light in optical instruments or in ocean waters [15]. Our objective was to simulate the propagation of sunlight underneath heterogeneous ice-covered ocean waters. Simulations were performed in an idealized ocean described by a cylinder of 120 m radius and 150 m depth (Figure 1). The water IOPs were selected to reflect pre-bloom conditions in the green–blue spectral region ($a = b = 0.05$ m^{-1}). These typical averaged values were measured during the GreenEdge 2016 campaign using an in situ spectrophotometer (ac-s from Sea-Bird Scientific) and represent the contribution of both pure water and the water constituents. The scattering phase function was described by a Fournier-Forand analytic form with a 3% backscatter fraction [16,17]. The inclusion of a 3D sea ice layer at the upper boundary of the ocean would require extensive computing power because of the high scattering properties of sea ice. Instead, sea ice was incorporated at the upper boundary of the ocean using a 2D light-emitting surface with a radius of 100 m. The angular distribution and magnitude of the light field emitted by the surface was chosen to mimic observed field data [18]. SimulO does not allow the use of arbitrary angular distribution for photon-emitting surfaces. To overcome this problem, two sources of photons were summed up in order to reproduce an observed under-ice light field (Figure 2). The first source was a regular Lambertian emitting surface while the second was a Lambertian emitting surface but restricted to an emission within 60 degrees of the zenith angle. A 5-m radius melt pond was set up at the center of the emitting surface (Figure 1). The melt pond had the same emitting angular distribution as the surrounding ice.

Its intensity was four times higher than the surrounding ice, which corresponds to typical conditions found in the Arctic during summer [19].

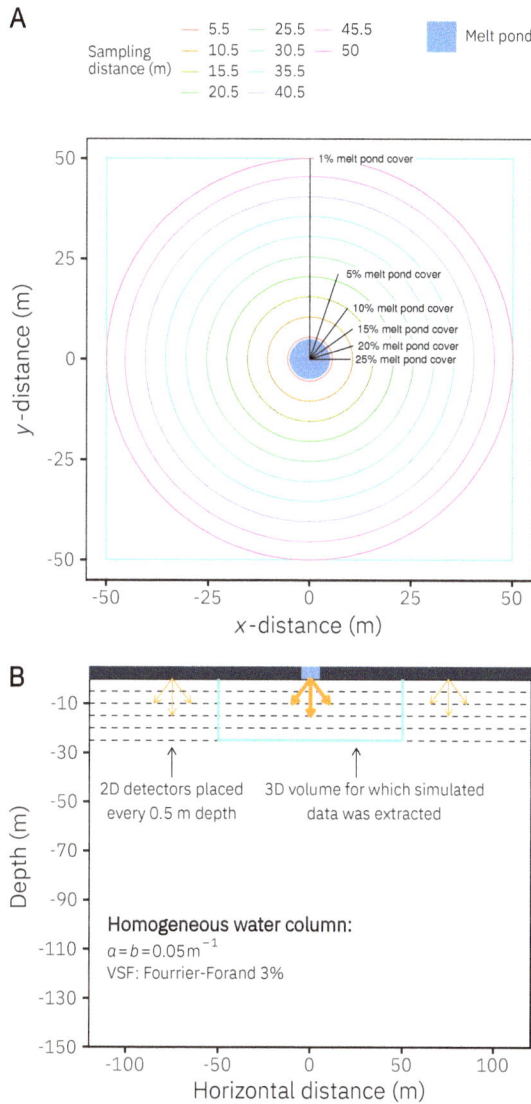

Figure 1. Spatial configuration used for the 3D Monte Carlo numerical simulations. (**A**) Surface view showing the percentage of the total area covered by the melt pond over the areas described by the black lines. For each of these areas, light profiles were averaged (see Figure 7). For visualization purpose, lines of the horizontal sampling distances from the centre of the melt pond have been plotted only at 5 m intervals. (**B**) 2D side view showing the 3D volume for which simulated data were extracted and how photon detectors were placed in the water column. Orange arrows indicate incident light sources.

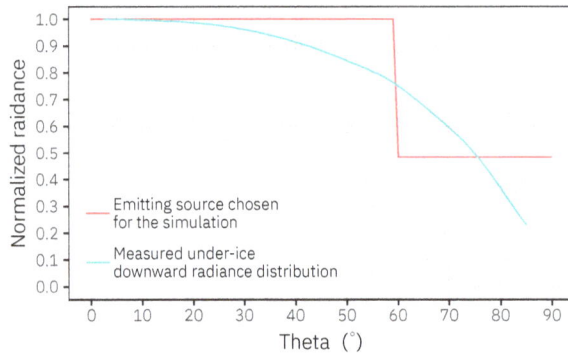

Figure 2. Comparison of the under-ice measured downward radiance distribution (the average cosine is ≈0.61, [18]) and the angular distribution of light-emitting source used in the paper.

Given our interest in surface light profiles, 2D horizontal software detectors were placed vertically every 0.5 m, from 0.5 m up to a depth of 25 m. Detectors include 1 m² pixels measuring downward irradiance and upward radiance (5-degree half angle of acceptance). In order to avoid the effect of the boundary (i.e., absorption by the side of the cylinder used to simulate the water column), data outside a radius of 50 m were not used (see the green box in Figure 1). A total number of 7.14×10^{10} photons were simulated to obtain a sufficient number of upwelling photons. The simulation took approximately 6000 h distributed over 2000 CPU cores. Because the geometry was symmetrical azimuthally, irradiance and radiance were averaged over the azimuth in order to increase the signal-to-noise ratio. Because of the low scattering coefficients used to reproduce in situ conditions observed during the sampling campaign, radiance profiles were noisy because a small number of upward photons could be captured. To address this issue, radiance profiles were smoothed using a Gaussian fit (Figure A2 included in the Appendix B).

2.3.2. Estimation of Reference and Local Light Profiles

To explore how the melt pond influences the averaged underwater irradiance and radiance profiles (Figure 1), data from the Monte Carlo simulation were averaged according to six different radii, corresponding to varying melt pond spatial proportions. The simulated light profiles were averaged within the following surface areas: (1) 10 m radius (25% melt pond cover), (2) 11.18 m radius (20% melt pond cover), (3) 12.91 m radius (15% melt pond cover), (4) 15.81 m radius (10% melt pond cover), (5) 22.36 m radius (5% melt pond cover) and (6) 50 m radius (1% melt pond cover). For each of these six configurations, the corresponding averaged light profile, $\overline{E_d}(z)$, was subsequently viewed as an adequate description of the average underwater light field. For the remainder of the text, these averaged profiles are referred to as reference light profiles. Furthermore, 50 light profiles, evenly spaced by 1 m from the melt pond centre, were extracted to mimic local measurements of light and to calculate associated diffuse attenuation coefficients.

2.4. Statistical Analysis

All statistical analyses and graphics were carried out with R 3.5.1 [20].

3. Results

3.1. Comparing In Situ Downward Irradiance (E_d) and Upward Radiance (L_u) Measurements

An example showing in situ downward irradiance (E_d) profiles and upward radiance (L_u) profiles at 16 visible wavelengths measured under-ice is presented in Figure 3. For the E_d profiles, subsurface light maxima at a depth of around 10 m are clearly visible between 400 and 560 nm. These peaks are not visible in the yellow/red region (580–700 nm). For the L_u profiles, no subsurface light maxima were found at any wavelength. To have a closer look at the shape of both E_d and L_u profiles, data below the 10 m depth were normalized to the value at 10 m (Figure 4). Below 10 m and between 400 and 580 nm, both E_d and L_u profiles presented the same shape (i.e., yield the same rate of attenuation with increasing depth). At longer wavelengths ($\geq$600 nm), differences between the shapes of E_d and L_u profiles increased. Irradiance and radiance diffuse attenuation coefficients (K_d and K_{Lu}) calculated for the layers of a 5 m thickness are compared in Figure 5 for all 83 profiles. In the blue/green/yellow regions (400–580 nm), the determination coefficients between K_{Lu} and K_d varied between 0.98 at the surface (10–15 m) and 0.64 at depth (75–80 m). For most of the surface layers, regression lines lined up with the 1:1 lines. Slight deviations from the 1:1 lines started to appear below 60 m where K_d was on average higher than K_{Lu}. The relationships including orange and red wavelengths are presented in Figure A3 included in the Appendix C. A linear regression analysis between all in situ normalized E_d and L_u profiles showed that determination coefficients (R^2) range between 0.75 and 1 (Figure A4). A sharp decrease and a high variability of calculated R^2 occurred beyond 575 nm. This suggests a gradual decoupling between E_d and L_u profiles at longer wavelengths, likely due to the effect of inelastic scattering (mostly Raman scattering).

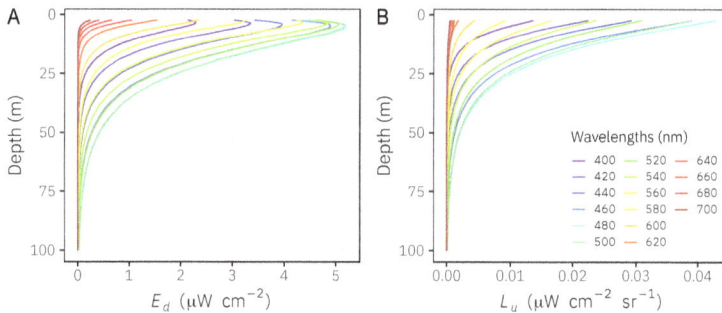

Figure 3. Examples of in situ downward irradiance ($E_d(z)$) and upward radiance ($L_u(z)$) profiles measured under-ice on 20 June 2016. Note the presence of subsurface maxima in the downward irradiance profiles and the absence of subsurface maxima in the upward radiance profiles.

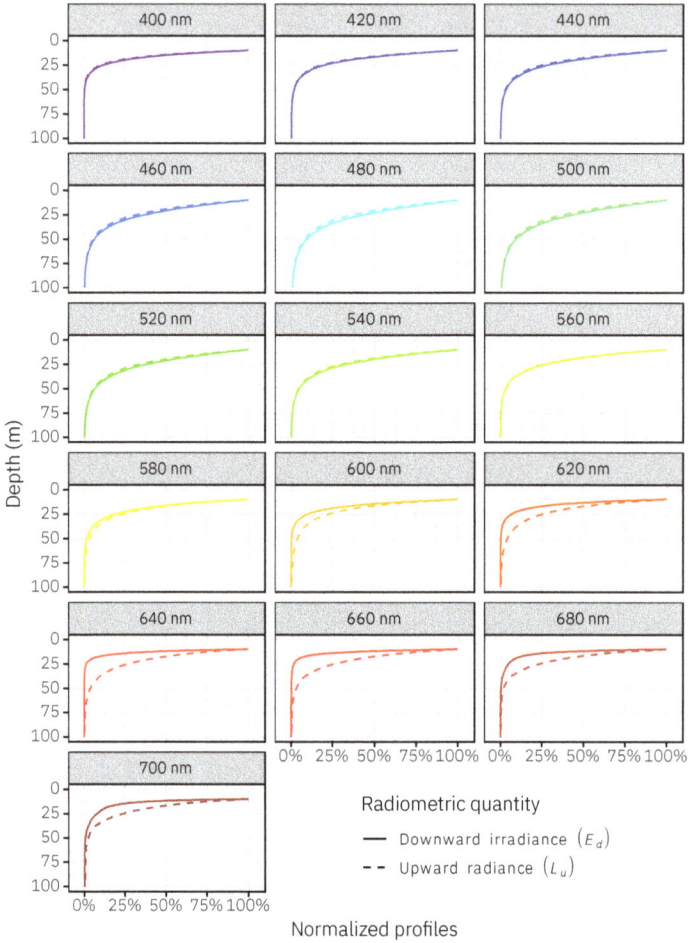

Figure 4. Comparison of downward irradiance ($E_d(z)$) and upward radiance ($L_u(z)$) for one example light profile measured under-ice. Profiles were normalized to the measured radiometric value at 10 m depth (under the subsurface light maximum) in order to emphasize the similar shape between $E_d(z)$ and $L_u(z)$.

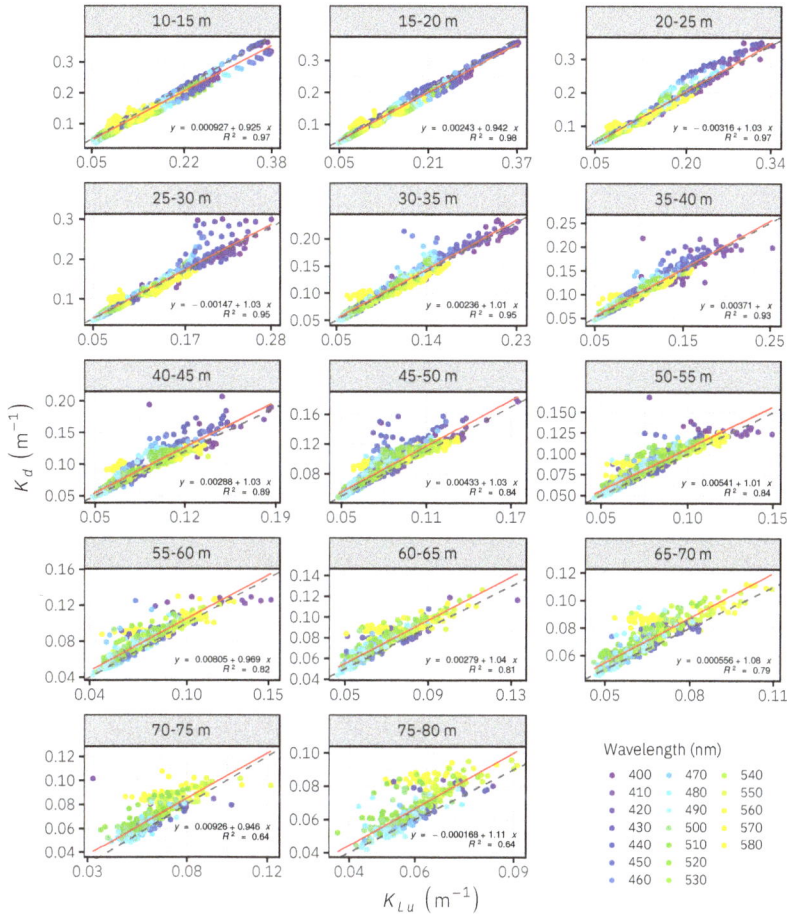

Figure 5. Scatter plots showing the relationships between the measured K_d and K_{Lu} in the spectral range between 400 and 580 nm at different depths (numbers in gray boxes). Red lines represent the regression lines of the fitted linear models. Regression equations and determination coefficients (R^2) are also provided in each plot. Dashed lines are the 1:1 lines.

3.2. 3D Monte Carlo Numerical Simulations

Figure 6 shows cross-sections of the simulated downward irradiance and upward radiance. A key difference for the upcoming discussion is that the simulated upward radiance was more homogeneous compared to the simulated downward irradiance. Figure 7 shows the reference irradiance, $E_d(z)$, and reference radiance, $L_u(z)$, profiles. The highest irradiance and radiance occurred when the melt pond occupied 25% of the sampling area, allowing for more light to propagate in the water column. None of the $E_d(z)$ and $L_u(z)$ reference profiles showed subsurface light maxima. Figure 8 shows the 50 simulated local downward irradiance and upward radiance profiles evenly spaced by 1 m in the horizontal distance from the melt pond centrer. Local downward irradiance profiles under the melt pond (0–5 m) showed a rapid decrease with increasing depth described by a monotonically exponential or quasi-exponential decrease. Local simulated downward irradiance profiles just outside the melt pond (5–10 m from the melt pond centre) were characterized with subsurface light maxima occurring at a depth of between approximately 5 and 10 m. Further away from the melt pond centre, downward irradiance profiles followed a monotonically exponential or quasi-exponential decrease. None of the simulated upward radiance profiles presented subsurface light maxima (Figure 8). From local simulated irradiance and radiance profiles (Figure 8), K_d and K_{Lu} were calculated by fitting Equation (1) between the depths of 0 m and 25 m. Results are presented in Figure 9. K_d varied between 0.065 and 0.157 m^{-1} and K_{Lu} between 0.079 and 0.116 m^{-1}. These K_d and K_{Lu} were used to propagate light downward from surface reference values $E_d(0^-)$. Figure 10 shows the profiles resulting from this calculation. A greater dispersion around the reference profiles (thick black lines in Figure 10) occurred when using K_d compared to the profiles generated with similarly derived K_{Lu} values. The relative differences between the depth-integrated values of each local profile (coloured lines in Figure 10) and the depth-integrated values of the reference profiles (thick black lines in Figure 10) were used to quantify the error of using either K_d or K_{Lu} as a proxy to predict downward irradiance in the water column (Figure 11). Below the melt pond, K_d overestimated the total downward irradiance by up to 40% when the melt pond occupied 1% of the surface area. In this region, the local K coefficients are inflated. In the transition region, at a horizontal distance of 5 and 10 m from the centre of the melt pond, where subsurface maxima are observed, K_d underestimated the downward irradiance by up to 35% when the melt pond occupied 25% of the surface area. Further away from the edge of the melt pond, the errors saturated to a maximum of -25%. The same behaviour is observed for K_{Lu} but with about two times less amplitude. The mean relative errors were lower by approximately a factor of two when using K_{Lu} (-7%) compared to K_d (-12%). Also, the prediction errors stabilized at a shorter horizontal distance from the centre of the melt pond when using K_{Lu} (≈ 10 m) compared with using K_d (≈ 20 m).

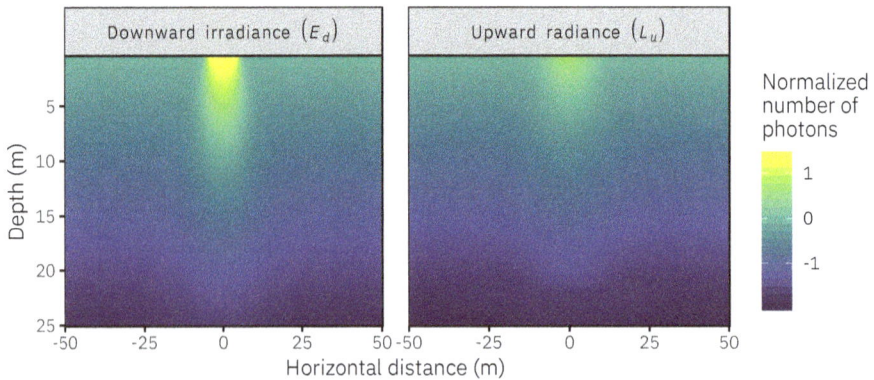

Figure 6. Cross-sections of simulated downward irradiance and upward radiance fields under a melt pond with a 5 m radius. The logarithm of the normalized number of photons has been used to create the scale for visualization. The normalization has been done using the values modelled at a 0.5 m depth and at a horizontal distance of 50 m from the centre of the melt pond.

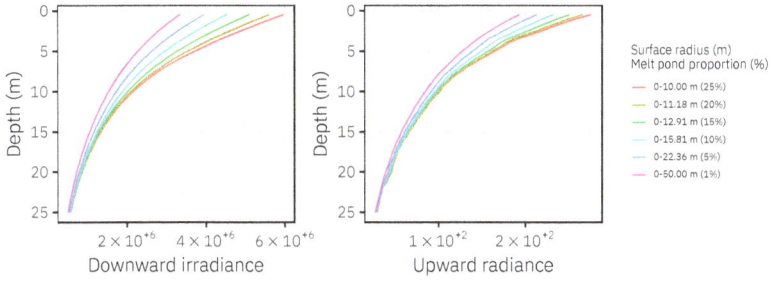

Figure 7. Simulated reference downward irradiance and upward radiance profiles ($\overline{E_d}(z)$, $\overline{L_u}(z)$ in relative units) for six different areas with varying proportions of the surface occupied by the melt pond (see Figure 1). Note that none of the averaged irradiance profiles show the same subsurface light maxima as observed with in situ data (see Figure 3).

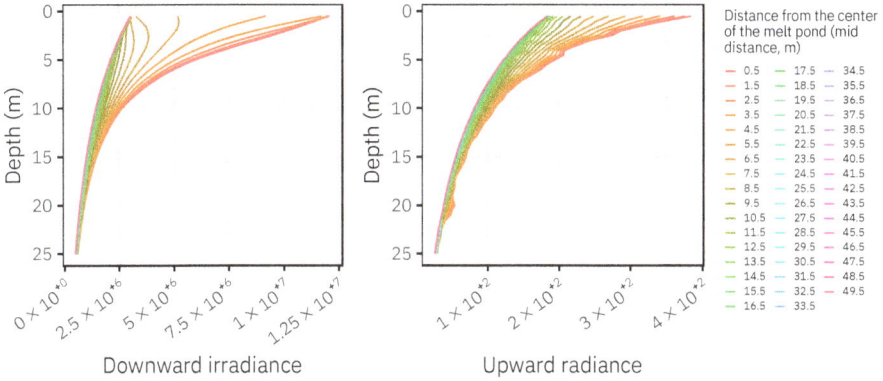

Figure 8. Simulated local downward irradiance and upward radiance profiles (expressed in relative units) at different horizontal distances from the centre of the melt pond (see Figure 1) used to compute K_d and K_{Lu}. These attenuation coefficients were used to propagate surface reference downward irradiance ($E_d(0^-)$, the surface values of the lines in Figure 7) through the water column.

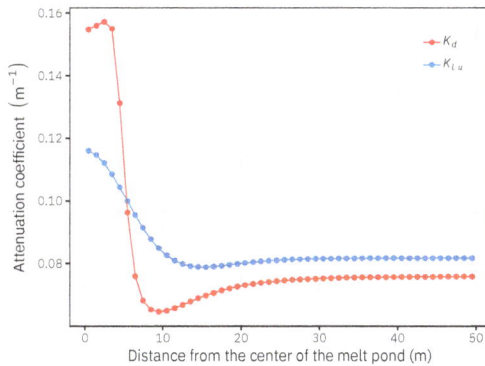

Figure 9. Diffuse attenuation coefficients calculated from local downward irradiance and upward radiance profiles simulated at different distances from the centre of the melt pond (see Figure 8).

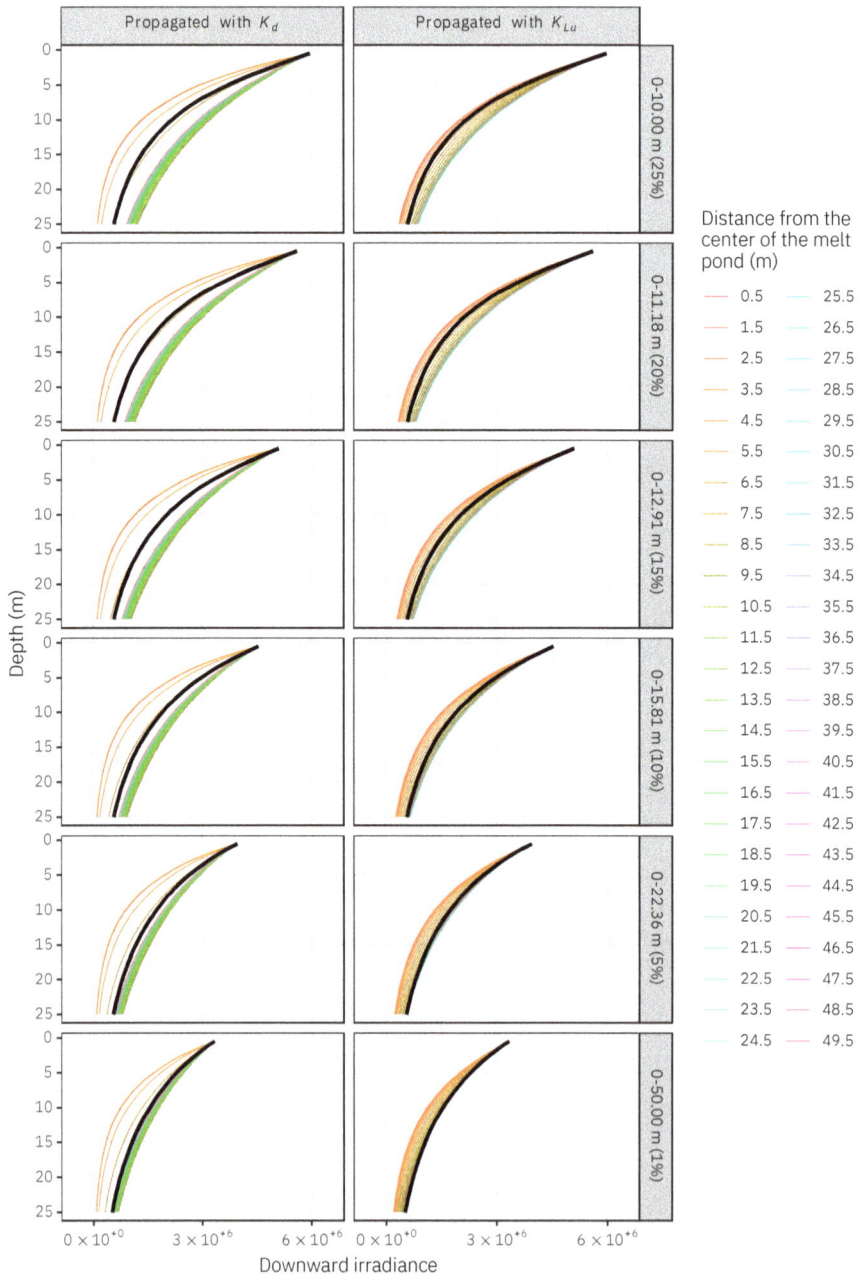

Figure 10. Reference downward irradiance profiles (thick black lines, in relative units) and propagated irradiance through the water column (coloured lines, in relative units) using local values of K_d and K_{Lu} (see Figure 8). Light was propagated using the surface reference downward irradiance.

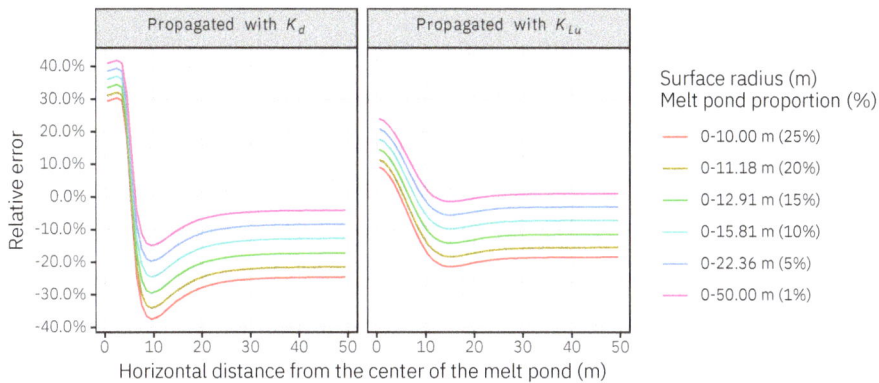

Figure 11. Relative errors of the predictions calculated as the relative differences between the depth integral of the reference and predicted irradiance profiles.

3.3. Inelastic Scattering

Based on in situ data, our results have pointed out that K_{Lu} is not a good proxy for K_d at longer wavelengths (Figures A3 and A4) because of the effect of Raman scattering. To validate this hypothesis, we used the HydroLight (Sequoia Scientific, Inc., Bellevue, WA, USA) radiative transfer numerical model to calculate theoretical downward irradiance and upward radiance and their associated vertical attenuation coefficients in an open water column in the presence of Raman scattering. The simulation was parametrised using IOPs measured during the field campaign (detailed information can be found in the supplementary section entitled Raman inelastic scattering included in Appendix A). The simulation was able to reproduce the observed decoupling between K_d and K_{Lu} observed at wavelengths $\geq$600 nm (Figure A5). These results are generally consistent with previous findings from radiative transfer simulations, which demonstrated the depth and spectral dependencies of diffuse attenuation coefficients as affected by Raman scattering [21,22].

4. Discussion

In the Arctic, melt pond coverage, lead coverage, and ice and snow thickness can vary greatly in both time and space [23,24]. Due to this sea ice heterogeneity, local under-ice measurements of downward irradiance are sometimes characterized by subsurface light maxima (Figure 3). To model such profiles, Laney et al. [9] proposed a semi-empirical parametrisation using two exponential terms (see Equation (3)). Whereas their method might provide adequate estimations of instantaneous downward diffuse attenuation coefficients at specific locations, fitting a double exponential might not be ideal because data are modelled locally and do not provide an adequate description of the average light field ($\overline{E_d}(z)$) as it would be seen, for example, by drifting phytoplankton cells. In such conditions, this paper argues that under-ice irradiance measurements should be analysed in the context of ice and surface properties within a radius of several meters over the horizontal distance because local measurements cannot be used as a proxy of the average light field.

Using in situ light measurements, it was found that E_d and L_u (and therefore K_d and K_{Lu}) were highly correlated below 10 m depth (Figures 4 and 5), even when subsurface light maxima were present (Figure 3). Furthermore, no subsurface light maxima were observed in the in situ upward radiance profiles. The reason is that the L_u radiometer measures upwelling photons coming from deeper depth, which have likely undergone more scattering. These photons thus originate from a larger surface area. This reinforces the idea that L_u is less influenced by sea ice surface heterogeneity.

Based on Monte Carlo simulations of radiative transfer, our results showed that the average downward irradiance profile, $\overline{E_d}(z)$, under heterogeneous sea ice cover follows a single-term exponential function, even when melt ponds occupy a large fraction of the study area (Figure 7). This is similar to what is observed under a wavy ice-free surface [10]. However, estimating $\overline{E_d}(z)$ for a given area is not straightforward, as it requires a large number of local profiles under the sea ice. An intuitive alternative to deriving the attenuation coefficient is to use upward radiance, which is less influenced by sea surface heterogeneity compared to downward irradiance (Figures 3–5). Monte Carlo simulations showed that a local estimation of K_{Lu} was a good proxy for $\overline{K_d}$ and that using K_{Lu} rather than K_d provided better estimations of the average downward profile by reducing the average error by approximately a factor of two (Figure 11).

There are at least two main factors influencing the quality of in situ downward irradiance measurements under heterogeneous sea ice. The first factor is the horizontal distance from the centre of the melt pond. Although the relative error of propagating $E_d(0^-)$ using both K_d and K_{Lu} showed the same pattern, the largest error occurred when using local estimations of K_d directly below the melt pond and up to 10 m from the melt pond edge (Figure 11). In contrast, the relative error associated with the use of K_{Lu} was much lower and stabilized just after approximately 10 m from the centre of the melt pond. The second factor driving the relative error of local measurements is the proportion occupied by melt ponds over the area of interest (Figure 11). Indeed, higher proportions of melt pond allow for more light to penetrate in the water column. Hence, local measurements made under surrounding ice are more likely to show subsurface light maxima (see Frey et al. [8]). Accordingly, when melt ponds accounted for 1% of the total area, averaged error in $E_d(z)$ using K_{Lu} was 1.33% but increased to 18% when the melt pond occupied 25% of the total area (Figure 11).

5. Conclusions

Our results show that under spatially heterogeneous sea ice at the surface (and for a homogeneous water column), the average irradiance profile, $\overline{E_d}(z)$, is well reproduced by a single exponential function. We also showed that propagating $E_d(0^-)$ using K_{Lu} is a better choice compared to K_d under heterogeneous sea ice. Nowadays, radiance measurements are becoming more routinely performed during field campaigns, so we argue that one should use K_{Lu} when available to propagate $E_d(0^-)$ through the water column under sea ice. The main difficulty remains in finding good estimates of averaged $E_d(0^-)$. In recent years, this has become easier with the development of remotely operated vehicles [3,4,25], remote sensing techniques, and drone imagery. In this study, we used a Monte Carlo approach to model an idealized surface with a single melt pond (Figures 1 and 6). Figure 11 shows that the effect of a melt pond with diameter 5 m is minimized at a horizontal distance of approximately 20 m or more. Therefore, when many melt ponds are characterizing an area, if one has to perform a single profile, measuring an upward radiance profile under bare ice as far away as possible from any melt pond would minimize the error in estimating the area-averaged downward irradiance profile using K_{Lu}. Although not representative of a complex Arctic sea ice surface, our simple surface geometry allowed to study the transition from a high to a low transmission sea ice. Further 3D Monte Carlo work could include a more complex geometry of heterogeneous surfaces.

Author Contributions: Conceptualization, P.M., G.B., S.L.-G., E.L. and M.B.; methodology, P.M., G.B., S.L.-G., E.L. and M.B.; field work, G.B., S.L.-G., and M.B.; writing—original draft preparation, P.M.; writing—review and editing, P.M., G.B., S.L.-G., E.L. and M.B.; supervision, M.B.; funding acquisition, M.B.

Funding: The GreenEdge project is funded by the following French and Canadian programs and agencies: ANR (contract #111112), CNES (project #131425), IPEV (project #1164), CSA, Fondation Total, ArcticNet, LEFE and the French Arctic Initiative (GreenEdge project).

Acknowledgments: This project would not have been possible without the support of the Hamlet of Qikiqtarjuaq and the members of the community as well as the Inuksuit School and its principal, Jacqueline Arsenault. The project is conducted under the scientific coordination of the Canada Excellence Research Chair on Remote Sensing of Canada's New Arctic Frontier and the CNRS and Université Laval Takuvik Joint International laboratory (UMI3376). The field campaign was successful thanks to the contributions of J. Ferland, G. Bécu, C. Marec, J. Lagunas-Morales, F. Bruyant, J. Larivière, E. Rehm, S. Lambert-Girard, C. Aubry, C. Lalande, A. LeBaron, C. Marty, J. Sansoulet, D. Christiansen-Stowe, A. Wells, M. Benoît-Gagné, E. Devred and M.-H. Forget from the Takuvik laboratory, C.J. Mundy and V. Galindo from University of Manitoba, and F. Pinczon du Sel and E. Brossier from Vagabond. We also thank Michel Gosselin, Québec-Océan, the CCGS Amundsen and the Polar Continental Shelf Program for their in-kind contribution in polar logistics and scientific equipment. We thank Laurent Oziel, Jade Larivière and Julien Laliberté for their contribution to radiometry measurements in 2016. This research was enabled in part by support provided by Calcul Québec (www.calculquebec.ca) and Compute Canada (www.computecanada.ca). S.L. Girard was supported by a postdoctoral fellowship from the Natural Sciences and Engineering Research Council of Canada (NSERC). We also acknowledge the Canada First Research Excellence Fund and the Sentinel North Strategy for their financial support. We thank Dariusz Stramski and one anonymous reviewer for their valuable comments which helped to greatly improve the manuscript.

Conflicts of Interest: The authors declare no conflict of interest.

Appendix A

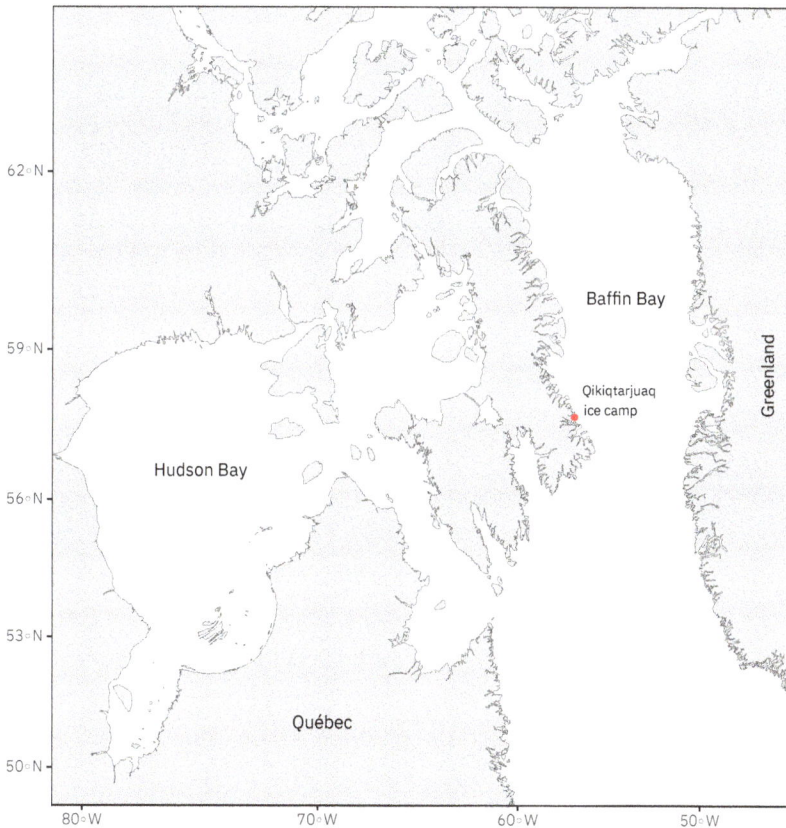

Figure A1. The field campaign was part of the GreenEdge project (www.greenedgeproject.info) which was conducted on landfast ice southeast of the Qikiqtarjuaq Island in the Baffin Bay (67.4797 N, 63.7895 W).

Appendix B. Smoothing Radiance Data

Due to the low scattering coefficients used to reproduce in situ conditions observed during the sampling campaign, radiance profiles were noisy because only few photons were scattered back in the upward direction (note the different y-scales). To overcome this problem, upward radiance data were smoothed using a Gaussian fit accordingly to Equation (A1):

$$f(x, \varphi, \mu, \sigma, k) = \varphi e^{-\frac{(x - \mu)^2}{2\sigma^2}} + k \tag{A1}$$

where x (m) is the horizontal distance from the center of the melt pond, σ (m) is the standard deviation controlling the width of the curve, φ is the height of the curve peak ($\varphi = \frac{1}{\sigma\sqrt{2\pi}}$), μ (m) is the position of the center of the peak, and k an offset coefficient.

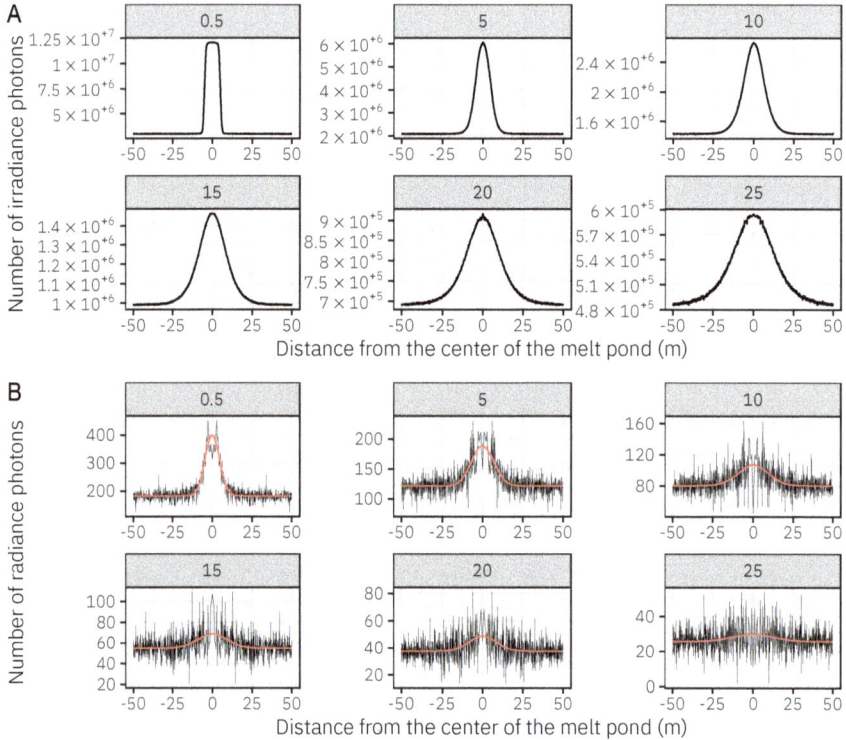

Figure A2. Examples showing the number of downward irradiance (**A**) and upward radiance (**B**) photons captured by the detectors of the Monte Carlo simulation at different depth ranges (numbers in gray boxes) as a function of the horizontal distance from the melt pond. The red lines represent the fitted Gaussian curves.

Appendix C. Raman Inelastic Scattering

Raman scattering is a process by which photons, interacting with water molecules, lose or gain energy and are scattered at a different wavelength than the one they were originating from. In Figures A3 and A4, one can observe a decoupling between K_d and K_{Lu} at longer wavelengths, possibly due to inelastic Raman scattering. To validate this hypothesis, we used the HydroLight radiative transfer numerical model to calculate downward irradiance and upward radiance and their associated attenuation coefficients in a water column.

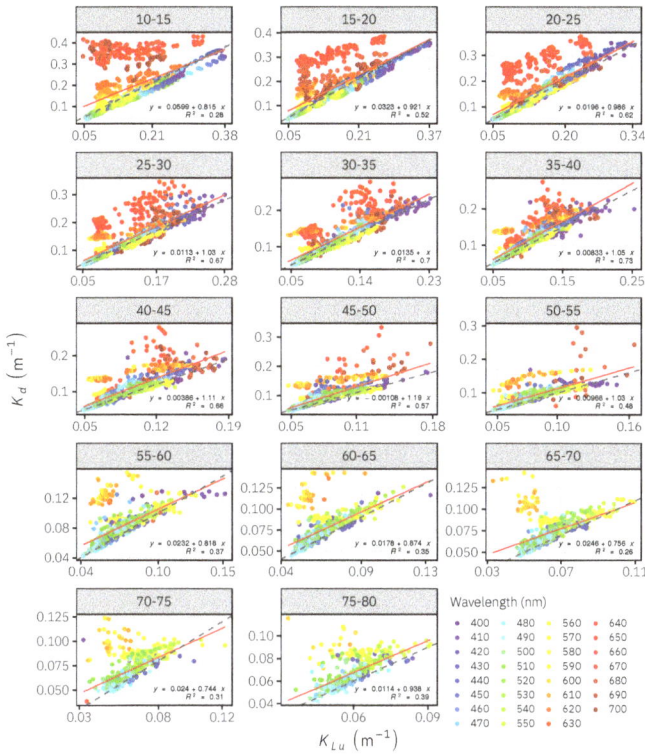

Figure A3. Scatter plots showing the relationships between downward irradiance ($E_d(z)$) and upward radiance ($L_u(z)$) between 400 and 700 nm at different depths (numbers in gray boxes). Red lines represent the regression lines of the fitted linear models. Dashed lines are the 1:1 lines. Note the large deviations between the data points and the 1:1 line occurring in the orange and red regions ($\geq$600 nm).

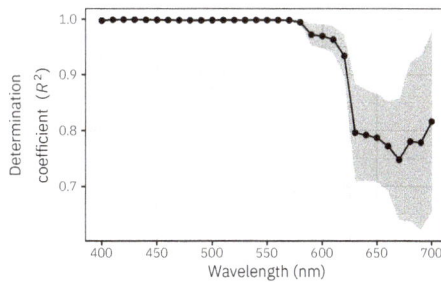

Figure A4. Average determination coefficient R^2 and standard deviation (shaded area) of the regressions between normalized (at 10 m depth) $E_d(z)$ and $L_u(z)$ profiles between 400 and 700 nm. At each wavelength, average values were computed from the 83 COPS measurements. A sharp decrease of R^2 occurred at wavelength longer than approximately 575 nm, suggesting a gradual decoupling between $E_d(z)$ and $L_u(z)$ profiles at longer wavelengths, possibly due to the effect of inelastic scattering.

HydroLight Simulations

Two HydroLight simulations were carried out to model downward irradiance and upward radiance with and without taking into account Raman inelastic scattering. The simulations were parameterized using an IOPs profile (ac-s from Sea-Bird Scientific) measured on the first of May 2015 in the Baffin Bay. Simulations were performed with the following characteristics:

- A surface free of ice.
- A surface without waves.
- Sun position at noon for May 1st (solar zenith angle = 45.39 degrees).
- A cloudless sky.
- No fluorescence.
- Using HydroLight default atmospheric parameters.
- The scattering phase function of water was described by a Fournier-Forand analytic form with a 3% backscatter fraction.
- EcoLight option was run.

The HydroLight simulations showed a decoupling between K_d and K_{Lu} starting at around 600 nm when Raman scattering was modelled (Figure A5). Similar decoupling was also observed with the in situ data (see Figure A3).

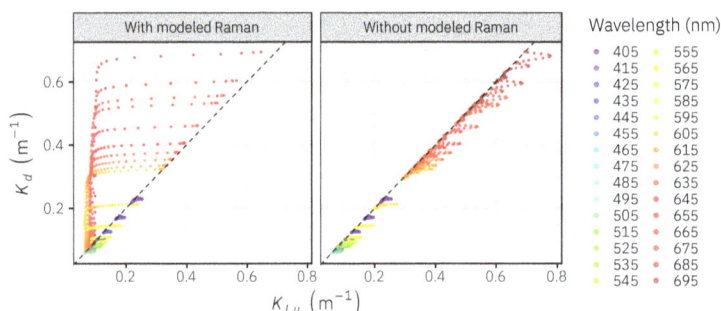

Figure A5. Scatter plots showing the relationships between K_d and K_{Lu} calculated from the downward irradiance and upward radiance profiles modelled with and without Raman scattering. The dashed lines represent the 1:1 lines.

References

1. Kirk, J.T.O. *Light and Photosynthesis in Aquatic Ecosystems*, 2nd ed.; Cambridge University Press: Cambridge, UK; New York, NY, USA, 1994.
2. Morel, A. An ocean flux study in eutrophic, mesotrophic and oligotrophic situations: The EUMELI program. *Deep Sea Res. Part I Oceanogr. Res. Pap.* **1996**, *43*, 1185–1190. [CrossRef]
3. Nicolaus, M.; Katlein, C. Mapping radiation transfer through sea ice using a remotely operated vehicle (ROV). *Cryosphere* **2013**, *7*, 763–777. [CrossRef]
4. Katlein, C.; Arndt, S.; Nicolaus, M.; Perovich, D.K.; Jakuba, M.V.; Suman, S.; Elliott, S.; Whitcomb, L.L.; McFarland, C.J.; Gerdes, R.; et al. Influence of ice thickness and surface properties on light transmission through Arctic sea ice. *J. Geophys. Res. Ocean.* **2015**, *120*, 5932–5944. [CrossRef] [PubMed]
5. Katlein, C.; Perovich, D.K.; Nicolaus, M. Geometric Effects of an Inhomogeneous Sea Ice Cover on the under Ice Light Field. *Front. Earth Sci.* **2016**, *4*. [CrossRef]
6. Lange, B.A.; Flores, H.; Michel, C.; Beckers, J.F.; Bublitz, A.; Casey, J.A.; Castellani, G.; Hatam, I.; Reppchen, A.; Rudolph, S.A.; et al. Pan-Arctic sea ice-algal chl a biomass and suitable habitat are largely underestimated for multiyear ice. *Glob. Chang. Biol.* **2017**, *23*, 4581–4597. [CrossRef] [PubMed]
7. Mundy, C.J.; Gosselin, M.; Ehn, J.; Gratton, Y.; Rossnagel, A.; Barber, D.G.; Martin, J.; Tremblay, J.É.; Palmer, M.; Arrigo, K.R.; et al. Contribution of under-ice primary production to an ice-edge upwelling phytoplankton bloom in the Canadian Beaufort Sea. *Geophys. Res. Lett.* **2009**, *36*, L17601. [CrossRef]
8. Frey, K.E.; Perovich, D.K.; Light, B. The spatial distribution of solar radiation under a melting Arctic sea ice cover. *Geophys. Res. Lett.* **2011**, *38*. [CrossRef]

9. Laney, S.R.; Krishfield, R.A.; Toole, J.M. The euphotic zone under Arctic Ocean sea ice: Vertical extents and seasonal trends. *Limnol. Oceanogr.* **2017**, *62*, 1910–1934. [CrossRef]

10. Zaneveld, J.R.V.; Boss, E.; Barnard, A. Influence of surface waves on measured and modeled irradiance profiles. *Appl. Opt.* **2001**, *40*, 1442. [CrossRef] [PubMed]

11. Smith, R.C.; Baker, K.S. Analysis of Ocean Optical Data II. In Proceedings of the 1986 Technical Symposium Southeast, Orlando, FL, USA, 7 August 1986; Volume 489, p. 95. [CrossRef]

12. Mobley, C.D. Measuring Radiant Energy, Ocean Optics Web Book. Available online: http://www.oceanopticsbook. info/view/light_and_radiometry/measuring_radiant_energy (accessed on 28 November 2018).

13. Petrich, C.; Nicolaus, M.; Gradinger, R. Sensitivity of the light field under sea ice to spatially inhomogeneous optical properties and incident light assessed with three-dimensional Monte Carlo radiative transfer simulations. *Cold Reg. Sci. Technol.* **2012**. [CrossRef]

14. Katlein, C.; Nicolaus, M.; Petrich, C. The anisotropic scattering coefficient of sea ice. *J. Geophys. Res. Ocean.* **2014**, *119*, 842–855. [CrossRef]

15. Leymarie, E.; Doxaran, D.; Babin, M. Uncertainties associated to measurements of inherent optical properties in natural waters. *Appl. Opt.* **2010**, *49*, 5415. [CrossRef] [PubMed]

16. Fournier, G.R.; Forand, J.L. Analytic phase function for ocean water. In Proceedings of the Ocean Optics XII, Bergen, Norway, 26 October 1994; Jaffe, J.S., Ed.; 1994; pp. 194–201. [CrossRef]

17. Mobley, C.D.; Sundman, L.K.; Boss, E. Phase function effects on oceanic light fields. *Appl. Opt.* **2002**, *41*, 1035. [CrossRef] [PubMed]

18. Girard, S.L.; Leymarie, E.; Marty, S.; Matthes, L.; Ehn, J.; Babin, M. High angular resolution measurements of the radiance distribution beneath Arctic landfast sea ice during the spring transition. *Earth Space Sci.* **2018**, *5*, 30–47.

19. Perovich, D.K. Sea ice and sunlight. In *Sea Ice*; John Wiley & Sons, Ltd.: Chichester, UK, 2016; Chapter 4, pp. 110–137. [CrossRef]

20. R Core Team. *R: A Language and Environment for Statistical Computing*; R Foundation for Statistical Computing: Vienna, Austria, 2018.

21. Li, L.; Stramski, D.; Reynolds, R.A. Effects of inelastic radiative processes on the determination of water-leaving spectral radiance from extrapolation of underwater near-surface measurements. *Appl. Opt.* **2016**, *55*, 7050. [CrossRef]

22. Berwald, J.; Stramski, D.; Mobley, C.D.; Kiefer, D.A. Effect of Raman scattering on the average cosine and diffuse attenuation coefficient of irradiance in the ocean. *Limnol. Oceanogr.* **1998**, *43*, 564–576. [CrossRef]

23. Landy, J.; Ehn, J.; Shields, M.; Barber, D. Surface and melt pond evolution on landfast first-year sea ice in the Canadian Arctic Archipelago. *J. Geophys. Res. Ocean.* **2014**, *119*, 3054–3075. [CrossRef]

24. Eicken, H.; Grenfell, T.C.; Perovich, D.K.; Richter-Menge, J.A.; Frey, K. Hydraulic controls of summer Arctic pack ice albedo. *J. Geophys. Res. Ocean.* **2004**, *109*, 1–12. [CrossRef]

25. Arndt, S.; Meiners, K.M.; Ricker, R.; Krumpen, T.; Katlein, C.; Nicolaus, M. Influence of snow depth and surface flooding on light transmission through Antarctic pack ice. *J. Geophys. Res. Ocean.* **2017**, *122*, 2108–2119. [CrossRef]

MDPI

St. Alban-Anlage 66

4052 Basel

Switzerland

Tel. +41 61 683 77 34

Fax +41 61 302 89 18

www.mdpi.com

Applied Sciences Editorial Office

E-mail: applsci@mdpi.com

www.mdpi.com/journal/applsci

www.ingramcontent.com/pod-product-compliance
Lightning Source LLC
Chambersburg PA
CBHW051703210326
41597CB00032B/5358